Experimental Methods in the Physical Sciences
Single-Photon Generation and Detection

VOLUME FORTY FIVE

Experimental Methods in the Physical Sciences

Thomas Lucatorto, Albert C. Parr and Kenneth Baldwin, *Editors in Chief*

Single-Photon Generation and Detection

Experimental Methods in the Physical Sciences

VOLUME FORTY FIVE

Editors

Alan Migdall
Sergey Polyakov
Jingyun Fan
Joshua Bienfang

AMSTERDAM • BOSTON • HEIDELBERG • LONDON
NEW YORK • OXFORD • PARIS • SAN DIEGO
SAN FRANCISCO • SINGAPORE • SYDNEY • TOKYO

Academic Press is an imprint of Elsevier

Academic Press is an imprint of Elsevier
225, Wyman Street, Waltham, MA 02451, USA
The Boulevard, Langford Lane, Kidlington, Oxford OX5 1GB, UK
Radarweg 29, PO Box 211, 1000 AE Amsterdam, The Netherlands

© 2013 Published by Elsevier Inc.

Portions of this Work were prepared by U.S. government employee in connection with their official duties, and therefore copyright protection is not available in the United States for such portions of the book pursuant to 17 U.S.C. Section 105.

No part of this publication may be reproduced, stored in a retrieval system or transmitted in any form or by any means electronic, mechanical, photocopying, recording or otherwise without the prior written permission of the publisher.

Permissions may be sought directly from Elsevier's Science & Technology Rights Department in Oxford, UK: phone (+44) (0) 1865 843830; fax (+44) (0) 1865 853333; permissions@elsevier.com. Alternatively you can submit your request online by visiting the Elsevier web site at http://elsevier.com/locate/permissions, and selecting Obtaining permission to use Elsevier material.

Notices
No responsibility is assumed by the publisher for any injury and/or damage to persons or property as a matter of products liability, negligence or otherwise, or from any use or operation of any methods, products, instructions or ideas contained in the material herein. Because of rapid advances in the medical sciences, in particular, independent verification of diagnoses and drug dosages should be made.

Library of Congress Cataloging-in-Publication Data
A catalog record for this book is available from the Library of Congress

British Library Cataloguing in Publication Data
A catalogue record for this book is available from the British Library

For information on all **Academic Press** publications
visit our web site at store.elsevier.com

Printed and bound in the United States of America
13 14 15 16 17   10 9 8 7 6 5 4 3 2 1

ISBN: 978-0-12-387695-9
ISSN: 1079-4042

Contents

Contributors	xiii
Volumes in series	xvii
Preface	xxi

1. Introduction — 1

- 1.1 Physics of Light—an Historical Perspective — 1
- 1.2 Quantum Light — 2
  - 1.2.1 What is Non-Classical Light? — 2
  - 1.2.2 What is a Photon? — 3
- 1.3 The Development of Single-Photon Technologies — 4
- 1.4 Some Applications of Single-Photon Technology — 8
- 1.5 This book — 9
  - 1.5.1 Single-Photon Detectors — 9
  - 1.5.2 Single-Photon Sources — 16
- 1.6 Conclusions — 17
- References — 18

2. Photon Statistics, Measurements, and Measurements Tools — 25

- 2.1 Quantized Electric Field & Operator Notation — 26
- 2.2 Source Characteristics — 28
  - 2.2.1 State Vector — 28
  - 2.2.2 Density Matrix and Photon Number Probabilities — 29
  - 2.2.3 Purity — 30
  - 2.2.4 Source Efficiency and Generation Rate — 31
  - 2.2.5 Second-Order Coherence, $g^{(2)}$ — 32
  - 2.2.6 Relating $g^{(2)}$ to $P(n)$ — 34
  - 2.2.7 Ideal and Non-Ideal Single-Photon Sources — 37
  - 2.2.8 To measure $P(n)$ or $g^{(2)}$? — 38
  - 2.2.9 Hanbury Brown-Twiss Interferometer — 38
  - 2.2.10 Bunching, Antibunching, and Poissonian Photon Statistics — 42
  - 2.2.11 High-Order Coherences — 44
  - 2.2.12 Indistinguishability — 45
  - 2.2.13 Other Sources — 47

	2.3	Detector Properties		52
		2.3.1	Detection Efficiency	53
		2.3.2	POVM Elements	55
		2.3.3	Photon-Number-Resolving (PNR) Capability	56
		2.3.4	Timing Latency and Rise Time	62
		2.3.5	Timing Jitter	62
		2.3.6	Dead Time, Reset Time, and Recovery Time	64
		2.3.7	Dark Count Rate	65
		2.3.8	Background Count Rate	65
		2.3.9	Afterpulse Probability	65
		2.3.10	Active Area	66
		2.3.11	Operating Temperature of Active Area	66
		References		66
3.	**Photomultiplier Tubes**			**69**
	3.1	Introduction		69
	3.2	Brief History		69
	3.3	Principle of Operation		71
		3.3.1	Photoelectron Emission and Photocathodes	72
		3.3.2	Secondary Emission, Dynodes	73
	3.4	Photon Counting with Photomultipliers		76
	3.5	Conclusion		82
		References		82
4.	**Semiconductor-Based Detectors**			**83**
	4.1	Photon Counting: When and Why		84
	4.2	Why Semiconductor Detectors for Photon Counting?		85
	4.3	Principle of Operation of Single-Photon Avalanche Diodes		85
	4.4	Performance Parameters and Features of SPAD Devices		87
		4.4.1	Photon Detection Efficiency	88
		4.4.2	Dark Count Rate (DCR)	88
		4.4.3	Afterpulsing	89
		4.4.4	Timing Jitter	90
		4.4.5	Crosstalk	92
		4.4.6	Fill-Factor	93
		4.4.7	Microelectronic Structure of a SPAD: Outline and Basic Features	93
	4.5	Circuit Principles for SPAD Operation		94
	4.6	Silicon SPAD Devices		98
		4.6.1	Planar SPAD Devices Fabricated in a Custom Technology	98
		4.6.2	Non-Planar SPAD Devices Fabricated in a Custom Technology	102
		4.6.3	High-Voltage, Complementary Metal-Oxide Semiconductor (HV-CMOS) SPADs	104
		4.6.4	Standard Deep-Submicron CMOS SPADs	106
	4.7	Silicon SPAD Array Detectors		108

Contents

	4.8	SPADS for the Infrared Spectral Range	113
		4.8.1 Infrared SPADs	113
		4.8.2 Basic InGaAs/InP SPAD Design Concepts	114
		4.8.3 DE and DCR Modeling and Performance	115
		4.8.4 Timing Jitter	117
		4.8.5 Afterpulsing	118
		4.8.6 Comparison of InGaAs/InP SPADs and Si SPADs	119
	4.9	Active Gating Techniques for InGaAs SPADs	120
		4.9.1 Introduction	120
		4.9.2 Sampling	122
		4.9.3 Cancellation	123
		4.9.4 Introduction to High-Speed Periodic Gating	125
		4.9.5 Sine-Wave Gating	127
		4.9.6 Self-Differencing	129
		4.9.7 Harmonic Subtraction	131
		4.9.8 Summary	132
	4.10	Future Prospects for Silicon SPADs	134
	4.11	Future Prospects for InGaAs SPADs	135
		References	137

5. Novel Semiconductor Single-Photon Detectors — 147

	5.1	Introduction	147
	5.2	Solid-State Photomultipliers and Visible-Light Photon Counters	148
		5.2.1 Introduction	148
		5.2.2 VLPC Structure and Operation	150
		5.2.3 SSPM and VLPC Performance	154
		5.2.4 Quantitative Model and its Current Limitations	161
		5.2.5 New Opportunities for VLPCs	163
		5.2.6 Conclusions	166
	5.3	Quantum-Dot-Based Detectors	166
		5.3.1 Detector Designs and Principles of Operation	167
		5.3.2 Photon-Number-Resolving Detection	172
		5.3.3 Modeling Photoconductive Gain	175
		5.3.4 Conclusions	179
		References	180

6. Detectors Based on Superconductors — 185

	6.1	Introduction	186
	6.2	Superconducting Nanowire Single-Photon Detectors	187
		6.2.1 Operating Principle	187
		6.2.2 Principal Strengths, Weaknesses	191
		6.2.3 Areas of Research	192
	6.3	Transition-Edge Sensors	194
		6.3.1 Operating Principle	195
		6.3.2 Principal Strengths and Weaknesses	199
		6.3.3 Research Areas	199

	6.4	Superconducting Tunnel Junction Detectors	201
		6.4.1 Operating Principle	201
		6.4.2 Strengths and Weaknesses	204
		6.4.3 Research Areas	204
	6.5	Microwave Kinetic-Inductance Detectors	204
		6.5.1 Operating Principle	205
		6.5.2 Strengths and Weaknesses	206
		6.5.3 Research Areas	207
	6.6	Conclusions and Perspective	208
		References	209

7. Hybrid Detectors — 217

- 7.1 Introduction — 218
- 7.2 Space-Multiplexed Detectors — 219
  - 7.2.1 Introduction — 219
  - 7.2.2 Theory of Operation — 220
  - 7.2.3 Experimental Implementations of Space-Multiplexed Detectors — 231
- 7.3 Time-Multiplexed Detectors — 236
  - 7.3.1 Introduction — 236
  - 7.3.2 Fiber-Loop Detectors — 237
  - 7.3.3 Weak-Homodyne Detection — 241
- 7.4 Up-Conversion Detectors — 243
  - 7.4.1 Introduction — 243
  - 7.4.2 Theory of Single-Photon Up-Conversion — 244
  - 7.4.3 Up-Conversion Techniques — 245
  - 7.4.4 Pulsed Up-Conversion — 249
  - 7.4.5 Ultrafast Up-Conversion — 250
- 7.5 Conclusion — 253
- References — 253

8. Single-Photon Detector Calibration — 257

- 8.1 Introduction — 257
- 8.2 Definitions — 259
- 8.3 Calibration Methods — 260
  - 8.3.1 Radiant Power Measurements (Substitution Method) — 261
  - 8.3.2 Correlated-Photon-Pair Calibration Method — 262
- 8.4 Practical Considerations — 263
  - 8.4.1 Semiconductor Single-Photon Avalanche Diodes — 264
  - 8.4.2 Transition Edge Sensors — 275
- 8.5 Conclusion — 279
- References — 280

9. Quantum Detector Tomography — 283

- 9.1 Introduction — 283
- 9.2 Quantum Tomography: Prelude — 286
  - 9.2.1 State Tomography — 287

		9.2.2	Process Tomography	288
	9.3	Detector Tomography		288
		9.3.1	General Introduction	289
		9.3.2	Photon-Number-Resolving Detectors	291
		9.3.3	Reconstruction without Phase-Sensitivity	293
		9.3.4	Reconstruction with Phase-Sensitivity: the Challenge	295
	9.4	Experimental Implementations of Detector Tomography		297
		9.4.1	Experimental Setup	298
		9.4.2	Q-Function	300
		9.4.3	Reconstructed POVM Elements	301
		9.4.4	Conditioning and Regularization	305
		9.4.5	Robustness of Detector Tomography	307
		9.4.6	Wigner Functions	308
	9.5	Conclusions		310
		References		311

10. The First Single-Photon Sources — 315

	10.1	Introduction		316
	10.2	Feeble Light vs. Single Photon		318
		10.2.1	In Search of Feeble Light's Wave-Like Properties: A Short Historical Review	318
		10.2.2	Quantum Optics in a Nutshell	319
		10.2.3	One-Photon Wavepacket	321
		10.2.4	Quasi-Classical Wavepacket	326
		10.2.5	The Possibility of an Experimental Distinction	328
		10.2.6	Attenuated Continuous Light Beams	329
		10.2.7	Light From a Discharge Lamp	331
		10.2.8	Conclusion: What is Single-Photon Light?	333
	10.3	Photon Pairs as a Resource for Single Photons		334
		10.3.1	Introduction	334
		10.3.2	Non-Classical Properties in an Atomic Cascade	335
		10.3.3	Anticorrelation for a Single Photon on a Beamsplitter	336
		10.3.4	The 1986 Anticorrelation Experiment	339
	10.4	Single-Photon Interferences		344
		10.4.1	Wave-Particle Duality in Textbooks	344
		10.4.2	Interferences with a Single Photon	344
	10.5	Further Developments		346
		10.5.1	Parametric Sources of Photon Pairs	346
		10.5.2	Other Heralded and "On-Demand" Single-Photon Sources	347
		10.5.3	"Delayed-Choice" Single-Photon Interference Experiments	348
		References		348

11. Parametric Down-Conversion — 351

	11.1	Introduction		352
	11.2	Single Photons from PDC: Theory		353
		11.2.1	Classical Description of PDC	354

	11.2.2	Quantum Mechanical Description of PDC	357
	11.2.3	Heralding Single Photons from PDC	360
	11.2.4	Heralding Pure Single-Photon Fock States	362
11.3	Bulk-Crystal PDC		367
	11.3.1	Birefringent Phase-Matching	367
	11.3.2	Heralded Single Photons from Triggered PDC	372
11.4	Periodically-Poled Crystal PDC		379
	11.4.1	Quasi-Phase-Matching	379
	11.4.2	Periodic Poling	383
	11.4.3	Optimal Focus Parameters for Heralding Efficiency	384
	11.4.4	Number Purity	388
	11.4.5	Spectral Purity	390
	11.4.6	Non-Uniform Periodic Poling	391
11.5	Waveguide-Crystal PDC		392
	11.5.1	History and Experimental Implementations	393
	11.5.2	Theory of PDC in Waveguides	394
	11.5.3	Heralding Single Photons from PDC in Waveguides	399
	11.5.4	Electric Field Modes in Waveguides	401
11.6	Comparison of Experimental Single-Photon Sources Using PDC		403
11.7	Overview of the Most Commonly Used Nonlinear Materials and Their Properties		404
11.8	Conclusion		404
	References		404

12. Four-Wave Mixing in Single-Mode Optical Fibers — 411

12.1	Introduction		412
12.2	Photon-Pair Generation in Optical Fibers		413
	12.2.1	Classical Four-Wave Mixing Theory and Phase-Matching Requirements	413
	12.2.2	Quantum Theory of Four-Wave Mixing	416
	12.2.3	Cross-Polarized Four-Wave Mixing in Birefringent Fibers	419
	12.2.4	Raman Scattering	420
12.3	Heralded Single-Photon Sources Based on sFWM		422
	12.3.1	Photon-Pair Generation in the Anomalous Dispersion Regime	425
	12.3.2	Photonic Crystal Fiber Sources in the Normal Dispersion Regime	427
12.4	Quantum Interference Between Separate Spectrally Filtered Fiber Sources		430
12.5	Intrinsically Pure-State Photons		436
	12.5.1	Generation of Spectrally Uncorrelated Two-Photon States Through Group Velocity Matching	436
	12.5.2	A Temporal Filtering Approach for Attaining Pure-State Photons	440
12.6	Entangled Photon-Pair Sources		444

Contents

12.7	Applications of Fiber Photon Sources—All-Fiber Quantum Logic Gates	454
12.8	Photonic Fusion in Fiber	458
12.9	Conclusion	460
	References	461

13. Single Emitters in Isolated Quantum Systems — 467

- 13.1 Introduction — 468
- 13.2 Single Photons from Atoms and Ions - A. Kuhn — 468
  - 13.2.1 Emission into Free Space — 469
  - 13.2.2 Cavity-Based Single-Photon Emitters — 471
  - 13.2.3 Photon Coherence, Amplitude, and Phase Control — 485
- 13.3 Single Photons from Semiconductor Quantum Dots - G. S. Solomon — 492
  - 13.3.1 Introduction — 492
  - 13.3.2 InAs-Based Quantum-Dot Formation — 493
  - 13.3.3 Exciton Energetics — 494
  - 13.3.4 Optically Accessing Single Quantum Dots — 497
  - 13.3.5 Single Photons From Single Quantum Dots — 499
  - 13.3.6 Weak QD-Cavity Coupling — 502
  - 13.3.7 Quantum-Dot Photon Indistinguishability — 505
- 13.4 Single Defects in Diamond - C. Santori — 511
  - 13.4.1 Introduction — 511
  - 13.4.2 The Nitrogen-Vacancy Center — 511
  - 13.4.3 Other Defects — 521
  - 13.4.4 Optical Structures in Diamond — 522
  - 13.4.5 Quantum Communication — 525
  - 13.4.6 Summary — 526
- 13.5 Future Directions — 526
- References — 527

14. Generation and Storage of Single Photons in Collectively Excited Atomic Ensembles — 541

- 14.1 Introduction — 541
- 14.2 Basic Concepts — 543
- 14.3 From Heralded to Deterministic Single-Photon Sources — 545
- 14.4 Interference of Photons from Independent Sources — 550
- 14.5 Conclusion and Outlook — 555
- Appendix — 556
  - A Write Process — 556
  - B Read Process — 559
- References — 560

Index — 563

Contributors

Numbers in parentheses indicate the pages on which the authors' contributions begin.

Alain Aspect (315), Laboratoire Charles Fabry, Institut d'Optique, CNRS, Univ Paris-Sud, 2 Avenue Augustin Fresnel, 91127 Palaiseau, France

Bryn Bell (411), Photonics Group, Merchant Venturers School of Engineering, University of Bristol, Bristol, BS8 1UB, UK

Karl K. Berggren (189), Department of Electrical Engineering and Computer Science, Massachusetts Institute of Technology, 77 Massachusetts Avenue, Cambridge, Massachusetts 02139, USA

Joshua C. Bienfang (1, 83), Joint Quantum Institute, University of Maryland and National Institute of Standards and Technology, Gaithersburg, MD 20899, USA

Alex Clark (411), Centre for Ultrahigh Bandwidth Devices for Optical Systems (CUDOS), Institute of Photonics and Optical Science (IPOS), School of Physics, University of Sydney, NSW 2006, Australia

Andreas Christ (351), Applied Physics, University of Paderborn, Warburger Straße 100, 33098 Paderborn, Germany

Hendrik B. Coldenstrodt-Ronge (217, 283), University of Oxford, Clarendon Laboratory, Parks Road, Oxford, OX1 3PU, United Kingdom

Sergio Cova (83), Politecnico di Milano, Dipartimento di Elettronica, Informazione e Bioingegneria, Piazza Leonardo da Vinci, 32 20133 Milano, Italy

Animesh Datta (283), Clarendon Laboratory, Department of Physics, University of Oxford, OX1 3PU, United Kingdom

Eric A. Dauler (185), Lincoln Laboratory, Massachusetts Institute of Technology, 244 Wood Street, Lexington, Massachusetts 02420, USA

Ivo Pietro Degiovanni (217), I.N.RI.M., Strada delle Cacce 91, 10135 Turin, Italy

Jingyun Fan (1), Joint Quantum Institute, University of Maryland and National Institute of Standards and Technology, Gaithersburg, MD 20899, USA

Alessandro Fedrizzi (351), Centre for Engineered Quantum Systems, Centre for Quantum Computer and Communication Technology, School of Mathematics and Physics, University of Queensland, Brisbane 4072, Australia

Eric J. Gansen (147), Department of Physics, University of Wisconsin-La Crosse, La Crosse, WI 54601, USA

Massimo Ghioni (83), Politecnico di Milano, Dipartimento di Elettronica, Informazione e Bioingegneria, Piazza Leonardo da Vinci, 32 20133 Milano, Italy

Philippe Grangier (315), Laboratoire Charles Fabry, Institut d'Optique, CNRS, Univ Paris-Sud, 2 Avenue Augustin Fresnel, 91127 Palaiseau, France

Yu-Ping Huang (411), Center for Photonic Communication and Computing, Department of Electrical Engineering and Computer Science & Department of Physics and Astronomy, Northwestern University, 2145 Sheridan Road, Evanston, Illinois 60208, USA

Hannes Hübel (351), Fysikum, University of Stockholm, Roslagstullsbacken 21, 10691 Stockholm, Sweden

Mark A. Itzler (83), Princeton Lightwave, Inc., 2555 US Route 130 S., Cranbury, NJ 08512, USA

Thomas Jennewein (351), Institute for Quantum Computing and Department of Physics & Astronomy, University of Waterloo, Waterloo, Canada N2L 3G1

Andrew J. Kerman (185), Lincoln Laboratory, Massachusetts Institute of Technology, 244 Wood Street, Lexington, Massachusetts 02420, USA

Jungsang Kim (147), Department of Electrical and Computer Engineering, Physics and Computer Science, Duke University, Durham, NC 27708, USA

Axel Kuhn (467), University of Oxford, Clarendon Laboratory, Parks Road, OX1 3PU, United Kingdom

Prem Kumar (411), Center for Photonic Communication and Computing, Department of Electrical Engineering and Computer Science & Department of Physics and Astronomy, Northwestern University, 2145 Sheridan Road, Evanston, Illinois 60208, USA

Paul G. Kwiat (147), Department of Physics, University of Illinois at Urbana-Champaign, Urbana, IL 61801, USA

Kyle S. McKay (147), National Institute of Standards and Technology, Boulder, CO 80303, USA

Alex McMillan (411), Photonics Group, Merchant Venturers School of Engineering, University of Bristol, Bristol, BS8 1UB, UK

Alan Migdall (1, 217), Joint Quantum Institute and National Institute of Standards and Technology, 100 Bureau Dr, Stop 8410, Gaithersburg, MD 20899, USA

Sae-Woo Nam (185), National Institute of Standards and Technology, 325 Broadway, Boulder, Colorado 80305, USA

Jian-Wei Pan (541), Hefei National Laboratory for Physical Sciences at Microscale and Department of Modern Physics, University of Science and Technology of China, Hefei, Anhui, 230026, PR China

Contributors

Sergey V. Polyakov (1, 69, 217, 257), National Institute of Standards and Technology, Gaithersburg, MD 20899, USA

John Rarity (411), Photonics Group, Merchant Venturers School of Engineering, University of Bristol, Bristol, BS8 1UB, UK

Alessandro Restelli (83), Joint Quantum Institute, University of Maryland, and The National Institute of Standards and Technology, Gaithersburg, MD 20899, USA

Danna Rosenberg (185), Lincoln Laboratory, Massachusetts Institute of Technology, 244 Wood Street, Lexington, Massachusetts 02420, USA

Charles Santori (467), Hewlett-Packard Laboratories, 1501 Page Mill Rd., Palo Alto, CA, 94304, USA

Christine Silberhorn (351), Applied Physics, University of Paderborn, Warburger Straße 100, 33098 Paderborn, Germany

Glenn S. Solomon (467), Joint Quantum Institute, National Institute of Standards and Technology, University of Maryland, Gaithersburg, MD, USA

Martin J. Stevens (25), National Institute of Standards and Technology, 325 Broadway, Boulder, CO 80305, USA

Ian A. Walmsley (217, 283), Clarendon Laboratory, Department of Physics, University of Oxford, Parks Road, Oxford, OX1 3PU, United Kingdom

Franco N. C. Wong (217), Massachusetts Institute of Technology, Research Laboratory of Electronics, Cambridge, MA 02139, USA

Lijian Zhang (283), Clarendon Laboratory, Department of Physics, University of Oxford, OX1 3PU, United Kingdom

Bo Zhao (541), Hefei National Laboratory for Physical Sciences at Microscale and Department of Modern Physics, University of Science and Technology of China, Hefei, Anhui, 230026, PR China

Kevin Zielnicki (147), Department of Physics, University of Illinois at Urbana-Champaign, Urbana, IL 61801, USA

Volumes in series

Editors-in-Chief
Thomas Lucatorto, Albert C. Parr and Kenneth Baldwin

Volume 1. Classical Methods
Edited by Immanuel Estermann

Volume 2. Electronic Methods, Second Edition (in two parts)
Edited by E. Bleuler and R. O. Haxby

Volume 3. Molecular Physics, Second Edition (in two parts)
Edited by Dudley Williams

Volume 4. Atomic and Electron Physics - Part A: Atomic Sources and Detectors; Part B: Free Atoms
Edited by Vernon W. Hughes and Howard L. Schultz

Volume 5. Nuclear Physics (in two parts)
Edited by Luke C. L. Yuan and Chien-Shiung Wu

Volume 6. Solid State Physics - Part A: Preparation, Structure, Mechanical and Thermal Properties; Part B: Electrical, Magnetic and Optical Properties
Edited by K. Lark-Horovitz and Vivian A. Johnson

Volume 7. Atomic and Electron Physics - Atomic Interactions (in two parts)
Edited by Benjamin Bederson and Wade L. Fite

Volume 8. Problems and Solutions for Students
Edited by L. Marton and W. F. Hornyak

Volume 9. Plasma Physics (in two parts)
Edited by Hans R. Griem and Ralph H. Lovberg

Volume 10. Physical Principles of Far-Infrared Radiation
Edited by L. C. Robinson

Volume 11. Solid State Physics
Edited by R. V. Coleman

Volume 12. Astrophysics - Part A: Optical and Infrared Astronomy
Edited by N. Carleton

Part B: Radio Telescopes; Part C: Radio Observations
Edited by M. L. Meeks

Volume 13. Spectroscopy (in two parts)
Edited by Dudley Williams

Volume 14. Vacuum Physics and Technology
Edited by G. L. Weissler and R. W. Carlson

Volume 15. Quantum Electronics (in two parts)
Edited by C. L. Tang

Volume 16. Polymers - Part A: Molecular Structure and Dynamics; Part B: Crystal Structure and Morphology; Part C: Physical Properties
Edited by R. A. Fava

Volume 17. Accelerators in Atomic Physics
Edited by P. Richard

Volume 18. Fluid Dynamics (in two parts)
Edited by R. J. Emrich

Volume 19. Ultrasonics
Edited by Peter D. Edmonds

Volume 20. Biophysics
Edited by Gerald Ehrenstein and Harold Lecar

Volume 21. Solid State Physics: Nuclear Methods
Edited by J. N. Mundy, S. J. Rothman, M. J. Fluss, and L. C. Smedskjaer

Volume 22. Solid State Physics: Surfaces
Edited by Robert L. Park and Max G. Lagally

Volume 23. Neutron Scattering (in three parts)
Edited by K. Skold and D. L. Price

Volume 24. Geophysics - Part A: Laboratory Measurements; Part B: Field Measurements
Edited by C. G. Sammis and T. L. Henyey

Volume 25. Geometrical and Instrumental Optics
Edited by Daniel Malacara

Volume 26. Physical Optics and Light Measurements
Edited by Daniel Malacara

Volume 27. Scanning Tunneling Microscopy
Edited by Joseph Stroscio and William Kaiser

Volume 28. Statistical Methods for Physical Science
Edited by John L. Stanford and Stephen B. Vardaman

Volume 29. Atomic, Molecular, and Optical Physics - Part A: Charged Particles; Part B: Atoms and Molecules; Part C: Electromagnetic Radiation
Edited by F. B. Dunning and Randall G. Hulet

Volume 30. Laser Ablation and Desorption
Edited by John C. Miller and Richard F. Haglund, Jr.

Volume 31. Vacuum Ultraviolet Spectroscopy I
Edited by J. A. R. Samson and D. L. Ederer

Volume 32. Vacuum Ultraviolet Spectroscopy II
Edited by J. A. R. Samson and D. L. Ederer

Volume 33. Cumulative Author Index and Tables of Contents, Volumes 1–32

Volume 34. Cumulative Subject Index

Volume 35. Methods in the Physics of Porous Media
Edited by Po-zen Wong

Volume 36. Magnetic Imaging and its Applications to Materials
Edited by Marc De Graef and Yimei Zhu

Volume 37. Characterization of Amorphous and Crystalline Rough Surface: Principles and Applications
Edited by Yi Ping Zhao, Gwo-Ching Wang, and Toh-Ming Lu

Volume 38. Advances in Surface Science
Edited by Hari Singh Nalwa

Volume 39. Modern Acoustical Techniques for the Measurement of Mechanical Properties
Edited by Moises Levy, Henry E. Bass, and Richard Stern

Volume 40. Cavity-Enhanced Spectroscopies
Edited by Roger D. van Zee and J. Patrick Looney

Volume 41. Optical Radiometry
Edited by A. C. Parr, R. U. Datla, and J. L. Gardner

Volume 42. Radiometric Temperature Measurements. I. Fundamentals
Edited by Z. M. Zhang, B. K. Tsai, and G. Machin

Volume 43. Radiometric Temperature Measurements. II. Applications
Edited by Z. M. Zhang, B. K. Tsai, and G. Machin

Volume 44. Neutron Scattering
Edited by Price and Fernandez-Alonso

Volume 45. Single-Photon Generation and Detection
Edited by Alan Migdall, Sergey Polyakov, Jingyun Fan, and Joshua Bienfang

Preface

Single-Photon Generation and Detection: Physics and Applications

Single-photon generation and detection is at the forefront of modern optical physics research. This book is intended to provide a comprehensive overview of the current status of single-photon techniques and research methods in the spectral region from the visible to the infrared.

The use of single photons, produced on demand with well defined quantum properties, offers an unprecedented set of capabilities that are central to the new area of quantum information, and are of revolutionary importance in areas that range from the traditional, such as high sensitivity detection for astronomy, remote sensing, and medical diagnostics, to the exotic, such as secretive surveillance and very long communication links for data transmission on interplanetary missions. Even some routine applications such as ellipsometry or refractive birefringence measurements can, by making use of a non-classical source's polarization correlations, be enhanced with these quantum sources and detectors.

One of the earliest examples in modern physics of the observable effect of the photon nature of light occurred in Hertz's experiments on the photoelectric effect, an effect that played a key role in Einstein's explanation of the nature of light in terms of Planck's quantum of energy. The photoelectric effect also played a key role in the development of the photomultiplier, a device now commonly used for single-photon detection over a broad range of energies. While people had been detecting energetic single particles and single photons in the X-ray and gamma-ray regions since close to the beginning of the 20th century using the Geiger-Muller tube, the detection of single photons at visible or lower energies was achieved only with the advent of the photomultiplier in the late 1930s. It is this much lower photon energy range, i.e., the visible and infrared, that will be the focus of this book.

Efforts to develop single-photon sources in the visible range go back to the use of two-photon atomic cascades in the 1950s in radiometry and lifetime measurements. These sources employed the two-step decay of a selected excited atomic state, with the first decay photon acting as a signal announcing the arrival of the second. Thus while the time of emission was not well defined, the pair emission allowed one photon to herald the existence of the other, effectively creating a single-photon source. (Later source development focused on this issue of controlling the time of emission.) Perhaps the most famous applications of such a cascade source occurred several decades later in the EPR experiments of Aspect *et al.*, in which the cascade was used to produce entanglement, a uniquely quantum mechanical property much beyond simple photon correlations, to explore our most fundamental concepts of reality.

There are presently four research areas in which advances in the capabilities and applications of single-photon sources and detectors are of high interest: (a) quantum communication, in which the performance of sources and detectors directly impacts the generation of, for example, secure cryptographic key, imposing stringent demands on source and detector development; (b) quantum computation, a technology offering exponential improvements over classical computing, but where requirements for photonic devices go well beyond what is currently available. Current challenges in this area include on-demand single-photon sources at specific narrow-band frequencies and detectors with very high count rates and very high efficiencies (e.g., 0.999); (c) metrology, where nonclassical control of light offers resolution beyond the classically allowable limits; and (d) fundamental tests of nature, where exquisitely controlled photonic signals are the key to making these schemes work. These fundamental tests include those connected with alternative theories to quantum mechanics and the search for gravitational waves.

The goal of this volume will be to provide a researcher with a comprehensive overview of single-photon science and technology to enable and enhance the design of the wide range of experimental research that makes use of these technologies, and to serve as a point of departure for those beginning experimental research in single- and correlated-photon-based science. The book will be broken into chapters focused specifically on the development and capabilities of the available detectors and sources to allow a comparative understanding to be developed by the reader along with an idea of how the field is progressing and what can be expected in the near future. With each technology, we will also provide a survey of the primary (and potential) applications that drive its development. We intend to make this the go-to reference for this field.

Chapter 1

Introduction

J.C. Bienfang, J. Fan, A. Migdall and S.V. Polyakov
Joint Quantum Institute, University of Maryland and National Institute of Standards and Technology, Gaithersburg, MD 20899, USA

Chapter Outline
1.1 Physics of Light—an Historical Perspective   1
1.2 Quantum Light   2
    1.2.1 What Is Non-Classical Light?   2
    1.2.2 What Is a Photon?   3
1.3 The Development of Single-Photon Technologies   4
1.4 Some Applications of Single-Photon Technology   8
1.5 This Book   9
    1.5.1 Single-Photon Detectors   9
    1.5.2 Single-Photon Sources   16
1.6 Conclusions   17
References   18

All these fifty years of conscious brooding have brought me no nearer to the answer to the question, 'What are light quanta?' Nowadays every Tom, Dick and Harry thinks he knows it, but he is mistaken.

Albert Einstein, 1954 [1]

1.1 PHYSICS OF LIGHT—AN HISTORICAL PERSPECTIVE

In the beginning there was light. And it was good. Not long thereafter people began to look for a comprehensive understanding of its nature. While the publication record starts off a little spotty, in the fifth century BC the Greek philosopher Empedocles concluded that light consists of rays that emanate from the eye. This point of view was questioned using what today might be characterized as a local realism argument by Euclid in his classic text on light propagation *Optica*. Euclid hypothesized that rays of light are emitted by external sources, but it was not until Ibn al-Haytham in 1000 AD that this view was put on a scientific footing.

The character of light itself was described by Descartes in the 17th century as "pressure" that was transmitted through space from a source to a detector. This idea was later developed by Huygens and Hooke into the wave theory of light. At about the same time, Gassendi put forward the contravening notion that light was a particle, an idea embraced and developed further by Newton. The differing perspectives of light as a particle versus a wave were generally considered resolved in favor of the wave picture by Young's double-slit experiment in 1803, and by Fresnel's experiments in diffraction. In the 1860s, further confirmation of this conclusion was framed in an elegant and deeply satisfying manner by the Maxwell equations: the prediction of polarized electromagnetic waves that propagate at what was understood to be the speed of light.

Problems with the waves-and-fluids view of electromagnetism arose in 1897, when J.J. Thomson discovered discrete particles carrying negative electric charge moving through vacuum. Then in 1900, in "an act of despair" Planck invoked quantized bundles of electromagnetic energy in the derivation of the blackbody radiation law [2,3], a step that not only embraced prior conjectures by Boltzmann in statistical mechanics, but also flew in the face of conventional understanding. It was originally considered an artifact of the derivation to be corrected later, but Einstein took the idea of light quanta more seriously in his 1905 description of the photoelectric effect [4]. Then in 1913 Bohr invoked the quantization of both energy and angular momentum to explain the discrete spectral emission lines observed in the Hydrogen-Balmer series. The wheels came completely off the wagon in 1924 when de Broglie hypothesized that not only light, but also particles of matter have wave-like properties. A flurry of subsequent discovery and advance that established the framework of quantum mechanics, most notably by Heisenberg, Born, Schrodinger, Pauli, and Dirac, culminated for the purposes of this book in 1927 when Dirac quantized the electromagnetic field, effectively developing a theory of light that encompassed the physical phenomenon that kicked off the entire revolution in the first place. The first direct detection of single photons was achieved in the 1930s. The atomic-cascade photon-pair source [5] developed in the 1950s and its use in the 1970s and 1980s [6–9] represents the first single-photon source. Then there was quantum light, and it was really good.

1.2 QUANTUM LIGHT

1.2.1 What is Non-Classical Light?

Before going too deeply into the details of single-photon technologies it is useful to at least provide a basic definition of what is meant by the term "quantum light." Quantum light, or "non-classical light," describes the broad class of states that cannot be emitted by "classical" sources such as discharge lamps or lasers.

Formally, the distinction between non-classical and classical light can be defined by writing the state in the Glauber-Sudarshan representation, in which the state is expanded in the basis of coherent states, weighted by a quasiprobability distribution [10]. If the quasiprobability is positive and bounded, then the light is considered classical, otherwise it is non-classical. Examples of classical light include thermal light emitted from a blackbody source, and coherent light emitted from a laser, while non-classical light includes squeezed states, photon-number (or Fock) states generally, and single-photon states specifically.

1.2.2 What is a Photon?

A photon is defined as an elementary excitation of a single mode of the quantized electromagnetic field [11]. The term "photon" was first introduced by Lewis in 1926 [12]. A mode k of the quantized electromagnetic field is labeled by its frequency v_k, and a single photon in that mode has energy equal to hv_k, where h is the Planck constant. While this monochromatic definition of a photon implies delocalization in time, it is common to talk about propagating "single-photon states" that are localized to some degree in time and space. Mathematically, such states can be described as superpositions of monochromatic modes [11]. While there is some discussion in the literature about the definition of a "photon wavefunction" [15], here we adopt the following operational definition of a single-photon state: given a detector that can determine the number of incident photons (in some finite-width frequency range) with 100% accuracy, a single-photon state is an excitation of the electromagnetic field (localized to some degree in both space and time) such that the detector measures exactly one photon for each incident state. In other words, a single-photon state is one for which the photon-number statistics have a mean value of one and a variance of zero. It should be noted that since the results of quantum measurements may depend on the procedure and apparatus used, the physics of the measurement process itself should also be considered [16].

It is also worth noting the distinction between single-photon sources and classical sources of light in terms of photon statistics. While the distinction between non-classical and classical light is defined formally, a qualitative description is that the statistical variation of classical sources is at least that of a Poisson process, while emission from a single-photon source has lower variance. This is easily understood from the fact that a single-photon source is, by definition, unable to emit a second photon for some time following a previous photon (antibunching), and this enforces some degree of order to the source's output. In contrast, a classical thermal source has an increased likelihood of emitting additional photons near an existing one (bunching), and this tendency goes hand-in-hand with increased statistical fluctuations. The low-noise nature of single-photon sources is an obvious advantage in many measurement situations.

1.3 THE DEVELOPMENT OF SINGLE-PHOTON TECHNOLOGIES

The investigation of quantum mechanics for the advance of mankind's understanding and technological capability remains one of the prime motivators of current research in the physical sciences. Of the many strange properties of quantum mechanics, those relating to the so-called coherent quantum effects take the widest departure from our daily (macroscopic) experience: the possibility for physical objects to appear to simultaneously hold mutually exclusive properties is a confounding fact of the natural world. This capability has the potential to be an extremely powerful technological tool, and the exploitation of coherent quantum effects is at the frontier of research. And yet, the feature that conveys these wondrous capabilities is also their primary impediment: coherence in quantum systems is difficult to preserve in the presence of interactions with the environment. Thus, while quantum effects are studied in a wide array of media, optical quantum-mechanical phenomena are particularly robust and accessible in an experimental setting, which makes them among the most widely investigated.

Today, single-photon technologies are an area of intense and sustained interest. They represent a bridge between the "classical" world of our daily experience and the quantum realm where we may access the extraordinary phenomena therein. Single-photon technologies are critical for Bell tests probing fundamental questions about the nature of reality, and to research and development in quantum information science. One emergent example in this field is the generation of verifiable random numbers [18–20]. In a more prosaic but no less important sense, single-photon technologies operate at the fundamental limit of electromagnetic signal strength, and are thus used to make the most sensitive measurements in a wide range of applications, notably: astrophysics, molecular biology, health and safety monitoring, environmental sensing, and imaging. Given the broad scope and magnitude of these applications, it is not surprising that while some form of single-photon technology has been available for nearly 80 years [21–23], the literature record shows continued and robust activity in this field; a database search for papers on "single-photon detection" shows sustained growth over the past four decades, and a similar growth, although one that started more recently, in papers on "single-photon sources," as can be seen in (Fig. 1.1). We also see that interest in single-photon sources coincided with the advent of quantum cryptography, and while that field may appear to be maturing, the growth in interest in single-photon measurement and quantum-enabled metrology appears to be robust.

Much of the research in single-photon technologies actually focuses on states comprised of two photons. While at first blush this may seem counterintuitive, correlated-photon pairs offer a wealth of possibilities in single-photon technologies. Although the generation of photon pairs is typically governed by a spontaneous random process, as in a classical light source,

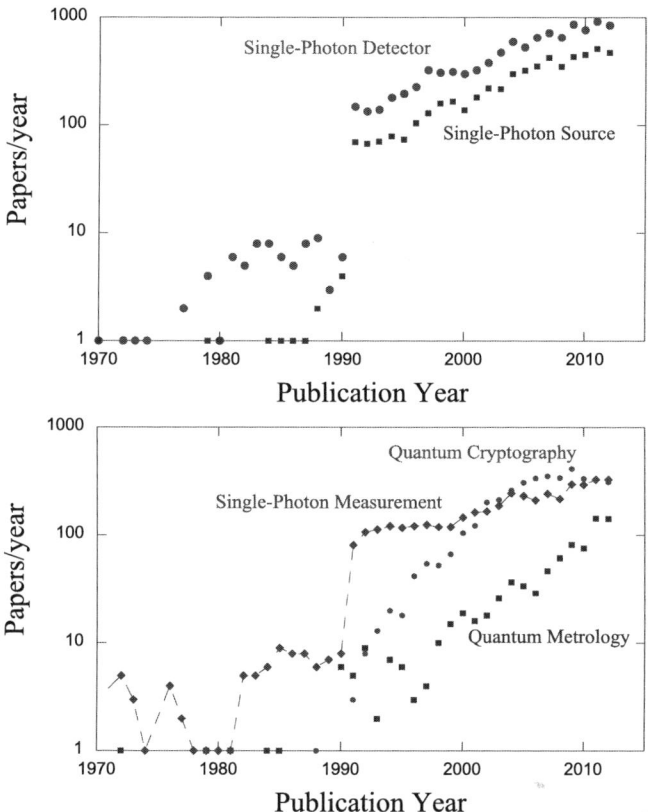

FIGURE 1.1 Growth of the field of single-photon source and detector technology as seen by the number of papers published each year using the indicated search terms in the publication database Web of Science [17]. Search terms: (a) "single-photon source" and "single-photon detector" and (b) "single-photon measurement," "quantum cryptography," and "quantum metrology." *Note:* The jump in the data at 1990 is an artifact due to the database methodology rather than an indication of a real change in the publication rates.

the pair-wise correlation of the two photons themselves is governed by strict conservation laws in a quantum-mechanical process. This fact allows one to produce light with non-classical properties by detecting one photon of the pair to indicate the existence of its correlated partner: a heralded single-photon state [24,25]. By extension, correlated pairs can be used as the basis for a primary standard technique [24,26] for calibrating single-photon detectors because they allow one to know when one and only one photon was incident on a detector under examination. Furthermore, such bipartite states provide a convenient way to study coherent quantum effects (e.g. entanglement) and have a variety of applications in quantum communication.

Figure 1.2 tells a nice story of the advance of photon-pair technology over the past five decades in terms of the detected-pair rate. The first pair sources were based on atomic cascades (c.f. Chapter 10), which improved from a few

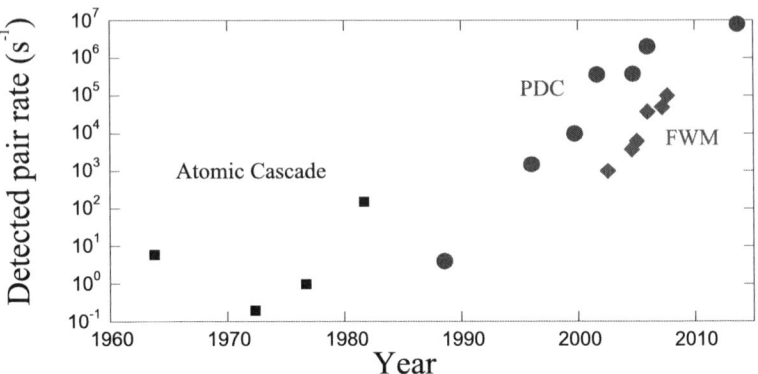

FIGURE 1.2 Detected photon-pair-rate progress. Detected photon-pair rates shown for atomic-cascade- (black squares), parametric-down-conversion- (blues dots), and fiber-four-wave-mixing-based sources (red diamonds) [6,7,9,27–35]. (For interpretation of the references to color in this figure legend, the reader is referred to the web version of this book.)

detected pairs/s in the 1960s to 150 pairs/s in the 1980s, and led to what is probably the most famous quantum-optics experiment to date: the test of Bell's theorem by Aspect et al. [9]. But those sources inherently emit into 4π steradians, meaning that the efficiency with which the emitted photons can be collected is severely limited. In contrast, the energy and momentum constraints governing the production of photon pairs in spontaneous parametric down-conversion (PDC) serve to concentrate the emission into a narrow range of angles, allowing for significantly more efficient collection of the pairs (c.f. Chapter 11). Indeed, although in the very first demonstration that PDC consisted of correlated pairs of photons, by Burham and Weinberg [36], the detection rate of ≈ 1 s^{-1} was actually lower than that of Aspect et al., the ratio of coincidence-to-accidentals (a measure of the overall collection efficiency, among other things) was more than 10 times higher. From that point there was exponential growth in the maximum reported detected-pair rate in PDC-based sources [33] until the late 2000s, where progress started to slow. Here it is important to note that Fig. 1.2 describes progress in both sources and single-photon detectors, and the apparent stall in progress in detected-pair rates was mostly due to detector limitations. By that time most photon-pair experiments used Si single-photon avalanche diodes (SPADs) for single-photon detection (c.f. Chapter 4). Initially SPADs were used in a passively quenched mode with a maximum count rate of $\approx 10^5$ s^{-1}, later they were used in the actively quenched mode, which allowed count rates up to $\approx 10^6$ s^{-1}, as can be seen from the graph, and recent reports make use of multiple actively quenched SPADs to achieve unprecedented detected-pair rates of $\approx 10^7$ s^{-1} [35]. In the past few years new detection technologies, such as multi-element and fast-gated SPADs (Chapters 4 and 7) and superconducting nanowire detectors (Chapter 6) capable of operating at rates approaching $\approx 10^9$ s^{-1} have been reported, and it is reasonable to expect corresponding advances in the detected-pair rate.

As mentioned above, collection efficiency continues to play, a critical limiting role in the overall performance of a correlated-pair source. Sources in which the emission is constrained to a well-defined spatial mode, as in four-wave-mixing (FWM) in single-mode fiber, have the potential to advance detected-pair rates even farther (Chapter 12). Figure 1.2 also shows exponential growth in rates from FWM sources, and these ideas have promoted the recent development of PDC in single-mode non-linear waveguides for correlated-pair generation (Chapter 11).

Heralded single-photon sources based on correlated photons are inherently probabilistic, and therefore have limited scalability in the context of real-world applications. Ideally, the next generation of sources would be deterministic on-demand sources that can produce a photon whenever required, and the most obvious approach is to use a single quantum emitter (e.g. a single two- or many-level system). Though seemingly straightforward, such sources are actually extremely challenging to implement. Using single atoms is a natural choice for such a system, but to address the collection problem seen by the original atom-based sources high-finesse cavities must be used to enhance the collection efficiency, and it is very difficult to maintain the coupling between a single emitter and the cavity. Nonetheless there have been a number of successes with both neutral alkali atoms, such as Cs and Rb [37–42], and Ca+ ions [43–45]. Solid-state single-photon emitters, such as quantum dots and color centers, are also attractive candidates. These systems are confined to a host material that can be used to form the optical cavity. For example, distributed-Bragg-reflection (DBR) mirrors on both sides of quantum dots can be used to define the emission mode, and these mirrors can be grown together with the dots. In more sophisticated designs, dots are integrated in pillars, micro-disks, or photonic-crystal cavities [46–52]. There are also efforts to build optical cavities around nitrogen-vacancy color centers in diamond [53,54] to improve coupling efficiency (Chapter 13). A common drawback of solid-state single-photon sources is their high rate of decoherence due to interactions between the emitter with host material. Current research is aimed at understanding and mitigating these effects [55].

Recently, the idea of using ensembles of emitters for generating single photons deterministically has been advanced (Chapter 14). In this scheme a collective excitation is generated in an ensemble of atoms, and later emitted when a read-out signal is applied to the ensemble, effectively acting as a quantum memory [56–59]. The use of an ensemble significantly enhances coupling of light to the emitters, meaning that an optical cavity is no longer necessary, while the deterministic character of the emission is preserved. Also note that a quantum memory used in conjunction with any photon-pair source is another path to deterministic single-photon generation [60–64]. The main distinction is whether the single photon is created externally and then stored in the memory, or created within the memory.

1.4 SOME APPLICATIONS OF SINGLE-PHOTON TECHNOLOGY

There are myriad applications that are enabled by, or that benefit from single-photon technologies. Currently quantum information is a particularly hot area, where examples of applications enabled by single-photon technology are quantum cryptography protocols [65–69], quantum communication and specifically quantum repeaters [70–74], and certain quantum computation protocols [75]. All these applications are being actively researched, in turn providing new and sometimes fundamental challenges for single-photon technology research. Clear demonstrations of how the performance of single-photon components affect quantum computation can be seen in recent work on boson sampling [76–79], an application in the field of linear-optics quantum computation. Another area of experimental research that relies on single-photon technology is the development of quantum receivers that can unambiguously discriminate between non-orthogonal states at error rates below the standard quantum limit, a fundamental measurement problem [80–82]. The latter is an example of a measurement that is not possible with a toolbox limited only to the classical world.

Single-photon technologies are also critical to fundamental tests of quantum mechanics that seek to determine whether or not non-local realism is absolutely required. Current efforts build on the Bell tests first implemented by Aspect, *et al.* [9] with atomic-cascade-based single-photon sources. However, the experimental implementation of tests of this type can have a number of so-called loopholes that may fail to conclusively exclude the possibility of alternate theories that offer a different perspective on the world than that suggested by quantum mechanics [83]. Currently there are intense efforts to close the remaining loopholes in such experiments, and this requires pushing the performance of single-photon devices to unprecedented levels [84–88]. Short of closing all the possible loopholes in a Bell test, other experimental tests can be performed to verify (or falsify) limited sub-classes of alternative theories, see for example [89]. Related to this are the emerging efforts to use the non-local realism of quantum mechanics and its indeterminism as a source of certifiable randomness [18–20], which may enable a whole new array of security related applications [90–94].

In addition to quantum-information, single-photon detectors are used for a wide variety of applications, including DNA sequencing [95–98], bioluminescence characterization [99], Förster resonance energy transfer for protein folding observation [100–102], light detection and ranging for remote sensing [103,104], and light ranging on shorter scales [105], optical time-domain reflectometry [106–112], picosecond imaging circuit analysis (PICA) [113–118], single-molecule spectroscopy [119–125] and fluorescence-lifetime measurements [126], medical applications such as diffuse optical tomography [127] and positron emission tomography [128], and finally, single-photon metrology [26,129–136].

1.5 THIS BOOK

This book attempts to provide a comprehensive overview of the current state of technology and techniques that are available to facilitate and advance the design of experiments involving single photons. The book is broken into chapters focused on the design, performance, and ongoing research of available detectors and sources, grouped by their underlying physical principles.

Most chapters were written by active researchers who have contributed significantly to the field. To bridge gaps between the sub-fields, common concepts were identified. Chapter 2 introduces those basic properties and measurement metrics that are generally accepted for the characterization of single-photon sources and single-photon detectors. The discussion begins with the statistics of light and presents how those statistics are reflected in the measurements of single-photon sources and detectors. This allows a comparative understanding to be developed, along with an impression of how the field is progressing and what may be expected in the near future. The rest of the book is divided into two parts: Chapters 3 through 9 describe single-photon detectors, and the ways to characterize them, and Chapters 10 through 14 present various types of sources.

1.5.1 Single-Photon Detectors

Single-photon detectors were invented approximately 80 years ago with the advent of the photomultiplier tube (PMT), and ever since then detectors have been an active area of research and development. While there are a limited number of material systems used for single-photon detection, generally: photomultipliers, semiconductors, and superconductors, within each of these families are nearly innumerable variations designed to enhance various properties. Of course, the ideal detector has 100 % detection efficiency, photon number-resolving (PNR) capability, arbitrarily accurate timing resolution, and no detrimental effects such as saturation, dark counts, etc. In fact, each of these ideal properties is nearly available in one type of detector or another, but not all in one detector at the same time. Unfortunately we are in little danger of ever achieving such a perfect detector. In reality, a detector represents a set of performance tradeoffs that must be selected based on the needs of a particular application. To aid with detector selection, Table 1.1 presents a survey of reported detector performance, and some relevant parameters. Refer to Chapter 2 for common definitions relevant to single-photon detectors.

TABLE 1.1 Comparison of Single-Photon Detectors [17] Based on Tables from [191,192] Using a Figure of Merit Given by the Ratio of the Detection Efficiency to the Product of the Dark-Count Rate and the Time Resolution (Assumed to be the Timing Jitter), $\eta/(D\delta t)$

Detector Type	Operation Temp. (K)	Detection Efficiency, Wavelength η (%), λ (nm)	Timing Jitter, δt (ns) (FWHM)	Dark-count Rate, D (ungated) (1/s)	Figure of Merit $\eta/(D\delta t)$	Max. Count Rate (10^6/s)	PNR Capability
PMT (visible–near-infrared) [137]	300	40 @ 500	0.3	100	1.3×10^7	10	some
PMT (infrared) [138]	200	2 @ 1550	0.3	200000	3.3×10^2	10	some
Si SPAD (thick junction) [139]	250	65 @ 650	0.4	25	6.5×10^7	10	none
Si SPAD (shallow junction) [140]	250	49 @ 550	0.035	25	5.6×10^8	10	none
Si SPAD (self-differencing) [141]	250	74 @ 600	–	2000	–	16	some
Si SPAD (linear-mode) [142]	78	56 @ 450	–	0.0008	–	0.01	full*
Si SPAD (cavity) [143]	78	42 @ 780	0.035	3500	3.4×10^6	10	none
Si SPAD (multipixel) [144,145]	290	40 @ 532	0.3	25000-500000	1×10^4	30	some
Hybrid PMT (PMT + APD) [146,147]	270	30 @ 1064	0.2	30000	5×10^4	200	none
Time multiplexed (Si SPAD) [148]	250	39 @ 680	0.4	200	5×10^6	0.5	some
Time multiplexed (Si SPAD) [149]	250	50 @ 825	0.5	150	7×10^6	2	some
Space multiplexed (InGaAs/InP SPAD) [150]	250	33 @ 1060	0.133	160000000	1.6×10^1	10	some
Space multiplexed (InGaAs/InP SPAD) [151]	250	2 @ 1550	–	–	–	0.3	none
InGaAs/InP SPAD (gated) [152]	220	20 @ 1545; 38 @ 1308	–	3300	–	5	none
InGaAs/InP SPAD (gated) [153]	200	10 @ 1550	0.370	91	3.0×10^5	0.01	none

Detector						
InGaAs/InP SPAD (self-differencing) [154]	240	10 @ 1550	0.055	16000	1.1×10^5	100	none
InGaAs/InP SPAD (self-differencing) [155]	240	10 @ 1550	–	–	–	–	full
InGaAs/InP SPAD (self-differencing) [156]	243	26 @ 1550	0.1	170000	4.5×10^6	1000	none
InGaAs/InP SPAD (sinusoidal gating) [157]	223	12 @ 1550	0.37	1570	5.8×10^7	43	none
InGaAs/InP SPAD (harmonic subtraction) [158]	251	25 @ 1310	0.077	313000	5.4×10^5	100	none
InGaAs/InP SPAD (discharge pulse counting) [159]	243	7 @ 1550	–	40000	–	10	none
InGaAs NFAD (monolithic negative feedback) [160,161]	243	6 @ 1550	0.4	28000	–	10	some
InGaAs (self-quenching & self-recovery) [162]	160	11.5 @ 1550	–	3300000	–	3	none
CIPD (InGaAs) [163]	4.2	80 @ 1310	–	–	–	0.001	full
Frequency up-conversion [164]	300	8.8 @ 1550	0.4	13000	1.7×10^4	10	none
Frequency up-conversion [165,156]	160	56 to 59 @ 1550	–	460000	–	5	none
Frequency up-conversion [167]	300	20 @ 1306	0.62	2200	1.5×10^5	10	none
VLPC [168]	7	88 @ 694	40	20000	1.1×10^3	10	some
VLPC [169]	7	40 @ 633	0.24	25000	6.7×10^4	10	some
SSPM [170]	6	76 @ 702	3.5	7000	3×10^4	30	full
TES(W) [171]	0.1	50 @ 1550	100	3	1.7×10^6	0.1	full
TES(W) [172,217]	0.1	95 @ 1556	4	–	–	0.1	full
TES(Ha) [173]	0.1	85 @ 850	28	–	–	0.1	full
TES (Ti) [174–176,216]	0.1	89 to 98 @ 850	100	–	–	1	full
Waveguide-SNSPD [177]	2	91(on-chip); 2 @ 1550	0.02	6000	1.7×10^5	1000	none

(Continued)

TABLE 1.1 (Continued)

Detector Type	Operation Temp. (K)	Detection Efficiency, Wavelength $\eta(\%)$, λ (nm)	Timing Jitter, δt(ns) (FWHM)	Dark-count Rate, D (ungated) (1/s)	Figure of Merit $\eta/(D\delta t)$	Max. Count Rate (10^6/s)	PNR Capability
NbTi-SNSPD [178]	2.3	74 @ 1550	0.07	100	1×10^8	–	none
WSi-SNSPD [179]	0.25	93 @ 1550	0.15	0.1(intrinsic); 1000	6.2×10^6	25	none
Parallel SNSPD [180]	2	2 (device) @ 1300	0.05	0.15	2.7×10^9 (device)	1000	some
Multi-element SNSPD [181]	2.5	76 @ 1550	0.07	10000	1.1×10^6	800	some
Series SNSPD [182]	1.2	1 @ 1300	0.08	5	2.5×10^7	100	some
STJ [183–185]	0.4	45 @ 350	2000	–	–	0.01	full
QD (resonant tunnel diode) [186]	4	12 @ 550	150	0.002	4×10^9	0.25	full
QDOGFET (field-effect transistor) [187–189]	4	2 @ 805	10000	150	10	0.05	full
SPT (single-photoelectron transistor) [190]	4	1 @ 1300	–	–	–	–	full

*PNR should be possible, but none has been demonstrated as of yet. While most parameter values are for the detector as a complete system, a few parameter values as indicated are given just for the intrinsic device.

Maximum count rate is a rough estimate from the detector's output pulse width or count rate that yields 100% dead time. The photon-number-resolving (PNR) capability is defined here as: none) for devices that are typically operated as a photon or no photon device; some) for devices that are made from multiple detectors that individually have no PNR capability and thus are limited in the photon number that can be resolved to the number of individual detectors; and full) for devices whose output is inherently proportional to the number of photons even if their proportional response ultimately saturates at high photon levels.

Chapter 3 reviews the PMT, the first detector able to detect single optical photons. Given its large active area, fast response, reliability, and response efficiency over a wide range of optical spectrum, the PMT remains one of the most widely used detectors. Two distinct disadvantages of the PMT, the bulk vacuum tube and the lower detection efficiency, particularly at near infrared and longer wavelengths, have motivated the development of solid-state alternatives.

Chapter 4 presents a comprehensive review, from fabrication to operation, of the most commonly used alternative to the PMT (certainly they are the most used in quantum information applications), the semiconductor-based detector. This is largely due to their higher detection efficiency, particularly in the red and near-infrared spectral regions, along with other advantages such as the compactness, low power dissipation, lower cost, and reliability.

Chapter 5 describes a variety of solid-state detectors that go beyond the mechanisms of typical semiconductor-based devices. These include the visible-light photon counter (VLPC) and the solid-state photomultiplier (SSPM), which have been in existence for some time, but have not been particularly accessible to many researchers. These devices have the advantage of high detection efficiency and high maximum rates, and some PNR capability. Also included in this chapter are devices that use quantum dots, photoconductive gain, and electrical readout of photo-generated carriers as a means to detect single photons. In addition to the potential for PNR detection, these latter schemes may even allow for the possibility of recording the spin of a incident photon. Such transfer of a photonic quantum state to a material system would represent a unique capability that could open the door to a whole new category of applications.

Chapter 6 reviews one of today's most promising single-photon-detection technologies, cryogenic-based detection, in the form of the superconducting transition-edge-sensor (TES) and the superconducting nanowire single-photon detector (SNSPD). Both of these detectors have demonstrated detection efficiencies approaching unity. In addition, TES detectors can provide information about the number of photons that arrive simultaneously, and can do so up to large numbers of photons. We refer to this as full PNR capability. The SNSPD, the younger of these two technologies and known for low timing jitter and high count rates, is now being used to provide PNR capability by means of multiplexing. And the advent of commercially available cryogen-free cooling systems has greatly improved the convenience of these detectors, albeit at high capital cost. SNSPDs generally operate at somewhat more accessible temperatures than TES (≈ 1 K rather than ≈ 0.1 K) and the development of devices with higher operating temperatures is an active area of research, offering the potential to move them out of the laboratory environment and introducing them to a wider range of applications.

Chapter 7 discusses the hybrid detection systems, whereby multiple technologies are combined to provide additional detection capabilities. These include using multiplexed detection techniques to provide PNR capability where the individual detector(s) has(have) none. In addition, these techniques can be

TABLE 1.2 Comparison of Single-Photon Sources.

Source Type	Prob. or Det.	Temp. (K)	Wavelength Range General	Wave-Length Tunability Specific	Inherent Bandwidth	Emission Efficiency	Output Spatial Mode	$g^{(2)}(0)$	Refs.
Faint laser Two photon	P	300	vis-IR	nm	GHz	1	single	1	
(heralded)– atomic cascade	P	–	vis-UV	MHz	atomic line	0.0001	multi	–	[9]
PDC bulk	P	300	vis-IR	nm	nm	0.6	multi	0.0014	[193–195]
periodically poled	P	300 to 400	vis-IR	nm	nm	0.84	multi/single*	–	[84–88]
waveguide (periodically poled)	P	300 to 400	vis-IR	nm	nm	0.8	single	–	[196]
gated	D	300	vis-IR	nm	nm	0.27	single	0.02	[197,198]
multiplexed (free-space, chip)	D	300	vis-IR	nm	nm	0.1	single	0.08	[199,200]
FWM									
DSF	P	4 to 300	IR	nm	nm	0.02	single	–	[201]
BSMF	P	300	vis-IR	nm	nm	0.26	single	0.022	[202]
PCF	P	300	vis-IR	10 nm	nm	0.18	single	0.01	[203]
SOI waveguide	P	300	IR	10 nm	nm	0.17	single	–	[204]
Laser-PDC hybrid	P	300	vis-IR	nm	nm	–	single	0.37	[205]
Isolated system–									
Single Molecule	D	300	500 nm to 750 nm	30 nm	30 nm	0.04	multi	0.09	[206–208]
Color center (NV)	D	300	640 nm to 800 nm	nm	nm	0.022	multi	0.07	[209]

Source	Type		Wavelength	Bandwidth				Ref
QD (GaN)	D	200	340 nm to 370 nm	nm	–	multi	0.4	[210]
QD (CdSe/ZnS)	D	300	500 nm to 900 nm	nm	0.05	multi	0.003	[211]
QD (InAs) in pillar cavity	D	5	920 nm to 950 nm	10 GHz	0.05	single	0.02	[212]
QD (InGaAs) in pillar cavity	D	10	932 nm	1 GHz	0.25	single	0.15	[54]
QD in tapered pillar	D	5	915, 950 nm	1 GHz	0.01	single	0.008	[52,213]
Single ion in cavity	D	≈0	atomic line	5 MHz	0.08	single	0.015	[44]
Single Atom in cavity	D	≈0	atomic line	10 MHz	0.05	single	0.05	[39,214]
Ensemble– Rb, Cs	D	10^{-4}	atomic line	10 MHz	0.2	single	0.25	[56,57]

Sources are characterized as probabilistic (P) or deterministic (D) (remembering the caveat that a deterministic source can in practice lose some or much of its determinism and operate in a more probabilistic fashion due to issues such as low emission efficiency)

*While bulk and periodically poled PDC sources are generally inherently multimode, they can be engineered to emit with high overlap to a single-spatial mode [215].

The wavelength range possible for each method is given, along with how far an individual source can be tuned. The inherent bandwidth indicates the typical spectral width of the emitted photons. The emission efficiency is the overall extraction efficiency of the source from generation of the photons to emission of the light, including any spectral filtering that would be necessary for typical quantum-information applications (the efficiency of the detector used to measure the source is not included). Note that for two-photon sources, the second-order correlation function $g^{(2)}(0)$ typically increases as the generation rate increases, so the values here are for the lower end of the generation ranges.

used to circumvent other deficiencies, such as dead time and maximum count rates. Another hybrid scheme is the use of frequency up-conversion, which transfers photons of an IR wavelength to the visible where Si SPADs have much better characteristics.

Because of the complexity of a detector's response, in many cases they need to be characterized by the end user. Chapter 8 addresses the important question of how one calibrates the detection efficiency of a single-photon detector and documents the issues relevant to achieving high accuracy. While knowing the absolute detection efficiency is critical both to metrology and to many quantum information applications, determining a device's efficiency can be challenging because many other properties of a detector must also be understood and characterized to achieve a highly accurate calibration. In addition to the usual challenges associated with high accuracy metrology, we should bear in mind that while each single-photon detection is a classical event, the measurement itself is a quantum process. The interface between these two regimes requires particular care and Chapter 9 analyzes single-photon detection, and the quantum process by which that measurement takes place.

1.5.2 Single-Photon Sources

Contemporary single-photon sources can be broadly categorized as either probabilistic or deterministic, with a probabilistic source being based on two-photon emission where one of the photons is used to herald the other as the "single-photon emission." Although many applications require an on-demand (i.e. deterministic) source of single photons, and this has led to intense research into developing truly deterministic single-photon sources, we note that photon-pair-based heralded single-photon sources are still the most widely used in applications such as quantum information, quantum-enabled measurement, and single-photon detector calibration. Table 1.2 provides a survey of reported performance for a variety of source technologies. The meaning of the relevant parameters used to describe a source is introduced in Chapter 2. The physics of different source types is discussed in Chapters 10 through 14.

Chapter 10 discusses the historic motivation and development of single-photon sources for testing the most fundamental of quantum mechanical features: wave-particle duality and entanglement. This application illustrates the fundamental distinction between a weak light pulse and a true single-photon source. The full appreciation of this distinction is key to understanding a broad range of applications in quantum information.

Spontaneous PDC is the most studied photon-pair production process and the most used for single-photon generation in quantum information applications. In particular, PDC-based single-photon sources offer the highest detected photon-pair rates (Fig. 1.2). Chapter 11 presents a comprehensive and detailed description of PDC-based single-photon sources over their 40 years development. An alternative and more recently developed type of photon-pair-based heralded single-photon source relies on four-wave mixing in single-mode optical fibers. The advantage of such a source is that the generation, extraction, encoding, delivery, and detection of photon pairs can be achieved all within the fiber, which greatly simplifies the design and operation of end-user applications. Such advantage of the four-wave mixing method has led to its use in a number of labs. Chapter 12 covers four-wave mixing sources and their applications in quantum information science.

"Single-emitter" quantum systems as on-demand sources of single photons are discussed in Chapter 13. While each of these single-emitter approaches uses a different material system (single atoms and ions, quantum dots, color centers, mesoscopic quantum wells, etc.), most rely on similar principles of operation. When single-photon emission is desired, some external control is used to put the system into an excited state that will emit a single photon upon relaxation to some lower energy state. Commonly, coupling emitters to optical cavities is employed to enforce high emission efficiency into a single spatial mode.

Chapter 14 introduces a different kind of on-demand single-photon source. This method uses collective excitations in ensembles of atoms with two metastable ground states and at least one excited state. All the atoms are first optically prepared in one of the ground states. A weak coupling pulse probabilistically transfers an ensemble into a superposition state, in which one atom of the ensemble ends up in the second ground state, but importantly, which atom was transferred cannot be determined, even in principle. This collective excitation can be stored in the ensemble until a single photon is required. Then, by applying a strong coupling pulse, the single ensemble excitation can be deterministically converted into a single photon. Sources of this type offer many advantages and conveniences over other on-demand sources.

1.6 CONCLUSIONS

Single-photon technology has had a short, but very eventful history. The concept of a photon was brought forth less than one century ago to address fundamental physics problems of the time. Since then, the field has experienced rapid and accelerating development. At the same time, the list of applications that require either single photons or single-photon detection has grown. In the following chapters the major technologies and applications are discussed in detail. Obviously, this book does not, and cannot, describe all related technologies and applications.

The development of single-photon sources and single-photon detectors have been closely interrelated, as a comprehensive understanding of detectors is not possible without studying sources at the same time, and vice versa. Our hope is that we have compiled a useful guide to the current state of the field, and one that would provide a curious reader enough background knowledge to aid their own research, as well as with the motivation to pursue these wonderful tools for the betterment of humanity.

ACKNOWLEDGMENTS

First, we thank all the colleagues who contributed to this book by co-authoring a chapter. We also thank our colleagues and friends who have supported our effort, by sharing their thoughts, suggestions, and advice. Below is an incomplete list in alphabetical order: Francisco E. Becerra, Matthew Eisaman, Elizabeth Goldschmidt, Mohammad Hafezi, Scott Glancy, Emmanuel Knill, Paul Lett, Richard Mirin, Joffrey Peters, Alessandro Restelli.

REFERENCES

[1] P. Speziali, "Albert Einstein-Michele Besso Correspondence 1903–1955," p. 453 (1972).
[2] H. Kragh, Physics World (2000).
[3] M. Planck, Ann. Phys. 4, 561 (1901).
[4] A. Einstein, Ann. Phys. 17, 132 (1905).
[5] S. Heron, R.W.P. McWirter, and E.H. Rhoderick, Nature 174, 564 (1954).
[6] S. Freedman and J. Clauser, Phys. Rev. Lett. 28, 938 (1972).
[7] E.S. Fry and R.C. Thompson, Phys. Rev. Lett. 37, 465 (1976).
[8] J.F. Clauser, Phys. Rev. D 9, 853 (1974).
[9] A. Aspect, P. Grangier, and G. Roger, Phys. Rev. Lett. 47, 460 (1981).
[10] L. Mandel, E. Wolf, "Optical Coherence and Quantum Optics," Cambridge University Press, New York, NY (1995).
[11] C. Cohen-Tannoudji, J. Dupont-Roc, and G. Grynberg, "Photons and Atoms: Introduction to Quantum Electrodynamics," Wiley-Interscience, New York (1997).
[12] G. Lewis, Nature 118, 874 (1926).
[13] P.A.M. Dirac, Proc. R. Soc. London: Ser. A 114, 243 (1927).
[14] E. Fermi, Rev. Mod. Phys. 4, 87 (1932).
[15] M.O. Scully and M.S. Zubairy, "Quantum Optics," Cambridge University Press, Cambridge (1997).
[16] R. Glauber, Rev. Mod. Phys. 78, 1267 (2006).
[17] Certain commercial equipment, instruments or materials are identified in this paper to foster understanding. Such identification does not imply recommendation or endorsement by the National Institute of Standards and Technology, nor does it imply that the materials or equipment are necessarily the best available for the purpose.
[18] T. Jennewein, U. Achleitner, G. Weihs, H. Weinfurter, and A. Zeilinger, Rev. Sci. Instrum. 71, 1675 (2000).
[19] M. Fiorentino, C. Santori, S.M. Spillane, R.G. Beausoleil, and W.J. Munro, Phys. Rev. A 75, 032334 (2007).
[20] I.J. Owens, R.J. Hughes, and J.E. Nordholt, Phys. Rev. A 78, 022307 (2008).
[21] L.A. Kubetsky, Multielement Electronic Device (USSR), Patent No. 24040 (1931).
[22] V. Zworykin, G. Morton, and L. Malter, Proc. IRE 24, 351 (1936).
[23] L.A. Kubetsky, Proc. Inst. Radio Eng. 254, 421 (1937).
[24] B.Y. Zeldovich and D.N. Klyshko, Sov. Phys. JETP. Lett. 9, 40 (1969).
[25] C.K. Hong and L. Mandel, Phys. Rev. Lett. 56, 58 (1986).
[26] S. V. Polyakov and A. L. Migdall, Opt. Express 15, 1390 (2007).
[27] F. Cristofori, P. Fenici, G. Frigerio, N. Molho, and P. Sona, Phys. Lett. 6, 171 (1963).
[28] Z. Ou and L. Mandel, Phys. Rev. Lett. 61, 54 (1988).
[29] P. G. Kwiat, K. Mattle, H. Weinfurter, A. Zeilinger, A.V. Sergienko, and Y. Shih, Phys. Rev. Lett. 75, 4337 (1995).
[30] P.G. Kwiat, E. Waks, A.G. White, I. Appelbaum, and P.H. Eberhard, Phys. Rev. A 60, R773 (1999).
[31] A.B. U'Ren, C. Silberhorn, K. Banaszek, and I.A. Walmsley, Phys. Rev. Lett. 93, 093601 (2004).

[32] C. Kurtsiefer, M. Oberparleiter, and H. Weinfurter, Phys. Rev. A 64, 023802 (2001).
[33] P. Kwiat, J. Altepeter, J. Barreiro, M. Goggin, E. Jeffrey, N. Peters, and A. Van Devender, in Quantum Communication, Measurement And Computing, edited by S. Barnett, E. Andersson, J. Jeffers, P. Ohberg, and O. Hirota (2004) AIP Conference Proceedings of vol. 734, pp. 337–341, 7th International Conference on Quantum Communication, Measurement and Computing, Glasgow, SCOTLAND, July 25–29, 2004.
[34] J. Altepeter, E. Jeffrey, and P. Kwiat, Opt. Express 13, 8951 (2005).
[35] P. Kwiat, Private commun. (2013).
[36] D. Burnham and D. Weinberg, Phys. Rev. Lett. 25, 84 (1970).
[37] A. Kuhn, M. Hennrich, and G. Rempe, Phys. Rev. Lett. 89, 067901 (2002).
[38] M. Hennrich, T. Legero, A. Kuhn, and G. Rempe, New J. Phys. 6, 86 (2004).
[39] M. Hijlkema, B. Weber, H.P. Specht, S.C. Webster, A. Kuhn, and G. Rempe, Nature Phys. 3, 253 (2007).
[40] T. Wilk, S.C. Webster, H.P. Specht, G. Rempe, and A. Kuhn, Phys. Rev. Lett. 98, 063601 (2007).
[41] B. Dayan, A.S. Parkins, T. Aoki, E.P. Ostby, K.J. Vahala, and H.J. Kimble, Science 319, 1062–1065 (2008).
[42] T. Aoki, A.S. Parkins, D.J. Alton, C.A. Regal, B. Dayan, E. Ostby, K.J. Vahala, and H.J. Kimble, Phys. Rev. Lett. 102, 083601 (2009).
[43] C. Maurer, C. Becher, C. Russo, J. Eschner, and R. Blatt, New J. Phys. 6, 94 (2004).
[44] M. Keller, B. Lange, K. Hayasaka, W. Lange, and H. Walther, Nature 431, 1075 (2004).
[45] H.G. Barros, A. Stute, T.E. Northup, C. Russo, P.O. Schmidt, and R. Blatt, New J. Phys. 11, 103004 (2009).
[46] A.J. Shields, Nat. Photonics 1, 215 (2007).
[47] E. Moreau, I. Robert, J. Gerard, I. Abram, L. Manin, and V. Thierry-Mieg, Appl. Phys. Lett. 79, 2865 (2001).
[48] M. Pelton, C. Santori, J. Vuckovic, B. Zhang, G. Solomon, J. Plant, and Y. Yamamoto, Phys. Rev. Lett. 89, 233602 (2002).
[49] D. Press, S. Goetzinger, S. Reitzenstein, C. Hofmann, A. Loeffler, M. Kamp, A. Forchel, and Y. Yamamoto, Phys. Rev. Lett. 98, 117402 (2007).
[50] A. Kress, F. Hofbauer, N. Reinelt, M. Kaniber, H. Krenner, R. Meyer, G. Bohm, and J. Finley, Phys. Rev. B 71, 241304 (2005).
[51] S. Laurent, S. Varoutsis, L. Le Gratiet, A. Lemaitre, I. Sagnes, F. Raineri, A. Levenson, I. Robert-Philip, and I. Abram, Appl. Phys. Lett. 87, 163107 (2005).
[52] M.E. Reimer, G. Bulgarini, N. Akopian, M. Hocevar, M.B. Bavinck, M.A. Verheijen, E.P. Bakkers, L.P. Kouwenhoven, and V. Zwiller, Nat. Commun. 3, 737 (2012).
[53] P.E. Barclay, K.-M. C. Fu, C. Santori, and R.G. Beausoleil, Appl. Phys. Lett. 95, 191115 (2009).
[54] O. Gazzano, S. Michaelis de Vasconcellos, C. Arnold, A. Nowak, E. Galopin, I. Sagnes, L. Lanco, A. Lemaitre, P. Senellart, and P. Senellart, Nat. Commun. 4, 1425 (2013).
[55] E. Flagg, S.V. Polyakov, T. Thomay, and G. Solomon, Phys. Rev. Lett. 109, 163601 (2012).
[56] C.W. Chou, S.V. Polyakov, A. Kuzmich, and H.J. Kimble, Phys. Rev. Lett. 92, 213601 (2004).
[57] D. N. Matsukevich, T. Chaneliere, S.D. Jenkins, S.-Y., Lan, T.A.B. Kennedy, and A. Kuzmich, Phys. Rev. Lett. 97, 013601 (2006).
[58] C.H. van der Wal, M.D. Eisaman, A. Andre, R.L. Walsworth, D.F. Philips, A.S. Zibrov, and M.D. Lukin, Science 301, 196 (2003).
[59] A. Kuzmich, W.P. Bowen, A.D. Boozer, A. Boca, C.W. Chou, L.-M. Duan, and H.J. Kimble, Nature 423, 731 (2003).
[60] T. Pittman, B. Jacobs, and J. Franson, Phys. Rev. A 66, 062302 (2002).
[61] E. Jeffrey, N.A. Peters, and P.G. Kwiat, New J. Phys. 6, 100 (2004).
[62] A.I. Lvovsky, B.C. Sanders, and W. Tittel, Nat. Photonics 3, 706 (2009).
[63] X.-S. Ma, S. Zotter, J. Kofler, T. Jennewein, and A. Zeilinger, Phys. Rev. A 83, 043814 (2011).
[64] E. Saglamyurek, N. Sinclair, J. Jin, J. A. Slater, D. Oblak, F. Bussières, M. George, R. Ricken, W. Sohler, and W. Tittel, Phys. Rev. Lett. 108, 083602 (2012).
[65] C. Bennett and G. Brassard, Proceedings of IEEE International Conference on Computers, Systems, and Signal Processing, Bangalore, India, IEEE, New York pp. 175–179 (1984).

[66] A.K. Ekert, Phys. Rev. Lett. 67, 661 (1991).
[67] C.H. Bennett, Phys. Rev. Lett. 68, 3121 (1992).
[68] Y. Zhao, B. Qi, X. Ma, H.-K. Lo, and L. Qian, Phys. Rev. Lett. 96, 070502 (2006).
[69] S. Pironio, A. Acn, N. Brunner, N. Gisin, S. Massar, and V. Scarani, New J. Phys. 11, 045021 (2009).
[70] H.J. Briegel, W. Dur, S.J. van Enk, J.I. Cirac, and P. Zoller, in The Physics of Quantum Information, edited by D. Bouwmeester, A. Ekert, and A. Zeilinger, Springer, Berlin (2000).
[71] L. Duan, M. Lukin, J. Cirac, and P. Zoller, Nature 414, 413–418 (2001).
[72] C. Simon, H. de Riedmatten, M. Afzelius, N. Sangouard, H. Zbinden, and N. Gisin, Phys. Rev. Lett. 98, 190503 (2007).
[73] N. Sangouard, C. Simon, H. de Riedmatten, and N. Gisin, Rev. Mod. Phys. 83, 33 (2011).
[74] N. Sangouard, C. Simon, B. Zhao, Y.-A. Chen, H. de Riedmatten, J.-W. Pan, and N. Gisin, Phys. Rev. A 77, 062301 (2008).
[75] C. H. Bennett, G. Brassard, C. Crépeau, R. Jozsa, A. Peres, and W. K. Wooters, Phys. Rev. Lett. 70, 1895 (1993).
[76] S. Aaronson and A. Arkhipov (2010), arXiv:1011.3245[quant-ph].
[77] M.A. Broome, A. Fedrizzi, S. Rahimi-Keshari, J. Dove, S. Aaronson, T.C. Ralph, and A.G. White, Science 339, 794 (2013).
[78] J.B. Spring, B.J. Metcalf, P.C. Humphreys, W.S. Kolthammer, X.-M. Jin, M. Barbieri, A. Datta, N. Thomas-Peter, N.K. Langford, D. Kundys, J.C. Gates, B.J. Smith, P.G.R. Smith, and I.A. Walmsley, Science 339, 798 (2013).
[79] K.R. Motes, J.P. Dowling, and P.P. Rohde (2013), arXiv:1307.8238V1[quant-ph].
[80] K. Tsujino, D. Fukuda, G. Fujii, S. Inoue, M. Fujiwara, M. Takeoka, and M. Sasaki, Phys. Rev. Lett. 106, 250503 (2011).
[81] F.E. Becerra, G.B.J. Fan, J. Goldhar, J.T. Kosloski, and A. Migdall, Nat. Photonics 7, 147 (2013).
[82] F.E. Becerra, J. Fan, and A. Migdall, Nat. Commun. 4, 2028 (2013).
[83] A. Khrennikov, AIP Conf. Proc. 1424, 160 (2012).
[84] A. Fedrizzi, T. Herbst, A. Poppe, T. Jennewein, and A. Zeilinger, Opt. Express 15, 15377 (2007).
[85] M. D. C. Pereira, F. E. Becerra, B. L. Glebov, J. Fan, S. W. Nam, and A. Migdall, Opts. Lett. 38, 1609 (2013).
[86] M. Giustina, A. Mech, S. Ramelow, B. Wittmann, J. Kofler, J. Beyer, A. Lita, B. Calkins, T. Gerrits, S. Nam, R. Ursin, and A. Zeilinger, Nature 497, 227 (2013).
[87] D.H. Smith, G. Gillett, M.P. de Almeida, C. Branciard, A. Fedrizzi, T.J. Weinhold, A. Lita, B. Calkins, T. Gerrits, H.M. Wiseman, S.W. Nam, and A.G. White, Nat. Commun. 3, 625 (2012).
[88] B.G. Christensen, K.T. McCusker, J.B. Altepeter, B. Calkins, T. Gerrits, A.E. Lita, A. Miller, L.K. Shalm, Y. Zhang, S.W. Nam, N. Brunner, C.C. W. Lim, et al. (2013), arXiv:1306.5772v1[quant-ph].
[89] G. Brida, I.P. Degiovanni, M. Genovese, A. Migdall, F. Piacentini, S.V. Polyakov, and P. Traina, J. Phys. Soc. Jpn. 82, 034004 (2013).
[90] M. Bellare and P. Rogaway, in Proceedings of the First ACM Conference on Computer and Communications Security, November 1993, pp. 62–73,(1993).
[91] J. Boyar, I. Damgard, and R. Peralta, J. Cryptology 13, 449 (2000).
[92] P. Bogetoft, D.L. Christensen, I. Damgard, M. Geisler, T. Jakobsen, M. Kroigaard, J. Da, Nielsen, J.B. Nielsen, K. Nielsen, J. Pagter, M. Schwartzback, et al. Financial Cryptography and Data Security, Lecture Notes in Computer Science vol. 5628, pp. 325–343, 13th International Conference, FC 2009, Accra Beach, Barbados, February 23–26, 2009.
[93] M.I. Michael J. Fischer and R. Peralta (2010), www.nist.gov/itl/csd/ct/upload/Public RandomnessService-2.pdf?
[94] M. Iorga, L. Bassham, J. Kelsey, R. Peralta, and M.J. Fischer (2013), http://www.nist.gov/itl/csd/ct/nist_beacon.cfm.
[95] U. Lieberwirth, J. Arden-Jacob, K.H. Drexhage, D.P. Herten, R. Muller, M. Neumann, A. Schulz, S. Siebert, G. Sagner, S. Klingel, M. Sauer, and J. Wolfrum, Anal. Chem. 70, 4771 (1998).
[96] J.-P. Knemeyer, N. Marme, and M. Sauer, Anal. Chem. 72, 3717 (2000).

[97] D.N. Gavrilov, B. Gorbovitski, M. Gouzman, G. Gudkov, A. Stepoukhovitch, V. Ruskovoloshin, A. Tsuprik, G. Tyshko, O. Bilenko, O. Kosobokova, S. Luryi, and V. Gorfinkel, Electrophoresis 24, 1184 (2003).
[98] I. Rech, A. Restelli, S. Cova, M. Ghioni, M. Chiari, and M. Cretich, Sen. Actuators B-Chemi. 100, 158 (2004).
[99] T. Isoshima, Y. Isojima, K. Kikuchi, K. Nagai, and H. Nakagawa, Rev. Sci. Instrum. 66, 2922 (1995).
[100] J.-P. Knemeyer, N. Marme, and M. Sauer, Science 283, 1676 (1999).
[101] A. Berglund, A. Doherty, and H. Mabuchi, Phys. Rev. Lett. 89, 068101 (2002).
[102] K. Suhling, P. French, and D. Phillips, Photochem. Photobiol. Sci. 4, 13 (2005).
[103] T. McIlrath, R. Hudson, A. Aikin, and T. Wilkerson, Appl. Opt. 18, 316 (1979).
[104] M. Vitebini, A. Adriani, and G. Didonfrancesco, Rev. Sci. Instrum. 58, 1833 (1987).
[105] S. Pellegrini, G. Buller, J. Smith, A. Wallace, and S. Cova, Meas. Sci. Technol. 11, 712 (2000).
[106] S. Personick, Bell Sys. Tech. J. 56, 355 (1977).
[107] B. Levine, C. Bethea, and J. Campbell, Appl. Phys. Lett. 46, 333 (1985).
[108] G. Ripamonti, M. Ghioni, and S. Vanoli, Electron. Lett. 26, 1569 (1990).
[109] A. Lacaita, P. Francese, S. Cova, and G. Riparmonti, Opt. Lett. 18, 1110 (1993).
[110] F. Scholder, J. Gautier, M. Wegmuller, and N. Gisin, Opt. Commun. 213, 57 (2002).
[111] A. Wegmuller, F. Scholder, and N. Gisin, J. Lightwave Technol. 22, 390 (2004).
[112] M. Legre, R. Thew, H. Zbinden, and N. Gisin, Opt. Express 15, 8237 (2007).
[113] in J. Kash, J. Tsang, D. Knebel, and D. Vallett, ISTFA 98: Proceedings Of The 24th International Symposium for Testing and Failure Analysis, pp. 483–488 (1998).
[114] J. Tsang, J. Kash, and D. Vallett, IBM J. Res. Dev. 44, 583 (2000).
[115] F. Stellari, F. Zappa, S. Cova, C. Porta, and J. Tsang, IEEE Trans. Electron Devices 48, 2830 (2001).
[116] N. Goldblatt, M. Leibowitz, and W. Lo, Microelectron. Reliab. 41, 1507 (2001).
[117] F. Stellari, A. Tosi, F. Zappa, and S. Cova, IEEE Trans. Instrum. Meas. 53, 163 (2004).
[118] S. Polonsky and K. Jenkins, IEEE Electron Device Lett. 25, 208 (2004).
[119] S. Soper, Q. Mattingly, and P. Vegunta, Anal. Chem. 65, 740 (1993).
[120] L.-Q. Li and L. Davis, Rev. Sci. Instrum. 64, 1524 (1993).
[121] I. Rech, G. Luo, M. Ghioni, H. Yang, X. S. Xie, and S. Cova, IEEE J. Sel. Top. Quant. Electron. 10, 788 (2004).
[122] M. Wahl, F. Koberling, M. Patting, H. Rahn, and R. Erdmann, Curr. Pharm. Biotechnol. 5, 299 (2004).
[123] M. Gosch, A. Serov, T. Anhut, T. Lasser, A. Rochas, P. Besse, R. Popovic, H. Blom, and R. Rigler, J. Biomed. Opt. 9, 913 (2004).
[124] X. Michalet, O.H.W. Siegmund, J.V. Vallerga, P. Jelinsky, J.E. Millaud, and S. Weiss, J. Mod. Opt. 54, 239 (2007).
[125] X. Michalet, R.A. Colyer, J. Antelman, O.H.W. Siegmund, A. Tremsin, J.V. Vallerga, and S. Weiss, Curr. Pharm. Biotechnol. 10, 543 (2009).
[126] S. Felekyan, R. Khnemuth, V. Kudryavtsev, C. Sandhagen, W. Becker, and C.A.M. Seidel, Rev. Sci. Instrum. 76, 083104 (2005).
[127] A. Pifferi, A. Torricelli, L. Spinelli, D. Contini, R. Cubeddu, F. Martelli, G. Zaccanti, A. Tosi, A. D. Mora, F. Zappa, and S. Cova, Phys. Rev. Lett. 100, 138101 (2008).
[128] V.C. Spanoudaki, A.B. Mann, A.N. Otte, I. Konorov, I. Torres-Espallardo, S. Paul, and S.I. Ziegler, J. Instrum. 2, 12002 (2007).
[129] D. Klyshko, Kvantovaya Elektron. 4, 1056 (1977).
[130] A. Malygin, A. Penin, and A. Sergienko, JETP Lett. 33, 477 (1981).
[131] A. Migdall, R. Datla, A. Sergienko, J. Orszak, and Y. Shih, Appl. Opt. 37, 3455 (1998).
[132] A. Migdall, E. Dauler, A. Muller, and A. Sergienko, Anal. Chim. Acta 380, 311 (1999).
[133] M. Ware and A. Migdall, J. Mod. Opt. 51, 1549 (2004).
[134] G. Brida, M. Genovese, M. Gramegna, M. Rastello, M. Chekhova, and L. Krivitsky, J. Opt. Soc. Am. B-Opt. Phys. 22, 488 (2005).
[135] S. Castelletto, I. P. Degiovanni, V. Schettini, and A. Migdall, Metrologia 43, S56 (2006).
[136] G. Brida, M. Genovese, and M. Gramegna, Laser Phys. Lett. 3, 115 (2006).
[137] http://jp.hamamatsu.com/resources/products/etd/pdf/m-h7422e.pdf (2010).

[138] http://jp.hamamatsu.com/resources/products/etd/pdf/NIR-PMT_APPLI_TPMO104 0E02.pdf (2010).
[139] http://www.perkinelmer.com/Category/Category/cat1-/IDSMI_TAXONOMY_DELE TIONS/cat2/IND_SE_CAT_SinglePhotonCountingModulesSPCM_001/key/10613 (2010).
[140] http://www.microphotondevices.com/media/pdf/PDM_v3_3.pdf (2010).
[141] O. Thomas, Z.L. Yuan, J.F. Dynes, A.W. Sharpe, and A.J. Shields, Appl. Rev. Lett. 97, 031102 (2010).
[142] M. Akiba, K. Tsujino, and M. Sasaki, Opt. Lett. 35, 2621 (2010).
[143] M. Ghioni, G. Armellini, P. Maccagnani, I. Rech, M.K. Emsley, and M.S. Unlu, J. Mod. Opt. 56, 309 (2009).
[144] D.A. Kalashnikov, S.H. Tan, M.V. Chekhova, and L.A. Krivitsky, Opt. Express 19, 9352 (2011).
[145] http://jp.hamamatsu.com/resources/products/ssd/pdf/s10362-11_series_kapd1022e05.pdf (2009).
[146] R.A. LaRue, G.A. Davis, D. Pudvay, K.A. Costello, and V.W. Aebi, IEEE Elect. Dev. Lett. 20, 126 (1999).
[147] N. Bertone, R. Biasi, and B. Dion, SPIE Proc. Semicond. Photodetectors II 5726, 153 (2005).
[148] M.J. Fitch, B.C. Jacobs, T.B. Pittman, and J.D. Franson, Phys. Rev. A 68, 043814 (2003).
[149] M. Micuda, O. Haderka, and M. Jezek, Phys. Rev. A 78, 025804 (2008).
[150] L.A. Jiang, E.A. Dauler, and J.T. Chang, Phys. Rev. A 75, 062325 (2007).
[151] G. Brida, I.P. Degiovanni, F. Piacentini, V. Schettini, S.V. Polyakov, and A. Migdall, Rev. Sci. Instrum. 80, 116103 (2009).
[152] D.S. Bethune, W.P. Risk, and G. Pabst, J. Mod. Opt. 51, 1359 (2004).
[153] C. Gobby, Z.L. Yuan, and A.J. Shields, Appl. Phys. Lett. 84, 3762 (2004).
[154] A.R. Dixon, Z.L. Yuan, J.F. Dynes, A.W. Sharpe, and A.J. Shields, Opt. Express 16, 18790 (2008).
[155] B.E. Kardynal, Z.L. Yuan, and A.J. Shields, Nat. Photonics 2, 425 (2008).
[156] Z. Yuan, A. Sharpe, J. Dynes, A. Dixon, and A. Shields, Appl. Phys. Lett. 96, 071101 (2010).
[157] Y. Nambu, S. Takahashi, K. Yoshino, A. Tanaka, M. Fujiwara, M. Sasaki, A. Tajima, S. Yorozu, and A. Tomita, Opt. Express 19, 20531–20541 (2011).
[158] A. Restelli, J. Bienfang, and A. Migdall, in Advanced Photon Counting Techniques VII, edited by M. A. Itzler and J. C. Campbell (2013), Proceedings Of The Society Of Photo-Optical Instrumentation Engineers (SPIE) of vol. 8727, p. 87270F.
[159] A. Yoshizawa, R. Kaji, and H. Tsuchida, Appl. Phys. Lett. 84, 3606 (2004).
[160] X. Jiang, M.A. Itzler, B. Nyman, and K. Slomkowski, Proc. SPIE Quantum Sensing and Nanophotonic Devices VI 7320, 732011 (2009).
[161] http://www.princetonlightwave.com/content/PNA-20XNFAD Datasheet_rv2.pdf (2010).
[162] K. Zhao, S. You, J. Cheng, and Y.-hua Lo, Appl. Phys. Lett. 93, 153504 (2008).
[163] M. Fujiwara and M. Sasaki, Appl. Opt. 46, 3069 (2007).
[164] H. Takesue, E. Diamanti, T. Honjo, C. Langrock, M.M. Fejer, K. Inoue, and Y. Yamamoto, New J. Phys. 7, 232 (2005).
[165] M.A. Albota and F.N.C. Wong, Opt. Lett. 29, 1449 (2004).
[166] A.P. Van Devender and P.G. Kwiat, J. Opt. Soc. Am. B 24, 295 (2007).
[167] H. Xu, L. Ma, A. Mink, B. Hershman, and X. Tang, Opt. Express 15, 7247 (2007).
[168] S. Takeuchi, J. Kim, Y. Yamamoto, and H.H. Hogue, Appl. Phys. Lett. 74, 1063 (1999).
[169] B. Baek, K. McKay, M. Stevens, J.K.H. Hogue, and S.W. Nam, IEEE J. of Quant. Electron. 46, 991 (2010).
[170] P.G. Kwiat, A.M. Steinberg, R.Y. Chiao, P.H. Eberhard, and M.D. Petroff, Appl. Opt. 33, 1844 (1994).
[171] D. Rosenberg, J.W. Harrington, P.R. Rice, P.A. Hiskett, C.G. Peterson, R.J. Hughes, A.E. Lita, S.W. Nam, and J.E. Nordholt, Phys. Rev. Lett. 98, 010503 (2007).
[172] A. E. Lita, A. J. Miller, and S. W. Nam, Opt. Express 16, 3032 (2008).
[173] A.E. Lita, B. Calkins, L.A. Pellochoud, A.J. Miller, and S. Nam, AIP Conf. Proc. 13th Int. Workshop Low Temp. Detectors LTD13 1185, 351 (2009).
[174] D. Fukuda, G. Fujii, T. Numata, A. Yoshizawa, H. Tsuchida, H. Fujino, H. Ishii, T. Itatani, S. Inoue, and T. Zama, Metrologia 46, S288 (2009).

[175] D. Fukuda, G. Fujii, T. Numata, A. Yoshizawa, H. Tsuchida, H. Fujino, H. Ishii, T. Itatani, S. Inoue, and T. Zama, Tenth International Conference on Quantum Communication, Measurement and Computation (QCMC), Brisbane, Queensland, Australia (2010).
[176] D. Fukuda, G. Fujii, T. Numata, K. Amemiya, A. Yoshizawa, H. Tsuchida, H. Fujino, H. Ishii, T. Itatani, S. Inoue, and T. Zama, Opt. Express 19, 870 (2011).
[177] C. Pernice, W.H.P. Schuck, O. Minaeva, M. Li, G. Goltsman, A. Sergienko, and H. Tang, Nat. Commun. 3, 10208 (2012).
[178] S. Miki, T. Yamashita, H. Terai, and Z. Wang, Opt. Express 21, 10208 (2013).
[179] F. Marsili, V.B. Verma, J.A. Stern, S. Harrington, A.E. Lita, T. Gerrits, I. Vayshenker, B. Baek, M.D. Shaw, R.P. Mirin, and S.W. Nam, Nat. Photonics 7, 210 (2013).
[180] A. Divochiy, F. Marsili, D. Bitauld, A. Gaggero, R.Leoni, F. Mattioli, A. Korneev, V. Seleznev, N. Kaurova, O. Minaeva, G. Gol'tsman, K.G. Lagoudakis, et al., Nat. Photonics 2, 302 (2008).
[181] D. Rosenberg, A.J. Kerman, R.J. Molnar, and E.A. Dauler, Opt. Express 21, 1440 (2013).
[182] S. Jahanmirinejad, G. Frucci, F. Mattioli, D. Sahin, A. Gaggero, R. Leoni, and A. Fiore, Appl. Phys. Lett. 101, 072602 (2012).
[183] A. Peacock, P. Verhoeve, N. Rando, A. van Dordrecht, B.G. Taylor, C. Erd, M.A.C. Perryman, R. Venn, J. Howlett, D.J. Goldie, J. Lumley, and M. Wallis, Nature 381, 135 (1996).
[184] A. Peacock, P. Verhoeve, N. Rando, A. van Dordrecht, B.G. Taylor, C. Erd, M.A.C. Perryman, R. Venn, J. Howlett, D.J. Goldie, J. Lumley, and M. Wallis, J. Appl. Phys. 81, 7641 (1997).
[185] T. Peacock, P. Verhoeve, N. Rando, C. Erd, M. Bavdaz, B. Taylor, and D. Perez, Astron. Astrophys. Supp. Ser. 127, 497 (1998).
[186] J.C. Blakesley, P. See, A.J. Shields, B.E. Kardyna, P. Atkinson, I. Farrer, and D. A. Ritchie, Phys. Rev. Lett. 94, 067401 (2005).
[187] M.A. Rowe, E.J. Gansen, M. Greene, R.H. Hadfield, T.E. Harvey, M.Y. Su, S.W. Nam, R.P. Mirin, and D. Rosenberg, Appl. Phys. Lett. 89, 253505 (2006).
[188] E.J. Gansen, M.A. Rowe, M.B. Greene, D. Rosenberg, T.E. Harvey, M.Y. Su, R.H. Hadfield, S.W. Nam, and R.P. Mirin, Nat. Photonics 1, 585 (2007).
[189] M.A. Rowe, G.M. Salley, E.J. Gansen, S.M. Etzel, S.W. Nam, and R.P. Mirin, J. Appl. Phys. 107, 063110 (2010).
[190] H. Kosaka, D.S. Rao, H.D. Robinson, P. Bandaru, and K. M.E. Yablonovitch, Phys. Rev. B 67, 045104 (2003).
[191] R. Hadfield, Nat. Photonics 3, 696 (2009).
[192] M.D. Eisaman, A. Migdall, J. Fan, and S.V. Polyakov, Rev. Sci. Instrum. 82, 071101 (2011).
[193] P.G. Kwiat and R.Y. Chiao, Phys. Rev. Lett. 66, 588 (1991).
[194] S. Fasel, O. Alibart, S. Tanzilli, P. Baldi, A. Beveratos, N. Gisin, and H. Zbinden, New J. Phys. 6, 163 (2004).
[195] Q. Wang, W. Chen, G. Xavier, M. Swillo, T. Zhang, S. Sauge, M. Tengner, Z.-F. Han, G.-C. Guo, and A. Karlsson, Phys. Rev. Lett. 100, 090501 (2008).
[196] G. Harder, V. Ansari, B. Brecht, T. Dirmeier, C. Marquardt, and C. Silberhorn, Opt. Express 21, 13975 (2013).
[197] S. Takeuchi, R. Okamoto, and K. Sasaki, Appl. Optics 43, 5708 (2004).
[198] G. Brida, I.P. Degiovanni, M. Genovese, A. Migdall, F. Piacentini, S.V. Polyakov, and I.R. Berchera, Opt. Express 19, 1484 (2011).
[199] X.-S. Ma, S. Zotter, J. Kofler, T. Jennewein, and A. Zeilinger (2010), arXiv: 1007.47981[quant-ph].
[200] M.J. Collins, C. Xiong, I.H. Rey, T.D. Vo, J. He, S. Shahnia, C. Reardon, M.J. Steel, T.F. Krauss, A.S. Clark, and B.J. Eggleton, arXiv: 1305.7278v1[quant-ph].
[201] S.D. Dyer, M.J. Stevens, B. Baek, and S.W. Nam, Opt. Express 16, 9966 (2008).
[202] B.J. Smith, P. Mahou, O. Cohen, J.S. Lundeen, and I.A. Walmsley, Opt. Express 17, 23589 (2009).
[203] A. Ling, J. Chen, J. Fan, and A. Migdall, Opt. Express 17, 21302 (2009).
[204] H. Takesue, Y. Tokura, H. Fukuda, T. Tsuchizawa, T. Watanabe, K. Yamada, and S.-i. Itabashi, Appl. Phys. Lett. 91, 201108 (2007).
[205] T.B. Pittman, J.D. Franson, and B.C. Jacobs, New J. Phys. 9, 195 (2007).
[206] S.G. Lukishova, A.W. Schmid, A.J. McNamara, R.W. Boyd, and J. Carlos R. Stroud, IEEE J. Sel. Top. Quant. Electron. 9, 1512 (2003).
[207] R. Alleaume, F. Treussart, J.-M. Courty, and J.-F. Roch, New J. Phys. 6, 85 (2004).

[208] S.G. Lukishova, A.W. Schmidz, C.M. Supranowitzy, N. Lippa, A.J. Mcnamara, R.W. Boyd, and J.C. R. Stroud, J. Mod. Opt. 51, 1535 (2004).
[209] A. Beveratos, R. Brouri, T. Gacoin, A. Villing, J.-P. Poizat, and P. Grangier, Phys. Rev. Lett. 89, 187901 (2002).
[210] S. Kako, C. Santori, K. Hoshino, S. Gotzinger, Y. Yamamato, and Y. Arakawa, Nat. Mater. 5, 887 (2006).
[211] X. Brokmann, E. Giacobino, M. Dahan, and J. Hermier, Appl. Phys. Lett. 85, 712 (2004).
[212] A.J. Bennett, D.C. Unitt, P. Atkinson, D.A. Ritchie, and A.J. Shields, Opt. Express 13, 50 (2005).
[213] J. Claudon, J. Bleuse, N.S. Malik, M. Bazin, P. Jaffrennou, N. Gregersen, C. Sauvan, P. Lalanne, and J.-M. Gerard, Nat. Photonics 4, 174 (2010).
[214] J. McKeever, A. Boca, A.D. Boozer, R. Miller, J.R. Buck, A. Kuzmich, and H. J. Kimble, Science 303, 1992 (2004).
[215] P.G. Evans, R.S. Bennink, W.P. Grice, T.S. Humble, and J. Schaake, Phys. Rev. Lett. 105, 253601 (2010).
[216] D. Fukuda, G. Fujii, T. Numata, K. Amemiya, A. Yoshizawa, H. Tsuchida, H. Fujino, H. Ishii, T. Itatani, S. Inoue, T. Zama, IEEE Trans. Appl. Supercond. 21 , 241 (2011).
[217] A. Lamas-Linares, B. Calkins, N.A. Tomlin, T. Gerrits, A.E. Lita, J. Beyer, R.P. Mirin and S.W. Nam, Appl. Phys. Lett. 102, 231117 (2013).

Chapter 2

Photon Statistics, Measurements, and Measurements Tools

Martin J. Stevens
National Institute of Standards and Technology, 325 Broadway, Boulder, CO 80305, USA

Chapter Outline

2.1	**Quantized Electric Field & Operator Notation**	26
2.2	**Source Characteristics**	28
	2.2.1 State Vector	28
	2.2.2 Density Matrix and Photon Number Probabilities	29
	2.2.3 Purity	30
	2.2.4 Source Efficiency and Generation Rate	31
	2.2.5 Second-Order Coherence, $g^{(2)}$	32
	2.2.6 Relating $g^{(2)}$ to $P(n)$	34
	2.2.7 Ideal and Non-Ideal Single-Photon Sources	37
	2.2.8 To Measure $P(n)$ or $g^{(2)}$?	38
	2.2.9 Hanbury Brown-Twiss Interferometer	38
	2.2.10 Bunching, Antibunching, and Poissonian Photon Statistics	42
	2.2.11 High-Order Coherences	44
	2.2.12 Indistinguishability	45
	2.2.13 Other Sources	47
2.3	**Detector Properties**	52
	2.3.1 Detection Efficiency	53
	2.3.2 POVM Elements	55
	2.3.3 Photon-Number-Resolving (PNR) Capability	56
	2.3.4 Timing Latency and Rise Time	62
	2.3.5 Timing Jitter	62
	2.3.6 Dead Time, Reset Time, and Recovery Time	64
	2.3.7 Dark Count Rate	65
	2.3.8 Background Count Rate	65

2.3.9 Afterpulse Probability 65
2.3.10 Active Area 66
2.3.11 Operating Temperature of Active Area 66
References 66

2.1 QUANTIZED ELECTRIC FIELD & OPERATOR NOTATION

This section introduces some of the basic quantum optics notation found in many textbooks, most closely following the treatments of Gerry and Knight [16], Loudon [25], and Walls and Milburn [47]. Sections 2.2 and 2.3 then detail the characteristics of single-photon sources, and detectors, respectively.

The quantized electric field at position $\vec{r}$ and time t can be written as an operator,

$$\hat{\mathbf{E}}(\vec{r},t) = \sum_j \left[\hat{\mathbf{E}}_j^{(+)}(\vec{r},t) + \hat{\mathbf{E}}_j^{(-)}(\vec{r},t) \right], \quad (2.1)$$

where the sum is over j orthogonal modes. Each mode j can be defined as having some particular spatial extent, propagation vector, center wavelength, polarization, and spectral and temporal profiles. $\hat{\mathbf{E}}_j^{(+)}(\vec{r},t)$ and $\hat{\mathbf{E}}_j^{(-)}(\vec{r},t)$ represent the positive and negative frequency components, respectively, of the field in mode j. They are related as

$$\hat{\mathbf{E}}_j^{(-)}(\vec{r},t) = \left[\hat{\mathbf{E}}_j^{(+)}(\vec{r},t) \right]^\dagger, \quad (2.2)$$

where $\dagger$ denotes the Hermitian conjugate.

The electromagnetic field is treated as a quantized harmonic oscillator with a Hamiltonian

$$\hat{H} = \sum_j \hbar \omega_j \left(\hat{a}_j^\dagger \hat{a}_j + \frac{1}{2} \right), \quad (2.3)$$

where $\hbar = h/2\pi$ and h is the Planck constant and $\frac{1}{2}\hbar\omega_j$ is the energy of the vacuum fluctuations in mode j. The single-mode creation operator $\hat{a}_j^\dagger$ denotes the addition of one photon to mode j:

$$\hat{a}_j^\dagger |n\rangle_j = \sqrt{n+1}|n+1\rangle_j, \quad (2.4)$$

where the photon number state $|n\rangle_j$ indicates that exactly n photons occupy mode j. Similarly, the annihilation operator $\hat{a}_j$ describes the removal of one photon from the mode:

$$\hat{a}_j |n\rangle_j = \sqrt{n}|n-1\rangle_j, \quad (2.5)$$

The number states are eigenstates of the single-mode number operator, $\hat{n}_j$, such that

$$\hat{n}_j |n\rangle_j = \hat{a}_j^\dagger \hat{a}_j |n\rangle_j = n|n\rangle_j, \quad (2.6)$$

In perhaps the most commonly cited example of a single-mode field, the mode j can be taken to indicate a plane-wave, monochromatic field with angular frequency ω_j, wave vector $\vec{k}_j$, and polarization described by the unit vector $\vec{e}_j$. The positive-frequency component of the field in this mode can be written as [16]

$$\hat{\mathbf{E}}_j^{(+)}(\vec{r},t) = \mathcal{E}_j \vec{e}_j \hat{a}_j e^{i(\vec{k}_j \cdot \vec{r} - \omega_j t)}, \tag{2.7}$$

with $\mathcal{E}_j = i\sqrt{\hbar\omega_j/2\varepsilon_0 V}$, where ε_0 is the vacuum permittivity and V is the mode volume.

To describe fields that are not strictly monochromatic, and thus have some non-zero spectral width, it is convenient to replace the single-mode creation and annihilation operators $\hat{a}_j^\dagger$ and $\hat{a}_j$ with their continuous-mode counterparts $\hat{a}_j^\dagger(\omega)$ and $\hat{a}_j(\omega)$. These operators obey the commutation relation [25]

$$[\hat{a}_j(\omega), \hat{a}_j^\dagger(\omega')] = \delta(\omega - \omega'). \tag{2.8}$$

The number operator in this case is rewritten as

$$\hat{n}_j = \int d\omega \, \hat{a}_j^\dagger(\omega) \hat{a}_j(\omega), \tag{2.9}$$

and the positive-frequency field operator becomes

$$\hat{\mathbf{E}}_j^{(+)}(\vec{r},t) = \int d\omega \mathcal{E}_j \vec{e}_j \hat{a}_j(\omega) e^{i(\vec{k}_j \cdot \vec{r} - \omega t)}. \tag{2.10}$$

Fourier transformation yields the continuous-mode operators in the time domain, [25]

$$\hat{a}_j^\dagger(t) = \frac{1}{\sqrt{2\pi}} \int d\omega \, \hat{a}_j^\dagger(\omega) e^{-i\omega t},$$
$$\hat{a}_j(t) = \frac{1}{\sqrt{2\pi}} \int d\omega \, \hat{a}_j(\omega) e^{-i\omega t}, \tag{2.11}$$
$$\hat{n}_j = \int dt \, \hat{a}_j^\dagger(t) \hat{a}_j(t),$$

with the commutation relation

$$[\hat{a}_j(t), \hat{a}_j^\dagger(t')] = \delta(t - t'). \tag{2.12}$$

To describe experiments with pulsed light, it is convenient to define a photon-wavepacket creation operator [25]

$$\hat{a}_{j,f}^\dagger = \int dt f_j(t) \hat{a}_j^\dagger(t) = \int d\omega F_j(\omega) \hat{a}_j^\dagger(\omega), \tag{2.13}$$

where $f_j(t)$ and $F_j(\omega)$ represent the temporal and spectral profile, respectively, of the photon wavepacket, and are normalized according to Eqs. (2.12) and (2.13) [25]

$$\int d\omega |F_j(\omega)|^2 = \int dt |f_j(t)|^2 = 1. \tag{2.14}$$

In this case, the positive-frequency component of the field can be written

$$\hat{\mathbf{E}}_j^{(+)}(\vec{\mathbf{r}},t) = \int d\omega \mathcal{E}_j \vec{\mathbf{e}}_j F_j^*(\omega) \hat{a}_j(\omega) e^{i(\vec{\mathbf{k}}_j \cdot \vec{\mathbf{r}} - \omega t)}. \tag{2.15}$$

Substituting $F_j(\omega) = \delta(\omega - \omega_j)$ into Eq. (2.15) yields the single-mode monochromatic case in Eq. (2.7). If the bandwidth over which $F_j^*(\omega)$ is non-zero is sufficiently narrow compared to the center frequency ω_j, one may make the approximation

$$\hat{\mathbf{E}}_j^{(+)}(\vec{\mathbf{r}},t) \simeq \mathcal{E}_j \vec{\mathbf{e}}_j e^{i(\vec{\mathbf{k}}_j \cdot \vec{\mathbf{r}} - \omega_j t)} \int d\omega F_j^*(\omega) \hat{a}_j(\omega). \tag{2.16}$$

Time-dependence is incorporated into quantum optics calculations using one of three representations: the Schrödinger picture, the Heisenberg picture, or the Interaction picture. In the Schrödinger picture, the state vector $|\psi^{(S)}(t)\rangle$ is time-dependent but the operators representing observables $\hat{O}^{(S)}$ are independent of time. In the Heisenberg picture, the state vectors $|\psi^{(H)}\rangle$ are time-independent, while the operators $\hat{O}^{(H)}(t)$ depend on time. The Heisenberg picture is often used to describe pulsed experiments where the duration of the pulse is short compared to the temporal resolution of the detectors. In this case it is common to use a time-independent density matrix, which describes the state of the field averaged over the duration of the pulse, and further averaged from pulse to pulse. It may not always be trivial to calculate the average over the pulse duration, but experimentally this is often what is measured given that single-photon detectors are relatively slow in comparison to the duration of ultrafast laser pulses. In the Interaction picture, both the state vectors $|\psi^{(I)}(t)\rangle$ and the operators $\hat{O}^{(I)}(t)$ are allowed to depend on time. The Interaction picture allows the computation of coherences that depend on time delay τ, such as $g^{(2)}(\tau)$. Each of these three representations are used in this book, with the choice of picture based on convenience for a given scenario. The superscript labels (S), (H), and (I) are used here for illustration, but are typically not included elsewhere.

2.2 SOURCE CHARACTERISTICS

2.2.1 State Vector

If the light emitted by a source is in a pure state, it can be described by a state vector. One can write a generalized multi-mode version of the state vector for a quantized electric field in a pure state, at position $\vec{\mathbf{r}}$ and time t, as [25]

$$|\psi(\vec{\mathbf{r}},t)\rangle = \sum_{n_1,n_2,n_3,\ldots} c_{n_1,n_2,n_3,\ldots}(\vec{\mathbf{r}},t)|n_1\rangle_1 |n_2\rangle_2 |n_3\rangle_3 \ldots, \tag{2.17}$$

where $c_{n_1,n_2,n_3,\ldots}(\vec{\mathbf{r}},t)$ is the complex coefficient of the multi-mode number state ket representing n_1 photons in mode 1, n_2 photons in mode 2, and so on.

This ket can be written as [15]

$$|n_1\rangle_1|n_2\rangle_2|n_3\rangle_3 \cdots = \prod_{j=1,2,3,\ldots} \frac{(\hat{a}_j^\dagger)^{n_j}}{\sqrt{n_j!}}|0\rangle, \qquad (2.18)$$

where $|0\rangle$ is the global vacuum, representing no photons in any mode. The infinite set of these multi-mode number states form a complete orthonormal set of basis states, so that [25]

$$[\hat{a}_j, \hat{a}_k^\dagger] = \delta_{j,k}. \qquad (2.19)$$

In many situations treated in this book, only one mode contains any photons, so the representation of the state can be simplified greatly. In the single-mode case, we can write

$$|\psi(\vec{r},t)\rangle = \sum_{n=0}^{\infty} c_n(\vec{r},t)|n\rangle, \qquad (2.20)$$

where we have dropped the mode subscript j on the number state. If the state vector does not depend on r or t over the region of interest, or if the information about time- and space-dependence is incorporated into the definition of the mode, we can use the simplified notation

$$|\psi\rangle = \sum_{n=0}^{\infty} c_n|n\rangle. \qquad (2.21)$$

In this case an *ideal* single-photon source will emit light in the state

$$|\psi\rangle = |1\rangle, \qquad (2.22)$$

which contains exactly one photon in exactly one optical mode.

2.2.2 Density Matrix and Photon Number Probabilities

A more general quantum representation of a state is the density matrix, which is valid for both pure and mixed states. The density-matrix operator for a mixed state can be written as an ensemble average of pure states,

$$\hat{\rho} = \sum_i p_i |\psi_i\rangle\langle\psi_i|, \qquad (2.23)$$

where p_i is the probability that the light field occupies state $|\psi_i\rangle$. For a pure state, only one of the coefficients p_i is non-zero, and is equal to one. The ensemble average of an observable $\hat{O}$ can be found from the trace of its product with the density matrix [25]:

$$\langle \hat{O} \rangle = \text{Tr}\{\hat{\rho}\hat{O}\}. \qquad (2.24)$$

When the photon number states are used as basis states, the diagonal elements of $\hat{\rho}$ contain the photon number probabilities,

$$P(n) = \text{Tr}\{\hat{\rho}|n\rangle\langle n|\} = \langle n|\hat{\rho}|n\rangle, \qquad (2.25)$$

where $P(n)$ is the probability that a source emits n photons. The photon number probabilities are normalized such that

$$\sum_{n=0}^{\infty} P(n) = \text{Tr}\{\hat{\rho}\} = 1, \qquad (2.26)$$

which satisfies the usual density-matrix normalization condition.

The notation $P(n > m)$ refers to the total probability that a source emits more than m photons

$$P(n > m) = \sum_{n=m+1}^{\infty} P(n). \qquad (2.27)$$

The probability of multi-photon emission, $P(n > 1)$, is a special case of this:

$$P(n > 1) = \sum_{n=2}^{\infty} P(n). \qquad (2.28)$$

For a pulsed source, $P(n)$ indicates the probability that the source emits n photons per pulse. The mean number of photons per pulse is

$$\mu = \text{Tr}\{\hat{\rho}\hat{n}\}. \qquad (2.29)$$

For typical photon sources, if $P(1) \gg P(n > 1)$, then it is usually the case that $\mu \simeq P(1)$.

For a continuous-wave (CW) source we define $P(n; T)$ as the probability that the source emits n photons in a time interval T, and $\mu(T)$ as the mean number of photons present over the same interval. If $P(1; T) \gg P(n > 1; T)$, then for typical sources $\mu(T) \simeq P(1; T)$. Care should be taken when choosing T, as there can be ambiguity in whether a source emits n photons simultaneously or in succession over the duration of T. This will be discussed in more detail in Section 2.2.5.

2.2.3 Purity

In many applications, particularly those that require indistinguishable photons, it is important that photons be emitted in pure states. The purity of a state can be quantified as

$$\mathcal{P} = \text{Tr}\{\hat{\rho}^2\}. \qquad (2.30)$$

Purity has an upper limit of unity for a pure state and a lower limit of $1/N$ for a completely mixed N-dimensional state. A simple example of a mixed state is a pulsed source whose state vector varies from pulse to pulse. If the variation is small, $\mathcal{P}$ may be approximately unity.

2.2.4 Source Efficiency and Generation Rate

For a pulsed source, the source efficiency η_{source} is defined as the probability that the source delivers one or more photons to an experiment for each pump pulse:

$$\eta_{\text{source}} = P(n > 0). \quad (2.31)$$

When characterizing a source, the experiment is often simply a detector or detection system (see Fig. 2.1). In the limit of low multi-photon probability, $P(1) \gg P(n > 1)$, the source efficiency can be estimated from the detected count rate and the detection efficiency as

$$\eta_{\text{source}} \simeq P(1) \simeq \frac{R_{\text{detect}}^{\text{correct}}}{\eta_{\text{DE}} R_{\text{pump}}}, \quad (2.32)$$

where η_{DE} is the detection efficiency defined in Section 2.3.1, R_{pump} is the repetition rate of the electrical or optical pump, and $R_{\text{detect}}^{\text{correct}}$ is the detected count rate, corrected as necessary to account for detector nonidealities such as dark counts, dead time, afterpulsing and differing response to single photons as compared to multiple photons. These corrections, which may or may not be important depending on the specific source and detector involved, are discussed in Chapter 8.

In some cases, it may be possible to decompose the source efficiency into two components,

$$\eta_{\text{source}} = \eta_{\text{gen}} \eta_{\text{extract}}, \quad (2.33)$$

where η_{gen} is the generation efficiency of the source itself and η_{extract} is the extraction efficiency. The generation efficiency is the probability that one or more photons are created within the source per pump pulse. The extraction efficiency includes all optical losses incurred in extracting the photons from where they are generated to where they are useful for an application. This can

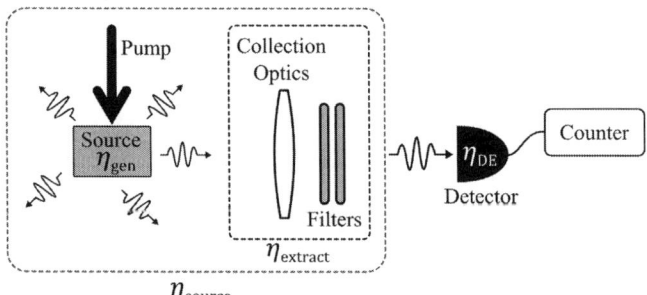

FIGURE 2.1 Schematic showing how the source efficiency, η_{source}, is defined for a pulsed source: as the total probability that a source can deliver one or more photons to an experiment for each pump pulse. In the case shown here, the "experiment" is simply a single-photon detector and a counter. The source efficiency is the product of the generation efficiency of the source itself, η_{gen}, with the efficiency with which photons are extracted from the source and delivered to the detector, η_{extract}.

include losses due to spectral filtering for defining the emission bandwidth and the rejection of any pump light, spatial filtering, geometric alignment, and beam shaping optics. Extraction efficiency is often characterized by measuring transmittance and mode overlap. If η_{gen} and η_{extract} are identified, their dependence on collection optics, spectral bandwidth of filters, pump power, and other critical parameters such as the physical interaction volume defined by the collection should also be given.

A pulsed source may also be characterized by the maximum repetition rate at which it could deliver photons to an experiment, $R_{\text{rep}}^{\max}$. If $R_{\text{rep}}^{\max}$ is identified, any effects that increased rate has on other characteristics, such as the multi-photon emission probability, should also be identified.

For a CW source, the generation rate R_{gen} is typically a more relevant quantity than efficiency. R_{gen} is the total rate at which photons are delivered to a detector, corrected, as in the pulsed case, for detection efficiency and other detector non-idealities:

$$R_{\text{gen}} = \frac{R_{\text{detect}}^{\text{correct}}}{\eta_{\text{DE}}}. \tag{2.34}$$

2.2.5 Second-Order Coherence, $g^{(2)}$

The second-order coherence, $g^{(2)}$, is by far the most common measurement used for determining the quality of a single-photon source, because it gives information about the source's multi-photon emission probability. $g^{(2)}$ is also referred to as the second-order correlation function or the normalized intensity correlation. The second-order coherence between mode j, measured at position $\vec{r_1}$ and time t_1, and mode k, measured at position $\vec{r_2}$ and time t_2 is [25]

$$g_{j,k}^{(2)}(\vec{r_1},t_1;\vec{r_2},t_2) = \frac{\langle \hat{E}_j^{(-)}(\vec{r_1},t_1)\hat{E}_k^{(-)}(\vec{r_2},t_2)\hat{E}_k^{(+)}(\vec{r_2},t_2)\hat{E}_j^{(+)}(\vec{r_1},t_1)\rangle}{\langle \hat{E}_j^{(-)}(\vec{r_1},t_1)\hat{E}_j^{(+)}(\vec{r_1},t_1)\rangle\langle \hat{E}_k^{(-)}(\vec{r_2},t_2)\hat{E}_k^{(+)}(\vec{r_2},t_2)\rangle}, \tag{2.35}$$

where the angled brackets denote an ensemble average; in a laboratory experiment, this typically entails an average over a large number of detected photons. Writing in terms of creation and annihilation operators, and cancelling common factors in the numerator and denominator, this becomes

$$g_{j,k}^{(2)}(\vec{r_1},t_1;\vec{r_2},t_2) = \frac{\langle \hat{a}_j^\dagger(\vec{r_1},t_1)\hat{a}_k^\dagger(\vec{r_2},t_2)\hat{a}_k(\vec{r_2},t_2)\hat{a}_j(\vec{r_1},t_1)\rangle}{\langle \hat{a}_j^\dagger(\vec{r_1},t_1)\hat{a}_j(\vec{r_1},t_1)\rangle\langle \hat{a}_k^\dagger(\vec{r_2},t_2)\hat{a}_k(\vec{r_2},t_2)\rangle}. \tag{2.36}$$

For measurements on a single mode ($j = k$), the j and k subscripts are typically omitted.

This can be written more compactly for a stationary source—one whose properties depend on a time delay $\tau = t_2 - t_1$ but not on specific values of t_1 and t_2. Stable CW sources are typically stationary, whereas pulsed sources are inherently not stationary. For a single-mode, stationary source measured

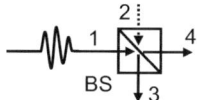

FIGURE 2.2 Illustration of the ports (modes) of an optical beamsplitter (BS). Here an arbitrary field is incident on port 1, and the vacuum field is incident on port 2. For a BS with a field reflection coefficient $\mathcal{R}$ and field transmission coefficient $\mathcal{T}$, the destruction operators can be related as $\hat{a}_3 = \mathcal{R}\hat{a}_1 + \mathcal{T}\hat{a}_2$ and $\hat{a}_4 = \mathcal{T}\hat{a}_1 + \mathcal{R}\hat{a}_2$.

at a single position ($\vec{r}_1 = \vec{r}_2$)—or two positions that can be considered as equivalent—this simplifies to the more familiar form:

$$g^{(2)}(\tau) = \frac{\langle \hat{a}^\dagger(t)\hat{a}^\dagger(t+\tau)\hat{a}(t+\tau)\hat{a}(t)\rangle}{\langle \hat{a}^\dagger(t)\hat{a}(t)\rangle^2}. \tag{2.37}$$

It is tempting to try to express $g^{(2)}(\tau)$ in terms of the time-dependent number operator, $\hat{n}(t) = \hat{a}^\dagger(t)\hat{a}(t)$. However, because $\hat{a}$ and $\hat{a}^\dagger$ do not commute, the normally ordered operators in this expression *cannot* be rearranged and

$$g^{(2)}(\tau) \neq \frac{\langle \hat{n}(t)\hat{n}(t+\tau)\rangle}{\langle \hat{n}(t)\rangle^2}. \tag{2.38}$$

In the special case of $\tau = 0$, on the other hand, it is possible to write $g^{(2)}$ in terms of $\hat{n}(t)$ as [25]

$$g^{(2)}(0) = \frac{\langle \hat{n}(t)(\hat{n}(t)-1)\rangle}{\langle \hat{n}(t)\rangle^2}. \tag{2.39}$$

Fortunately, this zero-delay value is typically the most relevant for characterizing a single-photon source. Using this expression, $g^{(2)}(0)$ can be related to the photon number probabilities $P(n)$; this will be discussed in the next section.

One important property of $g^{(2)}(\tau)$ for a light source is that its value does not change with loss as long as all modes experience the same loss. To show this, one need only verify that transmission through a beamsplitter (BS) does not change the second-order coherence, since loss can be modeled as transmission through a lossless BS with arbitrary reflectance and transmittance [26]. An example of such a BS is shown in Fig. 2.2. The input and output destruction operators can be related as [25]

$$\hat{a}_3 = \mathcal{R}\hat{a}_1 + \mathcal{T}\hat{a}_2,$$
$$\hat{a}_4 = \mathcal{T}\hat{a}_1 + \mathcal{R}\hat{a}_2. \tag{2.40}$$

The electric-field reflection and transmission coefficients, $\mathcal{R}$ and $\mathcal{T}$, are complex numbers that satisfy $|\mathcal{R}|^2 + |\mathcal{T}|^2 = 1$ and $\mathcal{R}\mathcal{T}^* + \mathcal{T}\mathcal{R}^* = 1$ for a lossless symmetric BS. For an arbitrary field input at port 1 and a vacuum field input at port 2, one can readily show that [25]

$$g^{(2)}_{44}(\tau) = \frac{|\mathcal{T}|^4 \langle \hat{a}_1^\dagger(t)\hat{a}_1^\dagger(t+\tau)\hat{a}_1(t+\tau)\hat{a}_1(t)\rangle}{|\mathcal{T}|^2 \langle \hat{a}_1^\dagger(t)\hat{a}_1(t)\rangle |\mathcal{T}|^2 \langle \hat{a}_1^\dagger(t+\tau)\hat{a}_1(t+\tau)\rangle}. \tag{2.41}$$

with a similar expression for $g_{33}^{(2)}(\tau)$. Thus the second-order coherence of the transmitted and reflected fields is equal to that of the incident light [25]:

$$g_{11}^{(2)}(\tau) = g_{33}^{(2)}(\tau) = g_{44}^{(2)}(\tau). \tag{2.42}$$

As a result, $g^{(2)}(\tau)$ can be measured accurately in an optical system that has low transmission or low extraction efficiency.

2.2.6 Relating $g^{(2)}$ to $P(n)$

To gain more insight into the information contained in the second-order coherence it is useful to see how it can be related to the photon number probabilities. Here we examine this relationship for both pulsed and CW sources.

2.2.6.1 Pulsed Sources

Pulsed sources are inherently not stationary, and Eq. (2.37) is not valid. In this case, it can be convenient to define a discrete version of $g^{(2)}$ [40]

$$g^{(2)}[m] = \frac{\langle \hat{a}^\dagger[l]\hat{a}^\dagger[l+m]\hat{a}[l+m]\hat{a}[l]\rangle}{\langle \hat{a}^\dagger[l]\hat{a}[l]\rangle\langle \hat{a}^\dagger[l+m]\hat{a}[l+m]\rangle}, \tag{2.43}$$

where l and m take on integer values denoting pulse number, and the angled brackets indicate an average over l. Thus $g^{(2)}[0]$ is the autocorrelation of the pulse train, while $g^{(2)}[1]$ is the cross-correlation of each pulse with its nearest subsequent neighbor (see Fig. 2.3).

In this chapter, square brackets are used to distinguish the discrete form $g^{(2)}[m]$, valid for pulsed sources, from the continuous form $g^{(2)}(\tau)$, valid for CW sources. In the discrete notation, $g^{(2)}[0]$ is the relevant metric for quantifying the multi-photon emission probability of pulsed single-photon sources, while $g^{(2)}(0)$ is relevant for continuously pumped sources. The remaining chapters of this book drop this distinction and follow the usual convention of using the notation $g^{(2)}(0)$ to characterize both pulsed and CW sources.

The zero-delay value can be written as

$$g^{(2)}[0] = \frac{\langle \hat{n}(\hat{n}-1)\rangle}{\langle \hat{n}\rangle^2}, \tag{2.44}$$

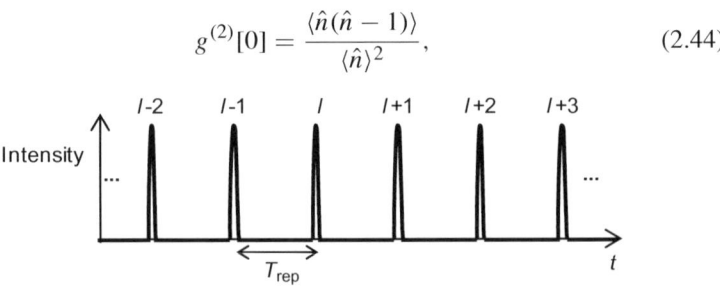

FIGURE 2.3 Optical pulse train illustrating the labeling scheme used in the discrete notation. T_{rep} is the repetition time of the system, which is typically set by the repetition time of the optical or electrical pump.

where the number operator $\hat{n} = \hat{a}^\dagger[l]\hat{a}[l]$ measures the number of photons in the lth pulse. The expectation values can be rewritten in terms of the density matrix as

$$g^{(2)}[0] = \frac{\text{Tr}\{\hat{\rho}\hat{n}(\hat{n}-1)\}}{(\text{Tr}\{\hat{\rho}\hat{n}\})^2}. \quad (2.45)$$

Only the diagonal terms of the density matrix (when written in the photon-number-state basis) contribute to $g^{(2)}[0]$. Thus we can rewrite this in terms of the per-pulse photon probabilities as

$$\begin{aligned} g^{(2)}[0] &= \frac{\sum_{n=0}^{\infty} n(n-1)P(n)}{[\sum_{n=0}^{\infty} nP(n)]^2} \\ &= \frac{2P(2) + 6P(3) + 12P(4) + \cdots}{[P(1) + 2P(2) + 3P(3) + \cdots]^2} \\ &= \frac{2P(2) + 6P(3) + 12P(4) + \cdots}{\mu^2}, \end{aligned} \quad (2.46)$$

where μ is the mean photon number per pulse. This expression can be used to bound the probability of multi-photon events by relating this probability to the numerator of $g^{(2)}[0]$ as [8,45]

$$P(n > 1) = \sum_{n=2}^{\infty} P(n) \leq \frac{1}{2} \sum_{n=0}^{\infty} n(n-1)P(n). \quad (2.47)$$

As a result, a measured $g^{(2)}[0]$ can be used to place an upper limit on the multi-photon probability:

$$P(n > 1) \leq \frac{1}{2}\mu^2 g^{(2)}[0]. \quad (2.48)$$

If $P(2) \gg P(n > 2)$, Eq. (2.46) can be approximated as

$$g^{(2)}[0] \simeq \frac{2P(2)}{\mu^2}. \quad (2.49)$$

In the special case where $P(1) \gg P(2) \gg P(n > 2)$, which holds for many low-efficiency sources, this simplifies to

$$g^{(2)}[0] \sim \frac{2P(2)}{[P(1)]^2}. \quad (2.50)$$

Note that $g^{(2)}[0]$ does not directly reflect the two:one photon ratio $P(2)/P(1)$, as one might naively expect. Although $g^{(2)}[0]$ is the most commonly cited metric of single-photon source quality, for some applications $P(2)/P(1)$ is the more relevant quantity.

The discussion above assumes each pulse is treated as a whole, and photon statistics are averaged over the pulse duration. More information about a source can be obtained by studying the detailed time evolution of $g^{(2)}$ during each pulse [1,12].

2.2.6.2 CW Sources

If $P(1; T) \gg P(2; T) \gg P(n > 2; T)$ for a time interval T over which $g^{(2)}(\tau)$ is approximately equal to the zero-delay value (*i.e.*, $g^{(2)}(T) \simeq g^{(2)}(0)$), then this can be approximated as

$$g^{(2)}(0) \simeq \frac{2P(2; T)}{[P(1; T)]^2}. \quad (2.51)$$

Care should be taken when choosing T, given that a source may be able to emit two photons simultaneously or in succession.

To illustrate this point, it is useful to consider an example. Suppose we have an approximation to a single-photon source that emits photons at a rate $R_{gen} = 10^6$ s^{-1} and has the $g^{(2)}(\tau)$ shown in Fig. 2.4, where $g^{(2)}(0) = 0.1$ and $g^{(2)}(\tau)$ is approximately constant over a time delay range >1 ns about the origin, as shown in Fig. 2.4b. Thus, over a time interval $T = 1$ ns, the probability of the source emitting one or more photons is

$$P(n > 0; 1 \text{ ns}) = 10^{-9} \text{ s} \times 10^6 \text{ s}^{-1} = 10^{-3}. \quad (2.52)$$

Given this very low probability, we can make the approximation $P(1; 1 \text{ ns}) \simeq P(n > 0; 1 \text{ ns})$. Solving Eq. (2.51) for the two-photon probability over a 1 ns interval yields

$$P(2; 1 \text{ ns}) \simeq \frac{1}{2} g^{(2)}(0)[P(1; 1 \text{ ns})]^2. \quad (2.53)$$

Using the numbers for this example source, we find

$$P(2; 1 \text{ ns}) \simeq 5 \times 10^{-8}. \quad (2.54)$$

The two:one photon ratio in this case is

$$\frac{P(2; 1 \text{ ns})}{P(1; 1 \text{ ns})} \simeq 5 \times 10^{-5}. \quad (2.55)$$

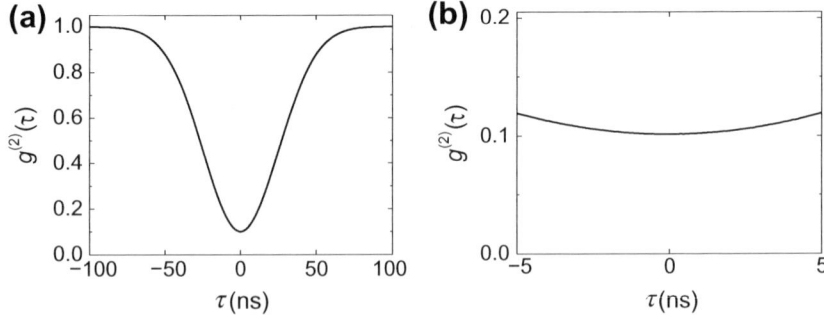

FIGURE 2.4 Second-order coherence of the example non-ideal single-photon source discussed in the text. Plot (b) is a close-up view near zero delay of the data in (a).

By contrast, a source with Poissonian photon statistics and the same generation rate would have

$$P(2; 1 \text{ ns}) \simeq 5 \times 10^{-7} \tag{2.56}$$

and

$$\frac{P(2; 1 \text{ ns})}{P(1; 1 \text{ ns})} \simeq 5 \times 10^{-4}. \tag{2.57}$$

The two:one photon ratio is an order of magnitude worse in the latter case, although neither is particularly large when considering such a short time duration. Judging from the width of the antibunching dip, however, the photons emitted by this source are likely to have wavepacket envelopes with durations much longer than 1 ns, leading to much higher two-photon probabilities. Calculations over time scales where $g^{(2)}(\tau)$ is not constant are outside the scope of the simple treatment outlined here.

2.2.7 Ideal and Non-Ideal Single-Photon Sources

As mentioned in Sections 2.2.1 and 2.2.4, an ideal single-photon source will emit exactly one photon into a well-defined optical mode that can be efficiently collected and delivered to an experiment. Two important and distinct features of an ideal source are: (1) it never emits more than one photon at a time, and (2) it emits single photons with unity source efficiency. An ideal pulsed source thus has $\eta_{\text{source}} = 1$ and $g^{(2)}[0] = 0$. For an ideal CW source, the generation rate is limited only by the temporal duration of each photon wavepacket (often referred to as temporal coherence) and $g^{(2)}(0) = 0$, with $g^{(2)}(\tau) \to 1$ as $\tau \to \infty$.

To illustrate the distinction between these two features, consider a non-ideal source in the form of a pulsed source that never emits two or more photons in the same pulse, yet has $\eta_{\text{source}} < 1$. Such a source could emit light into the pure state

$$|\psi\rangle = \sqrt{P(0)}|0\rangle + \sqrt{P(1)}|1\rangle, \tag{2.58}$$

or the mixed state

$$\hat{\rho} = P(0)|0\rangle\langle 0| + P(1)|1\rangle\langle 1|. \tag{2.59}$$

The multi-photon emission probability of either source is zero,

$$P(n > 1) = 0, \tag{2.60}$$

as is the second-order coherence at zero delay,

$$g^{(2)}[0] = 0. \tag{2.61}$$

The source efficiency, mean photon number, and probability of one-photon emission are all equal:

$$\eta_{\text{source}} = \mu = P(1). \tag{2.62}$$

In practice, no single-photon source is ideal in all ways—yet much can be done with non-ideal sources. To determine whether a source is useful in a given application it is necessary to establish whether the multi-photon emission probabilities are sufficiently small and whether the source efficiency (or generation rate) is sufficiently large for that application.

2.2.8 To measure $P(n)$ or $g^{(2)}$?

It might seem that measuring the photon number distribution $P(n)$ would be the most direct way to characterize a single-photon source. One could then compare the single-photon probability $P(1)$ with the multi-photon probabilities $P(2), P(3)$, etc. Photon-number-resolving detectors do exist (they are discussed in Section 2.3 and in subsequent chapters); to date, however, even the best devices do not yet combine very high efficiency with low timing jitter and short dead times [31,46,27]. Unless the detection efficiency of a number-resolving detector is 100%, the measured photon statistics will be skewed by loss. In principle, the loss matrix can be inverted to reconstruct the statistics of the optical state [46,27,5], but such an inversion works best with efficiencies as high as possible. By contrast, $g^{(2)}$ can be measured accurately even with click/no-click detectors with low detection efficiencies. This is one of the key reasons that $g^{(2)}$ is used to characterize the vast majority of single-photon sources.

2.2.9 Hanbury Brown-Twiss Interferometer

Typically, $g^{(2)}(\tau)$ is measured with a Hanbury Brown-Twiss (HBT) interferometer [19], such as the one shown in Fig. 2.5. This consists of a beamsplitter, two discrete single-photon detectors, and some form of timing or coincidence electronics. An incoming optical field in mode 1 is divided at the beamsplitter and sent to two output ports, modes 3 and 4, which are incident on two detectors, D_3 and D_4. One can appreciate how the HBT interferometer works qualitatively for an ideal single-photon source: if only one photon at a time is input in mode 1, there is no way for both detectors D_3 and D_4 to detect a photon simultaneously, and $g^{(2)}(0) = 0$.

The timing electronics record the relative time delay between the two detection events at D_3 and D_4. In the original HBT experiment [19], and for many years thereafter, the timing electronics consisted of a simple start-stop

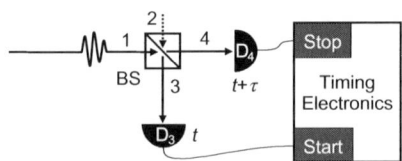

FIGURE 2.5 Hanbury Brown-Twiss interferometer. The light field to be measured is incident on input port 1 of the beamsplitter (BS), while a vacuum field is incident on input port 2. With appropriate normalization, a histogram of the correlations between photon numbers measured by detectors at output ports 3 and 4 can yield a good approximation to $g^{(2)}(\tau)$.

timing circuit. In this case, the timer starts when detector D$_3$ registers a photon and stops after a time delay τ, when detector D$_4$ records a photon. The timing electronics record a histogram of these start-stop delay times that is ideally proportional to $g^{(2)}(\tau)$. It is not critical that the optical path length from the BS to D$_3$ is equal to that from the BS to D$_4$, as any difference in photon arrival time can be compensated by changing the length of electrical cables or adjusting internal delays in the electronics. In the case of a simple start-stop timer, the stop pulse is often purposely delayed from the start, so that negative delay data ($\tau < 0$) can be recorded. In recent years, more sophisticated electronics have been employed. For example, time-tagging electronics can record the arrival times of all detected photons and data can be post-processed to form multi-start, multi-stop correlation histograms that more accurately resemble $g^{(2)}(\tau)$, especially at higher count rates.

In the remainder of this section we examine the HBT interferometer more quantitatively and explore how well real HBT experiments recover $g^{(2)}(\tau)$.

2.2.9.1 HBT with Photon-Number-Resolving Detectors

If D$_3$ and D$_4$ are ideal photon-number-resolving detectors with unity detection efficiency, they can accurately measure $\hat{n}_3(t)$, the number of photons in mode 3 at time t, and $\hat{n}_4(t+\tau)$, the number of photons in mode 4 at time $t+\tau$. Such detectors could be used to obtain a normalized correlation between the photon numbers in the two outputs,

$$\gamma_{\text{PNR}}^{(2)}(\tau) = \frac{\langle \hat{n}_3(t)\hat{n}_4(t+\tau)\rangle}{\langle \hat{n}_3(t)\rangle\langle \hat{n}_4(t+\tau)\rangle}, \tag{2.63}$$

where the subscript PNR denotes the use of photon-number-resolving detectors. Given that $[\hat{a}_3,\hat{a}_4^\dagger] = 0$, the operators in the numerator can be rearranged to show that this correlation is equal to the second-order coherence between these two modes [13,25]:

$$\gamma_{\text{PNR}}^{(2)}(\tau) = \frac{\langle \hat{a}_3^\dagger(t)\hat{a}_4^\dagger(t+\tau)\hat{a}_4(t+\tau)\hat{a}_3(t)\rangle}{\langle \hat{a}_3^\dagger(t)\hat{a}_3(t)\rangle\langle \hat{a}_4^\dagger(t+\tau)\hat{a}_4(t+\tau)\rangle} = g_{34}^{(2)}(\tau), \tag{2.64}$$

Using the beamsplitter transformations in Eq. (2.40) one can show that the measured photon number correlation between D$_3$ and D$_4$ is exactly equal to the second-order coherence of the incident light:

$$\gamma_{\text{PNR}}^{(2)}(\tau) = \frac{|\mathcal{R}|^2|\mathcal{T}|^2\langle \hat{a}_1^\dagger(t)\hat{a}_1^\dagger(t+\tau)\hat{a}_1(t+\tau)\hat{a}_1(t)\rangle}{|\mathcal{R}|^2\langle \hat{a}_1^\dagger(t)\hat{a}_1(t)\rangle|\mathcal{T}|^2\langle \hat{a}_1^\dagger(t+\tau)\hat{a}_1(t+\tau)\rangle} = g_{11}^{(2)}(\tau). \tag{2.65}$$

So far we have assumed 100% detection efficiency, but we can relax this requirement and still recover $g_{11}^{(2)}(\tau)$ exactly. To show this, we model the detection efficiencies, η_3 and η_4, as a beamsplitter with transmittance η_3 in front

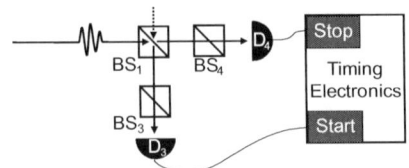

FIGURE 2.6 Model of a Hanbury Brown-Twiss interferometer employing detectors with nonunity detection efficiency. The detection efficiency of the detectors in modes 3 and 4 are taken to be η_3 and η_4. These efficiencies are modeled here as beamsplitters BS$_3$ and BS$_4$, with transmission coefficient $T_3 = \sqrt{\eta_3}$ and $T_4 = \sqrt{\eta_4}$, followed by detectors D$_3$ and D$_4$, respectively. D$_3$ and D$_4$ are both modeled as ideal photon-number-resolving detectors with $\eta_{DE} = 1$.

of D$_3$ and a beamsplitter with transmittance η_4 in front of D$_4$. The normalized correlation between the number of photons *detected* by D$_3$ and D$_4$ is also equal to $g^{(2)}(\tau)$ of the incident field [25,26] (see Fig. 2.6).

$$\gamma^{(2)}_{\text{PNR}(\eta_3,\eta_4)}(\tau) = \frac{\eta_3\eta_4\langle\hat{n}_3(t)\hat{n}_4(t+\tau)\rangle}{\eta_3\langle\hat{n}_3(t)\rangle\eta_4\langle\hat{n}_4(t+\tau)\rangle} = g^{(2)}_{11}(\tau). \tag{2.66}$$

2.2.9.2 HBT with "Click" Detectors

Since ideal PNR detectors with sufficient timing resolution are not readily available, HBT interferometers have typically employed single-photon detectors that can be described as "click," or "click/no-click," detectors. Such a detector will give an output response if it detects one or more photons, but has no way of distinguishing two or more photons from one photon. Click detectors are discussed in more detail in Section 2.3.3.

To explore the conditions under which an HBT interferometer with click detectors correctly measures the second-order coherence, we can define the measured temporal correlation function

$$\gamma^{(2)}_{\text{click}}(\tau) \equiv \left\langle \frac{p_{3,4}(\text{click}(t),\text{click}(t+\tau))}{p_3(\text{click}(t))p_4(\text{click}(t+\tau))} \right\rangle, \tag{2.67}$$

where $p_{3,4}(\text{click}(t),\text{click}(t+\tau))$ is the probability that detector 3 clicks at time t and detector 4 clicks at time $t+\tau$; $p_3(\text{click}(t))$ is the probability that detector 3 clicks at time t; $p_4(\text{click}(t+\tau))$ is the probability that detector 4 clicks at time $t+\tau$; and the average is taken over all time t. We can estimate this quantity from measurements by using

$$\hat{\gamma}^{(2)}_{\text{click}}(\tau) = \frac{N_c(\tau;\Delta\tau)}{R_3 R_4 \Delta\tau T_{\text{int}}}, \tag{2.68}$$

where $N_c(\tau;\Delta\tau)$ is the number of correlation events recorded by the timing electronics in the histogram bin centered at delay τ having width $\Delta\tau$. This quantity represents the number of times that a click on detector 3 is followed by a click on detector 4 after a delay ranging from $\tau - \Delta\tau/2$ to $\tau + \Delta\tau/2$. R_3 and

R_4 are the singles count rates on detectors D_3 and D_4, and T_{int} is the integration time or experiment run time. For a CW source, $\Delta\tau$ is often set at the minimum value allowed by the timing resolution of the detectors and electronics, but in some cases a wider bin can be chosen if $g^{(2)}(\tau)$ evolves more slowly with τ.

For pulsed sources, the measured correlation function can be normalized to find the area of each pulse in the histogram as

$$\hat{\gamma}^{(2)}_{click}[m] = \frac{N_c[m]}{R_3 R_4 T_{rep} T_{int}}, \tag{2.69}$$

where m takes on integer values and $\Delta\tau$ has been replaced by the pump repetition period T_{rep}. Assuming a lossless BS with arbitrary $\mathcal{R}$ and $\mathcal{T}$ and click detectors with detection efficiencies η_3 and η_4, the pulsed measurement outcome at zero delay can be written as a function of the photon probability distribution of the incident light by computing the joint probability distribution of photons detected by D_3 and D_4 [44].

In general, $\gamma^{(2)}_{click}[0]$ is not equal to $g^{(2)}[0]$, but under appropriate experimental conditions $\gamma^{(2)}_{click}[0]$ can approximate $g^{(2)}[0]$ quite well. It is illustrative to write out the first few terms of each factor in the numerator and denominator [44]:

$$\gamma^{(2)}_{click}[0] = \frac{\eta_3\eta_4|\mathcal{R}|^2|\mathcal{T}|^2\left[2P(2) + 6P(3)\left(1 - \frac{1}{2}\eta_3|\mathcal{R}|^2 - \frac{1}{2}\eta_4|\mathcal{T}|^2\right) + \cdots\right]}{\eta_3|\mathcal{R}|^2\left[P(1) + 2P(2)\left(1 - \frac{1}{2}\eta_3|\mathcal{R}|^2\right) + \cdots\right]\eta_4|\mathcal{T}|^2\left[P(1) + 2P(2)\left(1 - \frac{1}{2}\eta_4|\mathcal{T}|^2\right) + \cdots\right]} \tag{2.70}$$

where the $P(n)$'s represent the photon number probability distribution of the light input in mode 1. If we keep just the lowest order terms in Eq. (2.70), we can see that it reduces to Eq. (2.50). For higher photon numbers, $\mathcal{R}, \mathcal{T}, \eta_3$, and η_4 appear as correction factors, indicating that the higher photon number terms tend to be underestimated by the click detectors. In the limit of very low detection efficiencies, these corrections to the higher photon number terms become negligible. Applying l'Hôpital's rule to Eq. (2.70) yields

$$\lim_{\eta_3,\eta_4 \to 0} \gamma^{(2)}_{click}[0] = g^{(2)}[0], \tag{2.71}$$

regardless of the values of $\mathcal{R}$ and $\mathcal{T}$. Alternatively, if the source has very low multi-photon generation probabilities so that $P(1) \gg P(2) \gg P(n > 2)$, then

$$\gamma^{(2)}_{click}[0] \simeq \frac{2P(2)}{[P(1)]^2} \simeq g^{(2)}[0]. \tag{2.72}$$

Although the coincidence probability is maximized for $\mathcal{R} = \mathcal{T} = 1/2$, because of the way $\gamma^{(2)}_{click}[0]$ is normalized, it can still give a good approximation to $g^{(2)}[0]$ even when $\mathcal{R} \neq \mathcal{T}$ or when the detection efficiencies are not matched, $\eta_3 \neq \eta_4$. Because loss does not change $g^{(2)}$, even though it alters the photon probabilities $P(n)$, the measurement fidelity can be improved by introducing additional loss to the system. Care should be taken to ensure that any additional loss is the same for all modes; for example, any additional spatial or spectral filtering should be avoided.

2.2.9.3 Summary of HBT Interferometry in the Characterization of a Single-Photon Source

The HBT interferometer has several important properties that make it especially useful for accurate measurements of $g^{(2)}(\tau)$.

(1) The BS reflectance:transmittance ratio does not need to be perfectly 50:50.
(2) The detection efficiencies of D_3 and D_4 do not have to be 100 % or even matched to one another.
(3) Threshold or click detectors can be used without sacrificing accuracy, as long as $P(1) \gg P(2) \gg P(n > 2)$.
(4) Loss does not change $g^{(2)}(\tau)$, so it can be measured accurately even in a system with unknown and potentially large losses.

There are a number of key experimental limitations to the HBT interferometer. For example, if $g^{(2)}(\tau)$ varies significantly over time delays that are short compared to the timing jitter of detectors and/or electronics, then the measured $\gamma^{(2)}(\tau)$ will be a convolution of $g^{(2)}(\tau)$ and an instrument response function. While it may be possible in some cases to deconvolve these two components and obtain a better estimate of the source $g^{(2)}(\tau)$, it is generally not an easy task. In addition, normalizing by count rates, as in Eqs. (2.68) and (2.69), may prove inaccurate if the count rates vary during the acquisition time. In this case, it might be preferable to perform a piece-wise normalization over shorter time intervals, if possible.

Finally, it is worth noting that a beamsplitter is not necessarily required to measure the second-order coherence. Beamsplitters are used chiefly because readily available single-photon detectors have non-zero dead time and a limited ability to resolve photon number. Recent demonstrations have shown that $g^{(2)}(\tau)$ of single- or few-photon states can be measured with one detector, provided that the detector either has a very short dead time [42] or the ability to resolve photon number [51].

2.2.10 Bunching, Antibunching, and Poissonian Photon Statistics

A source exhibits **bunching** if photons are more likely to arrive closely spaced in time than they are to arrive further apart. Conversely, a source exhibits **antibunching** if photons are more likely to arrive far apart in time than close together. These conditions are typically expressed as $g^{(2)}(0) > g^{(2)}(\tau \neq 0)$ for bunching and $g^{(2)}(0) < g^{(2)}(\tau \neq 0)$ for antibunching [25,50]. (These relations break down for a source that is antibunched over some time scales and bunched over others [48].)

A source with **Poissonian** photon number statistics, such as the coherent state discussed in Section 2.2.13, is neither bunched nor antibunched, and has $g^{(2)}(\tau) = 1$ for all τ. In this case photon arrival times are distributed randomly, with two photons displaying no preference for arriving separated by short or long time delays [25,50]. Note however that $g^{(2)}(\tau) = 1$ does not necessarily imply that the photon number distribution follows a Poisson distribution. One example is a statistical mixture of three sources (single-photon, thermal, and coherent) in three modes that can add up in the right proportion to yield $g^{(2)}(\tau) = 1$ and yet have a probability distribution that cannot be described by a Poisson distribution [18].

Nevertheless, it is useful to define terms that delineate how $g^{(2)}$ varies from the value it takes for Poissonian sources. One way of looking at $g^{(2)}$ is as a quantity relating the mean photon number to the magnitude of the variation in photon number. To see this, we can examine the pulsed case and rewrite Eq. (2.44) as [25]

$$g^{(2)}[0] = 1 + \frac{(\Delta n)^2 - \mu}{\mu^2}, \qquad (2.73)$$

where $\mu = \langle \hat{n} \rangle$ is the mean photon number and the variance is

$$(\Delta n)^2 = \langle \hat{n}^2 \rangle - \langle \hat{n} \rangle^2. \qquad (2.74)$$

Thus $g^{(2)}$ is a measure of the relative magnitude of the mean and the variance. For a Poisson source or any other source with $g^{(2)} = 1$, the mean and variance are equal: $\mu = (\Delta n)^2$. If the variance in photon number is *larger* than the mean, then $g^{(2)} > 1$ and the source exhibits **super-Poissonian** behavior. If the variance is *smaller* than the mean, then $g^{(2)} < 1$ and the source exhibits **sub-Poissonian** behavior. Because classical light fields are constrained to $g^{(2)} \geq 1$, sub-Poissonian statistics can only be described using a quantized electromagnetic field [17,25].

Because of the timing jitter limitations of detectors and electronics, a measured $\gamma^{(2)}(\tau)$ may not fully capture the fastest temporal dynamics of $g^{(2)}(\tau)$. As a result, measuring $\gamma^{(2)}(0) = 1$ does not necessarily imply that $g^{(2)}(0) = 1$. Most thermal sources, for example, have coherence times far too short ($\ll 1$ ps) for the consequent bunching to be resolved in a typical $\gamma^{(2)}(\tau)$ measurement.

Most single-photon-source approximations exhibit both antibunching and sub-Poissonian photon statistics. However, an antibunched source does not necessarily have sub-Poissonian photon statistics [30,48]. One example is a source that exhibits "blinking," displaying antibunching on very short time scales and bunching on longer time scales [4]; an illustrative example of this is plotted in Fig. 2.7b. This highlights the importance of correctly normalizing $\gamma^{(2)}(\tau)$: if such a source is bunched over a much longer time scale than the antibunching (so that the bunching peak looks flat over the measured delay range), then a source may appear to have sub-Poissonian statistics, when in fact $g^{(2)}(\tau) > 1$.

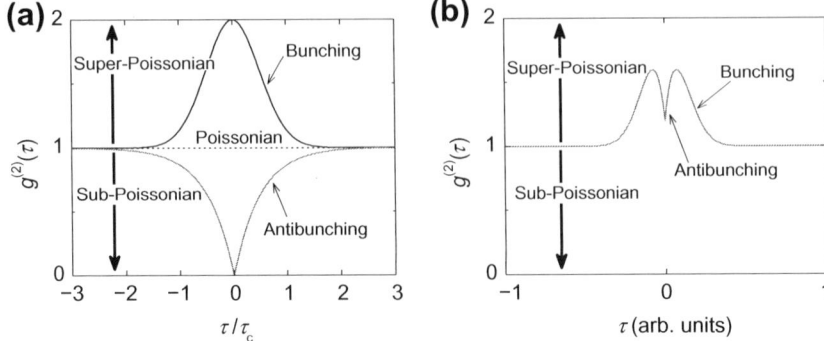

FIGURE 2.7 (a) Plot showing the relationships among bunching, antibunching, sub-, super- and Poissonian photon statistics. Plotted are three ideal sources: a chaotic, thermal-like source (solid black curve), a coherent source (dotted black curve), and a single-photon source (solid gray curve). τ_c is the coherence time of the bunched and antibunched sources. (b) Illustration of how a source could be antibunched without exhibiting sub-Poissonian photon statistics.

Likewise, sub-Poissonian photon statistics do not necessarily imply that a source will be antibunched, although this requires a light source that would be difficult to implement in a practical setting [30,48].

2.2.11 High-Order Coherences

In some cases, it may be relevant to measure coherences higher than second order, given that in principle one needs to know all orders to fully reconstruct the density matrix [17]. The third-order coherence for a single-mode stationary source can be written:

$$g^{(3)}(\tau_1,\tau_2) = \frac{\langle \hat{a}^\dagger(t)\hat{a}^\dagger(t+\tau_1)\hat{a}^\dagger(t+\tau_2)\hat{a}(t+\tau_2)\hat{a}(t+\tau_1)\hat{a}(t)\rangle}{\langle \hat{a}^\dagger(t)\hat{a}(t)\rangle^3}. \quad (2.75)$$

For a pulsed source, we can write a discrete version for the zero-delay value in terms of the density matrix:

$$g^{(3)}[0,0] = \frac{\mathrm{Tr}\{\hat{\rho}\hat{n}(\hat{n}-1)(\hat{n}-2)\}}{(\mathrm{Tr}\{\hat{\rho}\hat{n}\})^3}. \quad (2.76)$$

In terms of photon probabilities, this is

$$\begin{aligned}g^{(3)}[0,0] &= \frac{\sum_{n=0}^{\infty} n(n-1)(n-2)P(n)}{[\sum_{n=0}^{\infty} nP(n)]^3} \\ &= \frac{6P(3)+24P(4)+\cdots}{[P(1)+2P(2)+3P(3)+\cdots]^3} \\ &\simeq \frac{6P(3)}{[P(1)]^3},\end{aligned} \quad (2.77)$$

where the approximation in the last line holds provided $P(1) \gg P(n > 1)$ and $P(3) \gg P(n > 3)$. The temporal coherence can be generalized to arbitrary order m as

$$g^{(m)}(\vec{\tau}) = \frac{\langle \hat{a}^\dagger(t)[\prod_{i=1}^{m-1}\hat{a}^\dagger(t+\tau_i)][\prod_{i=m-1}^{1}\hat{a}(t+\tau_i)]\hat{a}(t)\rangle}{\langle \hat{a}^\dagger(t)\hat{a}(t)\rangle^m}, \qquad (2.78)$$

where $\vec{\tau} = (\tau_1, \tau_2, \ldots, \tau_{m-1})$. The leading-term approximation for a pulsed source when all delays are zero is

$$g^{(m)}[\vec{0}] \simeq \frac{m!P(m)}{[P(1)]^m}, \qquad (2.79)$$

assuming $P(1) \gg P(n > 1)$ and $P(m) \gg P(n > m)$. Thus, in sources with low mean photon number, to first order the mth-order coherence will tend to be dominated by the m-photon contribution. For an ideal single-photon source, all $P(n)$ are zero for $n \geq 2$, and so all higher-order coherences are also identically zero.

Because $g^{(2)}(\tau)$ has the form of an autocorrelation, it must be symmetric about τ. Higher-order coherences lack this constraint. It has been proposed that asymmetry in $g^{(3)}(\tau_1, \tau_2)$ can indicate irreversible processes, which could be used to distinguish a non-equilibrium steady state from true equilibrium in chemical reactions, for example [38].

2.2.12 Indistinguishability

As discussed above, while an ideal single-photon source will emit exactly one photon at a time, many applications further require indistinguishability of the emitted photons. This may mean that two or more successive photons from a single source are emitted into identical quantum states. It may also mean that multiple single-photon sources should emit photons that are indistinguishable from one another.

Two photons are perfectly indistinguishable if their density matrices $\hat{\rho}_1$ and $\hat{\rho}_2$ are equal. To quantify the difference between two unequal density matrices, we can define the indistinguishability of these two photons as

$$\mathcal{J}(\hat{\rho}_1, \hat{\rho}_2) = 1 - \frac{1}{2}\|\hat{\rho}_1 - \hat{\rho}_2\|^2, \qquad (2.80)$$

where $\|\hat{\rho}_1 - \hat{\rho}_2\|^2$ is the operational distance between $\hat{\rho}_1$ and $\hat{\rho}_2$ [23,35]. This operational distance has a maximum value of two, and $\mathcal{J}(\hat{\rho}_1, \hat{\rho}_2)$ has a minimum value of zero, if $\hat{\rho}_1$ and $\hat{\rho}_2$ are perfectly distinguishable. If $\hat{\rho}_1 = \hat{\rho}_2$, then $\|\hat{\rho}_1 - \hat{\rho}_2\|^2 = 0$ and $\mathcal{J}(\hat{\rho}_1, \hat{\rho}_2) = 1$. If the two input photons are in pure states, then the density matrices can be written as $\hat{\rho}_1 = |\psi_1\rangle\langle\psi_1|$, and $\hat{\rho}_2 = |\psi_2\rangle\langle\psi_2|$, and the indistinguishability is [23,35]

$$\mathcal{J}(\hat{\rho}_1, \hat{\rho}_2) = |\langle\psi_1|\psi_2\rangle|^2. \qquad (2.81)$$

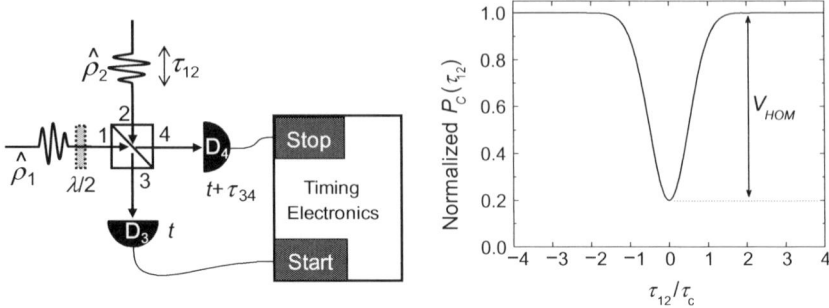

FIGURE 2.8 Left panel: Schematic for measuring HOM interference. The time delay τ_{12} is the elapsed time from when photon 1 arrives at the BS to when photon 2 arrives at the BS. This delay, which can be scanned by adjusting the optical path length of one of the input arms, can be varied in steps much smaller than the timing resolution of the single-photon detectors D_3 and D_4. Alternatively, a half-wave plate ($\lambda/2$) can be inserted into one input port (port 1 here) to rotate the polarization of $\hat{\rho}_1$ so that it is orthogonal to that of $\hat{\rho}_2$, making the two photons perfectly distinguishable. The time delay τ_{34} is the elapsed time from when a photon is detected by D_3 to when a photon is detected by D_4. Coincidences are events where $\tau_{34} = 0$: both D_3 and D_4 detect photons at the same time. Right panel: Theoretical plot of HOM dip for $V_{HOM} = 0.8$.

This is recognizable as the fidelity [49]. In principle, indistinguishability could be computed by separately measuring all the quantum properties of the two states, including the spectral, temporal, and spatial profiles, and the polarization state, and fully accounting for all measurement bias, such as mode selectivity.

In practice, the indistinguishability of two photons is instead typically quantified by observing Hong-Ou-Mandel (HOM) interference between them [20]. In an HOM interferometer, if two photons are incident on the two input ports of an ideal 50:50 beamsplitter, their paths can interfere such that both photons must exit the same output port of the BS.

A typical experimental setup is shown in Fig. 2.8. If two indistinguishable single photons in pure states are incident on the BS, one each at ports 1 and 2, they transform as $|1\rangle_1|1\rangle_2 \to (|2\rangle_3|0\rangle_4 + |0\rangle_3|2\rangle_4)/\sqrt{2}$. Both photons could be detected by either detector D_3 or detector D_4, but the probability of a coincidence between the two detectors, P_c, is zero. The visibility of the HOM interferometer is found by comparing this coincidence probability to the coincidence probability when the two input photons are made perfectly distinguishable.

One method of measuring the visibility is to vary the time delay, τ_{12}, between the photons incident to ports 1 and 2. The probability of coincidences (events where $\tau_{34} = 0$) between detected photons in output ports 3 and 4 is measured as a function of this time delay, $P_c(\tau_{12})$. In the ideal case, $P_c(0) = 0$. One can define a visibility [34]

$$V_{\text{HOM}} = \frac{P_c(\tau_{12} \gg \Delta\tau_{\text{dip}}) - P_c(0)}{P_c(\tau_{12} \gg \Delta\tau_{\text{dip}})}, \tag{2.82}$$

where the long-delay limit is measured at a delay τ_{12} much longer than the width of the HOM dip $\Delta\tau_{\text{dip}}$; in other words, the time delay is large enough that the photon wavepackets do not overlap in time.

As an alternative to scanning the delay, if all photons are linearly polarized, a half-wave plate can be inserted into one of the input ports of the beamsplitter. In this way, the two photons can be made perfectly distinguishable by setting the polarizations orthogonal to one another. The measured coincidence probabilities for parallel polarizations, $P_c(\|)$, and for orthogonal polarizations, $P_c(\perp)$, can then be used to determine the visibility as [11]

$$V_{\text{HOM}} = \frac{P_c(\perp) - P_c(\|)}{P_c(\perp)}. \tag{2.83}$$

The HOM visibility can reach its maximum value of unity only if three conditions are met: (1) there must never be more than one photon in either input port, (2) the two photons must be indistinguishable from one another, and (3) both photons must be in pure states. If there is exactly one photon at each input port and the BS has a 50:50 split ratio, V_{HOM} can be written in terms of the input density matrices $\hat{\rho}_1$ and $\hat{\rho}_2$ as [35]

$$V_{\text{HOM}} = \text{Tr}\{\hat{\rho}_1 \hat{\rho}_2\} = \frac{\text{Tr}\{\hat{\rho}_1^2\} + \text{Tr}\{\hat{\rho}_2^2\} - \|\hat{\rho}_1 - \hat{\rho}_2\|^2}{2}. \tag{2.84}$$

Because $\text{Tr}\{\hat{\rho}_1^2\} \leq 1$ and $\text{Tr}\{\hat{\rho}_2^2\} \leq 1$, this visibility serves as a lower limit to indistinguishability

$$V_{\text{HOM}} \leq \mathcal{J}(\hat{\rho}_1, \hat{\rho}_2). \tag{2.85}$$

If the two input photons are in pure states $|\psi_1\rangle$ and $|\psi_2\rangle$, then $\text{Tr}\{\hat{\rho}_1^2\} = \text{Tr}\{\hat{\rho}_2^2\} = 1$, and hence the visibility and indistinguishability are equivalent [23,35]:

$$V_{\text{HOM}} = \mathcal{J}(\hat{\rho}_1, \hat{\rho}_2) = |\langle\psi_1|\psi_2\rangle|^2. \tag{2.86}$$

V_{HOM} can be reduced below unity by many mechanisms, including background counts, multi-photon events, and other experimental imperfections. This makes HOM interferometry a powerful tool for identifying any unforeseen imperfections in a source or an experimental setup that need to be addressed.

2.2.13 Other Sources

In addition to single-photon sources, several other sources will be discussed throughout this book, both to highlight the ways they contrast with single-photon sources, and as sources that have important applications in their own rights.

2.2.13.1 Coherent Source
A coherent state is defined as [17]

$$|\alpha\rangle = \exp\left(-\frac{1}{2}|\alpha|^2\right) \sum_{n=0}^{\infty} \frac{\alpha^n}{\sqrt{n!}} |n\rangle. \tag{2.87}$$

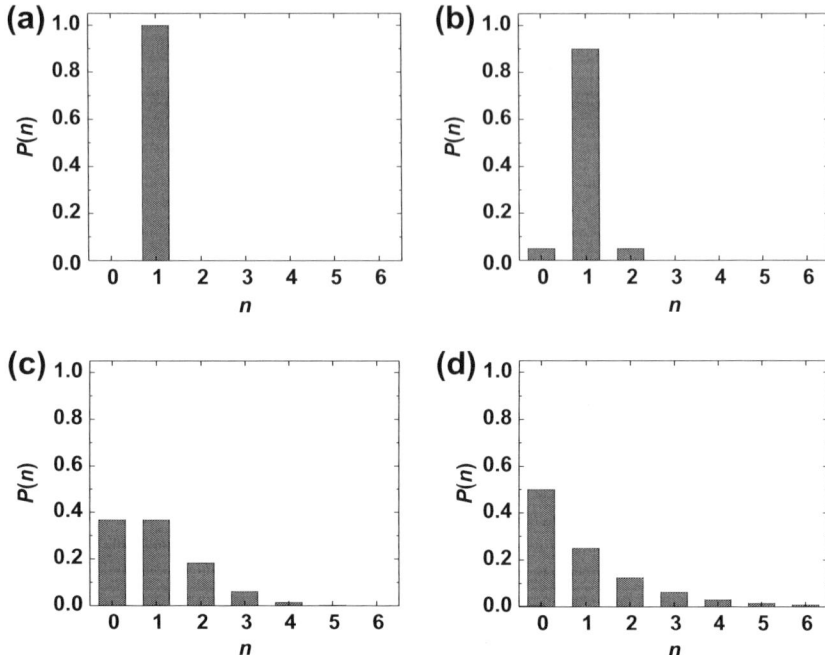

FIGURE 2.9 Photon number probability distributions of four pulsed sources, each with a mean photon number $\mu = 1$. (a) Ideal single-photon source, $g^{(2)}[0] = 0$. (b) Non-ideal single-photon source, $g^{(2)}[0] = 0.1$. (c) Coherent source, $g^{(2)}[0] = 1$. (d) Thermal source, $g^{(2)}[0] = 2$.

The photon statistics for a pulsed coherent source follow a Poisson probability distribution

$$P(n) = \frac{e^{-\mu}\mu^n}{n!}, \tag{2.88}$$

where $\mu = |\alpha|^2$ is the mean photon number per pulse. This distribution is shown in Fig. 2.9c for $\mu = 1$ and in Fig. 2.10c for $\mu = 0.1$. The second-order coherence of a coherent state is $g^{(2)}(\tau) = 1$ for all τ, as shown in Fig. 2.11c for a pulsed coherent source. Higher orders of coherence are also equal to unity: $g^{(m)}(\vec{\tau}) = 1$ for all $\vec{\tau}$. A stable single-mode laser operated well above threshold typically emits light that is a good approximation to a coherent state [25,15].

The photon probability distribution of a coherent state remains Poissonian after attenuation. Nonetheless, by attenuating a coherent state it is possible to make the two:one photon ratio

$$\frac{P(2)}{P(1)} = \frac{\mu}{2} \tag{2.89}$$

arbitrarily small; however, this improvement comes at the expense of a lowered mean photon number, μ, and hence a lowered source efficiency. As discussed in Sections 2.2.2 and 2.2.9, attenuation does not change the second-order coherence.

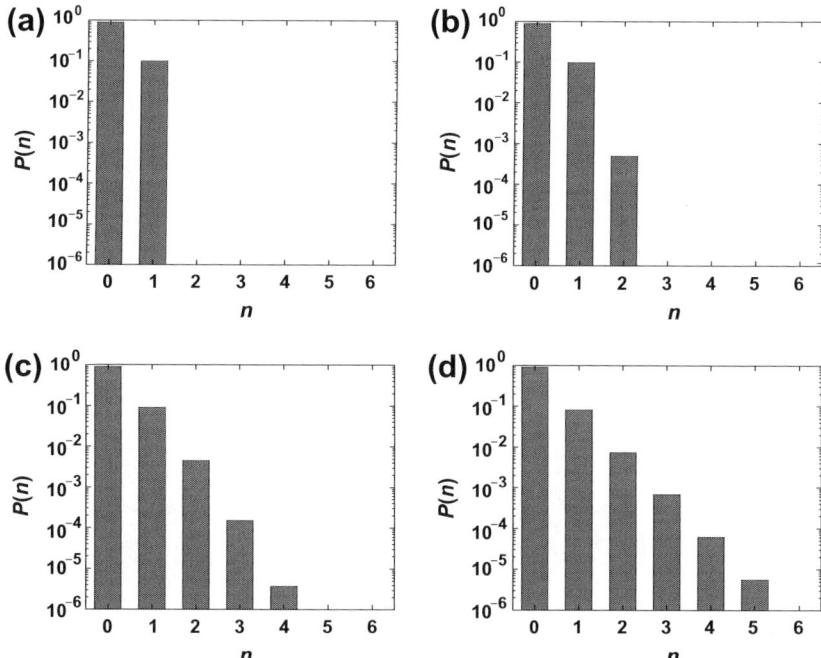

FIGURE 2.10 Photon number probability distributions of four pulsed sources with mean photon number $\mu = 0.1$. (a) Single-photon source with $g^{(2)}[0] = 0$. (b) Non-ideal single-photon source with $g^{(2)}[0] = 0.1$. (c) Coherent source, $g^{(2)}[0] = 1$. (d) Thermal source, $g^{(2)}[0] = 2$. Note the log scale on the $P(n)$ axis.

2.2.13.2 Thermal Source

The density matrix for a thermal source can be written [25]

$$\hat{\rho} = \sum_n \frac{\mu^n}{(1+\mu)^{n+1}} |n\rangle\langle n|. \tag{2.90}$$

This source has a photon probability distribution

$$P(n) = \frac{\mu^n}{(1+\mu)^{n+1}}. \tag{2.91}$$

The zero-delay values of the temporal coherences are $g^{(2)}(0) = 2$ and $g^{(m)}(\vec{0}) = m!$. Thus a thermal source is both *bunched*, since the zero-delay values of the coherences are higher than the values at non-zero delays, and *super-Poissonian*, since those values are greater than one.

In practice, a true thermal source is difficult to implement and characterize experimentally because coherence times of thermal sources are typically much shorter than the temporal resolution of single-photon detectors. As a substitute, one can make a pseudo-thermal source by scattering a CW laser off a time-dependent scattering medium, such as a rotating wheel of ground glass [2]. The

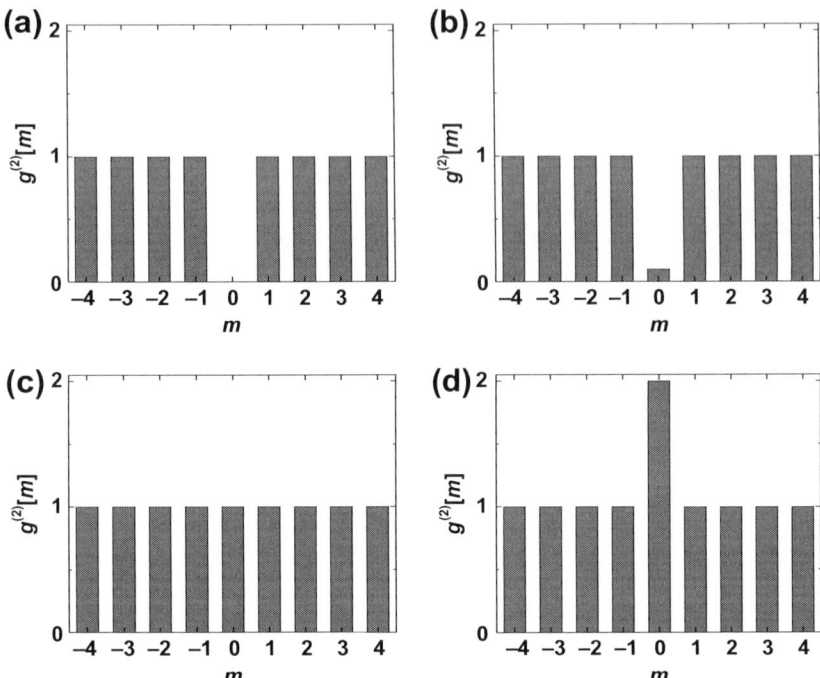

FIGURE 2.11 Second-order coherences of the four pulsed sources in Figs. 2.9 and 2.10: (a) single-photon source with $g^{(2)}[0] = 0$, (b) non-ideal single-photon source with $g^{(2)}[0] = 0.1$, (c) coherent source, and (d) thermal source. The height of each bar represents the total area of each pulse in the normalized correlation histogram.

resulting time-dependent speckle pattern is then sampled in the far field over an area small compared to the mean speckle size. The effective coherence time of the source can then be modified by adjusting the speed at which the scattering medium changes, for example, by adjusting the rotation speed of the ground glass wheel.

A thermal or pseudo-thermal source is sometimes referred to as a chaotic source, although it is chaotic only in the sense that it appears random, and has nothing to do with chaos theory. The bunching exhibited by a thermal source can be fully explained as intensity fluctuations of a classical electromagnetic field [25].

2.2.13.3 Pair Source

Photon pairs can be generated in many different ways. Common examples include atomic cascades (see Chapter 11), parametric downconversion (Chapter 12), and four-wave mixing (Chapter 13). The specifics of photon number distributions and coherences will depend on the details of the source, and are detailed for each type of source in subsequent chapters. Here, we focus on one example, a two-mode squeezed state, which is the model state for photon pairs generated via parametric downconversion [16]:

$$|\xi_2\rangle = \exp(\xi^* \hat{a}_s \hat{a}_i - \xi \hat{a}_s^\dagger \hat{a}_i^\dagger)|0\rangle_s |0\rangle_i, \qquad (2.92)$$

where $\xi = re^{i\theta}$ is the squeezing parameter and $\hat{a}_s^\dagger$ and $\hat{a}_i^\dagger$ denote creation operators acting on the signal and idler modes, respectively. For each photon created in the signal mode, a corresponding paired photon is created in the idler mode:

$$|\xi_2\rangle = \frac{1}{\cosh r} \sum_{n=0}^{\infty} e^{in\theta}(-\tanh r)^n |n\rangle_s |n\rangle_i. \qquad (2.93)$$

The photon number probabilities for signal and idler are identical, $P_s(n) = P_i(n) = P(n)$, and can be written in terms of the mean photon number $\mu_s = \mu_i = \mu$ as

$$P(n) = \frac{(\tanh r)^{2n}}{(\cosh r)^2} = \frac{\mu^n}{(1+\mu)^{n+1}}. \qquad (2.94)$$

Thus the number of photons in each follows a thermal distribution, with μ the average number of pairs generated.

One important application of a pair source is a heralded single-photon source, where detection of one member of the pair (idler) is used to herald the presence of the other (signal). Conditioning on detection of an idler photon decreases the probability of finding no photons in the signal field. This increases the effective (conditional) mean photon number $\mu_{\text{effective}}$ without significantly changing the two:one photon ratio $P(2)/P(1)$; as a result the ratio $2P(2)/[P(1)]^2$ decreases and $g^{(2)}[0]$ can drop below unity (see Eq. (2.50)). This can be seen by comparing Fig. 2.12a and b. If the mean pair number μ is lowered, this conditional $g^{(2)}[0]$ can be made arbitrarily close to zero. If there is loss in the heralding arm, the overall rate of heralded photons will drop, increasing the conditional $g^{(2)}[0]$ compared to lossless heralding detection, as illustrated in Fig. 2.12c. Channel losses in the heralded arm will cause a drop of only the overall rate of photons, but will not affect multi-photon suppression or $g^{(2)}[0]$ (see Fig. 2.12d). Practical issues related to heralded sources will be covered in greater detail in Chapters 11–13.

Taking into account the spectral composition of the signal and idler, and taking the low-squeezing limit, Eq. (2.92) can be approximated as [28]

$$|\xi_2\rangle \simeq \sqrt{1-\xi}|0\rangle_s|0\rangle_i + \sqrt{\xi} \int d\omega_s \int d\omega_i \Psi(\omega_s, \omega_i) \hat{a}_s^\dagger(\omega_s) \hat{a}_i^\dagger(\omega_i)|0\rangle_s|0\rangle_i, \qquad (2.95)$$

where $\Psi(\omega_s, \omega_i)$ is the signal-idler joint wavefunction, which describes the amplitude distribution for creating a photon pair with signal frequency ω_s and idler frequency ω_i. This joint wavefunction is determined by both energy conservation and phase-matching conditions, and can be characterized by measuring the joint spectral density. This will be discussed in Chapter 12.

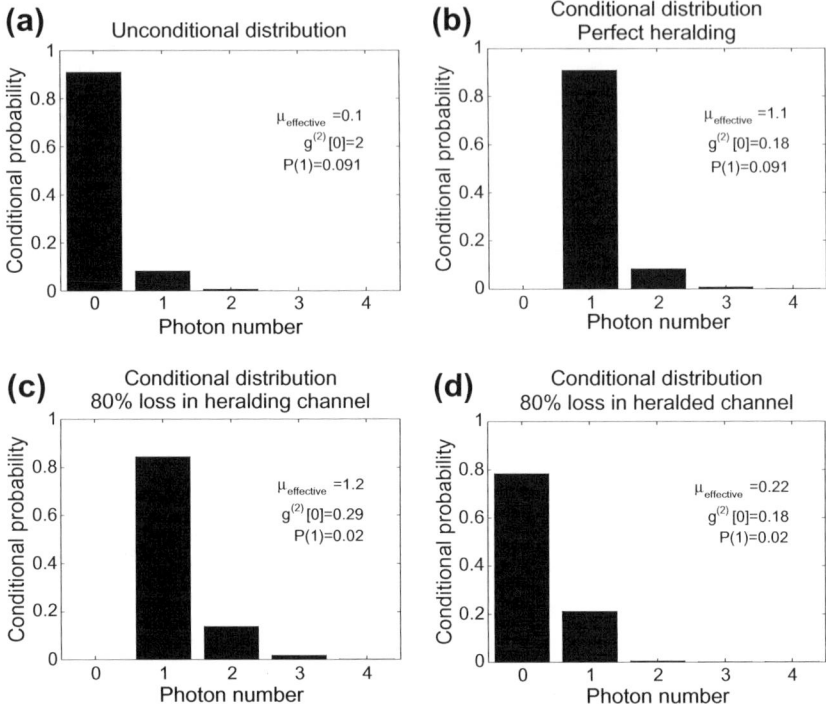

FIGURE 2.12 Basic principles of a heralded single-photon source based on a pair source, assuming a click/no-click single-photon detector (see Section 2.3.3.1) in the heralding arm. (a) Unconditional photon number distribution in each arm. Conditional photon number distribution of the heralded arm with (b) lossless heralding; (c) loss in heralding channel only; (d) loss in heralded channel only. $\mu_{\text{effective}}$ is the mean photon number in the heralded arm after conditioning on the heralding field and $P(1)$ is the overall, unconditional probability of producing one photon in the heralded arm. *Image courtesy of Elizabeth Goldschmidt.*

2.3 DETECTOR PROPERTIES

In its simplest, ideal form, a single-photon detector is a device that produces one electrical output pulse in response to a single input photon. The perfect detector would have a detection efficiency of 100 % and ideally would be able to resolve photon number. Timing latency, timing jitter, and dead time would all be zero. The ideal device would have no dark counts and zero probability of afterpulsing.

Not surprisingly, no existing detector satisfies all these criteria, and no one type of detector outperforms all others in all these metrics. For a given application, the choice of detector will typically involve trade-offs, where some of these characteristics are improved at the expense of others. This section will formally define the properties used to evaluate the performance of the detectors discussed in this book.

The single-photon detectors covered in this book can in general be categorized as photon counters. They give information only about the terms on the diagonal of the density matrix, when the density matrix is written in the photon-number-state basis. Photon-counting measurements, including $g^{(2)}$ measurements performed with photon-counting detectors, typically are sensitive only to certain characteristics of the modes of the photons under study. For example, photon-counting experiments are not typically sensitive to the details of the spatial, spectral and temporal mode profiles, except in the sense that detectors only count photons in the modes that reach the detectors, at wavelengths the detectors are sensitive to, and at times when the detectors are active. Hong-Ou-Mandel interference is a notable exception to this, as it is highly sensitive to overlap of the spatial, spectral, temporal, and polarization modes of the two incident photons. Other measurements, such as those of high-order coherences $g^{(n)}$, can yield information about the number and nature of multiple modes having different photon statistics [18].

Several other methods exist for measuring signals at the single-photon level, many of which involve mixing the single-photon field with a strong optical field and using a conventional (as opposed to single-photon) optical detector. These methods, which include homodyne tomography [29,36], heterodyne detection, and spectral interferometry [10], can yield information about the off-diagonal elements of $\hat{\rho}$. These techniques tend to be exquisitely sensitive to the mode, since they give information about only the modes of an optical state that overlap with the separately prepared strong optical field, which is often referred to as the local oscillator. These techniques are outside the scope of this book.

2.3.1 Detection Efficiency

One of the most important characteristics of a single-photon detector is its detection efficiency, η_{DE}. For a free-space coupled detector, η_{DE} represents the probability that a photon incident on the active area of the detector results in an electrical output of sufficient magnitude to be registered by the external electronics, provided that the detector and its electronics are armed and ready to sense an incoming photon. It should be noted that a detector often appears as a single package, which may have components required for its operation, such as a window on a hermetic package, or more sophisticated optics to guide light onto the active area of the detector. Because these components are inseparable parts of the device, their overall transmittance should be included in η_{DE}.

In a fiber-coupled detection system, the term detection efficiency can be somewhat more problematic to define because one must decide where the source to be measured ends and the detector begins. The preferred dividing line is within the portion of the detector's input fiber that is accessible for a connector or fusion splicing. Thus η_{DE} represents the fraction of photons that have been coupled into (or generated in) the optical fiber that yield a measurable electrical output, and is sometimes referred to as the "system detection efficiency" to

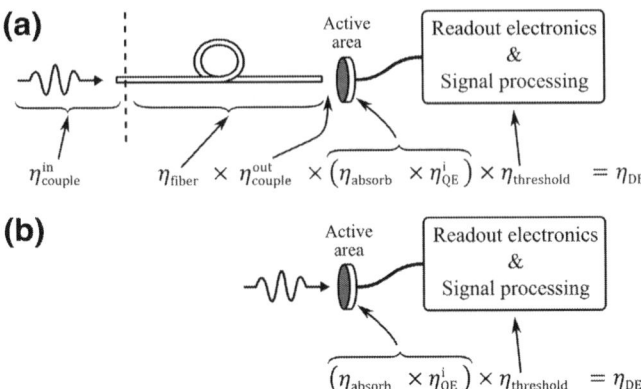

FIGURE 2.13 Illustration of decomposition of detection efficiency for (a) fiber-coupled and (b) free-space-coupled single-photon detector systems, where the dashed line indicates the demarcation between efficiencies associated with the detector and those associated with the experiment.

distinguish it from the detection efficiency of the bare detector. While this definition does not include the (invariably non-unity) efficiency of coupling from a free-space optical experiment into fiber, it does include losses within the fiber and losses associated with coupling light out of the fiber and onto the active area of the actual sensor. Thus it is important to remember that when using a fiber-coupled detector in a free-space experiment, η_{DE} must be reduced by the fiber input coupling efficiency to find the detection efficiency for that situation (see Fig. 2.13).

For a fiber-coupled detector, the system detection efficiency, η_{DE}, can be decomposed into a product of several discrete efficiencies [33]:

$$\eta_{DE} = \eta_{fiber} \eta_{couple}^{out} \eta_{absorb} \eta_{QE}^{i} \eta_{threshold}. \qquad (2.96)$$

The first two component efficiencies quantify the efficiency with which light reaches the sensitive region of the detector. The fiber transmission, η_{fiber}, is typically quite high for visible and near-infrared wavelengths, unless particularly long fiber is used. The output coupling efficiency, η_{couple}^{out}, is a measure of the spatial overlap of the optical mode exiting the fiber with the active area of the detector. η_{couple}^{out} accounts for any misalignment or size mismatch between the output mode of the optical fiber and the active area of the device.

The third and fourth component efficiencies are intrinsic to the detector element itself. The absorption efficiency, η_{absorb}, is the fraction of photons incident on the detector active area that are absorbed. The internal quantum efficiency, η_{QE}^{i}, is the fraction of absorbed photons that yield an output electrical signal. When a fiber-coupled detector is used with a free-space experiment, remember to include η_{couple}^{in}, the coupling efficiency of the input light into the fiber, as a prefactor.

The threshold efficiency, $\eta_{threshold}$, quantifies the efficiency with which the output electrical signal is registered by external counting or timing electronics.

While $\eta_{\text{threshold}}$ is often unity or very close to it, this may not always be the case. For example, if electrical pulses resulting from dark counts (or background counts arising from detected background blackbody radiation) have a comparable but somewhat smaller amplitude than signal pulses, then a threshold in the external electronics may be adjusted to reduce the background count rate. In some cases, a trade-off can be made between $\eta_{\text{threshold}}$—and hence detection efficiency—and this background count rate [32, 14].

For a free-space-coupled detector, the detection efficiency is the probability that a photon incident on the detector is recorded by the measurement electronics. In this case, the decomposition simplifies to

$$\eta_{\text{DE}} = \eta_{\text{absorb}} \eta_{\text{QE}}^i \eta_{\text{threshold}}. \tag{2.97}$$

There may be special cases for a free-space-coupled detector where the optical coupling system has a significant impact on the detector performance, for example if the detector's active area is especially small. In such a case, as for the fiber-coupled detector case, a coupling efficiency term, η_{couple}, should also be included in η_{DE}.

2.3.2 POVM Elements

The set of POVM (Positive-Operator-Valued-Measure) operators are useful for characterizing the outcome of measurements with a single-photon detector or detection system. The probability of obtaining result m, given an optical field described by the density matrix $\hat{\rho}$ and a detector with detection efficiency η_{DE} can be written as [37]

$$p_{\text{DE}}(m) = \text{Tr}\{\hat{\rho}\hat{\pi}_m\}, \tag{2.98}$$

where $\hat{\pi}_m$ is the detector POVM operator for outcome m and depends on η_{DE}.

The POVM operators of a detector that is only sensitive to the number of photons in an optical field (*i.e.*, the diagonal elements of $\hat{\rho}$ in the number state basis) can be written as

$$\hat{\pi}_m = \sum_{n=m}^{\infty} p(m|n)|n\rangle\langle n|, \tag{2.99}$$

where $p(m|n)$ is the conditional probability that detector records m photons at the output, given n photons at the input. In this case, we can write [9]

$$p_{\text{DE}}(m) = \sum_{n=m}^{\infty} p(m|n) P(n). \tag{2.100}$$

The form of the $\hat{\pi}_m$ matrices are given in the following section for two ideal detector types. In a real measurement system, the POVM elements will only approximate these ideal cases. Nonetheless, the $\hat{\pi}_m$ for each outcome m can be determined by performing detector tomography, as discussed in Chapter 9.

2.3.3 Photon-Number-Resolving (PNR) Capability

Single-photon detectors can be broadly grouped into three categories of PNR capability: "no," "some" or "full" PNR capability [7]. While these categories are of course somewhat arbitrary, they capture some essence of the fundamental operation of each type of device.

PNR capability can be useful in multiple distinct ways. For single-shot measurements, if the detector is to correctly identify the number of incident photons it is essential that η_{DE} be very close to 100%. For averaged or ensemble measurements, high detection efficiency is less critical: if one can determine the detector POVM elements with low uncertainty, then it should be possible to invert these POVM elements and recover the photon statistics of the incident light with high accuracy. Of course, this procedure works best with high detection efficiency.

2.3.3.1 No PNR Capability ("click/no-click" Detectors)

These are detectors that operate (or typically operate) as a one-or-more-photon or no-photon device. That is, they register only *whether* photons were detected, but do not provide information on the number of detected photons. These are variously referred to as "threshold," "click" or "click/no-click" detectors, and are the most readily available and widely used type of single-photon detectors. Examples discussed in this book include photomultiplier tubes (PMTs) in Chapter 3, single-photon avalanche diodes (SPADs) in Chapter 4, and superconducting nanowire single-photon detectors (SNSPDs) in Chapter 6.

For an ideal threshold detector, with no dark counts or afterpulsing and with detection efficiency η_{DE}, the POVM for no output is [22]:

$$\hat{\pi}_0 = \sum_{n=0}^{\infty} (1 - \eta_{DE})^n |n\rangle\langle n|. \tag{2.101}$$

The probability of the detector delivering no output for this detection efficiency, $p_{DE}(0)$, can be expressed in terms of the photon number probabilities of the incident light, $P(n)$, as

$$p_{DE}(0) = \sum_{n=0}^{\infty} (1 - \eta_{DE})^n P(n). \tag{2.102}$$

For this detector, the POVM for yielding a click is [21,22]

$$\hat{\pi}_{\text{click}} = \sum_{n=1}^{\infty} [1 - (1 - \eta_{DE})^n] |n\rangle\langle n|. \tag{2.103}$$

Thus the probability of a click at the output is

$$p_{DE}(\text{click}) = \sum_{n=1}^{\infty} [1 - (1 - \eta_{DE})^n] P(n). \tag{2.104}$$

2.3.3.2 Some PNR Capability (Number Resolving Through Multiplexing)

This category consists of devices that are constructed by multiplexing individual non-PNR detectors. The maximum number of photons that these devices can resolve is obviously limited by the number of individual detectors and/or multiplexing pathways. Characteristics of the device can change with the number of photons detected and the spatial distribution of the illumination, since some of the individual elements or areas become unresponsive after firing. Crosstalk between individual elements may also degrade these characteristics, so any crosstalk—or evidence for lack thereof—should be discussed. Time-multiplexed systems will necessarily operate with an increased dead time compared to the individual elements, thus limiting the overall repetition rate.

Despite the potential downsides, these detectors can be quite useful provided the crosstalk is minimal and the number of individual elements or pathways is large compared to the number of incident photons. In fact, some detectors in this category outperform many detectors considered to have full PNR capability, most notably because some detectors in this latter category have very low system detection efficiencies.

Examples of detectors classified as having some PNR capability include the multi-element SNSPDs discussed in Chapter 6 and the time-multiplexed and space-multiplexed detection schemes described in Chapter 7. The visible light photon counter (VLPC) covered in Chapter 5 is a somewhat unusual example in this category, since it does not consist of individual, discrete detection elements; nonetheless, we include it in this category because a region of the detector is effectively "dead" (or "blocked") after detection of one photon leads to a controlled avalanche in that region.

2.3.3.3 Full (or Intrinsic) PNR Capability

Detectors with full PNR capability are devices whose output is inherently proportional to the number of photons, excluding usual saturation limits to which all detectors are subject. It has been suggested that a detector with full PNR capability is a device whose probability of detecting n photons scales *only* as η_{DE} without any other n-dependence. Detectors satisfying this definition of full PNR capability include the solid-state photomultiplier (SSPM) and quantum dot optically gated field effect transistor (QDOGFET), which are discussed in Chapter 5, and the transition edge sensor (TES) described in Chapter 6.

In the absence of dark counts and afterpulsing, the POVM describing how an ideal photon number-resolving detector gives an output indicating m detected photons is [22,41]

$$\hat{\pi}_m = \sum_{n=m}^{\infty} \binom{n}{m} \eta_{DE}^m (1 - \eta_{DE})^{n-m} |n\rangle\langle n|. \qquad (2.105)$$

Thus, the probability of this PNR detector registering m photons, for a detection efficiency of η_{DE}, can be computed from the photon number probabilities $P(n)$ as

$$p_{\text{DE}}(m) = \sum_{n=m}^{\infty} \binom{n}{m} \eta_{\text{DE}}^m (1 - \eta_{\text{DE}})^{n-m} P(n). \tag{2.106}$$

The conditional probability that the detector records m photons at the output, given exactly n photons at the input, is

$$p(m|n) = \binom{n}{m} \eta_{\text{DE}}^m (1 - \eta_{\text{DE}})^{n-m}. \tag{2.107}$$

As an example, the probability of a PNR detector yielding a one-photon output is

$$p_{\text{DE}}(1) = \sum_{n=1}^{\infty} n \eta_{\text{DE}} (1 - \eta_{\text{DE}})^{n-1} P(n). \tag{2.108}$$

Expanding out the first few terms, we find

$$p_{\text{DE}}(1) = \eta_{\text{DE}} [P(1) + 2(1 - \eta_{\text{DE}}) P(2) + 3(1 - \eta_{\text{DE}})^2 P(3) + \cdots] \tag{2.109}$$

If $\eta_{\text{DE}} \simeq 1$, then $p_{\text{DE}}(1)$ will tend to be dominated by the one-photon component of the input light. By contrast, if $\eta_{\text{DE}} \ll 1$, higher photon number terms will play a more important role.

One way to characterize the number-resolving capability of a detector is to compute the fidelity of the measured $\hat{\pi}_m$ with the ideal $\hat{\pi}_m$ for each value of m, following the approach used by Feito et al. with classical input states sent to time-multiplexed detectors [9]. Brida et al. showed that this scheme can be improved by leveraging the nonclassical correlations of a parametric downconversion source [5].

An even more stringent test is to compute the fidelity with the ideal POVM elements for a PNR detector with $\eta_{\text{DE}} = 1$; in this case, each ideal POVM element has a single component: $\hat{\pi}_m = \delta_{m,n} |n\rangle\langle n|$. A PNR detector having high fidelity with these POVM elements would be immensely powerful for single-shot measurements of photon number.

Another way to characterize PNR capability is to invert the measured POVM and find the fidelity of the reconstructed photon probability distributions with the incident $P(n)$'s.

In the situation considered so far in this section, outcomes m will only be incorrect because not all photons are detected ($\eta_{\text{DE}} \neq 1$). In this case, $p(m|n)$ is zero for all $m > n$: one incident photon will never be misidentified as two photons, two will never be misidentified as three, and so on. Realistic PNR detectors can also misidentify the number of incident photons if there is significant variation in the amplitude of the output electrical pulses for a fixed number of incident photons. This variation is typically characterized by measuring a histogram of the distribution of pulse heights or areas, as is illustrated in the next section for the example of an energy-resolving detector.

2.3.3.4 Energy-Resolving Detectors

An energy-resolving detector is one whose output electrical pulse height (or area) is directly proportional to the energy of each absorbed photon. A notable example is the TES discussed in Chapter 6. The pulse-height distribution, when plotted with photon energy on the x-axis, can be used to determine the energy resolution, ΔE, which is the full-width-at-half-maximum (FWHM) of each peak in the distribution. For some detectors, the energy resolution depends on either the photon energy or the photon number; if so, $\Delta E(\hbar\omega)$ or $\Delta E(n)$ should be specified or plotted.

However, ΔE contains somewhat limited information about a detector's ability to resolve photon number. In many applications the most relevant quantity is the probability that detection events are associated with the wrong photon number. Unfortunately, it is typically not possible to independently measure the distributions for each individual photon number. Instead, a source such as an attenuated laser pulse with a Poisson distribution is typically used, and all the photon number peaks are measured at once. The overlap of adjacent peaks can be quantified by measuring the peak visibility, defined as [24]

$$V_{\text{peak}} = \frac{(max - min)}{(max + min)}, \qquad (2.110)$$

where *max* is the average maximum value of two adjacent peaks and the *min* is the lowest value between these two maxima. V_{peak} may be determined from the raw histogram data or from more sophisticated multi-peak fitting routines [24].

Figure 2.14 illustrates simulated pulse-height distribution histograms for detectors with three different energy resolutions and two different photon energies. Each individual photon peak in this conceptual model has an ideal Gaussian shape—something that cannot necessarily be assumed about practical detectors [24]. In all four panels, the source is assumed to follow Poissonian statistics with mean photon number $\mu = 2$. Panels (a)–(c) illustrate the effect of changing the energy resolution for a fixed photon energy of $\hbar\omega = 1$ eV. The individual photon peaks are clearly well separated from one another in (a), where $\Delta E = 0.2\hbar\omega$ and $V_{\text{peak}} \simeq 1$ for all peaks. In (b), each peak overlaps just slightly with its neighboring peaks, yet only ~1% of detection events will misidentify the photon number. In this case, V_{peak} ranges from ~0.88 to ~0.90 for the peaks shown. In (c), the resolution is further degraded such that $\Delta E = \hbar\omega = 1$ eV, and peak visibility is undefined for all minima but one. In this case, a large fraction of photon numbers will be misidentified.

Figure 2.14d shows what happens when ΔE is held fixed at 1 eV (the same value as in Fig. 2.14c), but the photon energy is increased to 2.2 eV. The result looks much the same as in Fig. 2.14b. This illustrates another feature of energy-resolving detectors: in addition to resolving the number of photons at a fixed wavelength, they can also be used to distinguish single photons of different wavelengths, provided the photon energies are separated well enough to be resolved [6]. This can also present experimental challenges. For example, in a

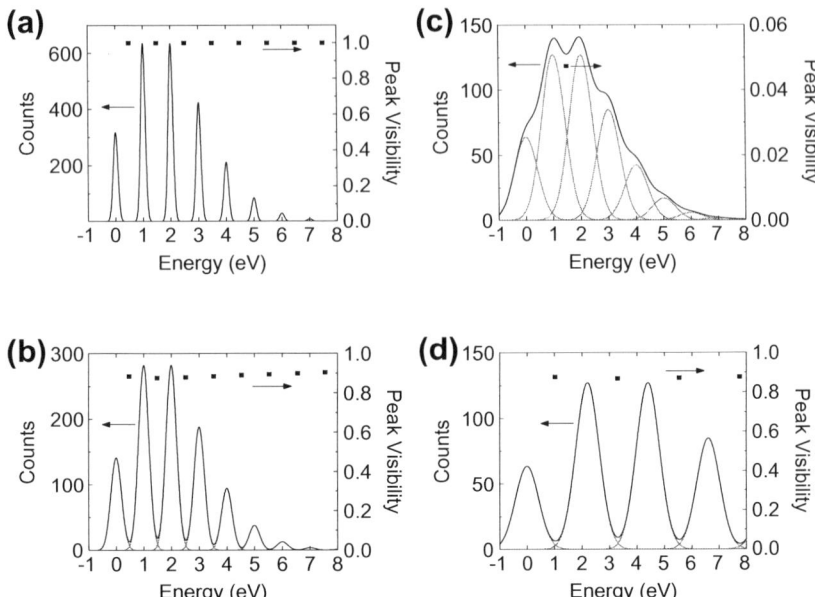

FIGURE 2.14 Simulated pulse energy distributions for an ideal energy-resolving detector, assuming Gaussian distributions for each photon number peak. Gray lines are individual photon number peaks and black lines are the sum of all the individual peaks. (a)–(c) Simulated histograms for a photon energy $\hbar\omega = 1$ eV for three different energy resolutions: (a) $\Delta E = 0.2$ eV, (b) $\Delta E = 0.45$ eV, (c) $\Delta E = 1$ eV $= \hbar\omega$. (d) Simulation for $\Delta E = 1$ eV, which is the same energy resolution as in (c), but with a higher photon energy of $\hbar\omega = 2.2$ eV. In these simulations, ΔE does not depend on energy or photon number. Visibilities (dots) are derived using Eq. (2.110).

parametric downconversion experiment, a single pump photon of energy $2\hbar\omega$ can be misidentified as two downconverted photons of energy $\hbar\omega + \hbar\omega$.

As well resolved as the individual photon peaks are in Fig. 2.14a, a detector with such a histogram is only useful for resolving photon number in a single-shot measurement if the detection efficiency is quite high. By contrast, a detector with the pulse-height distribution in Fig. 2.14c with $\eta_{\text{DE}} \simeq 1$ would in many respects outperform a detector with the distribution in Fig. 2.14a, even if the latter had a relatively high detection efficiency of $\eta_{\text{DE}} = 0.76$. This point is illustrated in Tables 2.1 and 2.2, which display the probability of outcome m given n input photons, $p(m|n)$, for the first few values of m and n. Table 2.1 shows these the probabilities for $\Delta E = \hbar\omega$ and $\eta_{\text{DE}} = 1$; Table 2.2 shows them for $\Delta E = 0.2\hbar\omega$ and $\eta_{\text{DE}} = 0.76$.

Note that in both cases, the probability of correctly identifying one photon at the input as one photon at the output is $p(1|1) \simeq 0.76$. As the incident photon number increases, the high-efficiency detector maintains $p(n|n) \simeq 0.76$, whereas the detector with better energy resolution but lower efficiency quickly degrades: $p(2|2) \simeq 0.58$ and $p(3|3) \simeq 0.44$.

TABLE 2.1 Expected values of $p(m|n)$, which is the probability of detector outcome m, given that n photons are incident on a detector with full (but non-ideal) number resolving capability ($\Delta E \simeq \hbar\omega$) that has ideal detection efficiency η_{DE}. In Tables 2.1–2.3, all values are rounded to the nearest 0.01 and values < 0.01 are omitted for clarity

		\multicolumn{4}{c}{n}			
		0	1	2	3
	0	0.88	0.12		
m	1	0.12	0.76	0.12	
	2		0.12	0.76	0.12
	3			0.12	0.76

TABLE 2.2 Expected values of $p(m|n)$ for a near-ideal number resolving detector ($\Delta E \simeq 0.2\hbar\omega$) with non-ideal detection efficiency $\eta_{DE} = 0.76$. Such a detector would have a pulse energy distribution shown in Fig. 2.14a

		\multicolumn{4}{c}{n}			
		0	1	2	3
	0	1	0.24	0.06	0.01
m	1		0.76	0.36	0.13
	2			0.58	0.42
	3				0.44

As a third example, Table 2.3 shows $p(m|n)$ for a detector that combines the sub-optimal energy resolution of the detector in Table 2.1 ($\Delta E = \hbar\omega$) with the efficiency of the detector in Table 2.2 ($\eta_{DE} = 0.76$). The matrix for in Table 2.3 is simply the product of the two matrices in Tables 2.1 and 2.2.

It should be emphasized that these tables have been computed for the conceptual model described in the text, about which we have much more information than we could typically obtain for a realistic PNR detector—unless it had very well-resolved photon number peaks like those in Fig. 2.14a. In nearly all cases, there will be some finite overlap of adjacent photon number peaks, and conclusions about the probability of photon number misidentification can only be drawn by making assumptions about the shape of the peaks in the overlapping region.

Also note that multiplexed detectors that operate by taking a digital summation of the number of elements that detect a photon will typically

TABLE 2.3 Expected values of $p(m|n)$ for a non-ideal number resolving detector ($\Delta E = \hbar\omega$) with non-ideal detection efficiency $\eta_{DE} = 0.76$

		\multicolumn{4}{c}{n}			
		0	1	2	3
m	0	0.88	0.30	0.10	0.03
	1	0.12	0.61	0.35	0.15
	2		0.09	0.48	0.38
	3			0.07	0.38

not suffer from the problem of overlapping peaks, and thus a pulse-height distribution plot is not necessary.

2.3.4 Timing Latency and Rise Time

The timing latency of a detector, t_{latency}, is the time that elapses between when a photon is incident on a detector or detector system and when the subsequent output electrical pulse crosses a given threshold level, as shown in Fig. 2.15. The choice of the threshold level is based on a variety of factors specific to a given detector: timing resolution, tradeoffs between signal strength and electronic noise, and the ability to trigger subsequent timing electronics are some examples of considerations relevant to the choice of threshold. The rise time of the output pulse, τ_{rise}, is often characterized as the time required for the electrical output signal to rise from 10 % to 90 % of its maximum value.

2.3.5 Timing Jitter

Timing jitter is a measure of the pulse-to-pulse variation in t_{latency}, and is typically determined by characterizing the instrument response function (IRF) of a detector [3]. A typical measurement scheme is shown in Fig. 2.16. A very short optical pulse from a laser is split at a beamsplitter. One beamsplitter

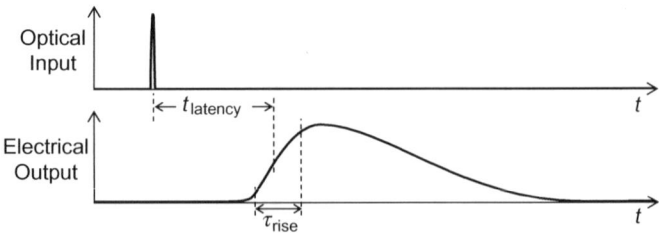

FIGURE 2.15 Illustration of definitions of timing latency, t_{latency}, and rise time, τ_{rise}.

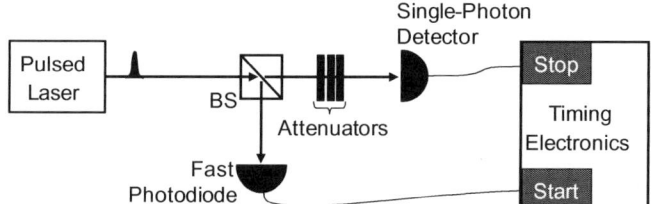

FIGURE 2.16 Typical setup for measuring a single-photon detector's instrument response function, which can be used to determine τ_{jitter}. The timing electronics here are operated in forward start-stop mode, where the fast photodiode (or alternatively the laser clock output, if available) starts a timer, and the signal from the single-photon detector stops the timer. For practical reasons, timing electronics are sometimes operated in reverse start-stop mode [3], where the single-photon detector starts the timer and the conventional fast photodiode stops it; this results in a time-reversed histogram, but will give the same value for τ_{jitter}.

output—comprised of a large, classical signal—is incident on a conventional fast photodiode. The other output of the beamsplitter is heavily attenuated before impinging on the single-photon detector. The timing electronics record a histogram of start-stop time delays that is proportional to the IRF.

The IRF measured using such a scheme will, in general, contain jitter contributions from several components, including the fast photodiode, the single-photon detector, and the timing electronics. The IRF may also be broadened due to the finite duration of the laser pulse. Ultrafast lasers with pulse durations < 1 ps and low-jitter (< 1 ps) conventional fast photodiodes are readily available; if these are used, the jitter is then largely dominated by contributions from the single-photon detector and the timing electronics, both of which are almost always greater than ~20 ps FWHM, as discussed in Chapter 1. In some cases, it may be possible to deconvolve the various contributions to the IRF to estimate the jitter of the single-photon detector alone, but deconvolution should be performed with due care.

Figure 2.17 shows a measured IRF of a silicon SPAD, illustrating the definition of timing jitter, τ_{jitter}, which is the FWHM of the IRF. One can also specify $\tau_{\text{jitter}}^{(1\%)}$, which is the full-width at 1% of maximum of the IRF. For single-photon avalanche diodes, $\tau_{\text{jitter}}^{(1\%)}$ is often significantly larger that the FWHM, and is relevant when the signals of interest may arrive at a comparable time scale, such as in the case of quantum communications at GHz rates.

In some cases, the timing performance of a detector can depend on the detected count rate. In SPADs, for example, both jitter and latency have been shown to vary significantly for count rates above $\sim 1 \times 10^6 /\text{s}$ [39]. Pile-up may also affect a jitter measurement. Pile-up occurs when the timing electronics record only the first photon from an optical pulse, and are unable (during the recovery time of the detector and/or electronics) to record a second or third photon from that same pulse [3]. This will typically lead to an underestimate of the timing jitter. Careful measurement of jitter typically requires attenuating

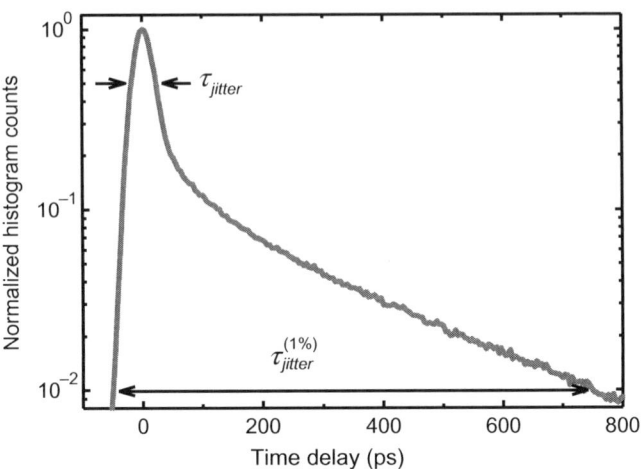

FIGURE 2.17 Example instrument response function illustrating the definitions of τ_{jitter} and $\tau_{jitter}^{(1\%)}$. For this device, τ_{jitter} is ~40 ps and $\tau_{jitter}^{(1\%)}$ is ~790 ps. *(Adapted from [43].)*

the optical pulse so that the count rate of the detector is much lower than the repetition rate of the laser. One can also test whether pile-up has affected a jitter measurement by observing whether τ_{jitter} is a function of the single-photon detection rate by decreasing the incident photon flux. Because of the potential for pile-up, and because it is important to know the effective timing jitter in a given measurement, the count rate and pump repetition rate (or the fraction of pump pulses in which a photon is detected), should accompany reported jitter measurements.

The timing jitter of detectors with partial or full number resolving capability may depend on the number of photons detected, so timing jitter may be specified separately for 1-photon events, 2-photon events, and so on.

2.3.6 Dead Time, Reset Time, and Recovery Time

The dead time, t_{dead}, is the duration of time, beginning at the start of a detection event, during which a detection system is incapable of producing an output electrical signal in response to additional incident photons. During the dead time, the detection efficiency is zero, as illustrated in Fig. 2.18. The dead time may be caused by intrinsic processes in the photosensitive system or it may be induced by external control systems in order to produce a particular performance characteristic.

The reset time, t_{reset}, is the time over which the detection efficiency increases from zero back to its initial value. If the detection efficiency approaches this initial value very slowly, it may be necessary to specify t_{reset} as the elapsed time after which the detection efficiency changes by less than some small percentage of its value.

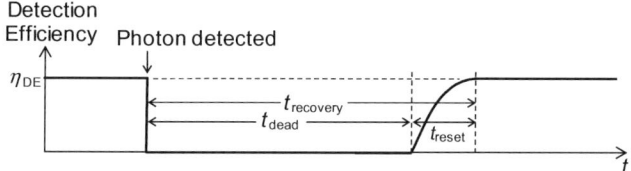

FIGURE 2.18 Example of detection efficiency plotted as a function of time. If the efficiency drops to zero after a photon is detected, then the dead time, t_{dead}, is the elapsed time until the efficiency is non-zero again. The reset time, t_{reset}, is the time required for the detection efficiency to recover to its initial value, η_{DE}. The recovery time, $t_{recovery}$, is the sum of t_{dead} and t_{reset}.

The total time required for the detection efficiency to recover to its steady-state value after a detection event is the recovery time, $t_{recovery} = t_{dead} + t_{reset}$. In detectors or systems with a very short reset time, $t_{dead} \simeq t_{recovery}$, and the terms can be used interchangeably.

The operation of the detector during its reset requires some special consideration. The fact that the detection efficiency is in transition during its reset can strongly affect measurements at high count rates. In addition, the reset action in some types of detectors (notably SPADs) can affect the ability of the electronics to sense a detection event that may have occurred during the reset. This is the origin of the so-called twilight events in some actively quenched SPAD detectors, and is discussed in Chapter 8.

2.3.7 Dark Count Rate

The dark count rate, R_{dark}, is the average number of counts registered by a detector per second when all input light to the detector is blocked.

2.3.8 Background Count Rate

In situations where not all background photons can be adequately blocked from reaching a detector, the background count rate, $R_{background}$, is sometimes quoted. While $R_{backgound}$ is not an intrinsic property of a detector, some detectors are more susceptible to it than others, especially those that are sensitive to mid- or far-infrared photons present in room-temperature blackbody radiation.

2.3.9 Afterpulse Probability

The afterpulse probability, $P_{afterpulse}$, is the excess probability for a detector to output an additional dark count due to a preceding detection event. The time interval over which the dark count probability is elevated should also be specified. Because $P_{afterpulse}$ can depend on the overall count rate, the afterpulse rate at the zero-count-rate limit should be distinguished from the afterpulse probability at higher count rates. In cases where the afterpulsing probability is especially high, it may be reduced by gating the detector. Afterpulsing is most commonly observed in SPADs.

2.3.10 Active Area

The active area of a detector, A_{active}, is the area of the absorbing region of the device, assuming normal incidence. It is typically fairly easy to couple all of the incident light to a device with $A_{\text{active}} > 100 \times 100 \ \mu m^2$. By contrast, for devices with active areas $10 \times 10 \ \mu m^2$ of or less, η_{couple} may be compromised and may have to be included in the definition of η_{DE}.

2.3.11 Operating Temperature of Active Area

The operating temperature of a detector can present obstacles to implementation. For example, superconducting detectors can exhibit many outstanding characteristics, but must operate in cryogenic systems. As a result, light must typically be coupled to the devices through an optical fiber.

ACKNOWLEDGMENTS

The author is indebted to J. Bienfang, B. Calkins, J. Fan, M. Förtsch, T. Gerrits, S. Glancy, E. Goldschmidt, A. Lamas-Linares, F. Marsili, A. Migdall, R. Mirin, S. Nam, S. Polyakov, K. Shalm, K. Silverman and N. Tomlin for many helpful conversations and comments on the chapter.

REFERENCES

[1] M. Aßmann, F. Veit, M. Bayer, C. Gies, F. Jahnke, S. Reitzenstein, S.Höfling, L. Worschech, and A. Forchel, "Ultrafast Tracking of Second-Order Photon Correlations in the Emission of Quantum-Dot Microresonator Lasers," Phys. Rev. B 81, 165314 (2010).
[2] F.T. Arecchi, "Measurement of the Statistical Distribution of Gaussian and Laser Sources," Phys. Rev. Lett. 15, 912 (1965).
[3] W. Becker, Advanced Time-Correlated Single Photon Counting Techniques, Vol. 81 (2005) (Springer Series in Chemical Physics).
[4] A. Beveratos, S.Kühn, R. Brouri, T. Gacoin, J.-P. Poizat, and P. Grangier, "Room Temperature Stable Single-Photon Source," Eur. Phys. J. D 18, 191–196 (2002).
[5] G. Brida, L. Ciavarella, I.P. Degiovanni, M. Genovese, A. Migdall, M.G. Mingolla, M.G.A. Paris, F. Piacentini, and S.V. Polyakov, "Ancilla-Assisted Calibration of a Measuring Apparatus," Phys. Rev. Lett. 108, 253601 (2012).
[6] B. Cabrera, R.M. Clarke, P. Colling, A.J. Miller, S. Nam, and R.W. Romani, "Detection of Single Infrared, Optical, and Ultraviolet Photons Using Superconducting Transition Edge Sensors," Appl. Phys. Lett. 73, 735 (1998).
[7] M.D. Eisaman, J. Fan, A. Migdall, and S.V. Polyakov, "Invited Review Article: Single-Photon Sources and Detectors," Rev. Sci. Instrum. 82, 071101 (2011).
[8] S. Fasel, O. Alibart, S. Tanzilli, P. Baldi, A. Beveratos, N. Gisin, and H. Zbinden, "High-Quality Asynchronous Heralded Single-Photon Source at Telecom Wavelength," New J. Phys. 6, 163 (2004).
[9] A. Feito, J.S. Lundeen, H. Coldenstrodt-Ronge, J. Eisert, M.B. Plenio, and I.A Walmsley, "Measuring Measurement: Theory and Practice," New J. Phys. 11, 093038 (2009).

[10] D.N. Fittinghoff, J.L. Bowie, J.N. Sweetser, R.T. Jennings, M.A. Krumbügel, K.W. DeLong, R. Trebino, and I.A. Walmsley, "Measurement of the Intensity and Phase of Ultraweak, Ultrashort Laser Pulses," Opt. Lett., 21, 884–886 (1996).
[11] E.B. Flagg, A. Muller, S.V. Polyakov, A. Ling, A. Migdall, and G.S. Solomon, "Interference of Single Photons from Two Separate Semiconductor Quantum Dots," Phys. Rev. Lett. 104, 137401 (2010).
[12] E.B. Flagg, S.V. Polyakov, T. Thomay, G.S. Solomon, "Dynamics of Nonclassical Light from a Single Solid-State Quantum Emitter," Phys. Rev. Lett. 109, 163601 (2012).
[13] A. M. Fox, Quantum optics: an introduction. Oxford University Press, Oxford, New York (2006).
[14] D. Fukuda, G. Fujii, T. Numata, K. Amemiya, A. Yoshizawa, H. Tsuchida, H. Fujino, H. Ishii, T. Itatani, S. Inoue, and T. Zama, IEEE Trans. Appl. Superdond. 21, 241–245 (2011).
[15] J.C. Garrison, and R.Y. Chiao, Quantum Optics. Oxford University Press, Oxford, New York (2008).
[16] C.C. Gerry and P.L. Knight, Introductory Quantum Optics. Cambridge University Press, New York (2005).
[17] R.J. Glauber, "Optical Coherence and Photon Statistics," in edited by C. deWitt, A. Blandin, and C. Cohen-Tannoudji, Quantum Optics and Electronics (Les Houches 1964), Gordon and Breach, New York, pp. 63–185 (1965).
[18] E.A. Goldschmidt, F. Piacentini, I.R. Berchera, S.V. Polyakov, S. Peters, S. Kück, G. Brida, I.P. Degiovanni, A. Migdall, M. Genovese, "Mode reconstruction of a light field by multi-photon statistics," Phys. Rev. A 88, 013822 (2013).
[19] R. Hanbury Brown and R.Q. Twiss, "Correlations between photons in two coherent beams of light," Nature 177, 27–29 (1956).
[20] C.K. Hong, Z.Y. Ou, and L. Mandel, "Measurement of subpicosecond time intervals between two photons by interference," Phys. Rev. Lett. 59 2044–2046, (1987).
[21] P. Kok and S.L. Braunstein, Phys. Rev. A 61, 042304 (2000).
[22] P. Kok, W.J. Munro, Kae Nemoto, T.C. Ralph, Jonathan P. Dowling, and G.J. Milburn, "Linear Optical Quantum Computing with Photonic Qubits," Rev. Modern Phys. 79, 135–174 (2007).
[23] J. Lee, M.S. Kim, and C. Brukner, Phys. Rev. Lett. 91, 087902 (2003).
[24] Z.H. Levine, T. Gerrits, A.L. Migdall, D.V. Samarov, B. Calkins, A.E. Lita, S.W. Nam, "Algorithm for Finding Clusters with a Known Distribution and its Application to Photon-Number Resolution Using a Superconducting Transition-Edge Sensor," J. Opt. Soc. Am. B 29, 2066 (2012).
[25] R. Loudon, The Quantum Theory of Light, 3rd Ed. Oxford University Press, Oxford, New York (2000).
[26] R. Loudon and P.L. Knight, J. Mod. Optics 34, 709–759 (1987).
[27] J.S. Lundeen, A. Feito, H. Coldenstrodt-Ronge1, K.L. Pregnell, Ch. Silberhorn, T.C. Ralph, J. Eisert, M.B. Plenio, and I.A. Walmsley, Tomography of quantum detectors. Nature Phys. 5, 27–30 (2009).
[28] A.I. Lvovsky, Wojciech Wasilewski, and Konrad Banaszek, "Decomposing a Pulsed Optical Parametric Amplifier into Independent Squeezers," Journal of Modern Optics 54, 721–733 (2007).
[29] A.I. Lvovsky and M.G. Raymer, Continuous-Variable Optical Quantum-State Tomography. Rev. Mod. Phys. 81, 299–332 (2009).
[30] L. Mandel, "Squeezed States and Sub-Poissonian Photon Statistics," Phys. Rev. Lett. 49, 136–138 (1982).
[31] A.J. Miller, S.W. Nam, J.M. Martinis, and A.V. Sergienko, Demonstration of a Low-Noise Near-Infrared Photon Counter with Multiphoton Discrimination. Appl. Phys. Lett. 83, 791–793 (2003).
[32] A.J. Miller *et al.* in Proc. of the 8th Int. Conf. Quantum Comm. Meas. Comp, 445 (2007).
[33] Aaron J. Miller, Adriana E. Lita, Brice Calkins, Igor Vayshenker, Steven M. Gruber, and Sae Woo Nam, "Compact Cryogenic Self-Aligning Fiber-to-Detector Coupling with Losses Below One Percent," Opt. Express 19, 9102–9210 (2011).
[34] P.J. Mosley, Generation of Heralded Single Photons in Pure Quantum States, Ph. D. Dissertation, University of Oxford (2007).
[35] P.J. Mosley, J.S. Lundeen, B.J. Smith, P. Wasylczyk, A.B. U'Ren, C. Silberhorn, and I.A. Walmsley, Phys. Rev. Lett. 100, 133601 (2008).

[36] C.R. Müller, B. Stoklasa, C. Peuntinger, C. Gabriel, J. Řeháček, Z. Hradil, A.B. Klimov, G. Leuchs, Ch. Marquardt, and L.L.Sánchez-Soto, "Quantum Polarization Tomography of Bright Squeezed Light," New J. Phys. 14, 085002 (2012).
[37] M.A. Nielsen and I.L. Chuang, Quantum Computation and Quantum Information. Cambridge University Press, Cambridge (2000).
[38] H. Qian and E.L. Elson, Fluorescence Correlation Spectroscopy with High-Order and Dual-Color Correlation to Probe Nonequilibrium Steady States," Proc. Natl. Acad. Sci. 101, 2828–2833 (2004).
[39] I. Rech, I. Labanca, M. Ghioni, S. Cova, "Modified Single Photon Counting Modules for Optimal Timing Performance," Rev. Sci. Instrum. 77, 033104 (2006).
[40] C. Santori, Generation of Nonclassical Light Using Semiconductor Quantum Dots, Ph.D. Dissertation, Stanford University (2003).
[41] M.O. Scully, and W.E. Lamb, Phys. Rev.179, 368 (1969).
[42] G.A. Steudle, S. Schietinger, D. Höckel, S.N. Dorenbos, I.E. Zadeh, V. Zwiller, O. Benson, "Measuring the Quantum Nature of Light with a Single Source and a Single Detector," Phys. Rev. A 86, 053814 (2012).
[43] M.J.Stevens, R.H. Hadfield, R.E. Schwall, S.W. Nam, R.P. Mirin, J.A. Gupta, "Fast lifetime measurements of infrared emitters using a low-jitter superconducting single-photon detector," Appl. Phys. Lett. 89, 031109 (2006).
[44] M.J. Stevens, S. Glancy, S.W. Nam, R.P. Mirin, "Third-Order Antibunching from an Imperfect Single-Photon Source," (2013) (To be submitted).
[45] E. Waks, C. Santori, Y. Yamamoto, "Security Aspects of Quantum Key Distribution with Sub-Poisson Light," Phys. Rev. A 66, 042315 (2002).
[46] E. Waks, E. Diamanti, B.C. Sanders, S.D. Bartlett, and Y. Yamamoto, "Direct Observation of Nonclassical Photon Statistics in Parametric Down-Conversion," Phys. Rev. Lett. 92, 113602 (2004).
[47] D.F. Walls, and G.J. Milburn, Quantum Optics, 2nd Ed., Heidelberg, Springer, Berlin (2008).
[48] X.T. Zou and L. Mandel, "Photon-antibunching and sub-Poissonian photon statistics," Phys. Rev. A 41, 475 (1990).
[49] H.-A. Bachor and T. C. Ralph, A Guide to Experiments in Quantum Optics, 2nd Edition, Wiley-VCH (2004).
[50] L. Mandel and E. Wolf, Optical Coherence and Quantum Optics, Cambridge University Press (1995).
[51] T. Gerrits, M.J. Stevens, B. Baek, B. Calkins, A. Lita, S. Glancy, E. Knill, S.W. Nam, R.P. Mirin, R.H. Hadfield, R.S. Bennink, W.P. Grice, S. Dorenbos, T. Zijlstra, T. Klapwijk, and V. Zwiller, "Generation of degenerate, factorizable, pulsed squeezed light at telecom wavelengths," Optics Exp. 19, 24434 (2011).

Chapter 3

Photomultiplier Tubes

Sergey V. Polyakov
National Institute of Standards and Technology, Gaithersburg, MD 20899, USA

Chapter Outline
3.1 Introduction   69
3.2 Brief History   69
3.3 Principle of Operation   71
    3.3.1 Photoelectron Emission and Photocathodes   72
    3.3.2 Secondary Emission, Dynodes   73
3.4 Photon Counting with Photomultipliers   76
3.5 Conclusion   82
References   82

3.1 INTRODUCTION

Photomultiplier tubes (PMTs), also known as photomultipliers, are remarkable devices. While a PMT was the first device to detect light at the single-photon level, invented more than 80 years ago, they are widely used to this day, particularly in biological and medical applications. Modern PMTs deliver low noise and low jitter detection over a wide dynamic range. However, they offer limited detection efficiency, especially for longer wavelengths. We discuss the physical mechanisms behind the photon detection in PMTs, the history of their development, and the key characteristics of PMTs in photon-counting mode.

3.2 BRIEF HISTORY

There are two distinct phenomena that are fundamental to the operation of a photomultiplier tube. The first is the photoelectric effect. A range of materials emit electrons when illuminated with light. The requirement is that the photons have energy that is equal to or exceeding the so-called workfunction of the photoelectric material. The second is the phenomenon of secondary emission. When an incident electron possesses sufficient energy, it can knock out multiple electrons when hitting a surface, or when passing through a medium. In a PMT, an incident photon excites a photoelectron from the photocathode; then the photoelectron is accelerated by an electric field between the photocathode and the first dynode, where it knocks out multiple electrons. Those electrons are subsequently accelerated to a second dynode where each electron knocks out a few more electrons. This acceleration and multiple electron emission process is repeated on subsequent dynodes, with the number of dynode stages typically ranging from about 10 to about 20, yielding a final pulse of $\approx 10^6$ electrons at the PMT anode.

Heinrich Hertz discovered the photoelectric effect in 1887 [1]. Later, Albert Einstein established that the mechanism behind this effect is unequivocally quantum, and that individual photons of light transfer their energy to single electrons [2]. The photoelectric effect therefore provides a way to turn photons into an electrical signal. Of course, the electrical signal from just one electron is too tiny to be detected, as it would be within the noise of an ordinary amplifier. Therefore, special steps for noise reduction are necessary. Luckily, single-photon detection does not require faithful amplification: so long as we can distinguish presence of an electrical signal due to a photon detection from the noise in the electrical circuit, the height and shape of an electrical signal can vary from one detection to another.

This problem was resolved using the mechanism of secondary emission, discovered by L. Austin and H. Starke in 1902. Secondary emission occurs only if incoming (primary) electrons possess enough energy to liberate the electrons in the target material. The electrons emitted in this process are called secondary electrons. The average number of secondary electrons emitted in the process (and hence an amplification coefficient) rapidly increases with the energy of primary electrons, and then saturates. Saturation properties depend on the secondary emission surface. This effect is easily understood. As the energy of primary electrons grows, the number of excited electrons in the secondary-emissive material increases, but that happens at a greater depth. The deeper the electrons are excited, the lower is the chance that they can reach the surface and escape. As a result, the fraction of electrons that can escape saturates, and then decreases. Most surfaces saturate with primary electrons energy of ≈ 1 keV.

While the noise properties of an amplifier based on secondary emission are compatible with the needs of single-photon detection (practically, no secondary electrons can be generated in the absence of primary electrons), the limited gain of this process is a significant obstacle. In 1930, Leonid Kubetsky found a way to overcome this gain limitation. He proposed a multistage secondary emission device, [3] where the output of one stage is sent onto a subsequent secondary emission stage. The first high-gain amplifier implementation of this scheme was reported by Vladimir Zworykin and colleagues [4]. They characterized a three-stage amplifier and reported a maximum measured gain of 60. Within less than a year, Kubetsky reported a measured amplification gain of "about 1000" [5].

Contemporary with these developments, a much improved photoelectric material was found, Cesium-antimony (Cs_3Sb) [6], which has much higher emission probability for the photoelectron (or *quantum efficiency*, discussed later in this Chapter in detail). Further commercial development of PMTs was led by RCA in the United States and Hamamatsu in Japan.

3.3 PRINCIPLE OF OPERATION

A typical PMT, as illustrated in Fig. 3.1, contains a photocathode, several dynodes, and an anode in a sealed glass envelope with a high vacuum inside. Photons incident on a PMT undergo the following steps:

1. Photons enter the tube through the input window.
2. Photons excite electrons, some of which are emitted from its surface into the vacuum. These emitted electrons are called photoelectrons.
3. Photoelectrons are focused with a focusing electrode onto the first secondary-electron emission surface, called a dynode. The number of photoelectrons is multiplied via the secondary-electron emission effect.
4. The secondary emission is repeated several times, on each of the dynodes in the device. Thus, high-gain amplification is achieved.
5. Secondary electrons from the last dynode are collected on the anode.
6. A current spike is detected by detection electronics.

In the following we discuss details of the two major elements of a PMT: photocathodes and dynodes. Further discussion of PMT engineering and usage can be found in a recent comprehensive handbook [7].

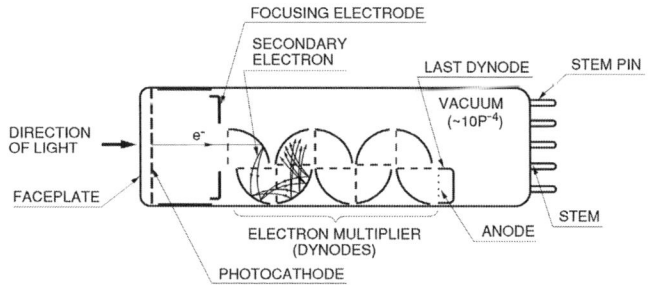

FIGURE 3.1 Typical photomultiplier tube (PMT). *Reproduced with permission from [7].*

3.3.1 Photoelectron Emission and Photocathodes

To be emitted to vacuum, an electron needs to overcome the vacuum level barrier, also known as the material workfunction. Such a barrier occurs at the surface of any solid. As a photon strikes a photocathode, electrons within the material can acquire some of the photons' energy and become excited. Then, as an excited electron diffuses through the material, it may come to the material's surface and escape to the vacuum provided that its energy is greater than the material's workfunction (i.e. sufficient to overcome the vacuum level barrier). The entire emission scenario can be quantitatively described as a series of probabilistic events. Thus, the *quantum efficiency*, or the ratio of photoelectrons to the incident photons, is a product of the probabilities of each of the steps. This quantum efficiency may be presented as:

$$\eta_{\text{QE}} = (1 - R)\frac{P_e(\lambda)}{k}\frac{1}{1 + 1/kl}P_s, \qquad (3.1)$$

where R is a reflection coefficient of incident light, k is the absorption coefficient, P_e is the probability that a photon can excite an electron of a sufficient energy, l is a mean escape length of the electrons, P_s is the probability that an electron that reaches the surface escapes into vacuum, and λ is the wavelength of the photon.

Another important factor contributing to the photocathode performance is its operation mode. Photocathodes may work in reflection or transmission, depending on the tube configuration. In reflection mode, the electrons are emitted from the illuminated side of the photocathode, and the input light normally comes through the side of the PMT's glass vacuum envelope. Transmission-mode photocathodes are used for head-on photomultiplier tubes. In this latter case, photoelectrons are emitted from the back side of the photocathode. In this configuration the light comes through the end of the cylinder-shaped glass envelope and often the photocathode is deposited directly on the inside of the glass.

Note that the parameters in the equation for quantum efficiency depend on both the choice of material for the photocathode and its operation mode. Photocathodes are generally made of compound alkali metals, or semiconductors activated with alkali, because these materials offer low vacuum level barriers. Cs-I and Cs-Te are used for ultraviolet light, for wavelengths as short as 115 nm. The short wavelength cutoff is due to limitations of glass used to seal the tube rather than the photocathode material. The photocathodes themselves can operate with even shorter wavelengths if the input window is removed. Both of these compounds are insensitive to visible and infrared light. Bialkali (Sb-Rb-Cs, Sb-K-Cs) and multialkali (Sb-Na-K-Cs) photocathodes are sensitive to the photons of visible light (up to 700 nm for bialkali and up to 850 nm for multialkali photocathodes). To extend sensitivity even further, to about 1000 nm, semiconductors, such as GaAsP, GaAs, and InGaAs activated with alkali (Cs), are used as photocathodes. Relatively recently, [8,9], spectral sensitivity has reached telecom wavelengths (wavelengths up to 1700 nm) in specially engineered semiconductor-based structures based on PN junctions (albeit with relatively low efficiencies).

Table 3.1 summarizes typical peak quantum efficiencies and sensitivity ranges of the PMTs with typical photocathode materials. Note that in most cases, the short wavelength sensitivity cutoff is due to window transmittance limitations, while the long wavelength sensitivity cutoff is due to properties of a photocathode material.

From Table 3.1, we see that quantum efficiency of photocathodes, and hence of photomultipliers is highest in the ultraviolet. And while spectral sensitivities have expanded into the near infrared, the low quantum efficiencies there make PMTs much less attractive than other technologies, many of which are discussed in the following chapters.

3.3.2 Secondary Emission, Dynodes

As we have outlined, the amplification of the photocurrent in photomultipliers is done via a secondary emission. To optimize secondary emission, an electrode made of stainless steel, copper-beryllium or nickel is coated by a so-called secondary-emissive material. These include alkali-antimonide, beryllium oxide, magnesium oxide, gallium phosphide, and arsenide-gallium phosphide. Figure 3.2 shows typical secondary emission ratios δ obtainable with different secondary emissive surfaces. δ is defined as an average number of secondary electrons per primary electron. Secondary emission ratios depend on the energy of the primary electron, therefore, to use secondary emission as an amplification mechanism for photoelectrons efficiently, primary electrons need to be accelerated. Acceleration is accomplished by applying an electric field between the dynodes. As follows from Fig. 3.2, δ grows proportionally to the accelerating potential V, and then it saturates after V reaches a certain value, typically on the order of 1000 volts. Therefore, typical amplification is 10–100 per an accelerating surface. As discussed in the historical section of this chapter, to arrange for a higher amplification, multiple secondary emission surfaces must be used. In modern photomultipliers up to 20 dynodes are used in a sequence.

In a PMT, the accelerating electric field is created by applying electrical potentials to dynodes. The dynode shape is an important engineering consideration because the electric field and thus electron trajectories depend on the dynode shape. The goal here is improving collection efficiency of primary electrons, and reducing the uncertainty in latency time of the device (electron transit time spread).

The collection efficiency of the first dynode (that is, the dynode that immediately follows the photocathode) is most critical, because any inefficiency at this stage adversely affects the detection efficiency and overall performance of the photomultiplier. Typically, the collection efficiency of the first dynode

TABLE 3.1 Properties of Typical Photocathodes

Photocathode Material	Reflection Mode		Transmission Mode	
	Spectral Range, (nm)	Peak Quantum Efficiency (%)	Spectral Range, (nm)	Peak Quantum Efficiency (%)
Cs-I	115 to 200	26 @ 125 nm	115 to 200	13 @ 130 nm
Cs-Te	115 to 320	37 @ 210 nm	115 to 320	14 @ 210 nm
Sb-Cs	185 to 750	25 @ 280 nm	–	–
Bialkali	185 to 750	30 @ 260 nm	160 to 650	27 @ 390 nm
Multialkali	185 to 900	30 @ 260 nm	160 to 850	25 @ 280 nm
Ag-O-Cs	–	–	400 to 1200	0.36 @ 740 nm
GaAs(Cs)	185 to 930	23 @ 300 nm	380 to 890	14 @ 760 nm
InGaAs(Cs)	300 to 1040	16 @ 370 nm	–	–
InP/InGaAs(Cs)	300 to 1700	1 @ 1200 nm	950 to 1700	2 @ 1550 nm

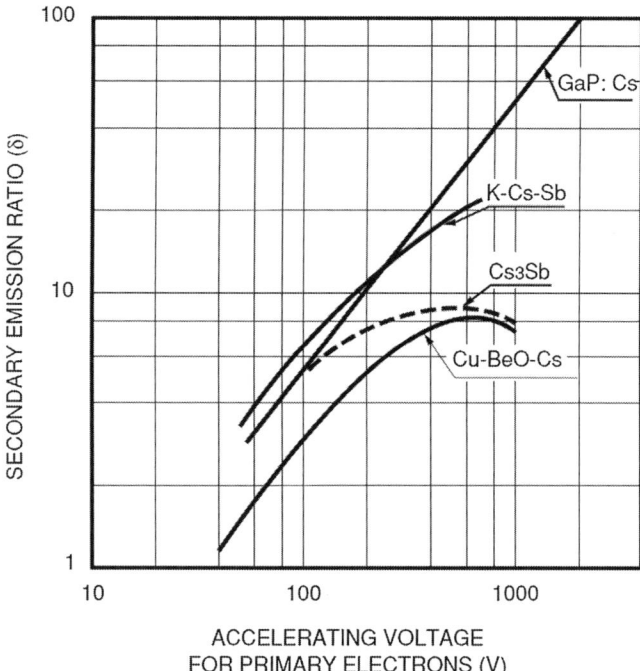

FIGURE 3.2 Secondary emission ratios as a function of an accelerating potential for typical dynodes. *Reproduced with permission from [7].*

is optimized to be better than 60–90%. It is common to engineer a better collection efficiency by focusing the trajectories of photoelectrons between a photocathode and the first dynode with a focusing electrode. If the goal is to minimize electron transit times, collection efficiency can be sacrificed in favor of shorter distances between dynodes and/or higher electric fields. A variety of dynode shapes are commercially available, with different designs leading to differences in device characteristics. In addition to temporal response and secondary-electron collection efficiency mentioned earlier, spatial and angular uniformity of detection efficiency depend on a dynode shape. Next we discuss timing characteristics of the PMTs, which are very important for many single-photon applications.

Let us define the relevant timing characteristics used to describe temporal performance of a photomultiplier, see Fig. 3.3. This figure depicts the typical electrical signal of a photomultiplier in response to a delta-function-like optical input (which could be a single photon). The latency time (also known as electron transit time) is determined by the time it takes electrons to traverse the tube from a photocathode to an anode. The latency time varies from shot to shot, because photoelectrons and secondary-electron beams may take different paths in a tube. Therefore it is relevant to not only characterize average latency but also

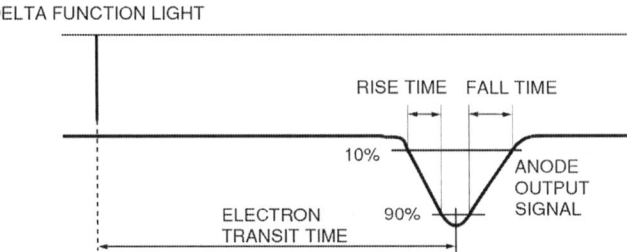

FIGURE 3.3 Temporal characterization of an output from a PMT. *Reproduced with permission from [7].*

its variation (or timing jitter), sometimes referred to as transit time spread (TTS) in photomultiplier literature. It is characterized by full width at half maximum (FWHM) of the latency time distribution histogram. This is not to be confused with pulse width, the overall width of the electrical output pulse, also measured by its FWHM. The rise time of the electrical signal is defined as the time it takes the output pulse to rise from 10 to 90 percent of the peak pulse height. The fall time is defined analogously, i.e. as time it takes for the electrical signal to change from 90 to 10 percent of the peak output pulse height.

In Table 3.2, we list most common dynode types and provide typical temporal, uniformity, and collection efficiency values. Figure 3.4 shows graphical depictions of photomultiplier tube geometries and dynode shapes. We see that PMT's timing characteristics depend strongly on dynode design. In addition, PMT characteristics may vary depending on the dynode's size and shape. We also see that there are tradeoffs between configurations that minimize TTS and maximizing collection efficiency or spatial response uniformity. Thus, the most suitable dynode shape should be chosen depending on a particular application.

3.4 PHOTON COUNTING WITH PHOTOMULTIPLIERS

Let us discuss how a photomultiplier device can be used for single-photon detection and photon counting. We have seen that a photocathode is sensitive to single photons, and that single photoelectrons can be sufficiently amplified with a sequence of dynodes. At the same time, current in the absence of the optical input (dark current) is small, because secondary emission occurs very seldom with no primary electrons present. These are the three most important prerequisites for photon counting. However, additional considerations are also required.

One obvious issue is a photomultiplier's response to a delta-function-like input. This is because a single photoelectronic event on a photocathode may be treated as such an input. The electrical current at the output is no longer a

TABLE 3.2 Characteristics of Photomultipliers vs Dynode Geometry

Dynode Type	Rise Time, (ns)	Fall Time, (ns)	Pulse Width, (ns)	Latency, (ns)	TTS, (ns)	Uniformity	Collection Efficiency
Linear-focused	0.7 to 3	1 to 10	1.3 to 5	16 to 50	0.37 to 1.1	Poor	Good
Circular-cage	3.4	10	7	31	3.6	Poor	Good
Box-and-grid	<7	25	13 to 20	57 to 70	<10	Good	Very good
Venetian blind	<7	25	25	60	<10	Good	Poor
Fine mesh	2.5 to 2.7	4 to 6	5	15	<0.45	Good	Poor
Metal channel	0.65 to 1.5	1 to 3	1.5 to 3	4.7 to 8.8	0.4	Good	Good

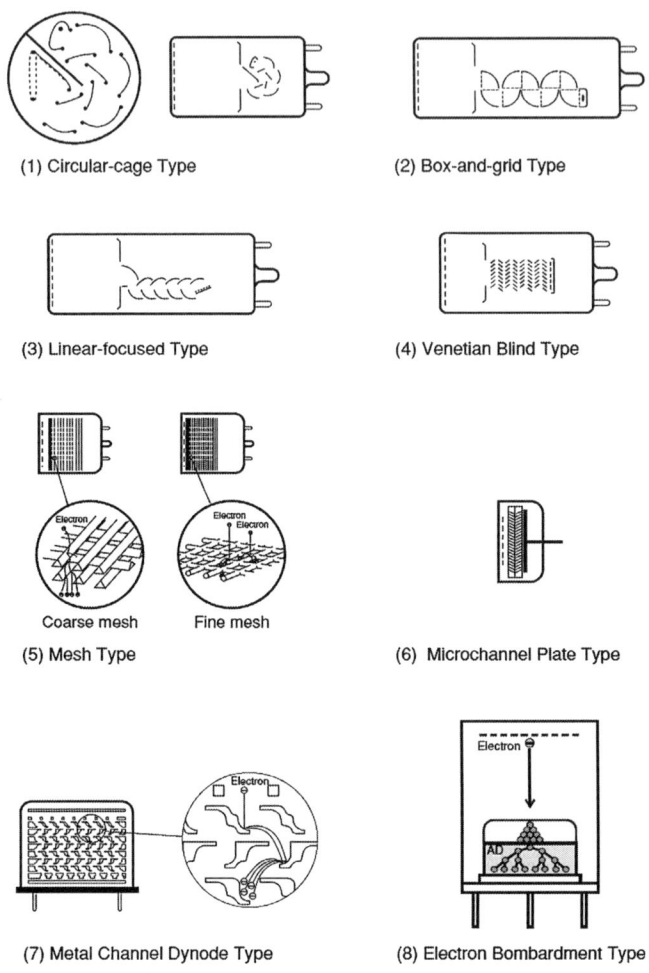

FIGURE 3.4 Typical dynode shapes. *Reproduced with permission from [7].*

delta-function, and has a finite width (sometimes referred to as an instrument response function), characterized by FWHM, $\Delta\tau$. It therefore becomes difficult to distinguish single photons that come at a rate close to or exceeding $1/\Delta\tau$, because the output pulses overlap in time. The second issue is that the overall gain of each photoelectron significantly varies. To first order, the number of secondary electrons released from a dynode obeys a Poisson distribution with an average number of secondary electrons equal to δ, the secondary-electron multiplication factor. This process is repeated as many times as there are dynodes. Thus, the uncertainty in the number of secondary electrons gives rise to a rather wide distribution of output pulse heights, making it impossible to reliably discriminate between a detection event due to absorption of just one photon versus that due to absorption of multiple photons. Given these complications, a typical use of photomultipliers (and many other detectors, as we will see in the following chapters of the book) is to distinguish between an input with no photons from that of one or more photons. Such detectors are referred to as *non-photon-number resolving* (or, sometimes, as *click/no-click* detectors). Thus, a click/no-click detector distinguishes events when one or more photons have been collected from when no photons have been detected with an exceptional signal-to-noise ratio. The leading edge of an output photoelectronic current provides a reliable (to within the TTS) "timestamp" for a detection event, useful in many applications, particularly for quantum information technology.

To characterize the performance of a photomultiplier in the photon-counting regime, a pulse-height distribution is measured. A typical pulse-height distribution of a photomultiplier is shown in Fig. 3.5a. This pulse-height distribution is measured with and without an input optical signal. The latter measurement can be used to characterize the so-called dark current. The origin of dark current is typically thermal-electron emission that occurs on the photocathode and the dynodes. Obviously, the thermal electrons from the dynodes go through fewer stages of amplification, resulting in pulses with lower amplitude than those due to photoelectrons.

Typically, a photomultiplier output is analyzed with a discriminator set at some threshold level L. Once the electrical output exceeds the threshold, a detection occurs. By setting an appropriate threshold, most spikes of current consistent with the amplification of thermal electrons that escaped from dynodes (especially those dynodes that are closer to the anode) may be filtered out. However, for any reasonable value of L, some dark-current events, particularly these originated from a photocathode and first few dynodes, will still exceed the threshold, and get counted. Detection events due to dark current are called *dark counts*. Obviously, given a detection event, it is generally impossible to determine for sure if that event is a dark count, or it is due to a true detection of a photon. One can only determine the probability that a detection event is (or is not) a dark count, based on the light level and the average number of dark counts per second. As we see in Fig. 3.5a, the dark count rate decreases as the level of threshold L increases. At the same time, a large fraction of legitimate counts are rejected when that threshold level is set too high.

The overall number of counted events for each measurement (with and without optical input) as a function of threshold level is presented in Fig. 3.5b. This figure is obtained by the integration of a histogram in Fig. 3.5a from L to ∞, as a direct consequence of a threshold detection. This trivial observation leads us to an important conclusion: while intrinsic *quantum efficiency* of the photo-sensitive material (in this case, a photocathode) is a constant value for a given material, *detection efficiency* of the device is not: by changing the detection threshold the overall detection efficiency is changed, as can be understood from

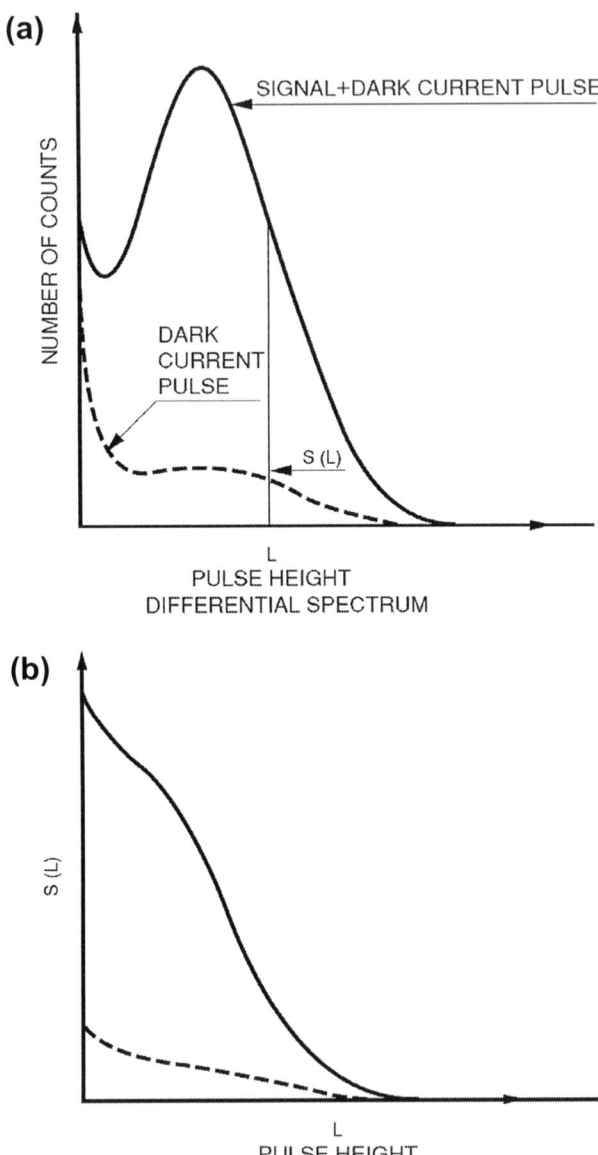

FIGURE 3.5 (a) Pulse-height histogram for characterization of a PMT; (b) total detected pulses of a PMT vs. a level for a discriminating threshold (integrated pulse-height histogram). Depending on where a discriminating threshold is set, a certain detection efficiency and dark count rates are achieved. *Reproduced with permission from [7].*

Fig. 3.5b. We see that with the same input optical power the total number of counts $S(L)$ decreases as the threshold L increases, which reduces detection efficiency.

Both dark-current and signal-current pulse-height distributions should be taken into account when selecting an optimal threshold. Obviously, the choice of L is application dependent. One possible application of photon counting is to establish the radiant power of a weak optical source. In this case, the signal-to-noise ratio should be maximized by selecting an appropriate L. Other applications call for maximum detection efficiency. In such cases, L can be lowered to improve detection efficiency at the expense of signal-to-noise ratio. The trade-offs between characteristics similar to those seen in these examples are quite common to other detector technologies, and will be addressed in the following chapters. Note that the ultimate goal for detector development is to find a technical solution which improves *all* characteristics of single-photon detection, as opposed to sacrificing one of the performance metrics in favor of the other.

Another important consequence of using a discriminator is the possibility of the output pulse overlap. If the two pulses are too close to be resolved, only one count will be recorded. That is, immediately after a single-photon detection, the detection system becomes effectively blind to the new incoming photons. The time after an initial detection during which no new input photons can be detected is the *dead time*. As explained in Chapter 2, dead time is a common issue with many different single-photon detectors, and the mechanisms behind dead time differ among the various technologies. In photomultipliers, both pulse length and discrimination threshold affect the dead time. A typical dead time of photomultipliers is on the order of 10 ns, but can be as short as 1 ns. This compares favorably with many other single-photon-detection technologies, whose dead times range from 10s of ns to microseconds. Short dead times are important for measurements at high photon rates. Note here that, in practice, dead time can also be limited by external factors such as pulse-counting electronics, rather than the detector itself.

Due to dead time, when measuring the intensity of a weak classical source (for example, an attenuated laser or an incoherent white light source), the true photon flux is underestimated. This effect becomes noticeable at photon detection rates $\approx 1\%$ of the maximal counting rate $f_{max} = 1/t_{dead}$, where t_{dead} is the dead time. This saturation effect makes single-photon detectors highly nonlinear. For an uncorrelated (Poisson-like) source, such as a weak coherent light beam, one can readily apply a simple correction factor, which allows estimation of the number of photons that were missed during the dead time: $N_{true} = N_{measured}/(1 - N_{measured} t_{dead})$. This correction is good (to within 1%) up to $f_{max}/10$. When the source of light is correlated (such as a single-mode thermal source or a non-classical source of light), the formula above cannot be used.

The last issue covered in this section is *afterpulsing*. Afterpulses are false detection events that occur after a prior detection event. Afterpulses are common to many single-photon detectors, but are caused by different underlying effects for different types of detectors. In PMTs, there are two effects that cause afterpulses. The first mechanism is elastic scattering of the electrons on the first dynode. Usually, the characteristic time delay for this kind of afterpulses is short: on the order of a nanosecond. For most photon-counting PMTs this is not a problem, because such an afterpulse is generated in the immediate vicinity of the event that induced it, and this afterpulse is not detected by the electronics because of the dead time. There are, however, afterpulses that are delayed with respect to the initial pulse long beyond the dead time. These are caused by the ionization of residual gas molecules in the tube. The positive ions may return to the photocathode and knock out electrons that are amplified as if they are legitimate events, and cause false counts. This process occurs over a time range scaling from several hundreds of nanoseconds to tens of microseconds, which may create serious problems for photon counting, and particularly to precision measurements with PMTs, because it is impossible to attribute particular detections to a true detection versus an afterpulse (c.f., dark counts). Instead, a statistical approach is used: the probability of an afterpulse may be estimated, and a statistical correction can be applied.

As with many single-photon detectors, the properties of PMTs vary not only among the different types of PMTs, but also from one device to another. Moreover, they may change depending on the age of the particular PMT, voltage used as an accelerating potential, history of previous light exposure, and other factors.

3.5 CONCLUSION

In this chapter, we have described the first photon-counting technology, invented over 80 years ago. Remarkably, this technology is still used widely today. In addition to reviewing how PMTs function, we saw how physical properties of the device (instrument response function, quantum efficiency, etc.) result in a set of properties that are specific to the photon-counting detectors (detection efficiency, dark count rate, dead time, afterpulsing). These properties are universal, applicable to photon-counting detectors with other technologies. In the following chapters, other technologies of single-photon detection will be introduced.

REFERENCES

[1] H. Hertz, Ann. Phys. 267, 983–1000 (1887).
[2] A. Einstein, Ann. Phys. 17, 132 (1905).
[3] L. Kubetsky, USSR, Author's Certificate No. 24040, Priority date 4 August 1930 (1930).
[4] V. Zworykin, G. Morton, and L. Malter, Proc. IRE 24, 351 (1936).
[5] L. Kubetsky, Proc. IRE 25, 421 (1937).
[6] P. Gorlich, Z. Angew. Phys. 101, 335 (1936).
[7] Photomultiplier Tubes, Basics and Applications, 3rd ed., Hamamatsu Photonics K.K. (2007).
[8] M. Niigaki, T. Hirohata, T. Suzuki, H. Kan, and T. Hiruma, Appl. Phys. Lett. 71, 27 (1997).
[9] K. Nakamura and H. Kyushima, Jpn. J. Appl. Phys. 67, 5 (1998).

Chapter 4

Semiconductor-Based Detectors

Sergio Cova[*], Massimo Ghioni[*], Mark A. Itzler[†], Joshua C. Bienfang[‡], and Alessandro Restelli[‡]

[*]Politecnico di Milano, Dipartimento di Elettronica, Informazione e Bioingegneria, Piazza Leonardo da Vinci, 32 20133 Milano, Italy
[†]Princeton Lightwave, Inc., 2555 US Route 130 S., Cranbury, NJ 08512, USA
[‡]Joint Quantum Institute, University of Maryland, and National Institute of Standards and Technology, Gaithersburg, MD 20899, USA

Chapter Outline

4.1	Photon Counting: When and Why	84
4.2	Why Semiconductor Detectors for Photon Counting?	85
4.3	Principle of Operation of Single-Photon Avalanche Diodes	85
4.4	Performance Parameters and Features of SPAD Devices	87
	4.4.1 Photon Detection Efficiency	88
	4.4.2 Dark Count Rate (DCR)	88
	4.4.3 Afterpulsing	89
	4.4.4 Timing Jitter	90
	4.4.5 Crosstalk	92
	4.4.6 Fill-Factor	93
	4.4.7 Microelectronic Structure of a SPAD: Outline and Basic Features	93
4.5	Circuit Principles for SPAD Operation	94
4.6	Silicon SPAD Devices	98
	4.6.1 Planar SPAD Devices Fabricated in a Custom Technology	98
	4.6.2 Non-Planar SPAD Devices Fabricated in a Custom Technology	102
	4.6.3 High-Voltage, Complementary Metal-Oxide Semiconductor (HV-CMOS) SPADs	104
	4.6.4 Standard Deep-Submicron CMOS SPADs	106
4.7	Silicon SPAD Array Detectors	108
4.8	SPADs for the Infrared Spectral Range	113
	4.8.1 Infrared SPADs	113
	4.8.2 Basic InGaAs/InP SPAD Design Concepts	114
	4.8.3 DE and DCR Modeling and Performance	115
	4.8.4 Timing Jitter	117

	4.8.5 Afterpulsing	118
	4.8.6 Comparison of InGaAs/InP SPADs and Si SPADs	119
4.9	**Active Gating Techniques for InGaAs SPADs**	**120**
	4.9.1 Introduction	120
	4.9.2 Sampling	122
	4.9.3 Cancellation	123
	4.9.4 Introduction to High-Speed Periodic Gating	125
	4.9.5 Sine-Wave Gating	127
	4.9.6 Self-Differencing	129
	4.9.7 Harmonic Subtraction	131
	4.9.8 Summary	132
4.10	**Future Prospects for Silicon SPADs**	**134**
4.11	**Future Prospects for InGaAs SPADs**	**135**
References		137

4.1 PHOTON COUNTING: WHEN AND WHY

There is nowadays a widespread and growing interest in low-level light detection and imaging. This interest is driven by the need for high sensitivity in various scientific and industrial applications such as fluorescence spectroscopy in life and material sciences, quantum computing and cryptography, profiling of remote objects with optical radar techniques, particle sizing, and more. In particular, the use of fluorescence-lifetime spectroscopy as both an analytical and research tool has increased markedly in recent years finding remarkable applications in chemistry, biochemistry, and biology.

Photon counting has long been recognized as the technique of choice for attaining the ultimate sensitivity in measurements of optical signals. However, advanced analog detectors (such as back-illuminated charge-coupled devices (CCDs)) with ultra-low dark current can also be used in some instances to measure very weak photon fluxes. A basic issue must therefore be clearly addressed: when and why are photon-counting detectors advantageous? For applications where the measurement time is very short or the arrival time of the optical signal must be known with high precision (e.g., high-frame-rate imaging, fluorescence correlation spectroscopy (FCS), or fast optical coincidences), photon-counting detectors have an advantage over analog detectors, which have electronic readout noise in addition to dark-current noise. For short measurement times the readout noise exceeds the dark-current noise and sets the sensitivity limit of analog detectors, whereas readout noise simply does not exist in photon-counting detectors. These photon-counting detectors exploit an internal amplification mechanism, which, in response to single photons, generates macroscopic electrical signals that are much larger than any electronic circuit noise.

4.2 WHY SEMICONDUCTOR DETECTORS FOR PHOTON COUNTING?

Photon counting and time-correlated single-photon counting (TCSPC) techniques were developed using photomultiplier tubes (PMT), that is, vacuum-tube detectors with high internal gain (see Chapter 3), and high-performance PMTs have been produced industrially since the 1940s. Commercially available devices can provide remarkable performance, even up to rates of millions of counts per second, and compact and rugged PMT devices have been developed to address the typical drawbacks of vacuum-tube devices. Among their advantages, the most significant and distinct is the PMT's large sensitive area ($\approx cm^2$), which greatly simplifies the design of the optical system. Micro-channel plate (MCP) PMTs also offer picosecond timing jitter. However, PMTs suffer from low detection efficiency (DE). In the visible, the DE of conventional bialkali and multialkali photocathodes reach 20 % to 25 % between 400 nm and 500 nm, whereas a DE up to ≈ 40 % can be achieved between 450 nm and 650 nm using a GaAsP photocathode [1,2]. In the infrared, PMTs have much lower DEs.

Semiconductor-based detectors are a valuable alternative to PMTs. Besides the well-known advantages of solid state versus vacuum-tube devices (small size, ruggedness, low power dissipation, low supply voltage, high reliability, low cost, etc.), semiconductor detectors provide inherently higher detection efficiency, particularly in the red and near-infrared spectral regions.

The development of semiconductor-based detectors of single photons has been slower than that of PMTs. Avalanche multiplication of carriers in reverse-biased p-n junctions is used in ordinary avalanche photodiodes (APDs) to obtain internal amplification in the detector similar to that in PMTs. However, in an ordinary APD, the multiplication of both holes and electrons causes an inherent positive feedback that produces strong fluctuations in the avalanche gain. Such fluctuations increase more steeply with the applied voltage than the average gain. In the best case (that is, in silicon diodes made with special structure and technology), the useful gain is limited to $\approx 5 \times 10^2$, as opposed to the $\approx 10^6$ gain easily reached by PMTs. Therefore, even in the best APDs the current pulses due to single-photon absorption have very small and wildly fluctuating amplitudes, hence it is only marginally possible to detect single-photon pulses and their timing is highly uncertain.

4.3 PRINCIPLE OF OPERATION OF SINGLE-PHOTON AVALANCHE DIODES

The positive feedback in the avalanche makes it possible to exploit a reverse-biased p-n junction in a different way for detecting single photons. In this operation mode, the p-n junction cannot be considered a detector with an amplifier inside, as is the case for APDs, but rather a detector with a digital flip-flop inside.

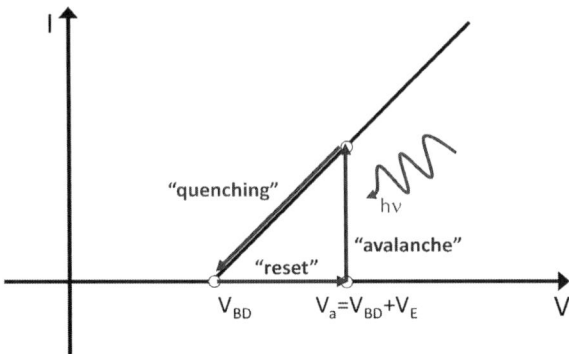

FIGURE 4.1 SPAD operation in the reverse I-V characteristics of a p-n junction.

As seen in Fig. 4.1, in the quiescent state the device is biased at voltage V_a above the breakdown voltage V_{BD}, and no current flows (the "OFF" state): in the junction depletion layer the electric field is very high, but no free carrier is present. When even a single charge carrier is injected in the high-field region it is strongly accelerated and can impact ionize and generate a secondary electron-hole pair, starting a self-sustaining avalanche multiplication process. The current then grows exponentially until the space-charge effect limits it to a constant level. This level is proportional to the excess bias voltage $V_E = V_a - V_{BD}$, hence an avalanche resistance can be defined. The p-n junction is thus switched to the "ON" state, where a constant macroscopic avalanche current flows (≈ 1 ma). The fast onset of the current marks the time of arrival of the photon that generated the initial charge carrier. The device remains in this ON state until the avalanche is quenched by an external circuit (quenching circuit), which drives the applied voltage down to V_{BD} (or lower). The quenching circuit then concludes the operation cycle by resetting the voltage to the original level above breakdown. The detector is insensitive to any subsequent photon arriving in the time interval from the avalanche onset to the voltage reset, which is the detector dead time.

This type of operation is called Geiger mode because of the analogy with the gas counters of ionizing radiation. The device operates in a way radically different from an ordinary APD and to avoid misunderstanding and confusion it was given a different name and acronym: single-photon avalanche diode (SPAD).

To be operated as a SPAD, a p-n junction must have a uniform breakdown over the entire active junction area in order to produce macroscopic current pulses with constant amplitude. That is, causes of localized breakdown must be avoided, such as edge effects and microplasmas within the active junction area. Besides this requirement, more stringent conditions must be fulfilled to attain acceptable SPAD performance, as discussed in Section 4.4.

Figure 4.2 outlines the structure of the silicon devices in which Geiger-mode operation was first observed. Haitz [3–5] developed these devices in a planar technology: a deeply diffused guard ring was used to avoid edge breakdown effects and to define a small sensitive area of diameter <10 μm.

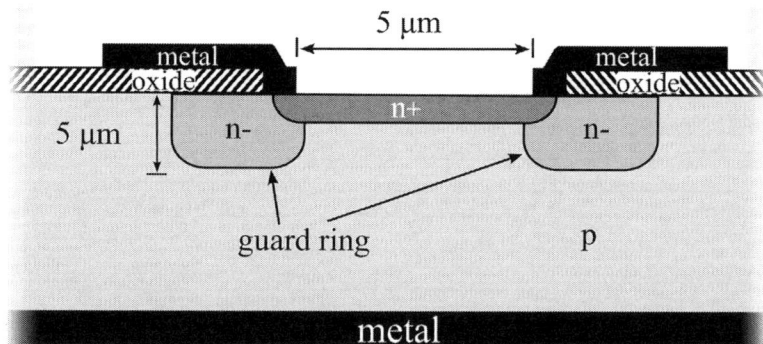

FIGURE 4.2 Cross-section of the p-n diode devised by Haitz [5] for investigating the physics of avalanche above breakdown.

Silicon SPADs have been extensively investigated and are now well developed. Considerable progress has been achieved in SPAD design and fabrication techniques, and devices with good characteristics are commercially available for applications over the visible spectral range, up to 1 μm. A thorough review of the available silicon SPADs and their state-of-the-art performance is given in Section 4.6. Recent applications, such as quantum cryptography (quantum key distribution, QKD), LADAR, and VLSI circuit characterization based on light emission from hot carriers in MOSFETs, require single-photon detectors with high efficiency, low noise, and picosecond timing jitter in the wavelength range above 1 μm. The DE in the near-infrared (NIR) range is extremely low for typical PMTs, and reaches only ≈1% for PMTs with photocathodes specifically designed for IR efficiency, but at the cost of high photon-timing jitter and high dark count rates [1]. The evolution of SPADs for extending the spectral range of photon counting beyond 1 μm started in the mid-1980s with studies carried out on commercially available APDs designed for fiber communications. Photon counting in the NIR range was first performed with Ge SPADs [6], and then extended with InGaAs/InP SPADs [7], which are now the workhorse for most experiments. A review of the state-of-the-art in InGaAs/InP SPADs is given in Section 4.8.

4.4 PERFORMANCE PARAMETERS AND FEATURES OF SPAD DEVICES

SPAD operation in Geiger mode is characterized by a number of basic performance parameters (see Chapter 2). The photon-detection efficiency is the probability that an incident photon triggers an avalanche (true detection event). The uncertainty in the photon arrival time is called timing jitter. The dark count rate (DCR) is the number of avalanches per unit time that occur in the absence of incident photons (i.e. false detection events, or "dark counts). Furthermore, physical phenomena specific to photon-counting devices can generate additional dark counts correlated to the occurrence of previous avalanche events, called afterpulses.

4.4.1 Photon Detection Efficiency

Besides the physical phenomena that determine the performance of semiconductor photodiodes in general (optical coupling, reflection, absorption, etc.), there are other physical effects that are important for SPAD operation. For a photon to be detected, it not only must be absorbed in the detector's active volume and generate a primary electron-hole pair, it is also necessary that the primary electron-hole pair succeeds in triggering an avalanche. The avalanche-triggering probability increases with the excess bias voltage, V_E, since it is enhanced by a higher electric field. Theoretical and experimental studies [8,9] have shown that this probability increases linearly at low V_E, and tends to saturate at high V_E. The detection efficiency behaves accordingly, as illustrated in Fig. 4.3.

4.4.2 Dark Count Rate (DCR)

Dark counts are due to carriers thermally generated within the SPAD junction, so the dark count rate increases with temperature. The DCR has the same role as the dark current in ordinary photodiodes, that is, its Poissonian fluctuations are the internal noise source of the detector. As shown in Fig. 4.4, the DCR of SPADs also increases with the excess bias voltage V_E. The rise is due not only to the avalanche-triggering probability, which also increases the DE, but also to the field enhancement of the carrier generation rate. It is well known that in silicon and other semiconductors thermal generation of carriers occurs through local energy levels located deep within the bandgap (levels closer to the midgap are the

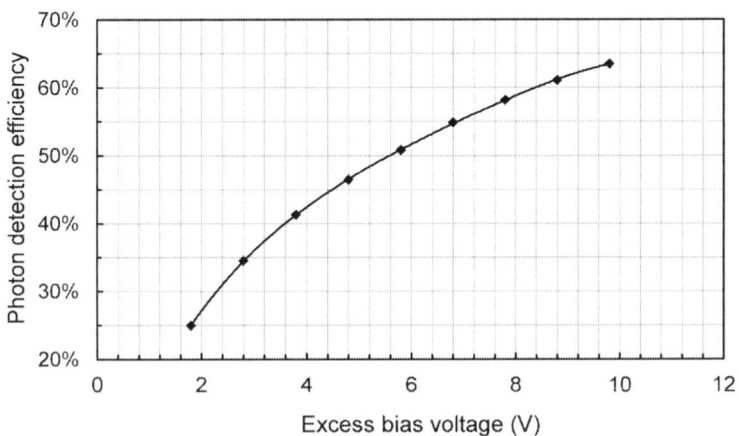

FIGURE 4.3 Photon detection efficiency versus excess bias voltage for SPAD devices reported in [16].

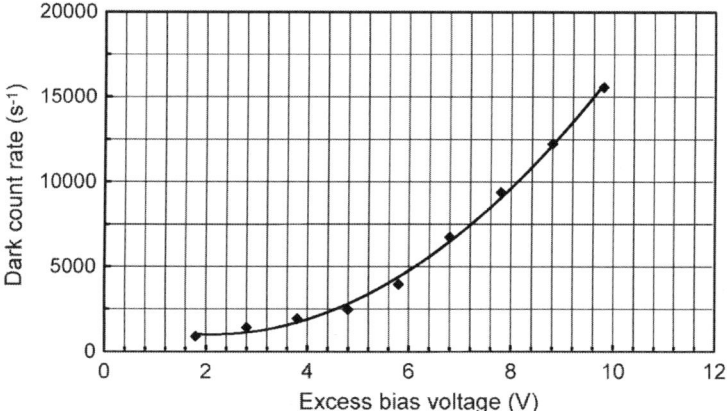

FIGURE 4.4 Dark count rate versus excess bias voltage for a 50 μm diameter SPAD reported in [16], operated at room temperature.

most efficient generation centers) [10]. Both the quality of the starting material and the technological processes used in the device fabrication have a strong impact on the density of deep energy levels and therefore on the generation rate. Transition-metal impurities are the most common source of deep levels. Metal contamination may occur during silicon handling, high-temperature heat treatments, or ion implantations. Unintentional contaminants, Fe, Cu, Ti, Ni are usually found in silicon in concentrations of $\approx 10^{11} \text{cm}^{-3}$ to 10^{12}cm^{-3} [11].

Poole-Frenkel and trap-assisted tunneling effects that occur at high electric fields ($>10^5$ V/cm) can greatly enhance the emission rate of deep energy levels (field-enhanced generation) [12,13]. At even higher field intensities ($>7 \times 10^5$ V/cm), direct band-to-band tunneling may take place, that is, strong generation of free carriers in the junction without the assistance of deep energy levels [14,15]. Tunnel-assisted generation is not reduced by lowering the temperature and therefore sets a limit to the reduction of the dark count rate obtained by cooling the detector. An important conclusion can thus be drawn for the design of SPADs: the electric field profile within the SPAD junction must be suitably tailored to avoid band-to-band tunneling and field-enhanced generation of carriers. This reduces the detector noise at room temperature and makes cooling more effective in reducing the DCR [16].

4.4.3 Afterpulsing

The noise in SPADs is further increased by an effect that does not play any role in ordinary APDs. Deep levels located at intermediate energies between mid-gap and band edge can act as minority carrier traps. During each avalanche pulse, a few carriers may be trapped in these levels and subsequently released, as outlined in Fig. 4.5a. The released carrier can retrigger the avalanche, thereby generating "afterpulses" correlated in time to the original avalanche triggered by the photon [5,17,18]. This release is statistical; the emission probability

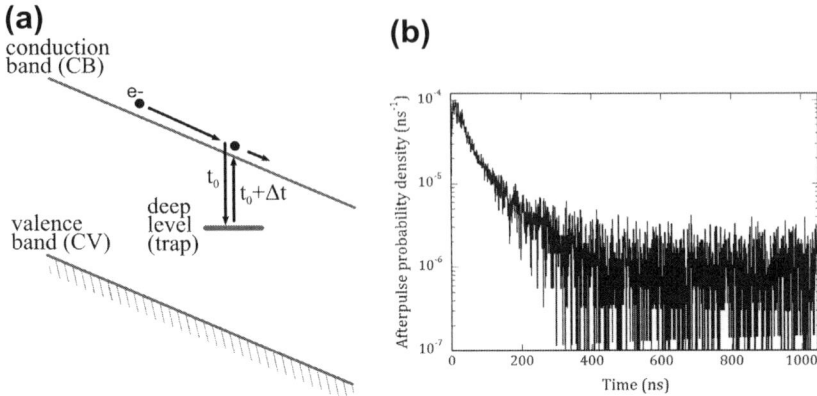

FIGURE 4.5 (a) Trapping and release of an electron by a deep level (CB is the conduction band; CV is the valence band); (b) Probability density of afterpulse generation after an avalanche in a 100 μm diameter silicon SPAD operating at room temperature.

per unit time has a characteristic value for each type of level involved, and the reciprocal of this emission probability is the exponential time constant (trap lifetime) for that level. Afterpulsing effects can be evaluated by using the time-correlated carrier counting technique described in Ref.[17]. Figure 4.5b shows the probability density as a function of time for the occurrence of an afterpulse after an initial avalanche pulse for a 100 μm Si SPAD operated at room temperature.

Afterpulsing introduces a positive feedback loop that can significantly increase the effective dark count rate [18]. Since traps are far from being saturated [18,19], their population increases linearly with the charge that flows during the avalanche pulse. Therefore, to reduce afterpulsing, the total avalanche charge should be reduced as far as possible.

If the trapped charge cannot be reduced to a sufficiently low level, a quenching procedure can be exploited to reduce the afterpulsing rate to a negligible, or at least acceptable level. After an avalanche, by deliberately maintaining the voltage at the quenching level (see Section 4.5) for an extended "hold-off" time, trapped carriers are given time to be released and thus will not retrigger the device when its voltage is returned above breakdown. While operating at lower temperatures improves noise performance of photodetectors, lower operating temperatures exacerbate the afterpulsing problem. This is because the trap-release process becomes much slower at lower temperatures [17], requiring a much longer hold-off time to achieve the same level of depopulation and seriously limiting the dynamic range in photon-counting measurements.

4.4.4 Timing Jitter

The onset of the avalanche pulse is correlated with the arrival time of the photon that generates the primary charge pair. Due to various physical effects,

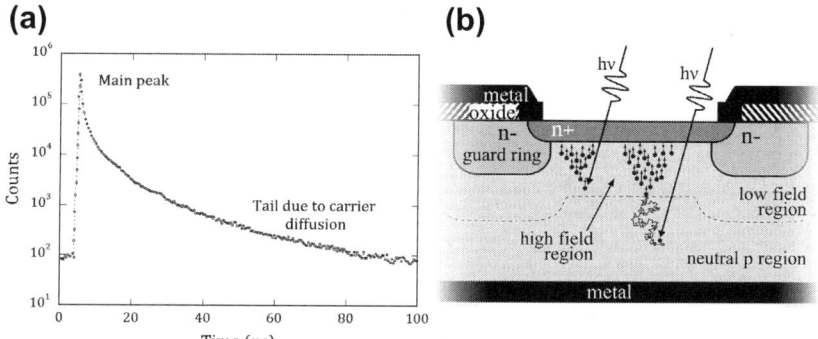

FIGURE 4.6 (a) Photon-arrival-time distribution measured in a standard TCSPC set up with laser-diode pulses of about 20 ps duration at 830 nm and a planar Si-SPAD device as in Fig. 4.2; (b) Outline of avalanche generation by photons absorbed in the depletion layer and in the neutral region.

however, the delay of the instant at which the onset is sensed with respect to the true arrival time of the photon is not constant, but subject to statistical fluctuations. The timing jitter is usually quoted as the full-width at half maximum (FWHM) of the photon arrival time distribution [20]. A typical photon-timing distribution is shown in Fig. 4.6a for the planar device structure developed by Haitz [3–5]. Two main components are evident. The narrow peak has a FWHM of about 60 ps and is due to carriers photogenerated in the junction depletion layer, which are immediately accelerated by the electric field (see Fig. 4.6b). The slow tail is due to carriers photogenerated in neutral regions near the depletion layer that migrate by diffusion, eventually reaching the edge of the depletion layer where they are accelerated by the electric field. The tail limits the photon-timing resolution. In QKD applications, diffusion tails can lead to inter-symbol error when the timing jitter exceeds the clock period, resulting in photon events being recorded in subsequent bit periods. In such applications the FW10%M and FW1%M are important parameters as well as the FWHM. Furthermore, the amplitude and duration of the tail are wavelength dependent due to the dependence of the optical absorption coefficient (i.e., of the penetration length) [21]. This is a significant drawback in applications where the light source is not monochromatic. It is clear that SPAD devices intended for photon-timing applications should have a structure designed to minimize any diffusion-tail effect.

The FWHM of the main peak exceeds the contribution due to the noise in the pulse-timing circuits, and can therefore be ascribed to statistical fluctuations in the build-up of the avalanche, from the first seed (the photogenerated electron-hole pair) to the macroscopic current level of the sensing threshold of the timing circuit. Various research groups have endeavored to explain and compute it in terms of a local statistical build-up with essentially one-dimensional models of the current rise. This approach, however, predicted FWHM values shorter than 10 ps. As a possible source of further fluctuations, the lateral propagation of the avalanche from the first seed to the whole detector area was then investigated. Two lateral diffusion mechanisms were highlighted in the literature: multiplication-assisted lateral diffusion of carriers [22], and photon emission from hot carriers in the avalanche [23]. The former mechanism is dominant in SPAD devices with thin depletion layers (≈ 1 μm), whereas the latter is dominant in SPAD devices having a thick depletion layer (≈ 20 μm), see Section 4.6. The efficiency of both processes is enhanced by a higher electric field; by increasing the excess bias voltage V_E, the photon-timing resolution is improved in all cases [24]. It has been recently demonstrated that remarkable timing performance is achievable by sensing the avalanche current at very low level (about a hundred μA), when the multiplication process is still confined within a small area around the photon absorption point [16,25]. To perform a true low-level sensing of the avalanche current it is critical to preserve the shape of the first part of the leading edge by minimizing any filtering action. To this end, a carefully designed current pick-up circuit must be used [25], as illustrated in Section 4.5.

In the context of SPAD arrays, a further set of performance parameters should be introduced besides the aforementioned ones. The most significant are crosstalk and fill-factor.

4.4.5 Crosstalk

Ideally, photons absorbed within the active volume of a pixel in a SPAD array are expected to contribute only to the signal of that pixel. In practice, however, such an event can cause detections in neighboring pixels, resulting in crosstalk. Crosstalk reduces the effective spatial resolution of an image sensor, leading to blurring. Two crosstalk mechanisms affect the performance of a SPAD array: optical crosstalk and electrical crosstalk.

4.4.5.1 Optical Crosstalk

Silicon p-n junctions emit photons when operated in the avalanche regime. The emission probability is very low: on the average, about one photon is emitted every 10^5 carriers crossing the junction [26]. In monolithic SPAD arrays, photons emitted from a SPAD can trigger an avalanche in another detector, thus causing optical crosstalk between the pixels of the array. The outcome is an incorrect evaluation of the optical signal detected by each pixel of the array. The crosstalk probability increases as the distance between pixels is reduced, and therefore sets a limit to the array density. Optical barriers placed between adjacent pixels (such as deep trenches coated with metals or heavily doped diffusions) cannot completely prevent the optical crosstalk [27] because photons can be reflected at the bottom surface of the chip, thus bypassing the optical barriers and contributing substantially to the crosstalk. A good strategy to minimize this contribution is to adopt thick and highly doped substrates to increase the free-carrier absorption of avalanche-generated photons [28].

4.4.5.2 Electrical Crosstalk

Carriers photogenerated in the quasi-neutral region below the p-n junction can diffuse laterally and trigger an avalanche in a neighboring pixel. Since the mean penetration depth strongly increases with wavelength, photons in the red and near-infrared ranges tend to induce more electrical crosstalk than short-wavelength photons. Electrical crosstalk may be effectively addressed by exploiting dielectric and/or junction isolation techniques, as discussed in Section 4.7.

4.4.6 Fill-Factor

Fill-factor is defined as the active-to-total area ratio of a single pixel. It is a key figure of merit, especially for applications involving diffuse illumination of the SPAD array (such as 3D imaging and profiling of remote objects). In general, the fill-factor is limited by the sizes of the guard ring structure, the isolation structure, and the quenching and counting/timing circuitry associated with each pixel; it is usually of the order of a few percent. Lenslet arrays can be used to improve the fill-factor [29], but at the cost of greater complexity of the system and lower flexibility in applications due to the reduced range of acceptance angles.

4.4.7 Microelectronic Structure of a SPAD: Outline and Basic Features

Taking into account the overview given above of SPAD performance and parameters, the essential elements of the structure of a SPAD device can now be highlighted. The device core is the *avalanche region*, where a high electric field provides carrier multiplication by impact ionization. For obtaining uniform DE, the material properties and the electric field intensity in this region must be as uniform as possible over the entire detector area. The thickness of the avalanche region has to be thin in order to limit the zone where the high electric field enhances the thermal generation of carriers, i.e., the DCR. To obtain high DE, carriers photo-generated in the contiguous *absorption and drift region* of the depletion layer must also be driven to the avalanche region, but by a lower electric field. In fact, moderate field intensity in this region is sufficient to ensure a saturated drift velocity (i.e., fast collection) and avoids enhancing thermal carrier generation. At the edge of the avalanche zone, deeply diffused *guard-rings* or other equivalent structures avoid local concentration of the electric field intensity. The depletion layer is sandwiched between neutral semiconductor layers at the top and bottom, which should be either transparent at the wavelength of interest, or at least very thin. Carriers photo-generated in these layers either recombine and do not contribute signals, or are collected at the avalanche region after diffusing in the neutral region, thus generating a diffusion tail in the photon-timing distribution (c.f. Section 4.6.1).

4.5 CIRCUIT PRINCIPLES FOR SPAD OPERATION

The circuit that quenches the avalanche and resets the bias voltage plays an integral role in the performance of SPADs [18]. In early studies on silicon avalanche diodes in Geiger mode the simple passive quenching circuit (PQC) outlined in Fig. 4.7a was used. The bias voltage is applied through a large ballast resistor R_L; a small resistor R_S is connected to the other terminal for observing the current pulse. The avalanche current discharges the total capacitance C_T at the diode terminal, which is the sum of the junction capacitance C_d and of the stray capacitance C_s. The voltage V_d across the diode decreases toward V_{BD} and the avalanche current decreases correspondingly. As the voltage V_d approaches V_{BD} the rate of decrease slows down. All the avalanche current flows through R_L and is reduced to the value $(V_a - V_{BD})/R_L$. If R_L is high enough to reduce the current below a few tens of μA, the number of avalanche carriers is small and the probability of interruption of the multiplication chain is high, and the avalanche can be finally quenched [5]. The voltage V_d then starts to recover slowly toward the bias voltage V_a (reset transition), as the small current in R_L recharges C_T with a long time constant $R_L C_T$. During the reset transition, the diode voltage is higher than V_{BD} and an avalanche can be triggered, but the diode fires at a voltage V_d lower than V_a (see Fig. 4.7b). It then operates with lower detection efficiency and impaired photon-timing resolution. In a passively quenched SPAD, after each avalanche the triggering probability has a continuous evolution, starting from practically nil and finally reaching the steady-state value determined by V_a. The behavior of the detector is thus peculiar: it is paralyzable, but with a time-dependent sensitivity to triggering events [18]. Furthermore, the voltage and current pulses produced during the reset transition have smaller amplitudes, as shown in Fig. 4.7b. Since a comparator is employed for sensing avalanches, pulses smaller than the threshold level are discarded, producing a dead time that is neither well known nor stable. The result is a loss of linearity at high

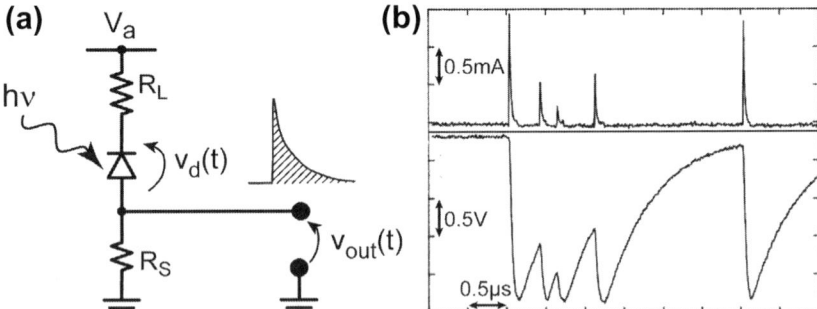

FIGURE 4.7 (a) Schematic circuit diagram of a SPAD in the passive-quenching arrangement (typical values $R_L = 500$ kΩ and $R_S = 50$ Ω); (b) Waveform of the avalanche current (upper trace) and of the voltage across the SPAD (lower trace).

counting rates that can be measured empirically, but is difficult to model and correct for accurately [18]. In summary, photon-counting measurements can be accurately performed only if the total count rate is low enough to make count losses negligible and correction unnecessary.

The drawbacks of the passive quenching can be reduced, though not eliminated, with modern circuit technologies that significantly reduce the stray capacitance C_s and thus shorten the duration of the reset transition. With surface mounting techniques and miniature components C_s can be reduced to a few picofarads. A monolithic integration of detector and ballast resistor, nowadays possible at least for silicon SPAD's, can reduce C_s well below 1 pF and the reset transition time well below 1 µs.

The solution that completely avoids the drawbacks of the passive quenching is the active quenching circuit (AQC), first devised in 1975 [30]. The principle is simple: to sense the rise of the avalanche and react back at the SPAD, forcing short quench and reset transitions with a controlled bias-voltage source. As outlined in Fig. 4.8a, the sensing comparator triggers a voltage driver to switch the bias voltage down to the breakdown voltage V_{BD} or below. After a controlled hold-off time, the bias voltage is then switched back to the operating level V_a. A standard pulse synchronous to the avalanche rise is derived from the comparator output, and can be used for photon counting and timing (Fig. 4.8b).

General principles and practical features of quenching circuits have been extensively dealt with in a tutorial paper [18]; this section concentrates on recent results and prospects for further development.

The matter may be better analyzed by focusing on the main requirements of the circuit, which concern the avalanche quenching transition, the subsequent hold-off time, and the reset transition.

Quenching should be as fast as possible to reduce the avalanche charge and related effects (power dissipation, carrier trapping, and light emission from hot carriers). To this end it is necessary to minimize both the delay T_d at

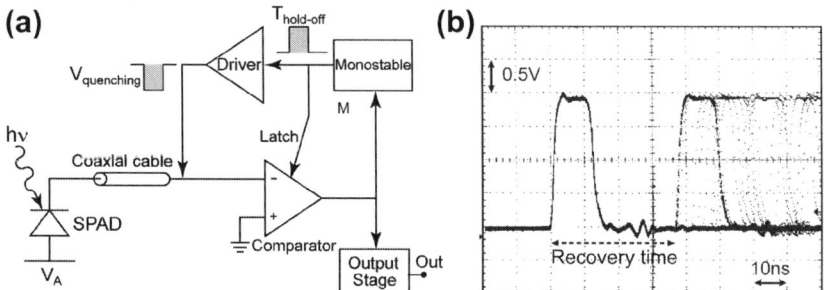

FIGURE 4.8 (a) Schematic diagram of a SPAD in an active-quenching circuit (AQC); (b) Output pulses from the AQC at a counting rate of ≈2 MHz (horizontal scale 10 ns/div; vertical scale 0.5 V/div).

which the quenching transition starts and the transition time T_q. With passive quenching circuits the transition is started immediately by the avalanche, but the transition time is determined by the time constant $R_D (C_d + C_s)$ (where R_D is the diode resistance during the avalanche). This time constant is fairly long in cases where R_D is not small and/or the capacitance $(C_d + C_s)$ is not minimal. Active-quenching circuits can ensure a very fast transition, but their feedback loop may produce a significant delay T_d, particularly in cases where the SPAD is not located near the quenching circuit [31,32]. A mixed passive-active quenching approach was developed [18], in which quenching is started with a passive transition and then sped up and completed by an active loop. A remarkable reduction of the avalanche charge can be obtained with discrete-component circuits [33], but further reductions can be achieved by integrating the load resistor R_L into the detector chip, and by integrating the complete quenching circuit on a chip [34]. A variant passive-active quenching circuit that was specifically devised for monolithic integration in CMOS technology cuts the avalanche current path instead of driving down the SPAD voltage [35].

The hold-off time, that is, the duration of the low-voltage fully quenched state, must be minimized to attain the highest counting rate. On the other hand, an adjustable hold-off time may be very useful for reducing the afterpulsing effect by allowing time for the release of the carriers trapped in deep levels [18]. This suggests developing circuits that have a negligible hold-off time and the capability of enforcing longer adjustable duration when needed.

The reset transition is a critical phase in all measurements and should therefore be very fast, at most a few nanoseconds. In photon counting it is highly desirable that the detector have a well-defined recovery time. That is, it is acceptable that the detector be totally insensitive for a given dead time (preferably a short time, of course), provided that it is then reset abruptly to its full efficiency. For this case, accurate equations for the correction of the count losses at high rates are available, and are based on well-known concepts of statistics. In reality, however, the situation is remarkably different: the recovery from zero efficiency is gradual and follows the evolution of the voltage with a non-linear dependence. Equations for accurately correcting the count losses due to such a variation have not been reported in the literature, and carrying out a quantitative statistical analysis of such a complex situation looks problematic. A gradual reset transition is also a significant drawback in photon-timing measurements, because the temporal resolution of SPADs depends strongly on the excess bias voltage. Therefore, a gradual reset transition causes a progressive degradation of the resolution as the count rate is increased. For both photon counting and photon timing, high-quality detector performance can be achieved at high count rates only if the duration of the reset transition, and thus the probability of detecting photons during the voltage recovery, is minimized. For this reason, active reset is explicitly advantageous. Passive quenching circuits have an inherent exponential reset transition, with time constant $R_L(C_d + C_s)$. In all cases where the circuit is not integrated in the detector chip, this time constant is at least a few hundred nanoseconds. It can be reduced to a few tens of nanoseconds in cases where the total capacitance ($C_d + C_s$) is in the range of 100 fF, as in fully integrated chips with small detector diameter (less than 20 μm). But even in such cases the reset transition is gradual over tens of nanoseconds and the drawbacks are reduced, but are not negligible.

The reset must not only be fast, but also very accurate. That is, for ensuring accurate photon counting and timing the bias voltage of the detector must be cleanly restored back to the final level. Therefore, in actively driven reset transitions care must be taken to avoid perturbations such as overshoots and ringing; a final part of the recovery that slowly approaches the steady-state voltage level must be avoided as well. To this end, it may be useful to enforce a reset action that pulls the voltage to the target level and keeps it there for a short time (typically 10 ns). that is, to introduce a final hold-on time. However, this approach can result in incorrect timing of photons that arrive during the hold-on time.

Obtaining accurate photon-timing sets further requirements that may conflict with the requirements for obtaining fast quenching with minimization of the avalanche charge. In particular, a marked dependence of the timing resolution on the circuit employed for timing the pulses is observed for SPAD devices with active area wider than 10 μm. Such dependence can be understood by taking into account the physical processes that determine the rise of the avalanche current [22,36–38]. During the initial stage of the avalanche, when the multiplication of carriers is localized within a small area around the seed point (the point of incidence of the photon), the avalanche current rises with relatively small statistical fluctuations. The subsequent rise to the avalanche signal's peak value corresponds to the progressive spreading of the multiplication over the sensitive area of the detector, and has stronger statistical fluctuations. Consequently, if the triggering threshold of the circuit that senses the avalanche is reached not in the very first part of the rise, but later during the spreading of the avalanche over the area, the timing jitter is remarkably larger than if the threshold is reached early in the avalanche growth. Any low-pass filtering that slows down the rise of the signal sensed by the timing circuit will have the effect of shifting the triggering instant to later times in the avalanche process, and will therefore degrade the timing resolution. The voltage waveform developed by the avalanche on the SPAD is inherently subject to a low-pass filtering action due to the charging of the diode and stray capacitances ($C_d + C_s$). Therefore, circuits that use this voltage waveform to sense the avalanche are not suitable for high-resolution timing with SPADs that have larger active areas. In fact, they provide good timing performance only in cases where the capacitance ($C_d + C_s$) is reduced to very low level, as in SPADs with less than 10 μm diameter and a low-capacitance quenching circuit. In larger area SPADs the lateral-propagation effect is stronger, and the larger intrinsic capacitance of the diode significantly contributes to the parasitic low-pass filtering action.

Research has demonstrated that the tradeoff between active area diameter and time resolution may be overcome by detecting the avalanche current during

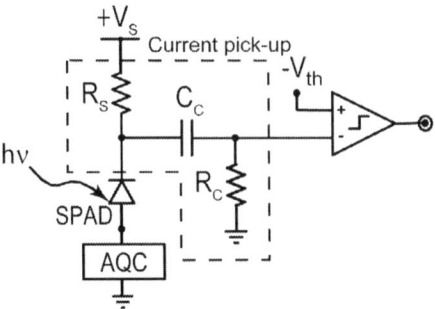

FIGURE 4.9 Simplified block diagram of a current pick-up circuit that can be added to any of the existing AQCs for improving the photon timing jitter.

the initial part of the rise, about at the 100 μA level [25]. By employing a separate current pick-up circuit (see Fig. 4.9), an unprecedented time resolution of 35 ps was obtained with a 100 μm diameter SPAD. This patented technique [39] employs AC coupling with a very fast time constant for extracting a short signal that reflects the rise of the avalanche current, which makes it possible to maintain excellent timing performance up to very high count rates. This technique enables the use of large-area SPADs in high-performance TCSPC measurements, as illustrated by results recently obtained with 200 μm diameter SPADs [16]. A monolithic integrated circuit that includes both the active quenching circuit and the current pick-up and timing circuit in a chip has recently been developed [40].

4.6 SILICON SPAD DEVICES

The silicon SPAD devices reported to date can be grouped in four categories, according to their fabrication technology:

- planar SPADs fabricated in a custom technology;
- non-planar SPADs fabricated in a custom technology;
- SPADs fabricated in a high-voltage complementary metal-oxide-semiconductor (HV-CMOS) technology;
- SPADs fabricated in a standard deep-submicron CMOS technology.

Their features and performance are reviewed in the following sections.

4.6.1 Planar SPAD Devices Fabricated in a Custom Technology

4.6.1.1 Early Planar SPAD Devices

The precursor of modern planar SPADs was introduced at the Shockley laboratory in the early 1960s, during studies of the physics of avalanche multiplication with high electric field intensities [3–5]. It was necessary to carry out experiments on avalanching junctions that had reasonably uniform electric field and were absolutely free from the so-called microplasmas (extended defects such as metal precipitates, dislocations, etc.). The approach was to fabricate many n+p junctions with very small diameter (a few microns) surrounded by a deeply diffused guard ring, and then select the few devices that did not contain a microplasma. The n+p junctions were fabricated by diffusing a shallow (<0.5 μm) n+ layer in a p-bulk substrate (Fig. 4.2). This simple structure has two key features: it operates at low voltage (about 30 V), resulting in limited power dissipation during the avalanche (a few hundred milliwatts), and it is fabricated in an ordinary silicon wafer with a planar technology, and thus is amenable to monolithic integration with other detectors and circuits. However, a thorough analysis reveals some weaknesses of the early planar structure as a SPAD device. The deep guard-ring diffusion causes the detection efficiency to be non-uniform in the active zone, giving it a dome-shaped distribution. This effect arises from the fact that the n-diffusion acts laterally from the edge of the device toward the center of the active junction over a distance almost equal to the diffusion depth, thereby decreasing the net p-doping and increasing the breakdown voltage from the center to the edge. Therefore the excess bias voltage progressively decreases from center to periphery, causing a decrease in the detection efficiency. This effect sets a strong limit to the minimum diameter of the active n+p junction, thus limiting the scalability of the structure for implementing arrays. Furthermore, due to the deep guard ring, the diffusion tail of the photon-timing distribution has a multi-exponential, wavelength-dependent shape. This makes the study of fast fluorescent decays more complex, especially when the spectral distribution of the incident light is not accurately known [21,41].

4.6.1.2 Planar SPAD Devices on Epitaxial Silicon Substrates

In the past three decades, a number of custom planar technologies have been developed for fabricating SPAD devices with optimized performance [42–47]. The planar epitaxial devices outlined in Fig. 4.10 were first introduced in 1988 [43,48] to overcome the drawbacks of early planar structures. These devices have undergone continuous improvement and are now exploited in

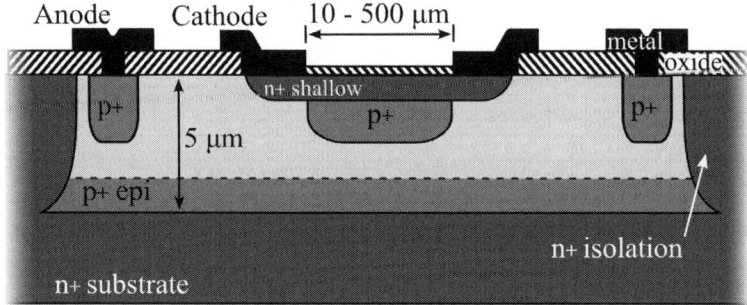

FIGURE 4.10 Schematic cross-section of the double-epitaxial SPAD device structure reported in [43].

commercially available photon detection modules from, for example, Micro Photon Devices [49][1].

The SPAD fabrication starts from an n-type substrate on top of which a p+/p− double-epitaxial layer is grown. The p-n junction formed between the epitaxial layer and the substrate limits the neutral region from which carriers are collected, thereby shortening the diffusion tail. The active n+p junction is built in the upper (2.5 μm thick), low-doped p-epilayer (10 Ω-cm). The buried p + epilayer (2 μm thick) establishes a low-resistivity path (0.3 Ω-cm) to the side ohmic contact. The extrinsic guard ring used in the early planar devices is replaced by a "virtual" guard ring structure. The concept of this virtual structure is straightforward: instead of using a lightly doped n-diffused guard ring to reduce the field in the peripheral region, the electric field in the central region of the shallow n + p junction is locally enhanced by means of a higher p-doping (*enrichment*). Implantation of boron followed by a drive-in diffusion is used to produce a ≈1 μm thick enrichment region, which defines the active area of the device. A highly doped, p-type diffusion (*sinker*) provides a low-resistivity path for the avalanche current flowing from the buried epilayer to the anode contact. Finally, a highly doped, n-type diffusion (*isolation*) region completely surrounds the detector. As a result the SPAD is enclosed in a p-well delimited by the isolation and by the substrate. This makes it possible to electrically isolate the detector from other SPADs or electronic devices fabricated on the same chip, thus allowing the fabrication of arrays (see Section 4.7) and monolithic integration of detectors and circuits in a chip.

Various design and fabrication parameters such as the boron-implanted dose, the conditions of the drive-in diffusion, the thickness and doping of the lightly doped epitaxial layer and of the buried layer can be easily customized for achieving a desired performance [16]. Continued improvement of the planar epitaxial technology, as described in [43], makes it possible to reliably fabricate SPAD devices with large active-area diameters (up to 500 μm) and exhibiting an excellent compromise between breakdown voltage (typically around 30 V), DE (50 % peak at 550 nm, decreasing to 25 % at 730 nm, and 12 % at 850 nm), DCR (from $10\ \text{s}^{-1}$ to $10^3\ \text{s}^{-1}$ at $-15\ °\text{C}$ for SPAD diameters ranging from 50 μm to 500 μm), total afterpulsing probability (about 1 % at $-15\ °\text{C}$), and timing jitter (better than 40 ps FWHM). Figure 4.11 (diamonds) shows the photon-timing distribution of a 200 μm SPAD detector illuminated with 10 ps FWHM optical pulses at 820 nm. The curve shows a prompt peak with a FWHM of 35 ps, and a clean exponential diffusion tail whose time constant is 280 ps [43].

Figure 4.11 (dots) demonstrates the advantage provided by fully custom SPAD technologies. By simply reducing the thickness of the p + epitaxial layer

---

[1] The identification of any commercial product or trade name does not imply endorsement or recommendation by the National Institute of Standards and Technology, nor does it imply that the materials or equipment implied are the best available for the purpose.

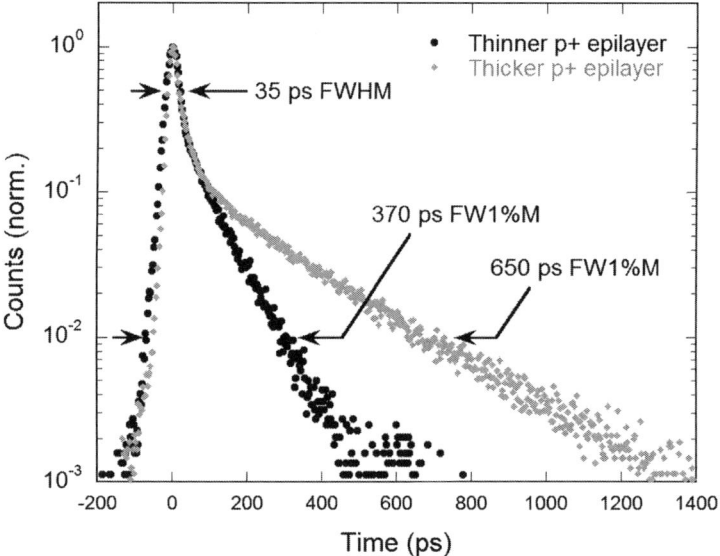

FIGURE 4.11 Photon-arrival-time distributions measured with two different 200 μm SPADs and a picosecond laser at 820 nm. The curves shows a prompt peak with a FWHM of 35 ps and a clean exponential diffusion tail with a time constant of 280 ps (diamonds), and 110 ps (dots), corresponding to a thickness of the neutral p+ buried layer of 2.4 μm, and 1.5 μm, respectively.

from 2.4 μm to 1.5 μm (in a different device), a reduction of the tail lifetime from 280 ps to 110 ps can be obtained.

Since the lifetime of the exponential tail does not depend on the photon wavelength, the timing resolution of the double-epitaxial SPAD is almost completely wavelength independent, which provides a remarkable advantage in reconvolution analysis of fast fluorescent decays [41,21]. To attain even better timing resolution completely free from diffusion tails, a fairly complex modification of the double-epitaxial device has been devised, fabricated, and tested [50]. The basic idea is to eliminate the neutral region beneath the active junction by exploiting a patterned buried layer. It has been verified that the diffusion tail can be completely eliminated in SPADs with diameters of up to about 10 μm. However, the performance of wider detectors remains less satisfactory, and the fabrication process is clearly more difficult than that of the previous double-epitaxial devices.

New developments in planar-epitaxial SPAD technology are mainly concerned with improving the DE in the red wavelength range (600 nm to 900 nm), either by incorporating a resonant cavity in the device structure or by increasing the thickness of the absorption region.

A resonant-cavity-enhanced (RCE) SPAD fabricated on a reflecting silicon-on-insulator (SOI) substrate has been reported by Ghioni et al. [51] and successfully exploited in three-dimensional imaging and QKD applications [52,53]. The RCE SPAD detectors have peak detection efficiencies ranging from 42% at 780 nm to 34% at 850 nm, and timing jitter of 35 ps FWHM. Typical dark count rates of 450 s^{-1}, 3500 s^{-1}, and 10^5 s^{-1} were measured at room temperature with RCE SPADs having 8 µm, 20 µm, and 50 µm diameters, respectively.

More recently, a red-enhanced (RE) SPAD device was devised, fabricated and characterized [54]. The key feature of RE-SPADs is a separate absorption and multiplication structure that provides a thick ($\approx$10 µm) absorption region with low electric field (hence no multiplication and negligible field-enhanced carrier generation), and a shallow multiplication region with a peaked electric field profile (designed to enhance the avalanche-triggering probability and to reduce the photon-timing jitter). The electric field profile in the two regions was designed to achieve the optimal tradeoff between operating voltage, avalanche-triggering probability (thus DE), DCR, and timing jitter. Experimental measurements on 50 µm RE-SPAD devices showed a significantly improved DE in the red region, reaching 40% at 800 nm (i.e., a factor 2.5 higher than the DE of standard planar SPADs) and 60% at 550 nm. The devices exhibit a remarkably low DCR, less than 10^3 s^{-1} at room temperature, decreasing to a $\approx$50 s^{-1} at 5 °C. Although the active volume of RE-SPADs is considerably larger than that of standard planar SPADs due to the increased thickness of the absorption layer, the DCR of the two detectors is comparable at room temperature. The thicker absorption region of the RE-SPAD does not significantly contribute to the DCR because the low electric field practically rules out field-enhanced generation of carriers. The dominant contribution to the DCR comes from the high-field multiplication region, whose design remained substantially unchanged. A timing jitter of 93 ps FWHM was obtained at room temperature, which is higher than the typical figure of standard planar SPADs (30–50 ps). This is due to the increased thickness of the absorption and drift region, resulting in increased transit times for photo-generated electrons of about 10 ps/µm at the saturated speed of 10^7 cm/s. Since photons are absorbed randomly in the drift region, timing jitter of $\approx$100 ps FWHM can be attributed to the 10 µm thick absorption region. Total afterpulsing probability lower than 1.5% was measured over the entire temperature range.

4.6.2 Non-Planar SPAD Devices Fabricated in a Custom Technology

The SLiK™ device sketched in Fig. 4.12 was devised by McIntyre and Webb at the former RCA Optoelectronics (now Excelitas Technologies Corp.), and employed to produce highly successful single-photon counting modules (SPCM) [55,56]. SLiK™ stands for "Super-low k," where k denotes the ratio of the ionization coefficient of holes to that of electrons. The device represents a remarkable evolution of an earlier reach-through avalanche diode structure pioneered by the same team [57,58]. It is built in special ultra-pure high-resistivity silicon wafers with a dedicated technological process; various device

Chapter | 4 Semiconductor-Based Detectors

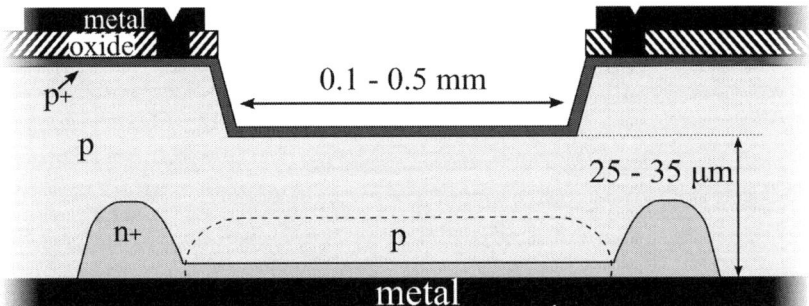

FIGURE 4.12 SLiK™ device developed at the former RCA ElectroOptics, now Excelitas Technologies Corp. [55,56].

features and processing steps are proprietary and covered by patents [59,60]. The active area of the detector is fairly wide ($\approx$180 μm). It is defined by a p + implant and deep diffusion in the central region of the bottom silicon surface and by a localized reduction of the wafer thickness to $\approx$30 μm, obtained by accurately etching the back of the wafer. A lightly diffused n guard ring around the shallow n + layer avoids edge breakdown. The device is illuminated from the back. Since the lightly doped p region (from 20 μm to 30 μm thick) is fully depleted, diffusion of photo-generated carriers takes place only in the surface p + layer (a few microns thick). The decrease of the electric field from its maximum at the n-p junction is gradual, hence the avalanche region is fairly extended. The breakdown voltage is high and strongly varies from sample to sample over a wide range from 250 V to 500 V. Thanks to the thick depletion layer, the DE is very high in the visible region and fairly good in the NIR up to about 1 μm. The typical value is significantly higher than 50 % over the entire range from 540 nm to 800 nm (the peak DE is 65 % at 650 nm), and is still about 3 % at 1064 nm [56]. Notwithstanding the remarkably large volume of the depletion layer, the DCR is very low, ranging from $\approx$100 s^{-1} to $\approx$10^3 s^{-1} at -10 °C. The afterpulsing probability is also strongly reduced, typically well below 1 %. The timing performance is moderate: with a broad illumination on the active area the timing distribution has a relatively broad peak with $\approx$450 ps FWHM, and an exponential diffusion tail that is one decade lower and has $\approx$160 ps lifetime. However, significant improvement in the timing performance can be gained by focusing the light at center of the active area and by using the current pick-up circuit discussed in Section 4.5 [61]. Devices similar to the SLiK™ have been recently developed by Laser Components, Inc., [62], having a larger active area (500 μm diameter) and a timing jitter $\leq$ 200 ps FWHM [63].

The SLiK™ devices have very good performance, but also a number of practical drawbacks. Due to the high breakdown voltage, the power dissipation during the avalanche is high, from 5 W to 10 W, and very effective cooling of the detector under normal operating conditions is mandatory (with Peltier

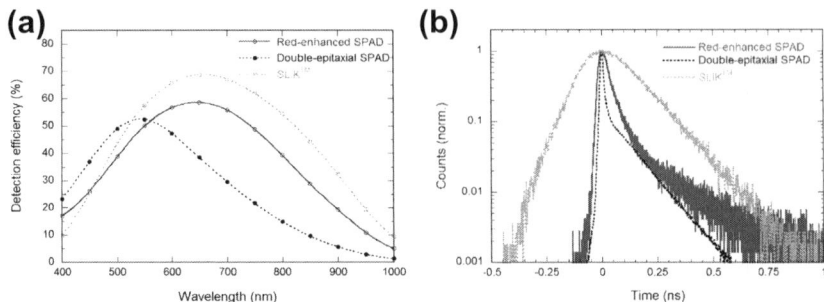

FIGURE 4.13 Detection efficiency (a) and photon timing distribution (b) of three different types of SPADs: SLiK™, planar epitaxial SPAD, and red-enhanced SPAD.

stages, or other means) [55]. The special fabrication technology is inherently complex, and the production yield of good devices is low and the cost is high. The devices are delicate and degradable, and integrating multiple detectors, or associated circuitry, is not possible. For reference, Fig. 4.13 compares the DE and timing jitter of three different devices, a SLiK™, a double-epitaxial SPAD [43] and a RE-SPAD [54].

4.6.3 High-Voltage, Complementary Metal-Oxide Semiconductor (HV-CMOS) SPADs

Foundry services for fabricating CMOS circuits have been available since the early 1990s, but for some years the available technologies have been far from meeting the requirements outlined above for fabricating SPADs, and all attempts gave poor results. The quality has been then steadily improving, but the progress in IC technologies is driven by the demands of circuits for consumer electronics, which are usually different from those of SPADs. Nevertheless, the monolithic integration of SPAD devices and CMOS circuits offers manifest advantages, from the availability of a fully supported, mature and reliable technology at reasonably low cost, to the possibility of developing complete systems on chip with a high degree of complexity.

The requirement for SPAD integration is that a suitable subset of fabrication steps can be specified within a complex CMOS process flow and used to build a planar p-n junction free from edge effects. However, some stringent technological requirements for SPAD fabrication must be carefully fulfilled. First of all, the quality of both the starting material and the fabrication process must guarantee very low concentrations of impurities (particularly transition-metal contaminants) that create deep energy levels within the silicon gap acting as generation-recombination centers and afterpulsing centers. Another key requirement is the ability to keep the electric field strength within the depletion region low enough to avoid band-to-band tunneling effects and to reduce the field-enhanced carrier generation as much as possible.

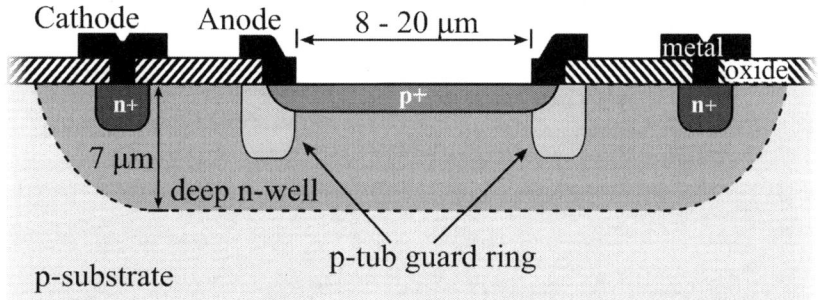

FIGURE 4.14 Schematic cross section of a typical HV-CMOS SPAD device, consisting of a shallow p+/deep n-well junction surrounded by a p-tub acting as an extrinsic guard ring to prevent edge breakdown.

In recent years, the development of high-voltage HV-CMOS technologies has been fueled by the rising demand for integrated circuits for automotive and control electronics (that is, circuits for the operation of motors, actuators, sensors, etc.). These circuits must fulfill more severe specifications than ordinary CMOS circuits; in particular they must operate with much higher voltage, and this imposes various requirements that are fairly consistent with those of SPAD devices. The main advantage of HV-CMOS technologies is the availability of a relatively low-doped deep n-well that provides up to 50 V isolation from the substrate [64]. HV-CMOS SPAD devices typically consist of a shallow p+/deep n-well junction surrounded by a p-tub acting as an extrinsic guard ring that prevents edge breakdown (Fig. 4.14). By relying on the p-tub guard ring, a breakdown voltage in the active area above 20 V can be obtained [64]. The deep n-well/p-substrate junction (a few microns deep) limits the depth of the neutral region from which minority carriers can be collected, thus reducing the length of the diffusion tail to less than 10 ns [64]. A number of small-area SPADs with diameter $\leq$20 μm and fairly low dark count rates were obtained using a 0.8 μm HV-CMOS technology by independent research groups [64–67]. Chips with more detectors and associated circuitry were implemented [65], as well as complete photon-counting modules (detectors and active quenching circuits) [66].

Due to the limitations of the 2-metal 0.8 μm HV-CMOS technology and its eventual obsolescence, there has been a push to migrate to more advanced technologies, and this has resulted in the first successful implementation of SPAD devices in a 0.35 μm HV-CMOS technology [35,68–70]. These 0.35 μm HV-CMOS SPADs exhibit a moderate DCR ($\approx 10^3$ s^{-1} for a 20 μm detector) and a maximum DE of 35% at 450 nm when biased 4 V above breakdown (Fig. 4.15). The DE however drops to $\approx$20% at 600 nm and it is <5% at 800 nm [35]. The reduction of the concentration of deep levels is not satisfactory in standard CMOS technologies, but the afterpulsing probability can nevertheless be reduced to fairly low levels by integrating the quenching circuit in the detector chip. In fact, the intrinsic capacitance of a small-area SPAD can be

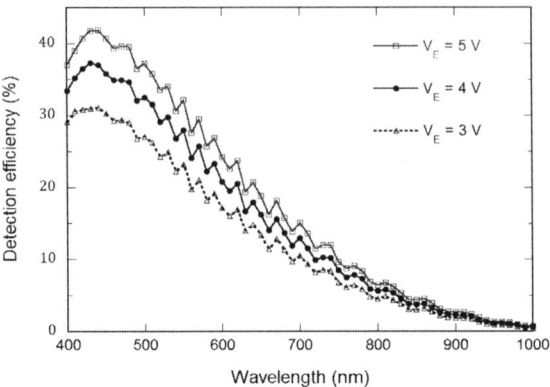

FIGURE 4.15 Detection efficiency of a 0.35 μm HV-CMOS SPAD device as a function of wavelength [35]. Measurements were performed at different excess bias voltages.

less than 100 fF. By integrating the quenching circuit, the stray capacitance can be brought to comparable level, thereby reducing the total capacitance by more than one order of magnitude with respect to an off-chip quenching circuit. The avalanche charge required for discharging a capacitance of about 100 fF is reduced to $\approx 6 \times 10^5$ electrons per Volt of excess bias. This minimization of the avalanche charge is also useful for reducing the optical crosstalk between adjacent SPAD detectors in a monolithic array, since the light emission from the hot avalanche carriers is proportional to the number of carriers that flow through the junction [26] (see Section 4.4.5).

A significant shortcoming of HV-CMOS SPADs arises from the p+n polarity of their active junction. Most of the depletion layer of this junction is accommodated in the n-well, so that the avalanche current is mainly triggered by holes, which are the minority carriers generated in this zone. In silicon, holes have a lower probability of avalanche initiation than electrons [8,9], and as a consequence the DE of a p+n SPAD is inherently lower than that of a complementary device structure with reversed n+p polarity. Another drawback of the currently available CMOS SPADs is the deeply diffused guard ring that causes non-uniform DE over the device active area, as discussed in Section 4.6.1.1.

4.6.4 Standard Deep-Submicron CMOS SPADs

CMOS technology with deep-submicron (DSM) resolution is mandatory for the fabrication of dense SPAD arrays with large numbers of pixels, adequate fill-factor, and smart pixels that include integrated electronics. However, a challenging basic issue must be faced: the inherent features of DSM CMOS technologies, namely the relentless trend toward higher doping levels, lower thermal budget, and thinner p- and n-well layers, conflict with SPAD detector performance. The smaller depth of carrier-collection layers limits the DE, and the high electric fields arising from higher doping result in strongly enhanced

DCRs due to band-to-band and trap-assisted tunneling effects. In addition, the reduced thermal budget and the lack of external gettering processes [71] also have adverse effects on afterpulsing.

Thus, the first major challenge for SPAD fabrication in DSM CMOS technologies is the choice of a suitable breakdown region that avoids trap-assisted and band-to-band tunneling. A second challenge that must be faced when scaling the SPAD size to a few micrometers is the design of a suitable guard-ring structure that is effective in preventing premature edge breakdown. To address these issues, designers have to cope with a number of design layers, models, and rules, without any flexibility in changing or adapting a process parameter to better match the requirements of a SPAD.

A number of SPAD structures in DSM CMOS have been proposed in recent years [72–77]. Most of them employ an explicit guard ring of low-doped p-well material around the central p+/n-well breakdown region similar to that shown in Fig. 4.14. Two SPADs with this structure have been reported at the 180 nm process node [72,73]. However, as discussed in [72], this device cannot be scaled much below 5 μm diameter because the p-well regions become so close that the active area of the SPAD is almost fully depleted (see also Section 4.6.1.1). A SPAD fabricated with 180 nm CMOS technology [74] proposes the use of shallow-trench isolation (STI) as the guard region, this approach being free from obvious limitations on minimum detector size and capable of fitting SPADs together compactly with other electronics. Unfortunately, an unacceptably high DCR of 10^6 s^{-1} was observed, which is likely due to the high density of deep-level carrier generation centers at the Si-SiO$_2$ interface [78]. If the active region of the SPAD is in direct contact with the STI walls, as in [74], the injection of free carriers into the avalanche region of the detector results in greatly increased DCR.

With 130 nm CMOS technology, various devices employing a p-well guard ring have been reported, invariably with high DCR (40×10^3 s^{-1} to 100×10^3 s^{-1}) [76]. Passivation of STI-interface traps failed to reduce DCR significantly, leading to the assumption that tunneling is the dominant mechanism [75].

Some improvements in compactness and scalability may be obtained by adopting the virtual guard-ring structure shown in Fig. 4.10. Richardson et al. [79] recently introduced three SPAD structures based on a novel retrograde buried n-well guard ring, capable of scaling from 32 μm to 2 μm in diameter. One of these structures is compatible with a standard 130 nm, triple-well CMOS technology. A remarkable sub-100 s^{-1} DCR for an 8 μm diameter SPAD was achieved at room temperature with 0.8 V excess bias, a maximum DE of 25 % at 560 nm, and negligible afterpulsing.

SPAD devices have been demonstrated in 90 nm CMOS technologies, but with significantly lower performance [80]. A notable exception is the 90 nm SPAD device reported by Webster et al. [81], where the deep n-well/p-substrate junction is used as the active junction, achieving a peak DE of 44 % at 690 nm and better than 20% at 850 nm. The 6.4 μm diameter SPAD also achieves a low DCR of 100 s^{-1} along with a low afterpulsing probability of 0.375% at 0.4 V excess bias voltage. Timing jitter as low as 50 ps FWHM was demonstrated, although the timing distribution had a relatively long diffusion tail. The key performance parameters reported for a variety of individual SPADs fabricated in standard deep-submicron technologies are summarized in Table 4.1.

4.7 SILICON SPAD ARRAY DETECTORS

Depending on the kind of application envisaged, two distinct directions are emerging for the fabrication of SPAD arrays. The first one is motivated by fast-growing applications like 3D imaging based on direct and indirect time-of-flight (TOF) techniques and low-light-level 2D imaging at high-frame rate, both of which require SPAD arrays with high pixel number and small pixel size, monolithically integrated in systems with electronics for information processing, that is, arrays with in-pixel electronics.

The integration of SPAD devices and associated electronics in submicron and deep-submicron CMOS technologies paved the way for the fabrication of large SPAD arrays that offer distinct advantages over charge-coupled device (CCD) and CMOS active pixel sensor (APS) imagers in these applications. Niclass et al.[82] first reported a large SPAD array implemented with 0.8 μm HV-CMOS technology. The array comprised 32 × 32 pixels, each with an independent SPAD device and a five-transistor digital circuit that provided quenching, pulse-shaping and column-access functions. The digital circuit occupies a square area of 54 × 54 μm^2, while the active area of the SPAD is 38 μm^2, resulting in a fill-factor of about 11%. Photon-timing operations were performed off-chip using a CMOS time-to-digital converter (TDC), and overall timing jitter of 115 ps FWHM was measured on a pixel. The main drawback of this design is that a sequential access is used, meaning that only one pixel can be processed at a time. While optical scanning is eliminated, frame rates remain relatively low (e.g., 250 frames per second for a pixel exposure time of 4 μs).

In-column array architectures were then introduced by the same research group, whereby processing is shared among clusters of pixels (for example, columns). To address the readout bottleneck an event-driven approach was devised, consisting of using the column as a bus that is addressed every time a photon is detected. The address of the relevant row is sent to the bottom of the column, where the photon time of arrival is evaluated, either off-chip [69,83,84], or on-chip [85]. The drawback of this approach is that multiple photon events cannot be detected simultaneously on the same column, which restricts the use of the event driven readout to applications where the expected photon flux hitting the sensor is low.

An alternative readout approach that allows the simultaneous detection of photons over an entire column is the latchless pipeline scheme [86]. In this approach, the absorption of a photon causes the SPAD to inject a digital signal

TABLE 4.1 Performance of a variety of SPADs fabricated with standard deep-submicron CMOS processes.

Reference	Technology Node	SPAD Diameter (µm)	V_E (V)	DCR (s^{-1}) @ Room Temp.	DE max	Total Afterpulsing Probability (%)	Dead Time (ns)	Timing Jitter FWHM (ps)
Faramarzpour et al. [72]	0.18 µm	20	2	70 k	5.5% @ 450 nm	–	30	–
Marwick and Andreou [73]	0.18 µm	10	0.5	100	–	–	–	–
Pancheri and Stoppa [77]	0.15 µm	10	5	230	32% @ 470 nm	2.1	30	170
Niclass et al. [76]	0.13 µm	10	1.7	100 k	34% @ 450 nm	–	450	144
Gersbach et al. [75]	0.13 µm	9	5	11 k	36% @ 480 nm	–	–	125
Richardson et al. [79]	0.13 µm	8	1.2	40 @ V_E = 0.8V	25% @ 560 nm	–	–	180
Karami et al. [80]	90 nm	8	0.13	8.1 k	9% @ 480 nm	32	1200	398
Webster et al. [81]	90 nm	6.4	0.4	100	37% @ 680 nm	0.375	15	82

into a delay line, which is then read externally. The timing of all injected pulses is evaluated to recover the time of arrival of the photon and to determine the pixel of origin.

The pixel access problem can also be overcome if time discrimination, photon counting, and any additional functionality (including local storage) is performed on-pixel. The advantage of this approach is the massive parallelism that can be achieved, potentially increasing the number of photons that can be detected and processed at the same time at reasonable power consumption. Many embodiments of this approach exist, depending on the level of complexity implemented at each pixel. The simplest one uses in-pixel information storage capability as realized by a counter [87–89]. Guerrieri *et al.* [89] designed and fabricated a high-speed single-photon camera based on a monolithic array of 32×32 smart pixels in a 0.35 µm HV-CMOS technology (Fig. 4.16). Each pixel (100 µm × 100 µm) is a completely independent photon-counting channel that includes a 20 µm diameter SPAD, analog sensing and avalanche quenching electronics, digital processing for counting the incoming photons, and memory and buffer stages for global shutter readout with no dead-time. Better than 35% peak DE is attained at 450 nm, decreasing to 8% at 800 nm, with DCR in the range of 10^3 s^{-1} for more than 75% of the SPAD devices. The 32×32 2D-imager can operate up to 10^5 frames per second with a dynamic range of 8 bits for counting. Noteworthy results have been obtained in challenging experiments [90], and work is in progress toward the in-pixel integration of a TDC [91].

With the implementation of the first SPADs in 130 nm CMOS technologies [75,76] it has been possible to integrate more functionality on a pixel, and remarkable results have been obtained [92]. A number of SPAD arrays were developed in which each pixel contains a multibit counter and a picosecond resolution TDC [93,94] or time-to-amplitude converter (TAC) [95]. Recently, a sensor was reported based on this concept, capable of detecting single photons

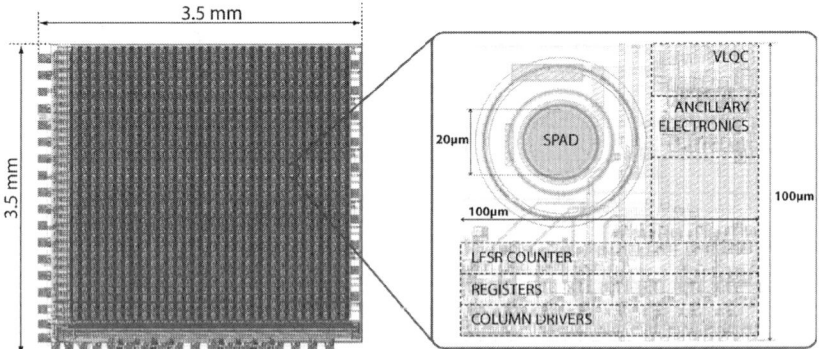

FIGURE 4.16 Layout of a monolithic array of 32×32 smart pixels fabricated in a 0.35 µm HV-CMOS technology [89]. The zoom shows the layout of a single pixel, including the SPAD detector, the analog sensing and quenching front-end (VLQC), and the digital counter and latch register.

over an array of 32 × 32 pixels, simultaneously evaluating their time-of-arrival with a time-bin width of 119 ps and a 10 bit range [96]. The array exploits the 8 μm low-noise SPAD devices described in [97], having a median DCR of 10^2 s^{-1} and a peak DE of 25 % at 460 nm when biased at 1 V above breakdown. Each channel operates independently, and contributes to an overall data rate from the chip of up to 10 Gb/s in time-correlated counting operation. The detector active area is 50 μm^2 and the total pixel area 2500 μm^2; hence the fill factor does not exceed 2 %. To mitigate this limitation, an array of microlenses based on a design described in [98] was used. The resulting concentration factor was characterized over all the pixels and showed strong variation across the array, with a median value of ≈5, corresponding to an effective fill factor of approximately 10 %. This SPAD array was successfully used in widefield fluorescence-lifetime imaging (FLIM) [99] experiments in the blue/green wavelength range [96]. More recently, an additional step was reported toward higher spatial and timing resolution with a new sensor of 160 × 128 pixels and 140 ps FWHM instrument response function [100]. The design includes a phase-locked-loop-stabilized 10 bit TDC array with 55 ps time bins.

The second direction in the fabrication of SPAD arrays is driven by applications in life sciences, such as fluorescence correlation spectroscopy (FCS) [101], multi-photon multifocal microscopy [102], spectrally resolved fluorescence-lifetime imaging (SFLIM) [1,99], luminescence/chemiluminescence detection in protein microarrays [103,104], fluorescence resonant energy transfer (FRET) [99]. In all of these applications the basic goal is to increase both throughput and miniaturization of the measurement system. These applications require large pixel sizes (50 μm to 100 μm diameter), high DE, and arrays of small or moderate pixel number (<100). An optimization of DE in the green/red region allows rapid and efficient detection of fluorescent emission from minimal quantities of biological material, i.e., from extremely small samples (down to single molecules of DNA and proteins). Large-area SPAD pixels are preferred to facilitate alignment of the detector array and to achieve good optical collection efficiency. Detectors used in multi-spot experiments (i.e. parallel excitation and detection of multiple spots in a sample) must be able to collect light from each individual spot with minimum contamination by emission from other spots. Although one could devise multiplexing schemes using a single detector to collect and disentangle signals originating from different locations, it is simpler and more effective to use multiple element detectors with a distinct element for each individual spot. It is also worth noting that a low ($\ll$1) fill factor is required in multi-spot detectors that must avoid optical cross-talk [105]. This requirement distinguishes these applications from the usual imaging applications, where a fill factor as close as possible to 100 % is generally preferred.

To meet these requirements, a research effort using the dedicated SPAD technology described in Section 4.6 was used to fabricate SPAD arrays with large-area elements. As an example, Fig. 4.17 shows a 6 × 8 SPAD array developed for chemiluminescent array detection and parallel FCS [106,107]. The

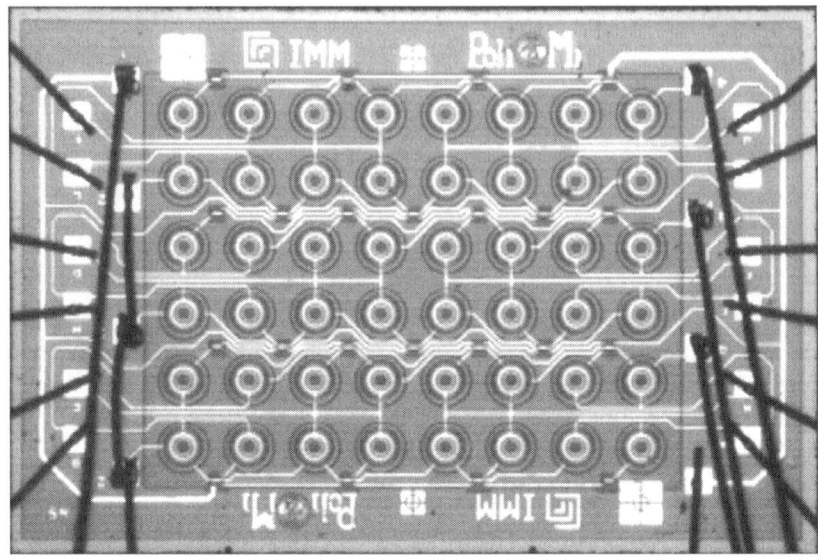

FIGURE 4.17 Microphotograph of a 6×8 SPAD array detector fabricated using a double-epitaxial silicon technology [106].

pixels have 50 μm diameter and 240 μm pitch. A low DCR was obtained at 5 V excess bias voltage at moderately low temperature (-15 °C with Peltier element cooling); the individual pixel DCR is 60 s^{-1} for about 40% of the elements and is below 5700 s^{-1} in the rest of the array. It was verified that the probability of optical crosstalk between elements is lower than 0.2%. Each pixel of the SPAD array shown above is connected to an integrated active-quenching circuit (i-AQC) that provides active-quenching and active reset pulses with a dead time of 65 ns, enabling a saturated photon counting rate of about 15×10^6 s^{-1}.

SPAD arrays with large-area elements and various geometries were successfully used in single-molecule FCS [108] and FRET [105] measurements and for wavefront sensing at high frame rate ($>10^4$ frames per second) in adaptive optics systems [109,110].

Due to the electrical coupling between adjacent pixels, the incorporation of high-performance photon timing capabilities into custom SPAD arrays is a challenging task even with small numbers of pixels. Fast and large voltage transients (≈ 1 V/ns) generated by the AQCs cause electrical disturbances on nearby pixels, which preclude low avalanche-sensing thresholds. As explained in Section 4.5, this increases timing jitter. An effective solution is monolithic integration of the avalanche pick-up circuit in close proximity to the SPAD. Thanks to the reduction of parasitic capacitances resulting from integration, it is possible to attain low timing jitter even with higher thresholds, reducing the issues due to electrical crosstalk. The lower parasitic capacitance also reduces the number of charge carriers flowing through the device during the avalanche, reducing both the afterpulsing probability and optical crosstalk.

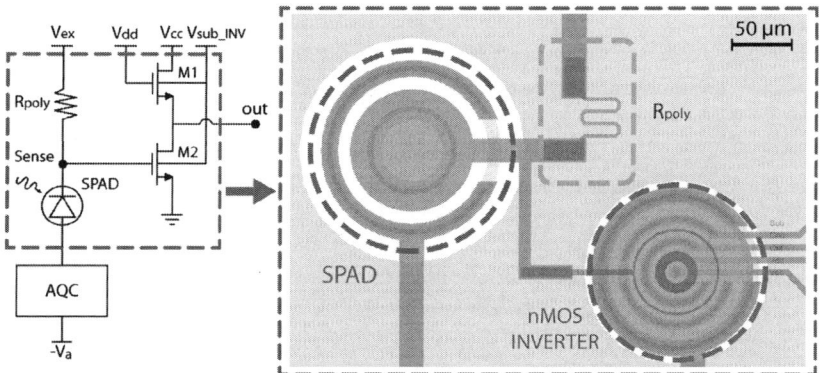

FIGURE 4.18 Schematic circuit (left) and layout (right) of a monolithic pixel including the SPAD, n-MOS inverter, and polysilicon resistance [111], fabricated in a full custom technology *"Reprinted with permission from [111]. Copyright 2012"*.

To enable the integration of MOS transistors in a custom SPAD technology, the dedicated SPAD process flow must be modified, and particular care must be taken to safeguard the structure and the performance of the detector. To this aim, only a few process steps necessary for the fabrication of a basic n-MOS transistor need to be added. In Ref. [111], a simple current pick-up circuit, including an n-MOS inverter and a polysilicon load resistance, were monolithically integrated near to the photodiode (see Fig. 4.18). The pixel was completed by an external standard-CMOS active quenching circuit, which provides stable timing performance up to high count rates ($>10^6$ s^{-1}).

In summary, research has demonstrated that SPAD arrays can attain performance comparable to that of state-of-the-art single-pixel detectors implemented in the same technology. CMOS-based SPAD arrays offer a significant functionality along with single-photon detection capability, whereas SPAD arrays manufactured in custom technologies represent a valuable tool for parallel high-throughput measurements of very low-light-level signals. There is a huge potential for improvement of SPAD arrays in terms of pixel number, detection efficiency and temporal resolution. As discussed in Section 4.10, progress will be mainly driven by user demands for detector performance in new and more diverse applications.

4.8 SPADS FOR THE INFRARED SPECTRAL RANGE

4.8.1 Infrared SPADs

While silicon provides excellent performance for the detection of photons at visible and near-infrared wavelengths, the rolloff in its optical absorption beyond ≈ 1 μm makes this material unsuitable for longer-wavelength detection. To serve applications in the wavelength range of 0.95 μm to 1.65 μm—particularly at the immensely important fiber-based telecommunications windows at 1.3 μm and 1.55 μm—photodetectors based on the InGaAsP compound semiconductor material system have been widely adopted. Avalanche diodes employing $In_{0.53}Ga_{0.47}As$ absorbers (which are lattice-matched to InP substrates) and InP multipliers have been technologically significant since they were first introduced in the late 1970s [112], and the enormous growth in fiber-optic communications during the late 1990s instigated dramatic improvements in the performance of InGaAs/InP APDs for use in fiber-optic receivers.

However, progress related to the telecom receiver-based "linear mode" operation of these devices, for which output photocurrent is proportional to input optical power, had essentially no impact on the performance and availability of SPADs based on a similar device design and material platform. In fact, up until the mid-2000s, there had been no InGaAs/InP devices designed specifically for photon-counting operation in Geiger mode, and system developers who had sought devices with good Geiger-mode performance at fiber-optic telecommunications wavelengths had been relegated to sampling the various commercially available telecom APDs to characterize their photon-counting performance [113–117].

Within the past decade, this situation has improved significantly. Researchers have found that the optimization of InP-based SPADs for detecting single photons requires design approaches that are quite distinct from those shown to be effective in optimizing APD linear-mode performance [116–122], primarily because the most critical performance attributes for linear-mode APDs (such as excess noise and gain-bandwidth product) are irrelevant for SPADs. This realization, and subsequent efforts expended on advancing the performance of InGaAs/InP SPADs, has led to significant progress for many of their properties [122,123]. For instance, there has been notable improvement in the fundamental tradeoff between single-photon detection efficiency and dark count rate, and high precision timing resolution has been demonstrated for these detectors. There has also been impressive scaling of these detectors to large format arrays [124,125] for emerging applications requiring single-photon imaging at short-wave infrared (SWIR) wavelengths.

4.8.2 Basic InGaAs/InP SPAD Design Concepts

All InP-based avalanche diodes deployed today are based on the separate absorption and multiplication (SAM) regions structure [112]. Figure 4.19 presents a schematic representation of a widely used InGaAs/InP device design platform [123]. This design entails a narrow bandgap $In_{0.53}Ga_{0.47}As$ layer (with bandgap $E_g \approx 0.75$ eV at 295 K), lattice-matched to InP, that provides efficient absorption of photons within a wavelength range between ≈0.9 μm and ≈1.6 μm. Adjacent to this absorption region is a wider bandgap InP region

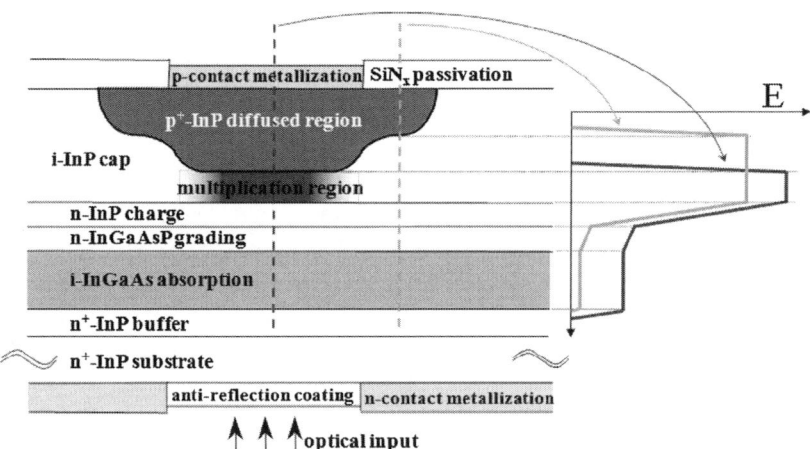

FIGURE 4.19 Schematic of a typical InGaAs/InP SPAD device structure and associated internal electric-field profiles (at right) [121].

($E_g \approx 1.35$ eV) in which avalanche multiplication occurs. A primary goal of the design is to maintain low electric field in the narrow bandgap absorber (to avoid dark carriers due to tunneling) while maintaining sufficiently high electric field in the multiplication region (so that impact ionization effects lead to significant avalanche multiplication). The inclusion of a charged layer between the absorption and multiplication regions (the SACM structure [126]) allows for more flexible tailoring of the internal electric field profile, along with the associated avalanche process, and is common to many InP-based avalanche diodes used today. The addition of grading layers between the InGaAs and InP layers in the structure is important to reduce hole trapping effects that result from the valence band offset that arises in an abrupt InGaAs/InP heterojunction [127].

As in any avalanche diode, the design of the lateral structure of the InGaAs/InP SPAD should achieve uniform gain in the high-field region of the device and suppress field enhancement due to p-n junction curvature at the device periphery that can lead to edge breakdown effects. The structure presented in Fig. 4.19 illustrates one commonly used scheme in which the device periphery is shaped using two concentric diffusions of different diameters [120,128], but other approaches are possible and have been demonstrated in the APD literature.

4.8.3 DE and DCR Modeling and Performance

One of the most fundamental design considerations for InGaAs/InP SPADs is managing the tradeoff between the single-photon DE and the DCR. As in Si SPADs, optical coupling issues are generally set aside and the DE is taken as the product of three probabilities: DE $= \eta_{QE} P_c P_a$, where η_{QE} is the quantum efficiency for carrier creation by absorption of an incident photon in the

InGaAs absorber; P_c is the probability that a photo-excited carrier is collected by injection into the InP multiplication region; and the avalanche probability P_a is the probability that a carrier injected into the multiplication region actually gives rise to a detectable avalanche. Although InGaAs/InP SPADs are often fiber pigtailed, the impact on the DE of optical coupling from the fiber to the SPAD active area is generally considered negligible in the context of the typical DE achieved with InGaAs/InP SPADs. In a well-designed device, the two dominant contributions to the DCR are thermal carrier generation in the narrow bandgap absorber and trap-assisted tunneling in the multiplier. The relative importance of these two mechanisms is determined by operating conditions; thermal generation will dominate at high temperature and low bias, while tunneling effects will dominate at low temperature and high bias.

DE and DCR calculations must include several dynamic processes, some of which are highly dependent on the local electric field intensity. Modeling of the avalanche probability P_a relies on a description of the avalanche process, and the adoption of appropriate expressions for the impact ionization coefficients in InP—particularly their temperature dependence [129]—is critical to the accuracy of the model. Dark-carrier creation can occur through field-dependent tunneling processes as well as thermally driven Shockley-Read-Hall processes. The first comprehensive description of a DE vs. DCR model for InP-based SPADs was developed by Donnelly *et al.* [118], and this formalism has been employed in additional work to treat both InGaAs/InP SPADs for 1.5 μm photon counting [121] as well as InGaAsP/InP SPADs for use at 1.06 μm [130]. One salient output of this model is that a wider multiplication region is highly beneficial for achieving a lower DCR at a given value of DE. Because a wider multiplication region reaches the avalanche-breakdown condition at lower electric field intensity than a narrower one, the former structure allows for Geiger-mode operation with reduced tunneling effects. (An alternative theoretical treatment of SPAD performance optimization with respect to multiplication-region width employs generalized breakdown probabilities [122] calculated using the recursive dead-space multiplication theory [131].) On the other hand, thermally generated dark counts originating in the InGaAs absorber are fairly independent of the width of the multiplication region, and are instead very sensitive to operating temperature. Therefore, reduction in operating temperature will improve DCR performance until temperatures are sufficiently low that tunneling effects dominate.

With the insight of the modeling described above, as well as continuous improvements related to the fabrication of these devices, the fundamental trade-off between DCR and DE has advanced to a performance level that is sufficient to serve many applications of photon counting at wavelengths near 1.5 μm. As illustrated in Fig. 4.20, devices with a size typical of fiber-coupled modules (e.g., 25 μm active area diameter) can provide DCR below 10^3 s^{-1} at 20% DE, and DCR below 10^4 s^{-1} for DE values of at least 40%, while operated with modest cooling provided by thermoelectric coolers (e.g., 223 K).

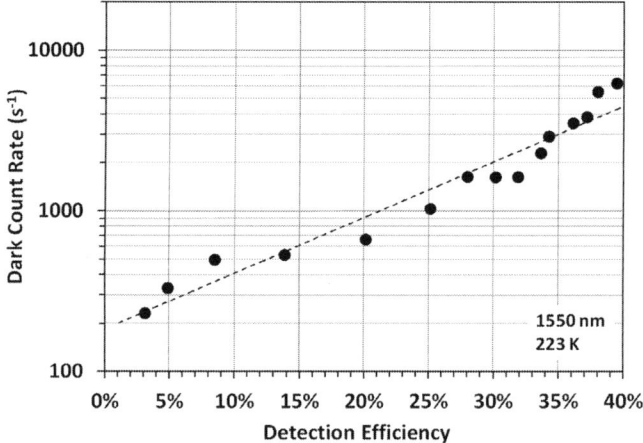

FIGURE 4.20 Dark count rate versus detection efficiency of an InGaAs/InP SPAD operating at 223 K in response to 1550 nm photons. The dashed line indicates the general trend in performance.

4.8.4 Timing Jitter

A number of physical mechanisms within any SPAD structure can contribute to timing jitter, i.e., uncertainty in the correlation between photon arrival time at the detector and the time of avalanche detection. In InGaAs/InP SPADs, these mechanisms include differences in the transit times of photoexcited carriers resulting from differences in the location of photon absorption, carrier propagation delay caused by the temporary trapping of carriers at heterojunctions formed by dissimilar semiconductor layers, and variations in the avalanche build-up time induced by the stochastic nature of the impact ionization process. Avalanche build-up time variation also includes effects related to the randomness of the spreading of the avalanche from an initially localized filament to a saturated avalanche process that fills the entire high-field active area of the device [24]. Aside from these stochastic processes, there is also the important consideration of local excess-bias non-uniformities resulting from non-uniform breakdown voltage across a given device's active area. If the excess bias exhibits considerable variation as a function of position in the device, the associated distribution of mean times to reach threshold further broadens the timing distribution and may increase the effective timing jitter significantly above that which would be found for a device with an ideally uniform excess bias.

For a well-designed InGaAs/InP SPAD, timing jitter of less than 50 ps FWHM has been demonstrated for an excess bias of ≈7 V [120] (see Fig. 4.21). This performance is comparable to the best results achieved with Si SPADs [25] and represents very good timing performance relative to other single-photon detector technologies. However, it should be noted that jitter is higher at lower excess bias (e.g., 100 ps at 3.5 V); obtaining very low jitter by using larger excess bias operation poses a tradeoff with DCR and afterpulsing. Additionally,

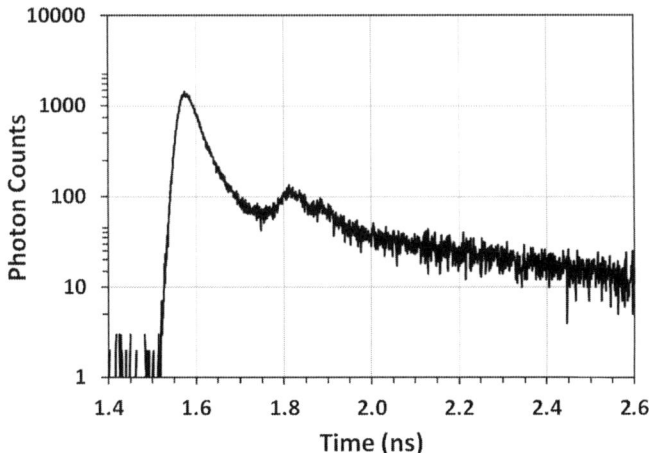

FIGURE 4.21 Photon-arrival-time distribution for an InGaAs/InP SPAD with 25 μm diameter operated with 7 V excess bias. The FWHM timing jitter is 46 ps. The influence of the distribution tail increases the root-mean-square (RMS) timing variation to 59 ps. (The minor peak at 1.8 ns is an artefact of the measurement apparatus.)

as discussed above in Section 4.4.4, low timing jitter is only possible with high-performance circuits that can accurately detect the onset of the avalanche response at very low signal levels without spurious detections caused by circuit transient response characteristics. The critical importance of the circuit in determining the jitter performance—as well as other SPAD performance parameters—are discussed further in the next section of this chapter.

4.8.5 Afterpulsing

Among the possible strategies for the mitigation of afterpulsing in InGaAs/InP SPADs, the most direct are (i) decreasing the density of material defects that act as potential charge traps and (ii) reducing the number of charges that are trapped in the first place by reducing the amount of charge that flows during each avalanche event. A dramatic improvement in the quality of the InGaAsP material system was driven by the enormous economic opportunities of the telecommunications boom beginning in the late 1990s. However, following the collapse of this market in 2002 and its relative maturation and commoditization over the past decade, there has been little indication of further improvement in InP material quality as relates to the density of defects in the multiplication region, defects that are understood to cause afterpulsing. A key problem in this area is the paucity of knowledge concerning what types of material defects could be acting as charge traps, and what causes their formation.

Therefore, essentially all recent efforts to mitigate afterpulsing have invoked the strategy of reducing the number of charges that are trapped by restricting avalanche events to having less charge flow. There have been experimental measurements to confirm that the amount of trapped charge and the consequent afterpulsing scale linearly with the charge flow per avalanche [132,133]. For situations in which gated-mode operation with very short (sub-ns scale) gates is appropriate, avalanche charge flow can be reduced dramatically because the falling edge of the gate acts to rapidly quench the avalanche. This basic concept has been implemented at high (GHz) gating frequencies, and are discussed in detail in Section 4.9. More general non-periodic solutions have focused on sensing the avalanche with as low a detection threshold as possible and then rapidly quenching it to minimize the charge flow [134]. Finally, there has been recent work on self-quenching InGaAs/InP SPADs [135–139] in which the monolithic integration of passive quenching elements can lead to reduced charge flow if the quench elements can be integrated with strictly minimized parasitic capacitance. (As discussed above in the context of Si SPADs, parasitic capacitive elements must be discharged and recharged with each avalanche event, and minimizing these capacitances can reduce the overall charge flow per avalanche, c.f. Section 4.5.) However, afterpulsing continues to pose the primary challenge to free-running and high-rate photon counting using InP-based SPADs.

4.8.6 Comparison of InGaAs/InP SPADs and Si SPADs

A comparison of InGaAs/InP SPAD performance with those of state-of-the-art Si SPADs [16] is useful as an indication of how far InP-based SPADs might progress if InGaAsP materials engineering can be brought to the level of Si materials engineering. The longer-wavelength InGaAs/InP detectors will always be at a performance disadvantage relative to Si detectors given the necessarily smaller bandgap of the InGaAs absorbers. However, the primary impact of the absorber bandgap on SPAD performance is its role in determining the contribution to the DCR of carriers generated thermally by generation-recombination via mid-gap states. This suggests that we can remove the bandgap disparity by comparing Si and InGaAs/InP device performance at different temperatures that compensate for the difference in bandgaps.

We first consider that dark-carrier thermal generation by Shockley-Read-Hall processes is proportional to $\exp(E_g/2k_BT)$ where E_g is the material bandgap, k_B is Boltzmann's constant, and T is temperature. Given that $E_g(\text{Si}) = 1.12$ eV at 20 °C, the exponent $E_g/2k_BT \approx 21.5$ for silicon. We then proceed to find the temperature for InGaAs at which $E_g(\text{InGaAs})/2k_BT$ gives the same value, which occurs at approximately -70 °C. By comparing Si SPAD and InGaAs/InP SPAD performance at these two respective temperatures, we remove the role of the material bandgap in thermal dark-carrier generation to allow a direct comparison of underlying material properties. This comparison is summarized in Table 4.2, assuming devices with a 50 μm active-region diameter.

Based on the rationale just described, Si SPADs exhibit superior material quality resulting in lower DCR, but only by a factor of 2 or 3 for a given value

TABLE 4.2 Comparison of State-of-the-Art Performance for Si and InGaAs/InP SPADs

	Si[1]	InGaAs/InP
Operating temperature	20 °C	−70 °C
Active region diameter	50 μm	
Wavelength	550 nm	1550 nm
DCR and DE[2]	10^4 s^{-1} at 60%	–
	2×10^3 s^{-1} at 40%	6×10^3 s^{-1} at 40%
	0.5×10^3 s^{-1} at 20%	10^3 s^{-1} at 20%
	–	0.5×10^3 s^{-1} at 10%
Jitter (FWHM)	30–50 ps	50–100 ps
Minimum hold-off for 1% afterpulsing[3]	≈10 ns	≈100 ns

[1] Si SPAD performance corresponds to thin Si SPAD structures as in [16]
[2] Si DE values are cited for 550 nm, for which the highest Si DE is obtained.
[3] Assumes 20% DE and free-running operation with fast active quenching of a few ns.

of DE. Thin Si SPADs have demonstrated somewhat lower timing jitter [25] than InGaAs/InP SPADs when operated with comparable electronic circuitry. For this parameter, the Si devices tend to operate over a range of timing jitter values that are about one-half the range exhibited by InP-based SPADs. Finally, a comparison of afterpulsing performance is complicated by the fact that it is highly circuit-dependent, so we rely on characterization in free-running operation with fast (i.e., a few ns) active quenching using the same backend electronics [140]. Si SPADs have the potential for an order of magnitude shorter hold-off times at 1% afterpulsing levels, but while this suggests lower trap densities in Si multiplication regions, at least some of this afterpulsing performance advantage is related to the much higher temperature operation of Si SPADs allowed by their larger bandgap absorber.

4.9 ACTIVE GATING TECHNIQUES FOR InGaAs SPADS

4.9.1 Introduction

As discussed in the preceding sections of this chapter, the efficiency, noise, resolution, and maximum count rate of any SPAD detection system derive from the co-operative performance of the SPAD and the circuitry used to control it. This connection is particularly strong for actively gated detection systems, in which the SPAD is biased in the linear-multiplication regime and raised above breakdown only during a short detection gate. Active bias gates are a useful means to improve the signal-to-noise (SNR) in applications with pulsed sources (e.g., ranging, fluorescence, communications), particularly (but not exclusively [141]) for devices with relatively high dark-count rates such as InGaAs or Ge SPADs [6,7], and can be simple to implement.

Over the past two decades, the demands of quantum information systems for low-noise high-efficiency single-photon detectors in the telecommunications windows [115,117,142] have motivated significant advances in active-gating techniques, resulting in dramatic improvements in detection efficiency and maximum count rate. The performance of these InGaAs detection systems can be so strongly influenced by the biasing scheme that it is common to identify them not by the type of SPAD, but by the type of gating and avalanche discrimination system used to control it. This section presents a comparative survey of these techniques and emphasizes the performance benefits various approaches can provide. While it is worthwhile to note the ongoing development of systems that combine gating with active and passive quenching circuits [18,113,115,143,144], this discussion is restricted to systems in which the avalanche current is terminated by the end of the gate itself, that is, quenching (as discussed above) is implemented by gate termination rather than some feedback mechanism [18].

Figure 4.22a shows one prototypical active-gating circuit. The bias on the cathode is the sum of the DC voltage, held at some value lower than the SPAD breakdown voltage, and the active gating signal AC-coupled to the SPAD. The avalanche current can be sensed as voltage across a resistive load, often chosen to be 50 Ω to keep the SPAD anode fast. As discussed in Section 4.8, InGaAs SPADs are generally devices that evolved from telecommunications applications [6], and as such tend to have low junction capacitance, typically of the order of 0.1 pF when biased close to breakdown. This capacitance acts as a high-pass filter on the spectral components of the applied gate pulse. Provided that the anode supports a wide bandwidth, the signal at the output is similar to that shown in Fig. 4.22b when a 5 V, 1.5 ns square gate pulse with $\approx$100 ps edges is applied to the cathode. These are the so-called gate transients. Avalanches due to single-photon absorption occur between these transient signals, as shown, and the discrimination of the avalanche signal from these gate transients, particularly when the gate duration is short ($\leq$1 ns), is the main subject of this section.

In the simple circuit shown in Fig. 4.22a, the SPAD presents an impedance discontinuity that may reflect spectral components of the gate signal back along any transmission line between the driving source and the SPAD. This effect can be used to increase the AC voltage experienced by the SPAD. On the other hand the transmission line may host multiple reflections that can complicate the gating waveform. A simple way to avoid this process is to provide an AC termination, as in [145].

If nothing more than a threshold comparator is used to detect avalanches, the discrimination threshold must be set at a level higher than the rising gate transient, and it was recognized early on that this has significant consequences

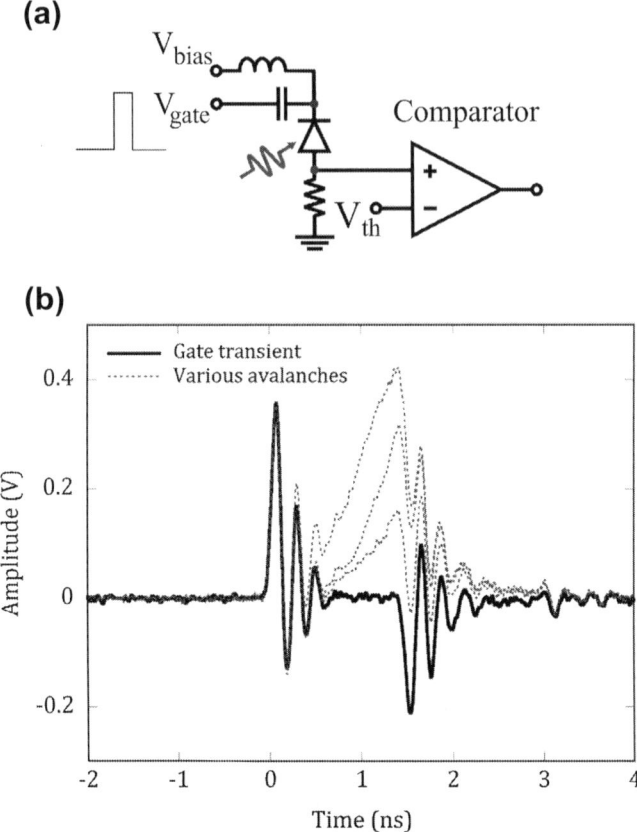

FIGURE 4.22 (a) A prototypical active gating circuit. (b) The gate transients at the SPAD anode, along with some avalanches, when a 5 V, 1.5 ns, square gate with 100 ps edges is applied to the cathode of an InGaAs/InP SPAD.

for the performance of the detection system. Specifically, larger avalanche signals correspond to larger amounts of charge, which populates more traps within the SPAD and increases the afterpulsing [114,146]. Better performance can be achieved with techniques that discriminate small avalanches from the gate transients, and a wide variety of techniques have been developed for this purpose, all representing different forms of filters. For classification we can categorize them as sampling schemes and cancellation schemes.

4.9.2 Sampling

Although not the first approach reported in the literature, sampling schemes are perhaps the most straightforward because they make use of the fact that the avalanche signal and the gate transients may be temporally distinguishable.

Sampling can therefore be implemented with modifications only to the discriminator electronics, for example, with an AND gate that samples the voltage between the gate transients [117,147], or with an analog-to-digital converter (ADC) that records the voltage between the gate transients for further discrimination in digital format [148]. Systems of this type have been used with 800 ps gates at repetition rates as high as 14 MHz with low afterpulsing at a detection efficiency of about 14%. Sampling becomes more challenging as the gate duration and avalanche amplitude become smaller, requiring higher bandwidth sampling systems. However, one alternative is to use the transient signal itself to sample the diode: if an avalanche occurs between the rising and falling edges of the gate, then the transient due to the falling edge of the gate becomes distorted, or even disappears, effectively reporting anything that may have occurred during the gate [149]. A comparator monitoring the falling-edge transient can then be used to identify when an avalanche occurred. This alleviates the need for high-speed sampling, and has been implemented with 1 ns gates at rates as high as 20 MHz [150].

4.9.3 Cancellation

There are a variety of approaches that all share the same basic idea: the SPAD biasing circuit is designed to generate matching replicas of the gate transients and subtract them to reveal an avalanche signal that may be obscured in one of the gate transients. One of the earliest techniques, developed at IBM in the late 1990s, is illustrated in Fig. 4.23 [151,152]. This circuit uses inverting and non-inverting reflections from the ends of coaxial cables to generate an opposing pair of gate transients that cancel in a passive network. Both signals pass through the SPAD, so the quality of the cancellation is determined primarily by the

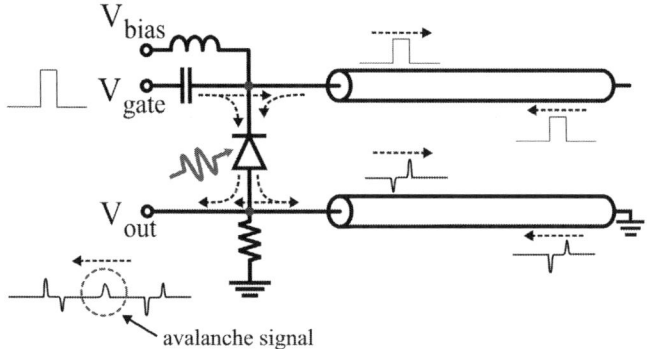

FIGURE 4.23 Gate-transient cancellation based on inverting and non-inverting reflections from matched transmission lines. The avalanche signal at the output must still be discriminated from the preceding and following uncancelled transients, as illustrated.

degree to which the properties of the coaxial lines (delay, attenuation, dispersion, reflection) match, which can be quite good. An additional means still must be applied to ignore the remaining gate transients, as illustrated in Fig. 4.23, and this limits the minimum interval between gates to at least twice the round-trip time of the transmission lines. It is also worthwhile to point out that the transmission line, particularly on the gate-driver side, can host multiple reflections that can affect the minimum threshold at short gate intervals. Nonetheless, this scheme is robust and has been applied at gate rates well above 1 MHz. When properly designed and implemented, this approach strongly suppresses the gate transient and allows the discrimination of small avalanche signals, on the order of 200 fC, reducing afterpulsing [153]. Another configuration of basically the same approach uses balun transformers, rather than open and shorted transmission lines, to generate the opposing gate transients, and was shown to support gate intervals as low as 5 ns [154].

An alternative scheme implements cancellation by applying the gate pulse to two separate SPADs, as in [155]. The resulting pair of gate transients can then be subtracted with active or passive circuit elements, for example, a 180° hybrid junction or a transformer. In this case the quality of the cancellation will be determined by how well the electrical response of the two diodes match, for which it is difficult, but not impossible, to control; Lu *et al.* used SPADs from adjacent locations on a wafer in a common-mode cancellation scheme with a sinusoidal gate signal (c.f. Section 4.9.5) and and achieved excellent transient suppression with an 80 MHz gate frequency [156,157]. They demonstrated detection efficiency as high as 43 % at 1310 nm.

As configured in Ref. [155], both SPADs can be used as detectors, and are distinguished by the orientation of the output avalanche signal (positive going or negative going), though simultaneous avalanches will result in distorted or completely missed detection events. Alternative configurations of this approach replace one of the SPADs with a "dummy" element, such as a diode [158], or a capacitor [143,159], whose electrical response is similar to that of a SPAD. These approaches are generally simple to implement and are effective in supressing the gate transient to improve (reduce) the avalanche discrimination threshold. However, the quality of the match between the SPAD and the reference element critically determines the minimum discrimination threshold, and hence the afterpulse performance and usable gate rate. Systems of this type have been demonstrated at rates up to 25 MHz [160].

Rather than generating a transient for cancellation, it is also possible to choose the gate waveform to facilitate the discrimination of avalanche signals. Zhang *et al.* [161] use a Gaussian gate waveform, which produces an anti-symmetric gate transient dominated by the first derivative of the Gaussian. Avalanches within this anti-symmetric structure are revealed by summing the transient with a delayed portion of the driving pulse, thereby creating a transient from which avalanche signals can be more effectively discriminated.

One of the most advantageous features of all these cancellation schemes is that a single gate pulse generates the reference signal used to supress the gate transient. For many of these schemes, this allows for essentially arbitrary gate waveforms, and more importantly, for asynchronous gating (up to the minimum supported repetition rate of the scheme). Asynchronous operation is particularly useful in conditional measurements, in which the detector is activated by an external event coincident with a signal of interest, as in correlated or heralded photon experiments.

4.9.4 Introduction to High-Speed Periodic Gating

The benefits demonstrated by reducing the avalanche charge motivated the investigation of methods that achieve even stronger gate-transient suppression and lower avalanche-discrimination thresholds. In 2006, Namekata *et al.* [162] demonstrated that by gating with a radio-frequency (RF) sine wave and using strong narrowband RF filters to supress the resulting gate transient, thresholds at unprecedented low levels (estimated to be 0.5 mV at the SPAD) could be used. The low threshold, in conjunction with short (sub-nanosecond) gate durations, efficiently detects avalanches with greatly reduced total charge; further developments have achieved average charge levels more than an order of magnitude lower than in sampling or cancellation systems, resulting in low afterpulsing even with gate frequencies of the order of 1 GHz. This section provides a survey of a variety of techniques that implement high-speed periodic gating, but first it is worthwhile to highlight some of their common characteristics.

The extremely strong transient suppression achieved in high-speed gating schemes is facilitated by both the periodicity and the high frequency of the gate. While this results in good sensitivity to minute avalanches, as is necessary for efficient high-speed operation, the periodicity also means that the device will be active every gate period, making asynchronous or triggered activation impossible unless further efforts are made. Therefore it is often the case that systems of this type employ a logical "hold-off," implemented after the output stage, that simply ignores outputs unlikely to be of interest. For example, such a hold-off is commonly applied immediately after a detection event to ignore the output from some number of subsequent gates that have high afterpulse probability (some typical hold-off times are included in Table 4.3).

Short bias-gate duration is also necessary for good performance in high-speed systems. For a fixed excess bias voltage, the total charge in an avalanche grows roughly exponentially with the gate duration [154], which means that the afterpulse performance can be significantly improved with even moderate reductions in the gate duration. However, the short bias gates, in some cases less than 200 ps, are on the order of the characteristic time scales for both the growth and the temporal jitter of the avalanche signal, and this has a significant impact on the temporal response of the detection system. One consequence is that the detection efficiency is strongly dependant on *when* in the gate a single

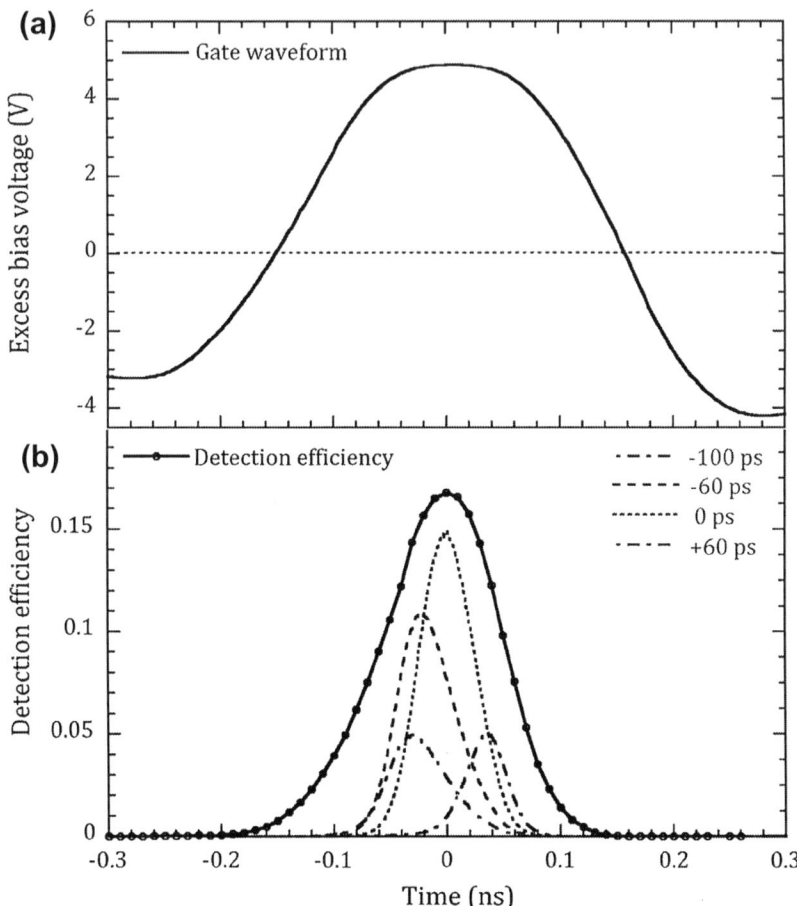

FIGURE 4.24 (a) A bias gate from a high-speed (1.25 GHz) periodically gated InGaAs detection system. The gate exceeds the breakdown voltage for roughly 320 ps. (b) Detection efficiency versus time, measured as a short (<30 ps) optical stimulus is stepped through the gate. Also shown are TCSPC histograms (counts normalized, 30 s acquisition, each) when the optical stimulus arrives at four different times relative to the peak of the detection efficiency. The position of peak detection efficiency in (b) relative to the gate voltage in (a) is unknown and arbitrarily aligned in this figure.

photon is absorbed. For example, Fig. 4.24a shows a bias gate from a high-speed gating system operating at 1.25 GHz, while Fig. 4.24b shows the detection efficiency of this gate as a function of the photon arrival time, as measured with a counter and an attenuated <30 ps optical stimulus whose temporal position is stepped through the gate. The bias gate exceeds breakdown for roughly 320 ps, but the detection system has a region of sensitivity with a FWHM of 140 ps. Obviously, the system is most efficiently used with ultra-short optical signals aligned to the peak of the distribution. Figure 4.24b also shows histograms of detection events for four different temporal positions of the optical stimulus with respect to the gate, as measured with a traditional start-stop TCSPC system. Although the FWHM of the detection-event histograms are narrower than the full region of sensitivity, they do not accurately represent the arrival time of the photon. For example, the stimulus that arrived at -100 ps (100 ps before the peak sensitivity) produced a histogram whose peak is located at -40 ps. Moreover, all the histograms overlap significantly, making it nearly impossible to even distinguish a photon that arrived early in the gate from one that arrived later. In general, high-speed periodically gated systems do not provide timing resolution other than that defined by the gate's temporal region of sensitivity.

Given the strong relationship between gate duration and afterpulsing, it is tempting to envision ever shorter electrical bias gates to improve performance. However, with a shorter bias gate, ensuring that initiated avalanches grow to a detectable level before the end of the gate requires increasing the excess bias voltage. The challenge is therefore to produce large amplitude GHz-rate gates, and maintain good suppression of the resulting gate transient. To date, good performance has been demonstrated with gate amplitudes up to 20 V using commercially available GHz amplifiers, though improvements in both components and techniques continue to be made [163].

4.9.5 Sine-Wave Gating

There are a variety of configurations that expand on the idea of sine-wave gating as introduced above: a narrow-band RF signal as the gate, and passive filters to suppress the gate transient; one setup is shown in Fig. 4.25. Systems of this type can be implemented with commercially available connectorized RF components and can achieve low-noise high-speed single-photon detection at gate frequencies in the GHz range.

Although the spectrum of the gate has a single RF component, the voltage dependence of the SPAD junction capacitance generates higher-order harmonics of the gate signal. Therefore, to achieve a low discrimination threshold the filters used to supress the gate transient must address not only the gate frequency, but its harmonics as well, as illustrated in Fig. 4.25. The avalanche signal has a broad spectrum that extends to roughly the inverse of the gate duration, and that

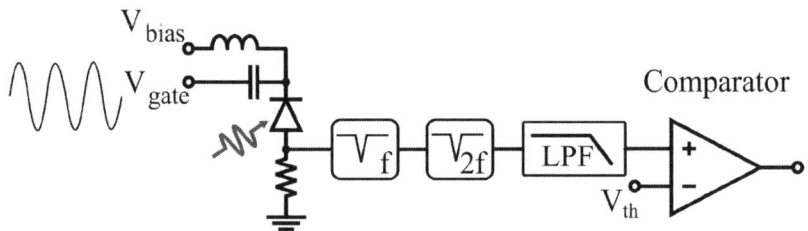

FIGURE 4.25 A sine-wave gating circuit. Notch filters at the gate frequency f, and 2f, and a low-pass filter (LPF) are used to suppress the harmonics of the gate.

necessarily spans the gate frequency. In sine-wave gating there is a fundamental tradeoff between supressing the gate signal and preserving the avalanche signal. Fortunately the gate harmonics lie at discrete frequencies, and a combination of narrow RF band-elimination (notch) filters and low-pass filters can be used to suppress them while efficiently passing the components of the avalanche signal that lie outside the filter bandwidths [164]. It has also been shown that solely low-pass filters with a corner below the gate frequency can be used to reject all the harmonics of the gate [165]. This approach affords more tuning of the gate frequency than is allowed by narrowband notch filters, but obviously low-pass filters the avalanche signal as well.

Regardless of which configuration is used, using passive filters to supress the gate transient distorts the avalanche signal. Low-pass filters limit the steepness of the rising edge of the avalanche, and multipole filters with sharp profiles induce strong dispersion, both of which change the shape of the avalanche waveform. The effect of the filters often appears as additional jitter in the distribution of detection events in a TCSPC measurement, as reported in [165–167]. However, it should be noted that the region of sensitivity (c.f. Fig. 4.24b) is not affected by the filter-induced distortions, and for this reason the importance of this additional jitter depends on the application. Alternatively, Liang et al.[168] demonstrated a variant of sine-wave gating in which the first harmonic is suppressed by cancellation with a reference sine wave, rather than with dispersive notch filters, and only low-pass filters are used to suppress the higher-order harmonics. This approach better preserves the avalanche spectrum and reduces the observed jitter in a TCSPC measurement, and is related to the harmonic subtraction technique discussed below. A consequence of using passive filters that may be more significant than jitter is that strong filtering may inhibit some avalanche signals from reaching the discrimination threshold before the end of the gate, either due to low-pass filtering or signal distortion. This is a particular concern for those avalanches that are initiated late in the gate, and may affect the detection efficiency of the system.

One of the main advantages of the sine-wave scheme is that the gate suppression with multiple filters can be strong (as high as -100 dB). These systems can therefore support extremely low discrimination thresholds, reducing avalanche charge and afterpulsing. The ultimate limit to the discrimination threshold is determined by thermally induced voltage fluctuations, or Johnson noise, at the output of the SPAD. The RMS voltage fluctuation across a resistor R, at temperature T, is given by $V_{Th} = (4k_B TRf)^{1/2}$, where f the measurement bandwidth. For a room temperature 50 Ω load in a 2 GHz bandwidth, the RMS thermal noise is 40 μV. A figure of merit for the actual usable threshold in the presence of such a Gaussian noise source is the 5σ level, or 0.2 mV, at which the probability that thermal noise would trigger an ideal comparator is in the 10^{-7} range, and thus below typical per-gate dark-count probabilities. Assuming a 1 V gate transient at the output of the SPAD, 74 dB of attenuation is required to suppress it below this thermal noise floor, an amount of attenuation that can be achieved fairly easily with RF filters. In practice, however, it is often the case that the noise floor is dominated by amplifiers in the output stage. Nonetheless, discrimination thresholds that correspond to total avalanche charges of the order of 10^4 electrons have been reported with sine-wave gating [164].

Sine-wave gating systems have an inherent inflexibility in the gate duration, as it is inextricably linked to the gate frequency and the excess bias voltage. For a given excess bias, the gate duration can be reduced by increasing the AC amplitude of the gate, making the sine wave more sharply peaked above the breakdown voltage. Following this approach Nambu *et al.* [166] used a 16 V gate signal at 1.244 GHz and report extremely low afterpulse probability. Sine-wave gating systems have been demonstrated at gate frequencies up to 2.23 GHz, and tend to operate most effectively at frequencies above 1 GHz given the link between the gate duration and frequency. Detection efficiencies up to 25 % have been reported [166]. Perhaps most noteworthy with respect to the schemes discussed earlier in this section, single-photon detection rates in the range of 10 MHz to 100 MHz can be achieved. This is the major advance enabled by high-speed periodically gated detection systems.

4.9.6 Self-Differencing

Self-differencing [169] is a high-speed periodic-gating scheme that, in contrast to sine-wave gating, supports arbitrary gate waveforms. A typical schematic is shown in Fig 4.26; in these systems the SPAD output is split evenly into two delay lines whose difference in propagation delay equals exactly one gate period, and the outputs of the delay lines are subtracted from each other. With a strictly periodic gate waveform, the gate transient from one period eliminates the transient produced in the previous period, revealing any avalanche that occurred in one of the gates. Along with the gate transient, the avalanche signal is also distributed to each delay line, resulting in the characteristic anti-symmetric signal shown in Fig. 4.26. While this is of little consequence at low count rates, it does mean that avalanches in adjacent gates will interfere, which can complicate measurements of afterpulsing [154].

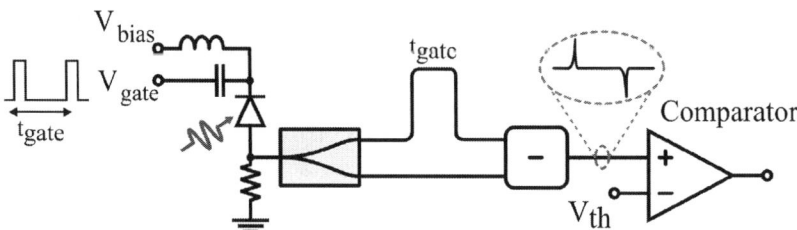

FIGURE 4.26 A self-differencing circuit. The SPAD output is split between two delay lines whose difference in propagation delay is equal to the period of the gate signal, t_{gate}. Taking the difference (−) of the outputs of the two delay lines subtracts the gate transients from successive gates, revealing avalanches as the anti-symmetric signal illustrated before the comparator.

The ability to use arbitrary periodic waveforms offers the flexibility to optimize the detection system to a given application, often for reducing the gate duration and minimizing the avalanche charge. Also, although the delay difference must equal one gate period for good cancellation, moderate changes in the gate frequency can be accommodated with coaxial line stretchers in each delay path. In addition, self-differencing does not require strongly dispersive RF filters, and can accurately report the avalanche waveform without undue distortion. Overall, self-differencing is a highly versatile approach to high-speed periodic gating.

Self-differencing has many similarities to the cancellation schemes discussed earlier [152]. However, in self-differencing the opposing transients that cancel with each other travel through delay lines that differ by a length equivalent to one gate period. The quality of the match of the attenuation and dispersion of these two different paths determines the gate-transient suppression, and therefor has a major impact on the overall performance of the detection system. Matching the attenuation and dispersion between the different delay lines is therefore critical, and becomes increasingly challenging as the signal bandwidth is increased, as with short square-wave bias gates with sharp edges (wide bandwidth). A variety of refinements to the self-differencing circuit have been developed to improve performance.

Yuan *et al.* [170] showed that by providing fine adjustable tuning to both the splitting ratio and the delay difference they were able to greatly improve the cancellation, and hence the overall performance. They demonstrated detection efficiency up to 23.5 % at 1550 nm, operating at 2 GHz with moderate afterpulse probability. Notably, they were able to show that the average charge per avalanche was as low as 35 fC. Restelli *et al.* [171] showed that the frequency-dependent losses in the coaxial delay lines could be well matched over the detection bandwidth by designing the long and short delay lines with different types of coaxial cable, thus requiring adjustment only to the delay difference (or the gate frequency) to optimize the cancellation.

Chen *et al.* [172] noted that imperfect transient suppression in the self-differencing output stage left a systematic periodic background signal. Being periodic, they demonstrated that it could be further suppressed by a second self-differencing stage. As described above, the self-differencing output stage distributes the avalanche signal between two gate periods, and thus reduces the SNR ratio. Interestingly, the double-self-differencing scheme presented by Chen *et al.* does not further reduce the amount of avalanche signal in the detection gate because two avalanche signals produced in the first stage combine to yield the same avalanche signal strength as with a single self-differencing stage. Unfortunately component losses cannot be ignored, and each power splitter or combiner imposes some loss to the avalanche signal. Nonetheless, Chen *et al.* were able to improve the transient suppression and better discriminate small avalanches from the transients by improving their discrimination threshold by 2 mV, and were able to demonstrate detection efficiency up to 30.5 % at 1550 nm with moderate afterpulsing.

One approach to improve transient cancellation in self-differencing systems is to relinquish some flexibility by using a narrowband sinusoidal gate in a self-differencing system [173]. As discussed above, the SPAD's voltage-dependant capacitance generates higher harmonics that can be removed with notch filters, in the same manner as in sine-wave schemes, and the remnant of the fundamental gate frequency can then be eliminated with the self-differencing circuit. In this case the broadband response of the delay lines is irrelevant, and care must only be given to match the attenuation of the two delay lines, which greatly simplifies the system and enhances the quality of the cancellation.

Although high-speed periodic gating schemes tend to operate more effectively at gate frequencies in the few-GHz range, the self-differencing scheme can be applied at lower gate frequencies by converting the output of the SPAD to an optical signal and using fiber-optic delay lines and balanced photodiode detection in the self-differencing output stage [174]. The low dispersion, low loss, and tunable power splitting available in fiber-optical components allow the gate repetition rates well into the MHz range with detection efficiency as high as 22% with moderate afterpulsing. It is worthwhile to note that while the fundamental signal to noise ratio may be exacerbated by the electrical-to-optical-to-electrical conversion, along with attendant amplification stages, some benefit is regained by the essentially lossless signal splitting.

4.9.7 Harmonic Subtraction

An alternative to sine-wave gating and self-differencing is shown in Fig. 4.27 [163,175]. Here, the gate waveform is synthesized from a discrete number of harmonics of the gate frequency, and the gate-transient suppression is achieved by destructive interference with reference signals at each harmonic that are

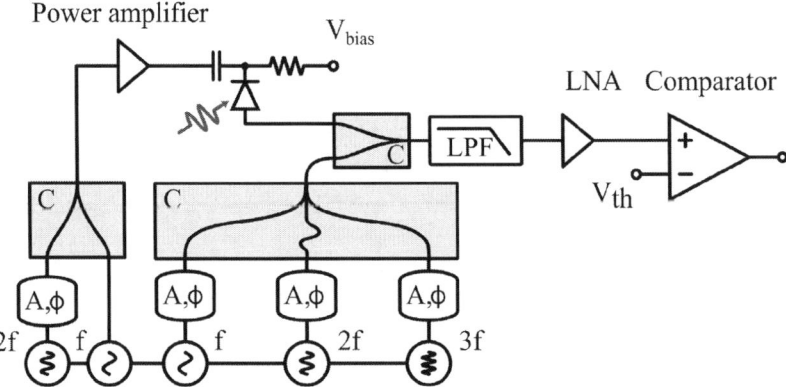

FIGURE 4.27 A harmonic-subtraction setup. The gate waveform is synthesized from harmonics of the gate frequency f, and the resulting harmonics in the gate transient are eliminated by destructive interference with reference signals from the source, with fine amplitude (A) and phase (ϕ) control, within the bandwidth of a low-pass filter (LPF). A low-loss combiner (C) and low-noise amplifier (LNA) efficiently preserve the avalanche signal.

generated directly at the RF source. The transient suppression in this scheme can approach what can be achieved with sine-wave gating, but without filtering or distorting the avalanche signal. This approach can support large-amplitude gate waveforms and excellent sensitivity to avalanche signals, both of which enhance the detection efficiency; Restelli *et al.* were able to reach detection efficiencies of 50% at 1310 nm. The use of multiple harmonics in the gate signal allows the gate duration to be reduced well below that of a sine-wave gate of the same frequency. The number of harmonics needed for transient suppression is determined solely by the detection bandwidth.

Harmonic subtraction has significant merits in the quality of the transient suppression, in the preservation of the undistorted avalanche signal, and in the ability to reduce the gate duration. It is also worthwhile to point out that the gate-synthesis in this approach allows the use of narrowband low-noise RF amplifiers. However, these benefits come at the expense of significant circuit complexity. Moreover, the quality of the transient suppression is determined by the ability to maintain nearly perfect destructive interference of multiple RF sinusoids; for a practical detection system, active stabilization of this interference is necessary. Nonetheless, this is a promising approach and has achieved the highest detection efficiency of any high-speed periodically gated system and afterpulsing comparable to the lowest levels reported to date.

4.9.8 Summary

A comparative survey of some of the high-speed (GHz) periodic gating schemes presented in this section is given in Table 4.3. Given the singular importance of afterpulsing in such schemes, and the difficulty in characterizing afterpulsing in a system that is gated on every nanosecond, the illumination rate, and the measurement technique and hold-off were used to characterize the system are specified (when available). The period of integration of the afterpulse-probability measurement is the inverse of the laser illumination rate. It should be noted that afterpulse events that occur during the hold-off immediately following a detection event are not counted. By extension, afterpulses measured within one hold-off time of an illuminated gate indicate that a detection event did *not* occur in that illuminated gate, meaning that the counted afterpulse was necessarily due to some earlier detection event. There is roughly an order of magnitude variation in the afterpulse probabilities, which reflects the strong (exponential) relationship between the gate duration and total charge.

Research in active-gating techniques continues to make significant contributions to single-photon detection technology. The benefits gating techniques have provided, particularly in improved count rates and detection efficiencies, come with tradeoffs, the most obvious being the reduction in detection duty cycle, or the time for which a detector is active. Both the benefits and tradeoffs underscore the importance of the bias and control circuitry to the

TABLE 4.3 Survey of a Variety of High-Speed (GHz) Periodic Gating Techniques, Along with the Various Refinements Applied, as Discussed in this Section. The Laser Illumination Rate Used During the Integrated Afterpulse-Probability (AP) Measurement, Along with the Type of Counter and Hold-off, are Specified Where Available.

Technique	Temp.	Gate Frequency	Detection Efficiency (λ)	Dark Count Probability (per gate)	Integrated AP	Laser Rate during AP Meas.	AP Meas. Method, Hold-off
Self-differencing, tunable difference [170]	243 K	1 GHz	27.8%	2.9×10^{-5}	8.8%	15.6 MHz	TCSPC, –
		2 GHz	23.5% (1550 nm)	1.32×10^{-5}	4.84%	31.25 MHz	
Self-differencing, with sine gate [173]	243 K	921 MHz	9.3% (1550 nm)	4.3×10^{-7}	3.4%	77 MHz	Gated counter, 10 ns hold-off
Sine-wave, notch filters [166]	223 K	1.244 GHz	11.6% (1550 nm)	5.8×10^{-7}	0.69%	9.7 MHz	TIA, 23 ns hold-off
Sine-wave, notch & low-pass filter [164]	223 K	1 GHz	10.4%	6.4×10^{-7}	1.6%	10 MHz	TCSPC, 50 ns hold-off
		2 GHz	10.5% (1550 nm)	6.1×10^{-7}	3.4%		
Sine-wave, cancel & low-pass filter [168]	240 K	1 GHz	10.4% (1550 nm)	6.1×10^{-6}	3.0%	10 MHz	–, 10 ns hold-off
Sine-wave, low-pass filter [165]	273 K	1.25 GHz	10% (1550 nm)	7×10^{-7}	1.6%	–	QKD performance, 8 ns hold-off
Harmonic subtraction [175]	251 K	1.25 GHz	25% (1310 nm)	2.4×10^{-5}	0.77%	19.5 MHz	Gated counter, 10 ns hold-off

performance of the detection system as a whole. The wide variety of gating techniques is a testament to the ingenuity of experimenters in extracting ever-better performance from imperfect devices.

4.10 FUTURE PROSPECTS FOR SILICON SPADs

The future progress of silicon SPADs will be mainly driven by user demands for detector performance, which naturally set requirements for the device design and fabrication technology.

A first basic request arising in many applications is to improve the detection efficiency. Enhanced DE is strongly desired (particularly in the red and near-infrared spectral ranges) in several life-science applications based on in vivo molecular imaging [176]. In order to achieve this, devices with thicker depletion layer must be designed and fabricated, which implies tailoring some steps in the fabrication process, or introducing new processing steps. In a dedicated fabrication technology this is quite natural. In standard CMOS technologies such flexibility is normally not offered; however, other applications may lead to develop new standard CMOS technologies with features suitable for producing SPADs with enhanced DE.

A second common request is a larger active area. In several techniques relying on confocal or near field microscopy (FCS, FLIM, combined FRET-FLIM), the illuminated spot size at the microscope image plane is small enough (a few tens of microns or less) to be easily covered by the SPAD active area, provided it has sufficiently large diameter (~ 100 μm). Fiber pigtailing of the detector, often employed for making the optical system more flexible, also benefits from a wider detector area because greater coupling efficiency can be achieved, and fibers with larger core diameters can be more easily accommodated. An increase of the detector diameter, however, sets stringent requirements on the quality of the starting material and of the fabrication process. A dedicated technology can exploit specific gettering steps performed as close as possible to the device active area. Such gettering is important not only because it is very difficult to obtain the required purity in the starting material, but also because contamination may be introduced later in the fabrication by unwanted marginal effects. Typical examples are faint residual contamination from furnaces previously employed in another fabrication, and side effects during the ion implantation. A further advantage of a dedicated technology is the capability of suitably shaping the electric field profile to minimize both band-to-band tunneling and field-assisted generation. Again, standard CMOS technologies do not offer this flexibility. Furthermore, the current trend of standard CMOS technologies toward low-thermal-budget processes and the lack of specific external gettering processes [71] that can be performed close to the SPAD raise some concerns about the evolution of these technologies toward fabrication of large-area devices.

A third quite specific requirement, arising from photon-timing applications, is to reduce the diffusion tail in the temporal response. For instance, a subnanosecond diffusion tail can benefit high-rate quantum key distribution (QKD) applications [53]. The diffusion tail can be reduced by keeping the neutral region very thin. Standard CMOS technologies do not allow any modification of the processing steps, whereas dedicated technologies are inherently flexible and fully customizable.

It is therefore likely that high-end applications requiring a combination of high DE, low DCR and low timing jitter will continue to rely on SPAD devices fabricated with dedicated technologies. Future developments in these technologies will likely focus on improving the DE in the red region of the spectrum. For example, a combination of red-enhanced [54] and resonant-cavity-enhanced [51] technologies might be used to develop frequency-selective SPAD devices with unprecedented DE at a desired wavelength.

There are no reasons to expect that the enhancement of sensitivity obtained with single SPAD detectors will not be extendable to array configurations used for multi-spot detection. The question to ascertain is how many SPADs such custom-technology arrays may eventually be able to contain without becoming cost-prohibitive, overly complex and, in the end, of little use to experimenters. A substantial increase in the pixel number can be achieved by resorting to more complex and sophisticated technologies, such as advanced multi-wafers and three-dimensional technologies that make possible the integration of custom SPAD arrays with high-performance CMOS electronics for quenching/timing. Recently, Aull *et al.*[177] reported a fully parallel laser radar imager based on a 64 × 64 SPAD array coupled to high-speed SOI CMOS circuits using 3D integration techniques.

On the other hand, CMOS integration has enabled progressively smaller feature sizes, to the point where it is now possible to envision extremely large imaging systems based on SPADs. Standard CMOS technologies will definitely outperform custom SPAD technology in all those applications where a high number (>1000) of pixels with small size, adequate fill factor ($>10\%$), and integrated electronics, are mandatory requirements. Future research activity in this area will be aimed at the development of denser arrays with larger formats ($>10^6$ pixels) by exploiting sub-100 nm CMOS technologies [80]. Significant development efforts will be necessary for achieving a satisfactory tradeoff between detector performance (e.g., active area, fill-factor, timing jitter, noise) and system complexity (in-pixel photon counting and timing circuitry, external readout electronics).

4.11 FUTURE PROSPECTS FOR InGaAs SPADs

As in the case of silicon SPADs, future improvements in the capabilities of InGaAs SPADs will be driven by the most pressing needs of applications that rely on these detectors. The tradeoff between DCR and DE described elsewhere in this chapter will continue to be a target for further progress, but this fundamental limitation poses significant challenges. DCR performance is intimately tied to materials properties, particularly with respect to bulk-material defects that lead to the thermal generation of dark carriers through Shockley-Read-Hall processes in the narrow-bandgap InGaAs absorption layer, as well as defects in the InP multiplication layer that lead to trap-assisted tunneling. Defects in the InP multiplier are also responsible for carrier trapping and detrapping that gives rise to afterpulsing effects. Dramatic improvements in epitaxial growth quality for the InGaAsP quaternary system were realized 10–15 years ago with the explosive growth in fiber-optic telecommunications applications that employed diode lasers and photodetectors based on this material system. Further progress on this front is likely to proceed much more modestly, especially in the absence of a similarly large new commercial market for these devices. Moreover, relative to the silicon material system, the InGaAsP material system serves vastly smaller markets and has considerably less technological maturity. Consequently, much less is known about the nature of InGaAsP materials limitations, and there is no comprehensive roadmap for materials improvement as there is in the silicon industry. The use of different III-V semiconductor materials with potentially favourable properties for SPAD devices may present interesting opportunities, but to the extent that these new materials will be even less technologically mature, they are likely to suffer from worse material quality. In light of these challenges, there is likely to be more rapid progress related to novel design approaches and implementation strategies, especially with regard to the electronic circuitry used to control SPAD functionality.

Beyond the fundamental DCR vs. DE tradeoff, the greatest recent focus for improvement of InGaAsP SPAD performance has been the effective photon counting rate of these devices. This need has been driven by the desire for GHz-scale bit rates for single-photon communications applications, especially in the context of quantum communications and quantum information processing. A similar requirement for higher-rate counting has also emerged in the context of applications of single-photon imaging such as 3D laser radar and low-light level imaging. While the inherent carrier dynamics of these devices can readily support response times well below 1 ns, afterpulsing effects pose a much more difficult challenge to high-rate counting. Because the elimination of defects that give rise to afterpulsing does not appear achievable as a near-term strategy, the reduction of afterpulsing effects is an example of the derivation of more viable improvements from clever circuit-based solutions. In particular, the dominant effective strategy among practitioners in the field has been to reduce the current flow per avalanche event to limit the amount of trapped charge that can potentially give rise to afterpulses, as discussed in Section 4.9.

Despite inherent materials challenges, the gradual maturing of the InGaAsP SPAD device platform has enabled the realization of imaging arrays with an evolution to successively larger formats. Figure 4.28 illustrates pixel maps for

Chapter | 4 Semiconductor-Based Detectors

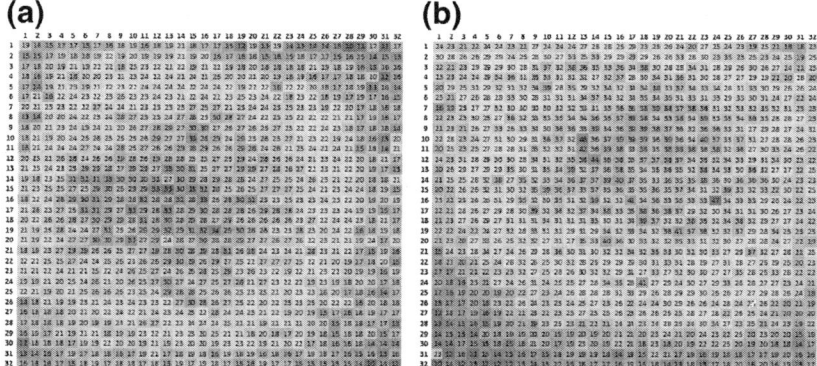

FIGURE 4.28 Performance maps of a 32 × 32 InGaAs/InP (1.55 μm) SPAD FPA operating with an excess bias of 3.25 V with modest cooling to 253 K. (a) The DCR is $<50 \times 10^3$ s^{-1} for *all* pixels. (b) Detection efficiency (in %) for all pixels, where the average pixel DE of 22 % includes all optical losses, including the microlens array for maintaining high fill factor.

the DCR and DE for a first-generation 32 x 32 array of InGaAs/InP SPADs on a 100 μm pitch designed for laser radar 3D imaging at 1.5 μm. Pixel yield is 100 % with well-behaved, fairly narrow distributions of pixel-level performance parameters [178]. Similar arrays with InGaAsP quaternary absorbers optimized for detection at 1.06 μm have also been commercially realized in 32 x 32 formats, as well as in larger 128 × 32 arrays with a 50 μm pitch [125]. The largest InGaAsP SPAD array demonstrated to date has a format of 256 × 64 pixels [179], and the progression to significantly larger formats seems inevitable.

REFERENCES

[1] W. Becker, "Advanced Time-Correlated Single-Photon Counting Techniques," Vol. 81, Springer (2005).
[2] Hamamatsu Photonics, MCP-PMT R3809U-50 Series Datasheet. Available online at: <http://www.hamamatsu.com/resources/pdf/etd/R3809U-50_TPMH1067E09.pdf>, 2011.
[3] A. Goetzberger, R. Scarlett, R. Haitz, and B. Mcdonald, "Avalanche Effects in Silicon P-N Junctions II. Structurally Perfect Junctions," J. Appl. Phys. 34, 1591–1600 (1963).
[4] R.H. Haitz, "Model for the Electrical Behavior of a Microplasma," J. Appl. Phys. 35, 1370–1376 (1964).
[5] R.H. Haitz, "Mechanisms Contributing to the Noise Pulse Rate of Avalanche Diodes," J. Appl. Phys. 36, 3123–3131 (1965).
[6] B. Levine and C. Bethea, "Single Photon Detection at 1.3 μm Using a Gated Avalanche Photodiode," Appl. Phys. Lett. 44 553–555 (1984).
[7] B. Levine, C. Bethea, and J. Campbell, "Near Room-Temperature 1.3 μm Single Photon-Counting with a Ingaas Avalanche Photodiode," Electron. Lett. 20, 596–598 (1984).
[8] R.J. McIntyre, "On the Avalanche Initiation Probability of Avalanche-Diodes Above the Breakdown Voltage," IEEE Trans. Electron Devices ED20, 637–641 (1973).
[9] W.G. Oldham, R.R. Samuelson, and P. Antognet, "Triggering Phenomena in Avalanche-Diodes," IEEE Trans. Electron Devices ED-19, 1056 (1972).
[10] S.M. Sze, "Physics of Semiconductor Devices," Wiley, New York, pp. 520–527 (1981).

[11] K. Graff, "Metal Impurities in Silicon-device Fabrication," Second ed., Springer-Verlag, Berlin (1995).
[12] G. Vincent, A. Chantre, and D. Bois, "Electric Field Effect on the Thermal Emission of Traps in Semiconductor Junctions," J. Appl. Phys. 50, 5484–5487 (1979).
[13] P.A. Martin, B.G. Streetman, and K. Hess, "Electric Field Enhanced Emission from Non-Coulombic Traps in Semiconductors," J. Appl. Phys. 52, 7409–7415 (1981).
[14] G.A.M. Hurkx, H.C. Degraaff, W.J. Kloosterman, and M.P.G. Knuvers, "A New Analytical Diode Model Including Tunneling and Avalanche Breakdown," IEEE Trans. Electron Devices 39, 2090–2098 (1992).
[15] G.A.M. Hurkx, D.B.M. Klaassen, and M.P.G. Knuvers, "A New Recombination Model for Device Simulation Including Tunneling," IEEE Trans. Electron Devices 39, 331–338 (1992).
[16] M. Ghioni, A. Gulinatti, I. Rech, F. Zappa, and S. Cova, "Progress in Silicon Single-Photon Avalanche Diodes," IEEE J. Sel. Top. Quantum Electron 13, 852–862 (2007).
[17] S. Cova, A. Lacaita, and G. Ripamonti, "Trapping Phenomena in Avalanche Photodiodes on Nanosecond Scale," IEEE Electron Device Lett. 12, 685–687 (1991).
[18] S. Cova, M. Ghioni, A. Lacaita, C. Samori, and F. Zappa, "Avalanche Photodiodes and Quenching Circuits for Single-Photon Detection," Appl. Opt. 35, 1956–1976 (1996).
[19] M. Ghioni, A. Giuduce, S. Cova, and F. Zappa, "High-Rate Quantum Key Distribution at Short Wavelength: Performance Analysis and Evaluation of Silicon Single Photon Avalanche Diodes," J. Mod. Opt. 50, 2251–2269 (2003).
[20] S. Cova, A. Longoni, and A. Andreoni, "Towards Picosecond Resolution with Single-Photon Avalanche Diodes," Rev. Sci. Instrum. 52, 408–412 (1981).
[21] G. Ripamonti and S. Cova, "Carrier Diffusion Effects in the Time-Response of a Fast Photodiode," Solid-State Electron 28, 925–931 (1985).
[22] A. Lacaita, M. Mastrapasqua, M. Ghioni, and S. Vanoli, "Observation of Avalanche Propagation by Multiplication Assisted Diffusion in p-n Junctions," Appl. Phys. Lett. 57, 489–491 (1990).
[23] P.P. Webb and R.J. McIntyre, "Recent Developments in Silicon Avalanche Photodiodes," RCS Eng. 27, 96–102 (1982).
[24] A. Spinelli and A.L. Lacaita, "Physics and Numerical Simulation of Single Photon Avalanche Diodes," IEEE Trans. Electron Devices 44, 1931–1943 (1997).
[25] A. Gulinatti, P. Maccagnani, I. Rech, M. Ghioni, and S. Cova, "35 ps Time Resolution at Room Temperature with Large Area Single Photon Avalanche Diodes," Electron. Lett. 41, 272–274 (2005).
[26] A. Lacaita, F. Zappa, S. Bigliardi, and M. Manfredi, "On the Bremsstrahlung Origin of Hot-Carrier-Induced Photons in Silicon Devices," IEEE Trans. Electron Devices 40, 577–582 (1993).
[27] I. Rech, A. Ingargiola, R. Spinelli, I. Labanca, S. Marangoni, M. Ghioni, and S. Cova, "A New Approach to Optical Crosstalk Modeling in Single-Photon Avalanche Diodes," IEEE Photonics Technol. Lett. 20, 330–332 (2008).
[28] I. Rech, A. Ingargiola, R. Spinelli, I. Labanca, S. Marangoni, M. Ghioni, and S. Cova, "Optical Crosstalk in Single Photon Avalanche Diode Arrays: A New Complete Model," Opt. Express 16 8381–8394 (2008).
[29] E. Charbon, "Towards Large Scale CMOS Single-Photon Detector Arrays for Lab-on-Chip Applications," J. Phys. -Appl. Phys. 41, (2008).
[30] P. Antognetti, S. Cova, and A. Longoni, "Study of the Operation and Performances of an Avalanche Diode as a Single Photon Detector," in Proceedings of the 2nd Ispra Nuclear Electronics Symposium, Stresa, Euratom Pub. 5370e, pp. 453–456, May 20–23 (1975) .
[31] S. Cova, "Active Quenching Circuit for Avalanche Photodiodes," US Patent No. 4,963,727; 16 Oct (1990) (priority date Oct. 20, 1988, Brevetto Italia 22367 A/88).
[32] A. Lacaita, S. Cova, C. Samori, and M. Ghioni, "Performance Optimization of Active Quenching Circuits for Picosecond Timing with Single Photon Avalanche Diodes," Rev. Sci. Instrum. 66, 4289–4295 (1995).
[33] M. Ghioni, S. Cova, F. Zappa, and C. Samori, "Compact Active Quenching Circuit for Fast Photon Counting with Avalanche Photodiodes," Rev. Sci. Instrum. 67, 3440–3448 (1996).
[34] F. Zappa, A. Lotito, A.C. Giudice, S. Cova, and M. Ghioni, "Monolithic Active-Quenching and Active-Reset Circuit for Single-Photon Avalanche Detectors," IEEE J. Solid-State Circuits 38, 1298–1301 (2003).

[35] S. Tisa, F. Guerrieri, and F. Zappa, "Variable-Load Quenching Circuit for Single-Photon Avalanche Diodes," Opt. Express 16, 2232–2244 (2008).
[36] A. Lacaita and M. Mastrapasqua, "Strong Dependence of Time Resolution on Detector Diameter in Single Photon Avalanche-Diodes," Electron. Lett. 26, 2053–2054 (1990).
[37] A. Lacaita, S. Cova, A. Spinelli, and F. Zappa, "Photon-Assisted Avalanche Spreading in Reach-Through Photodiodes," Appl. Phys. Lett. 62, 606–608 (1993).
[38] M. Assanelli, A. Ingargiola, I. Rech, A. Gulinatti, and M. Ghioni, "Photon-Timing Jitter Dependence on Injection Position in Single-Photon Avalanche Diodes," IEEE J. Quantum Electron 47, 151–159 (2011).
[39] S. Cova, M. Ghioni, and F. Zappa, "Circuit for High Precision Detection of the Time of Arrival of Photons Falling on Single Photon Avalanche Diodes," US Patent No. 6,384,663 B2; 07 May (2002) (priority date March 9, (2000)).
[40] A. Gallivanoni, I. Rech, D. Resnati, M. Ghioni, and S. Cova, "Monolithic Active Quenching and Picosecond Timing Circuit Suitable for Large-Area Single-Photon Avalanche Diodes," Opt. Express 14, 5021–5030 (2006).
[41] T.A. Louis, G. Ripamonti, and A. Lacaita, "Photoluminescence Lifetime Microscope Spectrometer Based on Time-Correlated Single-Photon Counting with an Avalanche Diode Detector," Rev. Sci. Instrum. 61, 11–22 (1990).
[42] M. Ghioni, S. Cova, A. Lacaita, and G. Ripamonti, "New Silicon Epitaxial Avalanche-Diode for Single-Photon Timing at Room-Temperature," Electron. Lett. 24, 1476–1477 (1988).
[43] A. Lacaita, M. Ghioni, and S. Cova, "Double Epitaxy Improves Single-Photon Avalanche-Diode Performance," Electron. Lett. 25, 841–843 (1989).
[44] W.J. Kindt and H.W. van Zeijl, "Modeling and Fabrication of Geiger Mode Avalanche Photodiodes," IEEE Trans. Nucl. Sci. 45, 715–719 (1998).
[45] J.C. Jackson, A.P. Morrison, D. Phelan, and A. Mathewson, "A Novel Silicon Geiger-Mode Avalanche Photodiode," in Digest of the International Electron Devices Meeting, New York (2002).
[46] B.F. Aull, A.H. Loomis, J.A. Gregory, and D. Young, "Geiger-Mode Avalanche Photodiode Arrays Integrated with CMOS Timing Circuits," in Digest of the 56th Annual Device Research Conference, pp. 58–59 (1998).
[47] E. Sciacca, A. C. Giudice, D. Sanfilippo, F. Zappa, S. Lombardo, R. Consentino, C. Di Franco, M. Ghioni, G. Fallica, G. Bonanno, S. Cova, and E. Rimini, "Silicon Planar Technology for Single-Photon Optical Detectors," IEEE Trans. Electron Devices 50, 918–925 (2003).
[48] A. Lacaita, M. Ghioni, and S. Cova, "Ultrafast Single-Photon Detector with Double Epitaxial Structure for Minimum Carrier Diffusion Effects," in Le Journal de Physique Colloques, Vol. 49–C4, Montpellier, France, pp. 633–636, September (1988).
[49] Micro-Photon-Devices, PDM Series Datasheet, Available online at: <http://www.micro-photon-devices.com/Docs/Datasheet/PDM.pdf>, 2013.
[50] A. Spinelli, M.A. Ghioni, S.D. Cova, and L.M. Davis, "Avalanche Detector with Ultraclean Response for Time-Resolved Photon Counting," IEEE J. Quantum Electron 34, 817–821 (1998).
[51] M. Ghioni, G. Armellini, P. Maccagnani, I. Rech, M.K. Emsley, and M.S. Unlu, "Resonant-Cavity-Enhanced Single-Photon Avalanche Diodes on Reflecting Silicon Substrates," IEEE Photonics Technol. Lett. 20, 413–415 (2008).
[52] N.J. Krichel, A. McCarthy, I. Rech, M. Ghioni, A. Gulinatti, and G.S. Buller, "Cumulative Data Acquisition in Comparative Photon-Counting Three-Dimensional Imaging," J. Mod. Opt. 58, 244–256 (2011).
[53] P.J. Clarke, R.J. Collins, P.A. Hiskett, M.-J. García-Martínez, N.J. Krichel, A. McCarthy, M.G. Tanner, J.A. O'Connor, C.M. Natarajan, S. Miki, M. Sasaki, Z. Wang, M. Fujiwara, I. Rech, M. Ghioni, A. Gulinatti, R.H. Hadfield, P.D. Townsend, and G.S. Buller, "Analysis of Detector Performance in a Gigahertz Clock Rate Quantum Key Distribution System," New J. Phys. 13, 075008 (2011).
[54] A. Gulinatti, I. Rech, F. Panzeri, C. Cammi, P. Maccagnani, M. Ghioni, and S. Cova, "New Silicon SPAD Technology for Enhanced Red-Sensitivity, High-Resolution Timing and System Integration," J. Mod. Opt. 59, 1489–1499 (2012).

[55] H. Dautet, P. Deschamps, B. Dion, A.D. MacGregor, D. MacSween, R.J. McIntyre, C. Trottier, and P.P. Webb, "Photon-Counting Techniques with Silicon Avalanche Photodiodes," Appl. Opt. 32, 3894–3900 (1993).

[56] Excelitas Technologies Corp., Single Photon Counting Module SPCM-AQRH Series Datasheet. Available online at: <http://www.excelitas.com/Downloads/DTS_SPCM-AQRH.pdf>, 2013.

[57] R.J. McIntyre, "Distribution of Gains in Uniformly Multiplying Avalanche Photodiodes – Theory," IEEE Trans. Electron Devices ED19, 703–713, 1972.

[58] P.P. Webb, R.J. McIntyre, and J. Conradi, "Properties of Avalanche Photodiodes," RCA Rev. 35, 234–278, 1974.

[59] R. J. McIntyre, "Silicon Avalanche Photodiode with Low Multiplication Noise," US Patent No. 4,972,242; 20 November (1990).

[60] R.J. McIntyre and P.P. Webb, "Low-Noise, Reach-Through, Avalanche Photodiodes," US Patent No. 5,583,352; 10 December (1996).

[61] I. Rech, I. Labanca, M. Ghioni, and S. Cova, "Modified Single Photon Counting Modules for Optimal Timing Performance," Rev. Sci. Instrum. 77, 033104-1–033104-5 (2006).

[62] Laser Components USA, Inc., SAP500-Series Datasheet, Available online at: <http://www.lasercomponents.com/fileadmin/user_upload/home/Datasheets/lcd/sap-series.pdf>, 2013.

[63] M. Stipcevic, H. Skenderovic, and D. Gracin, "Characterization of a Novel Avalanche Photodiode for Single Photon Detection in VIS-NIR Range," Opt. Express 18, 17448–17459 (2010).

[64] A. Rochas, M. Gani, B. Furrer, P.A. Besse, R. S. Popovic, G. Ribordy, and N. Gisin, "Single Photon Detector Fabricated in a Complementary Metal-Oxide-Semiconductor High-Voltage Technology," Rev. Sci. Instrum. 74, 3263–3270 (2003).

[65] A. Rochas, M. Gosch, A. Serov, P.A. Besse, R. S. Popovic, T. Lasser, and R. Rigler, "First Fully integrated 2-D array of single-photon detectors in standard CMOS technology," IEEE Photonics Technol. Lett. 15, 963–965, (2003).

[66] F. Zappa, S. Tisa, A. Gulinatti, A. Gallivanoni, and S. Cova, "Complete Single-Photon Counting and Timing Module in a Microchip," Opt. Lett. 30, 1327–1329 (2005).

[67] L. Pancheri and D. Stoppa, "Low-Noise CMOS Single-Photon Avalanche Diodes with 32 ns Dead Time," in ESSDERC 2007: Proceedings of the 37th European Solid-State Device Research Conference, edited by D. Schmitt Landsiedel and R. Thewes, IEEE, New York, pp. 362–365 (2007).

[68] A. Rochas, A. Pauchard, L. Monat, A. Matteo, P. Trinkler, R. Thew, and R. Ribordy, "Ultra-Compact CMOS Single Photon Detector," in Advanced Photon Counting Techniques, edited by W. Becker, Vol. 6372, SPIE-Int. Soc. Optical Engineering, Bellingham pp. U169–U176 (2006).

[69] C. Niclass, M. Sergio, and E. Charbon, "A Single Photon Avalanche Diode Array Fabricated in 0.3 μm CMOS and Based on an Event-Driven Readout for TCSPC Experiments," in Advanced Photon Counting Techniques, edited by W. Becker, Vol. 6372, SPIE-Int. Soc. Optical Engineering, Bellingham, pp. U216–U227 (2006).

[70] D. Mosconi, D. Stoppa, L. Pancheri, L. Gonzo, and A. Simoni, "CMOS Single-Photon Avalanche Diode Array for Time-Resolved Fluorescence Detection," in ESSCIRC 2006: Proceedings of the 32nd European Solid-State Circuits Conference, Montreaux, France, pp. 564–567, September (2006).

[71] M. J. Binns, S. Bertolini, R. Wise, D.J. Myers, and T. A. McKenna, "Effective Intrinsic Gettering for 200 mm and 300 mm P/P- Wafers in a Low Thermal Budget 0.13 μm Advanced CMOS Logic Process," in Semiconductor Silicon 2002, edited by H.R. Huff, L. Fabry, and S. Kishino, Electrochemical Society Series Volume: 2002, pp. 647–657 (2002).

[72] N. Faramarzpour, M.J. Deen, S. Shirani, and Q. Fang, "Fully Integrated Single Photon Avalanche Diode Detector in Standard CMOS 0.18 μm Technology," IEEE Trans. Electron Devices 55, 760–767 (2008).

[73] M.A. Marwick and A.G. Andreou, "Single Photon Avalanche Photodetector with Integrated Quenching Fabricated in TSMC 0.18 μm 1.8 VCMOS Process," Electron. Lett. 44, 643–644 (2008).

[74] H. Finkelstein, M.J. Hsu, and S.C. Esener, "STI-Bounded Single-Photon Avalanche Diode in a Deep-Submicrometer CMOS Technology," IEEE Electron Device Lett. 27, 887–889, (2006).

[75] M. Gersbach, J. Richardson, E. Mazaleyrat, S. Hardillier, C. Niclass, R. Henderson, L. Grant, and E. Charbon, "A Low-Noise Single-Photon Detector Implemented in a 130 nm CMOS Imaging Process," Solid-State Electron 53, 803–808 (2009).

[76] C. Niclass, M. Gersbach, R. Henderson, L. Grant, and E. Charbon, "A Single Photon Avalanche Diode Implemented in 130-nm CMOS Technology," IEEE J. Sel. Top. Quantum Electron 13, 863–869 (2007).

[77] L. Pancheri and D. Stoppa, "Low-Noise Single Photon Avalanche Diodes in 0.15 μm CMOS Technology," in Proceedings of the European Solid-State Device Research Conference (ESSDERC), Helsinki, Finland 2011, pp. 179–182 (2011).

[78] T. Hamamoto, "Sidewall Damage in a Silicon Substrate Caused by Trench Etching," Appl. Phys. Lett 58, 2942–2944 (1991).

[79] J.A. Richardson, E.A.G. Webster, L.A. Grant, and R.K. Henderson, "Scaleable Single-Photon Avalanche Diode Structures in Nanometer CMOS Technology," IEEE Trans. Electron Devices 58, 2028–2035 (2011).

[80] M.A. Karami, M. Gersbach, H.-J. Yoon, and E. Charbon, "A New Single-Photon Avalanche Diode in 90 nm Standard CMOS Technology," Opt. Express 18, 22158–22166 (2010).

[81] E.A.G. Webster, J.A. Richardson, L.A. Grant, D. Renshaw, and R.K. Henderson, "A Single-Photon Avalanche Diode in 90-nm CMOS Imaging Technology With 44% Photon Detection Efficiency at 690 nm," IEEE Electron Device Lett. 33, 694–696 2012.

[82] C. Niclass, A. Rochas, P.A. Besse, and E. Charbon, "Design and Characterization of a CMOS 3-D Image Sensor Based on Single Photon Avalanche Diodes," IEEE J. Solid-State Circuits 40, 1847–1854 (2005).

[83] C. Niclass, M. Sergio, and E. Charbon, "A CMOS 64 × 48 Single Photon Avalanche Diode Array with Event-Driven Readout," in ESSCIRC 2006: Proceedings of the 32nd European Solid-State Circuits Conference, edited by C. Enz, M. Declercq, and Y. Leblebici, IEEE, New York pp. 556–559, (2006).

[84] C. Niclass, M. Sergio, and E. Charbon, "A Single Photon Avalanche Diode Array Fabricated in Deep-Submicron CMOS Technology," in 2006 Design Automation and Test in Europe, Proceedings, Vols. 1–3, IEEE, New York, (2006) 79–84.

[85] C. Niclass, C. Favi, T. Kluter, M. Gersbach, and E. Charbon, "A 128 × 128 Single-Photon Image Sensor With Column-Level 10-Bit Time-to-Digital Converter Array," IEEE J. Solid-State Circuits 43, 2977–2989 (2008).

[86] M. Sergio, C. Niclass, and E. Charbon, "A 128 × 2 CMOS Single-Photon Streak Camera with Timing-Preserving Latchless Pipeline Readout," in Solid-State Circuits Conference - Digest of Technical Papers, ISSCC 2007. IEEE International, pp. 120–121 (2007).

[87] L. Carrara, C. Niclass, N. Scheidegger, H. Shea, and E. Charbon, "A Gamma, X-ray and High Energy Proton Radiation-Tolerant CIS for Space Applications," in Solid-State Circuits Conference - Digest of Technical Papers, ISSCC 2009. IEEE International, pp. 40–41 (2009).

[88] C. Niclass, C. Favi, T. Kluter, F. Monnier, and E. Charbon, "Single-Photon Synchronous Detection," IEEE J. Solid-State Circuits 44, 1977–1989 (2009).

[89] F. Guerrieri, S. Tisa, A. Tosi, and F. Zappa, "Two-Dimensional SPAD Imaging Camera for Photon Counting," IEEE Photonics J 2, 759–774 (2010).

[90] F. Guerrieri, L. Maccone, F.N.C. Wong, J.H. Shapiro, S. Tisa, and F. Zappa, "Sub-Rayleigh Imaging via N-Photon Detection," Phys. Rev. Lett. 105, 163602 (2010).

[91] B. Markovic, S. Tisa, A. Tosi, and F. Zappa, "Monolithic Single-Photon detectors and Time-to-Digital Converters for picoseconds Time-of-Flight ranging," in Sensors, Cameras, and Systems for Industrial, Scientific, and Consumer Applications XII, edited by R. Widenhorn and V. Nguyen, Vol. 7875, SPIE-Int. Soc. Optical Engineering, Bellingham, p. 78750P (2011).

[92] European Community Sixth Framework Programme, IST-FET Open, project MEGAFRAME Million Frame Per Second Time-Correlated Single Photon Camera. Available Online: <http://www.megaframe.eu>.

[93] M. Gersbach, Y. Maruyama, E. Labonne, J. Richardson, R. Walker, L. Grant, R. Henderson, F. Borghetti, D. Stoppa, and E. Charbon, "A Parallel 32 × 32 Time-To-Digital Converter Array Fabricated in a 130 nm Imaging CMOS Technology," in Proceedings of ESSCIRC, IEEE, New York, pp. 197–200 (2009).

[94] J. Richardson, R. Walker, L. Grant, D. Stoppa, F. Borghetti, E. Charbon, M. Gersbach, and R.K. Henderson, "A 32 × 32 50 ps Resolution 10 bit Time to Digital Converter Array in

130 nm CMOS for Time Correlated Imaging," in presented at the IEEE Custom Integrated Circuit Conference, New York, (2009).
[95] D. Stoppa, F. Borghetti, J. Richardson, R. Walker, L. Grant, R.K. Henderson, M. Gersbach, and E. Charbon, "A 32 × 32-Pixel Array with In-Pixel Photon Counting and Arrival Time Measurement in the Analog Domain," in 2009 Proceedings of ESSCIRC, IEEE, New York, pp. 205–208 (2009).
[96] M. Gersbach, Y. Maruyama, R. Trimananda, M.W. Fishburn, D. Stoppa, J.A. Richardson, R. Walker, R. Henderson, and E. Charbon, "A Time-Resolved, Low-Noise Single-Photon Image Sensor Fabricated in Deep-Submicron CMOS Technology," IEEE J. Solid-State Circuits 47, 1394–1407 (2012).
[97] J.A. Richardson, L.A. Grant, and R.K. Henderson, "Low Dark Count Single-Photon Avalanche Diode Structure Compatible With Standard Nanometer Scale CMOS Technology," IEEE Photonics Technol. Lett. 21, 1020–1022 (2009).
[98] S. Donati, G. Martini, and M. Norgia, "Microconcentrators to Recover Fill-Factor Inimage Photodetectors with Pixel on-Boardprocessing Circuits," Opt. Express 15, 18066–18075 (2007).
[99] J.R. Lakowicz, Principles of Fluorescence Spectroscopy, 3rd ed. Springer, (2006).
[100] C. Veerappan, J. Richardson, R. Walker, D.-U. Li, M. W. Fishburn, Y. Maruyama, D. Stoppa, F. Borghetti, M. Gersbach, R.K. Henderson, and E. Charbon, "A 160 × 128 Single-Photon Image Sensor with on-Pixel 55 ps 10b Time-to-Digital Converter," in Solid-State Circuits Conference Digest of Technical Papers (ISSCC), 2011 IEEE International pp. 312–314 (2011).
[101] R. Rigler and E.S. Elson, Fluorescence Correlation Spectroscopy: Theory and Applications, Springer-Verlag, Berlin, (2001).
[102] J. Bewersdorf, R. Pick, and S.W. Hell, "Multifocal Multiphoton Microscopy," Opt. Lett. 23, 655–657 (1998).
[103] G. MacBeath, "Protein Microarrays and Proteomics," Nat. Genet. 32, 526–532 (2002).
[104] T. Kodadek, "Protein Microarrays: Prospects and Problems," Chem. Biol. 8, 105–115 (2001).
[105] X. Michalet, R.A. Colyer, G. Scalia, A. Ingargiola, R. Lin, J.E. Millaud, S. Weiss, O.H.W. Siegmund, A.S. Tremsin, J.V. Vallerga, A. Cheng, M. Levi, D. Aharoni, K. Arisaka, F. Villa, F. Guerrieri, F. Panzeri, I. Rech, A. Gulinatti, F. Zappa, M. Ghioni, and S. Cova, "Development of New Photon-Counting Detectors for Single-Molecule Fluorescence Microscopy," Philos. Trans. R. Soc. B Biol. Sci. 368, 1611 (2013).
[106] A. Restelli, I. Rech, P. Maccagnani, M. Ghioni, and S. Cova, "Monolithic Silicon Matrix Detector with 50 μm Photon Counting Pixels," J. Mod. Opt. 54, 213–223 (2007).
[107] S. Marangoni, I. Rech, M. Ghioni, P. Maccagnani, M. Chiari, M. Cretich, F. Damin, G. Di Carlo, and S. Cova, "A 6 × 8 Photon-Counting Array Detector System for Fast and Sensitive Analysis of Protein Microarrays," Sensors Actuators B Chem. 149, 420–426, (2010).
[108] R.A. Colyer, G. Scalia, I. Rech, A. Gulinatti, M. Ghioni, S. Cova, S. Weiss, and X. Michalet, "High-Throughput FCS Using an LCOS Spatial Light Modulator and an 8 × 1 SPAD Array," Biomed. Opt. Express 1, 1408–1431 (2010).
[109] F. Zappa, A. Gulinatti, P. Maccagnani, S. Tisa, and S. Cova, "SPADA: Single-Photon Avalanche Diode Arrays," IEEE Photonics Technol. Lett. 17, 657–659 (2005).
[110] F. Zappa, S. Tisa, S. Cova, P. Maccagnani, D.B. Calia, R. Saletti, R. Roncella, G. Bonanno, and M. Belluso, "Single-Photon Avalanche Diode Arrays for Fast Transients and Adaptive Optics," IEEE Trans. Instrum. Meas. 55, 365–374 (2006).
[111] C. Cammi, F. Panzeri, A. Gulinatti, I. Rech, and M. Ghioni, "Custom Single-Photon Avalanche Diode with Integrated Front-End for Parallel photon Timing Applications," Rev. Sci. Instrum. 83, 033104 (2012).
[112] K. Nishida, K. Taguchi, and Y. Matsumoto, "InGaAsP Heterostructure Avalanche Photodiodes with High Avalanche Gain," Appl. Phys. Lett. 35, 251–253 (1979).
[113] A. Lacaita, F. Zappa, S. Cova, and P. Lovati, "Single-Photon Detection Beyond 1 μm: Performance of Commercially Available InGaAs/InP Detectors," Appl. Opt 35, 2986–2996 (1996).
[114] G. Ribordy, J.D. Gautier, H. Zbinden, and N. Gisin, "Performance of InGaAs/InP Avalanche Photodiodes as Gated-Mode Photon Counters," Appl. Opt. 37, 2272–2277 (1998).

[115] P.A. Hiskett, G.S. Buller, A.Y. Loudon, J.M. Smith, I. Gontijo, A.C. Walker, P.D. Townsend, and M.J. Robertson, "Performance and Design of InGaAs/InP Photodiodes for Single-Photon Counting at 1.55 μm," Appl. Opt. 39, 6818–6829 (2000).

[116] P.A. Hiskett, J.M. Smith, G.S. Buller, and P.D. Townsend, "Low-Noise Single-Photon Detection at Wavelength 1.55 μm," Electron. Lett. 37, 1081–1083 (2001).

[117] D. Stucki, G. Ribordy, A. Stefanov, H. Zbinden, J.G. Rarity, and T. Wall, "Photon Counting for Quantum Key Distribution with Peltier Cooled InGaAs/InP APDs," J. Mod. Opt. 48, 1967–1981 (2001).

[118] J.P. Donnelly, E.K. Duerr, K.A. McIntosh, E.A. Dauler, D.C. Oakley, S.H. Groves, C.J. Vineis, L.J. Mahoney, K.M. Molvar, P.I. Hopman, K.E. Jensen, G.M. Smith, S. Verghese, and D.C. Shaver, "Design Considerations for 1.06 μm InGaAsP-InP Geiger-Mode Avalanche Photodiodes," IEEE J. Quantum Electron 42, 797–809 (2006).

[119] S. Pellegrini, R.E. Warburton, L.J.J. Tan, J.S. Ng, A.B. Krysa, K. Groom, J.P.R. David, S. Cova, M. J. Robertson, and G.S. Buller, "Design and Performance of an InGaAs-InP Single-Photon Avalanche Diode Detector," IEEE J. Quantum Electron 42, 397–403 (2006).

[120] M.A. Itzler, R. Ben-Michael, C.-F. Hsu, K. Slomkowski, A. Tosi, S. Cova, F. Zappa, and R. Ispasoiu, "Single Photon Avalanche Diodes (SPADs) for 1.5 μm Photon Counting Applications," J. Mod. Opt. 54, 283–304 (2007).

[121] X. Jiang, M.A. Itzler, R. Ben-Michael, and K. Slomkowski, "InGaAsP-InP Avalanche Photodiodes for Single Photon Detection," IEEE J. Sel. Top. Quantum Electron 13, 895–905 (2007).

[122] D.A. Ramirez, M.M. Hayat, and M.A. Itzler, "Dependence of the Performance of Single Photon Avalanche Diodes on the Multiplication Region Width," IEEE J. Quantum Electron 44, 1188–1195, (2008).

[123] M.A. Itzler, X. Jiang, M. Entwistle, K. Slomkowski, A. Tosi, F. Acerbi, F. Zappa, and S. Cova, "Advances in InGaAsP-Based Avalanche Diode Single Photon Detectors," J. Mod. Opt, 58, 174–200 (2011).

[124] S. Verghese, J.P. Donnelly, E.K. Duerr, K.A. McIntosh, D.C. Chapman, C.J. Vineis, G.M. Smith, J.E. Funk, K.E. Jensen, P.I. Hopman, D.C. Shaver, B.F. Aull, J.C. Aversa, J.P. Frechette, J.B. Glettler, Z.-L. Liau, J.M. Mahan, L.J. Mahoney, K.M. Molvar, F.J. O'Donnell, D.C. Oakley, E.J. Ouellette, M.J. Renzi, and B.M. Tyrrell, "Arrays of InP-based Avalanche Photodiodes for Photon Counting," IEEE J. Sel. Top. Quantum Electron 13, 870–886 (2007).

[125] M.A. Itzler, M. Entwistle, M. Owens, K. Patel, X. Jiang, K. Slomkowski, S. Rangwala, P.F. Zalud, T. Senko, J. Tower, and J. Ferraro, "Comparison of 32 × 128 and 32 × 32 Geiger-Mode APD FPAs for Single Photon 3D LADAR Imaging," in Advanced Photon Counting Techniques V, edited by M. A. Itzler and J. C. Campbell," SPIE-Int. Soc. Optical Engineering, Bellingham, Vol. 8033, p. 80330G (2011).

[126] J.C. Campbell, A.G. Dentai, W.S. Holden, and B.L. Kasper, "High-Performance Avalanche Photodiode with Separate Absorption Grading and Multiplication Regions," Electron. Lett. 19, 818–820 1983.

[127] S.R. Forrest, O.K. Kim, and R.G. Smith, "Optical Response time of In0.53Ga0.47As/InP Avalanche Photodiodes," Appl. Phys. Lett. 41, 95–98 (1982).

[128] Y. Liu, S.R. Forrest, J. Hladky, M.J. Lange, G.H. Olsen, and D.E. Ackley, "A Planar InP/InGaAs Avalanche Photodiode with Floating Guard Ring and Double Diffused Junction," J. Light. Technol. 10, 182–193 (1992).

[129] F. Zappa, P. Lovati, and A. Lacaita, "Temperature Dependence of Electron and Hole Ionization Coefficients in InP," in Eighth International Conference on Indium Phosphide and Related Materials, 1996. IPRM '96, pp. 628–631 (1996).

[130] X. Jiang, M.A. Itzler, R. Ben-Michael, K. Slomkowski, M.A. Krainak, S. Wu, and X. Sun, "Afterpulsing Effects in Free-Running InGaAsP Single-Photon Avalanche Diodes," IEEE J. Quantum Electron 44, 3–11 (2008).

[131] D.A. Ramirez, M.M. Hayat, G. Karve, J.C. Campbell, S.N. Torres, B.E.A. Saleh, and M.C. Teich, "Detection Efficiencies and Generalized Breakdown Probabilities for Nanosecond-Gated Near Infrared Single-Photon Avalanche Photodiodes," IEEE J. Quantum Electron 42, 137–145 (2006).

[132] K.E. Jensen, P.I. Hopman, E.K. Duerr, E.A. Dauler, J.P. Donnelly, S.H. Groves, L.J. Mahoney, K.A. McIntosh, K.M. Molvar, A. Napoleone, D.C. Oakley, S. Verghese, C.J.

Vineis, and R.D. Younger, "Afterpulsing in Geiger-Mode Avalanche Photodiodes for 1.06 μm Wavelength," Appl. Phys. Lett. 88, 133503-1–33503-3 (2006).
[133] M. Liu, C. Hu, J.C. Campbell, Z. Pan, and M.M. Tashima, "Reduce Afterpulsing of Single Photon Avalanche Diodes Using Passive Quenching With Active Reset," IEEE J. Quantum Electron 44, 430–434 (2008).
[134] F. Zappa, A. Tosi, and S. Cova, "InGaAs SPAD and Electronics for Low Time Jitter and Low Noise," in Photon Counting Applications, Quantum Optics, and Quantum Cryptography, edited by I. Prochazka and A. L. Migdall, Bellingham: SPIE-Int. Soc. Optical Engineering, Vol. 6583, p. 65830E (2007).
[135] K. Zhao, A. Zhang, Y. Lo, and W. Farr, "InGaAs Single Photon Avalanche Detector with Ultralow Excess Noise," Appl. Phys. Lett. 91, 081107-1–081107-3 (2007).
[136] K. Zhao, S. You, J. Cheng, and Y. Lo, "Self-Quenching and Self-Recovering InGaAs/InAlAs Single Photon Avalanche Detector," Appl. Phys. Lett. 93, 153504-1–153504-3, (2008).
[137] M.A. Itzler, X. Jiang, B. Nyman, and K. Slomkowski, "InP-Based Negative Feedback Avalanche Diodes," in Quantum Sensing and Nanophotonic Devices VI, edited by M. Razeghi and R. Sudharsanan, SPIE-Int. Soc. Optical Engineering, Bellingham, Vol. 7222, p. 72221K, (2009).
[138] X. Jiang, M.A. Itzler, B. Nyman, and K. Slomkowski, "Negative Feedback Avalanche Diodes for Near-Infrared Single-Photon Detection," in Advanced Photon Counting Techniques III, edited by M. A. Itzler and J. C. Campbell, SPIE-Int. Soc. Optical Engineering, Bellingham, Vol. 7320, p. 732011, (2009).
[139] M. A. Itzler, X. Jiang, B. M. Onat, and K. Slomkowski, "Progress in Self-Quenching InP-Based Single Photon Detectors," in Quantum Sensing and Nanophotonic Devices VII, edited by M. Razeghi and R. Sudharsanan, SPIE-Int. Soc. Optical Engineering, Bellingham, Vol. 7608, p. 760829 (2010).
[140] A. Tosi, A.D. Mora, F. Zappa, and S. Cova, "Single-Photon Avalanche Diodes for the Near-Infrared Range: Detector and Circuit Issues," J. Mod. Opt. 56, 299–308, (2009).
[141] S. Cova, A. Longoni, and G. Ripamonti, "Active-Quenching and Gating Circuits for Single-Photon Avalanche-Diodes (SPADs)," IEEE Trans. Nucl. Sci. 29, 599–601 (1982).
[142] C. Marand and P. Townsend, "Quantum Key Distribution Over Distances as Long as 30 Km," Opt. Lett. 20, 1695–1697 (1995).
[143] A. Dalla Mora, A. Tosi, F. Zappa, S. Cova, D. Contini, A. Pifferi, L. Spinelli, A. Torricelli, and R. Cubeddu, "Fast-Gated Single-Photon Avalanche Diode for Wide Dynamic Range Near Infrared Spectroscopy," IEEE J. Sel. Top. Quantum Electron 16, 1023–1030 (2010).
[144] C. Hu, X. Zheng, J.C. Campbell, B.M. Onat, X. Jiang, and M.A. Itzler, "Characterization of an InGaAs/InP-Based Single-Photon Avalanche Diode with Gated-Passive Quenching with Active Reset Circuit," J. Mod. Opt. 58, 201–209 (2011).
[145] A. Yoshizawa and H. Tsuchida, "A 1550 nm Single-Photon Detector Using a Thermoelectrically Cooled InGaAs Avalanche Photodiode," Jpn. J. Appl. Phys. Part 1-Regul. Pap. Short Notes Rev. Pap. 40, 200–201, (2001).
[146] F. Zappa, A. Lacaita, S. Cova, and P. Webb, "Nanosecond Single-Photon Timing with Ingaas/Inp Photodiodes," Opt. Lett 19, 846–848 (1994).
[147] G. Ribordy, N. Gisin, O. Guinnard, D. Stucki, M. Wegmuller, and H. Zbinden, "Photon Counting at Telecom Wavelengths with Commercial InGaAs/InP Avalanche Photodiodes: Current Performance," J. Mod. Opt. 51, 1381–1398 (2004).
[148] P.L. Voss, K.G. Koprulu, S.K. Choi, S. Dugan, and P. Kumar, "14 MHz Rate Photon Counting with Room Temperature InGaAs/InP Avalanche Photodiodes," J. Mod. Opt. 51, 1369–1379, (2004).
[149] A. Yoshizawa, R. Kaji, and H. Tsuchida, "Gated-Mode Single-Photon Detection at 1550 nm by Discharge Pulse Counting," Appl. Phys. Lett. 84, 3606–3608 (2004).
[150] A. Yoshizawa, S. Odate, and H. Tsuchida, "Discharge Pulse Counting for Low-Noise Single-Photon Detection at 155 nm Using InGaAs Avalanche Photodiode Cooled to 130 K," Jpn. J. Appl. Phys. Part 1-Regul. Pap. Brief Commun. Rev. Pap. 46, 220–222 (2007).
[151] D.S. Bethune, R.G. Devoe, C. Kurtsiefer, C. T. Rettner, and W.P. Risk, "System for Gated Detection of Optical Pulses Containing a Small Number of Photons Using an Avalanche Photodiode," US 6,218,657 B1, 17April (2001).

[152] D.S. Bethune and W.P. Risk, "An Autocompensating Fiber-Optic Quantum Cryptography System Based on Polarization Splitting of Light," IEEE J. Quantum Electron 36, 340–347 (2000).
[153] D.S. Bethune, W.P. Risk, and G.W. Pabst, "A High-Performance Integrated Single-Photon Detector for Telecom Wavelengths," J. Mod. Opt. 51, 1359–1368 (2004).
[154] A. Restelli, J. C. Bienfang, and A. L. Migdall, "Time-Domain Measurements of Afterpulsing in InGaAs/InP SPAD Gated with Sub-Nanosecond Pulses," J. Mod. Opt. 59, 1465–1471 (2012).
[155] A. Tomita and K. Nakamura, "Balanced, Gated-Mode Photon Detector for Quantum-bit Discrimination at 1550 nm," Opt. Lett. 27, 1827–1829, Oct. 2002.
[156] Z. Lu, W. Sun, J.C. Campbell, X. Jiang, and M.A. Itzler, "Corrections to 'Common-Mode Cancellation in Sinusoidal Gating With Balanced InGaAs/InP Single Photon Avalanche Diodes'; [Dec 12 1505-1511]," IEEE J. Quantum Electron 49, 59–59 (2013).
[157] J.C. Campbell, W. Sun, Z. Lu, M.A. Itzler, and X. Jiang, "Common-Mode Cancellation in Sinusoidal Gating With Balanced InGaAs/InP Single Photon Avalanche Diodes," IEEE J. Quantum Electron 48, 1505–1511 (2012).
[158] G. Wu, C. Zhou, X. Chen, and H. Zeng, "High performance of gated-mode single-photon detector at 1.55 µm," Opt. Commun. 265, 126–131 (2006).
[159] C. Y. Zhou, G. Wu, and H. P. Zeng, "Multigate single-photon detection and timing discrimination with an InGaAs/InP avalanche photodiode," Appl. Opt. 45, 1773–1776 (2006).
[160] Y. Liang, Y. Jian, X. Chen, G. Wu, E. Wu, and H. Zeng, "Room-Temperature Single-Photon Detector Based on InGaAs/InP Avalanche Photodiode With Multichannel Counting Ability," IEEE Photonics Technol. Lett. 23, 115–117 (2011).
[161] Y. Zhang, X. Zhang, and S. Wang, "Gaussian Pulse Gated InGaAs/InP Avalanche Photodiode for Single Photon Detection," Opt. Lett. 38, 606–608 (2013).
[162] N. Namekata, S. Sasamori, and S. Inoue, "800 MHz Single-Photon Detection at 1550-nm Using an InGaAs/InP Avalanche Photodiode Operated with a sine Wave Gating," Opt. Express 14, 10043–10049 (2006).
[163] A. Restelli, J.C. Bienfang, and A.L. Migdall, "Single-Photon Detection Efficiency up to 50 % at 1310 nm with an InGaAs/InP Avalanche Diode Gated at 1.25 GHz," Appl. Phys. Lett. 102, 141104 (2013).
[164] N. Namekata, S. Adachi, and S. Inoue, "Ultra-Low-Noise Sinusoidally Gated Avalanche Photodiode for High-Speed Single-Photon Detection at Telecommunication Wavelengths," IEEE Photonics Technol. Lett. 22 529–531 (2010).
[165] N. Walenta, T. Lunghi, O. Guinnard, R. Houlmann, H. Zbinden, and N. Gisin, "Sine Gating Detector with Simple Filtering for Low-Noise Infra-Red Single Photon Detection at Room Temperature," J. Appl. Phys. 112, (2012).
[166] Y. Nambu, S. Takahashi, K. Yoshino, A. Tanaka, M. Fujiwara, M. Sasaki, A. Tajima, S. Yorozu, and A. Tomita, "Efficient and Low-Noise Single-Photon Avalanche Photodiode for 1.244-GHz Clocked Quantum Key Distribution," Opt. Express 19, 20531–20541 (2011).
[167] N. Namekata, S. Adachi, and S. Inoue, "1.5 GHz Single-Photon Detection at Telecommunication Wavelengths Using Sinusoidally Gated InGaAs/InP Avalanche Photodiode," Opt. Express 17, 6275–6282 (2009).
[168] Y. Liang, E. Wu, X. Chen, M. Ren, Y. Jian, G. Wu, and H. Zeng, "Low-Timing-Jitter Single-Photon Detection Using 1-GHz Sinusoidally Gated InGaAs/InP Avalanche Photodiode," IEEE Photonics Technol. Lett. 23, 887–889 (2011).
[169] Z.L. Yuan, B.E. Kardynal, A.W. Sharpe, and A.J. Shields, "High Speed Single Photon Detection in the Near Infrared," Appl. Phys. Lett 91, 041114 (2007).
[170] Z.L. Yuan, A.W. Sharpe, J.F. Dynes, A.R. Dixon, and A.J. Shields, "Multi-Gigahertz Operation of Photon Counting InGaAs Avalanche Photodiodes," Appl. Phys. Lett 96, (2010).
[171] A. Restelli and J. C. Bienfang, "Avalanche Discrimination and High-Speed Counting in Periodically Gated Single-Photon Avalanche Diodes," in Advanced Photon Counting Techniques VI, SPIE-Int. Soc. Optical Engineering, Vol. 8375, p. 83750Z (2012).
[172] X. Chen, E. Wu, G. Wu, and H. Zeng, "Low-Noise High-Speed InGaAs/InP-Based Single-Photon Detector," Opt. Express 18 7010–7018 (2010).
[173] J. Zhang, R. Thew, C. Barreiro, and H. Zbinden, "Practical Fast Gate Rate InGaAs/InP Single-Photon Avalanche Photodiodes," Appl. Phys. Lett. 95, 091103 (2009).

[174] Y. Jian, E. Wu, G. Wu, and H. Zeng, "Optically Self-Balanced InGaAs-InP Avalanche Photodiode for Infrared Single-Photon Detection," IEEE Photonics Technol. Lett. 22, 173–175 (2010).

[175] A. Restelli, J. C. Bienfang, and A. L. Migdall, "Gigahertz-gated InGaAs SPAD System with Avalanche Charge Sensitivity Approaching the Fundamental Limit," in Advanced Photon Counting Techniques VII, edited by M.A. Itzler and J.C. Campbell, SPIE-Int. Soc. Optical Engineering, Bell, Vol. 8727, p. 87270F (2013).

[176] V. Ntziachristos, J. Ripoll, L. V. Wang, and R. Weissleder, "Looking and Listening to Light: The Evolution of Whole-Body Photonic Imaging," Nat. Biotechnol 23, 313–320 (2005).

[177] B. Aull, J. Burns, C. Chen, B. Felton, H. Hanson, C. Keast, J. Knecht, A. Loomis, M. Renzi, A. Soares, V. Suntharalingam, K. Warner, D. Wolfson, D. Yost, and D. Young, "Laser Radar Imager Based on 3D Integration of Geiger-Mode Avalanche Photodiodes with Two SOI Timing Circuit Layers," in Solid-State Circuits Conference - Digest of Technical Papers. ISSCC 2006. IEEE International, pp. 1179–1188 (2006).

[178] M.A. Itzler, M. Entwistle, M. Owens, K. Patel, X. Jiang, K. Slomkowski, S. Rangwala, P.F. Zalud, T. Senko, J. Tower, and J. Ferraro, "Geiger-Mode Avalanche Photodiode Focal Plane Arrays for Three-Dimensional Imaging LADAR," in Infrared Remote Sensing and Instrumentation XVIII, edited by M. Strojnik and G. Paez, Eds. Bellingham: SPIE-Int. Soc. Optical Engineering, Vol. 7808, p. 78080C (2010).

[179] R.D. Younger, K.A. McIntosh, J.W. Chludzinski, D.C. Oakley, L.J. Mahoney, J.E. Funk, J.P. Donnelly, and S. Verghese, "Crosstalk Analysis of Integrated Geiger-Mode Avalanche Photodiode Focal Plane Arrays," in Advanced Photon Counting Techniques III edited by M. A. Itzler and J. C. Campbell, Bellingham: SPIE-Int. Soc. Optical Engineering, Vol. 7320, p. 73200Q (2009).

Chapter 5

Novel Semiconductor Single-Photon Detectors

Jungsang Kim[*], Kyle S. McKay[†], Paul G. Kwiat[‡], Kevin Zielnicki[‡], and Eric J. Gansen[††]

[*]*Department of Electrical and Computer Engineering, Physics and Computer Science, Duke University, Durham, NC 27708, USA*
[†]*National Institute of Standards and Technology, Boulder, CO 80303, USA*
[‡]*Department of Physics, University of Illinois at Urbana-Champaign, Urbana, IL 61801, USA*
[††]*Department of Physics, University of Wisconsin-La Crosse, La Crosse, WI 54601, USA*

Chapter Outline

5.1 Introduction	147
5.2 Solid-State Photomultipliers and Visible-Light Photon Counters	148
5.2.1 Introduction	148
5.2.2 VLPC Structure and Operation	150
5.2.3 SSPM and VLPC Performance	154
5.2.4 Quantitative Model and Its Current Limitations	161
5.2.5 New Opportunities for VLPCs	163
5.2.6 Conclusions	166
5.3 Quantum-Dot-Based Detectors	166
5.3.1 Detector Designs and Principles of Operation	167
5.3.2 Photon-Number-Resolving Detection	172
5.3.3 Modeling Photoconductive Gain	175
5.3.4 Conclusions	179
References	180

5.1 INTRODUCTION

The development of improved single-photon detectors is crucial to the advancement of quantum information technologies and measurement science.

Desired characteristics for detectors include high detection efficiency (DE), low dark count and afterpulsing probability, fast response time, and low timing jitter. In addition, certain applications require detectors that are not only sensitive to single photons, but that can also resolve the number of incident photons that arrive simultaneously. Such photon-number-resolving detectors are a key enabling technology for linear-optics quantum computing [1], impact the security of quantum communications [2], and provide crucial measurement tools for studying the quantum nature of light [3–7]. There are also demands for optical receivers that can do more than count photons. For instance, detectors capable of preserving the spin of photogenerated carriers are applicable to systems that encode information in the polarization states of photons. Such detectors could lead to the development of quantum repeaters [8], which enable quantum communication over distances far beyond the fiber attenuation length. Improving known detector technologies and searching for new methods of photodetection are both active areas of research aimed at realizing higher performing detection systems that are compact, robust, and easy to operate.

In this chapter, we describe some of the novel semiconductor-based single-photon detectors, including visible-light photon counters (VLPCs), solid-state photomultipliers (SSPMs), and quantum-dot-based detectors. We will cover the basic operation principles of these devices, experimental results that have demonstrated their unique attributes, concrete models to quantify their performance, and future prospects for these detector technologies.

5.2 SOLID-STATE PHOTOMULTIPLIERS AND VISIBLE-LIGHT PHOTON COUNTERS

5.2.1 Introduction

In a typical photodetector device, the incident photon is converted into an electrical signal, most commonly electrical current, to be sensed using amplifier circuits. In a photodiode, an incident photon is converted into an electron-hole pair generated across the bandgap in the host semiconductor material, which is then detected using a transimpedance amplifier. In a photoconductive detector, the absorption of a photon changes the conductance of the current-carrying channel, which is then detected by the readout circuit. When detecting a single photon, the device typically requires an internal amplification mechanism that brings the original (primary) signal generated by the photon to above the noise level of the readout amplifier circuit. Simple analysis shows that for those devices utilizing electrical current as the readout signal, the internal gain must be at least $\approx 10^4$ to overcome the thermal noise of the transimpedance amplifier operating at room temperature, when the total shunting capacitance of the device (plus any stray capacitance) is in the $\approx$pF range [9]. Examples of devices using this approach include single-photon avalanche diodes (SPADs) and photomultiplier tubes (PMTs). Use of low temperature devices or micron-scale devices connected to readout electronics without much stray capacitance can in principle overcome these constraints.

A sensitive photon detector utilizing impurity-band conduction (IBC) at low temperatures was invented in the 1980s [10]. This approach is especially suitable for detecting far-infrared photons (8–30 μm wavelength range), and extensive modeling and device development followed [11,12]. Soon it was recognized that a high performance single-photon detector could be constructed using impact-ionization of the impurity band as the gain mechanism, which led to the invention of the SSPM by Petroff and Stapelbroek, capable of detecting single photons in the wavelength range of 0.4–28 μm [13,14]. The VLPC is a variation of SSPM optimized for enhanced DE in the visible and suppressed response in the IR for applications in high-energy physics [15,16]. The absolute detection efficiency of SSPMs was first tested in the context of quantum optics experiments by Kwiat *et al.* at UC Berkeley in collaboration with the former Rockwell group that had developed the SSPMs and VLPCs [17–19]. The noise, detection efficiency, and multi-photon detection capabilities of VLPCs were characterized by Kim *et al.* at Stanford University in collaboration with the same researchers [20–22]. Both types of devices have demonstrated promising attributes desired in quantum information experiments, including high DE (up to ≈88% for the detector system, with >95% projected inferred device efficiency), relatively low timing jitter, and photon-number-resolving capability with <1% bit error rate (BER) for resolving a single-photon and two-photon states measured in the visible spectrum (702 nm).

While the VLPC has attractive performance features, it was designed for a specific target application (particle tracking in high-energy physics experiments) and leaves room for further optimization toward applications in quantum information processing. First, since the dominant photon absorption process is across the silicon bandgap, the sensitive response is limited to visible wavelength range (0.4–1.0 μm). Extending the operational spectrum of these devices into the ultraviolet and the infrared is interesting for ion-trap quantum information processing [9,23] and long-distance quantum communication experiments [24,25]. Second, the VLPC features relatively large dark counts when operated at maximum DE. Third, the timing jitter of the VLPC [26] is comparable to conventional SPADs [27,28]. Improved SPAD designs with a thinner active layer reduce the transit-time spread of the carriers and lead to much smaller timing jitter (as low as 20 ps, typically at the expense of lower DE) [29,30], and similar adjustments could be made for VLPCs.

In this section we provide the operating principles and demonstrated device performance of SSPMs and VLPCs as single-photon detectors. We summarize a physical model that describes the carrier transport and multiplication gain in the VLPCs as presented in Ref. [31], which can be utilized to improve the device metrics relevant for various applications in quantum information processing.

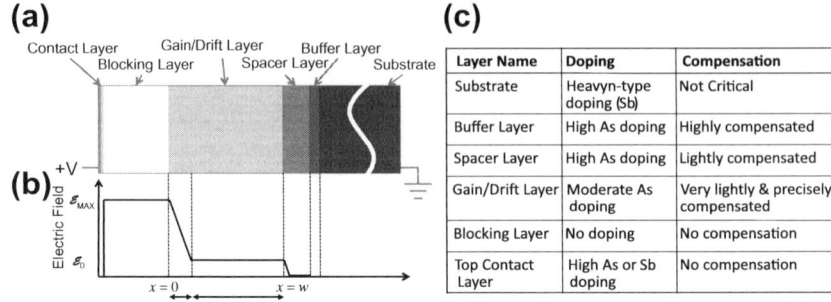

FIGURE 5.1 Schematic of (a) the VLPC layer structure and (b) resulting electric field profile under operating conditions. (c) Table describing the design requirement for each layer. *Reprinted with permission from [31]. Copyright 2011, SPIE.*

We will conclude with some proposals for new device geometries that might achieve high DE in the telecom wavelength range.

5.2.2 VLPC Structure and Operation

Brief descriptions of the VLPC operating principle are outlined in several publications [20, 26, 32]. Here, we present a more quantitative description of the carrier transport mechanism [33, 34], which allows us to construct a concrete model as described in detail in Ref. [31]. Based on this model, we can develop a software tool to analyze the device behavior that will aid in future optimization of the device design.

Figure 5.1a and table in Fig. 5.1c show the schematic structure of VLPC layers grown on a degenerate n-type silicon substrate, typically doped with antimony (Sb). The first layer is a thin n-type buffer layer heavily doped with arsenic (As) as donors, relatively highly compensated by adding p-type dopant boron (B) as acceptors. The second layer is a thicker n-type spacer layer, heavily doped with As and moderately compensated with B. The third layer is a thick n-type layer, moderately doped with As, that functions as both the gain and drift layer (described below). The compensation (using B) in this layer is very light, and the level must be precisely controlled for proper device operation. The doping and compensation levels of this layer determine many critical performance characteristics of the VLPC, such as gain, dark count rate, timing jitter, and maximum count rate. The next layer is an intrinsic blocking layer, without any doping or compensation. The final layer is a very thin, highly n-type doped top contact layer. An SSPM has similar device structure, but the blocking layer is substantially thinner.

The buffer layer and the top contact layer are degenerately doped, and remain highly conductive at all temperatures. At the device operating temperature of

<10 K, neither the donors (with doping density of N_D) nor the acceptors (density N_A, of B compensation) in the moderately doped gain/drift layer can be thermally excited so there are no free-carriers in the conduction or valence bands. In the absence of compensating acceptors, all donors would remain neutral and the material would behave as a perfect insulator. In the presence of compensating acceptors, some donors (N_A of them) lose their electrons to nearby acceptor atoms and become positively charged (these charges are called impurity-band holes, or D^+ charges). When the donor doping density N_D is sufficiently high, electrons from other nearby donors can hop or tunnel onto these ionized donors, effectively moving the D^+ charges from one donor atom to another. This conduction mechanism for D^+ charges is referred to as impurity-band conduction (IBC). The density of D^+ charges leading to the IBC is determined by the compensating acceptor density N_A, while the mobility of the D^+ charges is determined by the donor density N_D that determines the mean distance between the donors the charge has to hop/tunnel.

When a positive bias is applied to the top contact of the device, an electric field profile develops across the device as shown schematically in Fig. 5.1b. While the electric field in the intrinsic blocking layer is high ($\mathcal{E}_{MAX}$), there are only very low levels of (unintended) dopants here and all carriers generated into the conduction or valence band travel across this region with high mobility. In the gain/drift layer, the electric field from the bias will push the D^+ charges toward the substrate, leaving a negatively charged depletion layer (the negative charge is the excess electron, contributed by the donors and trapped at the compensating acceptors). The electric field decreases linearly through the depletion layer according to the Poisson equation, and reaches a steady-state value $\mathcal{E}_D$ (described below). The electric field in the depletion layer remains high, and any free-carriers in the conduction band can impact-ionize neutral donors to create additional conduction-band electrons. For this reason, the depletion layer is also referred to as the gain layer. The width of the depletion (gain) layer W_d can be estimated by solving the Poisson equation using N_A as the charge density, and results in the relationship $W_d \approx 1/\sqrt{N_A}$. The remainder of the gain/drift layer in Fig. 5.1a sustains a relatively low field $\mathcal{E}_D$, and is referred to as the drift layer. The electric field is quickly diminished once it reaches the spacer layer, as the level of compensation is higher (causing higher D^+ density).

Critical to the device operation is the mechanism of finite bias current generation in the device, which serves as the origin of the electric field $\mathcal{E}_D$ in the drift region. At zero bias, the only mechanism for the neutral donors (with electron energy level E_D below the conduction band minimum) to contribute carriers to the conduction band is through thermal activation (Fig. 5.2a). When a large field $\mathcal{E}$ is applied such that the bottom of the conduction band tilts substantially, the rate of carrier generation from the neutral donor to the conduction band is increased substantially due to the Poole-Frenkel effect

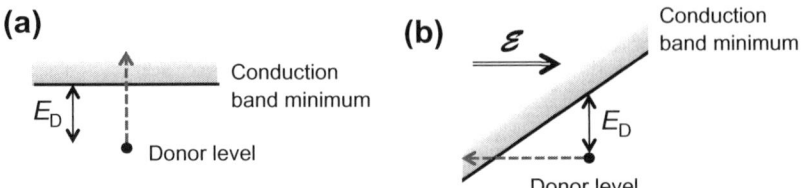

FIGURE 5.2 Mechanisms for bias current generation. (a) At zero field, a carrier in the conduction band can only be generated by thermal excitation of the donor atoms. (b) At finite bias field $\mathcal{E}$, a conduction-band electron can be generated by direct (or thermally assisted) field ionization. *Reprinted with permission from [31]. Copyright 2011, SPIE.*

(Fig. 5.2b). The enhancement factor is given by [34,35]

$$F = \left(\frac{k_B T}{\beta\sqrt{\mathcal{E}}}\right)^2 \left[1 + \left(\frac{\beta\sqrt{\mathcal{E}}}{k_B T} - 1\right)\exp\left(\frac{\beta\sqrt{\mathcal{E}}}{k_B T}\right)\right] + \frac{1}{2}, \quad (5.1)$$

where k_B is the Boltzmann constant, T is the temperature, $\beta \equiv \sqrt{e^3/\pi\epsilon}$, e is the electron charge (in cgs units), and ϵ is the electric permittivity of silicon. At the operating temperature of $T \approx 7$ K, one estimates that $F \approx 10^{11}$ for high-field region ($\mathcal{E} \approx 8000$ V/cm) and $F \approx 10^5$ for low-field region ($\mathcal{E} \approx 2000$ V/cm), although the details of the field ionization model depend strongly on local field distribution at high fields. It is reported that the observed activation energies are smaller than that predicted by this simple model by about a factor of 2, and the difference is attributed to non-Coulomb impurities and impurity-band effects [34].

Since the thermal generation rate is so much higher at the high-field regions, many electrons are generated in the gain layer close to the blocking layer while fewer electrons are generated close to and within the drift layer. The large number of conduction-band electrons generated near the top of the gain layer quickly reaches the blocking layer. The (fewer) electrons generated deep in the gain layer or in the drift layer accelerate substantially in the gain layer due to the high field, and impact-ionize other neutral donors. This process can quickly multiply the number of electrons in the conduction band, creating an avalanche of a substantial number of electrons ($>2 \times 10^4$) for each initiating electron. If such an avalanche process is triggered by a photogenerated primary carrier, the resulting electrical pulse corresponds to a photon detection signal. When the avalanche is triggered by a thermally generated carrier deep in the gain layer or in the drift layer, the pulse is indistinguishable from a photogenerated pulse and results in a dark count pulse.

One can calculate the impact-ionization gain experienced by each thermally generated electron. Given the electron-donor impact-ionization coefficient as a function of position $\xi(x)$ ($x = 0$ is chosen to be the boundary between the blocking layer and gain layer, as shown in Fig. 5.1b), the avalanche gain $M(x)$

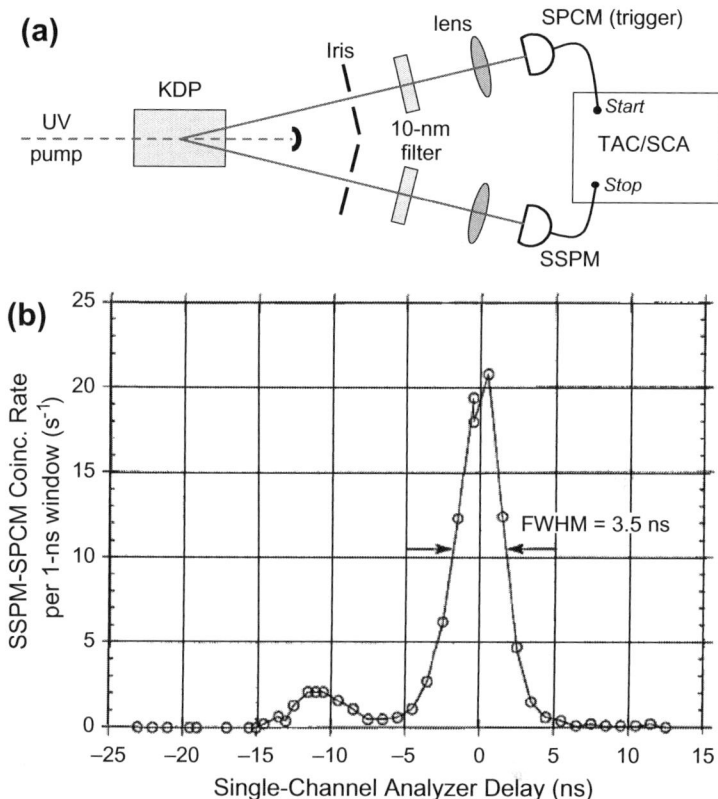

FIGURE 5.3 (a) Schematic of method to measure absolute detector efficiency and timing jitter. 351 nm laser light pumps a nonlinear KDP crystal, producing pairs of correlated photons. Detection of one photon by the "trigger" detector (here the SPCM) heralds the presence of the sister photon at the SSPM. (b) Time-correlation profile showing coincidences between an SSPM at a singles rate of 5×10^4 s^{-1} and a trigger SPCM at 255 s^{-1}. The single-channel analyzer window corresponds to 1.0 ns. UV: Ultraviolet, KDP: Potassium dihydrogen phosphate, SPCM: Single-photon counting module, TAC/SCA: Time-to-amplitude converter/Single-channel analyzer, FWHM: Full-width at half-maximum. *Reprinted with permission from [18]. Copyright 1994, Optical Society of America.*

for an electron generated at a position $x > 0$ can be calculated by [12,34]

$$M(x) = \exp\left(\int_0^x \xi(x')dx'\right). \quad (5.2)$$

The impact-ionization coefficient depends on parameters like the neutral donor density, local electric field, kinetic energy of the electron, donor energy level, and the cross-section of the impact ionization. A simple model for the impact-ionization coefficient is given by $\xi(x) = N_D \sigma_I \exp[-\mathcal{E}_c/\mathcal{E}(x)]$, where σ_I is the impact-ionization cross-section, and $\mathcal{E}_c$ is the critical field for impact ionization [12,34]. Parameters relevant for the description of the avalanche gain at low temperatures and high fields are not readily available for silicon, and further measurements are required to obtain an accurate model. The integral in Eq. (5.2) can be evaluated either numerically or analytically for the electric field distribution (such as shown in Fig. 5.1b), over the integration range of interest $0 \leq x \leq w$. Once $M(x)$ is evaluated, one can calculate the total current density of the device as [12]

$$\mathcal{J}_B = M(0)\mathcal{J}_p(0) + M(w)\mathcal{J}_n(w) + q \int_0^w g(x)M(x)dx, \tag{5.3}$$

where $\mathcal{J}_{n(p)}(x)$ denotes the current density due to conduction-band electrons (impurity-band holes, D^+) at location x, q is the absolute value of electron charge, and $g(x)$ is the generation rate of electrons at location x (that includes both thermal and photogeneration rates). This total current has to be constant throughout the device to satisfy the continuity equation. The bias current of the VLPC device is given by Eq. (5.3) in the absence of photogeneration, i.e., when $g(x) = G_0(x)$, the thermal generation rate. The current density $\mathcal{J}_B$ in the drift layer is mostly contributed by the D^+ holes, since the generation rate into the conduction band is low and the avalanche gain is small in this region due to low electric fields. The bias current in this region creates an electric field $\mathcal{E}_D$ via Ohm's law $\mathcal{J}_B \simeq \sigma_D \mathcal{E}_D$, where σ_D is the mobility of D^+ charges in the impurity band. One can estimate the carrier mobility σ_D of the IBC from the density of dopants N_D and N_A.

An electron generated by thermal activation or impact ionization leaves an ionized donor (D^+ charge) behind. Since the mobility of D^+ charges for the IBC is orders of magnitude lower than that of conduction-band electrons, they drift slowly toward the substrate contact. For the fast-moving electrons, the existence of D^+ charges is perceived as a spatial distribution of static positive charge that negates the applied field and reduces the local electric field. This space-charge effect is transient (i.e., disappears when the D^+ charges are eventually transported to the substrate), but saturates the avalanche gain given by Eq. (5.2). This effect is partially responsible for limiting the size of an avalanche in response to an incident photon (or dark count) in VLPCs, resulting in low multiplication noise [20].

5.2.3 SSPM and VLPC Performance

In this section, we summarize the reported performance characteristics of the SSPM and VLPC devices measured by the authors' research groups to date, as presented in Refs. [17–19,36,22,21,26]. While VPLCs have been made available to a select research groups in quantum optics over the past two decades, the availability of the original SSPM devices has been much more limited. Most of the device characterizations reported here are for high performance VLPCs; any measurements of the SSPM devices are noted explicitly.

5.2.3.1 Detection Efficiency, Dark Counts, and Count Rate

The potential to achieve high DE single-photon detectors using the SSPM [13,14] and VLPC [15] was highlighted in the early development of these devices, demonstrating the operating principles and initial applications. Using correlated photon pairs from a spontaneous parametric down-conversion source, one can make careful measurement of the absolute single-photon detection efficiency [37,38]. In this technique, the finite detection efficiency of a single detector (e.g., an SSPM) can be determined by normalizing the coincidence count rate ($\propto \eta_{\text{SSPM}} \eta_{\text{trigger}}$) between two detectors looking at the pairs to the singles count rate ($\propto \eta_{\text{trigger}}$) at the "trigger" detector (see Chapter 8). One of the main sources of inefficiency arose from reflection of the incident photons off the detector surface itself (and off the ends of the coupling fibers used to transport the incident light to the detectors). To minimize this, a retro-reflection mirror was used to give most of the light reflected back a second chance to be detected; this improved the detection efficiency by a relative 17%, yielding a system DE of 70.9(1.9)% [17,18]. In contrast to SPADs, no observable afterpulsing was seen in the SSPM devices.

Since SSPMs and VLPCs are sensitive to photons with wavelengths up to 28 μm (though the latter were specifically designed to be less sensitive at these longer wavelengths), one main challenge toward a detection system with high DE is to construct a set of radiation shields in the cryostat that filters out room-temperature thermal radiation in the 2–30 μm range, while maximizing the transmission of the photons targeted for detection. For the system used to operate the SSPM at UC Berkeley, the input photons were coupled to the detector through a poly methyl methacrylate (PMMA) fiber, with the detector end cooled to ≈7K. However, a careful post-measurement of these plastic input fibers revealed substantially more loss than expected, perhaps due to thermal-cycling-induced microcracks in the fibers; if this loss were eliminated, the inferred SSPM DE$_{\text{int}}$ = 96(3)% at 660 nm [19]. One method suggested to avoid such transmission losses through the in-coupling fiber is to use *glass* fibers (which have low loss in the visible wavelength ranges of interest) of sufficiently small core diameter so that longer wavelengths are simply not propagating modes, i.e., they leak out into the fiber cladding; preliminary investigations have shown some success with this technique.

A different option to filter out background thermal radiation is to switch to free-space coupling, incorporating acrylic plates, which strongly absorb photons beyond ≈2 μm. For the cryostat system used to characterize the VLPCs at an operating temperature ≈6.5 K at Stanford, a set of radiation shields at 77 K and 4 K was constructed with windows made out of acrylic plates antireflection (AR) coated near 700 nm [22]. The windows were secured to the metallic radiation shields using an indium seal, and any residual holes in the shields were stuffed using black felt to cut down on the light leakage. The throughput of the three acrylic windows was measured to be ≈97% at the testing wavelength of 694 nm. The rate of room-temperature photons impinging on the detector surface in

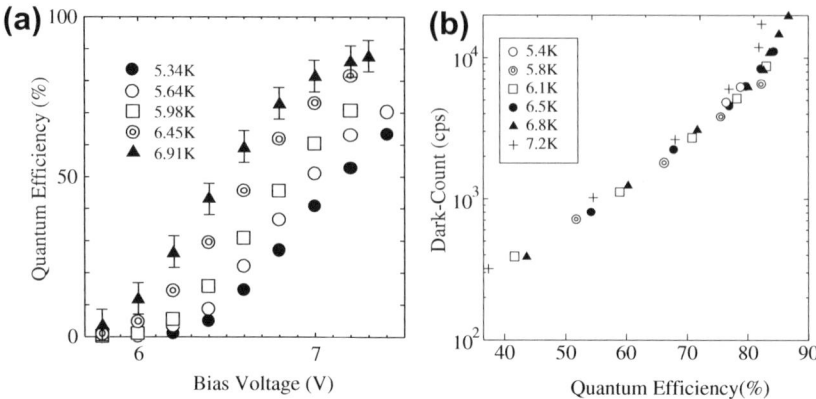

FIGURE 5.4 (a) DE (denoted as quantum efficiency) of a VLPC as a function of the operating temperature and bias voltage. (b) VLPC dark count as a function of DE (denoted as quantum efficiency), measured by varying the operating temperature and bias voltage. *Reprinted with permission from [22]. Copyright 1999, American Institute of Physics.*

the absence of the radiation shields was estimated to be $\approx 10^{15}$ photons per second in the relevant wavelength band (<28 μm), which was eliminated nearly completely in the setup, verified by the dark count measurements (see below).

The DE of the VLPC device was optimized near 550 nm, adequate for scintillating-fiber particle-tracking experiments [15], and featured $\approx 15.6\%$ reflection at 694 nm. As above, a spherical refocusing mirror was used to recollect the reflected photons back onto the detector surface for a second chance of being detected. Using this approach, the impact of inadequate AR coating was reduced to $\approx 2.5\%$ of the net DE reduction.

A weak coherent laser beam at 694 nm attenuated using calibrated neutral density filters (with uncertainty of $\pm 5\%$) was focused onto the VLPC surface through the windows in the cryostat system. The DE of the VLPC device was measured as a function of operating temperature and bias voltage by counting the number of detected photons (electrical signals) for a given input light level. Figure 5.4a shows the DE of the system as a function of the bias voltage and operating temperature. The detection efficiency decreases as the operating temperature is decreased, but it can be compensated by increasing the bias voltage. This is due to the increasing resistance of the drift layer at lower temperatures that acts as a series resistance, effectively reducing the bias voltage across the gain layer. A maximum detection efficiency of 88% was observed at a bias voltage of 7.3 V at 7 K. Upon further increase in the bias voltage, the device entered a breakdown state (large amount of current flowing through the device) and did not function as a single-photon detector. The breakdown voltage did not depend appreciably on the operating temperature.

The dark count rate of the device was measured as a function of the bias voltage and operating temperature. If the main contribution of the dark counts comes from the background photons from room-temperature thermal radiation, one would expect the dark counts to scale linearly with the detection efficiency as the incident thermal photons remain constant through the experiment. Instead, we find that all data points collapse onto a universal curve where the dark counts increase roughly exponentially with the detection efficiency (Fig. 5.4b), a strong indication that the major source of the dark counts is internal to the device. This observation suggests that the main source of dark counts is the thermal generation of electrons that trigger the avalanche multiplication, enhanced by the Poole-Frenkel effect described in Eq. (5.1). We see mild enhancement of this thermal generation at the highest operating temperatures at high voltage (7.2 K data points in Fig. 5.4b), that does not accompany DE enhancement.

After a photon detection event, the local area over which the electrical pulse is generated still contains a substantial number of D^+ charges. Since the recovery of these charges is slow, the region of the photon detection remains inactive for subsequent photon detection due to the reduction of the local electric field, until the concentrations of D^+ charges recover to the pre-avalanche levels. While this "dead time" remains local and other regions of the VLPC remain active for photon detection, it leads to saturation and limits the device DE at higher levels of incident photon flux. Experimentally, reduction of device DE by $\approx 10\%$ was seen at photon counting rates of $\approx 1 \times 10^6$ s^{-1}. Simple models that divide the VLPC into many parallel independent photon detectors with substantial dead times (corresponding to the D^+ charge recovery time) have been proposed to explain the saturation effects, although the estimates for the model parameters corresponding to the recovery times and the physical extent of the avalanche zone corresponding to a single-photon detection event still have substantial variations [32,39].

5.2.3.2 Timing Jitter

The SSPM timing jitter was determined using down-conversion pairs, whose intrinsic time correlation—typically <1 ps—allows for precise relative timing measurements. As shown in Fig. 5.3b, a jitter of ≈ 3.5 ns was reported by measuring coincidences between an SSPM and a commercial single-photon counting module (EG&G SPCM-200-PQ); the timing resolution in these early measurements was likely limited by the bandwidth in the readout electronics [17,18]. The VLPC timing jitter was investigated using the experimental setup shown in Fig. 5.5a [26]. To measure the jitter over a wide wavelength range, a supercontinuum light source was used, generated from a mode-locked Ti:sapphire femtosecond laser propagating through a highly nonlinear photonic crystal fiber. Part of the pump laser power was directed through a dichroic beam splitter to a fast photodiode, to provide a timing reference for the incident photon. The signal from the photodiode was used to trigger the start of a time interval measurement. The rest of the supercontinuum source is fed into a grating monochromator after a series of color filters and neutral density filters, to select a narrow spectral band of photons to be delivered to the VLPC in

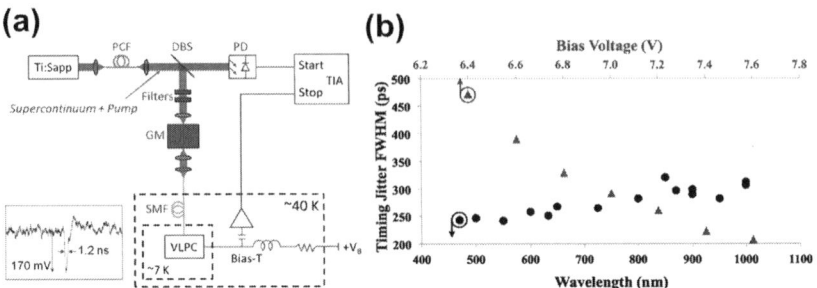

FIGURE 5.5 (a) Schematic of the experimental setup to measure the timing jitter of the VLPC. PCF: photonic crystal fiber, DBS: dichroic beamsplitter, PD: fast photodiode, Filters: color filters and neutral density filters, GM: grating monochromator, SMF: single-mode fiber, TIA: time-interval analyzer. (b) FWHM values of the timing jitter, measured as a function of the bias voltage at a fixed wavelength of 633 nm and temperature of 7.0 K (gray triangles), and as a function of incident photon wavelength at a fixed bias of 7.2 V and temperature of 7.0 K (black circles). *Reprinted with permission from [26]. Copyright 2010, IEEE.*

a cryostat through a single-mode fiber. The electrical output pulse from the VLPC generated by the incident photon is amplified using a cold ($\approx$40 K) broadband amplifier (shown in the inset of Fig. 5.5a), and provides the stop trigger to the time-interval measurement. The time-interval analyzer (TIA) takes the histogram of the time interval between the pump pulse and the photon detection event, providing a measure of the timing jitter in the VLPC photon detection process.

Figure 5.5b shows the full-width at half-maximum (FWHM) of the timing jitter measured as a function of the bias voltage and incident photon wavelength. The photon counting rate was kept at $\approx 50 \times 10^3$ s^{-1}, well below the saturation levels. The timing jitter monotonically decreases from $\approx$470 ps to 200 ps as the bias voltage is increased from 6.4 V to 7.6 V, at 7.0 K. It also shows mild dependence on the wavelength of the incident photos, increasing from below 250 ps at 400 nm to just above 300 ps at 1000 nm, when biased at a fixed voltage of 7.2 V.

The drift velocity of the conduction-band electrons in the drift layer decreases dramatically to $\approx 1.3 \times 10^6$ cm/s, (for a typical E-field of 1 keV/cm in the drift region) due to scattering with neutral donors, as explained by Erginsoy's formula [40]. Depending on where this secondary electron was generated in the device, the onset of avalanche gain can be delayed due to such transit-time variations leading to substantial timing jitter of $\approx$250 ps, which forms the "baseline" timing jitter shown in Fig. 5.5b [26]. The wavelength dependence can be explained by additional transit-time spread of the photogenerated holes to reach the drift layer. The absorption length of the photons near 400 nm is $\approx$200 nm, meaning that all photons are absorbed roughly within this depth from the surface [41]. A hole resulting from photon absorption will drift toward the drift layer where a secondary electron is generated; since the location of the photon absorption (and thus the hole generation) is concentrated in a narrow region of the blocking layer, the transit time is fairly uniform. Near 1000 nm, the absorption length of the photons increases to $\approx$100 μm, meaning the photon absorption is spread over the entire depth of blocking layer and the gain/drift layer. The hole drift velocity is $\approx 1 \times 10^7$ cm/s, so the transit-time spread of the holes over the thickness of the blocking/gain layer ($\approx$15 μm) is $\approx$150 ps. This is consistent with the increase in the FWHM of $\approx$70 ps over the wavelength range. If the photon is absorbed in the drift layer, the photogenerated primary electron triggers the avalanche as it propagates through the drift and gain layers toward the top contact. The slow drift velocity of these electrons leads to much longer transit times for these photon detection events, resulting in a long tail in the timing jitter measurement at longer wavelengths. As the bias voltage is lowered, the photo-generated hole will have to propagate further into the drift region before it can impact-ionize the secondary electron. The larger spread for the location of secondary electron generation leads to long timing jitter due to low electron drift velocity ($\approx$370 ps per 10 μm of travel at the drift velocity of $\approx 2.7 \times 10^6$ cm/s). The key to reducing the timing jitter further is to (1) increase the mobility of electron transport in the drift layer, (2) reduce the thickness of the drift layer, or (3) ensure that the hole generates the secondary electron within a well-defined region in the drift layer.

5.2.3.3 Low Multiplication Noise and Photon-Number-Resolving Capability

The gain mechanism for SSPMs and VLPCs fundamentally differs from the SPADs in two major ways. First, the impact-ionization process that produces the extra carriers occurs between the conduction band and the impurity band. This has several consequences: (1) The low mobility of the D^+ charges ensures that they do not gain sufficient kinetic energy to trigger an impact-ionization event. (2) Since the electrons initiating the impact-ionization process require much less kinetic energy, the device operates at much lower bias voltages than a typical SPAD. (3) An electron generated by a photon absorption or an impact-ionization event needs to accelerate for a while before gaining sufficient energy to initiate another impact-ionization event, so there is an effective "dead time" between impact-ionization events. This leads to non-Markovian statistics in the impact-ionization process [43], allowing for a low multiplication noise compared to the standard APD model [42]. Second, the D^+ charges generated as a result of the impact-ionization process relax very slowly compared to the electrons, leading to a space-charge effect that acts to compensate the local electric field as the avalanche grows. This leads to a self-quenching of the avalanche multiplication. Due to these two mechanisms, SSPMs and VLPCs feature near noise-free avalanche multiplication [20]. Figure 5.6 shows the measured pulse height distribution of a VLPC, which fits well to a gamma distribution using an

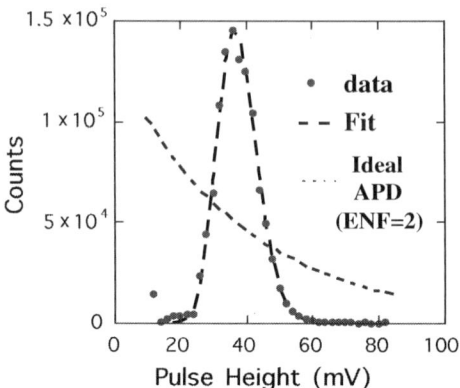

FIGURE 5.6 Pulse height distribution of the VLPC. Ideal avalanche photodiodes (APDs) following McIntyre's noise model [42] feature an excess noise factor (ENF) of 2, and the pulse height distribution follows an exponential distribution (blue dotted line). Similar behavior is expected for SPADs. The experimental data (red dots) fit very well to a gamma distribution corresponding to ENF = 1.025 (black dashed line). (For interpretation of the references to color in this figure legend, the reader is referred to the web version of this book.) *Reprinted with permission from [36]*.

excess noise factor (ENF) of 1.025. Ideal noiseless multiplication corresponds to ENF = 1, while the best possible performance of SPAD multiplication under Markovian noise statistics correspond to ENF = 2 (exponential distribution, shown in blue dotted line). This data indicates that the pulses generated by a photon detection event in a VLPC feature very narrow dispersion.

Unlike Geiger-mode SPADs where the entire device breaks down upon detection of a photon [44], we have seen that a photon detection event only creates a local avalanche in VLPCs, leaving the remainder of the device available for detection of other photons. Such local avalanche gain combined with near-zero multiplication noise results in photonnumber resolving capability of VLPCs [16,21,45], and similar behavior is to be expected from SSPMs. Figure 5.7 shows the experimental setup used to quantitatively measure the multiphoton detection capability of VLPCs. A nonlinear BBO crystal was pumped using a 351 nm Ar-ion laser in type-II phase matching configuration, and a pair of photons were generated at 702 nm via degenerate spontaneous parametric down-conversion process. The photons were spatially separated, and combined using a polarizing beam splitter after a finite time delay is introduced to one of the photons using a delay line. The two photons were incident on the VLPCs in the cryostat.

The pulse height distribution (in log scale) of the detected photons when one of the beam paths is blocked is shown in Fig. 5.8a. It features a narrow distribution similar to that shown in Fig. 5.6. When both beam paths are incident on the VLPC with no time delay, a second peak at twice the pulse height shows up as shown in Fig. 5.8b. This second peak indicates a two-photon detection event. By fitting each peak using a gamma distribution with adequate ENF and

Chapter | 5 Novel Semiconductor Single-Photon Detectors

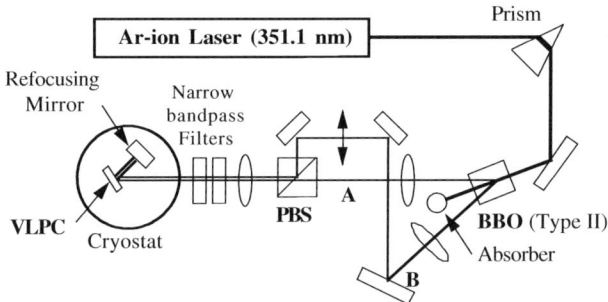

FIGURE 5.7 Experimental setup to measure the multi-photon detection capability of the VLPC. PBS: Polarizing beamsplitter. *Reprinted with permission from [21]. Copyright 1999, American Institute of Physics.*

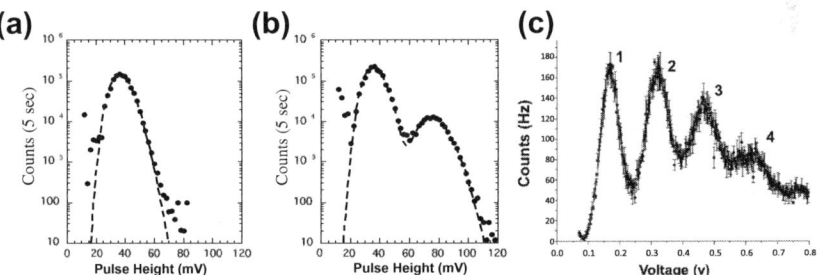

FIGURE 5.8 Pulse height distributions showing VLPC photon number resolution. (a) Only PDC photon path incident on detector, so only single-photon detection events are seen. (b) Both PDC photon paths are incident on detector with the same time delay so single- and two-photon detection events are seen (log-scale). *Reprinted with permission from [21]. Copyright 1999, American Institute of Physics.* (c) Pulse height distribution in linear scale for multi-photon detection events (1–4 photons).

considering the overlap of the two distributions, we estimated that the bit error rate of identifying a single-photon event as a two-photon event or *vice versa* is less than 1%, if the threshold for the photon-number discrimination is set at an adequate level. Figure 5.8c shows a pulse height distribution of measuring an attenuated laser beam (featuring Poisson distribution in photon number) in linear scale, demonstrating the VLPC's ability to detect multi-photon states.

5.2.4 Quantitative Model and its Current Limitations

Based on the carrier transport mechanisms discussed in Section 5.2.2, one can construct a simulator that will enable the estimation of various device performance characteristics, such as DE, gain, dark counts, and timing jitter at various photon wavelengths. In the presence of thermally generated carriers resulting in substantial bias current ($\approx$1–10 μA/cm^2 in VLPCs), we treat the photogenerated carriers and their multiplication gain as a perturbation to the steady-state operation. For a given doping profile, layer thicknesses, operating temperature, and bias voltage condition, we utilize the following procedure to find a steady-state solution:

Step 1. Solve Poisson's equation in the absence of bias current to calculate the initial electric field distribution.

Step 2. Based on the resulting field distribution, use the Poole-Frenkel enhancement factor [Eq. (5.1)] to calculate the thermal generation rate of conduction-band electrons at each location x.

Step 3. For each carrier generated, calculate the multiplication gain using Eq. (5.2) and the electric-field profile obtained in **Step 1**.

Step 4. Calculate the integrated current density using Eq. (5.3).

Step 5. Calculate the resulting electric field in the drift layer using its conductivity and Ohm's law.

Step 6. Using this electric field, solve Poisson's equation again to update the electric field distribution.

Step 7. Repeat **Step 2** through **Step 6** until the field distribution converges to a steady-state value.

At the end of this process, one would obtain the final electric-field distribution inside the device (Fig. 5.9a), avalanche gain (Fig. 5.9b) and generation rate of conduction-band electrons (Fig. 5.9c) as a function of the location x (of initial charge generation), as well as the total bias current. From this information, one can estimate all relevant performance characteristics of the VLPC device, and optimize the device design parameters (thickness and doping of each layer) to achieve desired device performance.

Currently, there are some limitations on this strategy that require further work. First, the multiplication gain calculated in **Step 3** saturates due to nonlinear space-charge effects, which have not been incorporated into the model. This is approximated to zeroth order by imposing an upper limit for the gain of each carrier. Second, the Poole-Frenkel enhancement factor calculated

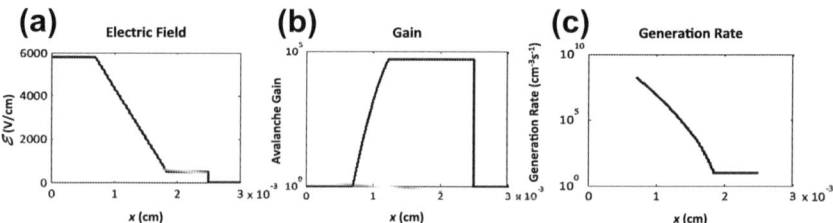

FIGURE 5.9 Example of calculated metrics of the VLPC device, such as (a) electric field profile, (b) gain of primary/secondary electron as a function of generation location x, and (c) thermal generation of carriers enhanced by Poole-Frenkel effect. *Reprinted with permission from [31]. Copyright 2011, SPIE.*

for an isolated donor may not describe the generation rate accurately. Other generation mechanisms such as Shockley-Read process have been suggested to bridge this discrepancy [34]. Third, the critical field $\mathcal{E}_c$ for impact ionization used in the calculation of gain [Eq. (5.2) in **Step 3**] has to be determined accurately as a function of the doping levels in the gain layer. Fourth, the conductivity of the drift layer necessary for calculating the field in this layer in **Step 5** has to be determined accurately. Of these, the Poole-Frenkel enhancement factor and the avalanche gain depend exponentially on the local electric field and the effective binding energy of the donors, and can dramatically affect the device operation characteristics. These parameters are extrapolated from the measurement of blocked impurity-band (BIB) far-infrared detectors operating at similar temperatures but at much lower electric fields (and low avalanche gain) [11,12,34]. Relevant measurement data in high fields do not exist today to establish a quantitatively accurate model for VLPC operation. Further experimental work on establishing these basic parameters is necessary to model the ideal operating condition.

5.2.5 New Opportunities for VLPCs

5.2.5.1 Extending Operating Wavelength Range

The DE of VLPC in the UV wavelength range is reduced due to the very large absorption coefficient of silicon below 400 nm. The photon is absorbed in the top 50 nm of the silicon material in the Ohmic contact layer, so the absorbed carriers cannot be injected into the gain/drift layers to trigger the electrical pulse. The DE of the VLPC can be enhanced by making the top contact layer very thin [41]. This can either be achieved by shallow implantation of dopants followed by UV laser annealing [46], or by growing highly doped ultra-thin contact layers using epitaxial growth techniques [47]. UV-enhanced VLPCs could find useful applications in fluorescence microscopy/spectroscopy and quantum information processing in trapped ion systems.

Although the SSPM features absorption in the infrared up to 28 μm wavelength via direct photoexcitation of the donor atoms, the detection efficiency in the 1–2 μm range remains too low to be of practical interest in a top (or bottom) illuminated device [14]. Even if the absorption coefficient is low, if one can make the depth of the absorption layer on the order of a millimeter, all photons will eventually be absorbed by the device layer. Such device geometry can be achieved by an edge illuminated device, as shown schematically in Fig. 5.10. In the proposed infrared photon counter (IRPC) device, the device layers are optimized to reduce the blocking, spacer and contact layers to minimum thicknesses, so that the bulk of the device layer can absorb the IR photons. The device can be grown on a silicon-on-insulator substrate to physically isolate the device layers from the handle substrate. Then, the active device layer is patterned into a waveguide shape of length ≈1 mm, and transferred onto an insulating transfer substrate with metallic contact pads already patterned on them. The

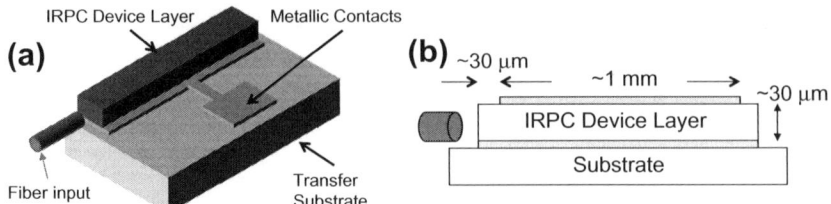

FIGURE 5.10 Schematic (a) perspective and (b) cross-sectional view of the proposed IRPC device. *Reprinted with permission from [31]. Copyright 2011, SPIE.*

incident photons are coupled at the edge of the waveguide IRPC device, and the photons remain inside the IRPC device layers due to total internal reflection. Over the length of the waveguide device, nearly all IR photons will be absorbed and the photogenerated carriers will each trigger an electrical pulse. A high DE single-photon detector in the telecom wavelength band can be constructed this way. Figure 5.10b shows how one can potentially suppress the response of this device to visible and far IR photons (wavelengths 3–30 μm). If one fabricates the metallic contact pads ≈30 μm from the illuminating edge of the detector to eliminate the bias fields there, the photons absorbed in this region will not trigger an electrical pulse; since most of the visible photons and far-IR photons will be absorbed in this region of the device, the device will not be sensitive to these photons. Ironically, the low absorption coefficient of the telecom photons will ensure that the number of signal photons absorbed in this low-response region is small, and provide the spectral selectivity of the IRPC.

5.2.5.2 Reducing Dark Counts and Balancing Maximum Count Rates

The dark counts of VLPCs can be reduced substantially by compromising the DE to some degree: while the device dark count rate is ≈20,000 s^{-1} at the operating condition that yields ≈88% DE, the dark count rate can be reduced to ≈500 s^{-1} when the device bias is reduced to operate at 50% DE level (Fig. 5.4b) [22]. Although these dark count rates may seem high compared to SPADs, one has to recall that the active area of a VLPC (≈1 mm in diameter) is much larger than a typical SPAD (≈50 μm in diameter). When scaled down to similar dimensions, a VLPC's typical dark count rate drops to about 50 s^{-1}, comparable to or less than most SPADs. However, unlike SPADs that can recover very quickly (≈50 ns) by actively quenching the avalanche breakdown following a photon detection event, the recovery time for the VLPC's active region from a photon detection event is much longer (estimated to be ≈ms) due to slow relaxation of D^+ carriers [39], leading to low saturation count rates. Thus, simply scaling the device to typical SPAD dimensions will reduce the saturation count rates to ≈2500 s^{-1}, substantially compromising the maximum count rate, as well as the photon-number-resolving capability. Figure 5.11 shows the results

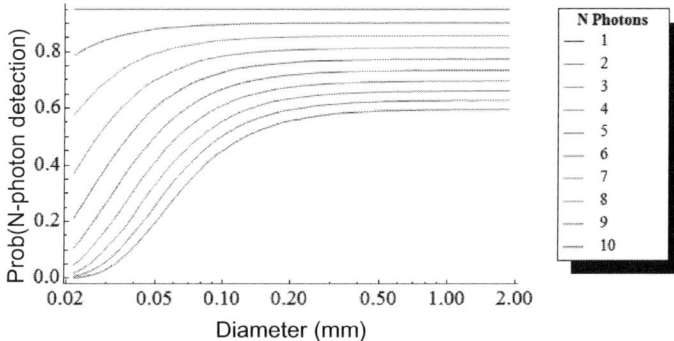

FIGURE 5.11 The calculated ideal probability of correctly identifying N input photons as a function of detector diameter, assuming a 95% detection efficiency and a 12 μm square effective avalanche filament area. This is determined from a simulation of photons in a Gaussian beam hitting random locations on the detector surface. A successful N-photon detection event is recorded if all photons create an avalanche, and no two avalanches overlap. Traces from bottom up are for $N = 10$ to 1, respectively.

of a Monte Carlo simulation of the probability of correctly identifying various numbers of input photons, as a function of detector diameter; even if the device diameter was reduced from 1 to 0.1 mm (presumably reducing the dark counts by a factor of 100), one could still reliably resolve several photons.

The recovery time of the D^+ carriers in the VLPC is determined by the mobility of these carriers, and can be improved substantially by increasing the donor and compensating acceptor densities N_D and N_A in the gain region. Unless done carefully, the increase in N_D could lead to higher dark count rates. One can optimize the device layers to include a gradient in the compensation charge profile in the gain region, so that larger gain, higher avalanche trigger probability, and fast relaxation of D^+ charge can be achieved while reducing the thermal generation rate of carriers that trigger dark counts.

5.2.5.3 Reducing Timing Jitter

The main source of timing jitter for VLPCs is the spread in transit time of secondary electrons through the drift layer toward the gain layer before it can trigger a full avalanche. The mobility of conduction-band electrons is reduced substantially in this region due to scattering with the high density of neutral donors, and the fact that the electric field is quite low [26]. To reduce the timing jitter by speeding up these electrons, one can reduce the thickness of the drift layer, reduce the donor density N_D, or increase the electric field in this region. However, all of these approaches are undesirable since they will either substantially increase the D^+ charge-recovery time or the dark count rate. An alternate strategy to reduce the timing jitter is to ensure that the secondary electron is generated before the primary hole reaches too deep into the drift region, thereby minimizing the distance the secondary electron has to travel before reaching the gain layer. This can be achieved by increasing the donor density and dramatically reducing the overall thickness of the drift layer. This must be accompanied by the reduction of the field to balance the increase in dark count rates.

5.2.6 Conclusions

In this Section, we described the operating principles of SSPMs and VLPCs in detail, summarized the reported performance characteristics, and outlined a systematic approach to quantitatively model the device behavior. The model has the potential to estimate important device performance metrics such as DE, gain, dark count rate, saturation levels, and timing jitter. Our ability to accurately model the dependence of device characteristics on various design variables could lead to optimized device design for target performance metrics. To improve the accuracy of the model, we need to measure the transport properties of conduction-band electrons and impurity-band holes (D^+ charges) under low temperature and high-field conditions, at various doping levels. We also discussed novel device geometries for extending the wavelength range over which a VLPC operates, including a waveguide detector device for high performance single- (and multiple-) photon detection in the telecommunication band.

5.3 QUANTUM-DOT-BASED DETECTORS

Within the past 15 years, a new method of detection has emerged that is well suited for photon-number-resolving detection and may lead to the development of spin-preserving detectors. This new method makes use of *photoconductive gain* [48] associated with persistent, three-dimensional confinement of photogenerated carriers in semiconductor diodes and field-effect transistors (FETs). In these devices, individual carriers excited by photons are confined to quantum dots (QDs) that function as optically addressable floating gates. The gain mechanism that provides these devices with single-photon sensitivity is not an avalanche or other multiplication process, but instead relies on a single trapped carrier influencing the transport of many mobile carriers flowing through a conductive channel. Because photoconductive gain can be produced independently and additively by multiple trapped carriers, this method of detection can be used for photon-number-resolving detection. Additionally, a number of aspects of QD-based structures make them promising spin-preserving single-photon detectors. The optical selection rules that govern the absorption of light in III–V semiconductor materials (e.g., GaAs) can be exploited to provide a direct connection between the polarization of the absorbed photon and the spin of the photoexcited carrier. Moreover, QDs have been shown to be effective storage centers for charge and spin, exhibiting spin lifetimes on the order of milliseconds [49,50].

5.3.1 Detector Designs and Principles of Operation

Here, we survey a number of the QD-based single-photon detectors that have been demonstrated to date. This is not meant to be an exhaustive listing, but is instead intended to highlight the various ways QDs can be used to detect individual photons. There has been an enormous amount of work aimed at using QDs for detectors of mid- and far-infrared radiation, and single-photon sensitivity has been reported (e.g., Ref. [51]). QD infrared photodetectors (QDiP) are the most prominent QD-based detectors in the literature, and there are many excellent review articles on these devices (e.g., Ref. [52]). In this section, we focus our attention on devices that use photoconductive gain to detect single photons in the visible and near-infrared spectral regions. We divide the devices into two groups, those that employ a single QD for photodetection and those that use an ensemble of QDs.

5.3.1.1 Single-Quantum-Dot Detectors

In 2003, Kosaka *et al.* demonstrated that a single-photoelectron transistor (SPT) that produces an electron trap using surface-mounted metal gates could be used to detect single photons with a wavelength of 1.3 μm. The composition of this detector is described in detail in Ref. [53] and is illustrated schematically in Fig. 5.12a. The transistor channel consists of an $In_{0.53}Ga_{0.47}As$ well bordered by $In_{0.52}Al_{0.48}As$ barriers. A Si-doped layer provides excess electrons to the conduction band that congregate in the $In_{0.53}Ga_{0.47}As$ well in the form of a two-dimensional electron gas (2DEG). The 2DEG provides conduction between the source and drain ohmic contacts of the transistor. The absorption region of the structure, where the electrostatically created trap resides, consists of a layer of $In_{0.53}Ga_{0.47}As$ sandwiched between two layers of InP. This design was chosen so that the effective g_e factor of the trap is approximately zero, making the two spin states nearly indistinguishable. The electron trap is produced when two semicircular gates on the surface of the structure are negatively biased, producing a two-dimensional potential minimum for conduction-band electrons in the absorption region. The barriers on either side of the absorption layer serve to confine electrons in the third dimension, effectively creating an artificial QD.

As illustrated in Fig. 5.12a, the device detects photons by trapping photoexcited electrons in the QD. The negatively charged gates play a double role in the operation of the detector; in addition to forming the 3D electron trap, they also pinch off the channel current by repelling the 2DEG electrons. When a photon is absorbed in the absorption region, it excites an electron from the valence band to the conduction band, leaving behind an empty state, or hole. The photogenerated electron is repelled by the negatively charged gates and

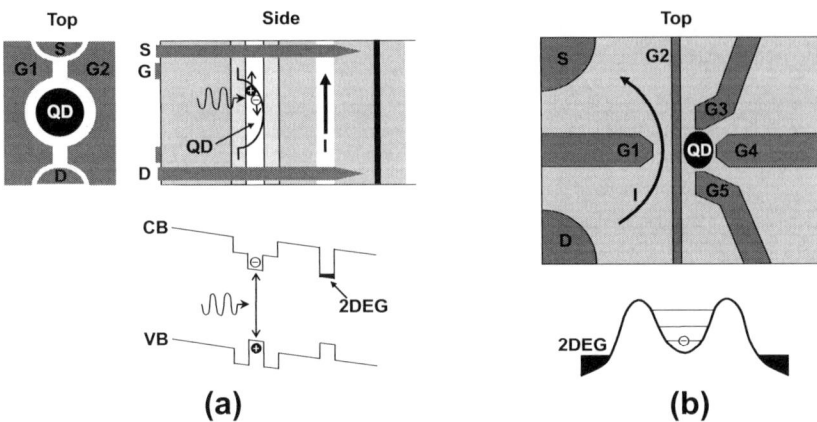

FIGURE 5.12 (a) Schematic diagrams of the composition and bandstructure of the SPT detector reported in Ref. [53]. The layers of the heterostructure are $In_{0.52}Al_{0.48}As$ (gray), InP (light gray), $In_{0.53}Ga_{0.47}As$ (white), and Si-doped $In_{0.52}Al_{0.48}As$ (black). (b) Schematic representations of the QPC device described in Ref. [54] and the electrostatically created QD. Source (S), drain (D), and gate (G) contacts are indicated in the diagrams, and CB and VB denote the conduction band and valence bands, respectively.

confined to the trap. Meanwhile, the hole is attracted to the gates and pulled laterally out of the trap area. While trapped in the dot, the electron further decreases the current flowing through the channel by enhancing the electric field produced by the gates. This condition of negative persistent photoconductivity is a distinguishing feature of the detector design. Kosaka et al. [53] demonstrated that a SPT with a 1 μm diameter circular active area could detect individual photons of light at the technologically relevant wavelength of 1.3 μm with DE of ≈1% when cooled to 4.2 K.

In 2005, a single-QD device with additional functionality was demonstrated [54] that used a $GaAs/Al_{0.3}Ga_{0.7}As$ quantum point contact (QPC) FET [Fig. 5.12b] to read out the charge trapped in the electrostatic dot. In this case, not only could a single photon be detected, but electrons could also be electrically emptied from the dot by modulating the voltage of one of the gates. The key layers of the device included a Si-doped n-$Al_{0.3}Ga_{0.7}As$ layer, and an undoped GaAs absorption layer, separated by a thin undoped $Al_{0.3}Ga_{0.7}As$ spacer. The 2DEG formed at the $GaAs/Al_{0.3}Ga_{0.7}As$ interface. As illustrated in Fig. 5.12b, a total of five metallic surface-mounted gates were used to form the QD at the interface and to produce the QPC (a constriction in the 2DEG) next to the dot. With a negative potential applied to the five gates, the QD is formed, trapping some of the 2DEG electrons, like a coastal pool on a beach. The QD can be emptied by using one of the gates (G4) as a plunger which can be activated to push electrons out of the dot one-by-one. As in the case of the SPT, when a photon is absorbed near the trap, the photogenerated electron is pushed into the center of the dot by the gate fields while the hole is pulled out. While trapped, the electron enhances the constriction of the QPC, reducing the amount of channel current. Once the photon is detected, the plunger gate can again be activated to empty the dot, ensuring that it is ready for another photon. Rao *et al.* reported a DE of ≈10% at an operating wavelength of 760 nm for a 400 nm diameter active area device cooled to 430 mK [54].

5.3.1.2 Multiple-Quantum-Dot Detectors

An alternative approach to electrostatically producing QDs using surface-mounted gates is to let the QDs form naturally during the semiconductor growth process. Self-assembled QDs are tiny islands of semiconductor material that can form during molecular beam epitaxy (MBE). In this process, layers of semiconductor material are deposited in an atomic-layer by atomic-layer fashion. QDs form when a thin layer of semiconductor is deposited on a thicker layer of differing lattice constant. The lattice mismatch between the two layers produces strain, which causes the thinner layer to separate into tiny islands, or dots. Created in this way, QDs can be strategically positioned in semiconductor devices with precisely controlled size, density, and material composition. Consequently, the allowed energy levels of the QDs can be tailored for specific applications.

In 2000, Shields *et al.* demonstrated a novel detector that consists of a $GaAs/Al_{0.33}Ga_{0.67}As$ modulation-doped FET that contained a single layer of self-assembled InAs QDs [55]. In this device, shown schematically in Fig. 5.13a, the QD layer is sandwiched between two layers of $Al_{0.33}Ga_{0.67}As$ and positioned in close proximity to the GaAs channel of the FET that contains the 2DEG. The device structure is specifically designed so that the channel conductance of the FET is extremely sensitive to the amount of charge contained in the QDs. Because the QDs exhibit bound states that lie below the bottom of the conduction band of the GaAs layer, the QDs are naturally populated with excess electrons. The negatively charged dots repel the electrons in the neighboring 2DEG in spatial regions adjacent to each QD. This causes areas of depletion in the 2DEG population and reduces the mobility of the carriers as a result of the spatial variation in the potential. In both ways, the local gating effects of the QDs inhibit current flow through the channel of the FET. The size of the active area plays a key role in the operation of the device, because the smaller the active area, the more influence each dot has on the total conductance of the channel. When a photon is absorbed in the $Al_{0.33}Ga_{0.67}As$, the photogenerated hole can become trapped by a dot due to Coulombic attraction between the hole and the negatively charged QDs. Once captured, it recombines with a trapped electron, discharging the QD. The reduction in the net charge of the dot reduces its influence on the channel 2DEG, resulting in *increased* current flow. The *positive persistent photoconductivity* associated with this process provides the devices with single-photon sensitivity as long as the active area is small enough.

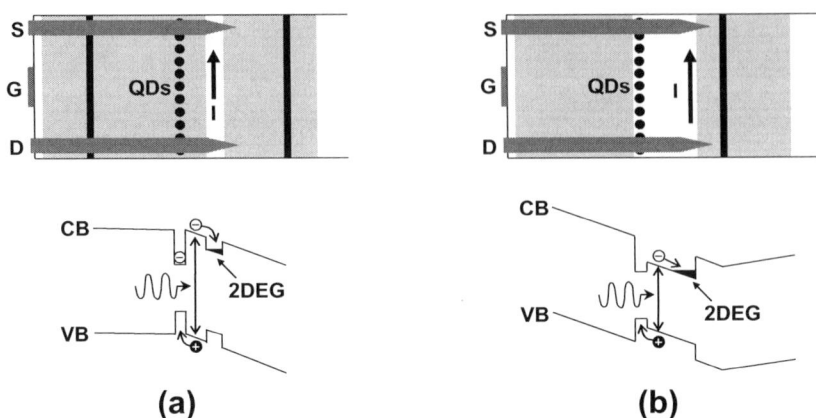

FIGURE 5.13 Schematic diagrams of the composition and bandstructure for transistor-based single-photon detectors that utilize self-assembled QDs, as reported in (a) Ref. [55] and (b) Ref. [56]. The layers of the heterostructures are $Al_yGa_{1-y}As$ (gray), GaAs (white), and Si-doped $Al_yGa_{1-y}As$ (black). Source (S), drain (D), and gate (G) contacts are indicated in the diagrams, and CB and VB denote the conduction band and valence bands, respectively.

Using self-assembled QDs in this way has been shown to be an effective method for detecting single photons, and there have been a number of reports on detectors based on this scheme. In their original letter, Shields *et al.* [55] showed that a device exhibiting an active area of ≈ 2 μm^2, containing approximately 100 QDs, could sense individual photons with DE of 0.48% when cooled to 4 K. It was subsequently demonstrated that the device could also detect photons at 77 K (with increased electrical noise) [57] and could be used in conjunction with a cryogenic radio-frequency amplifier to operate at a detection rate of 4×10^5 s^{-1} [58]. In these works, the detector was reset deterministically by periodically forward biasing the gate, which caused electrons from the 2DEG to flow toward the gate contact and repopulate the dots. Alternatively, Kardynal *et al.* showed that a probabilistic reset, based on electron tunneling, could be achieved in a similarly designed structure, allowing for continuous-mode operation with a constant gate voltage [59]. In this device, the QDs are recharged by electrons that tunnel from the channel well into the dots. The reset time can be controlled by tuning the voltage applied to the back gate of the device. Operated in this way, average reset times ranging from tens of microseconds to less than 200 ns were achieved, and photon-number discrimination was also reported [60].

Crucial to achieving high DE in a QD-based device is capturing a high percentage of the photogenerated carriers (either electrons or holes) excited in the structure. In 2006, Rowe *et al.* demonstrated that improved DE can be achieved by strategically placing the QDs and potential barriers [56]. The composition and principles of operation of the device, referred to as a QDOGFET (quantum dot, optically gated, field-effect transistor), are shown in Fig. 5.13b.

Here, a high density (400–500 μm^{-2}) layer of InGaAs QDs are embedded in a GaAs/Al$_{0.2}$Ga$_{0.8}$As high-electron mobility transistor; the QD layer and a 100 nm GaAs absorption layer sit side-by-side and are flanked by thick Al$_{0.2}$Ga$_{0.8}$As barriers. With a reverse bias applied to the gate, the electron and hole excited in the absorption region are sorted by the internal electric field and sent in opposite directions. The hole is directed toward the QDs, while the electron is sent to join the 2DEG. The Al$_{0.2}$Ga$_{0.8}$As barrier positioned next to the dots functions as a backstop, prohibiting the hole from traveling past the QD layer. Once trapped in a dot, the positively charged hole screens the internal field produced by the gate contact, increasing the amount of current flowing in the 2DEG. The dot can be emptied by temporarily forward biasing the gate. This floods the QDs with electrons, which subsequently recombine with any trapped holes.

The QDOGFET design is specifically tailored to efficiently detect photons with energy greater than the bandgap of GaAs, but less than that of Al$_{0.2}$Ga$_{0.8}$As. These photons are selectively absorbed by the GaAs region where the gate field and the barriers work together to direct the photogenerated holes to the QDs. In studies reported by Rowe et al. [56], it was demonstrated that a 2.7 μm^2 active area device cooled to 4 K could detect ≈70% of the 800 nm wavelength photons deposited in its absorption region. Subsequent work detailed the carrier dynamics involved in the operation of QDOGFETs [61,62], investigated how the sensitivity of the detectors depends on detection rate [63] and temperature [64], and demonstrated the photon-number-resolving capabilities of the devices [65].

In addition to using a lateral (in-plane) electrical current to read out photocharged QDs, single-photon detection can also be achieved using a vertical current that runs through the QDs. A single-photon detector of this type was demonstrated in 2005 by Blakesley et al. [66], who used self-assembled QDs embedded in a double-barrier resonant tunneling diode. As illustrated in Fig. 5.14, the emitter and collector regions of their QD resonant tunneling diode (QDRTD) are separated by intrinsic layers that include a 10 nm wide GaAs quantum well with 10 nm wide Al$_{0.33}$Ga$_{0.67}$As barriers, a single layer of a self-assembled InAs QDs, and a 310 nm thick GaAs absorption region.

Paramount to detecting photons with the device is that the amount of current that flows from the collector to the emitter of the structure depends very sensitively on the bias voltage. The current is strongly limited by tunneling through two Al$_{0.33}$Ga$_{0.67}$As barriers. Resonant tunneling occurs for bias voltages where the energies of the electrons on the emitter side of the barrier align with the discrete conduction-band energy level of the quantum well. A sharp peak in the current is produced for voltages where this resonance condition is met. To detect photons, the diode is biased near resonance, where the current is particularly sensitive to changes in the voltage. When a photon excites an electron-hole pair in the absorption region, the hole is directed toward the QDs by the internal electric field while the electron travels in the opposite toward

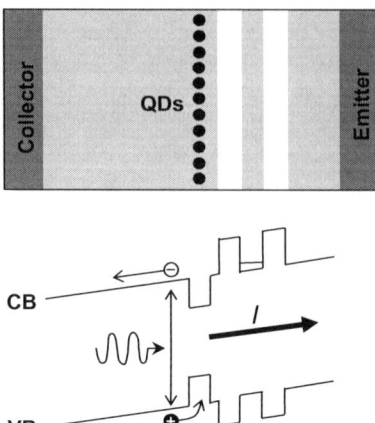

FIGURE 5.14 Schematic diagrams of the composition and bandstructure of the QDRTD detector reported in Ref. [66]. The diode consists of a GaAs emitter with n-type graded doping and a GaAs collector that is n^+-doped separated by an intrinsic region containing InAs QDs. The intrinsic layers of the heterostructures are GaAs (gray) and AlGaAs (white). CB and VB denote the conduction band and valence bands, respectively.

the collector. Once the hole is captured by a QD, it alters the internal field in the vicinity of the quantum well, shifting its energy levels and modifying the resonant tunneling condition. This results in a photo-induced change in the current, which persists for as long as the hole is trapped in the dot. Using a cross-wire geometry, a QDRTD with an active area of 1 μm^2 was demonstrated to exhibit DE of ≈11% for 550 nm wavelength photons when cooled to 5 K [66].

5.3.2 Photon-Number-Resolving Detection

Detectors that utilize QD-ensembles can trap many photogenerated carriers at once, making them well suited for photon-number-resolving detection. In a sense, the layer of QDs can be viewed as an array of individual charge detectors, where the 2DEG functions as an integrated analog read-out circuit that reveals the number of photo-activated elements via its current. However, in order for the number-resolving capabilities of the detectors to be realized, it is imperative that (i) each trapped carrier produce the same change in the channel current regardless of its location within the QD ensemble and that (ii) parasitic signals from charges trapped in defect states do not contaminate the signals associated with the QDs. In practice, defect states [67] and geometric limitations make satisfying these conditions a tall order; however, photon-number-resolution has been demonstrated in QD-based detectors [60,65].

In 2007, it was demonstrated that a QDOGFET can discriminate between the detection of 0, 1, 2, and 3 photons with relatively high fidelity [65].

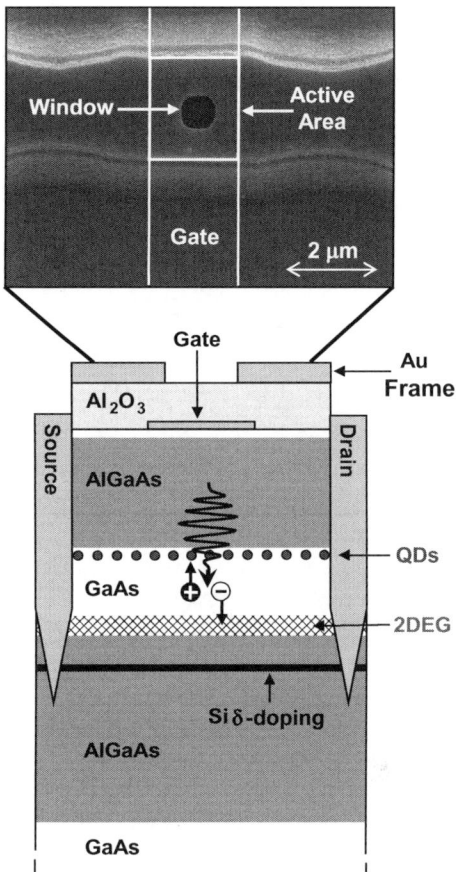

FIGURE 5.15 Composition of a QDOGFET with photon-number-resolving capability. The semiconductor heterostructure that forms the basis for the detector is comprised of [from bottom to top] a GaAs buffer layer (200 nm); an $Al_{0.20}Ga_{0.8}As$ spacer layer (2.5 μm); Si δ-doping; $Al_{0.20}Ga_{0.8}As$ (70 nm); a GaAs absorption layer (100 nm); InGaAs QDs; $Al_{0.20}Ga_{0.8}As$ (200 nm); and an n-doped GaAs cap layer. The gated portion of the channel mesa is 2.0 μm × 2.4 μm and masked by Au frame with a 0.7 μm × 0.7 μm transmission window. The Au frame is electrically isolated from the rest of the structure by a 100 nm thick layer of Al_2O_3. *Reprinted with permission from Ref. [65]. Copyright 2007, Nature Publishing Group.*

The composition of the QDOGFET that was the focus of that study is shown in Fig. 5.15. The detector was fabricated by depositing and annealing Ni/Au/Ge source and drain ohmic contacts on the structure surface, by etching a channel mesa between the contacts, and by depositing a 4 nm thick, semitransparent, Pt Schottky barrier gate midchannel. The fabrication process was completed by covering the channel mesa with a transparent layer of Al_2O_3 and by masking the active area of the detector with an opaque Au frame.

In theory, the condition that each trapped carrier needs to produce the same change in the channel current is inherent to the QDOGFET geometry. It can be shown [62] that in the small-signal limit, the increase in the channel current (I) caused by the addition of N photogenerated holes to the QD plane is given by

$$\Delta I = g_m \frac{eW}{\epsilon' A} N, \tag{5.4}$$

where e is the elementary charge (in SI units), W is the distance between the Pt gate and the QD layer, ϵ' is the electric permittivity of the host semiconductor material, A is the active area, and g_m is the transconductance of the detection system. Consequently, the channel current provides a direct measure of the number of detected photons.

In practice, however, spatial edge effects associated with the finite size of the gate contact can cause nonuniformities in the responses produced by different QDs. In addition, studies on unmasked QDOGFETs indicate that photons absorbed in the ungated portions of the channel mesa (near the edges of the active region) also affect the channel current [61]. These signals can contaminate the photoresponses produced by the charging of QDs within the active area. Masking the active area with an opaque Au aperture prevents the absorption of photons in the ungated portions of the channel mesa and promotes hole capture in QDs located within the interior of the active region, which are expected to produce the most uniform responses.

The ability of the QDOGFET to resolve different numbers of detected photons was investigated by illuminating the device with attenuated laser pulses and constructing histograms of the responses produced by the individual pulses of light. In Fig. 5.16, an example histogram is shown where on average $\overline{N} = 1.1$ photons were detected per pulse. The histogram is made up of a number of individual equally spaced peaks caused by the detection of discrete numbers of photons. The size and spacing of the peaks are in excellent agreement with the responses predicted by Eq. (5.4) for the geometry, material composition, and transconductance of the device used in the study and reflect the Poisson statistics of the laser light.

The resolving power of the detector is degraded by electrical noise and other mechanisms that broaden the histogram peaks. While the $N = 0$ peak is broadened solely by electrical noise, nonuniform contributions of holes trapped in different QDs of the ensemble will additionally broaden the peaks associated with higher number states. The change in current caused by N photons is the sum of N single-photon changes in the channel current. Consequently, as long as the signal changes caused by the charging of separate QDs are uncorrelated, the variances of the individual peaks should increase linearly with N. Quantitative analysis of the QDOGFET's number resolution was performed by fitting acquired histograms with a series of Gaussian functions (Fig. 5.16) [65]. The fits were constrained such that the areas of the Gaussians obey Poisson statistics, with $\overline{N}$ defined as a free fit parameter. The peak widths extracted

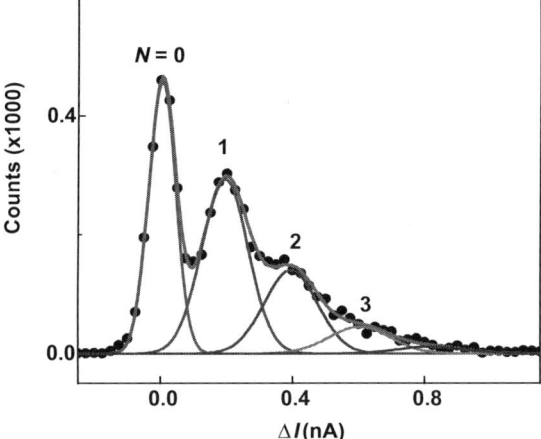

FIGURE 5.16 Histogram of current changes, ΔI, measured for a train of laser pulses, where on average 1.1 photons were detected per laser pulse. The black circles represent experimental data, and the gray curve is a Poisson fit to the data, where the detector response is taken into account by assuming that each photon-number peak follows a Gaussian distribution (colored curves) [65].

from the numerical fits were found to be consistent with those expected for uncorrelated signals. The Gaussian fits were further utilized to determine the probability that the QDOGFET could accurately determine the number of detected photons in a single shot given *a priori* knowledge that the photons were produced by a Poisson source and that $\overline{N} = 1.1$ [65]. In this analysis, the boundaries of decision regions for different number states were defined by the intersections of neighboring Gaussian peaks. The probability of accuracy was calculated for each number state by dividing the area of the Gaussian function contained within the corresponding decision region by the total area of all the Gaussian functions contained within that region. The resulting values indicated that N could be determined in a single-shot measurement with better than 83% confidence.

5.3.3 Modeling Photoconductive Gain

While QD-based detectors have been demonstrated using a number of different geometries, the one thing they all have in common is the use of photoconductive gain. To detect single photons, weak, photo-induced signals must be differentiated from random fluctuations associated with electrical noise. Electrical noise in two-dimensional electron systems has been studied extensively due to its impact on high-speed high-electron mobility transistors (HEMTs, e.g., Refs. [68–71]) and Hall-bar structures (Ref. [72] and references therein). The noise spectra of devices with relatively large active areas (dimensions >1 μm) are typically dominated by $1/f$-noise with Lorentzian-shaped contributions riding on top of the $1/f$ background. The temperature [68–70,73–75] and size [73] of the structure along with the existence of applied electric fields (typically produced by a gate contact) [71,73,75] are all known to have a strong influence on the spectral characteristics of the noise in two-dimensional electron systems. While photoconductive gain can be used to combat the effects of noise, a common question asked is, "does the persistent nature of photoconductive gain inherently limit the detection speeds of QD-based devices?" Here, we attempt to answer that question by describing the mechanics of photoconductive gain quantitatively, and detailing the relationship between electrical noise and detector sensitivity.

5.3.3.1 Basic Formalism
A basic mathematical formalism for detectors that utilize photoconductive gain was presented by Rowe et al. [63]. In this approach, a linear systems model is applied to the detection process in order to relate detector sensitivity, as quantified by the signal-to-noise ratio (S/σ), to the experimental parameters of the system in the specific case that the arrival times of the light pulses are known. As described previously in the chapter, when a photon is detected by a QD-based detector, it produces a persistent change, or step, in the channel current. A transimpedance amplifier is commonly used to convert these changes in current to changes in voltage, which can subsequently be monitored by additional electronics. An effective way to reduce the impact of electrical noise when the arrival times of the pulses of light are known is to apply an *average difference filter* (ADF) to the photo-induced steps.

An ADF integrates the signal over equal time intervals ($\tau/2$) before and after the arrival of the light pulse and then takes the difference of the two integrated values. From the viewpoint of linear systems, the impulse response function of the ADF filter is given by $w(t) = \theta(\tau/2-t)\theta(t) - \theta(\tau/2+t)\theta(-t)$, where $\theta(t)$ is the Heaviside step function. A pulse arriving at $t = 0$, producing a voltage change of the form $V(t) = \Delta V \theta(t) + V_0$, will result in a filtered signal,

$$S = \int_{-\infty}^{\infty} V(t)w(t)dt = \Delta V \tau/2, \quad (5.5)$$

that is directly proportional to the magnitude of the photo-induced step, ΔV. The linear dependence of S on τ underlines the utility of the QDs to produce persistent responses by storing photogenerated carriers, thus extending the allowable integration time of the system.

Although a longer measurement time will produce larger filtered signals, the impact τ has on the sensitivity of the detection system depends on the characteristics of the electrical noise. If the ADF is applied to a series of photo-induced steps of equal magnitude, the electrical noise in the signal will produce variation in the filtered signals with variance,

$$\sigma^2 = \int_0^\infty W^2(f)N(f)df, \quad (5.6)$$

where $W(f) = (2/i\pi f)\sin^2(\pi f\tau/2)$ is the unnormalized transfer function of the ADF, and $N(f)$ is the power spectral density (PSD) of the noise. With the ADF applied, only noise frequencies close to the measurement frequency, $f_m = 1/\tau$, substantially influence the sensitivity of the measurement. For frequencies below the corner frequency of the thermal noise floor in 2DEG systems, the PSD is typically dominated by a $1/f$ dependence: $N(f) = B^2/f$. In this case, $\sigma = B\tau\sqrt{\ln 2}$, and the signal-to-noise ratio of the filtered photoresponses,

$$S/\sigma = \frac{\Delta V}{2B\sqrt{\ln 2}}, \quad (5.7)$$

is independent of the measurement time.

This analysis indicates that there is not an inherent "speed limit" for detectors based on photoconductive gain as long as $1/f$ noise is the dominant noise contribution [63]. Any deviations from pure $1/f$ noise, however, will cause the sensitivity to vary with measurement time. For example, beyond the corner frequency of 2DEG systems, thermal noise becomes the dominate noise source. For measurement frequencies beyond the corner frequency, σ will grow nonlinearly with increased f_m, resulting in decreased sensitivity for higher detection rates. HEMT devices can have corner frequencies in the gigahertz range, indicating that high-speed operation should be possible with QD-based devices through proper device design.

5.3.3.2 Application to QDOGFET Detectors

The validity and usefulness of this quantitative treatment of photoconductive gain has been demonstrated in the context of QDOGFET detectors [63,64]. It has been shown to provide predictability in the performance of these devices and can be used to guide optimization of the overall detection system. A basic read-out scheme for QDOGFET detectors is shown in Fig. 5.17, where the conductive channel of the device is wired in series with a resistive load, R_L. The output voltage, V, is amplified, and monitored for photo-induced changes by additional electronics (not shown). A photon produces a change in the output voltage of magnitude $\Delta V = R_L \Delta I$, where ΔI depends on the composition and geometry of the QDOGFET, as indicated in Eq. (5.4). As a result, the voltage change produced by a photon can be expressed in terms of the experimental parameters as

$$\Delta V = \frac{g_m e W R_L}{\epsilon' A}. \quad (5.8)$$

This photo-induced signal competes with the electrical noise in the detection system.

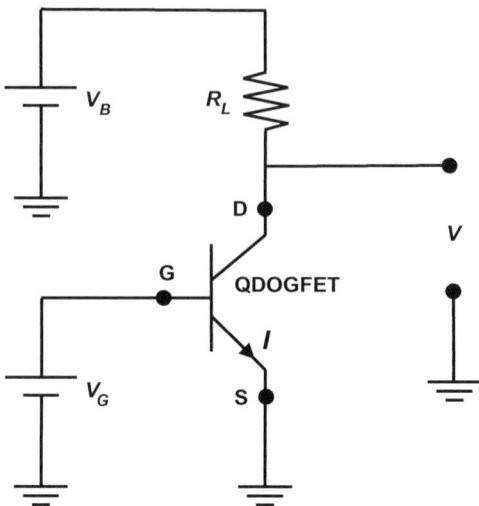

FIGURE 5.17 Schematic of circuitry commonly used for QDOGFET single-photon detectors. Voltage supplies, V_B and V_G, are used to produce the channel current, I, and set the gate voltage of the QDOGFET. The transconductance of the system is subsequently defined by $g_m = \Delta I / \Delta V_G$. D, G, and S denote the drain, gate, and source terminals, respectively.

In addition to the noise produce by the external amplifier, the total noise in the amplified signal contains two other dominant contributions, the noise produced by the QDOGFET and the thermal noise associated with the load resistance. The PSD (in units of V^2/Hz) of the noise produced by the detector and load is given by

$$N_V = (R_Q || R_L)^2 (N_Q + N_L), \tag{5.9}$$

where $R_Q || R_L$ is the parallel combination of the detector and load resistances; N_Q is the PSD (A^2/Hz) of the QDOGFET noise; and N_L is the thermal noise of the load resistor. Example noise spectra for a QDOGFET structure are shown in Fig. 5.18a. The noise spectra exhibit a high degree of $1/f$ character, consistent with the noise produced in other HEMT devices.

With knowledge of the noise spectrum and other electrical and geometrical parameters of the system, the sensitivity of the detector can be predicted for any measurement frequency using Eqs. (5.5) and (5.6). For instance, in Fig. 5.18b the signal-to-noise ratio predicted by the model is plotted as a function of operating temperature for a measurement frequency of $f_m = 50$ kHz. Also shown in the figure are values determined by performing optical measurements with the detection system. There is very good agreement between the optically determined values and those predicted by the model. The model can be further utilized to predict how the sensitivity of the detection system will vary with measurement frequency and with changes to the other experimental parameters

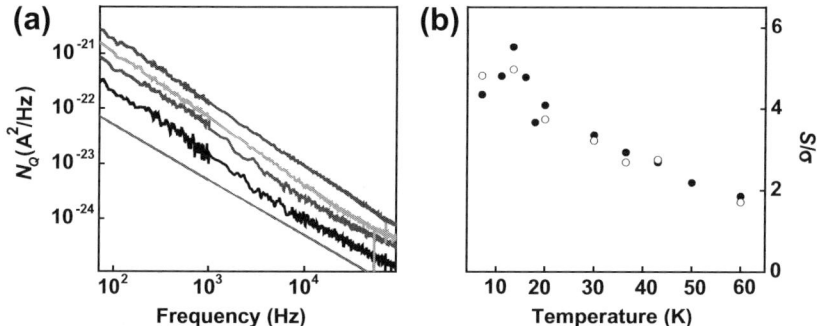

FIGURE 5.18 (a) PSD of the QDOGFET noise for select operating temperatures: from top to bottom- 60 K (red), 37 K (green), 18 K (blue), 11 K (black). The gray line, representing perfect $1/f$ dependence, is shown for comparison. (For interpretation of the references to color in this figure legend, the reader is referred to the web version of this book.) (b) The signal-to-noise ratio determined by optical measurements (open circles) and predicted using Eqs. (5.5) and (5.6) and experimentally measured noise spectra for $\tau = 20$ μs (solid circles) [64].

of the system [64]. Consequently, it can be used as a helpful tool for system optimization.

5.3.4 Conclusions

In this Section, we have provided an overview of QD-based single-photon detection. We surveyed a number of different device configurations and described the different ways that they utilize QDs and photoconductive gain to detect single photons of light. We also demonstrated how self-assembled QDs can be used to construct a photon-number-resolving detector and outlined a mathematical treatment of photoconductive gain as it applies to the sensitivity of single-photon detectors. The detection mechanisms used by QD-based devices are unique in that they rely on the electrical readout of an individual trapped carrier excited by a photon. This characteristic of the devices could potentially be put to use in future spintronic applications and could lead to the development of quantum repeaters [8].

Further development of the detectors described in this Section should yield significant improvements in performance. For instance, while these devices have exhibited low DE to date, improved performance can be achieved by optimizing the thickness of the absorbing layers and by incorporating the structures into optical cavities. The semiconductor makeup of the devices makes them particularly compatible with periodic Bragg mirrors that have been shown to enhance the efficiency of other semiconductor-based detectors [76]. Furthermore, while many of the devices described here operate at wavelengths below 1 μm, it should be possible to construct detectors sensitive to light at telecommunications wavelengths by modifying the material systems of the devices. Future device development should also focus on designing devices for high-speed operation (10^6–10^9 s^{-1} detection rates), increasing the active area of the detectors, and raising the operating temperature of the devices to ease their cooling requirements and cost of operation. Although persistent photoconductivity has been observed at much higher temperatures [77], experimental demonstrations of QD-based detectors have been limited to temperatures ranging from <1 K to 77 K. Increasing the operating temperature will be an important step to developing detectors suitable for commercial applications.

ACKNOWLEDGMENTS

J.K. and K.S.M. would like to thank Henry Hogue and Maryn Stapelbroek for providing VLPC detectors and help in developing the carrier transport model, Burm Baek, Martin Stevens, and Sae Woo Nam for experimental efforts in VLPC timing jitter measurements. J.K. also thanks Shigeki Takeuchi and Yoshihisa Yamamoto for early collaboration on VLPC characterization. P.G.K. would like to thank Michael Petroff for loan and eventual donation of and collaboration on the SSPM devices; Philippe Eberhard and Aephraim Steinberg for early experimental characterizations; and Alan Bross and Adam Para for consulting on our more recent VLPC samples. E.J.G. would like to acknowledge the significant contributions made by Richard Mirin and Mary Rowe who spearheaded the efforts to develop and study QDOGFET single-photon detectors. He also gratefully acknowledges Shelley Etzel, Marion Greene, and Todd Harvey who were instrumental in the growth and fabrication of the devices as well as Robert Hadfield, Sean Harrington, Sae Woo Nam, John Nehls, Danna Rosenberg, Mackay Salley, and Mark Su for their contributions to the QDOGFET research presented in this chapter.

REFERENCES

[1] E. Knill, R. Laflamme, and G.J. Milburn, "A Scheme for Efficient Quantum Computation with Linear Optics," Nature 409, 46–52 (2001).

[2] G. Brassard, N. Lütkenhaus, T. Mor, and B.C. Sanders, "Limitations on Practical Quantum Cryptography," Phys. Rev. Lett. 85, 1330 (2000).

[3] G. Di Giuseppe, M. Atatüre, M.D. Shaw, A.V. Sergienko, B.E. Saleh, M.C. Teich, A.J. Miller, S.W. Nam, and J. Martinis, "Direct Observation of Photon Pairs at a Single Output Port of a Beam-Splitter Interferometer," Phys. Rev. A 68, 063817 (2003).

[4] E. Waks, E. Diamanti, B.C. Sanders, S.D. Bartlett, and Y. Yamamoto, "Direct Observation of Non-Classical Photon Statistics in Parametric Downconversion," Phys. Rev. Lett. 92, 113602 (2004).

[5] E. Waks, B. Sanders, E. Diamanti, and Y. Yamamoto, "Highly Nonclassical Photon Statistics in Parametric Down Conversion," Phys. Rev. A 73, 33814 (2006).

[6] D. Achilles, C. Silberhorn, and I.A. Walmsley, "Direct, Loss-Tolerant Characterization of Nonclassical Photon Statistics," Phys. Rev. Lett. 97, 043602 (2006).

[7] E. Waks, E. Diamanti, and Y. Yamamoto, "Generation of Photon Number States," New J. Phys. 8, 4 (2006).

[8] E. Yablonovitch, H.W. Jiang, H. Kosaka, H.D. Robinson, D.S. Rao, and T. Szkopek, "Optoelectronics Quantum Telecommunications Based on Spins in Semiconductors," Proc. IEEE 91, 761–780 (2003).

[9] J. Kim and C. Kim, "Integrated Optical Approach to Trapped Ion Quantum Computation," Quant. Inf. Comput. 9, 181–202 (2009).

[10] M.D. Petroff and M.G. Stapelbroek, Blocked Impurity Band Detectors, US Patent No. 4,568,960, Filed 23 October 1980, Granted 4 February (1986).
[11] M.D. Petroff and M.G. Stapelbroek, "Responsivity and Noise Models of Blocked Impurity Band Detectors," in Proceedings of the IRIS Specialty Group on Infrared Detectors (1984).
[12] F. Szmulowicz and F.L. Madarasz, "Blocked Impurity band Detectors—An Analytical Model: Figures of merit," J. Appl. Phys. 62, 2533–2540 (1987).
[13] M.D. Petroff, M.G. Stapelbroek, and W.A. Kleinhans, "Detection of Individual 0.4–28 μm Wavelength Photons Via Impurity Impact Ionization in a Solid State Photomultiplier," Appl. Phys. Lett. 51, 406–408 (1987).
[14] M. Petroff and M. Stapelbroek, "Photon-counting Solid-State Photomultiplier," IEEE Trans. Nucl. Sci. 36, 158–162 (1989).
[15] M. Atac, J. Park, D. Cline, D. Chrisman, M. Petroff, and E. Anderson, "Scintillating Fiber Tracking for the ssc Using Visible Light Photon Counters," Nucl. Instrum. Methods Phys. Res. Sec. A Accelerat. Spectrom. Detect. Assoc. Equip. 314, 56–62 (1992).
[16] D. Adams, M. Adams, B. Baumbaugh, I. Bertram, A. Bross, D. Casey, S. Chang, M. Chung, C. Cooper, C. Cretsinger, R. Demina, G. Fanourakis, T. Ferbel, S. Grnendahl, J. Hinson, B. Howell, H. Johari, J. Kang, C. Kim, S. Kim, D. Koltick, F. Lobkowicz, S. Margulies, J. Moromisato, M. Narain, C. Park, S. Reucroft, Y. Park, R. Ruchti, J. Solomon, E. VonGoeler, J. Warchol, M. Wayne, E. Won, and Y. Yu, "First Large Sample Study of Visible Light Photon Counters (vlpc's)," Nucl. Phys. B Proc. Suppl. 44, 340–348 (1995).
[17] P.G. Kwiat, A.M. Steinberg, R.Y. Chiao, P.H. Eberhard, and M.D. Petroff, "High-Efficiency Single-Photon Detectors," Phys. Rev. A 48, R867–R870 (1993).
[18] P.G. Kwiat, A.M. Steinberg, R.Y. Chiao, P.H. Eberhard, and M.D. Petroff, "Absolute Efficiency and Time-Response Measurement of Single-Photon Detectors," Appl. Opt. 33, 1844–1853 (1994).
[19] P.H. Eberhard, P.G. Kwiat, M.D. Petroff, M.G. Stapelbroek, and H.H. Hogue, "Detection Efficiency and Dark Pulse Rate of Rockwell (sspm) Single Photon Counters," Appl. Photon. Technol. 1, 471–474 (1995).
[20] J. Kim, Y. Yamamoto, and H.H. Hogue, "Noise-Free Avalanche Multiplication in Si Solid State Photomultipliers," Appl. Phys. Lett. 70, 2852–2854 (1997).
[21] J. Kim, S. Takeuchi, Y. Yamamoto, H.H. Hogue, "Multiphoton Detection Using Visible Light Photon Counter," Appl. Phys. Lett. 74, 902–904 (1999).
[22] S. Takeuchi, J. Kim, Y. Yamamoto, and H.H. Hogue, "Development of a High-Quantum-Efficiency Single-Photon Counting System," Appl. Phys. Lett. 74, 1063–1065 (1999).
[23] D.L. Moehring, P. Maunz, S. Olmschenk, K.C. Younge, D.N. Matsukevich, L.M. Duan, and C. Monroe, "Entanglement of Single-Atom Quantum Bits at a Distance," Nature 449, 68–71 09 (2007).
[24] N. Lütkenhaus, "Security Against Individual Attacks for Realistic Quantum Key Distribution," Phys. Rev. A 61, 052304 (2000).
[25] E. Waks, C. Santori, and Y. Yamamoto, "Security Aspects of Quantum Key Distribution with Sub-Poisson Light," Phys. Rev. A 66, 042315 (2002).
[26] B. Baek, K. McKay, M. Stevens, J. Kim, H. Hogue, and S.W. Nam, "Single-Photon Detection Timing Jitter in a Visible Light Photon Counter," IEEE J. Quantum Electron. 46, 991–995 (2010).
[27] A. Lacaita, S. Cova, and M. Ghioni, "Four-Hundred-Picosecond Single-Photon Timing with Commercially Available Avalanche Photodiodes," Rev. Sci. Instrum. 59, 1115–1121 (1988).
[28] E. Moreau, I. Robert, J.M. Gerard, I. Abram, L. Manin, and V. Thierry-Mieg, "Single-Mode Solid-State Single Photon Source Based on Isolated Quantum Dots in Pillar Microcavities," Appl. Phys. Lett. 79, 2865–2867 (2001).
[29] S. Cova, A. Lacaita, M. Ghioni, G. Ripamonti, and T.A. Louis, "20-ps Timing Resolution with Single-Photon Avalanche Diodes," Rev. Sci. Instrum. 60, 1104–1110 (1989).
[30] H. Takesue, E. Diamanti, C. Langrock, M.M. Fejer, and Y. Yamamoto, "10-GHz Clock Differential Phase Shift Quantum Key Distribution Experiment," Opt. Express 14, 9522–9530 (2006).
[31] J. Kim, K.S. McKay, M.G. Stapelbroek, and H.H. Hogue, "Opportunities for Single-Photon Detection Using Visible Light Photon Counters," Proc. SPIE 8033, 80330Q (2011).
[32] J. Kim, S. Somani, and Y. Yamamoto, Nonclassical Light from Semiconductor Lasers and LEDs, Springer-Verlag, Berlin, Heidelberg (2001).

[33] G.B. Turner, M.G. Stapelbroek, M.D. Petroff, E.W. Atkins, and H.H. Hogue, "Visible Light Photon Counters for Scintillating Fiber Applications: I. Characteristics and Performance," SciFi 93-Workshop on Scintillating Fiber Detectors, World Scientific, New Jersey, pp. 613–620 (1993).

[34] M.G. Stapelbroek and M.D. Petroff, "Visible Light Photon Counters for Scintillating Fiber Applications: II. Principles of Operation," SciFi 93-Workshop on Scintillating Fiber Detectors, World Scientific, New Jersey, pp. 621–629 (1993).

[35] A.G. Milnes, "Deep Impurities in Semiconductors," Wiley, New York, (1973).

[36] J. Kim, Single Photonics: Generation and Detection of Heralded Single Photons, PhD Thesis, Stanford University (1999).

[37] D.N. Klyshko, "Utilisation of a Two-Photon Light for Absolute Calibration of Photoelectric Detectors," Sov. J. Quantum Electron. 10, 1112–1117 (1980).

[38] S.V. Polyakov and A.L. Migdall, "High Accuracy Verification of a Correlated-Photon-Based Method for Determining Photoncounting Detection Efficiency," Opt. Express 15, 1390–1407 (2007).

[39] A. Bross, J. Estrada, P. Rubinov, C. Garcia, and B. Hoeneisen, "Localized Field Reduction and Rate Limitation in Visible Light Photon Counters," Appl. Phys. Lett. 85, 6025–6027 (2004).

[40] C. Erginsoy, "Neutral Impurity Scattering in Semiconductors," Phys. Rev. 79, 1013–1014 (1950).

[41] K.S. McKay, J. Kim, and H.H. Hogue, "Enhanced Quantum Efficiency of the Visible Light Photon Counter in the Ultraviolet Wavelengths," Opt. Express 17, 7458–7464 (2009).

[42] R.J. McIntyre, "Multiplication Noise in Uniform Avalanche Diodes," IEEE Trans. Electron Devices ED-13, 164 (1966).

[43] R.A. LaViolette and M.G. Stapelbroek, "A Non-Markovian Model of Avalanche Gain Statistics for a Solid-State Photomultiplier," J. Appl. Phys. 65, 830–836 (1989).

[44] H. Dautet, P. Deschamps, B. Dion, A.D. MacGregor, D. MacSween, R.J. McIntyre, C. Trottier, and P.P. Webb, "Photon Counting Techniques with Silicon Avalanche Photodiodes," Appl. Opt. 32, 3894–3900 (1993).

[45] E. Waks, K. Inoue, E. Diamanti, and Y. Yamamoto, "High Efficiency Photon Number Detection for Quantum Information Processing," IEEE J. Quant. Electron. 9, 1502 (2003).

[46] I.W. Boyd and J.I.B. Wilson, "Laser Annealing for Semiconductor Devices," Nature 287, 278–278 (1980).

[47] J. Blacksberg, M.E. Hoenk, S.T. Elliott, S.E. Holland, and S. Nikzad, "Enhanced Quantum Efficiency of High-Purity Silicon Imaging Detectors by Ultralow Temperature Surface Modification Using sb Doping," Appl. Phys. Lett. 87, 254101 (2005).

[48] A. Rose, Concepts in Photoconductivity and Allied Problems, Interscience, New York (1963).

[49] M. Kroutvar, Y. Ducommun, D. Heiss, M. Bichler, D. Schuh, G. Abstreiter, and J.J. Finley, "Optically Programmable Electron Spin Memory Using Semiconductor Quantum Dots," Nature 432, 81 (2004).

[50] B.D. Gerardot, B. Brunner, P.A. Dalgarno, P. Ohberg, S. Seidl, M. Kroner, K. Karrai, N.G. Stoltz, P.M. Petroff, and R.J. Warburton, "Optical Pumping of a Single Hole Spin in a Quantum Dot," Nature 451, 441 (2008).

[51] S. Komiyama, O. Astaflev, V. Antonov, T. Kutsuwa, and H. Hirai, "A Single-Photon Detector in the Far-Infrared Range," Nature 403, 405–407 (2000).

[52] P. Martyniuk and A. Rogalski, "Quantum-Dot Infrared Photodetectors: Status and Outlook," Prog. Quantum Electron. 32, 89–120 (2008).

[53] H. Kosaka, H.D. Rao, D.S. Robinson, P. Bandaru, K. Makita, and E. Yablonovitch, "Single Photoelectron Trapping, Storage, and Detection in a Field Effect Transistor," Phys. Rev. B 67, 45104 (2003).

[54] D.S. Rao, T. Szkopek, H.D. Robinson, E. Yablonovitch, and H.W. Jiang, "Single Photoelectron Trapping, Storage, and Detection in a One-Electron Quantum Dot," J. Appl. Phys. 98, 114501 (2005).

[55] A.J. Shields, M.P. O'Sullivan, I. Farrer, D.A. Ritchie, R.A. Hogg, M.L. Leadbeater, C.E. Norman, and M. Pepper, "Detection of Single Photons Using a Field-Effect Transistor Gated by a Layer of Quantum Dots," Appl. Phys. Lett. 76, 3673–3675 (2000).

[56] M.A. Rowe, E.J. Gansen, M. Greene, R.H. Hadfield, T.E. Harvey, M.Y. Su, S.W. Nam, and R.P. Mirin, "Single-Photon Detection Using a Quantum Dot Optically Gated Field-Effect Transistor with High Internal Quantum Efficiency," Appl. Phys. Lett. 89, 253505 (2006).

[57] A.J. Shields, M.P. OSullivan, I. Farrer, D.A. Ritchie, M.L. Leadbeater, N.K. Patel, R.A. Hogg, C.E. Norman, N.J. Curson, and M. Pepper, "Single Photon Detection with a Quantum Dot Transistor," Jpn. J. Appl. Phys. 40, 2058–2064 (2001).
[58] B.E. Kardynal, A.J. Shields, N.S. Beattie, I. Farrer, K. Cooper, and D.A. Ritchie, "Low-Noise Photon Counting with a Radio-Frequency Quantum-Dot Field-Effect Transistor," Appl. Phys. Lett. 84, 419–421 (2004).
[59] B.E. Kardynal, A.J. Shields, I. Farrer, K. Cooper, and D.A. Ritchie, "Single Electron Dynamics in a Quantum Dot Field Effect Transistor," Appl. Phys. Lett. 89, 113503 (2006).
[60] B.E. Kardynal, S.S. Hees, A.J. Shields, C. Nicoll, I. Farrer, and D.A. Ritchie, "Photon Number Resolving Detector Based on a Quantum Dot Field Effect Transistor," Appl. Phys. Lett. 90, 181114 (2007).
[61] E.J. Gansen, M.A. Rowe, M.B. Greene, D. Rosenberg, T.E. Harvey, M.Y. Su, R.H. Hadfield, N.S.W., and R.P. Mirin, "Operational Analysis of a Quantum Dot Optically Gated Field-Effect Transistor as a Single-Photon Detector," IEEE. J. Sel. Top. Quantum Electron. 13, 1–11 (2007).
[62] M.A. Rowe, E.J. Gansen, M.B. Greene, D. Rosenberg, T.E. Harvey, M.Y. Su, R.H. Hadfield, S.W. Nam, and R.P. Mirin, "Designing High Electron Mobility Transistor Heterostructures with Quantum Dots for Efficient, Number-Resolving Photon Detection," J. Vac. Sci. Technol. B 26, 1174–1177 (2008).
[63] M.A. Rowe, M.G. Salley, E.J. Gansen, S.M. Etzel, S.W. Nam, and R.P. Mirin, "Analysis of Photoconductive Gain as it Applies to Single-Photon Detection," J. Appl. Phys. 107, 063110 (2010).
[64] E.J. Gansen, M.A. Rowe, S.D. Harrington, J.M. Nehls, S.M. Etzel, S.W. Nam, and R.P. Mirin, "Temperature Dependence of the Single-Photon Sensitivity of a Quantum Dot, Optically Gated, Field-Effect Transistor," J. Appl. Phys. 114, 93103 (2013).
[65] E.J. Gansen, M.A. Rowe, M.B. Greene, D. Rosenberg, T.E. Harvey, M.Y. Su, R.H. Hadfield, S.W. Nam, and R.P. Mirin, "Photon-Number-Discriminating Detection Using a Quantum-Dot, Optically Gated, Field-Effect Transistor," Nat. Photon. 1, 585–588 (2007).
[66] J.C. Blakesley, P. See, A.J. Shields, B.E. Kardynal, P. Atkinson, I. Farrer, and D.A. Ritchie, "Efficient Single Photon Detection by Quantum Dot Resonant Tunneling Diodes," Phys. Rev. Lett. 94, 067401 (2005).
[67] H. Kosaka, H.D. Rao, D.S. and Robinson, P. Bandaru, T. Sakamoto, and E. Yablonovitch, "Photoconductance Quantization in a Single-Photon Detector," Phys. Rev. B 65, 201307(R) (2002).
[68] Y. Chen, G.L. VanVliet, C.M. and Larkins, and H. Morko, "Generation-Recombination Noise in Nongated and Gated $Al_xGa_{1-x}As/GaAs$ Tegfets in the Range 1 Hz to 1 MHz," IEEE Trans. Electron Devices 47, 2045–2053 (2000).
[69] F. Hofman, R.J.J. Zijlstra, and J.M. Bettencourt de Freitas, "Voltage Noise in an $Al_xGa_{1-x}As$ – GaAs Heterostructure," J. Appl. Phys. 67, 2482–2487 (1990).
[70] J.R. Kirtley, T.N. Theis, P.M. Mooney, and S.L. Wright, "Noise Spectroscopy of Deep Level (dx) Centers in GaAs – $Al_xGa_{1-x}As$ Heterostructures," J. Appl. Phys. 63, 1541–1547 (1988).
[71] J.M. Peransin, P. Vignaud, D. Rigaud, and L.K.J. Vandamme, "$1/f$ Noise in Modfets at Low Drain Bias," IEEE Trans. Electron Devices 37, 2250–2253 (1990).
[72] J. Mller, Y. Li, S. von Molnr, Y. Ohno, and H. Ohno, "Single-Electron Switching in $Al_xGa_{1-x}As/GaAs$ Hall Devices," Phys. Rev. B 74, 125310 (2006).
[73] J. Mller, S. von Molnr, Y. Ohno, and H. Ohno, "Decomposition of $1/f$ Noise in $Al_xGa_{1-x}As/GaAs$ Hall Devices," Phys. Rev. Lett. 96, 186601 (20).
[74] R. Khlil, A.E. Hdiy, and Y. Jin, "Deep Levels and Low-Frequency Noise in Algaas/Gas Heterostructures," J. Appl. Phys. 98, 93709 (2005).
[75] Y. Li, C. Ren, P. Xiong, S. von Molnr, Y. Ohno, and H. Ohno, "Modulation of Noise in Submicron Gaas/Algaas Hall Devices by Gating," Phys. Rev. Lett. 93, 246602 (2004).
[76] I. Ozbay, E. and Kimukin, N. Biyikli, O. Aytur, M. Gokkavas, G. Ulu, M.S. Unlu, R.P. Mirin, K.A. Bertness, and D.H. Christensen, "High-speed > 90 Quantum-Efficiency p-i-n Photodiodes with a Resonance Wavelength Adjustable in the 795–835 nm Range," Appl. Phys. Lett. 74, 1072 (1999).
[77] J.J. Finley, M. Skalitz, M. Arzberger, A. Zrenner, G. Bhm, and G. Abstreiter, "Electrical Detection of Optically Induced Charge Storage in Self-Assembled Inas Quantum Dots," Appl. Phys. Lett. 73, 2618–2620 (1998).

Chapter 6

Detectors Based on Superconductors

Karl K. Berggren[*], **Eric A. Dauler**[†], **Andrew J. Kerman**[†], **Sae-Woo Nam**[‡], **and Danna Rosenberg**[†]

[*]Department of Electrical Engineering and Computer Science, Massachusetts Institute of Technology, 77 Massachusetts Avenue, Cambridge, Massachusetts 02139, USA
[†]Lincoln Laboratory, Massachusetts Institute of Technology, 244 Wood Street, Lexington, Massachusetts 02420, USA
[‡]National Institute of Standards and Technology, 325 Broadway St., Boulder, Colorado 80305, USA

Chapter Outline

6.1 Introduction	186
6.2 Superconducting Nanowire Single-Photon Detectors	187
6.2.1 Operating Principle	187
6.2.2 Principal Strengths, Weaknesses	191
6.2.3 Areas of Research	192
6.3 Transition-Edge Sensors	194
6.3.1 Operating Principle	195
6.3.2 Principal Strengths and Weaknesses	199
6.3.3 Research Areas	199
6.4 Superconducting Tunnel Junction Detectors	201
6.4.1 Operating Principle	201
6.4.2 Strengths and Weaknesses	204
6.4.3 Research Areas	204
6.5 Microwave Kinetic-Inductance Detectors	204
6.5.1 Operating Principle	205
6.5.2 Strengths and Weaknesses	206
6.5.3 Research Areas	207
6.6 Conclusions and Perspective	208
References	209

6.1 INTRODUCTION

Since the discovery of superconductivity in 1911 by Kamerlingh Onnes, superconducting devices have been an active area of research. In the early 1990s, it was recognized by many groups that it was possible to fabricate devices that were sufficiently sensitive to detect a single optical photon by a variety of superconductivity-based approaches [1–4]. Each of these proposed approaches was significantly different from each other and utilized different properties of superconducting materials or devices.

Understanding each of these approaches does not require a detailed microscopic understanding of superconductors; only a few basic concepts are required. The first is the basic notion that below a certain "critical" transition temperature T_c, the resistance of superconducting materials goes essentially to zero, as shown in Fig. 6.1a. Since this constitutes a thermodynamic phase transition, the dependence of resistance on temperature near T_c can be very steep [5]. Not surprisingly, researchers realized that this steep temperature sensitivity could be used to sense radiation [6,7]. Similarly, as shown in Fig. 6.1b, a phase transition occurs at fixed temperature as the current density (or equivalently, the magnetic field) is increased up to a critical value J_c. This sharp transition has also been exploited to implement a sensor [8]. Finally, associated with the superconducting phase transition is the opening of an energy gap between the ground state and low-lying electronic excited states. This gap corresponds to the binding energy of the superconducting charge carriers, the so-called "Cooper pairs" [9], and excitations out of the ground state, known as "quasiparticles," can be viewed as dissociation of these bound pairs. These quasiparticle excitations are created when energy is absorbed by the superconductor (e.g. from a sufficiently energetic photon). Typically their creation is indirect, mediated by the excitation of an electron-hole pair in the normal-metal states of the material (above the superconducting gap), which then relax to form the quasiparticles. However, the energy gap for quasiparticle excitation is significantly smaller than the energy of an optical photon (meV compared to eV), so that hundreds or thousands of these excitations are created when a single visible or near-infrared photon is absorbed.

Intrinsic to photon detection using superconductors is the need to operate significantly below room temperature. Although this can pose an engineering challenge, it has the advantage that many noise processes are strongly suppressed at low temperature, most notably those that can interfere with efficient detection or cause the generation of false detections, known as dark counts. Since the first demonstration of single-photon detection with superconductors by Peacock et al. [10], there has been active research to develop single-photon detectors with performance advantages over conventional semiconductor-based devices and photo-multiplier tubes. In this chapter, we will describe four types of superconducting devices: superconducting nanowire single-photon detectors, transition-edge sensors, superconducting

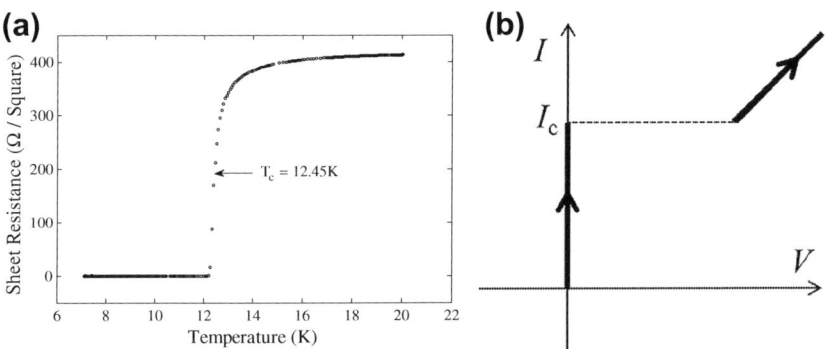

FIGURE 6.1 (a) Resistance vs. temperature of a superconducting thin film showing a transition between the superconducting, low-temperature, zero-resistance state, and the normal state at $T > T_c$ where T_c is the critical temperature of the material. (b) Schematic current-voltage characteristic of a superconducting wire, showing the superconducting branch, and a characteristic switching ("critical") current. Above this current, the material acts as a normal metal (i.e. a resistor).

tunnel junctions, and microwave kinetic-inductance detectors. Our survey supplements several existing review articles in the literature on these topics [11,12].

6.2 SUPERCONDUCTING NANOWIRE SINGLE-PHOTON DETECTORS

Superconducting nanowire single-photon detectors (SNSPDs) (also known as Superconducting single-photon detectors, or SSPDs), shown in Fig. 6.2, are a nanowire meander made from $\lesssim$10 nm think and $\lesssim$200 nm wide wires that locally switch between the superconducting and resistive state in response to the absorption of single photons [8].

6.2.1 Operating Principle

First, we briefly summarize the operating principle of the device. Later, we will go through each step in the process in more detail.

Due to the nanoscopic transverse dimension of these wires, in the presence of a bias current near the critical value, absorption of a photon results in local suppression or destruction of the superconducting state (by creation of quasiparticles) on picosecond timescales. Although initially only a nanoscopic length of the wire becomes resistive, the subsequent Joule heating (resulting from the current flowing through this now resistive length) causes this normal-state domain to grow to hundreds of nanometers in length, with a total effective resistance in the kΩ range. As a result, the current then diverts into the parallel load impedance across the detector (typically $\approx$50Ω), producing a measurable

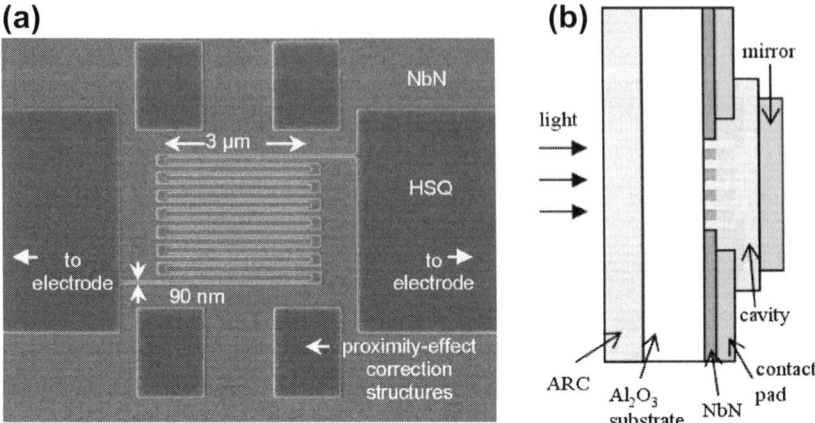

FIGURE 6.2 (a) Top-down scanning electron micrograph of a typical SNSPD pattern. The pattern shown consists of hydrogen silsesquioxane (HSQ) resist on a NbN film, prior to etching. As indicated, the active area of the device was ≈3 μm × 3 μm, and the wire width was ≈90 nm. Proximity-effect-correction structures (electrically isolated from the detector) were used to compensate for parasitic effects occurring during lithography. (b) Cross-section of SNSPD showing illumination through the device substrate. A thin anti-reflection layer can be applied to the substrate to reduce lost light, and an impedance-matching structure (essentially a small cavity) can be added to trap the light, enhance absorptance, and further suppress reflection. This image is not to scale: the NbN is typically only 3–5 nm thick, while the cavity is typically ≈200–300 nm thick, and the substrate is typically 100s of μm's thick. *Reprinted with Permission from [13].*

voltage pulse as illustrated in Fig. 6.3b [14]. Note that during this part of the response the detector is not sensitive to subsequent photons since there is no longer current flowing in it. With the diversion of current out of the wire the Joule heating is removed, and the wire cools and returns to the superconducting state. A crucial element for this cooling is a low thermal resistance between the nanowire and the substrate, which arises naturally from the extremely small film thickness, and in some cases also from a very good lattice match between the film and substrate. Once superconductivity is restored, the current recovers in the wire with a time constant arising from its kinetic inductance (associated with ballistic motion of Cooper pairs and much larger than the geometrical inductance due to magnetic fields) and load resistance. As the current recovers asymptotically toward the static bias current, the detection efficiency correspondingly approaches its initial static value. If another photon is absorbed during the recovery of the current, the instantaneous detection efficiency will be lower [14,15]. When properly designed and used, SNSPDs have demonstrated detection efficiencies above 90% for 1550 nm photons [16,17], timing jitter of ≈30 ps [18], reset times of a few nanoseconds [14], and dark count rates below 100 s^{-1} [19], though not all of these can be achieved simultaneously due to inherent performance tradeoffs [20]. Increasingly, SNSPDs have also been applied to detecting other quanta (including alpha particles [21] and ionic molecules [22]), but in this chapter we focus only on photonic

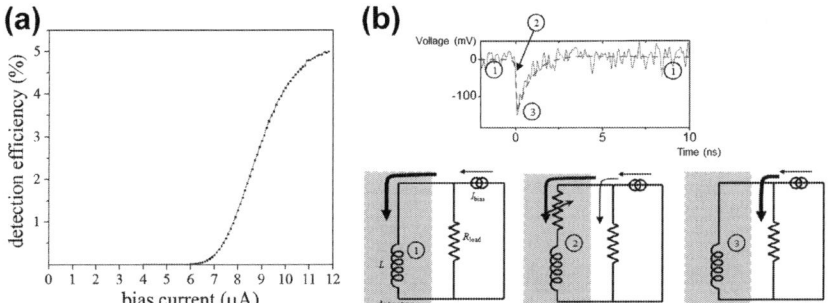

FIGURE 6.3 (a) Detection efficiency vs. bias current for an SNSPD. (b) Schematic showing switching and readout process for SNSPDs. The stage of each switching event in the photon detection cycle is labeled for (1) biased state prior to photon arrival; (2) photon hot spot resulting in resistive region in wire; and (3) restoration of superconductivity, after which the bias current eventually is restored.

applications. The design of SNSPDs and related detectors is constrained by factors detailed in the remainder of this section.

The detection process results from a sequence of steps: (1) photon absorption; (2) normal-state formation and readout of a current or voltage pulse; and (3) reset of the device to its starting state. Below, we go through these processes in further detail.

First, the photons must be absorbed by the superconducting nanowire. This absorption can be most efficiently accomplished with large active-area detectors in which a sizable fraction of the area is covered by wire, rather than by the gaps between wires. SNSPD active areas are typically a few tens of micrometers or smaller on a side, so either a waveguide (such as an optical fiber) must be brought in close proximity to the detector, or a focusing optic must be used. In either case, the waveguide or the optics must be precisely aligned with the detector active area, and this alignment must be maintained at cryogenic temperatures. While small detectors or poor alignment can result in low optical coupling efficiencies, well-designed systems can couple >95–99% of the light onto the SNSPD active area [16,17]. The superconducting materials used for the nanowires absorb over a very broad range of wavelengths from the ultraviolet into the infrared [23,24]. The nanowires are typically arranged with highly sub-wavelength spacing (for near-infrared and longer wavelengths), so that the meander pattern absorbs approximately uniformly, and thin-film multilayer structures can be used to optimize absorption of the incident light (c.f. Fig. 6.2b) [13,25]. Alternatively, nano-antenna structures [26] can be used to enhance detector absorption even when the wire spacing is comparable to the incident wavelength. Additionally, light polarized in the direction of the nanowire is more strongly absorbed, making the wires polarization sensitive (this sensitivity depends on device geometry, but for typical device dimensions the effect can result in a factor of 2 difference in absorption between the two polarizations). Absorption of the optimal polarization typically varies between 10% and 40% for single-pass structures [24] and can reach >90% for well-designed multilayer devices [27,28].

Second, to detect the photon absorption, the absorbed photon energy must result in the formation of a resistive region that spans the full width of the nanowire. The microscopic physics governing the formation of this resistive region is not fully understood, but the resistive region forms with higher probability for wires biased at higher currents, for higher-energy photons, for narrower and thinner wires, and for certain detector materials. In order to maximize the detection probability, usually the wire is biased close to the critical current because the probability of resistive-state formation approaches zero at low bias currents. This probability increases exponentially at intermediate bias levels and can saturate at sufficiently high bias (see Fig. 6.3a). The bias current can only approach the critical current density throughout the nanowire if the wire width, thickness, and superconducting properties are very spatially uniform along its length [29]. Sharp corners must also be avoided to achieve this uniformity, especially with wider wires [30–33], since the current tends to become non-uniform as it rounds a corner. Experimentally, the best materials for achieving a high probability of resistive-state-formation have been low-T_c, disordered, strongly type-II superconductors. There are several reasons for this: (i) low T_c means a small superconducting gap, which allows more quasiparticles to be created by each photon; (ii) strong disorder produces both a large normal-state resistivity (resulting in fast Joule heating), as well as a lower effective Cooper-pair density so that less thermal energy per volume is required to disrupt the superconducting state; and (iii) strong type-II implies both a short coherence length, such that a smaller volume must be initially heated, and a long magnetic penetration depth, which makes the current flow uniform throughout the wire cross-section. The resistive state forms with near-unity probability for visible-wavelength photons in uniform, sub-100 nm width wires at bias currents within 10–20% of the critical current for NbN. Narrower wires (20–50 nm) [34] and low-T_c materials (such as WSi [35]) have enabled high resistive-state-formation probabilities for currents even farther below critical, as well as for longer-wavelength photons. Figure 6.4 illustrates an example of the latter. Depending on the device design, fabrication, and material properties, substantial efficiency can be retained out to wavelengths in the $3 - 5$ μm range. In Fig. 6.4, the efficiency for a variety of NbN film thicknesses is shown, showing a general trend toward improved efficiency at lower thicknesses.

Finally, the thermal response of the detector and the electrical response of its readout circuit must be jointly chosen such that for all bias currents less than I_C, the static resistive state, known as a self-heating hotspot [37] (called the "latched" state in the SNSPD literature [38,40,41], and "phase separation" in the TES literature [39]), is *unstable*. This ensures that after a detection event the detector will return to the superconducting state and not remain in the latched state [38,40,41]. In general, the relatively large kinetic inductance of SNSPDs naturally ensures this condition is fulfilled. Cast in the time domain: if the

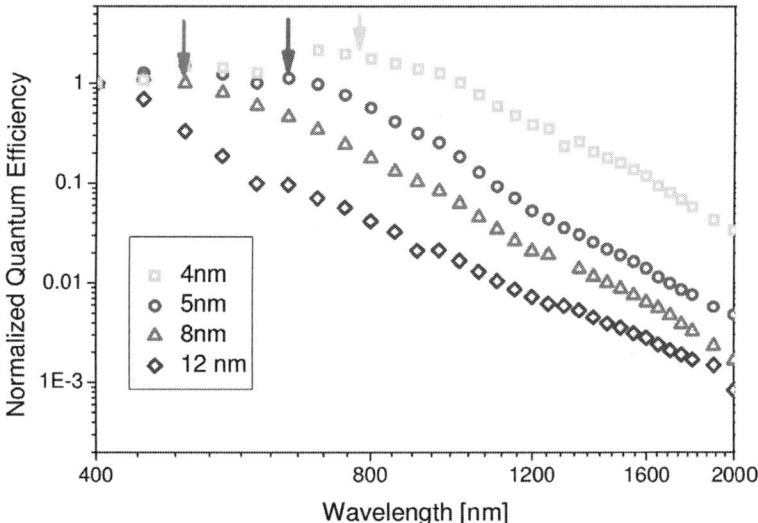

FIGURE 6.4 Efficiency normalized to the expected optical absorption as calculated theoretically at 400 nm optical wavelength, plotted as a function of wavelength for film thicknesses ranging from 4 nm to 12 nm, showing a roll-off at longer optical wavelength. In addition, thicker films (although they may in some cases be more absorptive) exhibit lower intrinsic efficiency, namely once a photon is absorbed, they are less likely to produce a voltage pulse[36].

electrical response of the system is sufficiently slower than the thermal response, the wire will have enough time to cool down and return to the superconducting state before the current in the wire begins to increase again.

Thus, the thermal response, which is governed by the superconductor, the surrounding material, and their interface properties, sets the ultimate limit on the achievable speed of the detector recovery, since the electrical time constant of the circuit must be kept longer than this to avoid latching [38,40,41]. The electrical time constant is determined by the load impedance and the detector's kinetic inductance. Typical SNSPDs have inductances of hundreds of nH, and a typical load impedance is $\approx 50\Omega$, resulting in reset times of ≈ 10 ns, which is much longer than the relevant thermal time constants. However, if the reset time is shortened sufficiently (either with smaller inductance or increased load impedance), the detector will latch at currents below I_C, and the maximum detection efficiency will no longer be accessible. Thus, latching can impose a tradeoff between minimizing reset time and maximizing detection efficiency [38,40].

6.2.2 Principal Strengths, Weaknesses

SNSPDs provide a combination of high detection efficiency, low timing jitter, fast reset times, and low dark count rates that cannot presently be matched by any other detector in the visible or short-wave infrared. The primary disadvantages of SNSPDs are their cryogenic operation (typically <3 K for the best performance from NbN devices; even lower temperatures for WSi [35]) and their small active area (at most a few tens of micrometers on a side, but typically ≈ 10 μm on a side). The low operating temperature also makes developing large arrays more challenging due to the restrictions that cryogenic cooling places on the thermal loads from readout electronics and optical access. The polarization dependence is also undesirable for some applications in which unpolarized light is to be detected, though it can be advantageous in other situations. SNSPDs are "click/no-click" detectors, and do not have an intrinsic power-sensing or calorimetric capability; an event consisting of two or more photons absorbed within a short time is indistinguishable from a single-photon event (though as with other click detectors, SNSPD arrays can approximate this capability [18]—more on this below).

Consequently, SNSPDs are currently best suited to applications that can tolerate the size, weight, and power associated with a sub-4 K cryocooler, that require at most a few tens of spatial modes, and which require very high speed, efficiency, and low noise (per resolvable time interval), such as quantum information processing and communication [42–44], high-sensitivity classical laser communications [45,46], biomedical imaging [47–49], and quantum-dot photonics [50–54].

6.2.3 Areas of Research

The precise, microscopic physics of the SNSPD photoresponse mechanism is still not well understood at present, and remains an active subject of research. Aside from general interest in the physics of superconductivity in a regime of both extreme spatial confinement and very close to the critical current, an understanding of this photoresponse could have substantial engineering impact. In particular, further improvements in the timing jitter and dark count rates of these devices, as well as their prospects for higher operating temperatures, will likely require such a detailed understanding. One way to begin to achieve this is to investigate the use of different superconducting materials, and to attempt to correlate microscopic material properties with detector behavior and performance. To this end, various superconducting materials have been investigated for use in SNSPDs, including NbN [8], NbTiN [55], TaN[33], MoRe [56], NbSi [57], WSi [35], MgB_2 [58], and Nb [59].

In addition, there is considerable practical interest in developing methods for coupling light more efficiently to SNSPDs, and this is also an ongoing area of research. There are several schemes reported in the literature using cryogenic optics or optical-fiber holders for SNSPD systems [60–64]. However, more recently there have been demonstrations of coupling light to nanowire detectors from on-chip waveguides or nano-antennae [16,26,65]; Fig. 6.5 highlights

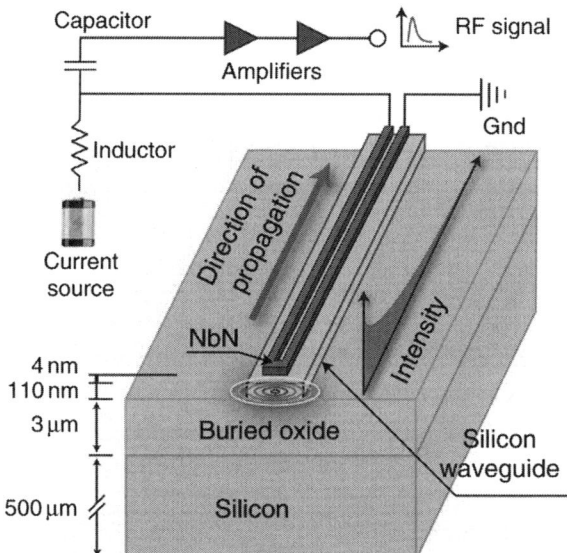

FIGURE 6.5 Diagram showing hairpin coupling of nanowire detector to an underlying waveguide. This traveling-wave-mode device can efficiently couple nearly 100% of the light in the waveguide into the niobium nitride. *Reprinted with permission from [16].*

one recent example in which an earlier design [66] was realized experimentally with high efficiency.

Optimized coupling of light into the nanowire system relies on accurate analytic and numerical modeling of the system. Generally, light polarized along the direction of the wires yields to analytic treatments [27,67], while transverse-polarized fields require numerical analysis due to the complexities of the surface effects [24,68]. All of these calculations require an estimate of the material index, which may vary depending on the details of growth [69], and thus needs to be extracted on a film-by-film basis. However, once a film is properly characterized, good agreement is achievable between the expected and observed optical performance [23,24,28].

Because patterning a wire is relatively straightforward, a variety of circuit topologies are under investigation to improve the performance by using multiple nanowires. For example, multiple nanowires can be integrated into a single active area (either as isolated, adjacent elements or interleaved elements). Independent electrical readout enables faster counting rates (lower inductance per element, and multiple elements counting in parallel) and a degree of photon-number resolution (multiple elements can detect photons-simultaneously, c.f. Chapter 7) [18,64]. Multiple nanowires can also be wired in parallel (with a single electrical readout) as shown in Fig. 6.6, either with [70] or without [71] resistors between the nanowires. This approach typically permits faster counting rates for a given active area. Wiring the detectors without resistors between them

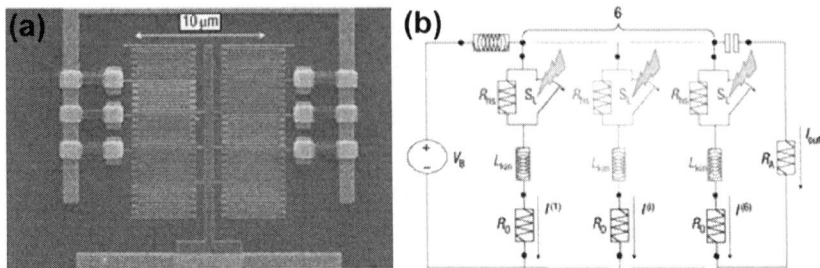

FIGURE 6.6 (a) Scanning electron micrograph colorized to illustrate the several parallel SNSPDs that constitute the device; (b) electrical schematic showing the expected mode of operation in which the signal current is summed onto the output amplifier (R_A). *Reprinted with permission from [73].*

can enable larger output signals (which is particularly useful for nanowires with very low critical currents), but provides no photon-number resolution. Using resistors can provide limited photon-number resolution, but the readout electronics to accomplish the required analog signal processing at high speeds have not yet been developed and these devices have only been demonstrated with relatively low detection efficiency. Recent work has shown that it may be possible to obtain photon-number resolution with a series connection of nanowire detectors [72].

In most applications to date, readout electronics for SNSPDs have been relatively simple, using a bias-tee and AC-coupled amplifiers at room temperature. However, as count rates, detection efficiencies, and array sizes are pushed closer to theoretical device limits, more sophisticated readout schemes become necessary, particularly in two areas. First, to maintain detection efficiency at high count rates, as well as to approach the maximum count rate dictated by the device's thermal time constant, a higher-impedance DC-coupled readout is required; this has been demonstrated recently using on-chip HEMT preamplifiers [20,74]. Second, larger array sizes will demand lower-power integrated wideband circuits that are compatible with available cooling capacities at these low temperatures. Such readout systems, based on both semiconducting (SiGe and CMOS) and superconducting (single flux quantum) logic [75,76], are currently being developed.

Since the first reports of single-photon detection using nanowires [77], the number of researchers and the number of papers in this field have grown rapidly. Even so, there are still open questions regarding the ultimate performance limits of these types of detectors. However, their demonstrated capabilities already make these devices ideal for many applications.

6.3 TRANSITION-EDGE SENSORS

Transition-edge sensors (TES) are highly sensitive microbolometers that can detect radiation from sub-mm wavelengths to gamma-rays. They typically consist of an absorber, a sensitive thermometer, and a weak thermal link to a cold bath. Here we focus on their application to detecting optical and near-optical photons. For a TES optimized to detect single photons, the temperature change caused by the absorption of a single photon is read out by the thermometer, and the system resets by cooling through the thermal link. Due to the low energies involved, detection of single photons at optical and near-infrared wavelengths requires low heat capacity and extremely sensitive thermometry.

6.3.1 Operating Principle

The most successful optical TES sensors use devices that operate below 1 K [78–80]. However, higher operating temperatures are being explored by sacrificing detector size [81]. For the devices operating below 1 K, the thermal isolation required is usually provided through the electron-phonon coupling, which can be weak at low temperatures. For example, in devices made using tungsten the sensor is cooled below the superconducting critical temperature T_c, typically $\approx$100 mK. A voltage bias is applied so that the resultant Joule heating raises the electron temperature of the detector to a temperature in the superconducting-to-normal transition, where the device has some non-zero resistance R_{TES}. The negative feedback inherent in voltage biasing a device with a positive temperature coefficient of resistance ($dR/dT > 0$) keeps the temperature stable; if the temperature increases (decreases), the resistance increases (decreases), resulting in a smaller (larger) amount of Joule heating. When a photon is absorbed the temperature of the sensor increases, resulting in a change in the electrical resistance. This change can be measured using a sensitive current amplifier such as a SQUID. The detector then resets as it cools through the weak thermal link. A typical example of a readout circuit is shown in Fig. 6.7. Because of the low operating temperature, and the need for a relatively stiff voltage bias, a small shunt resistor R_{bias} is placed in parallel with the TES detector, which is represented as R_{TES} in the schematic. The value of R_{bias} is chosen to be much smaller than R_{TES} at the operating point where the TES is properly biased to detect photons.

Transition-edge sensors are inherently energy resolving, and therefore can be engineered to resolve photon number for a monochromatic source of light. The response of the TES is determined by two coupled differential equations describing the electrical and thermal circuits. The differential equation that determines the current flowing in the devices depends upon the electrical circuit used to read it out. The thermal differential equation can be written as

$$C\frac{dT}{dt} = -P_{bath} + P_{joule} + P_{photons},$$

where C is the heat capacity of the device, P_{bath} is the power flowing from the electrons to the heat bath (or phonons), P_{Joule} is the Joule heating of the electrons, and $P_{photons}$ is the power from photons being detected. As an example,

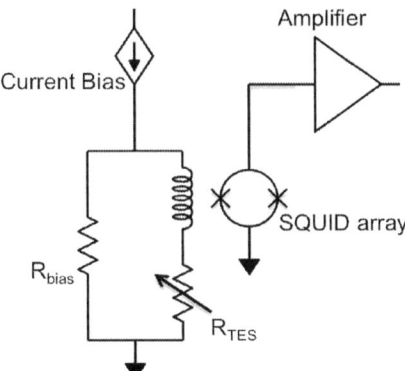

FIGURE 6.7 Electrical schematic of the bias and superconducting amplifier (SQUID) typically used to read out the signals from a TES detector. A voltage bias is formed by choosing a bias resistor that is much less than the operating resistance of the TES detector.

for tungsten-based TES detectors P_{bath} is given by the expression

$$P_{bath} = \kappa \times V \times (T_e^n - T_{ph}^n),$$

where κ is an electron-phonon decoupling constant, V is the volume of the device, and the exponent n is usually 5 for tungsten TES detectors. In general, the electrical and thermal differential equations are non-linear because of the non-linearity of the R vs. T curve in the superconducting transition, the temperature dependence of the thermal conductance to the heat bath, and the Joule power heating. However, reasonably good approximations can be obtained by taking the first-order terms in the power series expansion of these quantities in the temperature T. Using the equations above, and expressions for the Johnson noise of a resistor and for the phonon thermal noise, the energy resolution of a TES [82] is approximately given by

$$\Delta E_{FWHM} \approx 2\ln(2) \sqrt{\frac{4k_B T^2}{\alpha C}} \sqrt{\frac{n}{2}},$$

where α is a dimensionless parameter that characterizes the steepness of the R vs. T and k_B is the Boltzmann constant.

Another consequence of the coupled electrical and thermal differential equations is that the response of the TES can either be damped or oscillatory. Depending upon the parameters of the readout circuit and thermal circuit, the inductance of the readout and the heat capacity of the device can form an oscillator. The resistance of the device provides damping to this circuit. However, in some cases, this necessary damping can place a lower limit on the resistance usable for stable operation of the TES. It is usually preferable to bias the TES lower in its superconducting-to-normal transition to maximize the dynamic range for photon-number resolution. The optimum bias point for energy resolution depends upon details of the resistive transition and needs to be determined empirically for a given device. One challenge is that the noise contribution from the readout electronics can be significant when the TES is biased at high resistances because the intrinsic noise in the detector is proportional to the Johnson current noise.

As with SNSPDs, it is important to ensure that if the photon is absorbed by the superconductor it results in a measurable signal. For a TES, this requires that the entire sensor be biased in the superconducting transition uniformly so that the response to a photon will not vary spatially across the device. Consider the possibility of a temperature gradient along the direction of the current flow in the TES (perhaps because of a photon absorbed near one end). The hotter region will dissipate more Joule power, and tend to heat up even further. If this tendency is not overcome by thermal conduction within the device (which evens out the temperature across it), stable separation of the TES into a superconducting region and a resistive region can result. To avoid this the thermal conduction along/within the TES must be sufficiently greater than the thermal conductance between the device and the heat bath. The condition for stability against this type of phase separation [39] has been derived assuming that the thermal conduction within the TES is given by the Wiedemann-Franz law:

$$R_n < \frac{\pi^2 L T_c n}{G \alpha},$$

where R_n is the resistance of the TES when it is fully resistive, L is the Lorenz constant, T_c is the superconducting transition temperature, and G is the thermal conductance of the weak link. This phase separation effect is closely related to the 'latching' effect in SNSPDs described above, as the two systems are governed by similar electro-thermal equations [41].

A properly designed TES can have a quantum efficiency (i.e. the probability of generating a detectable signal given that a photon was absorbed) of essentially 100%. The external efficiency (i.e. including the probability of absorbing an incident photon) of TESs can also be increased close to unity by designing a stack of optical elements around the film, as discussed above in the case of SNSPDs. An example of a titanium TES embedded in an optical stack is shown in Fig. 6.8. Using numerical methods, the thicknesses of each layer can be determined to maximize the absorption in the TES material based on the optical properties of each layer.

An example of signals from a tungsten TES illuminated with an 805 nm pulsed laser is shown in Fig. 6.9. The rise times are typically <1 μs and are limited primarily by the inductance of the readout circuit divided by the resistance of the device. The recovery time is limited by the thermal properties of the device and the steepness of the resistive transition. Using different materials and different SQUID readout electronics, it is possible for TES detectors to

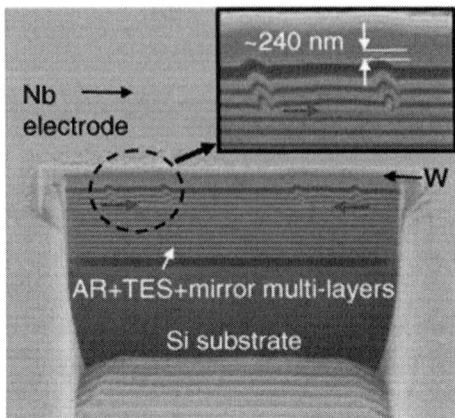

FIGURE 6.8 SEM image of a cross-section of a titanium TES embedded in a stack of dielectrics to enhance the absorption in the titanium sensor. The dark and bright layers represent SiO_2 and TiO_2, respectively. The titanium TES is located next to the red arrows. The entire structure is covered with a tungsten film used to protect the structure when fabricating the cross-section. (For interpretation of the references to color in this figure legend, the reader is referred to the web version of this book.) *Reprinted with permission from [83].*

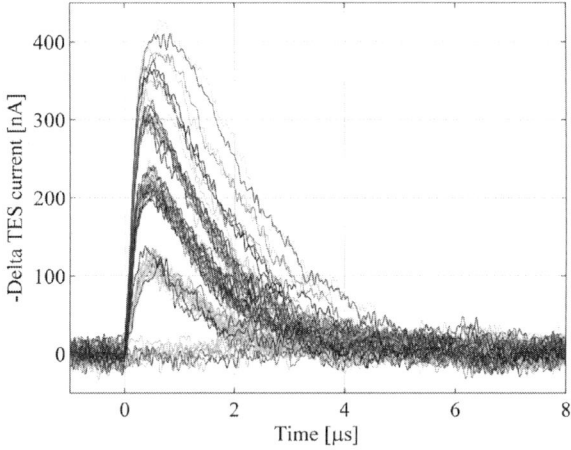

FIGURE 6.9 Response of a TES detector to an attenuated pulsed laser at 805 nm. Multiple trigger events from the pulsed laser are superimposed. Signals from no photon detected (flat traces), one, two, three, etc., photons are clearly distinguishable.

operate with faster rise times and hence lower jitter and faster recovery times [84,85].

From the traces in Fig. 6.9, it is easy to see the separation in pulse heights when zero, one, two, or three photons are absorbed. It is also easy to see in these traces that the accidental identification of background noise as photon counts is extremely unlikely, resulting in essentially no dark counts. However, in actual TES detector systems, there are still background counts from blackbody radiation that leaks into the fiber and is detected [86]. This is less of a concern in monochromatic systems, where as long as the energy of the source photons is sufficiently high, the blackbody background counts can be easily separated from signal counts based on pulse height.

6.3.2 Principal Strengths and Weaknesses

Transition-edge sensors have displayed the highest reported detection efficiencies for single-photon sensors. Following the discussion in Chapter 2 with respect to detection efficiency for a detector with optical input-coupling requirements such as a fiber-coupled detector in a cryostat, we define detection efficiency η_{DE} for TESs as the probability of detecting a photon that is inside the input fiber. Thus, η_{DE} includes the optical loss of the fiber within the cryostat and the coupling of the light from the fiber to the sensor, but not the coupling of light into the input end of the fiber. Published values of the system detection efficiency for fiber-coupled TES detectors are $\eta_{DE} > 95\%$ at 1550 nm [78] and $\eta_{DE} > 98\%$ at 850 nm [79]. The internal quantum efficiency of the TESs is believed to be extremely high, so the overall efficiency can likely reach values even closer to 100% through improvements in optical coupling and absorption in the detector.

The biggest weakness of the transition-edge sensor is the low operating temperature. The best results have been obtained using films with superconducting critical temperatures below 1 K. The most common cooling technologies used to achieve these temperatures are dilution refrigerators and adiabatic demagnetization refrigerators. Although both of these can be integrated with mechanical cryocoolers to enable cryogen-free operation, they are not yet sufficiently advanced to be considered "user-friendly," and their limited cooling power places restrictions on electrical and optical access, particularly in the context of detector arrays. The other weaknesses of the transition-edge sensor are the count rate and timing resolution. Recovery times are typically on the order of a microsecond, limiting the maximum count rate and effectively lowering the detection efficiency for high-flux sources. Furthermore, the timing resolution limits the maximum clock rate that could be used in a system to <100 MHz, and may make it harder to mitigate the effects of background counts in some applications.

6.3.3 Research Areas

The measured detection efficiency of a system with a TES sensor is often less than the expected device detection efficiency. Although TES detectors have the highest reported system detection efficiency, it is often limited by losses

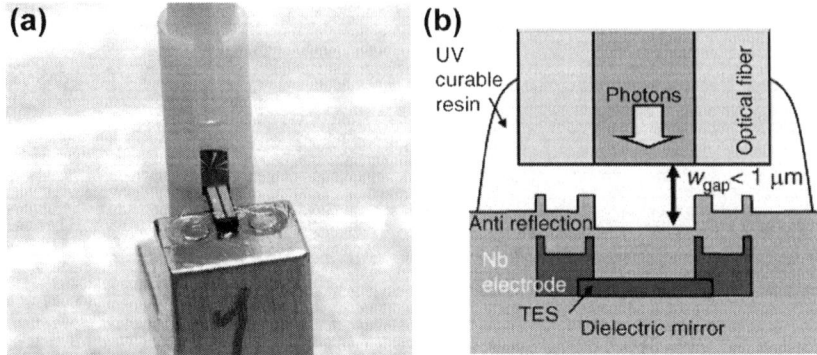

FIGURE 6.10 Examples of high efficiency methods of coupling of light from an optical fiber to a transition-edge sensor. (a) Micromachined silicon is used as a substrate to fit within a zirconia sleeve that aligns the ferrule with an optical fiber to the detector. The tip of a fiber is partially inserted into the zirconia sleeve. In the finished assembly, there is no discernible gap between the tip of the fiber and the detector chip. (b) Diagram (not drawn to scale) of an optical fiber aligned to a TES detector and held by a UV curable epoxy. *Reprinted with permission from [83].*

associated with coupling light from the optical fiber to the detectors in the system. A picture of the optical coupling scheme used in Miller *et al.* [61] is shown in Fig. 6.10a. In this scheme, the device is fabricated on a silicon substrate etched into a disk shape that can be placed inside a zirconia sleeve for a 2.5 mm-diameter fiber ferrule. The ferrule aligns the tip of the fiber to the center of the disk, where the TES is patterned. Another scheme is illustrated in Fig. 6.10b [79]. In this scheme, TES detectors are aligned to the optical fiber at room temperature and an index-matching epoxy is used to keep the fiber aligned and to reduce possible reflection losses between the tip of the fiber and the top surface of the TES. In addition to improving detection efficiency for fiber-coupled devices, there has been recent work to integrate TES detectors with photonic lightwave circuits [87]. One of the major limitations in this work is the relatively large mismatch in optical constants between the tungsten used for the TES and the silica used for the waveguide. This large mismatch would require a long length of tungsten to efficiently absorb all the photons; however, such a length is not possible because of the internal thermal conductance required for stable TES operation, as described earlier. Because of this, the efficiency achievable in this work was not close to 100%.

The relatively large timing jitter and recovery time has been an obstacle for more widespread use of the TES in quantum optics and quantum information experiments. However, there have been results with TES detectors indicating that they can be operated well within the timing resolution needed for MHz-clocked experiments. The simplest approach is to use a material with a higher operating temperature [84]. Another approach is to use gold cooling fins to introduce a second cooling path, along with more sophisticated SQUID electronics that reduce the inductance of the readout circuit and decrease the rise time [88]. The gold used to speed up the devices can also be used to extend the absorption area for TES detectors coupled to photonic lightwave circuits.

In contrast to conventional semiconductor-based detectors, PMTs, and SNSPDs, the TES is able to resolve the number of photons in a monochromatic optical pulse. However, processing the output signals to extract the photon number can be a non-trivial task for long data runs and multiple TES channels. There have been efforts to process signals with custom FPGA circuits [89] but nothing exists commercially that is straightforward to use. There also has been work [90,91] to extend the photon-number resolution of the TES to large photon numbers, where the device is extremely non-linear and in fact is saturated (i.e. the TES has reached its maximum resistance and stays hot). Extensions to the energy range for TES detectors may offer the potential to link classical approaches to optical power measurements to single-photon counting.

The ability of a TES detector to resolve photon number is a key feature that is still unmatched by other superconducting detector technologies. It also has been demonstrated to have high detection efficiency in many wavelength ranges commonly used for quantum optics experiments. TES detectors have not only been optimized to detect single visible and near-infrared wavelength photons but also sub-mm radiation, X-rays, and gamma-rays. In these applications, TESs can also be constructed in arrays with multiplexed SQUID readouts, and kilopixel arrays have been demonstrated for TESs operating at millimeter wavelengths [92]. In the future, more advanced versions of these multiplexing circuits may be integrated with the detectors to create even larger photon-counting arrays.

6.4 SUPERCONDUCTING TUNNEL JUNCTION DETECTORS

Superconducting tunnel junction detectors (STJs) were the pioneering technology in superconductor-based single-photon detection. These are superconducting devices that detect the quasiparticle excitations produced by an incident photon in a superconducting absorber by measuring the resulting charge tunneling through a junction barrier to ground [3,10].

6.4.1 Operating Principle

For most superconductors, the number of quasiparticles generated by the absorption of a photon can be written as

$$N_{\text{quasiparticles}} \approx \frac{E_{\text{photon}}}{1.74\Delta},$$

where E_{photon} is the energy of the photon and Δ is the superconducting gap energy. STJs are typically operated at temperatures below a tenth of T_c so that the thermal population of quasiparticles is negligible. Quasiparticles generated by a photon can be detected as net charge flowing across a voltage-biased Josephson tunnel junction between the absorber and a base electrode. This current is amplified by an intrinsic quasiparticle gain associated with Cooper-pair backtunneling through the barrier [93] and is typically measured with a charge amplifier. The best STJ results have also used a technique known as "quasiparticle trapping" to enhance this gain by confining the quasiparticles close to the tunnel barrier. In a quasiparticle-trapping structure, the photon is absorbed in a higher-T_c material that is in contact with a superconducting tunnel junction made from a lower-T_c material (typically aluminum). When the quasiparticle from the high-T_c absorber diffuses to the low-T_c material, it is extremely likely to scatter and relax to the gap energy of the lower-T_c material, so that it is effectively trapped in the lower-T_c electrode. This prevents diffusion away from the tunnel barrier, which can significantly enhance the gain when the quasiparticle lifetime is much longer than the tunneling time through the barrier. An example of the energy resolving capability is shown in Fig. 6.11 [83]. The detector clearly resolves the energies of single photons (produced by a lamp and a grating) whose energies are integer multiples of a 1183 nm photon, indicating that the energy from the absorption of a single 1183 nm photon can be distinguished from for example, two 1183 nm photons.

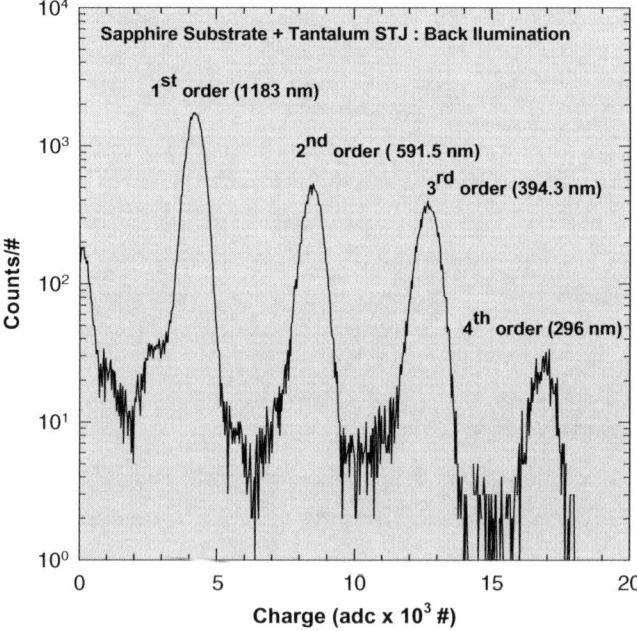

FIGURE 6.11 Histogram of the number of counts of the integrated charge (in analog-to-digital converter counts) of detection pulses generated by light sent through a monochromator set to 1183 nm. *Reproduced with permission [83].*

The theoretical limit to this energy resolution is associated with the statistics of the tunneling process, and can be approximated by

$$E_{\text{FWHM}} \approx 2.355\sqrt{1.74\Delta E_{\text{absorbed}}(F + G)},$$

where Δ is the energy gap, F is known as the Fano factor (typically ≈ 0.2), and G is a tunneling noise parameter used to describe the noise from the quasiparticle gain process. Usually, G is typically between 1 and 2, but recent work has shown that values less than 1 are possible [94,95].

Due to the difficulty of making reliable, high-quality (low-leakage) Josephson junctions, not many materials have been tested for use in STJs. The best devices use Al junctions and Ta, Nb, or Pb absorbers. Because the T_c of aluminum is ≈ 1.2 K, the best results require operating temperatures below 0.5 K. An additional complication when running STJ detectors is the need to use an applied magnetic field through the voltage-biased junction in order to suppress the DC supercurrent and Fiske modes [96]. Optimum operation of an STJ detector requires the user to adjust both the voltage bias and the magnetic field applied to the device. The diagram in Fig. 6.12 [83] shows an example of a detector with a tantalum-absorber and an aluminum junction. The best results were obtained by illuminating the detector through the sapphire substrate.

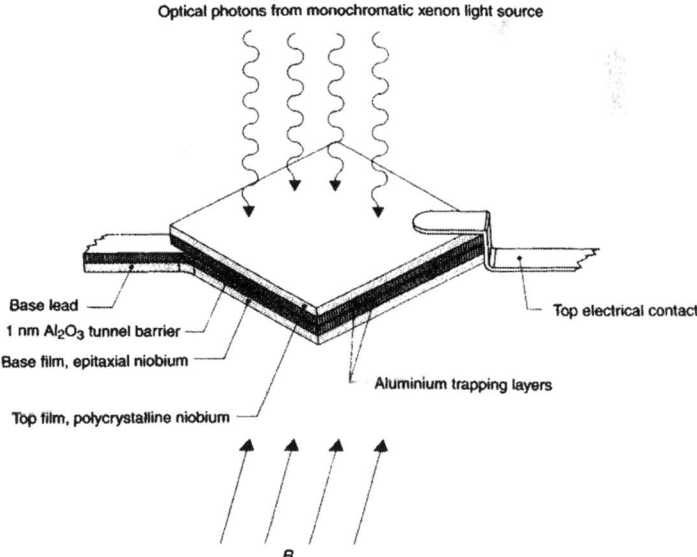

FIGURE 6.12 Illustration of an STJ device fabricated on sapphire using tantalum as the absorber and an aluminum tunnel junction. Also shown is the applied magnetic field used to suppress the Fiske modes. *Reproduced with permission [83].*

6.4.2 Strengths and Weaknesses

Like a TES, an STJ detector is capable of photon-number resolution. However, there are several issues that have limited its application in quantum optics and quantum information applications. In addition to the low operating temperatures needed (similar to the TES), it is difficult to optimize the STJ so that the absorption of the photons is near 100%. The best devices utilize highly crystalline superconducting absorbers, since imperfections in the crystal can shorten the quasiparticle lifetime resulting in a loss of signal. These defect sites can also cause variation in the measured pulse height depending upon the location where a photon is absorbed, which effectively reduces the energy resolution of the device. As a result, it is difficult to fabricate optical structures that simultaneously optimize the energy resolution and enhance the absorption so that it is nearly 100%.

6.4.3 Research Areas

Although STJs are not being optimized for quantum optics and quantum information applications today, breakthroughs in materials science and/or junction fabrication technology could yield energy-resolving detectors with high efficiency, higher operating temperatures, and higher speeds compared to TESs.

There is interest in the astrophysics and astronomy community to use STJs for wideband (UV to near-IR) imagers with moderate timing resolution (≈ 1 μs). For this application, the focus has been on making imaging arrays that may not have to deal with high count rates. One potential building block for such an imager would be a device consisting of two tunnel junctions that are located at the far ends of a superconducting absorber [97]. When a photon is absorbed, quasiparticles diffuse toward the two ends. The two resulting charge signals can give both the location of the absorbed photon and its total energy. Because of the recombination of quasiparticles in the superconducting absorber as they diffuse to the tunnel junctions, significant signal processing is needed to determine both photon position and energy, using prior calibration photon signals. One potential advantage for this device would be to effectively increase the area that can be covered by a given tunnel junction without completely sacrificing spatial resolution.

6.5 MICROWAVE KINETIC-INDUCTANCE DETECTORS

Microwave kinetic-inductance detectors (MKIDs) also detect the absorption of a photon by sensing the quasiparticles generated in a superconductor. However, in this kind of detector, the number of quasiparticles is measured via the small change in the superconductor's kinetic inductance [98].

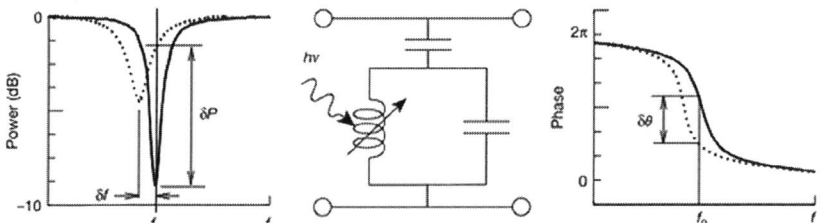

FIGURE 6.13 (left) Transmitted power as a function of applied microwave frequency showing the shift and increased loss associated with a resonator (center) resulting from the absorption of radiation. (right) Resulting phase change associated with the shift described in the left-hand-most frame. *Reprinted with permission from [99].*

6.5.1 Operating Principle

In an MKID, the change in kinetic inductance associated with photon-induced quasiparticle generation is measured by embedding the superconductor in a high-Q resonant microwave circuit. A schematic of one possible arrangement is shown in Fig. 6.13 [99]. In this figure, the MKID is drawn as a variable inductor that is wired in parallel with a capacitor. The LC tank circuit can be coupled to an electrical readout via a small coupling capacitor. Small changes in the inductance can be measured via a shift and/or broadening of the circuit's transmission or reflection resonance [100]. Since high-Q resonators with a variety of slightly different frequencies can be attached to the same feed line and read out in parallel on a single coaxial cable, these detectors are highly amenable to multiplexed readout of many pixels with minimal thermal load on the cryogenic refrigerator [100], as shown in Fig. 6.14. In addition, demultiplexing these signals can be done entirely at room temperature, where the required technology is readily available. State-of-the-art devices already contain thousands of pixels per feed line [101], and megapixel arrays are envisioned [102].

In principle, the MKID is an energy-resolving detector. Therefore, it is possible to resolve the number of photons absorbed by the detector from a monochromatic pulsed source. The fundamental energy resolution should ultimately be limited by the statistics of the quasiparticle generation (similar to the STJ, but absent the excess noise from the tunneling process):

$$E_{\text{FWHM}} \approx 2.355(1.74 \Delta E_{\text{absorbed}} F)^{1/2} \approx 45 \text{ meV}(1240 \text{ nm}/\lambda)^{1/2}(T_c/1 \text{ K})^{1/2}$$

However, amplifier noise and coupling to environmental noise [103–107] has prevented researchers from approaching the theoretical energy resolution. Very recently, photon detection at 1550 nm using an MKID has been demonstrated with photon-number resolution approaching that obtained with TES and STJ

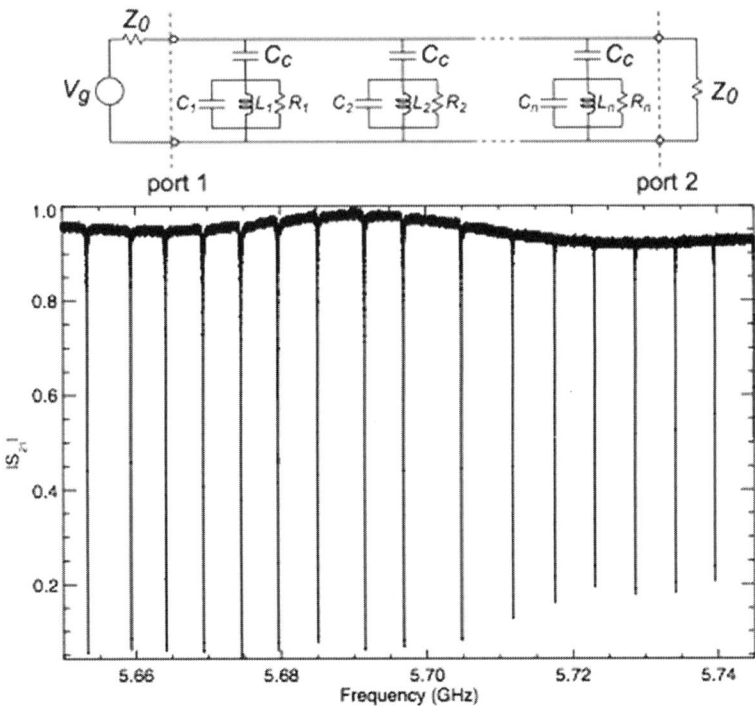

FIGURE 6.14 (top) Schematic of transmission-line geometry for MKID array in which a microwave signal is applied at port 1 and detected at port 2. (bottom) Transmission of microwave power through an MKID array in which the MKID elements resonate at unique frequencies. Each resonant dip in the S_{21} parameter corresponds to a unique element. *Reprinted with permission from [99].*

detectors [108]. A histogram of the output pulses from an MKID illuminated with a 1550-nm laser is shown in Fig. 6.15.

6.5.2 Strengths and Weaknesses

Like the TES and STJ detectors, the MKID resolves photon energy, which in pulsed monochromatic systems can be used both for photon-number resolution and for rejecting background counts at other wavelengths. Because MKIDs are read out using a resonant circuit, frequency-domain-multiplexing techniques can be used to read out many channels on a single wideband coaxial line, where the channel separation is determined by the desired signal bandwidth on each channel. For many astrophysical and astronomy applications, the required signal bandwidth is relatively small compared to the available $\approx$10 s of GHz frequency range, allowing substantial multiplexing. This is quite useful at the low operation temperature of these devices because one must minimize the number of wideband, low-loss readout lines to avoid excessive thermal loads.

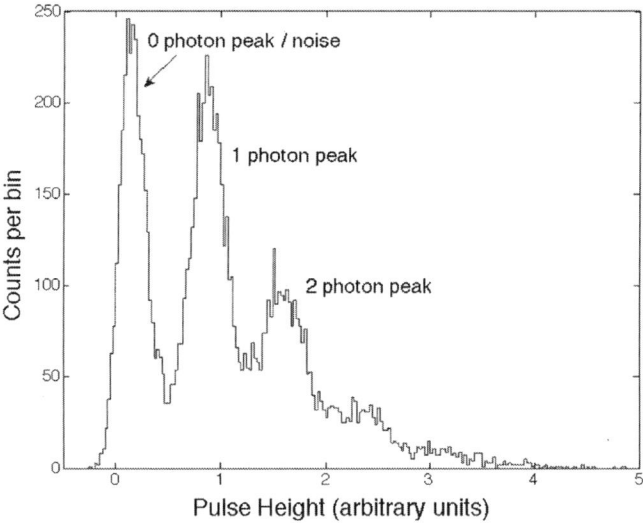

FIGURE 6.15 MKID output signal histogram from a pulsed 1550-nm laser showing evidence of quantization due to photon-number resolution.

However, for applications in which each channel requires a wide bandwidth, the frequency-domain-multiplexing technique becomes less useful.

As yet, not much effort has been devoted to optimizing the detection efficiency of these devices. There are two primary technical challenges faced in achieving close to 100% efficiency. The first is to ensure that the photons are absorbed in areas of the detector that are sensitive to changes in the inductance. Early MKID devices were planar microwave transmission-line resonators. As such, photons absorbed near a node of the resonator electric field produced a smaller change in inductance. This not only reduces the average sensitivity but also destroys the energy resolution. Two techniques have been developed to avoid this problem: quasiparticle traps can confine the photon-induced quasiparticles to a single region of the resonator with well-defined sensitivity, or lumped-element inductors with much smaller spatial variation of their sensitivity can be used [109,110]. The second problem is coupling of environmental noise to the device. Recent work has shown that the energy-resolving capability of MKIDs is extremely sensitive to unwanted microscopic two-level defects that have transition frequencies near the resonance of the MKID readout circuit. Common dielectrics that are typically used as spacers in multilayer structures (to improve absorption) tend to have a high density of these defects. This could substantially complicate the optimization of MKID efficiency.

6.5.3 Research Areas

MKIDs are an active area of research because of their potential in large imaging arrays for observational astrophysics and astronomy. There is significant ongoing effort to develop lower-noise cryogenic amplifiers optimized to readout MKID detectors [107, 111, 112] so that more detectors can be read out on one coaxial cable. In addition, there is a strong effort to determine and remove the sources of microscopic noise in the materials used with MKIDs.

6.6 CONCLUSIONS AND PERSPECTIVE

The performance of superconducting single-photon detectors often exceeds what is possible with their room-temperature counterparts. This additional performance has been used in a variety of quantum optics and quantum information applications [52, 113–120], as well as in other fields of research that require single-photon detection [121–125]. The majority of the technologies discussed above are either completely in the research domain, or else are in the very nascent stages of commercialization: acquisition of these technologies requires a significant expense and investment in infrastructure. It is thus worth considering the relative merits of these technologies and their potential when applied to various system needs.

There are two categories of systems that might be considered, which we will differentiate as narrow or broad in that they respectively have narrow performance requirements (e.g. focusing on a single metric like jitter or efficiency) or on a broad set of characteristics (e.g. simultaneously requiring high efficiency, low dark counts, fast reset time, low system size-weight-and-power). For narrow system requirements, the choice of detector is often trivially dictated by the system requirement, e.g. a system emphasizing low jitter would be naturally drawn to choose the SNSPD, which has the lowest jitter of any reported single-photon detector technology, while a system emphasizing photon-number resolution with a single detector element might prefer a TES detector. For broader system requirements, the SNSPDs tend to excel in that they combine high efficiency with fast reset, low dark counts, and low jitter. In addition, operation at 3 K permits a broader range of applications than many of the competing superconducting technologies, which require operation in the sub-1 K temperature range.

One often neglected advantage of all of the superconducting detector technologies is their relatively broad-band wavelength response. This advantage stems from the fundamental simplicity of the detection mechanism: photon absorption in metal is a relatively frequency-insensitive process (at least above the plasmon wavelength) and the superconducting gap of the materials used in these devices is so small that even relatively low-energy photons are often sufficient to precipitate a detection event. In some of the detector architectures (e.g. STJs and TESs) even radiation with photon energy below the superconducting gap can be detected (although practicalities prevent these detectors from achieving single-photon sensitivities at such long wavelengths).

Finally, one should of course mention the major disadvantage of all of these technologies—they require varying degrees of cryogenic cooling. This requirement necessitates power consumption on the order of kilowatts, and dimensions on the order of an electronics rack. However, such dimensions are not prohibitive for many applications where the detector element plays a crucial role or when there are other comparably large system components.

ACKNOWLEDGMENTS

In part, this work is sponsored by the National Aeronautics and Space Administration under Air Force Contract FA8721-05-C-0002. Opinions, interpretations, conclusions, and recommendations are those of the author and are not necessarily endorsed by the United States Government. This work was also sponsored by the National Science Foundation and by the National Institute of Standards and Technology. We thank Andrew Dane, and Faraz Najafi for assistance in preparing figures.

REFERENCES

[1] A.M. Kadin, M. Leung, A.D. Smith, and J.M. Murduck, "Nonbolometric Infrared Detection in Thin Superconducting Films via Photoproduction of Fluxon Pairs," IEEE Trans. Magn. 27, 1540–1543 (1991).
[2] A.M. Kadin and M.W. Johnson, "Nonequilibrium Photon-Induced Hotspot: A New Mechanism for Photodetection in Ultrathin Metallic Films," Appl. Phys. Lett. 69, 3938 (1996).
[3] M.A.C. Perryman, C.L. Foden, and A. Peacock, "A New Approach to Optical Photon Detection," In Proceedings of an ESA Symposium on Photon Detectors for Space Instrumentation (SEE N94-15025 03-19), pp. 21–26 (1992).
[4] K.D. Irwin, G.C. Hilton, D.A. Wollman, and J.M. Martinis, "X-ray Detection Using a Superconducting Transition-Edge Sensor Microcalorimeter with Electrothermal Feedback," 69, 1945–1947 (1996).
[5] M. Tinkham, Introduction to Superconductivity, Second ed., McGraw-Hill Science/Engineering/Math, p. 454 (1995).
[6] A. Goetz, "The Possible Use of Superconductivity for Radiometric Purposes," Phys. Rev. 55, 1270–1271 (1939).
[7] D. Andrews, W. Brucksch, W. Ziegler, and E. Blanchard, "Superconducting Films as Radiometric Receivers," Phys. Rev. 59, 1045–1046 (1941).
[8] G.N. Gol'tsman, O. Okunev, G. Chulkova, A. Lipatov, A. Semenov, K. Smirnov, B. Voronov, A. Dzardanov, C. Williams, R. Sobolewski, and G.N. Gol, "Picosecond Superconducting Single-Photon Optical Detector," Appl. Phys. Lett. 79, 705 (2001).
[9] J. Bardeen, L.N. Cooper, and J.R. Schrieffer, "Theory of Superconductivity," Phys. Rev. 108, 1175–1204 (1957).
[10] A. Peacock, P. Verhoeve, N. Rando, A. van Dordrecht, B.G. Taylor, C. Erd, M.A.C. Perryman, R. Venn, J. Howlett, D.J. Goldie, J. Lumley, and M. Wallis, "Single Optical Photon Detection with a Superconducting Tunnel Junction," Nature 381, 135–137, (1996).
[11] C.M. Natarajan, M.G. Tanner, and R.H. Hadfield, "Superconducting Nanowire Single-Photon Detectors: Physics and Applications," Supercond. Sci. Technol. 25, 063001 (2012).
[12] M.D. Eisaman, J. Fan, A. Migdall, and S.V Polyakov, "Invited Review Article: Single-Photon Sources and Detectors," Rev. Sci. Instrum. 82, 71101 (2011).
[13] K.M. Rosfjord, J.K.W. Yang, E.A. Dauler, A.J. Kerman, V. Anant, B.M. Voronov, G.N. Gol'tsman, and K.K. Berggren, "Nanowire Single-Photon Detector with an Integrated Optical Cavity and Anti-Reflection Coating," Opt. Express 14, 527–34 (2006).

[14] A. Kerman, E. Dauler, W. Keicher, J. Yang, K. Berggren, G. Gol'tsman, and B. Voronov, "Kinetic-Inductance-Limited Reset Time of Superconducting Nanowire Photon Counters," Appl. Phys. Lett. 88, (2006).
[15] R.H. Hadfield, A.J. Miller, S.W. Nam, R.L. Kautz, and R.E. Schwall, "Low-Frequency Phase Locking in High-Inductance Superconducting Nanowires," Appl. Phys. Lett. 87, 203505 (2005).
[16] W.H.P. Pernice, C. Schuck, O. Minaeva, M. Li, G.N. Goltsman, A.V. Sergienko, and H.X. Tang, "High-Speed and High-Efficiency Travelling Wave Single-Photon Detectors Embedded in Nanophotonic Circuits," Nat. Commun. 3, 1325 (2012).
[17] F. Marsili, V.B. Verma, J.A. Stern, S. Harrington, A.E. Lita, T. Gerrits, I. Vayshenker, B. Baek, M.D. Shaw, R.P. Mirin, and S.W. Nam, "Detecting Single Infrared Photons with 93% System Efficiency," Nat. Photonics 7, 210–214 (2013).
[18] E. Dauler, A. Kerman, B. Robinson, J. Yang, B. Voronov, G. Goltsman, S. Hamilton, and K. Berggren, "Photon-Number-Resolution with Sub-30-ps Timing Using Multi-Element Superconducting Nanowire Single Photon Detectors," J. Mod. Opt. 56, 364–373 (2009).
[19] T. Yamashita, S. Miki, K. Makise, W. Qiu, H. Terai, M. Fujiwara, M. Sasaki, and Z. Wang, "Origin of Intrinsic Dark Count in Superconducting Nanowire Single-Photon Detectors," Appl. Phys. Lett. 99, 161105 (2011).
[20] D. Rosenberg, A.J. Kerman, R.J. Molnar, and E.A. Dauler, "High-Speed and High-Efficiency Superconducting Nanowire Single Photon Detector Array," Opt. Express 21, 1440 (2013).
[21] H. Azzouz, S.N. Dorenbos, D. De Vries, E.B. Ureña, and V. Zwiller, "Efficient Single Particle Detection with a Superconducting Nanowire," AIP Adv. 2, 032124 (2012).
[22] A. Casaburi, N. Zen, K. Suzuki, M. Ejrnaes, S. Pagano, R. Cristiano, and M. Ohkubo, "Subnanosecond Time Response of Large-Area Superconducting Stripline Detectors for keV Molecular Ions," Appl. Phys. Lett. 94, 212502 (2009).
[23] L. Maingault, M. Tarkhov, I. Florya, A. Semenov, R. Espiau de Lamaëstre, P. Cavalier, G. Gol'tsman, J.-P. Poizat, and J.-C. Villégier, "Spectral Dependency of Superconducting Single Photon Detectors," J. Appl. Phys. 107, 116103 (2010).
[24] V. Anant, A.J. Kerman, E.A. Dauler, J.K.W. Yang, K.M. Rosfjord, and K.K. Berggren, "Optical Properties of Superconducting Nanowire Single-Photon Detectors," Opt. Express 16, 10750 (2008).
[25] F. Marsili, V.B. Verma, J.A. Stern, S. Harrington, A.E. Lita, T. Gerrits, I. Vayshenker, B. Baek, M.D. Shaw, R.P. Mirin, and S.W. Nam, "Detecting Single Infrared Photons with 93% System Efficiency," Nat. Photonics 7, 210–214 (2013).
[26] X. Hu, E.A. Dauler, R.J. Molnar, and K.K. Berggren, "Superconducting Nanowire Single-Photon Detectors Integrated with Optical Nano-Antennaes," Opt. Express 19, 17 (2010).
[27] E.F.C. Driessen, F.R. Braakman, E.M. Reiger, S.N. Dorenbos, V. Zwiller, and M.J.A. de Dood, "Impedance Model for the Polarization-Dependent Optical Absorption of Superconducting Single-Photon Detectors," Eur. Phys. J. Appl. Phys. 47, (2009).
[28] E.F.C. Driessen and M.J.A. de Dood, "The Perfect Absorber," Appl. Phys. Lett. 94, 171109 (2009).
[29] A. Kerman, E. Dauler, J. Yang, K. Rosfjord, V. Anant, K. Berggren, G. Gol'tsman, and B. Voronov, "Constriction-Limited Detection Efficiency of Superconducting Nanowire Single-Photon Detectors," Appl. Phys. Lett. 90, (2007).
[30] J.K.W. Yang, A.J. Kerman, E.A. Dauler, B. Cord, V. Anant, R.J. Molnar, and K.K. Berggren, "Suppressed Critical Current in Superconducting Nanowire Single-Photon Detectors With High Fill-Factors," IEEE Trans. Appl. Supercond. 19, 318–322 (2009).
[31] J. Clem and K. Berggren, "Geometry-Dependent Critical Currents in Superconducting Nanocircuits," Phys. Rev. B 84, 174510 (2011).
[32] H.L. Hortensius, E.F.C. Driessen, T.M. Klapwijk, K.K. Berggren, and J.R. Clem, "Critical-Current Reduction in Thin Superconducting Wires Due to Current Crowding," Appl. Phys. Lett. 100, 182602 (2012).
[33] A. Engel, A. Schilling, K. Il'in, and M. Siegel, "Dependence of Count Rate on Magnetic Field in Superconducting Thin-Film TaN Single-Photon Detectors," Phys. Rev. B 86, 140506 (2012).

[34] F. Marsili, F. Najafi, E. Dauler, F. Bellei, X. Hu, M. Csete, R.J. Molnar, and K.K. Berggren, "Single-Photon Detectors Based on Ultranarrow Superconducting Nanowires," Nano Lett. 11 2048–2053 (2011).
[35] B. Baek, A.E. Lita, V. Verma, and S.W. Nam, "Superconducting a-WxSi1-x Nanowire Single-Photon Detector with Saturated Internal Quantum Efficiency from Visible to 1850 nm," Appl. Phys. Lett. 98, 251105 (2011).
[36] M. Hofherr, D. Rall, K.S. Ilin, A. Semenov, N. Gippius, H.-W. Hübers, and M. Siegel, "Superconducting Nanowire Single-Photon Detectors: Quantum Efficiency vs. Film Thickness," J. Phys.: Conf. Ser. 234, 012017 (2010).
[37] W.J. Skocpol, M.R. Beasley, and M. Tinkham, "Phase-Slip Centers and Nonequilibrium Processes in Superconducting Tin Microbridges," J. Low Temp. Phys. 16, 145–167 (1974).
[38] J.K. W.Yang, A.J. Kerman, E.A. Dauler, V. Anant, K.M. Rosfjord, and K.K. Berggren, "Modeling the Electrical and Thermal Response of Superconducting Nanowire Single-Photon Detectors," IEEE Trans. Appl. Supercond. 17, 581–585, (2007).
[39] K.D. Irwin, G.C. Hilton, D.A. Wollman, and J.M. Martinis, "Thermal-Response Time of Superconducting Transition-Edge Microcalorimeters," J. Appl. Phys. 83, 3978 (1998).
[40] A. Semenov, P. Haas, H.-W. Hübers, K. Ilin, M. Siegel, A. Kirste, D. Drung, T. Schurig, and A. Engel, "Intrinsic Quantum Efficiency and Electro-Thermal Model of a Superconducting Nanowire Single-Photon Detector," J. Mod. Opt. 56, 345–351 (2009).
[41] A. Kerman, J. Yang, R. Molnar, E. Dauler, and K. Berggren, "Electrothermal Feedback in Superconducting Nanowire Single-Photon Detectors," Phys. Rev. B 79, 100509 (2009).
[42] A. Aspuru-Guzik and P. Walther, "Photonic Quantum Simulators," Nat. Phys. 8, 285–291 (2012).
[43] A. Acín, N. Brunner, N. Gisin, S. Massar, S. Pironio, and V. Scarani, "Device-Independent Security of Quantum Cryptography against Collective Attacks," Phys. Rev. Lett. 98, 230501 (2007).
[44] D. Bonneau, E. Engin, K. Ohira, N. Suzuki, H. Yoshida, N. Iizuka, M. Ezaki, C.M. Natarajan, M.G. Tanner, R.H. Hadfield, S.N. Dorenbos, V. Zwiller, J.L. O'Brien, and M. G. Thompson, "Quantum Interference and Manipulation of Entanglement in Silicon Wire Waveguide Quantum Circuits," New J. Phys. 14, 045003 (2012).
[45] M.E. Grein, A.J. Kerman, E.A. Dauler, O. Shatrovoy, R.J. Molnar, D. Rosenberg, J. Yoon, C.E. DeVoe, D.V. Murphy, B.S. Robinson, and D.M. Boroson, "Design of a Ground-Based Optical Receiver for the Lunar Laser Communications Demonstration," in 2011 International Conference on Space Optical Systems and Applications (ICSOS), pp. 78–82 (2011).
[46] B.S. Robinson, A.J. Kerman, E.A. Dauler, D.M. Boroson, S.A. Hamilton, J.K.W. Yang, V. Anant, and K.K. Berggren, "Demonstration of Gigabit-Per-Second and Higher Data Rates at Extremely High Efficiency Using Superconducting Nanowire Single Photon Detectors," in Optical Engineering + Applications, p. 67090Z–67090Z–8 (2007).
[47] S. Geissbuehler, N.L. Bocchio, C. Dellagiacoma, C. Berclaz, M. Leutenegger, and T. Lasser, "Mapping Molecular Statistics with Balanced Super-Resolution Optical Fluctuation Imaging (bSOFI)," Opt. Nanoscopy 1, 4 (2012).
[48] A.J. Nichols and C.L. Evans, "Video-Rate Scanning Confocal Microscopy and Microendoscopy," J. Vis. Exp.? JoVE , (2011).
[49] A. Crespi, M. Lobino, J.C.F. Matthews, A. Politi, C.R. Neal, R. Ramponi, R. Osellame, and J.L. O'Brien, "Measuring Protein Concentration with Entangled Photons," Appl. Phys. Lett. 100, 233704 (2012).
[50] E. Dauler, M. Stevens, B. Baek, R. Molnar, S. Hamilton, R. Mirin, S. Nam, and K. Berggren, "Measuring Intensity Correlations with A Two-Element Superconducting Nanowire Single-Photon Detector," Phys. Rev. A (At., Mol., and Opt. Phys.) 78, (2008).
[51] D. Elvira, X. Hachair, V.B. Verma, R. Braive, G. Beaudoin, I. Robert-Philip, I. Sagnes, B. Baek, S.W. Nam, E.A. Dauler, I. Abram, M.J. Stevens, and A. Beveratos, "Higher-Order Photon Correlations in Pulsed Photonic Crystal Nanolasers," Phys. Rev. A 84, 061802 (2011).

[52] M.J. Stevens, B. Baek, E.A. Dauler, A.J. Kerman, R.J. Molnar, S.A. Hamilton, K.K. Berggren, R.P. Mirin, S.W. Nam, Stevens, J. Martin, Burm, Dauler, and A. Eric, "High-Order Temporal Coherences of Chaotic and Laser Light," Opt. Express 18, 1430 (2010).

[53] R. Bose, D. Sridharan, H. Kim, G.S. Solomon, and E. Waks, "Low-Photon-Number Optical Switching with a Single Quantum Dot Coupled to a Photonic Crystal Cavity," Phys. Rev. Lett. 108 227402 (2012).

[54] R.E. Correa, E.A. Dauler, G. Nair, S.H. Pan, D. Rosenberg, A.J. Kerman, R.J. Molnar, X. Hu, F. Marsili, V. Anant, K.K. Berggren, and M.G. Bawendi, "Single Photon Counting from Individual Nanocrystals in the Infrared," Nano Lett. 12, 2953–2958 (2012).

[55] M.G. Tanner, C.M. Natarajan, V.K. Pottapenjara, J.A. O'Connor, R.J. Warburton, R.H. Hadfield, B. Baek, S. Nam, S.N. Dorenbos, E.B. Urenñ, T. Zijlstra, T.M. Klapwijk, and V. Zwiller, "Enhanced Telecom Wavelength Single-Photon Detection with NbTiN Superconducting Nanowires on Oxidized Silicon," Appl. Phys. Lett. 96, 221109 (2010).

[56] V.A. Seleznev, M.A. Tarkhov, B.M. Voronov, I.I. Milostnaya, V.Y. Lyakhno, A.S. Garbuz, M.Y. Mikhailov, O.M. Zhigalina, and G.N. Gol'tsman, "Deposition and Characterization of Few-Nanometers-Thick Superconducting Mo–Re Films," Supercond. Sci. Technol. 21, 115006 (2008).

[57] S.N. Dorenbos, P. Forn-Díiaz, T. Fuse, A.H. Verbruggen, T. Zijlstra, T.M. Klapwijk, and V. Zwiller, "Low Gap Superconducting Single Photon Detectors for Infrared Sensitivity," Appl. Phys. Lett. 98, 251102 (2011).

[58] H. Shibata, H. Takesue, T. Honjo, T. Akazaki, and Y. Tokura, "Single-Photon Detection Using Magnesium Diboride Superconducting Nanowires," Appl. Phys. Lett. 97, 212504 (2010).

[59] A.J. Annunziata, O. Quaranta, D.F. Santavicca, A. Casaburi, L. Frunzio, M. Ejrnaes, M.J. Rooks, R. Cristiano, S. Pagano, A. Frydman, and D.E. Prober, "Reset Dynamics and Latching in Niobium Superconducting Nanowire Single-Photon Detectors," J. Appl. Phys. 108, 084507 (2010).

[60] S. Miki, M. Takeda, M. Fujiwara, M. Sasaki, and Z. Wang, "Compactly Packaged Superconducting Nanowire Single-Photon Detector with an Optical Cavity for Multichannel System," (2010).

[61] A.J. Miller, A.E. Lita, B. Calkins, I. Vayshenker, S.M. Gruber, and S.W. Nam, "Compact Cryogenic Self-Aligning Fiber-to-Detector Coupling with Losses Below One Percent," Opt. Express 19, 9102–9110 (2011).

[62] J.-L.F.-X. Orgiazzi and A.H. Majedi, "Robust Packaging Technique and Characterization of Fiber-Pigtailed Superconducting NbN Nanowire Single Photon Detectors," Appl. Supercond. IEEE Trans. 19, 341–345 (2009).

[63] S. Miki, M. Fujiwara, M. Sasaki, and Z. Wang, "Development of SNSPD System With Gifford-McMahon Cryocooler," Appl. Supercond. IEEE Trans. 19, 332–335 (2009).

[64] D. Rosenberg, A.J. Kerman, R.J. Molnar, and E.A. Dauler, "High-Speed and High-Efficiency Superconducting Nanowire Single Photon Detector Array," Opt. Express 21, 1440 (2013).

[65] J.P. Sprengers, A. Gaggero, D. Sahin, S. Jahanmirinejad, G. Frucci, F. Mattioli, R. Leoni, J. Beetz, M. Lermer, M. Kamp, S. Höfling, R. Sanjines, and A. Fiore, "Waveguide Superconducting Single-Photon Detectors for Integrated Quantum Photonic Circuits," Appl. Phys. Lett. 99, 181110 (2011).

[66] X. Hu, C.W. Holzwarth, D. Masciarelli, E.A. Dauler, and K.K. Berggren, "Efficiently Coupling Light to Superconducting Nanowire Single-Photon Detectors," IEEE Trans. Appl. Supercond. (2009).

[67] A. Semenov, B. Günther, U. Böttger, H.-W. Hübers, H. Bartolf, A. Engel, A. Schilling, K. Ilin, M. Siegel, R. Schneider, D. Gerthsen, and N. Gippius, "Optical and Transport Properties of Ultrathin NbN Films and Nanostructures," Phys. Rev. B 80, 054510 (2009).

[68] M. Csete, Á. Sipos, F. Najafi, X. Hu, and K.K. Berggren, "Numerical Method to Optimize the Polar-Azimuthal Orientation of Infrared Superconducting-Nanowire Single-Photon Detectors," Appl. Opt. 50, 5949–56 (2011).

[69] D.D. Bacon, A.T. English, S. Nakahara, F.G. Peters, H. Schreiber, W.R. Sinclair, and R.B. van Dover, "Properties of NbN Thin Films Deposited on Ambient Temperature Substrates," J. Appl. Phys. 54, 6509 (1983).

[70] F. Marsili†, D. Bitauld, A. Fiore, A. Gaggero, R. Leoni, F. Mattioli, A. Divochiy, A. Korneev, V. Seleznev, N. Kaurova, O. Minaeva, and G. Goltsman, "Superconducting Parallel Nanowire Detector with Photon Number Resolving Functionality," J. Mod. Opt. 56, 334–344 (2009).

[71] M. Ejrnaes, R. Cristiano, O. Quaranta, S. Pagano, A. Gaggero, F. Mattioli, R. Leoni, B. Voronov, and G. Gol'tsman, "A Cascade Switching Superconducting Single Photon Detector," Appl. Phys. Lett. 91, 262509 (2007).

[72] S. Jahanmirinejad and A. Fiore, "Proposal for a Superconducting Photon Number Resolving Detector with Large Dynamic Range," Opt. Express 20, 5017 (2012).

[73] A. Divochiy, F. Marsili, D. Bitauld, A. Gaggero, R. Leoni, F. Mattioli, A. Korneev, V. Seleznev, N. Kaurova, O. Minaeva, G. Gol'tsman, K.G. Lagoudakis, M. Benkhaoul, F. Lévy, and A. Fiore, "Superconducting Nanowire Photon-Number-Resolving Detector at Telecommunication Wavelengths," Nat. Photonics 2, 302–306 (2008).

[74] A.J. Kerman, D. Rosenberg, R.J. Molnar, and E.A. Dauler, "Readout of Superconducting Nanowire Single-Photon Detectors at High Count Rates," J. Appl. Phys. 113, 144511 (2013).

[75] T. Ortlepp, M. Hofherr, L. Fritzsch, S. Engert, K. Ilin, D. Rall, H. Toepfer, H.-G.-G. Meyer, and M. Siegel, "Demonstration of Digital Readout Circuit for Superconducting Nanowire Single Photon Detector," Opt. Express 19, 18593–601 (2011).

[76] H. Terai and S. Miki, "Readout Electronics Using Single-Flux-Quantum Circuit Technology for Superconducting Single-Photon Detector Array," IEEE Trans. Appl. Supercond. 19, 350–353, (2009).

[77] A.D. Semenov, G.N. Gol'tsman, and A.A. Korneev, "Quantum Detection By Current Carrying Superconducting Film," Physica C: Supercond. 351, 349–356 (2001).

[78] A. E. Lita, A. J. Miller, and S. W. Nam, "Counting Near-Infrared Single-Photons with 95% Efficiency," Opt. Express 16, 3032–3040 (2008).

[79] D. Fukuda, G. Fujii, T. Numata, K. Amemiya, A. Yoshizawa, H. Tsuchida, H. Fujino, H. Ishii, T. Itatani, S. Inoue, and T. Zama, "Titanium-Based Transition-Edge Photon Number Resolving Detector with 98% Detection Efficiency with Index-Matched Small-Gap Fiber Coupling," Opt. Express 19, 870–875 (2011).

[80] L. Lolli, E. Taralli, and M. Rajteri, "Ti/Au TES to Discriminate Single Photons," J. Low Temp. Phys. 167, 803–808 (2012).

[81] D.F. Santavicca, F.W. Carter, and D.E. Prober, "Proposal for a GHz Count Rate Near-IR Single-Photon Detector Based on a Nanoscale Superconducting Transition Edge Sensor," in SPIE Defense, Security, and Sensing, p. 80330W–80330W–5 (2011).

[82] K.D. Irwin and G.C. Hilton, "Transition-Edge Sensors," in Topics in Applied Physics, Cryogenic Particle Detection, edited by C. Enss, 99th ed., vol. 149, Springer, pp. 63–150 (2005).

[83] T. Peacock, P. Verhoeve, N. Rando, C. Erd, M. Bavdaz, B.G. Taylor, and D. Perez, "Recent Developments in Superconducting Tunnel Junctions for Ultraviolet, Optical & Near Infrared Astronomy," Astron. Astrophys. Suppl. Ser. 127, 497–504 (1998).

[84] D. Fukuda, G. Fujii, T. Numata, A. Yoshizawa, H. Tsuchida, H. Fujino, H. Ishii, T. Itatani, S. Inoue, and T. Zama, "Photon Number Resolving Detection with High Speed and High Quantum Efficiency," Metrologia 46, S288–S292 (2009).

[85] A. Lamas-Linares, B. Calkins, N.A. Tomlin, T. Gerrits, A.E. Lita, J. Beyer, R.P. Mirin, and S. Woo Nam, "Nanosecond-Scale Timing Jitter for Single Photon Detection in Transition Edge Sensors," Appl. Phys. Lett. 102, 231117 (2013).

[86] A.J. Miller, A.E. Lita, D. Rosenberg, S. Gruber, and S. Nam, "Superconducting Photon Number Resolving Detectors: Performance and Promise," in Proceedings of the 8th International Conference on Quantum Communication, Measurement and Computing, pp. 445–450 (2007).

[87] T. Gerrits, N. Thomas-Peter, J. Gates, A. Lita, B. Metcalf, B. Calkins, N. Tomlin, A. Fox, A. Linares, J. Spring, N. Langford, R. Mirin, P.G. Smith, I. Walmsley, and S. Nam, "On-Chip, Photon-Number-Resolving, Telecommunication-Band Detectors for Scalable Photonic Information Processing," Phys. Rev. A 84, (2011).

[88] B. Calkins, A.E. Lita, A.E. Fox, and S. Woo Nam, "Faster Recovery Time of a Hot-Electron Transition-Edge Sensor By Use of Normal Metal Heat-Sinks," Appl. Phys. Lett. 99, 241114 (2011).
[89] S. Nam, J. Beyer, G. Hilton, K. Irwin, C. Reintsema, and J.M. Martinis, "Electronics for Arrays of Transition Edge Sensors Using Digital Signal Processing," IEEE Trans. Appl. Supercond. 13, 618–621 (2003).
[90] T. Gerrits, B. Calkins, N. Tomlin, A.E. Lita, A. Migdall, R. Mirin, and S.W. Nam, "Extending Single-Photon Optimized Superconducting Transition Edge Sensors Beyond the Single-Photon Counting Regime," Opt. Express 20, 23798–23810 (2012).
[91] Z.H. Levine, T. Gerrits, A.L. Migdall, D.V. Samarov, B. Calkins, A.E. Lita, and S.W. Nam, "Algorithm for Finding Clusters with a Known Distribution and its Application to Photon-Number Resolution Using a Superconducting Transition-Edge Sensor," J. Opt. Soc. Am. B 29 2066 (2012).
[92] D.S. Swetz, P.A.R. Ade, M. Amiri, J.W. Appel, E.S. Battistelli, B. Burger, J. Chervenak, M.J. Devlin, S.R. Dicker, W.B. Doriese, R. Dünner, T. Essinger-Hileman, R.P. Fisher, J.W. Fowler, M. Halpern, M. Hasselfield, G.C. Hilton, A.D. Hincks, K.D. Irwin, N. Jarosik, M. Kaul, J. Klein, J.M. Lau, M. Limon, T.A. Marriage, D. Marsden, K. Martocci, P. Mauskopf, H. Moseley, C.B. Netterfield, M.D. Niemack, M.R. Nolta, L.A. Page, L. Parker, S.T. Staggs, O. Stryzak, E.R. Switzer, R. Thornton, C. Tucker, E. Wollack, and Y. Zhao, "Overview of the Atacama Cosmology Telescope: Receiver, Instrumentation, and Telescope Systems," Astrophys. J. Suppl. Ser. 194, 41 (2011).
[93] K.E. Gray, "A Superconducting Transistor," Appl. Phys. Lett. 32, 392 (1978).
[94] D.D.E. Martin, P. Verhoeve, A. Peacock, A.G. Kozorezov, J.K. Wigmore, H. Rogalla, and R. Venn, "Resolution Limitation Due to Phonon Losses in Superconducting Tunnel Junctions," Appl. Phys. Lett. 88, 123510 (2006).
[95] P. Verhoeve, R. den Hartog, A. Kozorezov, D. Martin, A. van Dordrecht, J.K. Wigmore, and A. Peacock, "Time Dependence of Tunnel Statistics and the Energy Resolution of Superconducting Tunnel Junctions," J. Appl. Phys. 92, 6072 (2002).
[96] M. Fiske, "Temperature and Magnetic Field Dependencies of the Josephson Tunneling Current," Rev. Mod. Phys. 36, 221–222 (1964).
[97] C. Wilson, K. Segall, L. Frunzio, L. Li, D. Prober, D. Schiminovich, B. Mazin, C. Martin, and R. Vasquez, "Optical/UV Single-Photon Imaging Spectrometers Using Superconducting Tunnel Junctions," Nucl. Instrum. Methods Phys. Res. Sect. A: Accelerat. Spectrom. Detect. Assoc. Equip. 444, 449–452 (2000).
[98] D.G. McDonald, "Novel Superconducting Thermometer for Bolometric Applications," Appl. Phys. Lett. 50, 775 (1987).
[99] B.A. Mazin, B. Young, B. Cabrera, and A. Miller, "Microwave Kinetic Inductance Detectors: The First Decade," AIP Conf. Proc. 1185, 135–142 (2009).
[100] P.K. Day, H.G. LeDuc, B.A. Mazin, A. Vayonakis, and J. Zmuidzinas, "A Broadband Superconducting Detector Suitable for Use in Large Arrays," Nature 425, 817–821 (2003).
[101] B.A. Mazin, K. O'Brien, S. McHugh, B. Bumble, D. Moore, S. Golwala, and J. Zmuidzinas, "ARCHONS: A Highly Multiplexed Superconducting Optical to Near-IR Camera," in SPIE Astronomical Telescopes + Instrumentation, pp. 773518–773518–10 (2010).
[102] B.A. Mazin, B. Bumble, S.R. Meeker, K. O'Brien, S. McHugh, and E. Langman, "A Superconducting Focal Plane Array for Ultraviolet, Optical, and Near-Infrared Astrophysics," Opt. Express 20, 1503 (2012).
[103] J. Baselmans, S.J.C. Yates, R. Barends, Y.J.Y. Lankwarden, J.R. Gao, H. Hoevers, and T.M. Klapwijk, "Noise and Sensitivity of Aluminum Kinetic Inductance Detectors for Sub-mm Astronomy," J. Low Temp. Phys. 151, 524–529 (2008).
[104] R. Barends, N. Vercruyssen, A. Endo, P.J. de Visser, T. Zijlstra, T.M. Klapwijk, P. Diener, S.J.C. Yates, and J.J.A. Baselmans, "Minimal Resonator Loss for Circuit Quantum Electrodynamics," Appl. Phys. Lett. 97, 023508 (2010).
[105] R. Barends, N. Vercruyssen, A. Endo, P.J. de Visser, T. Zijlstra, T.M. Klapwijk, and J.J.A. Baselmans, "Reduced Frequency Noise in Superconducting Resonators," Appl. Phys. Lett. 97, 033507 (2010).

[106] R. Barends, J.J.A. Baselmans, S.J.C. Yates, J.R. Gao, J.N. Hovenier, and T.M. Klapwijk, "Quasiparticle Relaxation in Optically Excited High-Q Superconducting Resonators," Phys. Rev. Lett. 100, 257002 (2008).
[107] J. Gao, M. Daal, J.M. Martinis, A. Vayonakis, J. Zmuidzinas, B. Sadoulet, B.A. Mazin, P.K. Day, and H.G. Leduc, "A Semiempirical Model for Two-Level System Noise in Superconducting Microresonators," Appl. Phys. Lett. 92, 212504 (2008).
[108] J. Gao, M.R. Vissers, M.O. Sandberg, F.C.S. da Silva, S.W. Nam, D.P. Pappas, D.S. Wisbey, E.C. Langman, S.R. Meeker, B.A. Mazin, H.G. Leduc, J. Zmuidzinas, and K. D. Irwin, "A Titanium-Nitride Near-Infrared Kinetic Inductance Photon-Counting Detector and its Anomalous Electrodynamics," Appl. Phys. Lett. 101, 142602 (2012).
[109] B.A. Mazin, M. E. Eckart, B. Bumble, S. Golwala, P.K. Day, J. Gao, and J. Zmuidzinas, "Optical/UV and X-Ray Microwave Kinetic Inductance Strip Detectors," J. Low Temp. Phys. 151, 537–543 (2008).
[110] S. Doyle, J. Naylon, P. Mauskopf, A. Porch, and C. Dunscombe, "Lumped Element Kinetic Inductance Detectors," in International Space Terahertz Technology Conference Proceedings, pp. 170–177 (2007).
[111] M.A. Castellanos-Beltran, K.D. Irwin, G.C. Hilton, L.R. Vale, and K.W. Lehnert, "Amplification and Squeezing of Quantum Noise with a Tunable Josephson Metamaterial," Nat. Phys. 4, 929–931 (2008).
[112] B. Ho Eom, P.K. Day, H.G. LeDuc, and J. Zmuidzinas, "A Wideband, Low-Noise Superconducting Amplifier with High Dynamic Range," Nat. Phys. 8, 623–627 (2012).
[113] R.H. Hadfield, M.J. Stevens, S. S. Gruber, A.J. Miller, R.E. Schwall, R.P. Mirin, and S.W. Nam, "Single Photon Source Characterization with a Superconducting Single Photon Detector," Opt. Express 13, 10846–10853 (2005).
[114] M.B. Nasr, O. Minaeva, G.N. Goltsman, A.V. Sergienko, B.E. Saleh, and M.C. Teich, "Submicron Axial Resolution in an Ultrabroadband Two-Photon Interferometer Using Superconducting Single-Photon Detectors," Opt. Express 16, 15104 (2008).
[115] D. Elvira, A. Michon, B. Fain, G. Patriarche, G. Beaudoin, I. Robert-Philip, Y. Vachtomin, A.V. Divochiy, K.V. Smirnov, G.N. Gol'tsman, I. Sagnes, and A. Beveratos, "Time-Resolved Spectroscopy of InAsP/InP(001) Quantum Dots Emitting Near 2 μm," Appl. Phys. Lett. 97, 131907 (2010).
[116] S. Wang, W. Chen, J.-F. Guo, Z.-Q. Yin, H.-W. Li, Z. Zhou, G.-C. Guo, and Z.-F. Han, "2 GHz Clock Quantum Key Distribution Over 260 km of Standard Telecom Fiber," Opt. Lett. 37, 1008 (2012).
[117] D. Rosenberg, J.W. Harrington, P.R. Rice, P.A. Hiskett, C.G. Peterson, R.J. Hughes, A.E. Lita, S.W. Nam, and J.E. Nordholt, "Long-Distance Decoy-State Quantum Key Distribution in Optical Fiber," Phys. Rev. Lett. 98, 10503 (2007).
[118] T. Gerrits, S. Glancy, T. Clement, B. Calkins, A. Lita, A. Miller, A. Migdall, S. Nam, R. Mirin, and E. Knill, "Generation of Optical Coherent-State Superpositions By Number-Resolved Photon Subtraction from the Squeezed Vacuum," Phys. Rev. A 82, (2010).
[119] K. Tsujino, D. Fukuda, G. Fujii, S. Inoue, M. Fujiwara, M. Takeoka, and M. Sasaki, "Quantum Receiver Beyond the Standard Quantum Limit of Coherent Optical Communication," Phys. Rev. Lett. 106, (2011).
[120] D.H. Smith, G. Gillett, M.P. de Almeida, C. Branciard, A. Fedrizzi, T.J. Weinhold, A. Lita, B. Calkins, T. Gerrits, H.M. Wiseman, S.W. Nam, and A.G. White, "Conclusive Quantum Steering with Superconducting Transition-Edge Sensors," Nat. Commun. 3, 625 (2012).
[121] M.J. Stevens, R.H. Hadfield, R E. Schwall, S.W. Nam, and R.P. Mirin, "Time-Correlated Single-Photon Counting with Superconducting Single-Photon Detectors," in Optics East 2006, p. 63720U–63720U–10 (2006).
[122] G.S. Buller, N.J. Krichel, A. McCarthy, N.R. Gemmell, M.G. Tanner, C.M. Natarajan, X. Ren, and R.H. Hadfield, "Kilometer Range Depth Imaging Using Time-Correlated Single-Photon Counting," in SPIE Optical Engineering + Applications, vol. 8155, p. 81551I–81551I–8 (2011).
[123] B.S. Robinson, A.J. Kerman, E.A. Dauler, R.J. Barron, D.O. Caplan, M.L. Stevens, J.J. Carney, S.A. Hamilton, J.K.W. Yang, and K.K. Berggren, "781 Mbit/s Photon-Counting Optical Communications Using a Superconducting Nanowire Detector," Opt. Lett. 31, 444–446 (2006).

[124] M.G. Tanner, S.D. Dyer, B. Baek, R.H. Hadfield, and S. Woo Nam, "High-Resolution Single-Mode Fiber-Optic Distributed Raman Sensor for Absolute Temperature Measurement Using Superconducting Nanowire Single-Photon Detectors," Appl. Phys. Lett. 99, 201110 (2011).

[125] J. Hu, Q. Zhao, X. Zhang, L. Zhang, X. Zhao, L. Kang, and P. Wu, "Photon-Counting Optical Time-Domain Reflectometry Using a Superconducting Nanowire Single-Photon Detector," J. Lightwave Technol. 30, 2583–2588 (2012).

Chapter 7

Hybrid Detectors

Ivo Pietro Degiovanni[*], Sergey Polyakov[†], Alan Migdall[†],
Hendrik B. Coldenstrodt-Ronge[‡], Ian A. Walmsley[‡],
and Franco N.C. Wong[††]

[*]I.N.RI.M., Strada delle Cacce 91, 10135 Turin, Italy
[†]Joint Quantum Institute and National Institute of Standards and Technology, 100 Bureau Dr, Stop 8410, Gaithersburg, MD 20899, USA
[‡]University of Oxford, Clarendon Laboratory, Parks Road, Oxford, OX1 3PU, United Kingdom
[††]Massachusetts Institute of Technology, Research Laboratory of Electronics, Cambridge, MA 02139, USA

Chapter Outline

7.1 Introduction	218
7.2 Space-Multiplexed Detectors	219
7.2.1 Introduction	219
7.2.2 Theory of Operation	220
7.2.3 Experimental Implementations of Space-Multiplexed Detectors	231
7.3 Time-Multiplexed Detectors	236
7.3.1 Introduction	236
7.3.2 Fiber-Loop Detectors	237
7.3.3 Weak-Homodyne Detection	241
7.4 Up-Conversion Detectors	243
7.4.1 Introduction	243
7.4.2 Theory of Single-Photon Up-Conversion	244
7.4.3 Up-Conversion Techniques	245
7.4.4 Pulsed Up-Conversion	249
7.4.5 Ultrafast Up-Conversion	250
7.5 Conclusion	253
References	253

7.1 INTRODUCTION

We present an overview of efforts to improve photon-counting detection systems through the use of hybrid detection techniques such as spatial- and time-multiplexing of conventional detectors, and frequency up-conversion. We review the basic operation for these methods and illustrate their utility in a number of applications showing new or improved capabilities compared with conventional methods.

Photon-number-resolving (PNR) detectors are a critically important tool for many fields of optical science and technology [1] such as quantum metrology [2–4], quantum imaging [5], quantum information [6–8], and foundations of quantum mechanics [9]. Unfortunately, most conventional single-photon detectors can only distinguish between zero photons detected ("no-click") and one or more photons detected ("click"). What is generally meant by detectors with inherent (or full) PNR capability are those able to produce a signal proportional to the number of absorbed photons. However, there are currently few types of detectors with this intrinsic PNR ability [1], with the most promising type being the transition edge sensor (TES) [10] (discussed in Chapter 6). However, these PNR detectors require advanced cryogenic equipment and are hardly accessible or convenient for average laboratories or for end users. While future developments may increase the accessibility of these detectors, other technologies that promise more cost-effective PNR capability are being actively pursued as well. In the first two sections of this chapter, we examine the use of spatial [11–13] or temporal [14–17] multiplexing using conventional click/no-click detectors with no intrinsic PNR capability. In addition, multiplexing can also be used to address other deficiencies of detectors (such as accessing high order correlations of light [18, 19]).

Frequency up-conversion techniques address a different photon-counting issue: the challenges associated with the steep tradeoffs among efficiency, maximum count rates, dark-count rates, etc., at the longer wavelengths beyond 1 μm. Silicon single-photon avalanche diodes (SPADs) are well suited for continuous-wave (CW) operation with low dark counts, low afterpulsing, and reasonably high efficiency of greater than 50% at visible and near-infrared (near-IR) wavelengths. At the longer wavelengths, recent advances in InGaAs detector technologies (see Chapter 4) have closed the gap in these performance metrics (detection efficiencies up to 50% have been achieved [20]), but must still be operated in gated mode. Based on sum-frequency generation, up-conversion uses a strong pump to change the signal wavelength to the visible or near-IR for detection using standard Si SPADs. Additionally, using short pump pulses in the up-conversion process allows optical sampling of the signal photons to yield sub-picosecond temporal profiles of the signal.

In Section 7.2, we survey different types of spatially multiplexed detectors and compare their advantages and disadvantages in terms of important detector characteristics such as PNR capability, dark-count rates, and afterpulsing.

A spatially multiplexed detection system uses several conventional detectors connected with passive beamsplitters and/or active optical switches to provide measurement enhancements, particularly PNR capability and shorter effective dead times. A simple theory of the operation of a spatially multiplexed detector is described, and the detection statistics for both pulsed and continuous-wave photon sources are formulated. We also present a scheme using a tree of actively controlled optical switches that can surpass a passive tree configuration in achieving higher count rates, and we verify its operation in a proof-of-concept experiment.

In Section 7.3, we discuss time-multiplexed detection (TMD), which uses beamsplitters and time-delay loops to redistribute the photons in a signal pulse among multiple time slots. An advantage of TMD is that it reduces the number of detectors required for a multiplexing scheme. We present a theoretical description of TMD and discuss an experimental implementation. Finally we discuss phase-sensitive TMD and show proof-of-principle data obtained from such a system.

In Section 7.4, we review several experimental implementations of up-conversion-based single-photon counting including CW and pulsed modes of operation, and bulk and waveguide configurations. The basic theory of operation is described, and the special case in which the input quantum state is preserved is singled out as particularly useful for quantum information processing. Excessive background counts in the up-conversion process can be a problem in photon-counting applications and methods to mitigate this issue are discussed. Optical sampling using ultrashort pump pulses in up-conversion is shown to yield temporal information about single and entangled photons that would not have been possible with conventional detector technologies.

7.2 SPACE-MULTIPLEXED DETECTORS

7.2.1 Introduction

In this section we review the theory and practice of the spatially multiplexed detection approach and its application as a PNR detector. We present a simple general theory describing the operation of a spatially multiplexed detector. This theory models random and passive distribution of the photons toward each click/no-click (non-PNR) detector, for both pulsed and continuous photon sources. We discuss the various detectors developed in a spatially multiplexed form to date, such as the Si-photomultiplier, superconducting nanowire single-photon detector (SNSPD), or SPAD, and present the strategies behind them along with the technologies used to implement them. We compare the advantages and disadvantages of the different technologies in terms of important detector characteristics like PNR capability, as well as, dark-count-rates, afterpulsing, etc.

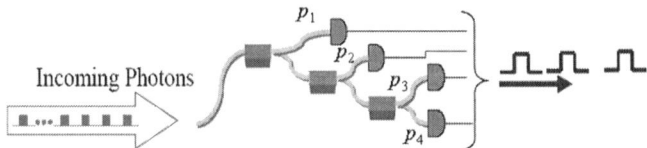

FIGURE 7.1 A passive spatially multiplexed detection assembly.

In its simplest form, a spatially multiplexed detection system is an assembly of detectors connected to an input optical field through a beam-splitter tree, see Fig. 7.1. This implementation will be referred to as a passive multiplexing scheme. A beam-splitter tree can be replaced with optical switches for better control over the multiplexing process and this approach will also be discussed using a scheme where the photons are distributed to the detectors through actively controlled optical switches. Active switching yields another interesting application of the spatially multiplexed detectors: the reduction of the effective dead time. It turns out that properly designed control of optical switches offers an advantage over a simple passive tree configuration in achieving high count rates (i.e., in reducing the effective dead time). We will summarize both theoretical considerations and proof-of principle experiments of this active-control approach.

7.2.2 Theory of Operation

Here we present some basic statistical models describing detection by spatially multiplexed detectors. Because in practice there are distinct differences in the detection of light from pulsed versus CW sources, we discuss these two cases separately, with the theory for a pulsed source presented in Section 7.2.2.1, followed by CW sources in Section 7.2.2.2.

7.2.2.1 Spatially Multiplexed PNR Detector with a Pulsed Source

Consider a state of N photons. The most obvious way to represent the distribution of photons from that state onto a spatially multiplexed array of $\mathcal{N}$ click/no-click detectors is through a multinomial distribution [21]

$$P(n_1, n_2, \ldots, n_{\mathcal{N}} | N, p_1, p_2, \ldots, p_{\mathcal{N}}) = \delta_{N, \sum_{i=1}^{\mathcal{N}} n_i} \delta_{1, \sum_{i=1}^{\mathcal{N}} p_i} N! \prod_{i=1}^{\mathcal{N}} \frac{p_i^{n_i}}{n_i!}, \quad (7.1)$$

where each photon has a probability p_i to impinge on the ith click/no-click detector, n_i is the number of photons impinging on the ith click/no-click detector, and δ_i is the Kronecker delta function. The conditional probability that n_i incident photons will produce a click on the ith detector is $p_i(\text{click}|n_i) = 1 - p_i(\text{no-click}|n_i)$ where $p_i(\text{no-click}|n_i) = (1 - \eta_i)^{n_i}$ and η_i is the detection efficiency of the ith detector (including the optical losses of the multiplexing system).

The probability that just the two detectors j and k click, while the others do not click when N photons are incident on the spatially multiplexed detector, is

$$\mathcal{P}(\text{click}_j, \text{click}_k | N) = \sum_{n_1,..n_j,...,n_k,...,n_{\mathcal{N}}} P(n_1,..n_j,...,n_k,...,n_{\mathcal{N}} | N, p_1,...,p_j,...,p_k,...,p_{\mathcal{N}})$$

$$\times p_j(\text{click}|n_j) p_k(\text{click}|n_k) \prod_{i=1 (i \neq j,k)}^{\mathcal{N}} p_i(\text{no-click}|n_i). \quad (7.2)$$

Thus, the probability of obtaining two clicks when N photons impinge on the detector array can be derived by summing over all the possible values of j and k:

$$Q_{\text{click}}(2|N) = \sum_{j=k+1}^{\mathcal{N}} \sum_{k=1}^{\mathcal{N}-1} \mathcal{P}(\text{click}_j, \text{click}_k | N). \quad (7.3)$$

An analogous argument holds for the probability $Q_{\text{click}}(k|N)$ of obtaining k clicks given N photons in the input field (with the obvious condition that $k \leq N$). Thus, the positive-operator-valued-measure (POVM) representing the detection process for a state with N photons is $\widehat{Q}_{\text{click}}(k) = \sum_{N=k}^{\infty} Q_{\text{click}}(k|N) |N\rangle \langle N|$.
The probability of observing k clicks per pulse is then

$$P_{\text{click}}(k) = \text{Tr}[\widehat{\rho} \widehat{Q}_{\text{click}}(k)] = \sum_{N=k}^{\infty} \rho_{N,N} Q_{\text{click}}(k|N), \quad (7.4)$$

where $\widehat{\rho}$ is the density matrix representing the quantum state emitted by the pulsed source and $\rho_{N,N}$ is the probability that the pulse contains N photons (the diagonal element of the density matrix $\widehat{\rho}$).

The situation is greatly simplified when the photons are distributed with equal probability among the click/no-click detectors (which, incidentally, is also the most efficient solution). In this case $p_i = 1/\mathcal{N}$, and all these detectors have the same efficiency $\eta_i = \eta$. Note that we use η instead of η_{DE} for each component detector's efficiency, η_{DE} is reserved to denote the instrument's *overall* detection efficiency. In this case the probability of having k clicks in the presence of N photons simplifies to

$$Q_{\text{click}}(k|N) = \frac{\mathcal{N}!}{k!(\mathcal{N}-k)!} \sum_{n_1,...,n_{\mathcal{N}}} \frac{N!}{\prod_{i=1}^{\mathcal{N}} n_i!} \left(\frac{1}{\mathcal{N}}\right)^N \left[\prod_{j=k+1}^{\mathcal{N}} (1-\eta)^{n_j}\right]$$

$$\times \left\{\prod_{j=1}^{k} [1-(1-\eta)^{n_j}]\right\} \delta_{N, \sum_{i=1}^{\mathcal{N}} n_i}. \quad (7.5)$$

An example of this probability function is shown in Fig. 7.2.

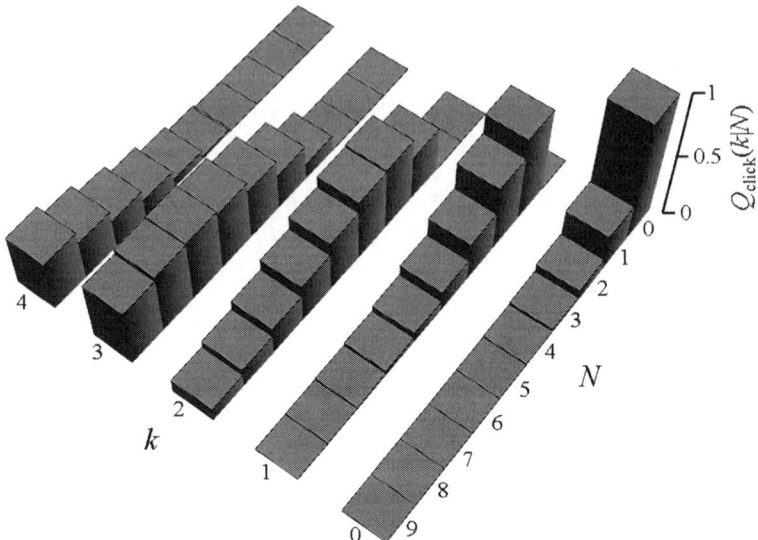

FIGURE 7.2 Probability $Q_{\text{click}}(k|N)$ of obtaining k clicks in the presence of N photons given by Eq. (7.5), and evaluated for a detection system of $\mathcal{N} = 4$ detectors with $\eta = 0.7$.

So far we have considered the case of perfect single-photon detectors. However, real detectors can have distorting effects, such as dead time, although if the dead time of the individual detectors is less than the inverse of the repetition rate of the source it will not be an issue. Similarly, if the maximum number of photons in a state is much smaller than the number of detectors, dead time effects may be negligible. Otherwise the statistical model must include the effects of dead time. Such treatment is beyond the scope of this book, but the investigation of some specific cases can be found in Ref. [22].

7.2.2.2 Spatially Multiplexed PNR Detector with a CW Source

The most traditional and well-established radiometric measurements involving macroscopic light levels and analog detectors are performed with continuous, stabilized light sources tied to radiometric scales. Operating any such radiometrically tied sources at photon-counting levels remains an outstanding technological challenge. The most commonly available photon-counting detectors exhibit nonlinearity due to dead time, limited PNR ability, etc. This nonlinear behavior is also strongly influenced by the temporal statistics of the emission source, e.g., dead time losses are different for sub-Poissonian and super-Poissonian light fields [23].

In this section we consider the simplest case: a Poissonian photon source [21]. The probability density function of the time interval Δ between the emission of two subsequent photons is $f_1(\Delta) = \mu \exp(-\mu \Delta)$, where μ is the mean photon rate of the emitted photons.

In this case, the probability of having m photons in the time interval T can be written in terms of the cumulative probability distribution function $F_m(T)$ as the difference $F_m(T) - F_{m+1}(T)$ [24,25]. $F_m(T)$ represents the probability that at least m photons impinge on a detector during a time interval from $t = 0$ to $t = T$:

$$F_m(T) = \int_0^T f_m(t) dt, \qquad (7.6)$$

where

$$f_m(t) = \mu(\mu t)^{m-1} \exp(-\mu t)/(m-1)! \qquad (7.7)$$

is the probability density function corresponding to an m photon Poissonian process, i.e., the convolution of m density functions $f_1(\Delta)$ describing m individual photons each arriving at some time during the interval T. Thus, one finds that the probability of having exactly m photons in the time interval T with a Poissonian photon source is:

$$F_m(T) - F_{m+1}(T) = \frac{(\mu T)^m}{m!} \exp(-\mu T) = \text{Poisson}(m|\mu T), \qquad (7.8)$$

which is the well-known Poissonian distribution with mean photon number μT.

We now describe the detection of photons from such a source by a spatially multiplexed detector. The first step is to find the distribution of photons for each single click/no-click detector. Following Section 7.2.2.1, this can be calculated from the discrete convolution of the multinomial distribution of Eq. (7.1) and the Poissonian distribution,

$$\sum_{m=0}^{\infty} P(n_1, n_2, \ldots, n_\mathcal{N}|m, p_1, p_2, \ldots, p_\mathcal{N}) \text{Poisson}(m|\mu T) = \prod_{i=1}^{\mathcal{N}} \text{Poisson}(n_i|p_i \mu T). \qquad (7.9)$$

Thus, we get $\mathcal{N}$ independent Poissonian processes, one for each of the detectors, and the mean photon number impinging on the ith detector in the time interval T is $p_i \mu T$.

This result can be used for both coherent and multimode thermal light. Indeed, in the limit of a measurement time T much longer than the correlation time of thermal light, the Poisson distribution of photons in Eq. (7.8) asymptotically holds for thermal light when the mean number of photons per mode of the thermal source is much smaller than one. Because for a Poissonian process the counting may be considered random (i.e., the events are equally distributed in time [21]), the photons impinge one at the time on the ith detector.

For a CW source, the distortion of the photon statistics caused by the ith detector's dead time t_{dead} can be neglected only when the mean time between two subsequent incoming photons $(p_i \mu)^{-1} \gg t_{\text{dead}}$ (where $p_i \mu$ is the photon rate impinging on the ith detector). When this condition is satisfied other detector characteristics, such as dark counts or dead time, may become important and should be included in the model. An example of such a more inclusive model

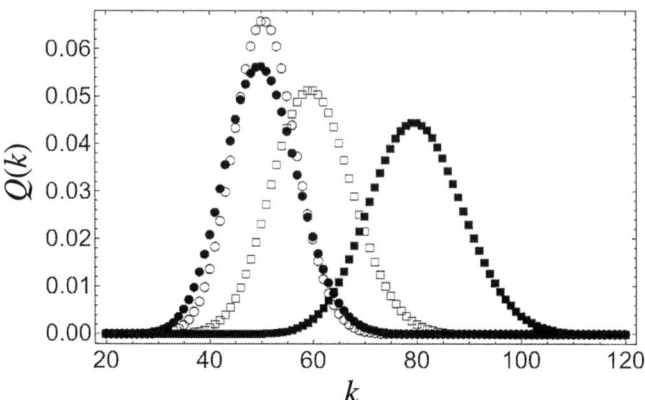

FIGURE 7.3 Distortion of incoming photon statistics induced by the detection efficiency and dead time. The incoming state with Poissonian statistics Poisson$(k|\mu T)$ with $\mu T = 80$ (filled squares) gets effectively attenuated due to a finite quantum efficiency $\eta = 0.7$ (open squares), see Eq. (7.10). The output distribution remains Poissonian with Poisson$(m|\eta\mu T)$. For the detectors with a paralyzable dead time t_{dead}, the output statistics is governed by Eq. (7.11), which gives a distribution Poisson$(k|\eta\mu T \exp(-\eta\mu t_{dead}))$, i.e., it also remains Poissonian with a new mean value of $\eta\mu T \exp(-\eta\mu t_{dead})$ (filled circles). For the detectors with a non-paralyzable dead time t_{dead}, the output distribution is no longer Poissonian, but rather sub-Poissonian, see Eq. (7.13) (open circles).

can be found in Ref. [26]. For simplicity, here we only include dead time in the model.

For negligible dead time (e.g., $p_i \mu t_{dead} \ll 10^{-2}$) and a detection efficiency of the ith detector, η_i, the probability distribution of counted events, $\mathcal{P}(m_i)$, can be written as a discrete convolution between the binomial distribution representing the counting process and the Poissonian distribution of the incoming photon

$$\mathcal{P}(m_i) = \sum_{n_i=m_i}^{\infty} \binom{n_i}{m_i} \eta_i^{m_i} (1-\eta_i)^{n_i-m_i} \text{Poisson}(n_i|p_i\mu T) = \text{Poisson}(m_i|\eta_i p_i \mu T).$$

(7.10)

The effect of a less-than-unity detection efficiency when measuring a Poissonian state is merely an attenuation of the input, with no other changes in detection statistics, see Fig. 7.3.

For a CW source neglecting the dead times is clearly too restrictive and all click/no-click detectors always exhibit some dead time. Since the time interval between subsequent arrivals of photons can be arbitrarily small in a Poissonian process with any mean number, dead time necessarily distorts the distribution of photon counts.

Due to differences in detector's electronics, it is usual to distinguish between two main types of dead time, i.e., non-paralyzable and paralyzable [27]. In the first, a dead time t_{dead} follows the detection of a photon, and all photons arriving within this fixed time interval are simply ignored. In the second, an incoming photon produces a dead time of t_{dead} if the detector was ready, or extends the dead time by the same interval t_{dead} if the detector was already dead. In both cases, all photons that arrive during the dead time are not counted.

For the case of paralyzable dead time, the distortion of the count statistics has been fully discussed in the literature [28–32]. The probability of counting a photon corresponds to the probability that nothing triggered the detector during the time interval t_{dead}, prior to the arrival of the photon in question. Based on Eq. (7.10), this probability can be expressed as $\pi = \exp(-\eta_i p_i \mu t_{\text{dead}})$. Thus, similar to Eq. (7.10), the probability of detecting k_i photons by the ith detector is simply

$$Q(k_i) = \sum_{m_i=k_i}^{\infty} \binom{m_i}{k_i} \pi^{k_i} (1-\pi)^{m_i-k_i} \mathcal{P}(m_i) = \text{Poisson}(k_i | \pi \eta_i p_i \mu T). \tag{7.11}$$

We use this result to estimate the true photon rate of a source for a given count rate measured by a multiplexed detector. As seen in Fig. 7.3, a paralyzable dead time also does not distort the Poissonian character of the input field.

For detectors with non-paralyzable dead time [33,34], a similar expression for the probability can be obtained by using a typical argument of renewal theory [24,25]. The detection circuit of such a detector is blocked for a fixed time t_{dead} after which it is able to detect a subsequent photon. This means that each time interval between adjacent photon arrivals is split into two parts. First, a fixed time interval t_{dead} after each successful detection during which the detector is inhibited (dead). Second, the remaining time (i.e., the interval between the end of the dead time t_{dead} and the succeeding detection). This gives rise to a Markovian stochastic process, in contrast to a Poissonian process that has no memory effects [21,24]. We treat this time interval between the two detections as the sum of the two time intervals introduced above.

Considering a measurement duration T, the probability of registering m counts given a fixed dead time t_{dead} is equal to $\Theta(T - mt_{\text{dead}})$, where Θ is the Heaviside function. The related probability density function is then $h_m(t) = \delta(t - mt_{\text{dead}})$ where δ is the Dirac delta function [35]. The probability density distribution of m counted photons by the ith detector is the convolution of the two probability density functions f_m (from Eq. (7.7)) and h_m,

$$\begin{aligned} g_m(t) &= \int f_m(t_1) h_m(t_2) \delta(t - t_1 - t_2) dt_1 dt_2 \\ &= \eta_i p_i \mu \left[\eta_i p_i \mu \left(t - mt_{\text{dead}}\right) \right]^{m-1} \frac{\exp\left[-\eta_i p_i \mu \left(t - mt_{\text{dead}}\right)\right]}{(m-1)!} \Theta(t - mt_{\text{dead}}). \end{aligned} \tag{7.12}$$

Following the same approach that led to Eq. (7.8), the probability of having exactly k_i photons counted in the time interval T in the presence of a Poissonian source and a detector with a non-paralyzable dead time is

$$Q(k_i) = \frac{\gamma\left[k_i; \eta_i p_i \mu \left(T - k_i t_{\text{dead}}\right)\right]}{(k_i - 1)!} - \frac{\gamma\left[k_i + 1; \eta_i p_i \mu \left(T - (k_i + 1) t_{\text{dead}}\right)\right]}{(k_i)!}, \tag{7.13}$$

where γ is the lower incomplete Gamma function [36]. Note that in this case of non-paralyzable dead times the shape of the statistics for the detected signal differs from the Poissonian shape of the input, see Fig. 7.3.

It can be shown [33] that the first moment of the probability distribution in Eq. (7.13) (i.e., the mean number of photon counts) $\mathcal{E}(k_i) = \sum_{k_i} k_i Q(k_i)$, in the asymptotic limit of $T \gg t_{\text{dead}}, (\eta_i p_i \mu)^{-1}$, corresponds to the well-known formula

$$\mathcal{E}(k_i) \simeq \frac{\eta_i p_i \mu T}{1 + \eta_i p_i \mu t_{\text{dead}}}, \tag{7.14}$$

and analogously it can be demonstrated that the variance is

$$\mathcal{E}(k_i^2) - \mathcal{E}(k_i)^2 \simeq \frac{\mathcal{E}(k_i)}{(1 + \eta_i p_i \mu t_{\text{dead}})^2}. \tag{7.15}$$

Thus, the average total number of counts measured by a spatially multiplexed detection system is

$$\mathcal{M} = \sum_{i=1}^{\mathcal{N}} \mathcal{E}(k_i). \tag{7.16}$$

When the photons are distributed with equal probability among the click/no-click detectors, $p_i = 1/\mathcal{N}$, and all these detectors have the same efficiency $\eta_i = \eta$, Eq. (7.16) simplifies to

$$\mathcal{M} = \frac{\eta \mu T}{1 + \eta \mu t_{\text{dead}}/\mathcal{N}}. \tag{7.17}$$

Note that Eq. (7.17) can also be interpreted as a reduced effective dead time (with a reduction factor $1/\mathcal{N}$). Therefore, a multiplexed system not only enables PNR capabilities from non-PNR detectors, but also can improve the dead time properties of a detector.

It turns out that when PNR capability is not required, dead time properties can be reduced beyond this $1/\mathcal{N}$ factor. We discuss this in the next section.

7.2.2.3 Dead Time Reduction

As we have seen, spatial multiplexing of detectors can not only enable PNR capabilities, but can also reduce the effective dead time, allowing for higher detection rates. Particularly, we saw that the effective dead time of a multiplexed system is reduced relative to a single detector by a factor of $\mathcal{N}$, the number of detectors in the passive "detector tree" arrangement.

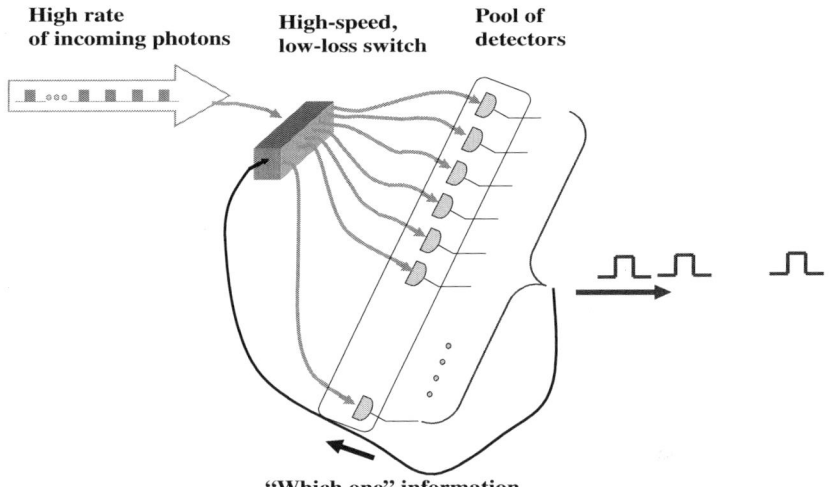

FIGURE 7.4 A pool of detectors and a fast switch are used to register a high rate of incoming photons. Incoming photons are switched to a ready detector. If that detector fires, it is switched out of the ready pool until recovery. If it does not fire, that detector remains active.

There exists a method to improve detection rates in a more efficient way than just randomly sending the photons toward the elements of the click/no-click detector array. This method requires a means to actively monitor the state of each detector to check if it is ready to register a photon or it is dead, and an optical switch to route subsequent incoming photons to a detector that is known to be ready [37–40], as shown in Fig. 7.4. We analyze this strategy analytically and numerically, and show that this scheme allows the $\mathcal{N}$-detector system to be operated at a detection rate significantly higher than $\mathcal{N}$-times the detection rate of an individual detector, while reducing the overall dead time.

The system's switching operation could be sequential, with each detector firing in order (the control system only switches to the next detector if the previous has fired), or it could be set up to switch the input to any live detector regardless of whether the previous detector had just fired. This latter implementation may allow for optimum use of an array of detectors where each detector may have a different dead time or when the switching time of the system is not negligible. In the simple model discussed here, we assume that all the detectors have the same dead time and switch transition time. The switch transition time includes any latency or other possible delays. While we include switch latency as part of the switch transition time, rather than as a separate parameter, we point out that latency may affect the choice of what firing order is used depending on what detection characteristic is most important for the particular application. For example, an application might have a repetitive pulsed photon source where the time to sense a detection is longer than the pulse period while the switch time by itself is less than the pulse period. In that case the detection system might benefit from operating in a mode where the input is immediately switched to another live detector regardless of whether the previous detector fired. This would reduce the effect of the long latency, although at a cost of a decreased likelihood of having at least one live detector available.

The switching strategy for the optical switch discussed here consists of simply re-routing photons to the next detector in the sequence of $\mathcal{N}$-detectors after the previous detector fires. This is the simplest implementation and is all that is required when the optical switching time is not a large fraction of each individual click/no-click detector's dead time. Even when the switching time is non-negligible, our assessment shows an advantage from using this scheme versus passive detector-tree schemes.

The relevant figure of merit in this context is dead time fraction (DTF), defined as the ratio of missed- to incident-events. A good device-independent benchmark for comparing different detection systems at high photon detection rates is the rate of incoming photons that results in a DTF of 10%, $R_{\text{DTF}=10\%}$. This is a practical limit for detector operation in many real-world applications.

We analytically estimate the DTF from the mean total count rate of the overall detector pool and effective dead time for each detector (which depends on their position in the switching system). We consider a Poissonian source, as described above, and a pool of identical detectors (with both equal detection efficiencies η and equal non-paralyzable dead times t_{dead}). From Eq. (7.17) we find the DTF for a detector tree to be

$$\text{DTF} = \frac{\eta\mu T - \mathcal{M}}{\eta\mu T} = 1 - \frac{1}{1 + \eta\mu t_{\text{dead}}/\mathcal{N}}. \qquad (7.18)$$

Analogously, for a detection system with an array of the same $\mathcal{N}$-detectors, and active multiplexing, an overall or "effective" dead time $T_{\text{dead}}(\mathcal{N})$ can be introduced, treating the whole system as a single detection unit:

$$\text{DTF} = 1 - \frac{1}{1 + \eta\mu T_{\text{dead}}(\mathcal{N})}. \qquad (7.19)$$

Therefore, the task reduces to calculating the effective dead time of the system. Because the optical switch only switches photons to a new detector after a count is registered, the effective dead time is given by the statistical contribution of the switching time, t_s and the single-detector dead time, t_{dead}, governed by the two cases illustrated in Fig. 7.5: (a) $\mathcal{N}$ events are counted in a time interval longer than $t_{\text{dead}} - t_s$, or (b) they occur in a time interval shorter than $t_{\text{dead}} - t_s$. In the latter case, a photon is switched back to a detector that is still dead, and the entire detector assembly saturates. Due to this saturation, the assembly dead time is longer than t_s by the remaining dead time of an individual

Chapter | 7 Hybrid Detectors

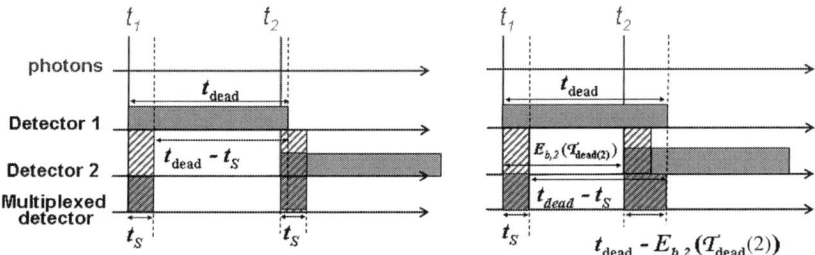

FIGURE 7.5 Detection of two consecutive photons by a multiplexed detector with $\mathcal{N} = 2$. The effective dead time $T_{\text{dead}}(2)$ is given by the statistical contribution of the two possible scenarios. (a) Full saturation is avoided: after a click of detector 1, the detector 2 clicks after a time interval greater than $t_{\text{dead}} - t_S$. By the time the detection assembly recovers from switching time t_S, the detector 1 will be ready to accept a new photon. (b) Full saturation of the assembly: after the click of detector 1, the detector 2 clicks after a time interval smaller than $t_{\text{dead}} - t_S$. In addition to switching dead time t_S, the assembly will remain fully saturated, until the first detector recovers.

detector. We write the effective dead time for an $\mathcal{N}$-detector assembly as:

$$T_{\text{dead}}(\mathcal{N}) = p_{a,\mathcal{N}}(T_{\text{dead}}(\mathcal{N}))t_S + p_{b,\mathcal{N}}(T_{\text{dead}}(\mathcal{N}))[t_{\text{dead}} - E_{b,\mathcal{N}}(T_{\text{dead}}(\mathcal{N}))], \quad (7.20)$$

where

$$p_{a,\mathcal{N}}(T_{\text{dead}}(\mathcal{N})) = \int_{t_{\text{dead}}-t_S}^{+\infty} g_{\mathcal{N}}(\Delta t | \eta \mu T, T_{\text{dead}}(\mathcal{N})) \mathrm{d}\Delta t, \quad (7.21)$$

and

$$p_{b,\mathcal{N}}(T_{\text{dead}}(\mathcal{N})) = \int_{0}^{t_{\text{dead}}-t_S} g_{\mathcal{N}}(\Delta t | \eta \mu T, T_{\text{dead}}(\mathcal{N})) \mathrm{d}\Delta t, \quad (7.22)$$

are the probabilities that case (a) or (b) occurs for $g_{\mathcal{N}}(\Delta t | \eta \mu T, T_{\text{dead}}(\mathcal{N}))$, which is the probability density distribution of the time interval Δt, between a count and the $(\mathcal{N}-1)^{\text{th}}$ preceding one. Note that the dependence of the above probabilities on $T_{\text{dead}}(\mathcal{N})$ requires solving an integral equation to obtain $T_{\text{dead}}(\mathcal{N})$. The mean time interval between a count and a $(\mathcal{N}-1)^{\text{th}}$ preceding count when case (b) occurs is given by:

$$E_{b,\mathcal{N}}(T_{\text{dead}}(\mathcal{N})) = \frac{\int_0^{t_{\text{dead}}-t_S} \Delta t\, g_{\mathcal{N}}(\Delta t | \eta \mu T, T_{\text{dead}}(\mathcal{N})) \mathrm{d}\Delta t}{\int_0^{t_{\text{dead}}-t_S} g_{\mathcal{N}}(\Delta t | \eta \mu T, T_{\text{dead}}(\mathcal{N})) \mathrm{d}\Delta t}. \quad (7.23)$$

Note that Eq. (7.12) allows writing an expression for the probability density distribution $g_{\mathcal{N}}(\Delta t | \eta \mu T, T_{\text{dead}}(\mathcal{N}))$, assuming a Poissonian input and introducing a constant effective dead time $T_{\text{dead}}(\mathcal{N})$ [34]:

$$g_{\mathcal{N}}(\Delta t | \eta \mu T, T_{\text{dead}}(\mathcal{N})) = \frac{(\eta \mu)^{\mathcal{N}-1}[\Delta t - (\mathcal{N}-1)T_{\text{dead}}(\mathcal{N})]^{\mathcal{N}-2}}{(\mathcal{N}-2)!} \\ \cdot \exp\{-\eta\mu[\Delta t - (\mathcal{N}-1)T_{\text{dead}}(\mathcal{N})]\}\Theta \\ \cdot [\Delta t - (\mathcal{N}-1)T_{\text{dead}}(\mathcal{N})]. \quad (7.24)$$

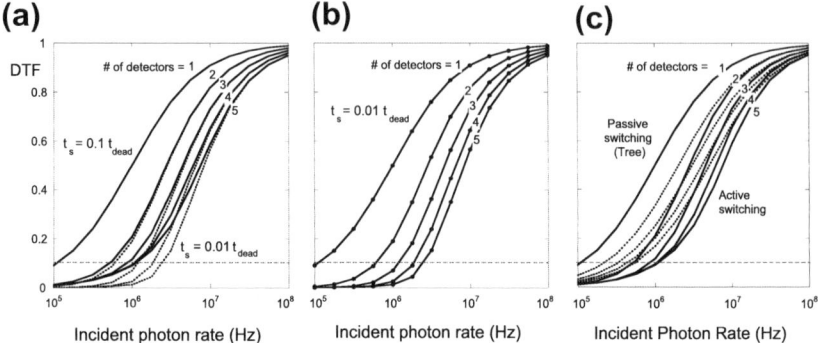

FIGURE 7.6 DTF versus the incoming photon rate for $\mathcal{N} = 1\text{--}5$ detectors with $t_{\text{dead}} = 1$ μs. (a) Analytically determined DTF for actively switched system with $t_s = 0.01\, t_{\text{dead}}$ (dotted lines) and $t_s = 0.1\, t_{\text{dead}}$ (solid lines). (b) DTF determined analytically (solid lines) compared to DTF determined using Monte-Carlo simulations (points) for $t_s = 0.01\, t_{\text{dead}}$. (c) DTFs with $t_s = 0.1\, t_{\text{dead}}$ for an actively switched scheme (solid lines) compared to a passive scheme (dotted lines). Horizontal dashed line indicates DTF = 10%, our benchmark level for practical detector operation.

An analytical formula for $\mathcal{T}_{\text{dead}}(\mathcal{N})$ exists only for $\mathcal{N} = 2$ detectors:

$$\mathcal{T}_{\text{dead}(2)} = \frac{t_{\text{dead}}}{2} - \frac{1 + 2W\left[\frac{(2t_s - t_{\text{dead}})\eta\mu - 1}{2}\right]}{2\eta\mu}, \quad (7.25)$$

where W is the principal value of the Lambert W-function [41]. For more detectors we use numerical methods to determine $\mathcal{T}_{\text{dead}}(\mathcal{N})$.

Interestingly, while neglecting the dynamic nature of the dead time and introducing a constant effective dead time $\mathcal{T}_{\text{dead}}(\mathcal{N})$ seems to be a restrictive assumption, the results obtained with this approach are in excellent agreement with experimental results, as well as Monte-Carlo simulations in all the regimes considered [38]. This is not surprising, because such a treatment merely swaps the order of integration when computing the averages.

Figure 7.6a shows the dead time fraction versus the incoming photon rate for systems with $\mathcal{N} = 1\text{--}5$ detectors with single-detector dead times of 1 μs and switching times equal to 1% and 10% of the single-detector dead time. For $t_s = 0.01\, t_{\text{dead}}$ the effect of switching time on the system is negligible, while for $t_s = 0.1\, t_{\text{dead}}$, the multiplexed scheme shows much less increase of the $R_{\text{DTF}=10\%}$ points with increasing detector number. Figure 7.6b compares the analytic theory with the Monte-Carlo results, showing good agreement with the simulation for all switching times t_s. Figure 7.6c compares the active multiplexed scheme just described with a passive scheme (detector/beam-splitter tree configuration) for $t_s = 0.1\, t_{\text{dead}}$. As judged by the $R_{\text{DTF}=10\%}$ points, the active multiplexed scheme surpasses the passive arrangement for relatively few detectors, $\mathcal{N} = 4, 5$.

Chapter | 7 Hybrid Detectors

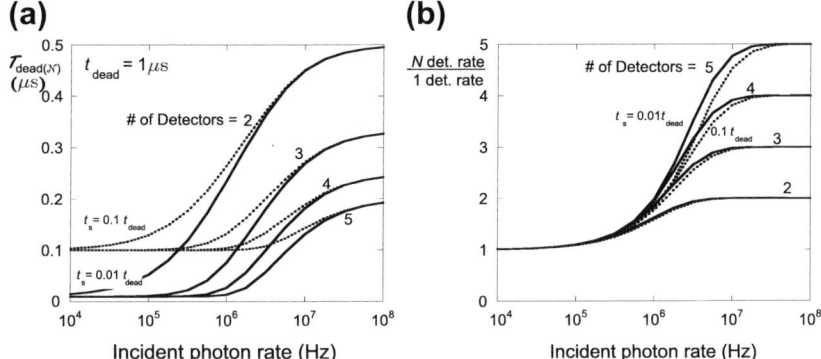

FIGURE 7.7 (a) Plots of $T_{\text{dead}}(\mathcal{N})$ versus incident photon rate for systems of 2–5 detectors (all with $t_{\text{dead}} = 1$ μs) with $t_s = 0.1\, t_{\text{dead}}$ (dotted lines) and $t_s = 0.01\, t_{\text{dead}}$ (solid lines). Note that the system dead time saturates at $t_{\text{dead}}/\mathcal{N}$ for high photon rates, while for low photon rates it is limited by t_s. (b) Ratio of the number of photons counted by multiplexed systems to that counted by a single detector, all for $t_{\text{dead}} = 1$ μs, and $t_s = 0.1\, t_{\text{dead}}$ (dotted lines), $t_s = 0.01\, t_{\text{dead}}$ (solid lines). The ratio limit for a high incoming photon rate is $\mathcal{N}$.

Figure 7.7a shows the mean effective dead time $T_{\text{dead}}(\mathcal{N})$ for $\mathcal{N}$ up to 5, versus the mean incident photon rate $(\eta\mu)$, for $t_s = 0.1\, t_{\text{dead}}$ and $0.01\, t_{\text{dead}}$. The effective dead time clearly satisfies the condition $t_s \leq T_{\text{dead}}(\mathcal{N}) \leq t_d/\mathcal{N}$. We see that the maximum effective dead time of the multiplexed scheme coincides with the detector-tree dead time. This means that for an optical switch with $t_s < t_{\text{dead}}/N$, the active scheme surpasses what is possible with a passive scheme. Figure 7.7b shows the ratio of the mean count rate for the multiplexed scheme to the count rate of a single detector, versus the mean incoming photon rate. We see that, as expected for high count rates, the maximum gain is $\mathcal{N}$-times the rate that would be obtained by a single detector.

Figure 7.8 shows $R_{\text{DTF}=10\%}$ versus the number of detectors for the active switching system at several switching times. The $t_s = 0.001\, t_{\text{dead}}$ result differs little from the case when the switching time is neglected. Up to $t_s = 0.02\, t_{\text{dead}}$ the results show significant advantage of the active-switch scheme over a passive beam-splitter tree for all numbers of detectors shown. Above $t_s = 0.2\, t_{\text{dead}}$ the advantage of the active system is significantly reduced until ultimately its figure of merit falls below that of the passive scheme for just a few detectors.

7.2.3 Experimental Implementations of Space-Multiplexed Detectors

7.2.3.1 Detector-Tree Arrangements

A spatially separated detector arrangement where detectors are connected by a series of beamsplitters is the simplest example of a space-multiplexed detector. The first experiment of this kind, where the two detectors were used

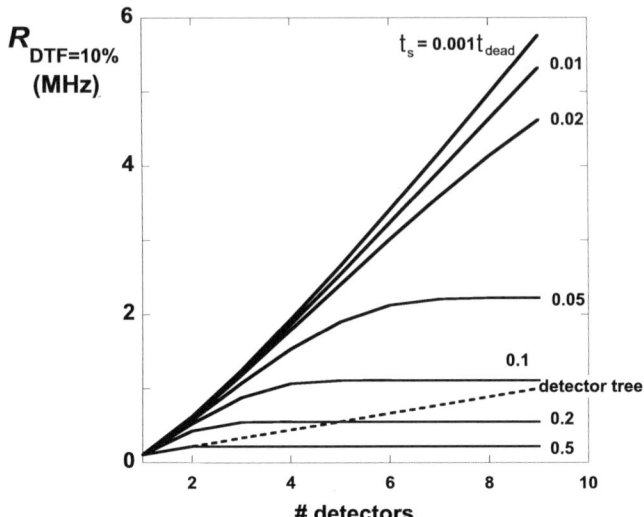

FIGURE 7.8 Incoming rate giving a DTF = 10% ($R_{DTF=10\%}$) versus the number of detectors, for $t_s = 0.001, 0.01, 0.02, 0.05, 0.1, 0.2, 0.5$ of t_{dead} (solid lines). The detector-tree configuration result is shown for comparison (dashed line).

to characterize multi-photon character of an input state of light, was that by Hanbury Brown and Twiss [42]. They used that system to show the difference in the photon number statistics of single-mode thermal light from that of multi-mode thermal light. Similar setups have been used in numerous experiments since, most notably to characterize second-order correlation properties of nonclassical states of light (with two-detector arrangements) and their multi-order correlation properties (with three or more detectors connected via a beam-splitter tree). The underlying theory and experimental techniques of this measurement are described in Chapter 2.

Because a detector-tree arrangement reduces the dead time of a detection system, multiplexing can also be used for counting photons at rates higher than that of an individual detector [39]. One example where this technique may be useful is in high-speed quantum communications, although in applications where security is the paramount concern, the subtleties of the detection system, such as detection efficiency variation with count rate, are often critical.

7.2.3.2 Photon-Number-Resolving Detector Arrays

As integration techniques develop, integration of many detectors into a single device becomes possible. New technology is allowing the integration of hundreds of click/no-click detectors, making it possible to achieve significant photon-number-resolving capacity, and making the effects of the dead time of each individual detector negligibly small.

One of the practical implementations of spatially multiplexed detectors uses a microlens array to image light from a fiber onto a two-dimensional SPAD array [43–45]. This is a simple and effective approach to achieving photon-number resolution with non-photon-resolving detectors by having the optical mode geometrically split among the detectors in an array. Each SPAD's output can be either read individually [43] or summed to give a single output pulse with amplitude proportional to the number of detected photons [44,45]. The drawbacks of these approaches usually include low overall detection efficiency, which means only a lower limit estimate of the photon number in an input pulse is possible, and uneven splitting of the input mode between the pixels so that saturation levels of different regions of a 2D array can vary significantly. In addition, there may be crosstalk between adjacent pixels, where the firing of one pixel causes one or more of its neighbor to fire, see Chapter 4.

SNSPDs can also be used in array formats. For example, the meanders of SNSPDs can be interleaved in such a way that each covers an equal part of the input mode. However, the number of multiplexed SNSPD devices used to detect the output of one mode has been low, with implementations so far reporting less than 10 separate individual meanders. A significant advantage of this arrangement is that it offers the potential of even faster operation speed than the already-fast single SNSPD, because the inductance of the individual wires is much lower than that of the longer single-wire meander of the original SNSPD, whose recovery time is inductance limited [46]. On the other hand, the need for cryogenic equipment to operate these detectors is a significant drawback to their use.

As with SPADs, there are two possible implementations for multiplexing. The first of these arrangements, the parallel-SNSPD [13], uses nanowires connected electrically in parallel. The currents through the parallel wires are summed so that the single analog output signal is proportional to the number of wires that have gone normal due to absorption of incident photons. This scheme was demonstrated with niobium nitride (NbN) nanowires 100 nm wide with a capability of counting up to four photons, with a dark-count rate of 0.15 Hz and a repetition rate of 80 MHz. A drawback of this parallel nanowire arrangement is that the analog output signals corresponding to different photon number states are not always well separated, leading to an uncertainty in interpreting the output. The demonstrated parallel-SNSPD performs well relative to other photon number-resolving detectors in regard to dark-count rate and maximum count rate.

The second scheme also runs parallel wires, but does so as completely separate detectors with individual outputs, thus the result is a digital output, i.e., the number of output pulses gives the number of photons detected. This scheme was demonstrated in a system of four separate wires [47]. In a recent publication [48], a system detection efficiency as high as 76% is reported. Therefore, this detector arrangement offers a combination of high detection efficiency, high temporal resolution, photon-number resolution, as well as reduced dead time through multiplexing, although the additional thermal loading of many output lines may be a challenge for such cryogenic detectors.

7.2.3.3 Dead Time Fraction Reduction via Active Detector Multiplexing

As shown from theory, active multiplexing of the detectors can boost detection rates while maintaining low saturation of the assembly by a factor exceeding the number of detectors in the assembly. This was verified in a proof-of-principle experiment with an assembly of just two detectors and a fast optical switch [39,40].

The experimental setup, presented in Fig. 7.9a, is built around a parametric down-conversion crystal that produces photon pairs at two different frequencies. The photon at 810 nm is detected by a silicon SPAD D_1 (with a dead time of 50 ns, that is negligible compared to the dead time of the infrared detectors in the system under test). The detection of an 810 nm photon heralds a 1550 nm photon in the signal arm, where different detector arrangements were tested. Two InGaAs detectors (D_{2A}, D_{2B}) connected through a fast optical switch were used to implement the multiplexing arrangements. The real-time logic for the active detector arrangements was built using a Field Programmable Gate Array (FPGA) (for some background on this, see for example [49]). For dead time reduction as discussed earlier in this chapter, the design for two detectors was built around an asynchronous Set-Reset Flip-Flop (RS-Trigger), see Fig. 7.9b. Note that even though FPGAs usually operate synchronously, asynchronous codes of low complexity can be implemented, although extensive testing of the asynchronous gate solutions is required. In particular, devices built using asynchronous gates may exhibit dead time. Fortunately, when SPAD outputs are used, single-photon detector dead time is usually longer than that of the FPGA logic. Another complication is the need to carefully consider timing (latency) and duration of the single-photon detector's output, which may differ from one detector to the other (even if the detectors are from the same batch) and be ready to modify the logical circuit accordingly.

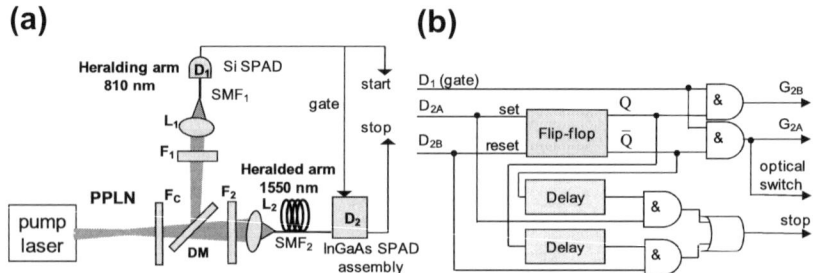

FIGURE 7.9 (a) Experimental setup for measuring DTF of multiplexed detectors. (b) Electronics schematic for detection dead time reduction.

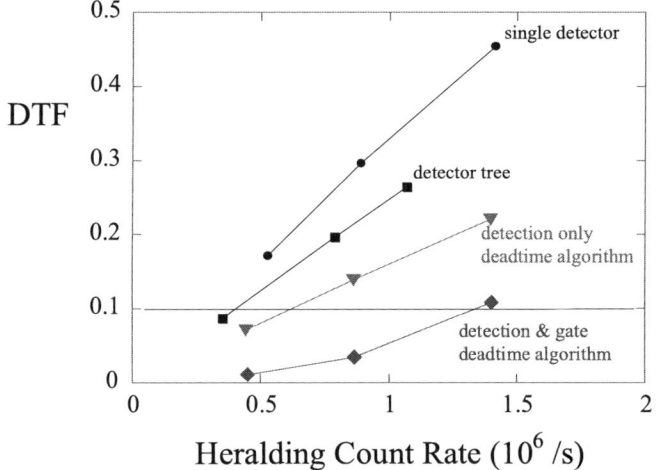

FIGURE 7.10 Measured DTF vs. heralding (D_1) count rate for a single detector, a detector tree, and multiplexed detector arrays with different dead time reduction algorithms: one considering dead time arising only from a detection as described in this chapter, and one that includes dead time due to detection of a photon and dead time due to gate electronics even if no detection occurs. Such gate electronics is specifically required for infrared SPADS [40]. Horizontal line is the benchmark level DTF = 10%.

Several detector configurations were compared: (i) a single detector, (ii) a detector-tree arrangement, and (iii) a multiplexed arrangement that was designed to reduce detection dead time, as discussed above (Fig. 7.10). The count rate for configuration (iii) is the highest for all values of DTF. In particular, for our chosen threshold of DTF = 10% we see a heralding count rate 2.3 times higher for the actively multiplexed scheme (iii) as compared to a single detector. Note that this improvement factor is achieved with just two detectors in the assembly.

Another interesting feature of active multiplexing is that other properties of detectors can be improved along with the dead time reduction. Because the output of only one of the detectors is monitored at any point of the protocol, the overall dark-count rate is just that of a single detector instead of scaling linearly with the number of detectors, as happens with passive multiplexing arrangements [39]. Another benefit of active multiplexing is that the afterpulse probability of an active arrangement will be always lower than that of a single detector or a detector tree and will depend on count rate [39].

While analyzing the performance limitations of active multiplexing, it was found that the maximum count rate increase is limited by a feature often found in gated InGaAs detectors that is not commonly appreciated. In addition to detection dead time, these detectors have a dead time of its gate input (i.e., in the case when a detector is gated on but no photon is detected, the gate circuitry requires some time to be ready to respond to a subsequent gate), thus a strategy for suppressing the impact of this gate dead time was developed, requiring only a modification of the FPGA firmware. A measurement using this modification with a two-detector multiplexed assembly in configuration (iii) resulted in a dead time reduction factor of nearly 5 when compared with a single InGaAs detector (Fig. 7.10). The details of this modification are beyond the scope of this chapter, but can be found in [40].

7.3 TIME-MULTIPLEXED DETECTORS

7.3.1 Introduction

In a time-multiplexed detector the pulse of photons impinging on the detector is split into several temporal modes as depicted on the left in Fig. 7.11. The resulting temporal modes can subsequently be detected with a non-photon-number-resolving, or click/no-click detector, such as a standard SPAD. A simple, however very powerful time-multiplexing detection scheme was suggested by Banaszek and Walmsley [14] in 2003 and is depicted on the right of Fig. 7.11. It is centered around a fiber loop used to store the light pulse under investigation. An electro-optic switch (S) can be used to couple the light pulse to be measured into the storage loop. A highly asymmetric coupler (C) is used to keep most of the pulse in the storage loop in each pass, but each time on average less than one photon per pulse leaves the cavity. A single-photon detector can be used to detect the output of this coupler and reconstruct the number of photons in the input pulse.

Depending on the fiber delay and source used, this time-multiplexed detector requires a fast switch to couple the pulse into the cavity and to ensure only a single pulse is transmitted into the cavity. The strength of the output coupler and

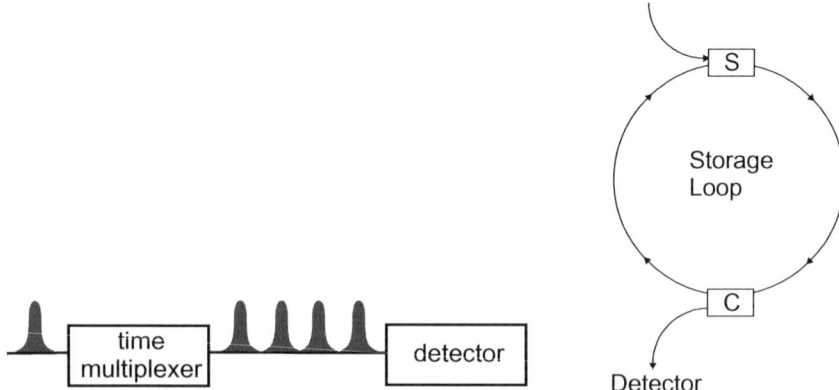

FIGURE 7.11 Left: Basic principle of time-multiplexed detection. Right: Simple storage loop TMD [14].

the time required between two measurements depend on the number of photons in the input pulse. When a pulse is coupled into the loop while a residual amount from the previous pulse is still present the overlap can prevent the correct estimation of the incoming photon statistics. In the following subsection we show an approximation of this active scheme using only passive splitters, though at the expense of limited photon-number-resolving capabilities.

7.3.2 Fiber-Loop Detectors

Figure 7.12 shows the schematic of a time-multiplexed detector based on fiber loops. An incoming pulse (I) is split at a fiber beamsplitter with a 50:50 splitting ratio resulting in two spatial output modes (II). One of these outputs is sent through additional optical fiber which realizes a delay of ΔT between them (III). At a second fiber beamsplitter the two temporal modes are recombined and split again, which leads to two temporal modes in each spatial mode (IV). In a second iteration more delay $2\Delta T$ is added to one of the two spatial modes shifting the two pulses yet again (V). The next recombination beamsplitter finally produces four temporal modes in two spatial outputs, which are detected by two single-photon detectors, for example SPADs.

When using non-photon-number-resolving detectors, the photon-number resolution of this network is constrained by the eight output modes of the beam-splitter network. In principle more delay and recombination stages could be added to the detector, increasing the number of output modes and thus the level of photon-number resolution. However this reduces the maximum possible repetition rate of the detector. If the mode that acquired the highest delay in the network could, in principle, overlap or interfere with the leading, undelayed mode of the next pulse entering the detector, the photon numbers in each pulse cannot be retrieved accurately. The maximum possible repetition rate of the detector is thus given by

$$f = \frac{1}{\sum \Delta T}. \qquad (7.26)$$

The time delays must be chosen according to the photon detectors used, taking into account the dead time of the detector, as well as its afterpulsing probability.

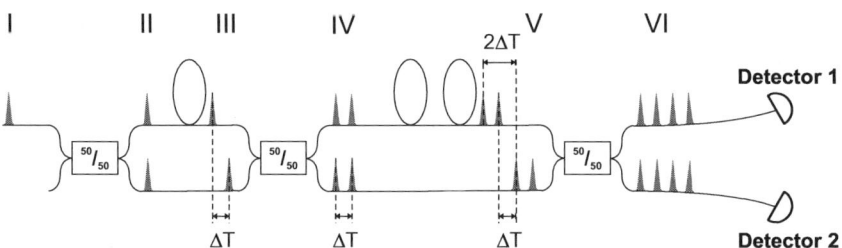

FIGURE 7.12 Time-multiplexed detector based on two fiber loops.

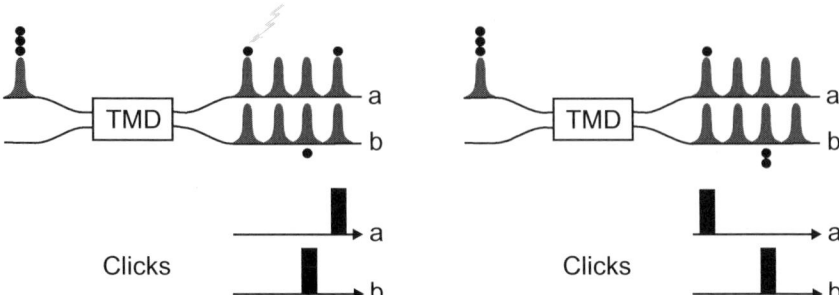

FIGURE 7.13 Left: Influence of losses (the first photon in channel a is lost). Right: Influence of the fiber loop network.

Current off-the-shelf SPADs have dead times of around 50 ns; however, due to afterpulsing, a base delay of 100 ns is typical. For two delay stages this sums to a maximum delay of 300 ns, which after 200 ns is added to separate two consecutive pulse trains, allowing a maximum detector repetition rate of 2 MHz. Data acquisition can be controlled by an FPGA, reading out the number of clicks from the detector for an incoming pulse. Time gating can be implemented with the FPGA to reduce effects from dark counts and afterpulsing.

The measured number of clicks, however, does not correspond directly to the number of photons in the input pulse. In Fig. 7.13 the effects of losses and the beam-splitter network are illustrated. Coupling into, as well as at splices within the fiber network, and beamsplitter imperfections are sources of loss, in addition to the intrinsic losses of the detector inefficiency. These may lead to only k clicks from an incoming pulse with n photons. A second effect is the distribution of the photons into the spatio-temporal modes, the convolution effect. Several photons can end up in the same mode, resulting in only one click event. For example, in Fig. 7.13 (right) two out of the three incoming photons end up in the same mode, causing only two clicks. Both these effects make the use of TMDs challenging for single-shot measurements. For ensemble measurements the photon number statistics can be determined.

We consider the photon number statistics $\vec{\sigma}$ of the input states, which are given by the diagonal elements of the density matrix $\hat{\rho}$ in the photon number basis. Most photon-number-resolving detectors cannot resolve phase information of the incoming state, making this a valid model. In Section 7.3.3 we will introduce the concept of weak-homodyne detection for extracting phase information from the measurement. Similar to photon-number statistics, we define the click statistics measured with the TMD $\vec{p}$, which are simply the probabilities p_n of measuring n clicks. The click statistics $\vec{p}$ for a set of input photon number statistics $\vec{\sigma}$ is given by [17]:

$$\vec{p} = \mathbf{C} \cdot \mathbf{L} \cdot \vec{\sigma}, \tag{7.27}$$

where the matrices **L** and **C** represent the effects of losses and convolution of several photons into a single bin, respectively. A single beamsplitter with reflectivity ϵ_{loss} in front of the detector can be used to model all losses with a single loss parameter [50,51]. The loss matrix contains the probabilities of k out of l photons being transmitted as

$$\mathbf{L}_{kl} = \begin{cases} \binom{l}{k} (1 - \epsilon_{\text{loss}})^k \epsilon_{\text{loss}}^{l-k} & \text{if: } l \geqslant k, \\ 0 & \text{otherwise,} \end{cases} \quad (7.28)$$

where $\binom{l}{k}$ is the binomial coefficient.

The convolution matrix is best calculated by considering all possible routes photons could travel through the beam-splitter network [17,52]. The probabilities for these photons ending up in a certain spatio-temporal mode after the fiber network are given by a set of N probabilities $b_1,\ldots,b_N$, where N is the number of spatio-temporal modes of the TMD. b_j then denotes the probabilities of a photon ending up in the jth bin, which are a characteristic of a given TMD. Several simplifications, for example, for perfect (50:50) splitting ratios [15] exist, but the most general form in which to calculate the convolution matrix are the probabilities of l incident photons distributed into k spatio-temporal modes. These k clicks can be caused by all possible combinations of k out of the N modes of the TMD, which can be described by the k-tuples $c = (c_1, c_2, \ldots, c_k)$ with $c_i \in [1, N] \cap \mathbb{N}$ and $c_1 \neq c_2 \neq \cdots \neq c_k$. The convolution matrix is then given by

$$\mathbf{C}_{kl} = \begin{cases} 0 & k > l \\ \sum_c b_{c_1} b_{c_2} \ldots b_{c_l} & k = l \\ \sum_c \left[\sum_d \frac{1}{id(d)!} \prod_{j=1}^{k} \binom{l - \sum_{i=0}^{j-1} d_i}{d_j} b_{c_1}^{d_1} b_{c_2}^{d_2} \ldots b_{c_k}^{d_k} \right] & k < l \end{cases}, \quad (7.29)$$

where the first two cases are trivial. In the third case more than one photon can end up in a single mode. l photons can be distributed into k bins according to the k-tuples $d = (d_1, d_2, \ldots, d_k)$ with $d_i \in [1, n-k+1] \cap \mathbb{N}$ and $d_1 \geqslant d_2 \geqslant \cdots \geqslant d_k$, with $\sum_k d_k = n$ and the definition $d_0 = 0$. $\prod_{j=1}^{k} \binom{l - \sum_{i=0}^{j-1} d_i}{d_j}$ accounts for the different ways the photons which are distributed into a single bin can be chosen and $id(d)!$ denotes the number of permutations with bins filled with the same number of photons to prevent overestimation by the binomial coefficient. Several illustrative examples for use of this formula are given in the literature [52].

The splitting ratios b_j required to calculate the convolution matrix of a TMD can be measured with bright light techniques [53] or photon counting [17]. The convolution matrix can also be determined by performing detector tomography, which is introduced in Chapter 9. Once measured, the convolution matrix for a given fiber network remains fixed, the loss matrix, however, may still change, for example, due to variations in coupling efficiency. Calibration techniques, introduced in Chapter 8, are therefore crucial for characterizing a TMD detector.

Many applications and experiments require the use of two detectors, such as the measurement of coincidences in down-conversion experiments (c.f. Chapter 11). In these experiments we can define the joint photon number statistics [54] σ of the two modes entering the two detectors. The entries of this matrix $(\sigma)_{m,n}$ are the probabilities of simultaneously having m photons in mode 1 and n photons in mode 2. Similar to the click statistics, we define the joint click statistics $(\mathbf{p})_{m,n}$ of getting m clicks in mode 1 simultaneously with n clicks in mode 2. Equation (7.27) then reads

$$\mathbf{p} = \mathbf{C}_1 \cdot \mathbf{L}(\eta_1) \cdot \sigma \cdot \mathbf{L}^{\mathrm{T}}(\eta_2) \cdot \mathbf{C}_2^{\mathrm{T}}, \tag{7.30}$$

where matrix transposition is denoted by the superscript "T" and subscripts indicate the properties of detector 1 and detector 2, respectively. Figure 7.14 shows examples of reconstructed joint photon number statistics.

The direct inversion of Eqs. (7.27) and (7.30) is possible; however, for low efficiencies in particular, it may lead to unphysical results such as negative probabilities. The use of optimization algorithms, such as least squares or maximum likelihood in the reconstruction, incorporates the constraint of obtaining physical states and enables state reconstruction even for overall efficiencies as low as 5% [54].

Background processes such as fluorescence can mix with the process under investigation and obscure the input photon number distribution. Assuming a background photon number distribution σ_{Bg}, the photon number distribution entering the detector σ_{Det} is given by the convolution [54]

$$\sigma_{\mathrm{Det}} = \sigma * \sigma_{\mathrm{Bg}}, \tag{7.31}$$

where σ denotes the joint photon number statistics of the processes under investigation. However, the joint click statistics $\mathbf{p}_{\mathrm{Det}}$, which are the entity

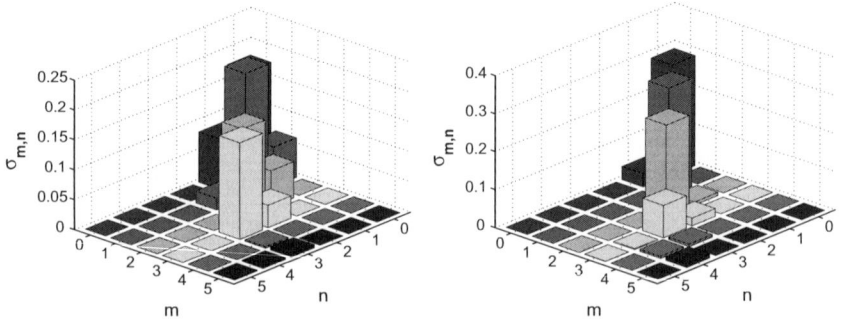

FIGURE 7.14 Joint statistics without (Left) and with (Right) background subtraction [54].

measured by the detector, cannot be written as a direct convolution of background click statistics $\mathbf{p}_{Bg}$ and clicks caused by the process under investigation $\mathbf{p}$. Photons originating in the actual process and the background are mixed in the fiber network, and then detected by SPADs that saturate at one photon. This effective interaction is described by the convolution matrix in Eqs. (7.27) and (7.30).

With the convolution matrices $\mathbf{C}_{1(2)}$ for the detector in mode 1 (mode 2) we define the reduced joint click statistics $\mathbf{r}_{Det}$ [54], which can be obtained by applying the inverse convolution matrix to Eq. (7.30)

$$\mathbf{r}_{Det} = \mathbf{C}_1^{-1} \cdot \mathbf{p}_{Det} \cdot \left(\mathbf{C}_2^T\right)^{-1} = \mathbf{L}(\eta_1) \cdot \boldsymbol{\sigma}_{Det} \cdot \mathbf{L}^T(\eta_2). \tag{7.32}$$

These reduced statistics can be assumed to be independent processes without interaction and we can rewrite Eq. (7.30) as

$$\mathbf{r}_{Det} = \mathbf{r} * \mathbf{r}_{Bg} = \left[\mathbf{C}_1^{-1} \cdot \mathbf{p} \cdot \left(\mathbf{C}_2^T\right)^{-1}\right] * \mathbf{r}_{Bg}. \tag{7.33}$$

Using the convolution theorem, the Fourier transform $\mathcal{F}$, and element-by-element matrix division, we find the desired reduced joint click statistics of the process under investigation [54]

$$\mathbf{p} = \mathbf{C}_1 \mathcal{F}^{-1} \left\{ \frac{\mathcal{F}\{\mathbf{r}_{Det}\}}{\mathcal{F}\{\mathbf{r}_{Bg}\}} \right\} \mathbf{C}_2^T, \tag{7.34}$$

which can then be used to reconstruct its joint photon number statistics $\boldsymbol{\sigma}$. The concept of reduced statistics remains valid with only one detector mode, described by Eq. (7.27). Figure 7.14 shows the effects of background on the measurement using the joint statistics of the two modes produced by spontaneous parametric down-conversion. This process, introduced in much greater detail in Chapter 11, produces photons in pairs – its joint statistics should therefore be diagonal. In the left of Fig. 7.14 the data shows significant off-diagonal components, indicating background contributions. After measuring the background, Eq. (7.34) was used to determine the reduced joint statistics and plot the reconstructed state on the right of Fig. 7.14, showing the success of background subtraction.

7.3.3 Weak-Homodyne Detection

Detectors with PNR capability usually work in the photon-number basis without any phase reference and thus only provide access to the diagonal elements of the density matrix in the number basis. However in many quantum experiments, for example entanglement witnessing [55,56], more information about the state is required. A standard way to do this is full state reconstruction by means of strong-field homodyne tomography [57,58], which mixes a local oscillator

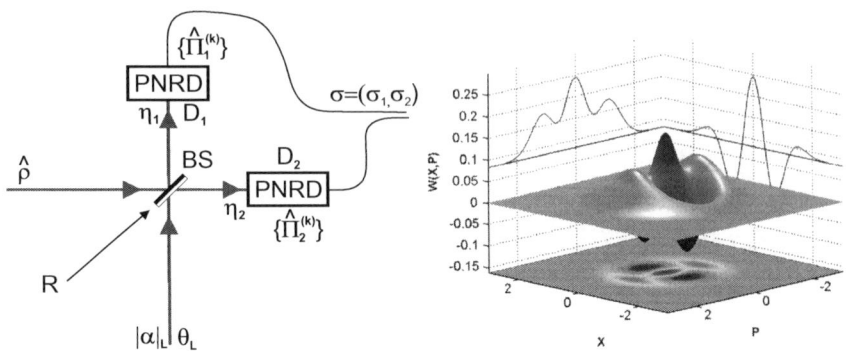

FIGURE 7.15 Left: Schematic of a phase-sensitive PNR detectors (PNRD). Right: Example of the exotic POVM elements [59] achievable with a weak-homodyne detector, illustrated by its Wigner function.

(LO) with the signal under investigation and subsequently takes measurements for different phase settings between the LO and the state. On the other hand, detectors with PNR capabilities cannot be used in strong-field homodyne detection as they do not resolve the number of photons usually present in the strong LO and might even be damaged by the high LO field strengths.

Based on the same concept, the weak-field homodyne detector mixes the probe state with a LO beam. However the LO strength is comparable to the signal to be detected and thus compatible with PNR detectors. Weak-field homodyne detectors have been suggested and analyzed theoretically [59], and demonstrated in recent experiments [59–62]. In Fig. 7.15 we show the principle of a weak-field homodyne detector: an input state $\hat{\rho}$ is mixed with a weak local oscillator at a beamsplitter of reflectivity R. Both outputs of the beamsplitter are subsequently detected by PNR detectors **D1** and **D2** and the click distribution $\sigma = (\sigma_1, \sigma_2)$ recorded. We describe the detectors using their POVM sets $\left\{\hat{\Pi}_1^{(k)}\right\}$ and $\left\{\hat{\Pi}_2^{(k)}\right\}$. Further parameters required for a full description of the detector are the amplitude $|\alpha_{LO}|$ and phase Θ_{LO} of the LO.

Analysis and rigorous mathematical treatment of the detector reveals the POVM elements for obtaining joint clicks $\sigma = (\sigma_1, \sigma_2)$

$$\hat{\Pi}_{\gamma,\text{theo}}^{(\sigma)} = \sum_{k,l=0}^{M} \left[\mathbf{C_1} \cdot \mathbf{L_1}\right]_{\sigma_1 k} \left[\mathbf{C_2} \cdot \mathbf{L_2}\right]_{\sigma_2 l} \hat{\Pi}_{\gamma,\text{ideal}}^{(k,l)}, \quad (7.35)$$

where we use the loss and convolution matrices for the TMD and $\hat{\Pi}_{\gamma,\text{ideal}}^{(k,l)}$ is the POVM element of a perfect photon-number-resolving detector, with the unitary transformation $\hat{U}_{\text{BS}}$ describing the beamsplitter

$$\hat{\Pi}_{\gamma,\text{ideal}}^{(k,l)} = \left(\hat{\mathbb{1}}_{in} \otimes |\alpha\rangle_L \langle\alpha|_L\right) \hat{U}_{\text{BS}}^\dagger \left(|k\rangle_1 |l\rangle_2 \otimes \langle k|_1 \langle l|_2\right) \hat{U}_{\text{BS}}. \quad (7.36)$$

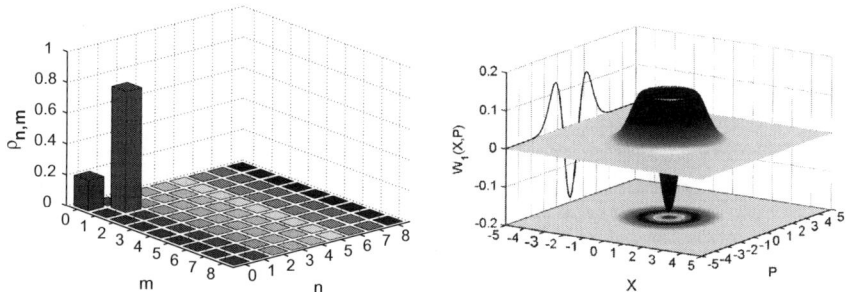

FIGURE 7.16 Results with the phase-sensitive TMD. Left: Joint photon number statistics of the heralded single-photon state. Right: Wigner function of the single-photon state.

In Fig. 7.16 we show the reconstruction of a heralded single-photon state, produced by spontaneous parametric down-conversion with a phase-sensitive TMD.

7.4 UP-CONVERSION DETECTORS

7.4.1 Introduction

Optical frequency up-conversion at the single-photon level [63–65] enables an alterative method for detecting single photons at wavelengths longer than 1 μm. Based on sum-frequency mixing in a nonlinear crystal, followed by detection in the visible or near-IR range, the technique provides a convenient way to detect single photons at wavelengths for which the performance of relatively inexpensive or commercially available single-photon counters is lacking. This is especially true for wavelengths greater than 1 μm where Si SPADs are not suitable. With a strong pump at an appropriate wavelength, the long-wavelength photon can be frequency translated to the visible or near-IR for detection by Si SPADs that are easier to use and have more favorable characteristics than InGaAs SPADs. For telecom-band photons, up-conversion and subsequent detection by Si SPADs can lead to significantly improved performance with higher detection efficiency, CW operation without gating, and lower dark counts compared with InGaAs SPADs. Even as this performance gap has narrowed somewhat in recent years with steady improvements of InGaAs SPAD operation (See Chapter 4 for details), detection via up-conversion remains a viable option. Up-conversion can also be utilized to provide unique single-photon measurement capabilities. In ultrafast up-conversion [66,67], ultrashort pump pulses are used to time-sample single-photon pulses and yield a single-photon temporal profile with sub-picosecond resolution that is orders of magnitude better than any state-of-the-art single-photon detector technologies.

Up-conversion of telecom-band photons at 1.3 μm and 1.55 μm for efficient detection is well suited to various tasks in optical and quantum communications. Communications-based applications include optical time-domain reflectometry [68], quantum key distribution [69–71], and photon-counting pulse-position modulation communication [72]. Up-conversion serves another important purpose in quantum optical information processing. Unity-efficient up-conversion preserves the quantum state of the input photon so that the output photon is identical to that of the input photon except it is at a different wavelength. Therefore, up-conversion offers wavelength freedom of choice in quantum optical information processing. For example, two frequency-nondegenerate photons can be made to produce a Hong-Ou-Mandel (HOM) interference signature [73] if one of the photons is frequency translated to match the wavelength of the other photon, or if both photons are up-converted to the same frequency [74]. A more practical application in a multi-wavelength quantum network [75] is to generate the entanglement at a convenient wavelength for high flux and high quality entanglement, followed by frequency translation to the desired wavelength without losing the entanglement [76].

7.4.2 Theory of Single-Photon Up-conversion

Consider the generation of an up-converted photon at frequency ω_2 from an input photon at frequency ω_1 through interaction with a strong monochromatic pump field at frequency ω_p in a second-order ($\chi^{(2)}$) nonlinear crystal. Energy conservation constrains these fields' frequencies to satisfy $\omega_1 + \omega_p = \omega_2$, and momentum conservation at the photon level requires that their wave vectors obey $\vec{k}_1 + \vec{k}_p = \vec{k}_2$. Assuming perfect phase matching and operation well within the phase-matching bandwidth, the input and up-converted output fields are related by a simple beam-splitter relationship [77]. In particular, the output at ω_2 is given by

$$\hat{A}_2(L) = \cos(\kappa|A_p|L)\hat{A}_2(0) + i\frac{A_p}{|A_p|}\sin(\kappa|A_p|L)\hat{A}_1(0), \quad (7.37)$$

where $\hat{A}_i(x)$ is the quantum field operator for frequency ω_i at the input ($x = 0$) or output ($x = L$) of the crystal, A_p is the classical pump field, κ is the nonlinear coupling coefficient, and L is the length of the crystal.

For vacuum input at ω_2 the up-converted output power, in photon units, is proportional to the input power

$$\langle \hat{N}_2(L) \rangle = \langle \hat{A}_2^\dagger(L)\hat{A}_2(L) \rangle = \sin^2(\kappa|A_p|L)\langle \hat{N}_1(0) \rangle. \quad (7.38)$$

The up-conversion efficiency is given by $\sin^2(\kappa|A_p|L)$ and reaches a maximum 100% when the pump field $|A_p| = A_{\pi/2} = \pi/2\kappa L$ (at a pump power $P_{\pi/2}$). We note that for $|A_p| = A_{\pi/2}$ Eq. (7.37) shows that

$$\hat{A}_2(L) = i\frac{A_p}{|A_p|}\hat{A}_1(0). \quad (7.39)$$

Except for an unimportant absolute phase, the output field operator will be in the same quantum state as the input field operator, and hence represents the ideal case of converting a photon from one frequency to another frequency without altering its quantum state. Quantum-state frequency conversion also applies to the field amplitude and was first demonstrated in squeezed states of light by Huang and Kumar [78]. Such state-preserving conversion is useful in many quantum information processing applications. For example, entangled photons can be generated at a convenient wavelength with high purity and then converted to a different wavelength that is optimized for a specific application. With efficient state-preserving frequency conversion a quantum key distribution (QKD) network can operate seamlessly between free-space operation at 800 nm and fiber-based system at 1.55 μm. We note that for nonzero detuning from perfect phase matching the quantum-state frequency translation is imperfect, even when $|A_p| = A_{\pi/2}$, because of dispersion [77].

For single-photon detection, frequency translation is primarily used to shift the photon wavelength to a region where good detector performance is readily available. Frequency conversion is not restricted to sum-frequency generation as we have discussed so far. Difference-frequency generation can also be used for frequency translation, satisfying the relation $\omega_1 - \omega_p = \omega_2$. In this case, the input photon has the highest frequency ω_1 and is converted to a lower frequency ω_2. For example, a visible photon can be mixed with a strong near-IR (say 800 nm) pump for conversion to the telecom wavelength, which can then be transmitted in a low-loss single-mode optical fiber to a remote location for subsequent detection by a SNSPD or TES, or for frequency up-conversion back to the visible, thus allowing states of visible photons to be transmitted over long distances. In difference-frequency conversion, the input quantum state can also be maintained at the output under appropriate pump and phase-matching conditions [79,80]. In general, frequency conversion works well for single-photon detection as long as the strong pump does not have the highest frequency, in which case the parametric down-conversion process would dominate and result in a large background of spontaneously emitted "noise" photons.

7.4.3 Up-Conversion Techniques

Single-photon frequency up-conversion can be implemented in several ways depending on the choice of nonlinear material, pump source, and the type of application. Below we briefly discuss several successful experimental demonstrations to illustrate the advantages and disadvantages of specific choices such as continuous-wave or pulsed pumping, and nonlinear bulk crystal or nonlinear waveguide.

Figure 7.17 shows the setup of one of the earliest single-photon up-conversion experiments by Albota and Wong [63]. Weak laser light at 1560 nm was converted to the visible at 633 nm using a 4 cm long bulk periodically poled lithium niobate (PPLN) crystal inside an enhancement ring cavity for

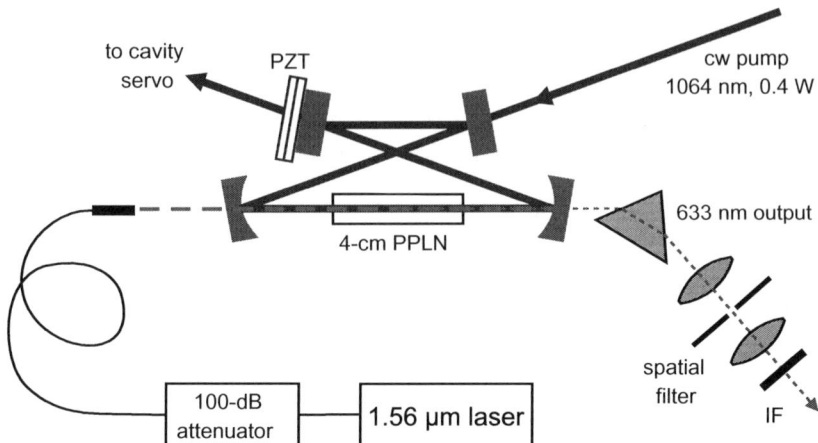

FIGURE 7.17 Experimental setup for cw single-photon up-conversion of 1.56 μm light using a bulk PPLN crystal and a pump enhancement cavity to reduce the required input pump power [63]. Spectral filtering with a dispersing prism and a 10 nm interference filter (IF) centered at 633 nm and spatial filtering with a pinhole are used to reduce pump-induced background counts. *Adapted from [63], Fig. 1.*

the 1064 nm pump. The cavity, which was single-pass for both the input and output up-converted photons, produced a circulating pump power of over 25 W at 1064 nm to reach near-unity conversion efficiency. With properly coated cavity mirrors and PPLN crystal there was very little loss for the input and output light, which is important for achieving a high overall detection efficiency that was primarily limited, in this case, by the spectral and spatial filters and by the Si SPAD detection efficiency. Intracavity up-conversion has also been demonstrated by placing the up-conversion crystal inside the pump laser cavity, which removes the need for phase locking the cavity to an external field [81].

Figure 7.18 shows the results of the single-photon up-conversion measurements that are in good agreement with the functional power dependence of Eq. (7.37). The up-conversion efficiency reaches 90%, and is limited by insufficient circulating pump power. It also plots the number of detected background photons (measured without input) and up-converted photons (measured difference with and without input) showing that the amount of noise photons can be significant and would adversely affect photon-counting applications such as QKD. The background photons were thought to have originated from a two-step cascaded process: pump-induced fluorescence and non-phase-matched parametric fluorescence that are in the same spatial mode and at the same wavelength λ_1 as the input, followed by efficient up-conversion from λ_1 to λ_2. The parametric fluorescence can be suppressed if the pump wavelength is chosen to be longer than the input wavelength, $\lambda_p > \lambda_1$, such that the energy of a pump photon is lower than the energy of the input photon.

Chapter | 7 Hybrid Detectors

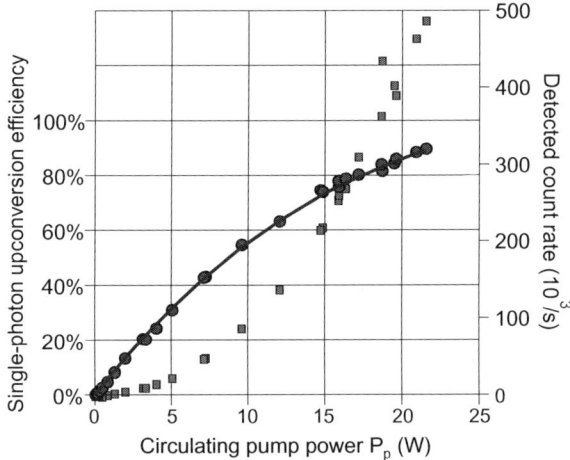

FIGURE 7.18 Intrinsic single-photon up-conversion efficiency (filled circles) of Ref. [63] using Fig. 7.17 setup as a function of circulating pump power, in accord with the functional form (solid curve) of Eq. (7.37). Up-converted signal counts (filled circles) and background counts (filled squares) are also plotted (right axis) indicating the significant background noise at high up-conversion efficiency. At the highest pump level, the overall system detection efficiency is 34%. *Adapted from [63], Fig. 3.*

The idea of using a pump wavelength longer than the input wavelength was first implemented by Langrock et al. [82]. Moreover, in view of the high pump power required for a bulk nonlinear crystal, a significantly more efficient nonlinear waveguide was used as the up-conversion medium [65], thus reducing the required pump power. Figure 7.19 shows the up-conversion results with a fiber-coupled PPLN waveguide and an input signal at 1.32 μm and a pump at 1.55 μm [82]. Two improvements are evident: only ≈100 mW of pump was required to achieve maximum intrinsic up-conversion efficiency, and a dark-count rate of ≈1.6×10^4/s was measured at that pump level. The use of a longer pump wavelength suppresses the dark-count rate by a factor of 50 compared with the same waveguide setup but with the pump and signal wavelengths interchanged. The dark count rate can be further reduced with additional filtering to enable up-conversion-based applications such as 1310 nm QKD [71].

In designing an up-conversion detection system, ultimately it is the overall system detection efficiency that should be maximized. One can relate the overall detection efficiency η_{DE} to various processes by $\eta_{DE} = \eta_u \eta_c \eta_f \eta_p \eta_s$, where η_u is the intrinsic up-conversion efficiency, η_c is the in- and out-coupling efficiency through the nonlinear medium, η_f is the filtering efficiency for spatial or spectral filtering that is needed to block background photons, η_p is the propagation efficiency through the up-conversion system other than those we have specifically noted, and η_s is the Si SPAD detection efficiency. It is straightforward to find a judicious combination of a highly nonlinear crystal,

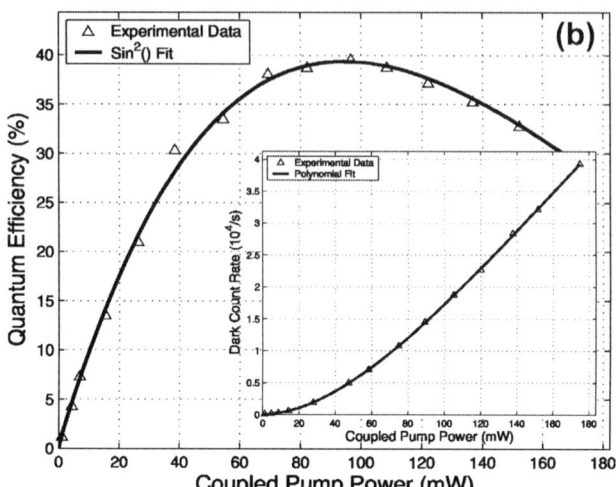

FIGURE 7.19 System detection efficiency and dark-counts measurements made with a fiber-coupled PPLN waveguide with a 1.55 μm pump and an input signal wavelength of 1.32 μm [82]. At ≈100 mW of pump, unity intrinsic up-conversion efficiency is reached with significantly lower dark counts than results shown in Fig. 7.18 that uses a pump wavelength shorter than the signal wavelength. *Reproduced from [82], Fig. 2(b).*

adequate pump power, and up-conversion configuration that produces near-unity intrinsic up-conversion efficiency. Extra efforts are usually needed to reduce other losses to achieve high system efficiency. For example, high-efficiency spectral filtering can be obtained using a reflective Bragg grating instead of a more lossy interference filter. Waveguide-based up-converters are convenient and require low pump powers, but coupling and waveguide losses can be a problem especially for free-space propagating signal photons. It should be noted that for certain quantum information processing tasks in which coincidences of two or more photons are measured, some amount of background counts can be tolerated because they are unlikely to be correlated with the two or more detected outputs.

For $\chi^{(2)}$ nonlinear crystals the up-conversion process is polarization specific and therefore the polarization of the input photon must be set accordingly. It is useful to be able to up-convert a photon with unknown polarization efficiently [83] and, better yet, retain its polarization in the output state after up-conversion which is essential for coherent up-conversion of polarization qubits. Figure 7.20 shows a simple two-up-converter scheme that can accomplish polarization-preserving up-conversion. A polarization beamsplitter separates the incoming photon into orthogonally polarized components that are up-converted by the two nonlinear crystals and then recombined at a second polarization beamsplitter for output. For any polarization-preserving up-conversion scheme it is important

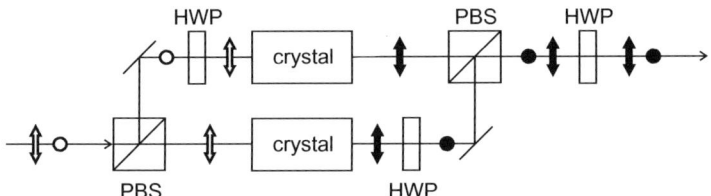

FIGURE 7.20 Schematic of a polarization-preserving up-converter. Input with an arbitrary polarization (open circle and arrow) is separated by a polarization beamsplitter (PBS) into horizontal and vertical polarizations. The vertical component in the upper arm is rotated into horizontal polarization before up-conversion. The lower-arm up-converted horizontal component is rotated into vertical polarization for recombination with the upper-arm horizontal component at the second PBS. A final output half-wave plate (HWP) restores the output state (filled circle and arrow) to its original input polarization state.

to maintain the relative phase between the two components by stabilizing the two paths interferometrically. Polarization-preserving up-conversion has been implemented using a single nonlinear crystal with a double-pass configuration [77]. In the first pass it up-converts one polarization component and in the return pass, after the light undergoes a 90° rotation for both input and output wavelengths, the orthogonal component is up-converted and then combined with the other polarization component for output. At the single-photon level, up-conversion of a polarization qubit while retaining the entanglement property with another photon has been demonstrated by Ramelow et al. [84].

7.4.4 Pulsed Up-Conversion

For cw sources of single photons and bi-photons the arrival times of the photons are random and the up-conversion apparatus must be continuously on, as in the examples discussed in the previous subsection. For pulsed sources, one can take advantage of the periodic nature of the photon arrival times by using a pulsed pump that is synchronized with the photon arrival times. To optimize the up-conversion efficiency the peak power of the pulsed pump should be equal to $P_{\pi/2}$, which can be easily obtained with a low average-power pump without the need for an enhancement cavity. The pulse width of the pump should also be larger than the input photon coherence time and the temporal jitters of the system.

Kwiat et al. first demonstrated pulsed up-conversion using a pulsed 1064 nm pump with an attenuated cw laser source at 1550 nm [64]. True synchronized pulsed up-conversion was later demonstrated using ≈600 ps pump pulses at 1064 nm and 200 ps signal pulses at 1550 nm showing near-unity intrinsic up-conversion efficiency [85]. Pulsed up-conversion using two or more spectrally and temporally distinct pump pulses can be utilized to increase the single-photon detection rate by spectrally separating the up-converted pulses and sending them to different single-photon detectors, thus overcoming the dead time of a single SPAD [86]. For pulsed up-conversion it is more appropriate to consider background count probability per pump pulse. A background count of 3×10^{-4} per 1 ns measurement time (comparable to pump pulse duration) was measured [64], which is comparable to that obtained in cw measurements [63] when normalized to 1 s. The same strategy of using a longer-wavelength pump to reduce the background counts was adopted by Xu et al. [71] in pulsed up-conversion of 1.32 μm photons showing a low dark-count rate of $\approx 2.2 \times 10^3$/s for a system detection efficiency of 20%. When the overall detection efficiency and the duty cycle of the pulsed system are taken into account, the dark-count rates for pulsed [71] and cw [82] up-conversion using a 1.55 μm pump are similar.

7.4.5 Ultrafast Up-Conversion

We can take advantage of the temporal degree of freedom in pulsed up-conversion in a different way. Instead of employing a pump pulse that is wider than the input photon's pulse width to ensure maximum temporal coverage and hence efficiency, the pump pulse can be much shorter to reveal temporal information of the incoming photons that is absent from typical up-conversion measurements and with a time resolution that is far better than that offered by any currently available single-photon counters. The idea is to use sub-picosecond pump pulses to optically sample the input photons which have typical coherence times of 1 ps. The input photon is up-converted only if the sub-ps pump pulse is present, and by scanning the arrival time of the pump pulse relative to the input photon, the temporal profile of the input photon can be mapped.

Kuzucu et al. utilized this time-resolved measurement technique in the setup of Fig. 7.21 to map the spatial variation of SPDC generation probability in a PPKTP crystal with a resolution of ≈ 1 mm [66]. A 790 nm mode-locked Ti:sapphire laser with 150 fs pulses was used to pump a 1 mm long periodically poled MgO-doped stoichiometric lithium tantalate (PPMgSLT) crystal that served as the up-conversion nonlinear medium. The type-0 phase-matched PPMgSLT crystal length was much shorter than the typical 40-mm length used in other up-conversion experiments because it is necessary to have a large phase-matching bandwidth (which is inversely proportional to the crystal length) for the short signal and pump pulses. The high peak power of the 150 fs pump pulse provided adequate up-conversion efficiency for the short crystal. To reduce timing jitter between the pump and the SPDC outputs, the same pump pulse was used to drive both the PPKTP down-conversion and the PPMgSLT up-conversion processes. Three optical delay lines provided separate timing control for the pump and the two input IR beams to the up-converter. Figure 7.21b shows that the two independent up-converters are implemented with a single crystal and a single pump by arranging the two IR inputs centered at 1580 nm from the PPKTP crystal in a noncollinear configuration. The three beams are in a

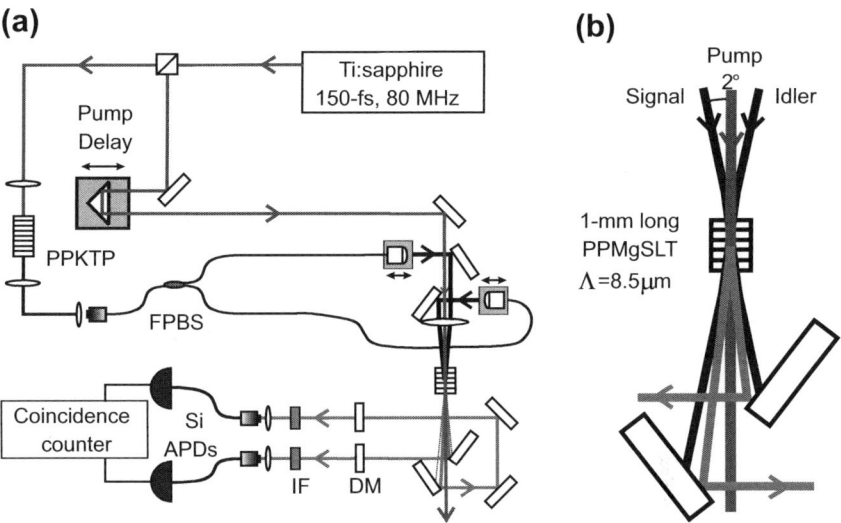

FIGURE 7.21 Schematic of sub-picosecond up-conversion experiments for measuring spatial variation in generation efficiencies along the length of a nonlinear crystal and for two-photon joint temporal measurements. *Reproduced from [66], Fig. 1.*

non-planar geometry to avoid coincident detection of any non-phase-matched down-converted photons that could be generated in PPMgSLT by the strong pump.

Because the pump and the SPDC outputs propagate at different group velocities in PPKTP, the location of the photon-pair emission in the crystal can be inferred from their arrival times relative to the pump pulse, with no (maximum) difference if they were emitted at the exit (entrance) facet of the crystal. The spatial resolution was set by the pulse width of the pump. Figure 7.22 displays the temporal profiles of signal and idler outputs from the PPKTP crystal, clearly showing that the generation probabilities in the front and back halves of the crystal are different. The time-resolved technique enables one to monitor the quality of the crystal's periodic grating structure and can be utilized as a diagnostic tool in the poling process.

Ultrafast up-conversion can be used as a new tool for characterizing two-photon entanglement. Tunable narrowband spectral filters are traditionally used for making joint spectral measurements of two correlated photons, but they can be quite lossy for very narrow bandwidths. Time-frequency Fourier duality suggests that one can learn as much from joint temporal measurements by optical sampling based on ultrafast up-conversion, whose timing resolution is easily adjustable and can be as fine as a few femtoseconds.

Kuzucu *et al.* applied the ultrafast up-conversion technique to directly measure the time anticorrelation characteristics of coincident-frequency entangled photon pairs generated by ultrafast SPDC in type-II phase-matched

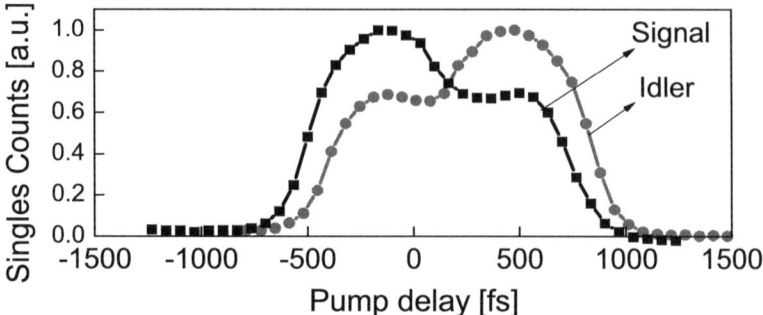

FIGURE 7.22 Normalized singles histogram for signal and idler outputs of PPKTP as a function of the pump delay. Signal and idler profiles are mirror images of each other because under extended phase-matching conditions the signal-pump and idler-pump time delays are equal with opposite signs [67]. *Reproduced from [66], Fig. 2(b).*

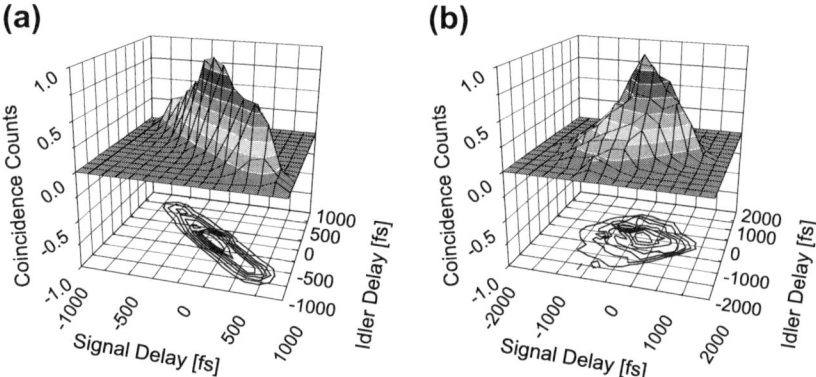

FIGURE 7.23 Two-photon joint temporal density for pump 3-dB bandwidth of (a) 6 nm and (b) 1.1 nm as measured by ultrafast up-conversion with 150 fs pulses [67]. *Reproduced from [67], Fig. 4.*

PPKTP under extended phase-matching conditions [87]. These specially phase-matched SPDC signal and idler photons are positively correlated in their frequencies and therefore, by Fourier duality, they should be anticorrelated in the time domain (relative to the pump excitation pulse for the PPKTP down-converter). That is, if the signal photon arrives a certain amount of time after the pump pulse, then the idler photon would arrive the same amount of time before the pump pulse. Using the same setup of Fig. 7.21 a background-free two-photon-coincidence temporal profile was obtained for a pump bandwidth of 6 nm as a function of the signal and idler delay times [67], as shown in Fig. 7.23a. It verifies that the arrival times of the two photons were indeed anticorrelated. Moreover, the joint temporal measurement capability makes it possible to monitor the result when one modifies the temporal correlation characteristics. Figure 7.23b shows a joint temporal density of a two-photon state that is nearly temporally unentangled when the pump bandwidth was reduced from 6 nm in Fig. 7.23a to 1.1 nm in (b) [67].

7.5 CONCLUSION

This chapter has covered a variety of hybrid detectors aimed at improving and expanding the capabilities of photon-counting measurements beyond those of a single non-PNR detector, particularly a SPAD. Multiplexing in space involving multiple SPADs, or in time with delay loops and a single SPAD, makes it possible to use conventional detectors to achieve a certain level of photon-number-resolving capability essential to many applications. Spatial multiplexing is well suited to overcoming single-SPAD limitations in maximum count rates, afterpulsing, and dead times. Multiplexing also enables new capabilities such as the weak-field homodyne detector based on the phase-sensitive time-multiplexed detection system.

Frequency up-conversion removes the need to operate at wavelengths within the range of sensitivity of Si SPADs. Instead, the photon can be translated from another wavelength of interest, such as the telecom band for low-loss transmission to a remote location via optical fibers. One particularly novel capability is sub-picosecond up-conversion that serves to optically sample the longer-duration photons to probe their rich temporal characteristics.

The future for hybrid detectors is bright because the demands of newly developed applications for high-performance photon-counting measurement capabilities cannot be met by existing single-photon detectors in a compact and convenient package. These hybrid detection techniques are expected to be further refined and utilized in an expanding range of applications, some of which have been developed during the time this book is being written. Hybrid measurement capabilities will form part of an essential detection toolbox for the growing fields of quantum information processing and low-light detection and imaging.

REFERENCES

[1] R.H. Hadfield, Nat. Photon. 3, 696 (2009); C. Silberhorn, Contemp. Phys. 48, 143 (2007) (*and refs. therein*).
[2] V. Giovannetti, S. Lloyd, and L. Maccone, Phys. Rev. Lett. 96, 010401 (2006).
[3] Y. Gao, P.M. Anisimov, C.F. Wildfeuer, J. Luine, H. Lee, and J.P. Dowling, J. Opt. Soc. Am. B 27, A170 (2010).
[4] V. Giovannetti, S. Lloyd, and L. Maccone, Nature Photon. 5, 222 (2011).
[5] G. Brida, M. Genovese, and I. Ruo Berchera, Nature Photon. 4, 227 (2010).
[6] E. Knill, R. Laflamme, and G.J. Milburn, Nature 409, 46 (2001).
[7] L.M. Duan, M.D. Lukin, J.I. Cirac, and P. Zoller, Nature 414, 413 (2001).
[8] T. Langer and G. Lenhart, New J. Phys. 11, 055051 (2009); J.L. O'Brien, A. Furusawa, and J. Vuckovic, Nature Photon. 3, 687 (2009); N. Gisin and R. Thew, Nat. Photon. 1, 165 (2007) (*and refs. therein*).
[9] M. Genovese, Phys. Rep. 413, 319 (2005) *and refs. therein*.

[10] K.D. Irwin and G.C. Hilton, "Cryogenic Particle Detection," in Topics in Applied Physics, edited by C. Enss, Vol. 99, Springer-Verlag, Berlin, p. 63 (2005).
[11] H. Paul, P.Törmä, T. Kiss, and I. Jex, Phys. Rev. Lett. 76, 2464 (1996).
[12] L.A. Jiang, E.A. Dauler, and J.T. Chang, Phys. Rev. A 75, 062325 (2007).
[13] A. Divochiy, F. Marsili, D. Bitauld, A. Gaggero, R. Leoni, F. Mattioli, A. Korneev, V. Seleznev, N. Kaurova, O. Minaeva, G. Gol'tsman, K.G. Lagoudakis, M. Benkhaoul, F. Lvy, and A. Fiore, Nat. Photon. 2, 302 (2008).
[14] K. Banaszek and I.A. Walmsley, Opt. Lett. 28, 52 (2003).
[15] M.J. Fitch, B.C. Jacobs, T.B. Pittman, and J.D. Franson, Phys. Rev. A 68, 043814 (2003).
[16] D. Achilles, C. Silberhorn, C. Śliwa, K. Banaszek, and I.A. Walmsley, Opt. Lett. 28, 2387 (2003).
[17] D. Achilles, C. Silberhorn, C. Śliwa, K. Banaszek, I.A. Walmsley, M.J. Fitch, B.C. Jacobs, T.B. Pittman, and J.D. Franson, J. Mod. Optics 51, 1499 (2004).
[18] M. Avenhaus, K. Laiho, M.V. Chekhova, and C. Silberhorn, Phys. Rev. Lett. 104, 063602 (2010).
[19] E.A. Goldschmidt, F. Piacentini, I.R. Berchera, S.V. Polyakov, S. Peters, S. Kueck, G. Brida, I.P. Degiovanni, A. Migdall, and M. Genovese, Phys. Rev. A 88, 013822 (2013).
[20] A. Restelli, J.C. Bienfang, and A.L. Migdall, Appl. Phys. Lett. 102, 141104 (2013).
[21] A. Papoulis, "Probability, Random Variables, and Stochastic Process," McGraw-Hill-International Edition (1991).
[22] I.P. Degiovanni, S. Polyakov, A. Migdall, Private Communication.
[23] R. Loudon, "The Quantum Theory of Light," Oxford Science Publications (2000).
[24] D.R. Cox and H.D. Miller, "The Theory of Stochastic Processes," Chapman and Hall Ed. (1978).
[25] D.R. Cox, "Renewal Theory," Chapman and Hall Ed. (1982).
[26] S. Castelletto, I.P. Degiovanni, and M.L. Rastello, Phys. Rev. A 67, 022305 (2003).
[27] J.W. Müller, Nucl. Instr. Meth. Phys. Res. A301, 543 (1991).
[28] J. Libert, Nucl. Instr. Meth. Phys. Res. 126, 589 (1975).
[29] J. Libert, Nucl. Instr. Meth. Phys. Res. 136, 563 (1976).
[30] J.W. Müller, Report BIPM-89/1.
[31] J.W. Müller, "Applied Modulo Counting, Advanced Mathematical Tools in Metrology," World Scientific, p.133 (1994).
[32] F. Esposito, N. Spinelli, R. Velotta, and V. Berardi, Rev. Sci. Instrum. 62, 2822 (1991).
[33] F. Carloni, A. Corberi, M. Marseguerra, and C.M. Porceddu, Nucl. Instr. Meth. Phys. Res. 78, 70 (1970).
[34] S. Castelletto, I.P. Degiovanni, and M.L. Rastello, Metrologia 37, 613 (2000).
[35] I.M. Gel'fand and G.E. Shilov, "Generalized Functions," vol.1, Academic Press (1964).
[36] M. Abramowitz and I.A. Stegun, "Handbook of Mathematical Functions," Dover Publications Inc. (1965).
[37] S.A. Castelletto, I.P. Degiovanni, V. Schettini, and A.L. Migdall, J. Mod. Opt., 54, 337 (2007).
[38] V. Schettini, S.V. Polyakov, I.P. Degiovanni, G. Brida, S. Castelletto, and A.L. Migdall, IEEE Quant. Electron. Sel. Top. 13, 978 (2007).
[39] G. Brida, I.P. Degiovanni, V. Schettini, S.V. Polyakov, and A. Migdall, J. Mod. Opt. 56, 405 (2009).
[40] G. Brida, I.P. Degiovanni, F. Piacentini, V. Schettini, S.V. Polyakov, and A. Migdall, Rev. Sci. Instr. 80, 116103 (2009).
[41] http://mathworld.wolfram.com/LambertW-Function.html.
[42] R. Hanbury Brown and R.Q. Twiss, Nature 177, 27 (1956).
[43] L.A. Jiang, E.A. Dauler, and J.T. Chang, Phys. Rev. A 75, 062325 (2007).
[44] D.A. Kalashnikov, S.-H. Tan, M.V. Chekhova, and L.A. Krivitsky, Opt. Express 19, 9352 (2011).
[45] http://jp.hamamatsu.com/resources/products/ssd/pdf/s10362-11_series_kapd1022e05.pdf(2009).
[46] A.J. Kerman, E.A. Dauler, W.E. Keicher, J.K.W. Yang, K.K. Berggren, G. Gol'tsman, and B. Voronov, Appl. Phys. Lett. 88, 111116 (2006).
[47] E.A. Dauler, A.J. Kerman, B.S. Robinson, J.K.W. Yang, B. Voronov, G. Goltsman, S.A. Hamilton, and K.K. Berggren, J. Mod. Opt. 56, 364 (2009).
[48] D. Rosenberg, A.J. Kerman, R.J. Molnar, and E.A. Dauler, Opt. Exp. 21, 1440 (2013).

[49] S.V. Polyakov, A. Migdall, S.W. Nam, AIP Conf. Proc. 1327, 505–519 (2011). http://www.nist.gov/pml/div684/grp03/multicoincidence.cfm.
[50] C. Silberhorn, D. Achilles, A.B. U'Ren, K. Banaszek, and I.A. Walmsley, "Characterization and Preparation of Higher Photon Number States," in Proceedings of 7th International Conference on Quantum Communication, Measurement and Computing, edited by S.M. Barnett, E. Andersson, J. Jefferes, P. Ohberg, and O. Hirota, pp. 342–345 (2004).
[51] D. Achilles, C. Silberhorn, and I.A. Walmsley, Phys. Rev. Lett. 94, 043602 (2006).
[52] H.B. Coldenstrodt-Ronge, J. S. Lundeen, K.L. Pregnell, A. Feito, B.J. Smith, W. Mauerer, C. Silberhorn, J. Eisert, M.B. Plenio, and I.A. Walmsley, J. Mod. Opt. 56, 432 (2009).
[53] H.B. Coldenstrodt-Ronge, "Photon Number Resolved Characterization of Quantum States," Diploma Thesis, Max Planck Research Group, Institute of Optics, Information and Photonics, University Erlangen-Nuremberg, Germany (March 2006).
[54] A.P. Worsley, H.B. Coldenstrodt-Ronge, J.S. Lundeen, P.J. Mosley, B.J. Smith, G. Puentes, N. Thomas-Peter, and I.A. Walmsley, Opt. Express 17, 4397 (2009).
[55] J. Eisert, F.G.S.L. Brandão and K.M.R. Audenaert, New J. Phys. 9, 46 (2007).
[56] G. Puentes, A. Datta, A. Feito, J. Eisert, M.B. Plenio, and I.A. Walmsley, New J. Phys. 12, 033042 (2010).
[57] D.T. Smithey, M. Beck, M.G. Raymer, and A. Faridani, Phys. Rev. Lett. 70, 1244 (1993).
[58] U. Leonhardt, "Measuring the Quantum State of Light," Cambridge University Press (2005).
[59] G. Puentes, J.S. Lundeen, M.P.A. Branderhorst, H.B. Coldenstrodt-Ronge, B.J. Smith, and I.A. Walmsley, Phys. Rev. Lett. 102, 080404 (2009).
[60] K. Laiho, K.N. Cassemiro, D. Gross, and C. Silberhorn, Phys. Rev. Lett. 105, 253603 (2010).
[61] L. Zhang, H.B. Coldenstrodt-Ronge, A. Datta, G. Puentes, J.S. Lundeen, X.-M. Jin, B.J. Smith, M.B. Plenio, and I.A. Walmsley, Nature Photon. 6, 364 (2012).
[62] L. Zhang, A. Datta, H.B. Coldenstrodt-Ronge, X.-M. Jin, J. Eisert, M.B. Plenio, and I.A. Walmsley, New J. Phys. 14, 115005 (2012).
[63] M.A. Albota and F.N.C. Wong, Opt. Lett. 29, 1449 (2004).
[64] A.P. Vandevender and P.G. Kwiat, J. Mod. Opt. 51, 1433 (2004).
[65] R.V. Roussev, C. Langrock, J.R. Kurz, and M.M. Fejer, Opt. Lett. 29, 1518 (2004).
[66] O. Kuzucu, F.N.C. Wong, S. Kurimura, and S. Tovstonog, Opt. Lett. 33, 2257 (2008).
[67] O. Kuzucu, F.N.C. Wong, S. Kurimura, and S. Tovstonog, Phys. Rev. Lett. 101, 153602 (2008).
[68] M. Legré, R. Thew, H. Zbinden, and N. Gisin, Opt. Express 15, 8237 (2007).
[69] R.T. Thew, S. Tanzilli, L. Krainer, S.C. Zeller, A. Rochas, I. Rech, S. Cova, H. Zbinden, and N. Gisin, New J. Phys. 8, 32 (2006).
[70] H. Takesue, T. Honjo, and H. Kamada, Jpn. J. Appl. Phys. 45, 5757 (2006).
[71] H. Xu, L. Ma, A. Mink, B. Hershman, and X. Tang, Opt. Express 15, 7247 (2007).
[72] M.A. Albota and B.S. Robinson, Opt. Lett. 35, 2627 (2010).
[73] C.K. Hong, Z.Y. Ou, and L. Mandel, Phys. Rev. Lett. 59, 2044 (1987).
[74] H. Takesue, Phys. Rev. Lett. 101, 173901 (2008).
[75] J.H. Shapiro, New J. Phys. 4, 47.1 (2002).
[76] S. Tanzilli, W. Tittel, M. Halder, O. Alibart, P. Baldi, N. Gisin, and H. Zbinden, Nature 437, 116 (2005).
[77] M.A. Albota, F.N.C. Wong, and J.H. Shapiro, J. Opt. Soc. Am. B 23, 918 (2006).
[78] J. Huang and P.R. Kumar, Phys. Rev. Lett. 68, 2153 (1992).
[79] Y. Ding and Z.Y. Ou, Opt. Lett. 35, 2591 (2010).
[80] N Curtz, R. Thew, C. Simon, N. Gisin, and H. Zbinden, Opt. Express 18, 22099 (2010).
[81] H. Pan and H. Zeng, Opt. Lett. 31, 793 (2006).
[82] C. Langrock, E. Diamanti, R.V. Roussev, Y. Yamamoto, M.M. Fejer, and H. Takesue, Opt. Lett. 30, 1725 (2005).
[83] H. Takesue, E. Diamanti, C. Langrock, M.M. Fejer, and Y. Yamamoto, Opt. Express 14, 13067 (2006).
[84] S. Ramelow, A. Fedrizzi, A. Poppe, N.K. Langford, and A. Zeilinger, Phys. Rev. A 85, 013845 (2012).
[85] A.P. VanDevender and P.G. Kwiat, J. Opt. Soc. Am. B 24, 295 (2007).
[86] L. Ma, J.C. Bienfang, O. Slattery, and X. Tang, Opt. Express 19, 5470 (2011).
[87] O. Kuzucu, M. Fiorentino, M.A. Albota, F.N. C. Wong, and F.X. Kärtner, Phys. Rev. Lett. 94, 083601 (2005).

Chapter 8

Single-Photon Detector Calibration

Sergey V. Polyakov
National Institute of Standards and Technology, Gaithersburg, MD 20899, USA

Chapter Outline
8.1 Introduction	257
8.2 Definitions	259
8.3 Calibration Methods	260
8.3.1 Radiant Power Measurements (Substitution Method)	261
8.3.2 Correlated-Photon-Pair Calibration Method	262
8.4 Practical Considerations	263
8.4.1 Semiconductor Single-Photon Avalanche Diodes	264
8.4.2 Transition Edge Sensors	275
8.5 Conclusion	279
References	280

8.1 INTRODUCTION

As we have seen thus far, single-photon detectors are based on a variety of nontrivial physical phenomena and often employ complex circuitry. Therefore, thorough characterization of a detector's properties is necessary before it can be relied on for accurate measurements. To this end, it is important to establish a uniform way to characterize single-photon detectors as accurately as possible regardless of the differences in their underlying technology. Such an undertaking reveals the interconnectedness of a detector's properties. As an example, detection efficiency is often considered the most critical parameter of a detector. However, it is by far not the only relevant parameter, and its usefulness is greatly reduced when presented without considerable additional information. Due to the inherent nonlinearity (saturation) of many single-photon detectors, the apparent detection efficiency depends critically on the count rate and photon statistics for which the efficiency is measured (or defined). Even when one attempts to more stringently define detection efficiency, the complex response of single-photon detectors makes it impossible to extrapolate it to the wide range of input conditions to which the device may be subjected. Characterization of detectors, therefore, requires much more than a simple measurement of the detection efficiency. A set of additional parameters, such as the dead and reset times (c.f. Chapter 2) and afterpulsing characteristics, are needed to describe a typical non-photon-number-resolving detector. Photon-number-resolving (PNR) detectors (with the notable exception of detectors with full PNR capability, such as transition edge sensors (TES)) often call for a more thorough characterization that includes a measurement of the positive-operator-valued measure (POVM) (see Chapter 2 for definition of PNR capability).

Unless a detector's response is truly linear with respect to the input photon number (i.e. unless a detector is a true PNR detector), the statistical properties of the input light field further complicate the measurement situation. Because input sources may have different auto-correlations (poissonian, bunched (thermal), or antibunched), and because, in general, a single-photon detector's efficiency is time-dependent, different fractions of the incoming light will be detected, missed, or counted more than once. In addition, detectors can produce counts in the absence of an input field, resulting in dark counts whose statistical properties are not related to those of the input field. There can also be a background field whose photon statistics are uncorrelated with the input light. Thus care should be taken to understand the impact of statistical properties of the actual input field on the detector parameters measured during characterization and calibration.

Given all these complications, a model of the detector's response should be developed and verified before any characterization takes place. To prove that the model for a detector is consistent, an independent verification of the calibration results obtained for a particular type of detector is necessary.

In this chapter we introduce the set of detector properties common to most modern detectors that should be determined for a complete characterization. Then we introduce methods for detector characterization, and finally we present practical recipes for calibration of non-PNR detectors and PNR detectors with full PNR capability. Although the recipes are based on specific underlying technologies (we use as examples single-photon avalanche diodes (SPADs) and TES detectors), the measurement algorithms are in general applicable to detectors with different principles of operation. A radically different approach is to treat the detector as a black box, i.e. to pay no attention to the underlying principles of operation and approach the problem from the viewpoint of quantum measurement theory to characterize the detection of arbitrary input quantum states (that is, find a detector's POVM). Although in practice even with such an approach some assumptions about the detector's operation are necessary to make the measurement problem tractable. Such a characterization is particularly useful for PNR detectors that exhibit saturation. POVM measurements are addressed in Chapter 9.

8.2 DEFINITIONS

Key properties of detectors based on a range of underlying principles have been introduced in the preceding chapters. Here we establish a generalized list of parameters that are relevant to calibrations with accuracies down to the $\approx 0.1\%$ level. More accurate calibrations may require considering additional parameters.

Recovery Time, Dead Time, and Reset Time. Many detectors, such as SPADs and single SNSPDs, can detect only one photon of the input field at a time, after which they become insensitive to incoming radiation for a certain amount of time. This period of time is called the device's dead time. Note also that returning to operation from this insensitive (dead) state may be a complex process, during which both the ability to detect photons and the device's detection latency (i.e. the time between a photon entering the detector and a signal output, see Chapter 2) can differ significantly from that of normal operation. This mode of a detector is called reset time. The above two modes of a detector, when the detector's operation is abnormal, are called the recovery time. Although the instantaneous detection efficiency at any point during this transient period may be a function of time elapsed since the detection event, one possible simplification useful for characterization with continuous wave sources is assuming a constant detection efficiency when the detector is alive and zero detection efficiency when the detector is dead, and adjusting the effective dead time accordingly.

Detection Efficiency. Detection efficiency is commonly defined as the probability of a detector to produce a successful detection given a single-photon input, when the detector is operating normally (far from its previous recovery), η_{DE}. Because of dead time (and assuming that both afterpulsing and dark count rates are zero), the actual rate of detection R_{out} will be bound by $R_{out} \leq \eta_{DE} R_{in}$, where R_{in} is the rate of incoming photons. Obviously, just knowing η_{DE} is not enough to describe the operation of a detector with realistic input fields. Alternatively, detection efficiency can be defined as a proportionality coefficient η_{cond} relating the number of detection signals to the number of photons incident on a detector, the auto-correlation properties of the input field, etc.

The detection efficiency of a full PNR detector is defined in a similar fashion. This probability should scale with the number of incident photons. That is, if η_{DE} is the probability to detect a single photon given a single-photon input, the η_{DE}^2 is the probability to detect a photon pair given two input photons, and in general η_{DE}^n is the probability to detect n photons given an n-photon state input. Any deviation of a real-life device from the full PNR assumption can significantly limit the accuracy of a precision measurement.

Dark Counts. Dark counts are output signals produced by the detector that do not correspond to actual photon detections. In most cases, nothing about the processes that cause dark counts makes them distinguishable from photon detection events. It is often assumed that dark counts obey Poissonian statistics (although at very high count rates that may no longer be true because of dead time effects). For an accurate calibration, dark count rates need to be characterized and included in the analysis.

Afterpulsing. Afterpulsing counts occur some time after a detection, sometimes even during the recovery of the detector. These counts originate from within the detector, and not from incoming photons. Particularly, in semiconductor avalanche detectors (see Chapter 4) afterpulses are triggered by carriers that were trapped and subsequently released from localized trap sites in the avalanche region after a previous avalanche. In a calibration the afterpulsing probability needs to be characterized as well.

8.3 CALIBRATION METHODS

In metrology, *primary standard methods* stand out because they depend on parameters of the physical system that can be independently verified by measurement and do not depend on implementation. In radiometry, a blackbody is a good electromagnetic source standard because there is a direct and universal dependence between electromagnetic radiation and temperature, as established by Planck [1]. A synchrotron is another such source, because its radiation is related only to the velocity, charge, number of particles, and magnetic field strength. That is, because the properties of the electromagnetic radiation can be fully described by these few parameters, one can reproduce these sources and independently measure (or calibrate) the response of radiometric equipment. This procedure independently reproduces a measurement scale for radiant power. Both synchrotron and blackbody-based calibration are examples of so-called fundamentally absolute calibration techniques or primary standard methods.

It turns out that with photon-counting detectors, an additional primary standard method can be implemented. This method is based on quantum sources that produce photons in pairs, so that detection of one photon of the pair indicates the presence of another. As we established, practical calibration techniques may need to be adapted to the specifics of the photon-counting device under consideration. In this regard, because the photon-pair-based method is inherently based on direct photon counting, adapting this method to a particular type of photon-counting detectors may be more straightforward than adapting a precision radiant-power-measurement-based method. In addition, the photon-pair-based calibration method may help with a more complete detector characterization, as will be seen later in the chapter.

We now outline the calibration methods as they apply to idealized detectors. By ideal, we mean that there are no competing mechanisms causing detectors to fire other than the input signal used for calibration. Then, we will introduce operational models of non-ideal detectors and study the consequences.

8.3.1 Radiant Power Measurements (Substitution Method)

Optical detectors are calibrated by measuring the radiant power emitted by a calibrated source. That source may be a primary standard, or a transfer standard (a detector whose calibration is tied ultimately to either a primary standard source or a primary standard detector). Primary standard sources are typically one of two types: blackbodies or synchrotrons. A blackbody source is used as a primary standard for the visible and near-infrared spectral bands. Synchrotron radiation can be used as a primary standard source over a broad spectral range, particularly in the ultraviolet band. Currently all transfer detectors are classical detectors, whose output is proportional to radiant power. Using one or more transfer detectors (or directly using a primary standard detector if one has access to such a device) is referred to as the substitution method, see Fig. 8.1, because the transfer detector is swapped with the detector under test (DUT) during calibration. Often, employing a transfer detector is advantageous to using a primary standard detector, despite the added systematic uncertainties, because the established procedures for calibrating transfer standards (classical detectors) are mature and well characterized. From the standpoint of a photon-counting detector, power measurements are often more problematic. The main difficulty is converting radiant power into photon flux. And even if detectors were perfect, an additional accurate measurement of the source's spectrum is needed. Other issues with single-photon detectors further complicate power measurements. These issues will be discussed later in our case study of substitution calibration.

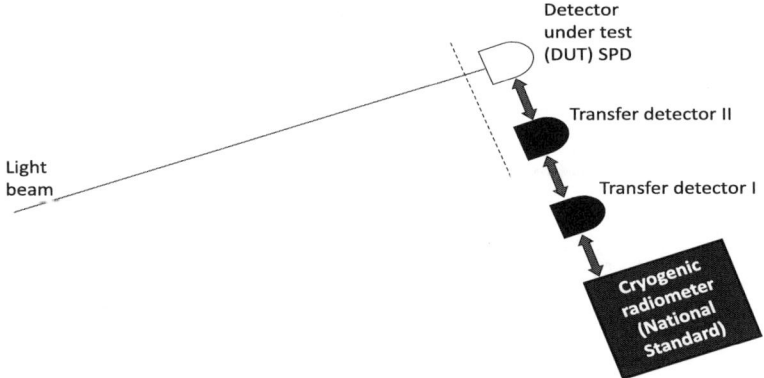

FIGURE 8.1 Conventional detection-efficiency calibration based on radiant power measurements (detector substitution). SPD—single-photon detector.

8.3.2 Correlated-Photon-Pair Calibration Method

The photon pairs produced in spontaneous parametric down-conversion (PDC), c.f. Chapter 11, provide a fundamentally absolute way to calibrate single-photon detectors [2–19]. In PDC, photons are produced in pairs whose frequencies and wavevectors are governed by energy and momentum conservation. Therefore the detection of one photon provides spatial, spectral, and temporal location information about (or *heralds*) the other photon of the pair with certainty. The absolute nature of the method derives from this quantum property of a PDC light source. Therefore, this method is a primary standard method [20]. Because the method relies on individual photon detections as a trigger, the scheme operates directly in the photon-counting regime and is thus well suited for photon-counting devices. To measure detection efficiency, a trigger detection system is placed to intercept some of the down-converted light. A DUT is then arranged to collect all the photons correlated to those seen by the trigger detector. The DUT channel detection efficiency is the ratio of the number of coincidence events to the number of trigger detection events in a given time interval. Coincidences are usually measured with start-stop time-to-digital systems, with the heralding detector output connected to the start, and the DUT output to the stop, Fig. 8.2. If we specify the total efficiency of the DUT and trigger channels by η_{DUT} and η_{trig}, respectively, then the total number of trigger counts is

$$N_{\text{trig}} = \eta_{\text{trig}} N_{\text{p}} \tag{8.1}$$

and the total number of coincidence counts is

$$N_{\text{c}} = \eta_{\text{DUT}} \eta_{\text{trig}} N_{\text{p}}, \tag{8.2}$$

where N_{p} is the total number of down-converted photons emitted into the trigger channel during the counting period. The absolute detection efficiency of the DUT channel is then simply

$$\eta_{\text{DUT}} = \frac{N_{\text{c}}}{N_{\text{trig}}}. \tag{8.3}$$

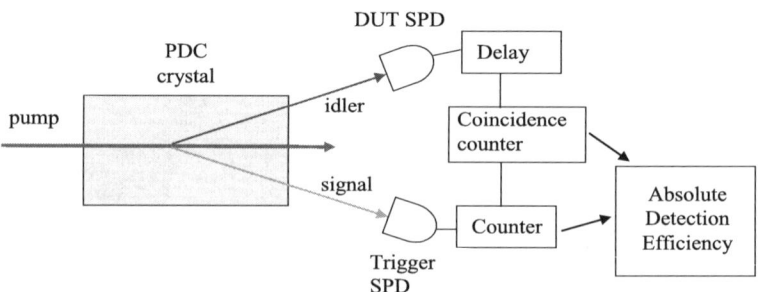

FIGURE 8.2 Detection-efficiency measurement based on photon-pair generation via parametric down-conversion.

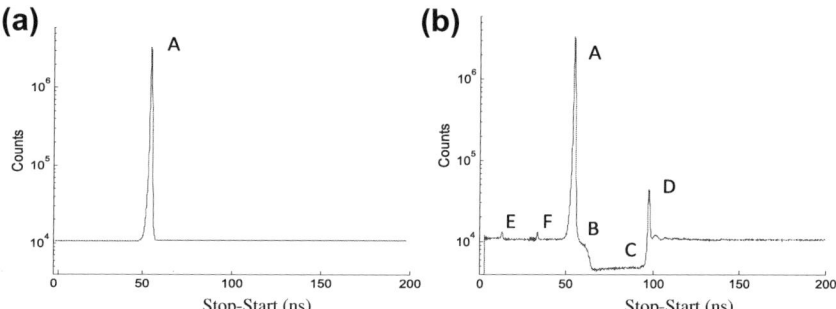

FIGURE 8.3 Start-Stop histograms that correspond to a correlated-pair calibration method with idealized detectors (a) and real detectors (b). In both cases the background is due to finite detection efficiency of the heralding detector (with real detectors, dark counts will also contribute to a background, but such contribution is small in this example). Features are identified as follows: A—main correlation peak, B—correlated photons that have arrived simultaneously with those in peak A during the reset mode of the detector exhibiting longer latency times. C—reduction of stop counts due to the recovery time of a detector. D—peak due to afterpulses and delayed counts in the reset mode (twilight counts). E and F—peaks specific to the setup, and not the DUT.

Note that this is the efficiency of the entire detection channel (including collection optics, etc.) and not just the efficiency of the DUT alone [21,22]. Therefore, all losses in the DUT path before reaching a detector must be determined. This includes losses within the source, usually a nonlinear crystal. Another important possible source of loss is the overlap between heralded photons that are correlated with trigger photons and the field of view of the DUT channel. Total channel efficiency is the product of the transmittances of individual optical elements in a DUT path and any beam-matching efficiencies, like the collection mode overlap. Uncertainties in characterizing these losses contribute to the overall uncertainty budget. Other uncertainties may also be present, but they are detector-specific.

With ideal detectors, the correlation function between the firing of the herald detector and the DUT would appear as shown in Fig. 8.3a. Here, the tallest correlation peak corresponds to the correlated signal due to PDC. It sits on a background that is generated by accidental or uncorrelated events, due to finite detection efficiency in a heralding channel. Before applying formula (8.3), this background should be removed. Then, the remaining number of correlated events gives N_c.

8.4 PRACTICAL CONSIDERATIONS

The basic theory discussed above assumes an ideal single-photon detector. Namely, the DUT produces a count with constant probability (for a single-photon input) that depends only on the detector efficiency and does not depend on any other factors; the detector should never produce a count if no photons were present; the single-photon detector is assumed to have no recovery time, etc. Below we discuss practical approaches to calibrating a real-world non-photon-number-resolving detector (a SPAD) and a photon-number-resolving detector (a TES).

8.4.1 Semiconductor Single-Photon Avalanche Diodes

8.4.1.1 Model of a Single-Photon Detector

Real-world detectors suffer from multiple features that complicate their output. To achieve high accuracy, all the features of single-photon detectors must be characterized first. Many of these features leave a trace on the correlation function as shown in Fig. 8.3b. Such histograms can be readily used to build a model of a real-world detector. Here we demonstrate how to build a model of a typical click/no-click single-photon detector based on such a correlation measurement, using a silicon SPAD as an example [23]. The recipe presented here can be modified to characterize any non-PNR detector. Building a detector model is important for both correlated measurements and radiant power measurements. On one hand, to find N_c for correlated measurements it is necessary to separate the correlated signal, due to true detections of twin photons, from background photons, with a high degree of accuracy. Measurements of radiant power on the other hand should be corrected for dark counts and afterpulsing, events that are not due to a photon detection.

We start with a correlated measurement, and obtain a histogram. We first note all features (peaks, valleys, etc.) of the histogram (see Fig. 8.3b) and determine their origins. In this example, features are labeled with letters A through F. The main coincidence peak (A) is the most prominent feature of the histogram and represent correlated events.

The trench (labeled C) to the right of the peak is due to detector recovery time. Because the DUT often fires due to detection of a photon from a correlated pair (feature A), it enters its recovery time, during which it is unable to fire again. Thus, the number of background events shortly after A is reduced. When subtracting the background, this trench needs to be properly modeled (to be correctly removed).

The origin of the small "shoulder" (labeled B) on the right of the main coincidence peak is less obvious. A study of this feature shows that the "shoulder" events are valid photon detections for which the detector latency is unusually long. This happens when a photon arrives while the detector is in the midst of its reset from a prior detection (c.f. Chapter 2). In this case, the detector may still manage to output an electric signal, but with an unusually long latency. Due to the inherent uncertainty associated with characterization of this additional complexity, this effect should be avoided when possible by reducing the photon flux. (In the comparison experiment discussed here, high photon rates are unavoidable, because of the use of classical transfer-standard

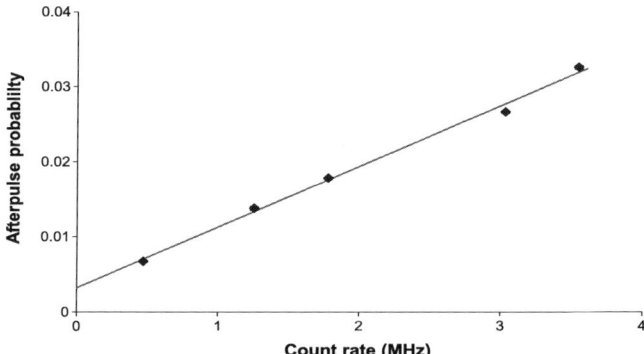

FIGURE 8.4 Afterpulsing probability of the DUT at various count rates. The apparent increase in afterpulsing probability with count rate is due to the twilight sensitivity of the detector, as photons arriving during the reset time (c.f. Chapter 2) trigger an avalanche with increased latency. The probability of true afterpulsing (due to trapped carriers) remains constant, and can be found by extrapolation of the linear dependence to low count rates.

detectors.) Because this behavior occurs during the device's return from its dead state, we refer to these detections as "twilight" events.

The afterpulse feature (D) is also affected by this mechanism. That is, events that occur due to variations in detector sensitivity and latency during the reset time can pile up in the vicinity of the afterpulse peak, causing the perceived probability of an afterpulse to be count-rate dependent. However, the probability of a purely electronic afterpulse (that is, an afterpulse due to trapped charge that was released during the reset time) should not depend on count rate. The contribution from twilight events is due to real photons. We verify this by measuring the number of events in the afterpulse peak as a function of count rate, and observe that, indeed, the afterpulse probability is a linear function of count rate with a positive offset, Fig. 8.4. This offset is the count-rate independent probability of an electronic afterpulse. Understanding this feature is very important for radiant power measurements because only proper afterpulses should be accounted for when calculating the number of photons detected. Also, due to the same effect, the true dead time of a real detector (i.e. when its detection efficiency is zero) is shorter than what is suggested by the duration of the dip C.

To demonstrate the twilight effect, and measure its properties directly, one can send pairs of attenuated optical pulses separated by a controllable delay to a detector, and record the output with a coincidence board. Figure 8.5 shows a number of histograms for varying delay times between the two optical pulses detected by a SPAD with a nominal recovery time of 49.5 ns. We see that if the photon comes during the first ≈ 40 ns of the recovery time, i.e. while the active quench circuit has the bias below breakdown (c.f. Chapter 4) it has no effect on the afterpulse (so the dead time is just ≈ 40 ns!). In this case, the afterpulse has a shape shown by the thick black line in Fig. 8.5 (previously observed in

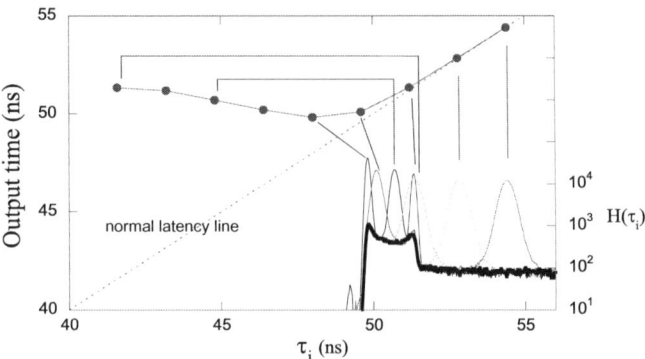

FIGURE 8.5 High resolution histogram of an afterpulse feature (c.f. D of Fig. 8.3) taken with photons arriving with a range of delays after a previous detection event, i.e. during dead time and the twilight zone. Thick line shows an afterpulse feature with no photons arriving in the twilight zone. Peaks (thin lines and right axis) show histogram counts (in 4 ps bins) due to a photon arriving in the twilight zone, when the detector is recovering from previous event. Upper curve (dots and left axis) shows variations in latency (output with normal latency would fall on the dashed line; an extra delay is indicated by the difference between the data and the dashed line).

Ref. [24]). If the two light pulses are separated by more than ≈ 40 ns, so the second pulse arrives while the SPAD bias is rising back to its original value, the counts in the afterpulse feature increase and we see a distinct new peak in the afterpulse shape, which can be delayed by up to 10 ns from the normal latency time. The data shows that the closer the photon arrives to the end of the recovery time, the shorter the observed delay (see the upper curve in Fig. 8.5), [23]. We conclude that the time after a detection when the SPAD does not put out any signal (c.f. feature C in Fig. 8.3b) is not equal to the dead time (and does not have to be equal to recovery time). The electronic pulses that represent these counts are delayed by up to the duration of the reset time, contributing to a higher-than-normal afterpulse peak D in Fig. 8.3b. The same mechanism is behind feature B. It appears due to correlated counts, i.e. due to photons that arrive at the same time as those of peak A. The only difference is that the detector is in its reset time, recovering from an earlier background detection. Note here that the extent of the reset time and latency profile of twilight counts depend strongly on the detector's underlying technology and its electronics.

Other, tiny peaks in Fig. 8.3b (E and F) may also be present and are specific to our particular setup. Most notably, peak E is produced by afterpulsing of the trigger detector. This is a very important point, because ideally, correlated-photon calibration techniques should not depend on the properties of the trigger detector. In this case, an afterpulse of the trigger detector re-starts the histogram acquisition, resulting in detecting correlated photons ≈ 55 ns earlier than the main correlation peak. Peak F is due to another common issue: there is a small back-reflection at either end of the fiber in the trigger arm and photons that experience a back-reflection at both ends of the fiber arrive at the trigger SPAD after traveling two extra fiber lengths. In the case of Fig. 8.3 the fiber is 2 m long and this extra double-pass corresponds to a delay of ≈ 20 ns. Again, the coincidences in this peak are valid and should be included in our determination of the true number of coincidences.

As shown, the use of just the correlation histogram allows one to identify a broad range of effects, even those whose contribution is minute (features E and F contain less than 0.5% of relevant counts). One remaining concern is that a click/no-click detector is generally not able to discriminate the instantaneous arrival of more than one photon from the arrival of a single photon. This needs to be addressed in the calibration model, particularly for a pulsed photon source [25].

The above study gives rise to the following model of a photon-counting detector:

- The detector has a constant detection efficiency η_{DE} when it is counting, and 0 when it is dead. This is a simplification, because the detection efficiency is not constant during the reset time.
- The detector has a recovery time, $t_{recovery}$, that is longer than its actual dead time t_{dead} due to the device activity during the reset. With our model, we do not need to accurately measure $t_{recovery}$, t_{dead}. To do that, many extra measurements (leading to extra uncertainties) would be needed. Instead, to find the effective t_{reset} we make use of the integrated number of twilight counts, readily found from the slope of Fig. 8.4 and assuming a constant detection efficiency η_{DE} during the entire reset time. This simplification does not affect accuracy of the calibration. The effective $t_{recovery}$ can be estimated from the histogram as the distance between the peak of features A and D in Fig. 8.3b. The effective dead time is then simply ($t_{recovery} - t_{reset}$). Note that this simplification allows us to skip integrating the actual variable detection efficiency over the nominal duration of the reset time.
- A true afterpulse is an event that is not due to a photon detection (i.e. purely due to charge trapping in SPADs). Its probability is also found from Fig. 8.4.

8.4.1.2 Correlated Calibration Technique

The model introduced above helps separate the background and correlated signals seen in Fig. 8.3, which is key to an accurate characterization. For simplicity, we require that the background and dark count rates on the trigger detector are negligible. We begin with a measured histogram denoted by $H(\tau_i)$, where τ_i is the time delay from the start event for the ith bin. Our goal is to separate the events represented by $H(\tau_i)$ into two categories: events correlated to the trigger pulse $C(\tau_i)$, such that $N_c = \sum C(\tau_j)$, and background events $B_m(\tau_i)$:

$$H(\tau_i) = C(\tau_i) + B(\tau_i). \tag{8.4}$$

Note that when we measure η_{DE}, we would have to not only find C, but also estimate how many correlated events were *missed* since the detector may have been dead when the correlated photon arrived. Similarly, the detector might be dead when a background event would otherwise have been recorded due to either a previous background event or a previous correlated event.

Following our model of the detector, we can express the recovery time $t_{recovery}$ in terms of the number of time bins $d = t_{recovery}/(\tau_{i+1} - \tau_i)$. For simplification, here d is assumed to be an integer. We also assume that the number of background events in a given bin is relatively constant before correlated events arrive, and we define B_0 as the average number of background events recorded per bin. Then, the probability b of an *alive* detector to register a background count is given by:

$$b = B_0/(N_{trig} - dB_0). \tag{8.5}$$

For ease of discussion, we will assume that the delay of the DUT channel is long enough that all of the correlated signals appear after the dth histogram bin. Then, the background in the presence of the correlated signal is:

$$B(\tau_i) = H(\tau_j) = B_0 \quad (i < d);$$
$$B(\tau_i) = b \times \left(N_{trig} - \sum_{j=i-d}^{i-1} H(\tau_j) - \Delta(\tau_i) \right) \quad (i \geq d). \tag{8.6}$$

Note that the number in parentheses is the number of scans where it was possible to record a background event in the bin. The sum gives the number of times that the detector was in recovery in the ith bin due to events in one of the previous d bins. The function $\Delta(\tau_i)$ accounts for the situations where a subsequent, closely spaced trigger pulse resets the histogram before it reaches the ith bin, a feature of some start-stop boards. (This accounts for the background level decrease for large js that can be seen in Fig. 8.3b).

We can use Eq. (8.6) together with Eq. (8.4) to determine C. Recall that $N_c = \sum C(\tau_j)$, summing over the main correlation peak (feature A), its shoulder (feature B), and the small features E and F (Fig. 8.3b). We find η_{cond} for the DUT, using Eq. (8.3), and accounting for losses in the DUT channel:

$$\eta_{cond} = \frac{\sum C(\tau_j)}{N_{trig}}. \tag{8.7}$$

The sum over the feature D gives the total number of afterpulses plus twilight counts. The probability of an afterpulse-like event is then simply the ratio of the two sums. Performing this measurement at different DUT count rates, and measuring this probability gives rise to Fig. 8.4. A linear regression helps separate the true afterpulse probability ($p_{afterpulse}$) from the probability of twilight counts, $p_{twilight}|_{cond}$. The ratio $d' = p_{twilight}|_{cond}/b$ gives the effective t_{reset}, measured in histogram bins.

To find the detection efficiency of the DUT η_{DE} from the data above, we calculate the corrected histogram $Z(\tau_i)$ that accounts for the occasional situation when DUT is dead because of a background detection during the arrival of a correlated photon as:

$$Z(\tau_i) = \frac{C(\tau_i)}{N_{\text{trig}} - \sum_{j=i-d}^{i-d'} B(\tau_j) - \Delta(\tau_i)} N_{\text{trig}}. \qquad (8.8)$$

Here we implicitly assume that the probability of the source to generate more than one correlated pair of photons per a histogram bin is negligible. In a last step, we use Eq. (8.3), where now $N_c = \sum Z(\tau_j)$. Note, that because we applied all the corrections to N_c, we should use η_{DE} instead of η_{DUT} (because corrections are designed to eliminate dependence on any prior events and extract a detection efficiency of an "always-active" detector). We see that the series of correlated measurements coupled with the detector's model introduced above provide a characterization of the detector. Particularly, we can extract detection efficiencies η_{DE} and η_{cond}, and find $p_{\text{afterpulse}}$ and t_{recovery} and an effective t_{dead}. In addition, measuring dark counts is required, though this measurement is trivial for most detectors. (Note that there exist detectors whose dark count rates are count-rate dependent, characterization of such detectors is beyond the scope of this algorithm.) We still need to prove, however, that this characterization is complete and adequate. This issue will be addressed by comparing a correlated-photon-based calibration to one based on a radiant power measurement.

8.4.1.3 Radiant Power Measurements with Single-Photon Detectors

To use a radiant power approach for calibration of single-photon detectors, one needs a method that turns a photon count into radiant power. This seems straightforward, as the energy per photon is given by $E = h\nu$, where ν is a weighted average frequency for a given source. However, proper corrections to the raw data are needed for an accurate conversion.

Consider a photon-counting measurement (during the time t) of a Poissonian input field by a single-photon detector described by the model introduced above. Such a source produces photons with a constant probability. Specifically, a Poissonian source's output does not depend on the history of previous emissions. On the other hand, a detector's response strongly depends on the history of previous photon detection. Therefore, the total number of counts N_{total} must be corrected so that the discrepancy introduced by the detector's behavior is removed. Corrections must include dark counts N_{dark}, effective dead time ($t_{\text{recovery}} - t_{\text{reset}}$), and real afterpulsing (whose probability is $p_{\text{afterpulse}}$). We write:

$$P = \eta_{\text{DE}} E \frac{N_{\text{total}} - N_{\text{dark}} - p_{\text{afterpulse}} N_{\text{total}}}{t - N_{\text{total}}(t_{\text{recovery}} - t_{\text{reset}})}. \qquad (8.9)$$

This expression shows that to accurately measure radiant power much more information about the detector than just η_{DE} is needed. A separate set of measurements aimed at characterizing the temporal response of the detector is necessary. For a stationary Poissonian input, however, it is possible to avoid some of this characterization if we are only interested in observing that stationary process, we simply use η_{cond}, which takes care of saturation effects:

$$P = \eta_{\text{cond}} E(N_{\text{total}} - N_{\text{dark}} - p_{\text{afterpulse}} N_{\text{total}})/t. \quad (8.10)$$

Finally, if we neglect the effect of dark counts, we can further simplify this formula:

$$P \approx \tilde{\eta}_{\text{cond}} E N_{\text{total}}/t. \quad (8.11)$$

Values of both η_{cond} and $\tilde{\eta}_{\text{cond}}$ strongly depend on the input conditions, i.e. input count rate and photon number statistics. Note that in the framework of the detector model η_{cond} in Eq. (8.10) is exactly the same as the detection efficiency that is given by analysis of the start-stop histogram, Eq. (8.7). This is a very useful observation if detector-substitution method results are to be directly compared to the correlated-photon results.

This measurement can be applied to direct measurements with a primary standard source or calibration against a classical transfer detector (using the detector-substitution method). The above formulae can be used to obtain radiant power for Poissonian sources. Sources with other statistical properties may require slight modification.

8.4.1.4 Experimental Implementation: Calibration and Verification

As we have seen, calibrating single-photon detectors requires adopting a detector model and characterizing the detector within the model. Because the end result critically depends on these assumptions, calibration should not only be performed with one method, but its results should be verified against another independent method. It is possible to combine the two absolute calibration techniques discussed: correlation- and substitution based, in a single experimental setup, as shown in Fig. 8.6. Because the coincidence events used for the two-photon calibration and the single-photon count rate of the DUT used for the conventional calibration can be recorded simultaneously, the two types of calibrations can be effectively conducted simultaneously. To obtain additional data for the substitution method, the DUT is replaced with a calibrated detector. In one example of such a comparison, [19], a 351 nm Ar$^+$ laser pumps a LiIO$_3$ crystal. The pump power is monitored by a dedicated detector. Nearly degenerate downconverted photons at $\approx$ 702 nm were sent to the two single-photon detectors (here SPADs were used): trigger and DUT. The DUT detector is mounted behind an aperture to reduce PDC light not correlated to that seen by the trigger, and a lens to collect the correlated light onto the DUT active area with $\approx$ 0.2 mm diameter. The independently calibrated transfer-standard photodetector is a cooled high-shunt resistance Si PIN photodiode. A pinhole mounted close to the Si PIN photodiode's surface was used to restrict the wings

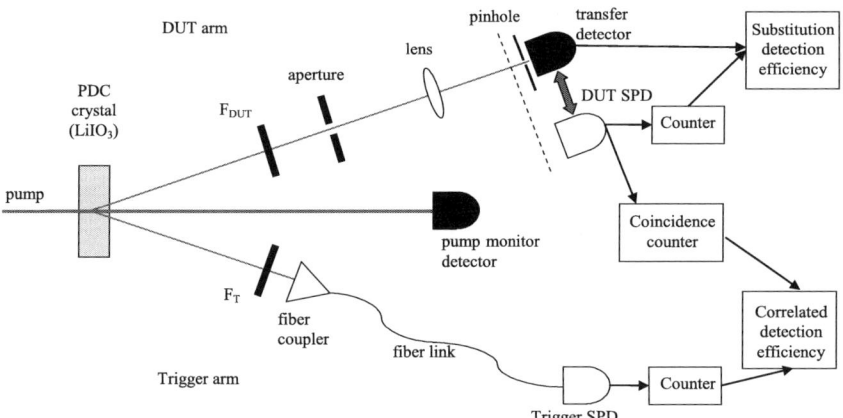

FIGURE 8.6 Setup used for a calibration and verification experiment. This setup combines the two independent absolute calibration methods. SPD—single-photon detector. In the experiment described here SPADs were used.

of the illuminating beam. In that way, the apertures of the SPAD and Si PIN photodiode are equal. In addition, narrow spectral filters are introduced in both trigger and DUT channels. The spectral band of the trigger-arm filter F_T is narrow enough to select only a small portion of the photons created by the downconversion process. The spectral bandwidth of F_{DUT} is significantly wider and completely encompasses the correlated band defined by F_T. The signals from the trigger and DUT single-photon detectors are collected by a circuit that records both the overall number of trigger and DUT events, and the correlation between trigger and DUT events in the form of a histogram with 0.1 ns temporal resolution. A typical histogram is shown in Fig. 8.3b.

Because the correlated method calibrates the entire optical path from within the nonlinear crystal to the DUT, as opposed to just the DUT detection efficiency, the transmittance of each optical element should be independently characterized. The uncertainties associated with those transmittance measurements will contribute to the overall uncertainty budget. An uncertainty budget with the list of physical properties typical for the radiant power calibration method is presented in Table 8.1, and the uncertainty budget for the correlated method is presented in Table 8.2. The values for the uncertainties are taken from Ref. [19]. They are presented here to aid with estimating potential trade-offs, and identifying the properties that could potentially limit a final calibration. For example, it is evident that the uncertainty associated with photon statistics is not the limiting factor in these measurements. Because the number of single-photon detections can be increased to reach better accuracy, the accuracy of auxiliary measurements determines the limit. In particular, for the substitution calibration method, the major source of uncertainty is the calibration of the transfer detector, which accounts for 0.1% in the overall uncertainty figure.

TABLE 8.1 Uncertainty Budget of a Substitution Calibration Method

Physical Property	Contribution to Relative DE Uncertainty (%)
Analog transfer-standard calibration (QE equivalent)	0.10
Spatial nonuniformity of photodiode at 700 nm, (standard deviation of central responsivity)	0.025
Analog measurement statistics and drift	0.06
Analog amplifier gain calibration at 10^{10} V/A	0.05
Pinhole backside reflection	0.1
DUT signal and background statistics	0.003
DUT afterpulsing	0.04
DUT dead time (due to rate changes with time)	0.02
Total	0.17

TABLE 8.2 Uncertainty Budget of a Correlated Calibration Method

Physical Property	Contribution to Relative DE Uncertainty (%)
Crystal reflectance	0.02
Crystal transmittance	0.009
Lens transmittance	0.02
Geometric collection	0.04
DUT filter transmittance	0.09
Trigger bandpass to virtual bandpass wavelength	0.07
Histogram background subtraction	0.03
Coincidence circuit correction	0.08
Counting statistics	0.07
dead time (due to rate changes with time)	0.022
Trigger afterpulsing	0.06
Trigger background and statistics	0.01
Trigger signal due to uncorrelated photons	0.033
Trigger signal due to fiber back-reflection	0.003
Total	0.18

Traceable calibration of conventional detectors to a lower uncertainty is difficult [26]. For the correlation method, a major source of uncertainty is calibration of the filters F_{DUT}, F_T. To measure transmittance properties of the filters a dedicated SIRCUS [27] facility was used, whose accuracy is among the best worldwide.

One interesting property of parametric down-conversion is energy conservation. That is, the sum of energy of the pair of photons produced in the parametric interaction is equal to that of the single pump photon. Therefore, if the central frequency of the trigger channel frequency filter is ω_T, then the center frequency of conjugate pairs heralded by the trigger is $\omega_{DUT} = \omega_{pump} - \omega_T$, where ω_{pump} is the frequency of pump. Thus, F_T filter in the trigger arm is conjugated with a certain range of frequencies in the DUT channel. This effect produces a so-called virtual bandpass filter in the DUT channel, whose spectral shape and transmittance can be found from the spectral shape and transmittance of F_T and the spectral properties of the pump. Therefore, uncertainty in this virtual bandpass filter's properties is related to both pump and F_T characterizations. Refer to Chapter 11 for more information on parametric down-conversion.

8.4.1.5 Comparison of Correlated Photon Pair and Substitution Calibration methods

To verify if both the detector model and measurements are valid, a comparison of the two independent primary standard methods is necessary. Such a comparison experiment was performed [19], consisting of a set of measurements where the DUT SPAD and the calibrated photodiode were alternately swapped into the setup. The experimental setup, shown in Fig. 8.6, allows the collection of data on the SPAD for the two calibration methods simultaneously by recording both the correlated signal (i.e. a histogram seen in Fig. 8.3b) and the total rate of uncorrelated events, thus calibrating the SPAD with the two methods under the same conditions. After each detector swap, the alignment of the DUT active area was verified. Due to power variations of the pump laser, it was necessary to normalize the conventional calibrations. This normalization allowed the correlation-based calibration data to be compared against all the conventional runs (and not just adjacent conventional runs), because conventional data taking became insensitive to power fluctuations in the setup. To compare calibrations performed by the two independent methods, the DE of the SPAD was determined from each dataset. Because the experimental conditions were slightly different each time, the DE values η_{cond} for each SPAD trial vary within 2%. The difference in DE from run to run of the DUT is due to two reasons. First, the DUT SPAD has significant non-uniformity of response across its detection area making it difficult to exactly reproduce the DUT position between swaps.

Second, because we use η_{cond}, rather than η_{DE}, its value depends on the count rate. The relative uncertainty of the detector-substitution DE measurement, and the relative uncertainty of the correlated method allows for a single run comparison of the two methods to $\approx 0.25\%$.

We define the agreement between the two independent calibration methods as

$$\Delta = \frac{\overline{\eta}_{\text{cond, substitution}} - \eta_{\text{cond, correlated}}}{(\overline{\eta}_{\text{cond, substitution}} + \eta_{\text{cond, correlated}})/2}, \quad (8.12)$$

where $\overline{\eta}_{\text{cond,substitution}}$ is the average DE over all calibrated detector runs for a fixed position of a SPAD. Ideally, if the two methods yield identical $\eta_{cond}, \Delta = 0$. Figure 8.7 shows Δ along with confidence bands at one and two standard deviations of a single comparison (0.25%) for all trials drawn against a zero-bias condition, $\Delta = 0$. We see that the differences between the two methods of calibration presented in Fig. 8.7 are distributed between the confidence bands, as would be expected for a normal distribution with six out of the nine measurements falling within one standard deviation of zero. We combine all trials to find an average measured Δ, and its uncertainty. The mean difference between the two methods is $\overline{\Delta} = 0.14(0.14)\%$, [19]. Thus, the mean difference between the two methods is consistent with the uncertainty of comparison, supporting the equivalence of the two absolute calibration methods, and limiting any residual bias to the level of the uncertainty of the comparison.

This measurement validates the detector model introduced earlier in the text to within experimental uncertainty. We expect that the detector model and the

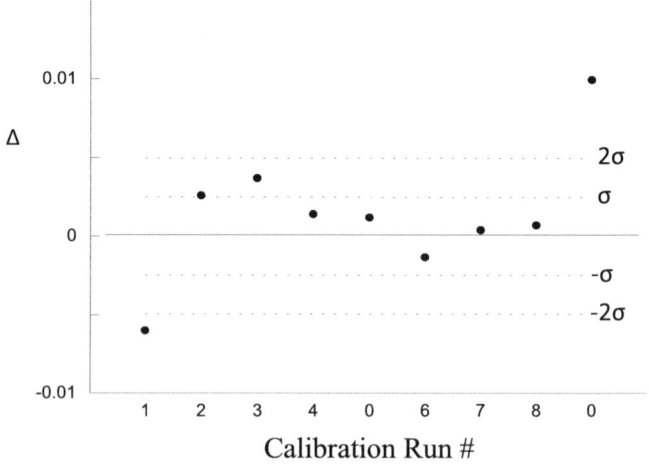

FIGURE 8.7 Comparison between two absolute calibration methods: correlated-photon pair and substitution (traceable to NIST scale). The size of the confidence bands reflects the uncertainty of each individual comparison and indicates the consistency of the overall comparison with zero bias between the two methods.

Chapter | 8 Single-Photon Detector Calibration   275

procedure for measuring its parameters would work for any non-PNR detector, regardless of its underlying physical principle of operation.

8.4.2 Transition Edge Sensors

Until this point, we have only dealt with non-photon-number-resolving detectors. Because building and verifying a methrologically accurate model of a photon-number-resolving detector may be very difficult, such a detector should be characterized by its POVM [28]. The POVM formalism is introduced in Chapter 9. If a detector's efficiency does not depend on the number of incoming photons, a POVM treatment may not be necessary; such detectors can be characterized solely by their detection efficiency. We discuss here the calibration of a detector with full photon-number-resolving capability using classification introduced in [29]. As an example, we focus on the specifics of calibrating transition edge sensor detectors (TES), c.f. Chapter 6. Calibration techniques discussed earlier may be adapted for a TES. Having full photon-number-resolving capability is advantageous for radiant power measurements. On the other hand, the signal-to-noise ratio of a TESs raw analog output may make it difficult to distinguish between certain detection events, particularly when the analog signal can be attributed to detection of both n and $n+1$ photon states. The modified calibration protocols described below are designed to take advantage of new capabilities and to circumvent the shortcomings of TES technology.

8.4.2.1 Radiant Power Calibration

A TES detector outputs an analog signal proportional to its temperature. When a photon or photons are absorbed, the detector's temperature slightly fluctuates as it heats up and then dissipates the heat. The peak temperature change is approximately proportional to the number of photons absorbed (assuming narrow bandwidth light). A typical response waveform spans about 1 μs (see Chapter 6). A power measurement consists of collecting raw analog waveforms and extracting detection events and photon numbers for each waveform. Because a TESs temporal response is relatively slow and the raw analog signal may be noisy, calibration with a pulsed source is advantageous. In this way the precise time of detection is known, which helps with the convergence of the waveform-fitting algorithm and allows the number of detected photons to be extracted with higher accuracy. Because the detector is photon-number resolving, coherent states of light with an average number of photons per pulse $\lambda \geq 1$ can be used. In practice, this number should not exceed more than a few photons per pulse, as the accuracy with which photon number can be extracted from the response waveforms decreases significantly beyond 10 or 20 photons. In contrast, for click/no-click detectors an average number of photons per pulse should be as close to zero as practically possible. For a Poissonian input, the probability $p(k)$ of k photons arriving in any single pulse is

$$p_{\text{in}}(k) = \lambda^k e^{-\lambda}/k! \qquad (8.13)$$

For a photon-number-resolving detector with DE of η, the probability to detect k photons is:

$$p(k) = (\eta\lambda)^k e^{-\eta\lambda}/k! \qquad (8.14)$$

We see that the distribution remains Poissonian, with a different effective mean number. As before, TES calibration results should be compared with an independent measurement of λ, which can be provided by using a transfer detector and an uncalibrated source, or by using a primary standard source.

Direct use of the substitution method with a classical detector is difficult because the slow response of a TES sets a relatively low limit on the incoming photon flux. Instead, a set of calibrated linear optical attenuators can be used to reach the required power levels. The optical attenuators must be switchable and repeatable and the power meter must be calibrated at a range of input power levels. The total number of attenuators is determined by the dynamic range of the power meter calibration and the required attenuation, because each attenuator should be independently calibrated using the power meter. When all attenuators are applied, the optical power should be suitable for a TES. The dominant uncertainty is usually a systematic error in the calibration of the power meter, combined with the uncertainties in the attenuators. A simplified diagram of a calibration setup is presented in Fig. 8.8. This calibration technique was implemented by Miller et al. [30]. In their setup, they used three attenuators. A similar version of this calibration setup was employed by [31], however their variable attenuator was pre-calibrated based on national standards, hence they could skip the *in situ* attenuator calibration.

The optical sensor used in the measurement depicted in Fig. 8.8 is calibrated in two steps. First, the power meter is compared to a transfer-standard detector at the wavelength of interest in a substitution setup [32]. Secondly, the linearity of the power meter response is measured relative to the calibration power level. Consideration of the power meter's linearity is necessary for attenuator calibration. To characterize the linearity, the method of additive power measurements can be used, also known as a triplet measurement [33]. A setup for such a measurement is shown in Fig. 8.9. The sum of the power measured

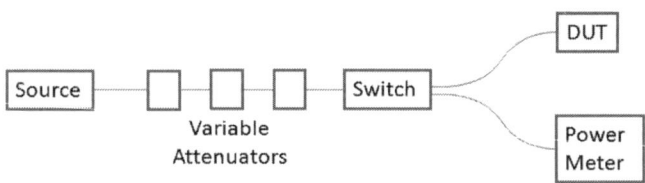

FIGURE 8.8 Multi-attenuator setup used for a TES calibration (DUT). *Reproduced from [30]*.

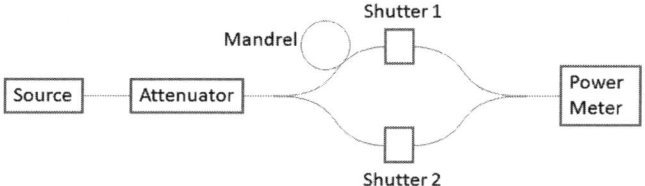

FIGURE 8.9 Linearity calibration of a power meter with a pulsed source. Three power measurements are taken: P_1 when only shutter 1 is open; P_2 when only shutter 2 is open and P_3 when both shutters are open. For a truly linear power meter, $P_3 = P_1 + P_2$. An extra fiber mandrel is added to one of the paths to prevent interference. *Reproduced from [30].*

in each of the two beamsplitter arms alone is compared to the power measured when the two arms are combined. To prevent interference on the last beam splitter before the power meter, a pulsed source and an extra fiber mandrel are used. If the response is nonlinear, there will be a difference between the sum and the full-power reading of the power meter. By making many of similar measurements over the entire dynamic range of the power meter, a dynamic calibration curve is obtained.

After applying the corrections to the measured count rate for a given source flux, a high-accuracy measurement of efficiency can proceed. Reference [30] claims an excellent reproducibility of detection-efficiency measurements across a batch of samples: $\approx 2\%$ (one standard deviation), and points to fiber alignment as a major source of uncertainty.

8.4.2.2 Correlated Calibration Technique

The calibration technique based on correlated-photon pairs (see Fig. 8.2) can be modified for calibrating TES detectors. In [34], a SPAD was used as a trigger and a TES detector as a DUT. Because TES detectors are photon-number resolving, there is room for improvement of the traditional technique. Specifically, it is possible to obtain independent estimates of detection efficiency for different photon-number states. Let $p_{\text{yes}}(n)$ be the probability of detecting n photons by the TES when the heralding channel fires, and $p_{\text{no}}(n)$ when the heralding channel does not. Then, for an ideal trigger channel, following the derivation of [34] one can write:

$$p_{\text{yes}}(0) = (1 - \eta) p_{\text{no}}(0), \tag{8.15}$$

$$p_{\text{yes}}(i) = (1 - \eta) p_{\text{no}}(i) + \eta p_{\text{no}}(i - 1), \tag{8.16}$$

where η is a detection efficiency of the entire TES channel. As before, inseparability of losses from detection efficiency presents a substantial difficulty because all the elements of the DUT optical path must be characterized separately and an overlap factor for the correlated photons with the DUT collection must also be determined. Because in many cases the TES's optical input is a single-mode fiber, this overlap is a significant factor. An accurate characterization of the overlap is challenging.

As we have shown, the trigger channel and the SPAD detector also need to be characterized to account for detections that are not related to photon pairs. Table 8.2 gives a list of important parameters in the trigger channel and their typical contribution to the overall uncertainty. The above formulae should be modified to account for false trigger events. Let ξ be the fraction of heralding events that are due to detecting a photon from a PDC pair (i.e. $(1-\xi)$ represents the number of false clicks in the heralding channel). To find ξ an additional experiment is needed in which the parametric down-conversion process is switched on and off (for example, by controlling phase matching). This method only works if SPAD afterpulses are identified and rejected. The authors of [34] explicitly take advantage of a pulsed source to reject detections that fall outside of a short temporal window defined by the source, therefore afterpulsing is not an issue. It is thus possible to write:

$$p_{\text{yes}}(0) = \xi(1-\eta)p_{\text{no}}(0) + (1-\xi)p_{\text{no}}(0), \qquad (8.17)$$

$$p_{\text{yes}}(i) = \xi[(1-\eta)p_{\text{no}}(i) + \eta p_{\text{no}}(i-1)] + (1-\xi)p_{\text{no}}(i). \qquad (8.18)$$

The above equations can be used to obtain independent values for η, based on events of different photon number. In [34], η values are found for $i = 0, 1$, and 2. These values agree to within an experimental uncertainty.

It is obviously possible to replace the SPAD in the heralding arm with another TES detector, and observe the corresponding joint photon-number distributions. A typical joint photon-number distribution obtained in this configuration is shown in Fig. 8.10 [35]. For each analog waveform a maximum pulse height is measured and plotted. The pulse height is proportional to total energy absorbed by the TES. Pulse heights cluster in accordance with photon number, with some dispersion due to analog noise. When the joint distribution is plotted for two detectors, a strong correlation between simultaneously detected photon numbers appears. Indeed if, for example, a 4-photon state was detected by TES 1, the probability for TES 2 to detect zero photons is very low. A joint photon distribution of this kind contains information about both detection efficiency (with all channel losses included) and a more complete measurement—the POVM. While approximate values for both the detection efficiency and the POVM can be inferred from this graph, additional measurements (such as background counts) are required to reach meteorological accuracy.

Note that if the overlap of correlated-photon modes is duly characterized, this calibration scheme is symmetric: in principle, either of the two detectors can be considered the trigger detector or the DUT. One can modify the setup further to take advantage of photon-number sensitivity by collecting both photons of the pair into one optical path to a single detector. Then, ideally, and in a regime of low probability of pair generation (so we can ignore multi-pair generation), we have:

$$p(2) = \eta^2 q(2), \qquad (8.19)$$

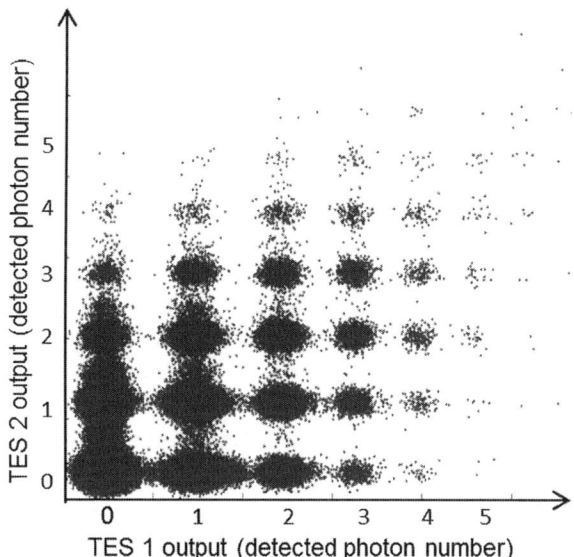

FIGURE 8.10 An output of a PDC source measured with two TES detectors (based on the pulse heights of the TES output waveforms). Photon number is assigned to each cluster peak. Strong photon-number correlation is evident: when a high-photon-number state is detected in one arm, the probability of detecting no photons in the other arm is lower than that of detecting one, two, or more. *Image courtesy of Thomas Gerrits [35].*

$$p(1) = 2\eta(1-\eta)q(2), \qquad (8.20)$$

$$p(0) = (1-\eta)q(2) + q(0). \qquad (8.21)$$

Here $p(n)$ are probabilities to detect n-photons in a trial. $q(n)$ are probabilities to generate an n-photon state, and η includes all coupling losses and detection efficiency for each of the two photons in the pair. These losses are assumed to be symmetric, but they may be asymmetric, because of the mode overlap. Because photons are generated in pairs, $q(1) = 0$. This leaves us with a set of three equations and three unknowns.

We have explored a few variants of a correlated technique suitable for TES detectors. We note that other variants exist, such as one that uses the DUT as its own trigger [36,37]. The above results are directly applicable to other detectors with full PNR capability.

8.5 CONCLUSION

We have presented the most common methods for calibrating single-photon detectors. We covered both radiant-power-based and correlated-photon-based calibration techniques, showcasing SPAD and TES calibration. It is evident that to achieve true meteorological accuracy these seemingly simple methods require careful consideration and attention. Even a basic measurement of radiant power with such detectors requires additional knowledge of both the source (spectrum, statistics), and detector (photon-number-resolving capabilities, recovery time, afterpulsing, etc.), as well as an appropriate detector model. SPAD calibration procedures can be adapted for most non-PNR detectors. Similarly, TES calibration procedures can be adapted to other detectors with full PNR capability. Other photon-number-resolving detectors, and most notably, multiplexed arrangements of non-number resolving detectors (c.f. Chapter 7), cannot be properly calibrated with these methods without substantial modifications. A better approach is to apply a POVM formalism, as described in the next chapter.

REFERENCES

[1] M. Planck, Verhandlungen der Deutschen Physikalischen Gesellschaft 2, 237 (1900).
[2] W.H. Louisell, A. Yariv, and A.E. Siegman, Phys. Rev. 124, 1646 (1961).
[3] B.Y. Zeldovich and D.N. Klyshko, Sov. Phys. JETP Lett. 9, 40 (1969).
[4] F. Zernike and J.E. Midwinter, "Applied Nonlinear Optics," Wiley, New York (1973).
[5] D.C. Burnham and D.L. Weinberg, Phys. Rev. Lett. 25, 84 (1970).
[6] D.N. Klyshko, Sov. J. Quantum Electron. 7, 591 (1977).
[7] D.N. Klyshko, Sov. J. Quantum Electron. 10, 1112 (1981).
[8] A.A. Malygin, A.N. Penin, and A.V. Sergienko, Sov. J. Quantum Electron. 11, 939 (1981).
[9] S.R. Bowman, Y.H. Shih, and C.O.Alley, "Laser Radar Technology and Applications I," in edited by J.M. Cruickskank and R.C. Harney, Proc. SPIE, vol. 663, pp. 24–29 (1986).
[10] K.D. Rarity, J.G. Ridley, and P.R. Tapster, Appl. Opt. 26, 4616 (1987).
[11] A.N. Penin and A.V. Sergienko, Appl. Opt. 30, 3582 (1991).
[12] V.M. Ginzburg, N. Keratishvili, E.L. Korzhenevich, G.V. Lunev, A.N. Penin, and V. Sapritsky, Opt. Eng. 32, 2911 (1993).
[13] P.G. Kwiat, A.M. Steinberg, R.Y. Chiao, P.H. Eberhard, and M.D. Petroff, Appl. Opt. 33, 1844 (1994).
[14] A. Migdall, R. Datla, A. Sergienko, J.S. Orszak, and Y.H. Shih, Metrologia 32, 479 (1995).
[15] G. Brida, S. Castelletto, I.P. Degiovanni, C. Novero, and M.L. Rastello, Metrologia 37, 625 (2000).
[16] G. Brida, S. Castelletto, I.P. Degiovanni, M. Genovese, C. Novero, and M.L. Rastello, Metrologia 37, 629 (2000).
[17] A. Ghazi-Bellouati, A. Razet, J. Bastie, M. Himbert, I. Degiovanni, S. Castelletto, and M. Rastello, Metrologia 42, 271 (2005).
[18] X.-H. Chen, Y.-H. Zhai, D. Zhang, and L.-A. Wu, Opt. Lett. 31, 2441 (2006).
[19] S.V. Polyakov and A.L. Migdall, Opt. Express 15, 1390 (2007).
[20] T.J. Quinn, Metrologia 34, 61 (1997).
[21] M. Ware and A.L. Migdall, J. Mod. Opt. 15, 1549 (2004).
[22] A.L. Migdall, IEEE Trans. Instrum. Meas. 50, 478 (2001).
[23] M. Ware, A. Migdall, J.C. Bienfang, and S.V. Polyakov, J. Mod. Opt. 54, 361 (2007).
[24] A. Spinelli, L.M. Davis, and H. Dautet, Rev. Sci. Instrum. 67, 55 (1996).
[25] W. Schmunk, M. Rodenberger, S. Peters, H. Hofer, and S. Kuck, J. Mod. Opt. 58, 1252 (2011).
[26] T.C. Larason, S.S. Bruce, and A.C. Parr, Spectroradiometric Detector Measurements: Part I – Ultraviolet Detectors and Part II – Visible to Near-Infrared Detectors, NIST Special Publication 250–41 (1998).
[27] S.W. Brown, G.P. Eppeldauer, and K.R. Lykke, Appl. Opt. 45, 8218 (2006).
[28] J.S. Lundeen, A. Feito, H. Coldenstrodt-Ronge, K.L. Pregnell, C. Silberhorn, T.C. Ralph, J. Eisert, M.B. Plenio, and I.A. Walmsley, Nat. Phys. 5, 27 (2009).
[29] M.D. Eisaman, J. Fan, A. Migdall, and S.V. Polyakov, Appl. Opt. 82, 071101 (2011).

[30] A.J. Miller, A.E. Lita, B. Calkins, I. Vayshenker, S.M. Gruber, and S.W. Nam, Opt. Express 19, 9102 (2011).
[31] D. Fukuda, G. Fujii, T. Numata, K. Amemiya, A. Yoshizawa, H. Tsuchida, H. Fujino, H. Ishii, T. Itatani, S. Inoue, and T. Zama, Opt. Express 19, 870 (2011).
[32] J.H. Lehman and X. Li, Appl. Opt. 38, 7164 (1999).
[33] I. Vayshenker, S. Yang, X.X. Li, T. Scott, C.L. Cromer, Optical Fiber Power Meter Nonlinearity Calibrations at NIST, NIST Special Publication 250–256 (2000).
[34] A. Avella, G. Brida, I.P. Degiovanni, M. Genovese, M. Gramegna, L. Lolli, E. Monticone, C. Portesi, M. Rajteri, M. L. Rastello, E. Taralli, P. Traina, and M. White, Opt. Express 19, 23249 (2011).
[35] T. Gerrits, Private Communication (2011).
[36] A. Czitrovszky, A. Sergienko, P. Jani, and A. Nagy, in Proc. SPIE, Vol. 3749, p. 422 (1999).
[37] A. Czitrovszky, A. Sergienko, P. Jani, and A. Nagy, Laser Phys. 10, 86 (2000).

Chapter 9

Quantum Detector Tomography

Lijian Zhang, Hendrik Coldenstrodt-Ronge, Animesh Datta, and Ian A. Walmsley
Clarendon Laboratory, Department of Physics, University of Oxford, OX1 3PU, United Kingdom

Chapter Outline
9.1 Introduction 283
9.2 Quantum Tomography: Prelude 286
    9.2.1 State Tomography 287
    9.2.2 Process Tomography 288
9.3 Detector Tomography 288
    9.3.1 General Introduction 289
    9.3.2 Photon-Number-Resolving Detectors 291
    9.3.3 Reconstruction without Phase-Sensitivity 293
    9.3.4 Reconstruction with Phase-Sensitivity: the Challenge 295
9.4 Experimental Implementations of Detector Tomography 297
    9.4.1 Experimental Setup 298
    9.4.2 Q-Function 300
    9.4.3 Reconstructed POVM Elements 301
    9.4.4 Conditioning and Regularization 305
    9.4.5 Robustness of Detector Tomography 307
    9.4.6 Wigner Functions 308
9.5 Conclusions 310
References 311

9.1 INTRODUCTION

Accurate knowledge of a quantum optical detector is essential for its fruitful utilization. Photodetectors are normally characterized by several parameters, including detection efficiency, spectral sensitivity, and noise-equivalent power (see Chapter 8). To be valid, such approach relies on a model of a particular detector. However, these parameters are insufficient to provide a satisfactory specification of a quantum detector. A quantum measurement apparatus is characterized by a set of operators $\{\hat{\Pi}^{(k)}\}$ that relate the input quantum system to the classical detector output. The link is given by the Born rule, which determines the probability p_k of obtaining the measurement result k, by Eq. (9.1)

$$p_k = \text{Tr}\left[\hat{\Pi}^{(k)}\hat{\rho}\right], \qquad (9.1)$$

where $\hat{\rho}$ is the density operator of the ensemble under investigation. The operators $\{\hat{\Pi}^{(k)}\}$, with completeness ($\sum_k \hat{\Pi}^{(k)} = 1$) and positivity ($\hat{\Pi}^{(k)} \geq 0$), are known as a positive-operator-valued measure (POVM). This set of operators must therefore be known if the full quantum power of the detector is to be used. This capability is central to realizing the potential of quantum technologies, such as enhanced precision metering, quantum communication, and quantum computation. The realization that non-classical states of light are a critical resource for enabling these technologies has brought to the fore the notion that measurement also plays a crucial role in these tasks and for some can even eliminate the need for quantum correlations in the input light beams. In enhanced precision measurement, appropriate measurements alone can give rise to super-resolution [1] and Heisenberg-limited sensitivity [2]. In communication, measurement completes the teleportation procedure, and thus, is central to quantum repeaters. And for computing, measurement-based schemes enable quantum computation [3,4]. It follows that measurement should also be considered a resource for quantum protocols. Further, most quantum information applications rely on a certain knowledge of the measurement apparatuses involved. In the framework discussed here, traditional methods of characterization (Chapter 8) can be thought of recovering only a partial knowledge of the POVM, which is particularly disadvantageous for photon-number-resolving detectors (PNRD) and/or detectors that can project the input light into exotic, purely quantum states.

In the framework of quantum mechanics, there is no general recipe for designing a detector that measures a given observable. This is perhaps surprising in the light of the central status of measurement in the theory [5]. In optics, recipes are known for building measurement devices from elementary components such as beam splitters and various kinds of photodetectors, along with ancillary classical light beams. However, the observables implemented by these detectors are in general taken to be known *a priori* based on a knowledge of the properties of the detectors, rather than inferred by a direct means. This characterization is typically based on partial calibrations or elaborate models that invoke several assumptions.

As photodetector technology advances, models that accurately describe their operation in terms of a set of parameters are often unknown. The advent of single-carbon-nanotube detectors [6], charge integration photon detectors (CIPD) [7], Visible Light Photon Counters (VLPC) [8], quantum dot arrays [9], superconducting transition-edge and picosecond nanowire sensors [10,11], or time-bin-multiplexed detectors based on commercial silicon single-photon avalanche diodes (SPADs) [12,13] calls for a new systematic approach. Certainly understanding in full detail, from first principles, the noise, loss and coherence characteristics of these technologies is not trivial. These more complex detectors call for an experimental approach to their characterization that does not rely on a model—indeed that can be used to verify models—and provides an identification of the measurement implemented by the detector that can then be used in experiments.

The problem may be addressed by using tools from quantum information processing, namely the idea of *quantum tomography*. This framework has been developed in order to characterize completely two essential resources: quantum states and quantum dynamics. The corresponding tools of quantum state tomography (QST) and quantum process tomography (QPT) are underpinned by the assumption of a fully characterized detector.

A distinct omission is that of the experimental tomography of detectors. This completes the triad of state [14–23], process [24–29], and detector tomography, required to fully specify an experiment. It thereby enables more accurate classification of measurement types, objective comparison of competing devices, and precise design of new detectors.

Characterizing a detector consists of determining its corresponding POVMs. These may be estimated by means of quantum detector tomography (QDT). In detector tomography, an unknown $\hat{\Pi}^{(k)}$ is characterized by performing a set of known measurements, based on a set of known input probe states $\{\hat{\rho}\}$. Many repetitions of the experiment on each of many identical copies of the probe states enable p_n to be estimated. From this estimate one can invert Eq. (9.1) to find $\hat{\Pi}^{(k)}$. A key requirement is that the set of reference states must be *tomographically complete*. That is, they must span the space of the POVM set. For a Hilbert space of dimension N, the set must have at least N^2 elements, depending on the number of measurement outcomes that are possible. In principle this is sufficient to enable the direct inversion of Eq. (9.1). However, it is also important that the set of reference states are experimentally feasible. That means that the states should themselves be well characterized, and that a large variety should be available. For photodetector measurements, this turns out to be straightforward. The set of coherent states $|\alpha\rangle$ of an optical beam are ideal candidates. They are overcomplete, and can thus form a tomographically complete set by transforming their amplitude (by means of attenuation, for example with a beam splitter) and their phase (with a simple delay line). Importantly, they are generated very easily by a laser. The set of input density operators give directly the probabilities

$$p_{k,\alpha} = \langle \alpha | \hat{\Pi}^{(k)} | \alpha \rangle = \pi Q^{(k)}(\alpha), \qquad (9.2)$$

where $Q^{(k)}(\alpha)$ is the Q-function of the detector POVM $\hat{\Pi}^{(k)}$. This is equal to the Husimi representation [30] of the POVM, and is uniquely related to the POVM. Therefore the set $\{Q^{(k)}(\alpha)\}$ contains a complete specification of the detector. Although simple in principle, QDT requires care in practice. The set of probe states must be carefully prepared with the correct modes, and adjustable amplitude and phase. The inversion of the experimental data requires conditioning on the POVM form (such as a requirement that the estimated operators be positive) and regularization is often needed to cope with noise in the data itself. This complexity should be set against the fact that traditional calibration of detectors has been used extensively with great success. Nonetheless, the increasing complexity of quantum detectors and the more elaborate tasks for which they are used call for a more complete and accurate method of characterization. In some key situations, such as entanglement witnessing, state and process tomography and state initialization, failure to understand the detector operation can result in erroneous inference about the overall system performance. To date, QDT has been successfully applied to SPADs [31], time-multiplexed detectors [32], transition-edge sensors [33], and superconducting nanowire detectors [34].

9.2 QUANTUM TOMOGRAPHY: PRELUDE

A crucial prerequisite for a successful measurement is information about the workings of the detector. The POVM set of a detector completely describes the detector and measurement, but so far, determination of the POVM set has been based on assumptions about the detector and theoretical models of the detector. The other two stages of a quantum experiment, state preparation and evolution (processing), have been experimentally investigated with state and process tomography respectively. Detector tomography is the missing link in this triad. In this chapter, we will start off with a brief outline of state and process tomography and develop the theoretical concept of detector tomography.

In tomography, several lower-dimensional images of a higher-dimensional object are used to eventually reconstruct the full object. It is a well-established technique in many sciences and technologies. Most familiar to the general public is its use in computerized tomography (CT) scanners for medical imaging [35–37]. In a CT scanner, the attenuation of X-rays by the human body is measured. By rotating the emitter and detector around the body, we can record the attenuation for different angles and finally the spatial distribution of attenuation in the body is reconstructed using a computer analysis. The mapping of a higher dimension distribution to a lower-dimensional subspace can be mathematically described by Radon transforms [38, 39].

In quantum optics, tomographic techniques were first used in state tomography by homodyne detection, reconstructing the Wigner function of a quantum state [15].

9.2.1 State Tomography

The ultimate goal of state tomography is the reconstruction of a quantum state, *i.e.* its density matrix. This is not as straightforward as it might seem, since it is not possible to perform state tomography with just a single measurement. Described by its density matrix, a quantum state is defined by all statistical information about the possible outcomes of all possible measurements that could be performed on the state.

Performing a measurement on a quantum state changes the state of the original system, which makes it impossible to determine the state of a "single copy" of a quantum state [40]. To obtain all statistical information about a quantum system, more than a single measurement is necessary, but due to the back-action induced by a measurement, only a single measurement can be performed on any single quantum state. Producing several "clones" of a quantum state is impossible as elucidated by the non-cloning theorem [41]. However, we can experimentally determine the quantum state of an ensemble of identically prepared systems. For a set of measurements $\{\hat{\Pi}^{(k)}\}$ we can record the probabilities p_k of obtaining results k, given by Eq. (9.1). Experimentally, quantum state tomography was first demonstrated for optical fields using homodyne detection [14,15] in the early 1990s to reconstruct the state of a single electromagnetic field mode. The quantum state to be measured was mixed at a beam splitter with a strong coherent field, the reference beam. Its outputs were detected with photodiodes and different marginal distributions of the quantum state were measured by changing the phase of the strong reference beam. An inverse Radon transform [15,38,39,42] or optimization techniques, like maximum likelihood [19,20], can be used to reconstruct the full Wigner function. This technique is related to the phase-sensitive detector tomography and we will give a more detailed description later. Quantum state tomography is a well-understood technique and it has been performed in different experimental realizations on a variety of quantum optical states [20–23,43,44] and other systems, such as molecules [16] and ions [17,45]. The experiences gained in performing the tomography experiments and their analysis will prove helpful when performing detector tomography. Although Eq. (9.1) is in principle directly invertible, small fluctuations in the experimentally measured statistics introduced, for example, by noise of the experimental equipment and finite number of trials, can have significant effects on the reconstructed density matrix [15]. Noise in the measured probabilities p_k can lead to cases where $\sum_k p_k \neq 1$, which leads to unphysical reconstructed states. These issues with direct reconstruction make the use of constrained optimization algorithms in state tomography favorable. For these reasons, we also favor an approach using constrained optimization over direct reconstruction. It is obvious that precise knowledge of the detector POVM set is crucial for performing state tomography. This emphasizes the need for techniques to determine the POVM sets of unknown detectors and verify our established models for existing detectors.

9.2.2 Process Tomography

The second stage of a quantum experiment is the evolution of the state. The aim of process tomography is to fully characterize the process generating the evolution. Process tomography was first proposed to reconstruct a process described by its Kraus operators [24,25,29,46,47], but other approaches include Ancilla-Assisted Process Tomography [28,48–50] and Direct Characterization of Quantum Dynamics [51,52]. A simple example for determining the process uses the Kraus operators $\{\hat{K}_j\}$ to describe the mapping $\hat{\rho} \mapsto \hat{\rho}'$ induced by the process $\varepsilon(\hat{\rho})$

$$\rho' \equiv \varepsilon(\hat{\rho}) = \sum_j \hat{K}_j \hat{\rho} \hat{K}_j^\dagger. \tag{9.3}$$

The Kraus operators can be reconstructed from the measurements made on $\hat{\rho}'$ and knowledge of the initial state $\hat{\rho}$. However to reconstruct all Kraus operators describing the complete quantum process, a suitable set of initial states needs to be chosen in order to cover the whole process.

Process tomography is based on precise knowledge of the initial and final quantum states, which is achieved by state tomography. This clearly requires well-characterized detectors as well, making detector tomography a key technique for all three stages of a quantum experiment.

9.3 DETECTOR TOMOGRAPHY

The effects of a quantum detector are encapsulated in its POVM set, and knowledge of the POVM set is necessary for predicting and interpreting measurement results using it.

It is usual to describe the action of the detector in terms of its operation when illuminated by classical light, with the ancillary assumption that it will operate the same way when illuminated by quantum light. Based on this idea, a POVM set for a SPAD has been developed [53]. More complicated detectors might still be modeled in terms of well-understood optical elements, such as beam splitters. Assuming these elements act in the same manner on the electromagnetic field operators as on the classical electromagnetic field, we may model POVM sets for more complicated detectors, such as time-multiplexed detectors (TMDs) [54]. As successful as this approach has been, it is not applicable to all detectors, for example, the superconducting nanowire detector in the nonlinear region [34].

Apart from the unproven fundamental assumption of working in the quantum regime the same way as for classical states, detectors could also become too complicated to be modeled classically. One could even imagine detectors without classical analogs, for example Mandel's quantum counter [55] or using ancilla-based measurement [56]. Detector tomography is a technique for characterizing the POVM set of an unknown detector which was first proposed as an extension of process tomography [57]. Optimization techniques were suggested [53] to potentially improve the estimation of the POVM elements from experimental data. Proposals for the experimental demonstration of detector tomography were first based on entangled photons [58], similar to which has been done for detector calibration [59]. However, the first complete analysis of a testbed was based on coherent states [54] which are easy to produce and control. This approach was used in the first experimental demonstration [32], which we will describe next.

9.3.1 General Introduction

In detector tomography the measurement outcomes of a detector under test (DUT) for a set of known input states are used to reconstruct the POVM elements of the DUT. In Fig. 9.1 we illustrate the process. Sets of probe states $\hat{\rho}_n$ are prepared and measured by the DUT with the unknown POVM set $\{\hat{\Pi}^{(m)}\}$. The probability of outcome m when probe state $\hat{\rho}_n$ is sent is given by Eq. (9.1)

$$p_m(\hat{\rho}_n) = \text{Tr}\left[\hat{\rho}_n \hat{\Pi}^{(m)}\right]. \tag{9.4}$$

Each of these measurement probabilities $p_m(\hat{\rho}_n)$ and probe state $\hat{\rho}_n$ combinations provides some information about the POVM element $\hat{\Pi}^{(m)}$ describing the measurement process. Therefore we could infer the POVM set $\{\hat{\Pi}^{(m)}\}$ if all the measurement probabilities are known for suitably chosen set of probe states. To reconstruct all POVM elements of a detector the set of probe states $\{\hat{\rho}_n\}$ must be tomographically complete, *i.e.* all states the detector might be sensitive to could be constructed from the probe state set. Mathematically that translates to the input states spanning the Hilbert space of the detector.

The POVM elements could be obtained by directly inverting the linear equations arising from Eq. (9.4) if enough probe states were used. However this can lead to unphysical POVM elements [53], violating, for example, the completeness condition $\sum_k \hat{\Pi}^{(k)} = \hat{1}$. Equation (9.4) may be inverted using maximum-likelihood techniques incorporating the constraints $\sum_k \hat{\Pi}^{(k)} = \hat{1}$ and that $\hat{\Pi}^{(k)}$ is positive, *i.e.* all eigenvalues of $\hat{\Pi}^{(k)}$ are ≥ 0 [53]. Convex optimization techniques, such as maximum likelihood, use all available experimental data and

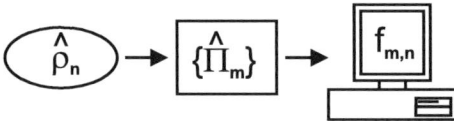

FIGURE 9.1 Schematic of detector tomography—an ensemble of known states $\hat{\rho}_n$ is sent to a detector with unknown POVM elements $\hat{\Pi}^{(m)}$ for measurements outcomes m. The frequencies $f_{m,n}$ of obtaining measurement outcome m with input state n are recorded and used to reconstruct the POVM elements in Eq. (9.4).

still yield physical results. Further optimization techniques have been developed [60,61] for the analysis of the experimental data presented in Ref. [32].

In an actual experimental realization we record the frequency $f_{m,n}$ for each measurement outcome m occurring when probe state n is used. For sufficiently large sample size we can replace the probabilities with the relative frequencies

$$p_m(\hat{\rho}_n) = \frac{f_{m,n}}{\sum_n f_{m,n}}, \tag{9.5}$$

and reconstruct the POVM elements. Two aspects of detector tomography are particularly important when planning the experimental realization: assumptions about the detector and choice of the set of probe states. Ideally one would like to perform detector tomography with the finest constraining assumptions possible, preparing the set of quantum probe states $\hat{\rho}_n$ with classical control over n and just measuring the detection outcomes m.

An implicit assumption made in the description of detector tomography so far is that of a memory-less detector, *i.e.* POVM elements are not influenced by previously performed measurements. Imagine a detector with a recovery time (c.f. Chapter 2), during this period of time it will be inhibited and the probability of no detection will be artificially higher (up to even unity) than during normal operation. Most, but not all, detectors can be assumed to operate in a memory-less regime, at least when certain operational parameters, such as dead time and afterpulsing, are respected.

Working with the most general set of assumptions, namely N positive POVM elements $\{\hat{\Pi}^{(k)}\}$ with $\sum_k \hat{\Pi}^{(k)} = \mathbf{1}$, and a d-dimensional state space, we have to estimate $(N-1)d^2$ parameters for the POVM elements. A further $d^2 - 1$ parameters arise from the probe state. Considering the large number of parameters, it becomes apparent that further assumptions, if valid, can make the process of reconstructing the POVM elements of an unknown detector much easier. In the following subsections we will develop reconstruction techniques for a phase-insensitive and a phase-sensitive PNRD. The first measures only the photon-number distribution, which is a diagonal operator in the photon-number basis, allowing further constraints on the POVM elements. However for the phase-sensitive PNRD, there is no such constraint, and full operators need to be reconstructed.

PNRDs operate in infinite dimensional Hilbert spaces. A basis of the states of the electromagnetic field are the eigenstates of the number operator $\hat{n} = \hat{a}^\dagger \hat{a}$. To reconstruct the POVM set of a detector that operates in such an infinite dimensional space it is infeasible to choose an infinite number of probe states, in a real experimental implementation. This emphasizes the choice of probe states for detector tomography. The main criterion in choosing the probe states for detector tomography is that they span the Hilbert space of the detector. To reconstruct all POVM elements $\{\hat{\Pi}^{(k)}\}$ the span of the probe states $\hat{\rho}_n$ must cover the space from which the POVM elements are taken. Even though operating in infinite dimensional Hilbert spaces, detectors are usually only sensitive to a finite subset of the infinite space. A classical analog is the different sensitivity regimes of a classical power meter. Despite operating on the space of power, a given power meter in a given configuration will only detect powers between a lower and an upper limit. For higher powers the detector saturates, *i.e.* it cannot differentiate between powers above a certain threshold. We call the power at which this occurs the saturation threshold.

This example also indicates how to handle infinite dimensional Hilbert spaces—they need to be truncated in experimental implementations, while making sure that the probe states from the truncated space still span the POVM's space. For the example of PNRDs, we can observe detector saturation when increasing the photon number of probe states over a detector-dependent threshold. For probe states with higher mean photon number we will not gain any additional information about the detector. Thus we can truncate the probe state space at the saturation threshold.

In the following subsections we will use PNRDs as an example to further develop the theory of detector tomography. In the experimental demonstration we use both phase-sensitive and phase-insensitive PNRDs. We start off by describing methods for non-phase-sensitive PNRDs, which can be extended for use with the phase-sensitive version.

9.3.2 Photon-Number-Resolving Detectors

In contrast to SPADs, a PNRD has more than two possible measurement outcomes, each of which can be assigned to an input photon number, making the detector capable of partially resolving the number of photons hitting it. Due to the finite detection efficiency and noise, there is generally not a one-to-one relation between the input photon number and the measurement outcome, and a PNRD normally measures the input photon-number statistics. Thus a natural way of expressing their POVM elements $\{\hat{\Pi}^{(k)}\}$ is in the photon-number basis

$$\hat{\Pi}^{(k)} = \sum_{m,n=0}^{\infty} \langle m|\hat{\Pi}^{(k)}|n\rangle |m\rangle\langle n|, \qquad (9.6)$$

where $\langle m|\hat{\Pi}^{(k)}|n\rangle = \Pi_{m,n}^{(k)}$ are the matrix elements of the POVM element in the photon-number basis [53]. A corresponding natural choice of probe states for a photon-number-resolving detector would be the Fock states $\{|n\rangle\}$. They span the photon-number space, however—especially with photon numbers higher than one—Fock states are difficult to produce. An alternative are the coherent states $\{|\alpha\rangle\}$, which can be directly produced with a laser and are easily expanded in the photon-number basis

$$|\alpha\rangle = \sum_{n=0}^{\infty} e^{(-\frac{|\alpha|^2}{2})} \frac{\alpha^n}{\sqrt{n!}} |n\rangle, \qquad (9.7)$$

following a Poissonian photon-number distribution. When attenuating the state $|\alpha\rangle$ with attenuation $1 - \eta$, the resulting state $|\eta\alpha\rangle$ is just another coherent state with lower mean photon number. Therefore we can produce any coherent state with lower mean photon number from an initial coherent state by means of variable attenuation which can be monitored with only classical devices. It still remains to be shown that we can use a set of coherent states to span the POVM space [54]. Coherent states form a basis for the state space of the electromagnetic field mode

$$\int \frac{d^2\alpha}{\pi} |\alpha\rangle\langle\alpha| = \hat{1}, \tag{9.8}$$

and any quantum state $|\psi\rangle$ can be expanded in the basis of coherent states.

Yet we cannot use all possible coherent states in a practical experiment, because that involves an infinite number of states. At the same time we cannot reconstruct an infinite number of matrix elements describing the POVM. PNRDs show saturation, *i.e.* when the number of photons entering the detector reaches a detector-dependent level, they cannot resolve further increases of photon number. We can therefore restrict the expansion of the POVM element in Eq. (9.6) to M photons

$$\hat{\Pi}^{(k)} = \sum_{m,n=0}^{M} \Pi_{m,n}^{(k)} |m\rangle\langle n|. \tag{9.9}$$

We only use a discrete subset $\{|\alpha_j\rangle\}$ of n_α coherent states and need to verify tomographic completeness of that set, *i.e.* that they span the M^2-dimensional subspace $\{|0\rangle\langle 1|, |0\rangle\langle 2|, \ldots, |M\rangle\langle M-1|, |M\rangle\langle M|\}$ we used to expand the POVM elements. Mathematically, this corresponds to

$$\mathrm{span}\left(\{|\alpha_j\rangle\langle\alpha_j|\}\right) = \mathrm{span}\left(|0\rangle\langle 1|, |0\rangle\langle 2|, \ldots, |M\rangle\langle M-1|, |M\rangle\langle M|\right). \tag{9.10}$$

This can be shown by truncating Eq. (9.7) at photon-number M as well and writing the probe states

$$\hat{\rho}_j = |\alpha_j\rangle\langle\alpha_j| = e^{-|\alpha|^2} \sum_{m,n=0}^{M} \frac{\alpha^m \alpha^{*n}}{\sqrt{m!n!}} |m\rangle\langle n|. \tag{9.11}$$

For each probe state $\hat{\rho}_j$ we get M^2 expansion coefficients $\langle m|\hat{\rho}_j|n\rangle$. We can relabel the basis elements $|m\rangle\langle n|$ with an index parameter $l \in \mathbb{N} \cap [1, M^2]$ to

$$\hat{\mathcal{P}}_l = |\tilde{m}(l)\rangle\langle\tilde{n}(l)|, \tag{9.12}$$

where we use the *modulo operation* to define $\tilde{m}(l) = [(l-1) \pmod{M}]$, $\tilde{n}(l) = [l - \tilde{m}(l) - 1]/M$. With this new index the matrix elements of all probe states can be written in a $n_\alpha \times M^2$ matrix $\mathbf{V}$ given by

$$V_{j,l} = \langle \tilde{m}(l)|\hat{\rho}_j|\tilde{n}(l)\rangle. \tag{9.13}$$

The rank of this matrix is the dimension of the vector space spanned by the set of coherent probe states $\{|\alpha_j\rangle\}$. Since we truncate the expansion of the states at the same photon-number M as the expansion of the POVM elements, we require $\mathbf{V}$ to have full rank M^2 in order to span the space of the detector. We calculate the determinant of $\mathbf{V}$. A non-zero determinant shows full rank and thus implies tomographic completeness of the chosen subset [54].

In the next section, we will show an experimental implementation of detector tomography with PNRDs. Completing the characterization of all three stages of a quantum experiment, detector tomography is a crucial element of quantum experiments since well-characterized detectors are necessary for state and process tomography. Considering the interdependencies of the techniques, starting off with detector tomography, followed by state tomography and finishing with the actual process is a recommended strategy for experiments.

9.3.3 Reconstruction without Phase-Sensitivity

For a phase-insensitive PNRD, we can simplify the detector tomography substantially. Since only the number of photons entering the detector (but no phase information) is registered, we can assume the off-diagonal POVM elements to be zero, that is, $\Pi_{m,n}^{(k)} = c_m \cdot \delta_{m,n}$, and Eq. (9.9) simplifies to

$$\hat{\Pi}^{(k)} = \sum_{n=0}^{M} \Pi_n^{(k)} |n\rangle\langle n|, \tag{9.14}$$

where $\Pi_n^{(k)} = \Pi_{n,n}^{(k)}$ are the diagonal elements of the matrix elements of $\hat{\Pi}^{(k)}$.

We will now find relations between the measured probabilities $p_{j,k}$ of getting detector outcome k for probe input states $\hat{\rho}_j = |\alpha_j\rangle\langle\alpha_j|$ and the set of POVM elements of the detector $\{\hat{\Pi}^{(k)}\}$. We use Eq. (9.4)

$$p_{j,k} = \operatorname{Tr} \hat{\rho}_j \hat{\Pi}^{(k)}, \tag{9.15}$$

and expand probe states and POVM elements in the photon-number basis according to Eqs. (9.7) and (9.14) respectively. Since $\Pi_{m,n}^{(k)} = c_m \cdot \delta_{m,n}$ we only have to consider the diagonal elements of the probe states, we can write

Eq. (9.15) for the probability $p_{j,k}$ as follows

$$p_{j,k} = \operatorname{Tr} \hat{\rho}_j \hat{\Pi}^{(k)} = \operatorname{Tr} e^{-|\alpha_j|^2} \sum_{n=0}^{M} \frac{\alpha_j^n \alpha_j^{*n}}{n!} |n\rangle\langle n| \sum_{m=0}^{M} \Pi_m^{(k)} |m\rangle\langle m|$$

$$= \operatorname{Tr} e^{-|\alpha_j|^2} \sum_{n=0}^{M} \sum_{m=0}^{M} \frac{|\alpha_j|^{2n}}{n!} \Pi_m^{(k)} |n\rangle\langle n|m\rangle\langle m|$$

$$= \operatorname{Tr} e^{-|\alpha_j|^2} \sum_{n=0}^{M} \frac{|\alpha_j|^{2n}}{n!} \Pi_n^{(k)} |n\rangle\langle n| = e^{-|\alpha_j|^2} \sum_{n=0}^{M} \frac{|\alpha_j|^{2n}}{n!} \Pi_n^{(k)}$$

$$= \sum_{n=0}^{M} F_{j,n} \Pi_{n,k}, \tag{9.16}$$

where we defined the matrices $\Pi_{n,k} = \Pi_n^{(k)}$ and

$$F_{j,n} = e^{-|\alpha_j|^2} \frac{|\alpha_j|^{2n}}{n!}. \tag{9.17}$$

This splits the equation into known contributions from the probe state in the matrix **F** and the unknown matrix elements of the POVM set in **Π**. The measured probabilities $p_{j,k}$ also define a matrix **P** with the entries $P_{j,k} = p_{j,k}$, and we can write Eq. (9.16) as a matrix equation

$$\mathbf{P} = \mathbf{F} \cdot \mathbf{\Pi}, \tag{9.18}$$

which could be directly inverted to obtain the matrix elements of the POVM set. However this is prone to noise and may give unphysical results, *e.g.* POVM sets which do not sum to unity. Thus we use optimization techniques to reconstruct the POVM set.

In an optimization problem, one tries to find the solution which minimizes a so-called "cost function" under given constraints. To illustrate this, we assume we have a trial set of POVM elements given by the matrix $\mathbf{\Pi}^{\text{trial}}$. For the input states **F** used, we expect to get measured probabilities according to $\mathbf{F} \cdot \mathbf{\Pi}^{\text{trial}}$. However we know the measured probabilities, which are given in **P**, thus we need to quantify how much they differ from $\mathbf{F} \cdot \mathbf{\Pi}^{\text{trial}}$. A measure for the difference in matrices is given by the Frobenius norm $||\mathbf{M}||_2$ of a matrix **M**, defined by $||\mathbf{M}||_2 = \sqrt{\sum_{\alpha,\beta} |M_{\alpha,\beta}|^2}$. A suitable "cost function" for retrieving the POVM elements would be given by

$$||\mathbf{P} - \mathbf{F} \cdot \mathbf{\Pi}||_2. \tag{9.19}$$

However it turns out that due to numeric instabilities this function is not ideal and can lead to spurious features, for example jumps between the matrix elements, in the reconstructed POVM [60,61]. We therefore introduce a filter function $g(\mathbf{\Pi})$ into the "cost function," favoring smooth POVM elements, leading to the final optimization problem

$$\min \left\{ ||\mathbf{P} - \mathbf{F} \cdot \mathbf{\Pi}||_2 - g\left(\mathbf{\Pi}\right) \right\}, \text{ such that } \hat{\Pi}_n \geq 0, \sum_{n=1}^{N} \hat{\Pi}_n = \hat{\mathbb{1}}. \quad (9.20)$$

One example of the filter function is given in Section 9.4.4.

9.3.4 Reconstruction with Phase-Sensitivity: the Challenge

The reconstruction of the POVM for the phase-sensitive photon-number-resolving detector starts again with Eq. (9.4) for the probabilities $p_{j,k}$ of a measurement outcome k with input probe state $\rho_j = |\alpha_j\rangle\langle\alpha_j|$

$$p_{j,k} = \text{Tr}\,\hat{\rho}_j \hat{\Pi}^{(k)}. \quad (9.21)$$

We use the exponential polar form $\alpha_j = |\alpha_j| e^{-i\theta_j}$ of complex numbers to rewrite Eq. (9.11)

$$\hat{\rho}_j = |\alpha_j\rangle\langle\alpha_j| = e^{-|\alpha_j|^2} \sum_{m,n=0}^{M} \frac{\alpha_j^m \alpha_j^{*n}}{\sqrt{m!n!}} |m\rangle\langle n|$$

$$= e^{-|\alpha_j|^2} \sum_{m,n=0}^{M} \frac{\left(|\alpha_j| e^{-i\theta_j}\right)^m \left(\alpha_j e^{-i\theta_j}\right)^{*n}}{\sqrt{m!n!}} |m\rangle\langle n|$$

$$= e^{-|\alpha_j|^2} \sum_{m,n=0}^{M} \frac{\left(|\alpha_j|\right)^{m+n}}{\sqrt{m!n!}} e^{-i(m-n)\theta_j} |m\rangle\langle n|, \quad (9.22)$$

which expands the probe states in the photon-number states. This form will prove more convenient for reconstructing the POVM elements, since we cannot ignore off-diagonal elements for the non-phase-sensitive detector.

We now substitute this representation of the probe state and the expanded POVM elements given by Eq. (9.9) into Eq. (9.21) for the probabilities $p_{j,k}$,

$$p_{j,k} = \text{Tr}\left(\hat{\rho}_j \hat{\Pi}^{(k)}\right)$$

$$= \text{Tr}\left(e^{-|\alpha_j|^2} \sum_{m,n=0}^{M} \frac{\left(|\alpha_j|\right)^{m+n}}{\sqrt{m!n!}} e^{-i(m-n)\theta_j} |m\rangle\langle n| \sum_{s,t=0}^{M} \Pi_{s,t}^{(k)} |s\rangle\langle t|\right)$$

$$= e^{-|\alpha_j|^2} \text{Tr}\left(\sum_{m,n=0}^{M} \sum_{s,t=0}^{M} \frac{\left(|\alpha_j|\right)^{m+n}}{\sqrt{m!n!}} e^{-i(m-n)\theta_j} \Pi_{s,t}^{(k)} |m\rangle\langle n|s\rangle\langle t|\right)$$

$$= e^{-|\alpha_j|^2} \text{Tr} \left(\sum_{m,n,t=0}^{M} \frac{(|\alpha_j|)^{m+n}}{\sqrt{m!n!}} e^{-i(m-n)\theta_j} \Pi_{n,t}^{(k)} |m\rangle\langle t| \right)$$

$$= e^{-|\alpha_j|^2} \sum_{m,n=0}^{M} \frac{(|\alpha_j|)^{m+n}}{\sqrt{m!n!}} e^{-i(m-n)\theta_j} \Pi_{n,m}^{(k)}. \qquad (9.23)$$

We use the index parameter $l \in \mathbb{N} \cap [1, M^2]$, defined after Eq. (9.12), to transform Eq. (9.23) into a matrix equation

$$\begin{aligned} p_{j,k} &= \sum_{m,n=0}^{M} e^{-|\alpha_j|^2} \frac{(|\alpha_j|)^{m+n}}{\sqrt{m!n!}} e^{-i(m-n)\theta_j} \Pi_{n,m}^{(k)} \\ &= \sum_{l=0}^{M^2} e^{-|\alpha_j|^2} \frac{(|\alpha_j|)^{(\tilde{m}(l)+\tilde{n}(l))}}{\sqrt{\tilde{m}(l)!\tilde{n}(l)!}} e^{-i(\tilde{m}(l)-\tilde{n}(l))\theta_j} \Pi_{\tilde{m}(l),\tilde{n}(l)}^{(k)} = \sum_{l=0}^{M^2} \tilde{F}_{j,l} \tilde{\Pi}_{l,k}, \end{aligned}$$
$$(9.24)$$

where we defined $\tilde{\Pi}_{l,k} = \Pi_{\tilde{m}(l),\tilde{n}(l)}^{(k)}$ and $\tilde{F}_{j,l} = e^{-|\alpha_j|^2} \frac{(|\alpha_j|)^{(\tilde{m}(l)+\tilde{n}(l))}}{\sqrt{\tilde{m}(l)!\tilde{n}(l)!}} e^{-i(\tilde{m}(l)-\tilde{n}(l))\theta_j}$. We can write Eq. (9.24) in matrix form similar to Eq. (9.18)

$$\mathbf{P} = \tilde{\mathbf{F}} \cdot \tilde{\mathbf{\Pi}}, \qquad (9.25)$$

with the entries of $\mathbf{P}$ defined through the measurement results $P_{j,k} = p_{j,k}$.

As in the phase-insensitive case, the POVM elements could be reconstructed using the convex optimization (c.f. Eq. (9.20))

$$\min \left\{ ||\mathbf{P} - \tilde{\mathbf{F}} \cdot \tilde{\mathbf{\Pi}}||_2 - g\left(\tilde{\mathbf{\Pi}}\right) \right\}; \text{ with the constraints } \hat{\tilde{\Pi}}_n \geq 0, \sum_{n=1}^{N} \hat{\tilde{\Pi}}_n = \hat{\mathbb{1}}. \qquad (9.26)$$

The problem is computationally considerably more demanding in the phase-sensitive case. In Table 9.1, we show a comparison of the dimensionality of the matrices involved in a phase-sensitive and non-phase-sensitive case. If n_{α_j} probe states are taken with a detector providing N clicks resolution, the click statistics $\mathbf{P}$ for both cases are represented by an $n_{\alpha_j} \times N$-matrix.

The probe state matrix $\tilde{\mathbf{F}}$ contains M^2 entries for each of the n_{α_j} probe states in the phase-sensitive case, while the matrix $\mathbf{F}$ contains $n_{\alpha_j} \times M$ entries in the phase-insensitive case. The same holds for the POVM matrices $\mathbf{\Pi}$ and $\tilde{\mathbf{\Pi}}$ which are $N \times M$ and $N \times M^2$ for phase-insensitive and phase-sensitive version respectively. In Table 9.1, we also give typical values used in the experimental tomography of a 8-bin TMD and a weak-homodyne version of this detector (see Chapter 7 for the introduction of these detectors).

TABLE 9.1 Comparison of the Dimensionality of the Matrices involved in Detector Tomography. The Phase-Sensitive Case Requires Significantly Longer Computation. We also Give Typical Sizes for the Matrices Used in a Practical Experiment

Matrix	P	F and $\tilde{F}$	Π and $\tilde{\Pi}$
Non-phase-sensitive	$N \times n_{\alpha_j}$ $= 9 \times 300$ $= 2700$	$n_{\alpha_j} \times M$ $= 300 \times 100$ $= 3 \cdot 10^4$	$M \times N$ $= 100 \times 9$ $= 900$
Phase-sensitive	$N \times n_{\alpha_j}$ 9×3500 $= 31500$	$n_{\alpha_j} \times M^2$ 3500×100^2 $= 3.5 \cdot 10^7$	$M^2 \times N$ $100^2 \times 9$ $= 9 \cdot 10^4$

The number of matrix elements for the direct inversion rises not only due to the increased size of POVM and probe state matrix, since we need to take into account the off-diagonal elements now, but also due to the increased number of probe states for a tomographically complete set (experimentally this means an additional modulation, the phase modulation, of the probe state). Another problem for the phase-sensitive detector tomography is to find a suitable regularization function $g\left(\tilde{\Pi}\right)$. Specially designed algorithms are required to overcome these difficulties. For one example we refer to [62,63], in which a recursive algorithm is developed that reduces the computational complexity from M^2 to M per recursion, making the phase-sensitive detector tomography feasible.

9.4 EXPERIMENTAL IMPLEMENTATIONS OF DETECTOR TOMOGRAPHY

To date QDT has been successfully applied to SPADs [31,64], TMDs [32,65], transition edge sensors [33], and superconducting nanowire single-photon detectors (SNSPD) [34]. In this section we present some of the results and discuss the experimental requirements.

Here we consider the tomography of a silicon-based SPAD and a 8-bin TMD. As given in Chapter 4, for a SPAD there are two possible measurement outcomes—click and no-click. In the linear region, the corresponding POVM elements in the photon-number basis are

$$\hat{\Pi}^{(0)} = \sum_{n=0}^{\infty} \left(1 - \eta_{\text{SPAD}}\right)^n |n\rangle \langle n|,$$

$$\hat{\Pi}^{(1)} = \hat{I} - \sum_{n=0}^{\infty} \left(1 - \eta_{\text{SPAD}}\right)^n |n\rangle \langle n|, \qquad (9.27)$$

where $\hat{\Pi}^{(0)}$ is the POVM element for the no-click event and $\hat{\Pi}^{(1)}$ for click, respectively, and η_{SPAD} is the detection efficiency.

For a 8-bin TMD, there are nine possible outcomes—from no-click to eight-click, and the corresponding POVM elements are

$$\hat{\Pi}^{(k)} = \sum_{j=0}^{N} [\mathbf{C} \cdot \mathbf{L}]_{kj} |j\rangle\langle j|, \quad (9.28)$$

where $\mathbf{C}$ and $\mathbf{L}$ are the convolution matrix and loss matrix, respectively, given in Chapter 7.

9.4.1 Experimental Setup

For the actual experimental implementation of SPAD and TMD detector tomography, precise control and knowledge of the probe beam is crucial. In Fig. 9.2 we show the setup used for tomography of the SPAD and TMD. The probe states are generated with a cavity dumped mode-locked titanium-sapphire (Ti:Sapphire) laser, generating femtosecond pulses in the 780 nm regime. The exact values for wavelength, bandwidth, and repetition rate which varied with the detector under test can be found in [65]. In the description of detector tomography in Section 9.3.2 we found that spanning the entire detector space is crucial for performing successful detector tomography. For PNRs we truncate the probe states given by Eq. (9.11) at the photon-number M, which is in the saturation regime of the detector. This saturation cut-off is determined by increasing the power with a variable attenuator until the click statistics did not change with further increases in power anymore.

Tomography of the 8-bin TMD requires average mean photon numbers ranging from far less than one photon per pulse to more than 100 photons per pulse for saturation. This dynamic range is obtained by rotating a half-wave plate (HWP) followed by a Glan-Thompson (GT) polarizer. Several filters are

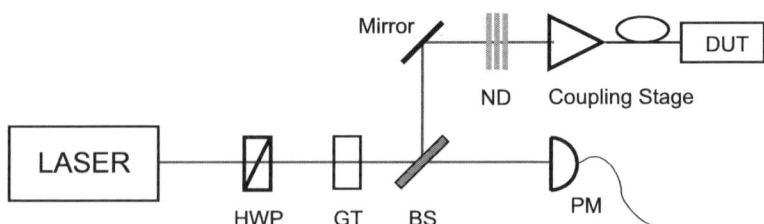

FIGURE 9.2 Schematic of the experimental setup used for detector tomography of phase-insensitive photon-number-resolving detectors: the probe beam is generated by a mode-locked Ti:Sapphire laser and subsequently sent through a half-wave plate (HWP) followed by a Glan-Thompson polarizer (GT). At a beam sampler (BS) about 95% of the beam is transmitted to a power meter (PM), while the reflected beam is further attenuated with calibrated neutral density filters (ND) and then coupled into a single mode fiber connected to the DUT.

required to attenuate the bright coherent beam coming from the laser down to the few photon level. The first attenuation stage is a beam sampler (*i.e.* a beam splitter with about 5% reflectivity), which also splits off a portion of the beam for power monitoring. The beam transmitted by the beam sampler is measured with a power meter. The reflection of the beam sampler is sent through several neutral density (ND) filters into single-mode fiber which is connected to the detector under test (DUT).

Determination of the POVM elements in Eq. (9.1) requires precise knowledge of the probe state $\hat{\rho}_j$. For coherent states this corresponds to knowledge of the mean number of photons $\langle \hat{n} \rangle = |\alpha|^2$, which corresponds to the time averaged power

$$\bar{P} = \langle \hat{n} \rangle \frac{hcf_{\text{rep}}}{\lambda} = \frac{|\alpha|^2 hcf_{\text{rep}}}{\lambda}, \tag{9.29}$$

of the probe state, where f_{rep} is the repetition rate of the laser pulses, λ is the optical wavelength of the probe state, h is Planck's constant, and c is the speed of light. Measuring the time averaged power $\bar{P}_j$ of a probe state allows estimation of the average photon number per pulse

$$|\alpha_j|^2 = \frac{\bar{P}_j \lambda}{hcf_{\text{rep}}}. \tag{9.30}$$

The low photon numbers required in detector tomography correspond to powers far outside the sensitivity regime of classical power meters, and swapping the DUT for a calibrated power meter for each probe state would also introduce coupling errors. However we can relate the power transmitted through the beam sampler to the power coupled to the DUT with an attenuation parameter γ which includes all losses in the system arising from the beam sampler, fiber coupling and additional neutral density filters attenuating the bright coherent state down to the low photon-number level.

For each experimental run we can measure the optical power coupled into the fiber without the additional ND filters and the power at the monitoring point behind the beam sampler. This ratio gives an estimate for the attenuation due to coupling and the beam sampler γ_{BC}. Additional, calibrated ND filters were added to further attenuate the coupled power by γ_{ND_i}, where i indicates the ith ND filter. The overall attenuation γ can be calculated with $\gamma = \gamma_{BC} \prod_i \gamma_{\text{ND}_i}$, leading to an estimate for the mean number of photons $|\alpha_j|^2 = \gamma \bar{P}_{\text{PM}j} \lambda/hcf$, where $\bar{P}_{\text{PM}j}$ is the time averaged power measured at the monitoring power meter for probe state j. Any errors in the calibration of the attenuator and power meter will propagate through the detector tomography process, so accurate calibrations of both are crucial.

State tomography experiments have shown that small experimental errors can lead to unphysical states in the reconstruction and make the reconstruction process more difficult. Stability of the probe states generated by the laser and reliability of the probe state $|\alpha_j|^2$ estimation are important parameters for detector tomography. For different settings of the HWP, we measured the relation between the power coupled to the fiber without additional ND filters and the power measured at the power monitor. This shows a nonlinearity less than 1%, which indicates that beam steering effects from the HWP are negligible. We also measured the intensity and spectrum (taken at $\lambda_0 = 789$ nm) of the laser used for tomography of the TMD. The result indicates a low fluctuation in the laser output. A further measurement of a histogram of the pulse power confirms that the relative uncertainty of the probe state amplitude is 2%.

9.4.2 Q-Function

To determine the POVM elements for the SPAD and TMD, coherent states with known amplitudes are launched and the detector outputs in the form of click statistics are recorded. For each amplitude setting this is repeated T times. The click probabilities with the measured click statistics can be calculated using Eq. (9.5).

For the tomography of a commercially available SPAD module, probe states are produced with mean photon number from $|\alpha_{\min}|^2 = 0.01$ up to $|\alpha_{\max}|^2 = 18.81$ and an additional measurement for $|\alpha_0|^2 = 0$ is taken with the probe beam completely blocked. With the lowest power showing less than 1% click rate and measurements from $|\alpha_{\text{sat}}|^2 = 9.26$ with click rates higher than 99%, the chosen ranges cover the entire detector space from low photon numbers up to detector saturation. Variable steps in $|\alpha_j|^2$ are chosen with the smallest graining down to $\Delta|\alpha_j|^2 = 0.01$ for low photons numbers, increasing up to $\Delta|\alpha|^2 > 0.6$ for the highest photon numbers in the saturation regime.

For each probe state amplitude α_j the probabilities $p_{j,k}$ of the click ($k = 1$) and no-click ($k = 0$) are calculated from the measurement statistics. This samples the Q-function given by Eq. (9.2) of the click and no-click POVM elements $Q_k(\alpha_j) = p_{j,k}/\pi$. In Fig. 9.3 the estimated probabilities are plotted against the mean number of photons in the coherent probe state. We note that the errors are too small to be resolved in this plot. For the click event POVM element we observe low probabilities for lower photon numbers, increasing exponentially toward higher photon numbers. For the no-click event POVM element the probabilities show the opposite behavior, decreasing exponentially toward higher photon numbers. This is in agreement with the qualitative analysis of the theoretical POVM elements and probabilities introduced in Eq. (9.27).

For the tomography of the TMD (an 8-bin realization described in Chapter 7, with two SPAD modules detecting the photons after the fiber network), the probe states are adjusted with mean photon number ranging from $|\alpha_{\min}|^2 = 0.19$ up to $|\alpha_{\max}|^2 = 124.47$. The estimated probabilities $p_{j,k}$ of obtaining k clicks are plotted in Fig. 9.4 against the mean number of photons in the probe beam. We restrict the plot to $|\alpha_{\max}|^2 = 60$ to keep the features at lower mean photon

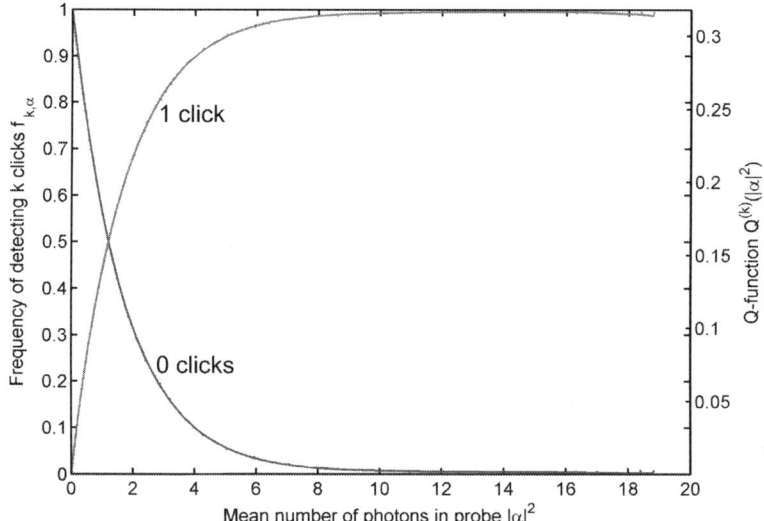

FIGURE 9.3 Estimated probabilities for the click and no-click event of a SPAD against the mean number of photons $|\alpha|^2$ in the coherent probe beam. The errors are too small to be resolved in the plot. The no-click event probability drops exponentially from one to almost zero, while the probabilities for the click event start at zero and increase to saturation.

numbers visible. For higher photon number the probability of recording a $k = 8$ click increased toward unity.

The no-click element shows similar behavior to the SPAD case with high probabilities at low photon numbers and a quick exponential decay. The intermediate elements $p_{j,k}$ with $k = 1, 2, \ldots, 7$ show peaks in their distinct sensitivity regimes, with the peak position moving toward higher photons for higher click number k. The width of the peaks is due to sensitivity to higher photon numbers because of losses in the detector, as well as the convolution, described by Eq. (9.28). We also observe insensitivity to mean photon numbers below the click number, i.e. $p_{j(|\alpha_j|^2 < k), k} = 0$, which indicates the detection noise (dark counts and afterpulsing) is negligible. The highest ($k = 8$) click element of the TMD shows similarity to the SPAD again: its probability rises from zero for low photon numbers toward unity for the highest photon numbers. The rise toward unity displays the saturation of the TMD for highest photon numbers. However, due to the resolution of clicks in between zero and eight this saturation happens much more slowly than for the SPAD case.

9.4.3 Reconstructed POVM Elements

With the method outlined in Section 9.3.3, the POVM elements of the SPAD and TMD can be reconstructed from the Q-functions. For phase-insensitive detectors like SPAD and TMD, the POVM elements $\hat{\Pi}^{(k)}$ for obtaining k clicks

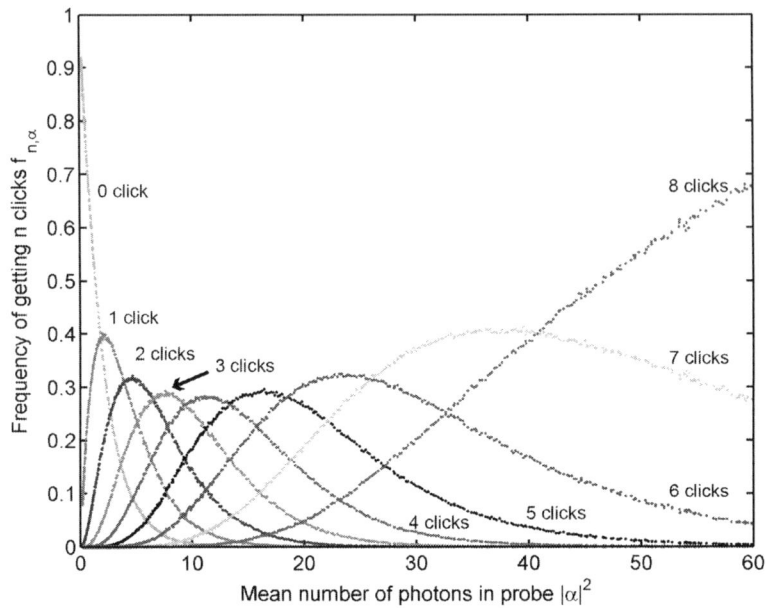

FIGURE 9.4 We show the estimated probabilities p_k for getting k clicks in the TMD plotted against the mean number of photons in the probe beam $|\alpha|^2$. Probe states up to $|\alpha|^2 \approx 124$ are used, which is sufficient for saturating the detector, restrict this plot to values up to $|\alpha|^2 = 60$ for clarity—the main features of the probability distribution are visible for lower $|\alpha|^2$ and the higher values only show saturation. Distinct sensitivity regimes are visible for the individual click elements, with the peak sensitivity higher for higher click numbers.

can be written as

$$\hat{\Pi}^{(k)} = \sum_{n=0}^{M} \Pi_n^{(k)} |n\rangle \langle n|, \qquad (9.31)$$

where $\Pi_n^{(k)}$ are the expansion coefficients of the POVM element in the photon-number basis and we truncated the expansion at photon-number M. The truncation was chosen such that it exceeds the saturation photon number, which is the lowest level for which the only probable detection event is the highest click event. For the SPAD that is the click event, while for the TMD it is the $k = 8$ clicks event. From this point increasing the probe state amplitude does not change the estimated probabilities significantly.

A benchmark for the reconstructed POVM elements are the theoretical models of the SPAD and TMD introduced in Eqs. (9.27) and (9.28). Based on the efficiencies obtained from the data sets one can calculate the expansion coefficients $\Pi_n^{(k)\text{th}}$ of the POVM elements given by the model. The difference between reconstructed and modeled expansion coefficients

$$\Delta_n^{(k)} = |\Pi_n^{(k)\text{th}} - \Pi_n^{(k)}|, \qquad (9.32)$$

can be used as a quantitative measure for the quality of the reconstruction.

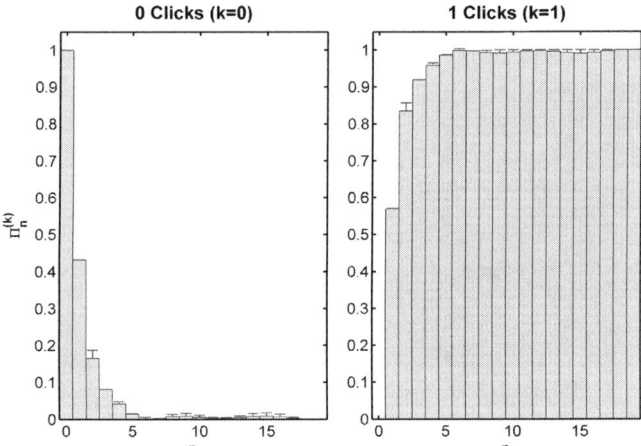

FIGURE 9.5 The reconstructed POVM elements for the SPAD no-click and click events in the photon-number basis. On top of each bar we plot the absolute deviation of that element from a theoretical model according to Eq. (9.32), showing the generally good agreement of the reconstructed POVM elements with the theoretical model. The sensitivity regimes for the two events are visible, with no contribution by zero photons to the click event and a quick decrease of the no-click event with photon number.

On the left of Fig. 9.5 we show the expansion coefficients of the POVM elements reconstructed for the SPAD against the number of photons. The deviation $\Delta_n^{(k)}$ from the theoretical POVM element (detection efficiency 56.8%) is plotted on top of each bar. The $k = 0$ POVM element shows high sensitivity for the low photon numbers and drops off quickly for increasing photon numbers. For the $k = 1$ click POVM element we observe a rapid increase from zero for no photons to almost maximum for already five photons. Zero photons always result in a $k = 0$ no-click event, since no input photons can only lead to a detection event due to dark counts and afterpulsing, the probability of which is suppressed by time gating. On the right of Fig. 9.5 we show the Q-functions for the reconstructed POVM, which show a good agreement with the measured statistics.

The reconstructed POVM elements for the TMD are shown in Fig. 9.6, with the deviation $\Delta_n^{(k)}$ from the theoretical model (detection efficiency 47.8%, three sequential fiber beam splitters with reflectivities, 50.18%, 50.60%, and 41.92%) plotted on top of each bar. Influences from dark counts and afterpulsing on the TMD POVM elements are also suppressed by gating and again the only element sensitive to zero photons is $k = 0$. The coefficients of the $k = 0$ element fall exponentially due to the detection efficiency. The higher $k > 0$ elements show their characteristic sensitivity regimes, with peak sensitivity in higher photon numbers for higher click numbers. The insensitivity to events with fewer photons than click number is clear for the POVM elements in Fig. 9.6

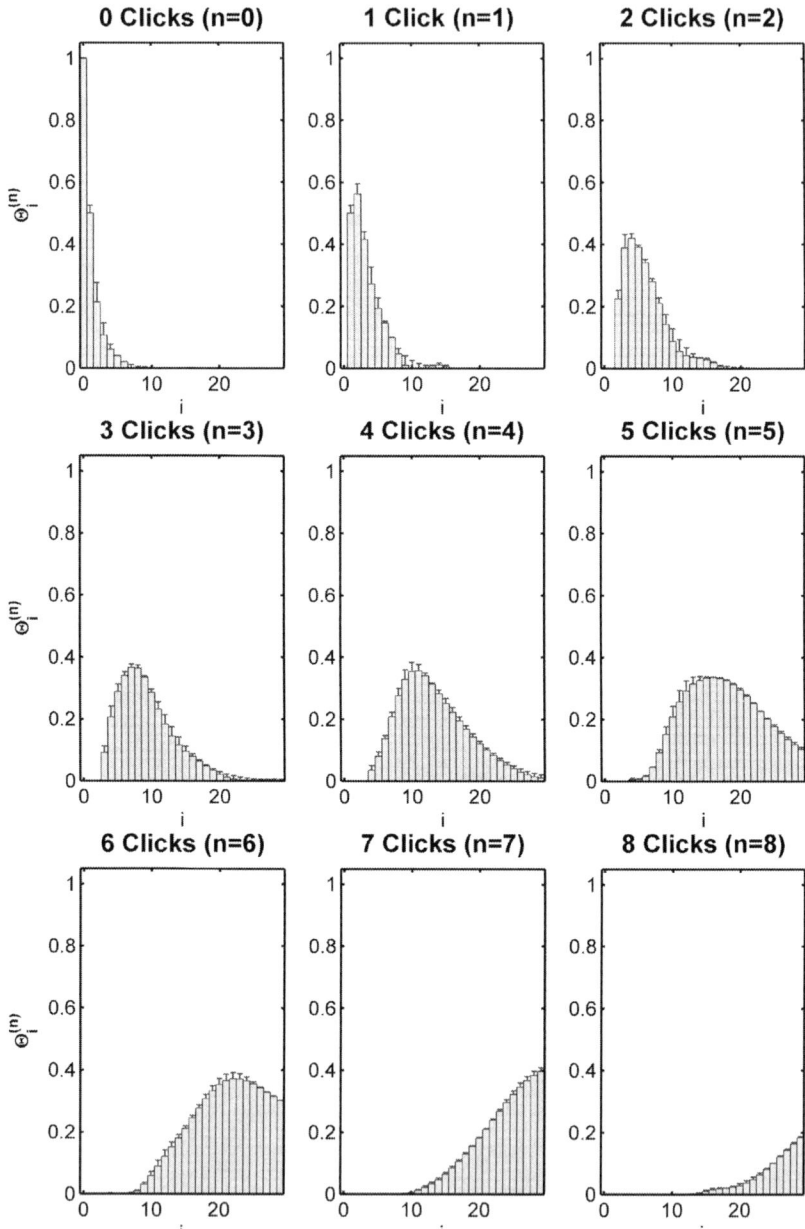

FIGURE 9.6 The reconstructed POVM elements for the $k = 0, \ldots, 8$ click events of the TMD. They show individual sensitivity regimes, with higher peak sensitivity are for higher click numbers. Insensitivity to photon numbers $|\alpha|^2 \leq k$ can be observed which shows negligible contributions from dark counts and afterpulsing. On top of each bar we plot the absolute deviation of each element from a theoretical model according to Eq. (9.32). Good agreement with the model developed in Section 9.4 is observed.

with $\hat{\Pi}^{(k)}_{n<k} \approx 0$. The low click number elements $k = 1$ and $k = 2$ have their peaks for photon numbers close to their click number, while for the higher $k > 2$ elements the areas of highest sensitivity are significantly shifted toward higher photon numbers, e.g. for $k = 4$ the peak sensitivity is for more than 10 incoming photons.

Another widely used benchmark for the similarity between $\hat{\Pi}^{(k)}$ and $\hat{\Pi}^{(k)\text{th}}$ is the fidelity

$$F = \left(\text{Tr} \left(\left(\sqrt{\hat{\Pi}^{(k)\text{th}}} \hat{\Pi}^{(k)} \sqrt{\hat{\Pi}^{(k)\text{th}}} \right)^{1/2} \right)^2 \right) \Big/ \text{Tr}(\hat{\Pi}^{(k)\text{th}}) \text{Tr}(\hat{\Pi}^{(k)}), \quad (9.33)$$

which is higher than 98.7% for all k, confirming the accuracy of the tomography results.

9.4.4 Conditioning and Regularization

One of the main problems encountered in the tomographic characterization of the detectors has to do with the numerical stability of the reconstruction. Such problems are common in tomography [66,67]. For instance, the transformations involved in the inverse Radon transform have inherent instabilities. In principle all the detection probabilities can be predicted by an accurate Glauber-Sudarshan P-function representation of the POVM element, which is given in Eq. (9.34).

$$\hat{\Pi}^{(k)} = \int P^{(k)}(\alpha) |\alpha\rangle\langle\alpha| d^2\alpha. \quad (9.34)$$

Yet the transformation from the Q-function to the P-function is not always well defined [68]. Multiple tools exist to bridge the link between homodyne tomography and the density matrix description [69]. One of them involves the use of pattern functions [70–72], that is, identifying some functions $G_n(\alpha)$ such that

$$\int Q^{(k)}(\alpha) G_n(\alpha) d^2\alpha = \theta_n^{(k)}. \quad (9.35)$$

However, finding the appropriate G_n involves the irregular wave functions (particular unbounded solutions of the Schrödinger equation) and proving them to be appropriate is typically as hard as estimating the error [73]. The use of maximum likelihood has also been explored and particularly for detector tomography [53,58]. However, the speed of the convergence is not generally guaranteed to be high, becoming exponential for certain problems. Our approach, following the spirit of maximum likelihood, translates the problem into a quadratic optimization one allowing for efficient semi-definite programming (cf. Eqs. (9.20) and (9.26)). The details of the approximations and filters that lead to our solution of the problem are provided in Ref. [61].

As most quantum detectors, especially those discussed here are lossy, there is a particular plausibility that their POVMs will be smooth. Indeed, if an optical detector has a POVM element with non-zero amplitude in $|n\rangle\langle n|$, then if it is lossy, it will have a positive amplitude in $|n+1\rangle\langle n+1|, |n+2\rangle\langle n+2|, \ldots, |n+K\rangle\langle n+K|$, decreasing with K but different from zero. In general, if the detector has a finite efficiency η which can be modeled with a BS, it will impose some smoothness on the distribution $\theta_n^{(k)}$. That is because if $G(n)$ is the probability of registering n photons and $H(n')$ is the probability that n' were present, then the loss process will impose

$$G(n) = \sum_{n'} \binom{n'}{n} \eta^n (1-\eta)^{n'-n} H(n). \tag{9.36}$$

Consequently, if $\theta_n \neq 0$, then $\theta_{n+1}, \theta_{n+2}$, etc., cannot be zero, but will have some relatively smooth distribution. This simple physical argument makes a certain smoothness plausible (but still should allow sharp transitions for $m < n$).

Let us first define what we mean by smooth. Smooth will mean in this context that the difference $|\theta_n^{(k)} - \theta_{n+1}^{(k)}|$ is small for all k and n. Known as smoothing regularization or more generally as Tikhonov regularization [74], the optimization in this context will mean that our minimization is defined as follows

$$\min \{\|\mathbf{P} - \mathbf{F}\mathbf{\Pi}\|_2 + yS\} \tag{9.37}$$

with

$$S = \sum_{k,n} (\theta_n^{(k)} - \theta_{n+1}^{(k)})^2 \tag{9.38}$$

for some fixed value of y. The smoothing function S will be independent of the detector, and will mildly penalize non-smooth POVM elements. This approach is further substantiated by the observation that the resulting POVMs are largely independent of the weight y that is given to the smoothness penalty.

For this detector (and for any photodiode-based detector) assuming loss is reasonable and can make the "smoothness" requirement appropriate. Let us however see if, without looking at the specific shape of our POVM, we can find an optimal smoothing coefficient y and justify further the use of the smoothing regularization.

One way to test this method is to quantify how resilient it is to noise in the data. To do so we introduce additional noise in $x = |\alpha|^2$ to the measured data. For example, we can alter x in $P_{i,n} = P(x_i(1+\delta_i), n)$ where $\delta = (\delta_1, \ldots, \delta_D)$ is again a vector of random variables distributed with a Gaussian distribution with zero mean. This simulates a statistical uncertainty in the measurement of the coherent state. To see its effect on the reconstruction we use the figure of merit $\|\Pi_\delta - \Pi_{\delta=0}\|_2$. This quantity should evaluate how POVMs differ from the one without noise. It is seen that the additional smoothing penalty makes the optimization more robust, largely independent of the value of y (we can multiply y by 100 and stay in the same regime). Using this smoothing regularization with noisy data seems therefore a good choice.

TABLE 9.2 Sensitivity of Optimal POVM Estimation to Smoothing

y	y Variation	Π Relative Error (%)
0.0001	×	27.3
0.001	×/1000	12.2
0.01	×/10	4
0.05	×/2	1
0.5	× 5	3
1	× 10	5

We have seen how smoothing makes the optimization more robust against noise, but we should also ask how sensitive this optimization is to the exact choice of y. To do so we may use the following procedure, comparing the POVM obtained using $y = 0.1$ with that obtained varying y over four orders of magnitude (cf. Ref. [61]). Remarkably doubling the value of y results in an overall relative error in the POVM of less than 1%. Multiplying (or dividing) y by 10 gives a variation below 5% and 100-fold variation results in a 12% variation. If we compare how this differs from the $y = 0$ case which is 110% different, then we can conclude that the optimization is quite insensitive to the exact choice of the *smoothing parameter y*. Table 9.2 provides some values for reference.

9.4.5 Robustness of Detector Tomography

A remaining problem is the robustness of the shown reconstruction to fluctuations in the input parameters, in particular the amplitude of the probe state. As shown in Section 9.4.1, the probe pulse energy jitter to be 1.88(2)%. For each probe state amplitude α_j we can model the fluctuation in pulse energy by a Gaussian distribution around α_j.

$$f_{\alpha_j}(\beta) = \frac{1}{\sigma\sqrt{2\pi}} e^{-\frac{(\beta-\alpha_j)^2}{2\sigma^2}}, \quad (9.39)$$

where σ is the standard deviation of the Gaussian distribution with $\sigma = 0.02|\alpha_j|^2$. Instead of the pure state density matrix $|\alpha_j\rangle\langle\alpha_j|$, the fluctuation leads to the use of the following probe state density matrix in the reconstruction of the POVM element

$$\hat{\rho}_j = \int d^2\beta f_{\alpha_j}(\beta) |\beta\rangle\langle\beta| = \sum_{m,n=0}^{\infty} E_{m,n,\alpha_j} |m\rangle\langle n|, \quad (9.40)$$

where we expanded the coherent state $|\beta\rangle$ in the photon-number basis, and combined the integral part in the factor

$$E_{m,n,\alpha_j} = \frac{1}{\sigma\sqrt{2\pi}\sqrt{m!n!}} \int d\beta \beta^{m+n} e^{-|\beta|^2 - \frac{(\beta-\alpha_j)^2}{2\sigma^2}}. \tag{9.41}$$

For the tomography of non-phase-sensitive detectors, only the diagonal elements $m = n$ in the density matrix are relevant, and according to the treatment in Section 9.3.3 we truncate the probe state at photon-number M, leading to the probe state

$$\hat{\rho}_j = \sum_{n=0}^{M} E_{n,n,\alpha_j} |n\rangle\langle n|, \tag{9.42}$$

which effectively changes the definition of matrix $F_{j,n}$ in Eq. (9.17), to

$$F_{j,n} = E_{n,n,\alpha_j} = \frac{1}{\sigma\sqrt{2\pi}n!} \int d\beta \beta^{2n} e^{-|\beta|^2 - \frac{(\beta-\alpha_j)^2}{2\sigma^2}}. \tag{9.43}$$

Reconstruction of the POVM elements using Eq. (9.43) instead of Eq. (9.17) in the convex optimization given by Eq. (9.20) yielded a set of POVM elements $\left\{\tilde{\hat{\Pi}}^{(k)}\right\}$ which incorporated the effects of probe state amplitude fluctuations. The relative difference between the POVM elements obtained with and without probe state amplitude fluctuations is calculated

$$\frac{||\hat{\Pi}^{(k)} - \tilde{\hat{\Pi}}^{(k)}||}{||\tilde{\hat{\Pi}}^{(k)}||} \leq 0.7\%, \tag{9.44}$$

which is negligibly small. The relative difference of individual matrix elements of the two POVMs obtained is less than 1.3% for any matrix element. This leads to the conclusion that technical noise from the laser is indeed less detrimental for detector tomography and the results shown in Section 9.4.3 are good estimations of the detector POVM elements even when neglecting fluctuations of the probe state amplitude.

9.4.6 Wigner Functions

The Wigner function is an intuitive way of visualizing quantum operations in phase space. The probability for obtaining a certain measurement result is the overlap of the state Wigner function with the Wigner function of the POVM element. The Wigner function of $\hat{\Pi}^{(k)}$ is given by

$$W^{(k)}(x,p) = W_{\hat{\Pi}^{(k)}}(x,p) = \frac{1}{\pi\hbar} \int_{-\infty}^{\infty} dy \langle x-y|\hat{\Pi}^{(k)}|x+y\rangle e^{\frac{2ipy}{\hbar}}. \tag{9.45}$$

Chapter | 9 Quantum Detector Tomography

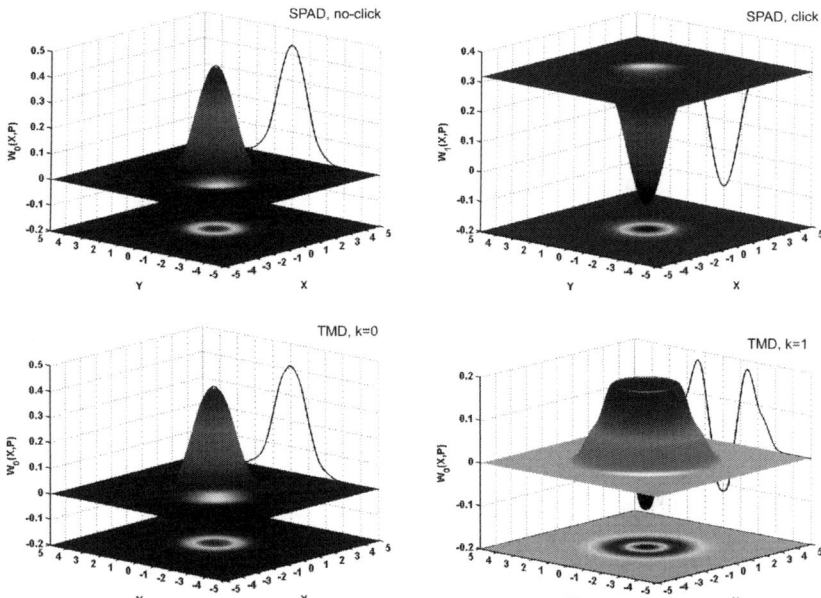

FIGURE 9.7 **Top:** The Wigner functions of the click and no-click POVM elements for the SPAD, calculated using Eq. (9.45). The functions are rotationally symmetric and we plot a cross-section in the back plane. The sensitivity regimes of the two POVM elements can be observed. **Top Left:** no-click has similarity with the vacuum Wigner function, broadened due to losses. **Top Right:** The click Wigner function shows no sensitivity in the center and higher sensitivity for increasing amplitude $|\alpha|$. However, the click Wigner function shows no similarity with the single-photon state, due to the non-photon-number-resolving nature of SPADs. **Bottom:** The Wigner functions of the $k = 0$ and $k = 1$ POVM elements for the TMD, calculated with Eq. (9.45). The functions are rotationally symmetric and we plot a cross-section in the back plane. **Left:** $k = 0$ clicks show similarity with the vacuum state. Losses broaden the Wigner function, making the $k = 0$ POVM element partially sensitive to higher photon numbers, too. **Right:** The $k = 1$ shows similarities to a single-photon Wigner function, broadened by loss, showing the partial photon-number resolution of the TMD. In contrast to the SPAD, higher amplitudes $|\alpha|^2$ are only marginally detected by this POVM element due to losses and the convolution effect.

In Fig. 9.7 we show the reconstructed Wigner functions for the SPAD for $k = 0$ and $k = 1$. Both Wigner functions are rotationally symmetric, due to the lack of phase sensitivity of the SPAD detector. The zero-click function shows high sensitivity for low photon number, with its maximum at the origin and has similarities with coherent vacuum state. Losses increase the sensitivity of the zero-click elements to higher photon numbers. This broadens the $k = 0$ click element Wigner function in comparison with the projection onto the vacuum.

The one-click event on the other hand is insensitive to the vacuum, with negative values in the center of phase space. It shows increasing sensitivity with increasing photon number. Often the SPAD is referred to as a single-photon detector, but the Wigner function of the click event shows no qualitative similarity with a single-photon Wigner function. Indeed, when using SPADs as single-photon detectors, one must keep in mind that their discriminative capabilities are between zero and more than zero photons, but not between the photon numbers. Calling a SPAD a single-photon detector only holds for non-significant contributions of multi-photon events in the probe beam.

In Fig. 9.7 we also show the Wigner functions of the zero-click and one-click POVM elements for the TMD. Similar to that of the SPAD, the zero-click element shows good overlap with the vacuum. The one-click Wigner function is similar to a single-photon Wigner function, again with some broadening due to losses and also the convolution. This confirms the capability of the TMD as a photon-number-resolving detector in the non-saturation region.

One particular interesting feature of the one-click Wigner function is its negativity at the origin, which is known as the non-classical signature of an operator. This reveals the non-classicality of the quantum measurement process, and confirms the capability of the TMD for detecting, verifying, and engineering non-classical optical states.

9.5 CONCLUSIONS

In this chapter we presented the theoretical framework of quantum detector tomography and two experimental implementations. QDT is a black-box, or device-independent, approach for the complete characterization of quantum detectors. By eliminating assumptions, full characterization enables more flexible design and use of detectors, be they noisy, nonlinear, inefficient, or operating outside their normal range. To demonstrate the accuracy of QDT, the experimental results given here show good agreement with the predictions of simple theoretical models. Yet the genuine power of QDT lies in the situations when the theoretical model becomes more complicated, or even fails due to too ideal assumptions. QDT of a weak-homodyne TMD has yielded the largest parametrization to date in quantum tomography experiments [62,63]. Tomography of superconducting nanowire detectors, on the other hand, shows the nonlinearity of single-photon detectors which makes the linear model in Eq. (9.27) not valid any more [34].

Another advantage of QDT relies on the fact that it characterizes the set of operators that link the input quantum state of the light field to the classical detector output. This allows us to ask precise quantitative questions about our power to verify and prepare non-classical states or herald quantum operations. It opens a path for the experimental study of concepts such as the non-classicality of detectors. For example, with the tomography of a single-photon detector with adjustable noise level, the effect of quantum decoherence in an optical detector has been demonstrated through the transition of a quantum measurement operator into a semi-classical regime [31].

REFERENCES

[1] K.J. Resch, K.L. Pregnell, R. Prevedel, A. Gilchrist, G.J. Pryde, J.L. O'Brien, and A.G. White, Phys. Rev. Lett. 98, 223601 (2007).
[2] B.L. Higgins, D.W. Berry, S.D. Bartlett, H.M. Wiseman, and G.J. Pryde, Nature 49, 393 (2007).
[3] M.A. Nielsen, Phys. Lett. A 308, 96 (2003).
[4] R. Raussendorf and H. Briegel, Quant. Inf. Comp. 6, 433 (2002).
[5] V.B. Braginsky, Y.I. Vorontsov, and K.S. Thorne, Science 209, 547 (1980).
[6] M. Freitag, Y. Martin, J.A. Misewich, R. Martel, and P. Avouris, Nano Lett. 3, 1067 (2003).
[7] M. Sasaki, K. Wakui, J. Mizuno, M. Fujiwara, and M. Akiba, in AIP Conference Proceedings, The Seventh International Conference on Quantum Communication, Measurement and Computing, 25-29 July 2004, Glasgow, UK, pp. 44–47 (2004).
[8] J. Kim, S. Takeuchi, Y. Yamamoto, and H. Hogue, Appl. Phys. Lett. 74, 902 (1999).
[9] A. Shields, M. O'Sullivan, I. Farrer, D. Ritchie, R. Hogg, M. Leadbeater, C. Norman, and M. Pepper, Appl. Phys. Lett. 76, 3673 (2000).
[10] G. Gol'tsman, O. Okunev, G. Chulkova, A. Lipatov, A. Semenov, K. Smirnov, B. Voronov, A. Dzardanov, C. Williams, and R. Sobolewski, Appl. Phys. Lett. 79, 705 (2001).
[11] A. Miller, S. Nam, J. Martinis, and A. Sergienko, Appl. Phys. Lett. 83, 791 (2003).
[12] D. Achilles, C. Silberhorn, C. Sliwa, K. Banaszek, and I.A. Walmsley, Opt. Lett. 28, 2387 (2003).
[13] D. Achilles, C. Silberhorn, C. Sliwa, K. Banaszek, I.A. Walmsley, M.J. Fitch, B.C. Jacobs, T.B. Pittman, and J.D. Franson, J. Mod. Opt. 51, 1499 (2004).
[14] K. Vogel and H. Risken, Phys. Rev. A 40, 2847 (1989).
[15] D.T. Smithey, M. Beck, M.G. Raymer, and A. Faridani, Phys. Rev. Lett. 70, 1244 (1993).
[16] T.J. Dunn, I.A. Walmsley, and S. Mukamel, Phys. Rev. Lett. 74, 884 (1995).
[17] C. Monroe, D.M. Meekhof, B.E. King, W.M. Itano, and D.J. Wineland, Phys. Rev. Lett. 75, 4714 (1995).
[18] S. Wallentowitz and W. Vogel, Phys. Rev. A 53, 4528 (1996).
[19] K. Banaszek, G.M. D'Ariano, M.G.A. Paris, and M.F. Sacchi, Phys. Rev. A 61, 010304 (1999).
[20] K. Banaszek, C. Radzewicz, K. Wódkiewicz, and J.S. Krasiński, Phys. Rev. A 60, 674 (1999).
[21] A.G. White, D.F.V. James, W.J. Munro, and P.G. Kwiat, Phys. Rev. A 65, 012301 (2001).
[22] A. Ourjoumtsev, H. Jeong, R. Tualle-Brouri, and P. Grangier, Nature 448, 784 (2007).
[23] J.S. Neergaard-Nielsen, B.M. Nielsen, C. Hettich, K. Mølmer, and E.S. Polzik, Phys. Rev. Lett. 97, 083604 (2006).
[24] I.L. Chuang and M.A. Nielsen, J. Mod. Opt. 44, 2455 (1997).
[25] J.F. Poyatos, J.I. Cirac, and P. Zoller, Phys. Rev. Lett. 78, 390 (1997).
[26] G.M. D'Ariano and L. Maccone, Phys. Rev. Lett. 80, 5465 (1998).
[27] M.A. Nielsen, E. Knill, and R. Laflamme, Nature 395, 52 (1998).
[28] J.B. Altepeter, D. Branning, E. Jeffrey, T.C. Wei, P.G. Kwiat, R.T. Thew, J.L. O'Brien, M.A. Nielsen, and A.G. White, Phys. Rev. Lett. 90, 193601 (2003).
[29] M.W. Mitchell, C.W. Ellenor, S. Schneider, and A.M. Steinberg, Phys. Rev. Lett. 91, 120402 (2003).

[30] M. Hillery, R.F. O'Connell, M.O. Scully, and E.P. Wigner, Phys. Rep. 106, 121 (1984).
[31] V.D'Auria, N. Lee, T. Amri, C. Fabre, and J. Laurat, Phys. Rev. Lett. 107, 050504 (2011).
[32] J. S. Lundeen, A. Feito, H. Coldenstrodt-Ronge, K.L. Pregnell, C. Silberhorn, T.C. Ralph, J. Eisert, M.B. Plenio, and I. Walmsley, Nat. Phys. 5, 27 (2009).
[33] G. Brida, L. Ciavarella, I.P. Degiovanni, M. Genovese, L. Lolli, M.G. Mingolla, F. Piacentini, M. Rajteri, E. Taralli, and M.G.A. Paris, New J. Phys. 14, 085001 (2011).
[34] M.K. Akhlaghi, A.H. Majedi, and J.S. Lundeen, Opt. Express 19, 21305 (2011).
[35] G.N. Hounsfield, Br. J. Radiol. 46, 1016 (1973).
[36] J. Ambrose, Br. J. Radiol. 46, 1023 (1973).
[37] E.C. Beckmann, Br. J. Radiol. 79, 5 (2006).
[38] J. Radon, Berichte über die Verhandlungen der Sächsische Akademie der Wissenschaften 69, 262 (1917).
[39] J. Radon, IEEE Trans. Med. Imag. 5, 170 (1986).
[40] A. Royer, Found. Phys. 19, 3 (1989).
[41] W.K. Wootters and W.H. Zurek, Nature 299, 802 (1982).
[42] U. Leonhardt, "Measuring the Quantum State of Light," Cambridge University Press (2005).
[43] K. Laiho, M. Avenhaus, K.N. Cassemiro, and C. Silberhorn, New J. Phys. 11, 043012 (2009).
[44] T. Gerrits, S. Glancy, T.S. Clement, B. Calkins, A.E. Lita, A.J. Miller, A.L. Migdall, S.W. Nam, R.P. Mirin, and E. Knill, Phys. Rev. A 82, 031802 (2010).
[45] J.P. Home, M.J. McDonnell, D.M. Lucas, G. Imreh, B.C. Keitch, D.J. Szwer, N.R. Thomas, S. Webster, D.N. Stacey, and A.M. Steane, New J. Phys. 8, 188 (2006).
[46] J.L. O'Brien, G.J. Pryde, A. Gilchrist, D.F.V. James, N.K. Langford, T.C. Ralph, and A.G. White, Phys. Rev. Lett. 93, 080502 (2004).
[47] M. Lobino, D. Korystov, C. Kupchak, E. Figueroa, B.C. Sanders, and A.I. Lvovsky, Science 322, 563 (2008).
[48] G.M. D'Ariano and P. Lo Presti, Phys. Rev. Lett. 86, 4195 (2001).
[49] G.M. D'Ariano and P. Lo Presti, Phys. Rev. Lett. 91, 047902 (2003).
[50] M. Riebe, K. Kim, P. Schindler, T. Monz, P.O. Schmidt, T.K. Körber, W. Hänsel, H. Häffner, C.F. Roos, and R. Blatt, Phys. Rev. Lett. 97, 220407 (2006).
[51] M. Mohseni and D.A. Lidar, Phys. Rev. Lett. 97, 170501 (2006).
[52] M. Mohseni and D.A. Lidar, Phys. Rev. A 75, 062331 (2007).
[53] J. Fiurášek, Phys. Rev. A 64, 024102 (2001).
[54] H.B. Coldenstrodt-Ronge, J.S. Lundeen, K.L. Pregnell, A. Feito, B.J. Smith, W. Mauerer, C. Silberhorn, J. Eisert, M.B. Plenio, and I.A. Walmsley, J. Mod. Opt. 56, 432 (2009).
[55] L. Mandel and E. Wolf, "Optical Coherence and Quantum Optics," Cambridge University Press (1995).
[56] A. Peres, Found. Phys. 20, 1441 (1990).
[57] A. Luis and L.L. Sánchez-Soto, Phys. Rev. Lett. 83, 3573 (1999).
[58] G.M. D'Ariano, L. Maccone, and P.L. Presti, Phys. Rev. Lett. 93, 250407 (2004).
[59] A.P. Worsley, H.B. Coldenstrodt-Ronge, J.S. Lundeen, P.J. Mosley, B.J. Smith, G. Puentes, N. Thomas-Peter, and I.A. Walmsley, Opt. Express 17, 4397 (2009).
[60] A.F. Feito, Ph.D. Thesis, Imperial College London (2008).
[61] A. Feito, J.S. Lundeen, H. Coldenstrodt-Ronge, J. Eisert, M.B. Plenio, and I.A. Walmsley, New J. Phys. 11, 093038 (2009).
[62] L. Zhang, H.B. Coldenstrodt-Ronge, A. Datta, G. Puentes, J.S. Lundeen, X.-M. Jin, B.J. Smith, M.B. Plenio, and I.A. Walmsley, Nat. Photonics 6, 364 (2012).
[63] L. Zhang, A. Datta, H.B. Coldenstrodt-Ronge, X.-M. Jin, J. Eisert, M.B. Plenio, and I.A. Walmsley, New J. Phys. 14, 115005 (2012).
[64] G. Brida, L. Ciavarella, I.P. Degiovanni, M. Genovese, A. Migdall, M.G. Mingolla, M.G. A. Paris, F. Piacentini, and S.V. Polyakov, Phys. Rev. Lett. 108, 253601 (2012).
[65] H.B. Coldenstrodt-Ronge, J.S. Lundeen, K.L. Pregnell, A. Feito, B.J. Smith, W. Mauerer, C. Silberhorn, J. Eisert, M.B. Plenio, and I.A. Walmsley, J. Mod. Opt. 56, 432 (2009).
[66] N. Boulant, T.F. Havel, M.A. Pravia, and D.G. Cory, Phys. Rev. A 67, 042322 (2003).
[67] M. Ježek, J. Fiurášek, and Z. Hradil, Phys. Rev. A 68, 012305 (2003).
[68] W. P. Schleich, "Quantum Optics in Phase Space," Wiley-VCH, Berlin (2001).
[69] A.I. Lvovsky and M.G. Raymer, Rev. Mod. Phys. 81, 299 (2009).

[70] U. Leonhardt, H. Paul, and G.M. D'Ariano, Phys. Rev. A 52, 4899 (1995).
[71] G.M. D'Ariano, U. Leonhardt, and H. Paul, Phys. Rev. A 52, R1801 (1995).
[72] A. Wünsche, J. Mod. Opt. 44, 2293 (1997).
[73] U. Leonhardt, M. Munroe, T. Kiss, T. Richter, and M. Raymer, Opt. Commun. 127, 144 (1996).
[74] S. Boyd and L. Vandenberghe, "Convex Optimization," Cambridge University Press, (2004).

Chapter 10

The First Single-Photon Sources

Alain Aspect and Philippe Grangier
Laboratoire Charles Fabry, Institut d'Optique, CNRS, Univ Paris-Sud, 2 Avenue Augustin Fresnel, 91127 Palaiseau, France

Chapter Outline

10.1	Introduction	316
10.2	Feeble Light Vs. Single Photon	318
	10.2.1 In Search of Feeble Light's Wave-Like Properties: A Short Historical Review	318
	10.2.2 Quantum Optics in a Nutshell	319
	10.2.3 One-Photon Wavepacket	321
	10.2.4 Quasi-Classical Wavepacket	326
	10.2.5 The Possibility of an Experimental Distinction	328
	10.2.6 Attenuated Continuous Light Beams	329
	10.2.7 Light From a Discharge Lamp	331
	10.2.8 Conclusion: What Is Single-Photon Light?	333
10.3	Photon Pairs As a Resource for Single Photons	334
	10.3.1 Introduction	334
	10.3.2 Non-Classical Properties in an Atomic Cascade	335
	10.3.3 Anticorrelation for a Single Photon on a Beamsplitter	336
	10.3.4 The 1986 Anticorrelation Experiment	339
10.4	Single-Photon Interferences	344
	10.4.1 Wave-Particle Duality in Textbooks	344
	10.4.2 Interferences with a Single Photon	344
10.5	Further Developments	346
	10.5.1 Parametric Sources of Photon Pairs	346
	10.5.2 Other Heralded and "On-Demand" Single-Photon Sources	347
	10.5.3 "Delayed-Choice" Single-Photon Interference Experiments	348
References		348

10.1 INTRODUCTION

This chapter shows how the concept of single-photon sources has emerged, starting in the early 1980s. It presents the quantum optics approach to "single-photon states" and "single-photon wavepackets." The quantum behavior of such states—a single photon yields one photodetection only—is contrasted with the behavior of attenuated classical lights, which always yield some possibility of a joint detection on both sides of a beam splitter. We describe the single-photon source that we developed in the early 1980s at Institut d'Optique, as well as the quantitative criterion (*"anticorrelation"*) that we introduced and used in a real experiment to show that it was indeed a single-photon source. We contrast these results with the ones that we obtained with a source of classical light pulses produced by a strongly attenuated light-emitting diode driven by nanosecond electric pulses. Such light pulses do not pass the anticorrelation test, and are definitely not single-photon pulses. We also describe the interference experiment we carried out with our single-photon source, which illustrates the notion of wave-particle duality. We conclude with a brief overview of further developments in sources of single photons, heralded or on-demand, as well as in wave-particle duality experiments, in particular Wheeler's delayed-choice experiments.

The rapidly developing field of quantum information [1] makes wide use of two types of sources of quantum light: sources of single photons on the one hand, and sources of pairs of entangled photons on the other hand. One might think that single-photon sources were developed first, but it turns out that the history is just the opposite: in the optical domain, sources of pairs of entangled photons were invented first, and only later came single-photon sources. This happened first with the source of entangled photons of Clauser and Freedman [2], about which a property related to the behavior of single photons was demonstrated two years later [3]. In the same vein, it took five years for the more efficient source of pairs of entangled photons of Aspect *et al.* [4], to be explicitly used and characterized, by Grangier and Aspect, as the first source of single photons [5]. Similarly, the first source of pairs of correlated photons produced by parametric down-conversion [6,7] preceded the use of that source to produce single photons by Mandel *et al.* [8]. Actually, all these single-photon sources were what is called, in modern quantum optics language, "heralded single-photon sources," i.e., single-photon wavepackets whose leading-edge time—or peak time in the case of a bell-shaped pulse—is known by the observation of the other photon of a pair [9]. This is why their development obviously demanded the existence of a source of pairs of photons correlated in time. It took almost another two decades until the first source of single photons "on-demand" appeared [10], i.e., a source of single-photon wavepackets whose leading-edge time can be chosen at will.

Although the question of single photons had been raised at the beginning of the 20th century in the context of single-photon interference (see Section 10.2.1), the question remained confused until the early 1980s, when we realized that none of the so-called "single-photon interference experiments" had been carried out with "one-photon states of light." Indeed, all these experiments had been performed with feeble-light beams issued from standard sources (such as discharge lamps), and it was clear from the formalism of quantum optics, that however weak, such lights could be described by quasi-classical states [11,12]. Therefore, their properties could be understood by the semi-classical model of matter-light interaction, in which light is described as a classical electromagnetic wave, and the notion of a single photon has no meaning. Inspired by the experiment of Clauser [3], and by the celebrated antibunching experiment of Kimble-Dagenais-Mandel [13], we found a simple quantitative criterion to test a characteristic property of a single photon, *anticorrelation*: when sent to a beam splitter, a single photon (i.e., a one-photon state of the quantized electromagnetic field) can be detected either on one side or on the other side of the beamsplitter, but never jointly on both sides. This is in contrast to the behavior of light that can be described by a classical wave, which is split on the beam splitter and always yields some possibility of a joint detection on both sides of the beam splitter. We thus had a criterion that could be used for a test of the single-photon character of the light emitted by a source, not only theoretically, but also experimentally.

Section 10.2 of this chapter is devoted to a detailed theoretical presentation, in the formalism of quantum optics (kept as simple as possible), of the difference between light emitted by true sources of single photons, and light emitted by any other source, as feeble as it may be. The main conclusion is that light from other sources, no matter how weak, does not have the same characteristic property as single photons. Even in the case of a strongly attenuated discharge lamp where it is tempting to describe the light as made of single-photon wavepackets separated from each other, it does not pass the single-photon test since we miss the information about the time at which each individual single-photon wavepacket is emitted.

In Section 10.3, we give some details about the single-photon source that we developed in the early 1980s, and about the precise quantitative anticorrelation criterion that we introduced and used in an experiment to show that it was indeed a single-photon source. We contrast these results with the ones that we obtained with a source of classical light pulses produced by a light-emitting diode (LED) driven by nanosecond electric pulses, and attenuated to an average level of 10^{-2} photons per pulse. Such light pulses do not pass the single-photon test, and are definitely not single-photon pulses.

In Section 10.4, we give some details about the interference experiment we carried out with our single-photon source. Combined with the experiment of Section 10.3 that uses the same source, it yields a striking demonstration of the so-called "wave-particle duality," one of the two "great mysteries" of quantum mechanics according to Feynman [14], and it can be used for an introductory course in quantum optics [15–17] (see also [18]).

Section 10.5 sketches further developments of modern sources of single photons, either heralded or on-demand, without many details, since these details can be found in other chapters of this book. We also mention further experiments on wave-particle duality, and in particular on Wheeler's delayed-choice experiment [19], which has been performed not only in its original form [20], but also in a more refined version [21,22].

Remark on Vocabulary
In this chapter, we use the wordings "single photon," "single-photon wavepacket," "single-photon pulse" on the one hand, and "one-photon wavepacket" on the other hand. Although these wordings are almost equivalent, we tend to use "single photon" as it would be used generically in the common language, or in the language of a general physicist, while we give to "one-photon wavepacket" a more technical meaning, i.e., a state of the light that is an eigenstate of the quantum observable "Number of Photons" of the formalism of Quantum Optics (Section 10.2.2).

10.2 FEEBLE LIGHT VS. SINGLE PHOTON

10.2.1 In Search of Feeble Light's Wave-Like Properties: A Short Historical Review

Almost as soon as Einstein introduced the notion of a quanta of light [23], i.e., a relativistic particle [24] of energy $\hbar\omega$ and momentum $\hbar\omega/c$, the question of the wave-like behavior of the corresponding particle became a major concern among physicists, including Einstein himself [25]. The first attempt to investigate the question experimentally [26] consisted of registering on a photographic plate the diffraction pattern of a needle illuminated with extremely attenuated light, so that the energy flux expressed in the number of photons per second would correspond to an average distance between the photons much larger than the size of the apparatus. This pioneering experiment was followed by a long series of diffraction [27] and interference [28–33] experiments with light emitted by strongly attenuated ordinary light sources, mostly discharge lamps, so that the average rate of photons entering the interferometric device, estimated as the power divided by the energy of a photon, ranged between 10^2 and 10^7 s^{-1}. Even at the largest of these rates, the average distance between photons was more than 10 m, much larger than the size of the interferometric device used in the corresponding experiment. It was thus concluded that "there was only one photon at a time in the interferometer," and the observation of fringes was then considered a demonstration that "a photon interferes with itself." Actually, one experiment [30] failed to observe the interference pattern expected for a wave, but it was soon repeated by other scientists who found the expected interference pattern [32]. There is thus little doubt that diffraction or interference phenomena can be observed even in conditions of very weak light intensity.

In the 1970s, the general wisdom was then that "single-photon wave-like behavior" had been experimentally demonstrated. However, revisiting that question in the early 1980s, we realized that, according to the formalism of modern quantum optics as developed by Glauber [12,34,35], none of the experiments cited above could be considered a demonstration of single particle interference, because in none of these experiments the light used could be considered as a single-photon wavepacket. This led us to perform the experiments of [5], presented in Sections 10.3 and 10.4. In the rest of this section, we use the formalism of quantum optics to highlight the difference between single-photon wavepackets and all the types of light used in the experiments above.

10.2.2 Quantum Optics in a Nutshell

We describe light in the standard formalism of quantum optics [17,36,37], in the Heisenberg representation. A particular light field is represented by a state vector independent of time $|\Psi\rangle$. When fluctuations must be accounted for, taking the statistical average will be sufficient, so we will not resort to the density matrix formalism nor to the notion of mixed states. The field observables depend on time (and position). The electric-field operator is decomposed into two adjoint operators, $\hat{\mathbf{E}}^{(-)}(\mathbf{r},t)$ and $\hat{\mathbf{E}}^{(+)}(\mathbf{r},t)$, corresponding respectively to positive and negative frequencies. These operators can be expanded on any set of modes of the electromagnetic field. A frequent choice is polarized homogeneous traveling waves, and the electric-field operator expansion then reads

$$\hat{\mathbf{E}}^{(+)}(\mathbf{r},t) = i \sum_{\ell} \mathcal{E}_{\ell}^{(1)} \vec{\varepsilon_{\ell}} \hat{a}_{\ell} \exp[i(\mathbf{k}_{\ell} \cdot \mathbf{r} - \omega_{\ell} t)], \qquad (10.1)$$

$$\hat{\mathbf{E}}^{(-)}(\mathbf{r},t) = [\hat{E}^{(+)}(\mathbf{r},t)]^{\dagger}. \qquad (10.2)$$

The mode ℓ is characterized by a wave-vector $\mathbf{k}_\ell$, an angular frequency $\omega_\ell = c|\mathbf{k}_\ell|$, and a polarization $\vec{\varepsilon_\ell}$ orthogonal to $\mathbf{k}_\ell$. The quantity

$$\mathcal{E}_{\ell}^{(1)} = \sqrt{\frac{\hbar \omega_\ell}{2\varepsilon_0 L^3}} \qquad (10.3)$$

is the "one-photon amplitude." It depends on the volume of quantization L^3, which is usually arbitrary, so that L should not appear explicitly in the final results of the calculations.

The adjoint operators $\hat{a}_\ell$, and $\hat{a}_\ell^{\dagger}$ are the destruction and creation operators for photons of the mode ℓ. They obey the fundamental commutation relations

$$[\hat{a}_\ell, \hat{a}_{\ell'}^{\dagger}] = \delta_{\ell\ell'} \qquad (10.4)$$

with $\delta_{\ell\ell'}$, the Kronecker symbol. They allow one to build a complete basis $\{|n_\ell\rangle; n_\ell = 0, 1 \ldots\}$ of the state space associated with the mode ℓ:

$$\hat{a}_\ell^\dagger |n_\ell\rangle = \sqrt{n_\ell + 1}|n_\ell + 1\rangle, \tag{10.5}$$

$$\hat{a}_\ell |n_\ell\rangle = \sqrt{n_\ell}|n_\ell - 1\rangle, \tag{10.6}$$

$$a_\ell |0_\ell\rangle = 0. \tag{10.7}$$

States $|n_\ell\rangle$, the so-called *number states*, are eigenstates of the operator "number of photons in the mode ℓ ":

$$\hat{N}_\ell = \hat{a}_\ell^\dagger \hat{a}_\ell, \tag{10.8}$$

the corresponding eigenvalue being precisely the number of photons n_ℓ:

$$\hat{N}_\ell |n_\ell\rangle = n_\ell |n_\ell\rangle. \tag{10.9}$$

One also defines the operator "total number of photons"

$$\hat{N} = \sum_\ell \hat{N}_\ell, \tag{10.10}$$

which can be measured with a wide-band photodetector operating in the photon-counting regime ("click detector").

There is no position operator for the photon, so one cannot define a density of probability of presence, as in the quantum mechanical description of a single massive particle. However, there is a very useful quantity that allows one to link theory to experiments with a click detector: the probability of a photodetection per unit of surface and time at the point $\mathbf{r}$ and time t. For a field in the state $|\Psi\rangle$, that quantity (also called the rate of single photodetections) is

$$w^{(1)}(\mathbf{r},t) = s\langle\Psi|\hat{\mathbf{E}}^{(-)}(\mathbf{r},t)\hat{\mathbf{E}}^{(+)}(\mathbf{r},t)|\Psi\rangle, \tag{10.11}$$

where s is the sensitivity of the detector. A most important quantity for modern quantum optics relates to the rate of double photodetections at $(\mathbf{r},t)$ and $(\mathbf{r}',t')$, which is defined by

$$d^2\mathcal{P} = w^{(2)}(\mathbf{r},t;\mathbf{r}',t')dt\,dt', \tag{10.12}$$

where $d^2\mathcal{P}$ is the probability of a double photodetection per unit surface around $\mathbf{r}$ during the time interval $[t, t + dt]$ and per unit surface around $\mathbf{r}'$ during $[t', t' + dt']$, with

$$w^{(2)}(\mathbf{r},t;\mathbf{r}',t') = s^2\langle\Psi|\hat{\mathbf{E}}^{(-)}(\mathbf{r},t)\hat{\mathbf{E}}^{(-)}(\mathbf{r}',t')\hat{\mathbf{E}}^{(+)}(\mathbf{r}',t')\hat{\mathbf{E}}^{(+)}(\mathbf{r},t)|\Psi\rangle. \tag{10.13}$$

The formalism above, and in particular the rates of single or double detections, will allow us to compare the properties of one-photon pulses with other types of lights: attenuated classical light pulses, attenuated laser beams, and light emitted from discharge lamps, attenuated or not.

Remark. Formulae (10.11) and (10.13) look similar to the semi-classical expressions for a classical electromagnetic field

$$\mathbf{E}_{\text{cl}}(\mathbf{r},t) = \mathbf{E}^{(-)}(\mathbf{r},t) + \mathbf{E}^{(+)}(\mathbf{r},t), \tag{10.14}$$

where $\mathbf{E}^{(+)}(\mathbf{r},t)$ is the complex amplitude of the field, and $\mathbf{E}^{(-)}(\mathbf{r},t)$ its complex conjugate. The rates of single and double photodetections are indeed, in the semi-classical point of view,

$$w^{(1)}(\mathbf{r},t) = s\mathbf{E}^{(-)}(\mathbf{r},t) \cdot \mathbf{E}^{(+)}(\mathbf{r},t) = \eta |\mathbf{E}^{(+)}(\mathbf{r},t)|^2 \tag{10.15}$$

and

$$w^{(2)}(\mathbf{r},t;\mathbf{r}',t') = s^2 |\mathbf{E}^{(+)}(\mathbf{r},t)|^2 |\mathbf{E}^{(+)}(\mathbf{r}',t')|^2. \tag{10.16}$$

The semi-classical and quantum expressions are, however, dramatically different both technically and conceptually. In the quantum formalism, the non-commutation of $\hat{\mathbf{E}}^{(-)}$ and $\hat{\mathbf{E}}^{(+)}$ entails the fact that the probability of a double detection is null for a single photon, as can be seen in Section 10.2.5. Such a statement does not hold in the semi-classical point of view. More generally, in the fully quantum point of view, observation of a photoelectron at time t and location $\mathbf{r}$ is associated with a photon being detected at $(\mathbf{r},t)$. If the photodetector is perfect, each detected photoelectron is therefore associated with a photon. In other words, the statistical distribution of the photoelectrons reflects the statistical distribution of the photons in the beam. This is in contrast to the semi-classical point of view, where there is no photon, and the discrete and probabilistic character of the photodetection signals stems from the discretization of the electric charge, or equivalently from the discontinuous and probabilistic character of the photodetection process itself, while the classical light intensity $|\mathbf{E}^{(+)}(\mathbf{r},t)|^2$ is a continuous quantity.

10.2.3 One-Photon Wavepacket

Any light state of the form

$$|1\rangle = \sum_{\ell} c_\ell |n_\ell = 1\rangle \tag{10.17}$$

is an eigenstate of $\hat{N}$ (see Eq. (10.10)) corresponding to the eigenvalue 1. It is a one-photon state. As a model of such a state in a collimated beam, we consider a one-photon state consisting of modes all propagating along the same direction defined by the unit vector $\mathbf{u}$, i.e.,

$$\mathbf{k}_\ell = \mathbf{u}\frac{\omega_\ell}{c}. \tag{10.18}$$

Equation (10.11) then gives the rate of photodetections:

$$w^{(1)}(\mathbf{r},t) = \eta \left\| \sum_\ell \vec{\varepsilon}_\ell \mathcal{E}_\ell^{(1)} c_\ell \exp\left[-i\omega_\ell \left(t - \frac{\mathbf{r}\cdot\mathbf{u}}{c}\right)\right] |0\rangle \right\|^2$$

$$= \eta \left| \sum_\ell \vec{\varepsilon}_\ell \mathcal{E}_\ell^{(1)} c_\ell \exp\left[-i\omega_\ell \left(t - \frac{\mathbf{r}\cdot\mathbf{u}}{c}\right)\right] \right|^2, \qquad (10.19)$$

which suggests a propagation along $\mathbf{u}$ at velocity c.

To simplify formulae, we write most often such quantities at $\mathbf{r} = 0$. The expression at $\mathbf{r}$ is readily obtained replacing t by $t - \mathbf{r}\cdot\mathbf{u}/c$.

To be more specific, let us consider the case of a Lorentzian distribution for $|c_\ell|^2$, which happens to describe light emitted by two-level-like single emitters, such as single atoms. More precisely, we take the form

$$c_\ell(t_j) = \frac{K_1}{\omega_\ell - \omega_0 + i\Gamma/2}, \qquad (10.20)$$

with

$$K_1 = |K_1| \exp\left(i\omega_\ell t_j\right), \qquad (10.21)$$

such that the state vector is normalized. For simplification, we take all modes to have the same polarization,

$$\vec{\varepsilon}_\ell = \vec{\varepsilon}. \qquad (10.22)$$

For L large enough, the sum in (10.19) can be transformed into an integral using the density of modes $\rho(\omega)$. If Γ is small compared to ω_0, the quantities $\mathcal{E}_\ell^{(1)}, \rho(\omega)$, and $|K_1|$ can be considered constant in the integral, with their values at ω_0. The remaining integral can be calculated by integration in the complex plane, yielding:

$$E_{t_j}^{(1)}(t) = \sum_\ell \mathcal{E}_\ell^{(1)} c_\ell e^{-i\omega_\ell t} = \rho(\omega_0)\mathcal{E}_{\omega_0}^{(1)}|K_1| \int d\omega \frac{\exp[-i\omega(t-t_j)]}{\omega - \omega_0 + i\Gamma/2}$$

$$= -2i\pi\rho(\omega_0)\mathcal{E}_{\omega_0}^{(1)} K_1 \mathrm{H}(\tau) \exp\left[\left(-\frac{\Gamma}{2} - i\omega_0\right)(t-t_j)\right]$$

$$= E_0 \mathrm{H}(t-t_j) \exp\left[\left(-\frac{\Gamma}{2} - i\omega_0\right)(t-t_j)\right], \qquad (10.23)$$

where $\mathrm{H}(t)$ is the Heaviside step function. The rate of photodetection at $\mathbf{r} = 0$ is then

$$w^{(1)}(0,t) = \eta |E_{t_j}^{(1)}(t)|^2 = \eta |E_0|^2 \mathrm{H}(t-t_j) \exp[-\Gamma(t-t_j)]. \qquad (10.24)$$

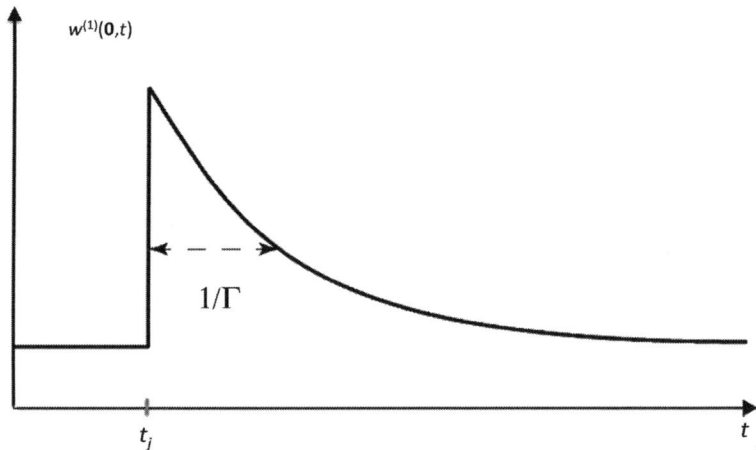

FIGURE 10.1 Average rate of photodetection at point at $\mathbf{r} = 0$ as a function of time, as given by Eq. (10.24) for a one-photon wavepacket with a leading edge at t_j. The rate of photodetection at point at $\mathbf{r}$ would be similar, with t_j replaced by $t_j + \mathbf{r} \cdot \mathbf{u}/c$.

Normalization of the state (10.17) with the coefficients c_ℓ given by (10.20) yields the condition

$$1 = \sum_\ell |c_\ell|^2 = \int d\omega\, \rho(\omega) \frac{|K_1|^2}{(\omega - \omega_0)^2 + \Gamma^2/4}$$
$$= \frac{2\pi}{\Gamma} \rho(\omega_0)|K_1|^2. \quad (10.25)$$

Hence

$$E_0 = -i\mathcal{E}_{\omega_0}^{(1)}[2\pi\rho(\omega_0)\Gamma]^{1/2}. \quad (10.26)$$

The rate of photodetection (10.24) at point $\mathbf{r} = 0$ is represented as a function of time in Fig. 10.1. It clearly suggests a wavepacket with a leading-edge at $t = t_j$, exponentially damped with a time constant Γ^{-1}. The result (10.24), however, must be understood in a statistical sense. One prepares a field in the form defined by Eqs. 10.17–10.20 at time $t = 0$, and one looks for photodetection by a detector at position $\mathbf{r}$. When the photodetection happens, its time is recorded. The experiment is repeated a great number of times, and the histogram of the results looks as shown in Fig. 10.1.

The one-photon state (10.17), of the form defined by (10.20), with (10.25), can be called a "one-photon wavepacket" with a leading-edge at t_j. We thus introduce the notation

$$|1(t_j)\rangle = \sum_\ell \frac{|K_1| \exp\left(i\omega_\ell t_j\right)}{\omega_\ell - \omega_0 + i\Gamma/2} |1_\ell\rangle. \quad (10.27)$$

The ensemble of these states, for all possible leading-edge times t_j, has properties that allow us to use them as a basis for all single-photon states. To show this, we establish a closure relation. We first introduce a *constant* density of states ρ_j, which has the dimension of an inverse time. We can then write

$$\sum_j |1(t_j)\rangle\langle 1(t_j)| = \rho_j \int dt_j |1(t_j)\rangle\langle 1(t_j)|. \tag{10.28}$$

Let us now express $|1(t_j)\rangle$ replacing $\sum_\ell$ by an integral in (10.17), with (10.20) and the density of states $\rho(\omega_\ell)$. We obtain

$$\sum_j |1(t_j)\rangle\langle 1(t_j)| = \rho_j \int dt_j \iint d\omega_\ell\, d\omega_{\ell'}\, \rho(\omega_\ell)\rho(\omega_{\ell'})$$

$$\times \frac{|K_1|^2 \exp i(\omega_\ell - \omega_{\ell'})t_j}{\left(\omega_\ell - \omega_0 + i\frac{\Gamma}{2}\right)\left(\omega_{\ell'} - \omega_0 - i\frac{\Gamma}{2}\right)}. \tag{10.29}$$

Using the fact that

$$\int dt_j\, e^{i(\omega_\ell - \omega_{\ell'})t_j} = 2\pi\, \delta(\omega_\ell - \omega_{\ell'}), \tag{10.30}$$

we obtain

$$\sum_j |1(t_j)\rangle\langle 1(t_j)| = \rho_j \int d\omega_\ell \frac{[\rho(\omega_\ell)]^2 |K_1|^2}{(\omega_\ell - \omega_0)^2 + \Gamma^2/4}$$

$$= \rho_j [\rho(\omega_0)]^2 |K_1|^2 \frac{2\pi}{\Gamma} \tag{10.31}$$

(as above, we take $\rho(\omega_\ell)$ constant over the bandwidth Γ around ω_0). Recalling (10.25), we finally have

$$\frac{1}{\rho_j \rho(\omega_0)} \sum_j |1(t_j)\rangle\langle 1(t_j)| = 1, \tag{10.32}$$

which can be used as a closure relation to expand single-photon states.

On the other hand, the $|1(t_j)\rangle$ only obey an approximate orthogonality relation:

$$\langle 1(t_j)|1(t_{j'})\rangle = \sum_\ell \sum_{\ell'} \frac{|K_1|^2 \exp(i\omega_\ell t_j) \exp(-i\omega_{\ell'} t_{j'})}{\left(\omega_\ell - \omega_0 + i\frac{\Gamma}{2}\right)\left(\omega_\ell - \omega_0 - i\frac{\Gamma}{2}\right)} \langle 1_\ell | 1_{\ell'}\rangle$$

$$= \sum_\ell \frac{|K_1|^2 \exp[i\omega_\ell(t_j - t_{j'})]}{(\omega_\ell - \omega_0)^2 + \Gamma^2/4} = |K_1|^2 \rho(\omega_0)$$

$$\times \int d\omega_\ell \frac{e^{i\omega_\ell(t_j - t_{j'})}}{(\omega_\ell - \omega_0)^2 + \Gamma^2/4}$$

$$= \frac{2\pi}{\Gamma} |K_1|^2 \rho(\omega_0) e^{i\omega_0(t_j - t_{j'})} e^{-\frac{\Gamma}{2}|t_j - t_{j'}|}. \quad (10.33)$$

Using (10.25) once more, we find

$$\langle 1(t_j)|1(t_{j'})\rangle = e^{i\omega_0(t_j - t_{j'})} e^{-\frac{\Gamma}{2}|t_j - t_{j'}|}. \quad (10.34)$$

This relation, as well as the closure relation (10.32), show that the ensemble of states $|1(t_j)\rangle$ is an overcomplete basis [12]. This is used in Section 10.2.7.

Remark. To obtain a result that is meaningful for a real experiment, we need to take into account the transverse profile of the light beam. A simple model that captures all the necessary features makes use of "top hat" modes, transversely homogeneous over a surface S, and with an arbitrary length L (along $\mathbf{u}$) (see for instance [17, Section 5B.1.2]). We have then

$$\left[\mathcal{E}_\omega^{(1)}\right]^2 = \frac{\hbar\omega}{2\varepsilon_0 LS} \quad (10.35)$$

$$\rho(\omega) = \frac{L}{2\pi c}. \quad (10.36)$$

Substituting in (10.26), we obtain an expression independent of L

$$E_0 = \sqrt{\frac{\hbar\omega_0}{2\varepsilon_0 Sc\Gamma^{-1}}}. \quad (10.37)$$

Note that this is the amplitude for a single photon in a volume $Sc\Gamma^{-1}$.

If the detector covers the whole beam, we must integrate $w^{(1)}$ [Eq. (10.24)] over S to obtain the probability of detection per unit time at position $\mathbf{r} = 0$, and we get

$$\frac{d\mathcal{P}^{(1)}}{dt} = \eta \frac{\hbar\omega_0}{2\varepsilon_0 c\Gamma^{-1}} \mathrm{H}\left(t - t_j\right) \exp[-\Gamma(t - t_j)]. \quad (10.38)$$

A perfect photodetector should detect a single photon with a probability of 1, i.e.,

$$\int dt \frac{d\mathcal{P}^{(1)}}{dt} = s \frac{\hbar\omega}{2\epsilon_0 c} = 1. \quad (10.39)$$

Hence, its sensitivity (in units of [electric field]$^{-2}$) per unit surface is

$$s_{\text{perfect}} = \frac{2\varepsilon_0 c}{\hbar\omega}. \quad (10.40)$$

10.2.4 Quasi-Classical Wavepacket

A fundamental reason for the success of the semi-classical model of matter-light interaction is the fact that most of the light sources available in everyday life, or even in laboratories, deliver light beams whose behavior can be fully described by the semi-classical model. In particular, the rates of single and joint photodetections can be expressed in terms of Eqs. (10.15) and (10.16). This can be understood, in the fully quantum optics formalism, by the fact that such light beams can be described by quantum states of radiation called *coherent states* or *quasi-classical states* [11,12]. A quasi-classical state $|\alpha_\ell\rangle$ of the mode ℓ is an eigenstate of the destruction operator $\hat{a}_\ell$

$$\hat{a}_\ell |\alpha_\ell\rangle = \alpha_\ell |\alpha_\ell\rangle, \tag{10.41}$$

with α_ℓ a complex number. A multimode quasi-classical state is

$$|\Psi_{qc}\rangle = |\alpha_{\ell=1}\rangle \otimes |\alpha_{\ell=2}\rangle \otimes \cdots \otimes |\alpha_\ell\rangle \otimes \cdots \tag{10.42}$$

This state is an eigenstate of the positive-frequency electric-field operator (10.1):

$$\hat{\mathbf{E}}^{(+)}(\mathbf{r},t)|\Psi_{qc}\rangle = \mathbf{E}_{cl}^{(+)}(\mathbf{r},t)|\Psi_{qc}\rangle \tag{10.43}$$

with the eigenvalue

$$\mathbf{E}_{cl}^{(+)}(\mathbf{r},t) = i \sum_\ell \mathcal{E}_\ell^{(1)} \alpha_\ell \vec{\varepsilon}_\ell \exp\{i(\mathbf{k}_\ell \cdot \mathbf{r} - \omega_\ell t)\}. \tag{10.44}$$

It turns out that $\mathbf{E}_{cl}^{(+)}(\mathbf{r},t)$ is the positive frequencies part of a classical field that we can associate with $|\Psi_{qc}\rangle$. One can then check by simple inspection that the rates of simple or double photodetection (10.11) or (10.13) obtained for the state (10.42) are identical to the ones obtained using the semi-classical expressions (10.15) and (10.16) with the classical field (10.44).

Such quasi-classical states—or, more generally, a statistical ensemble of states of the form (10.42)—allow one to describe, in the quantum optics formalism, the light emitted by what we will thus call *classical sources*, for instance a thermal source, or a laser operated well above threshold (see Section 10.2.6). But they also allow us to build quasi-classical wavepackets that lead to the same probability of single detections as the one-photon wavepackets considered in Section 10.2.3. To show this, we take again the case of propagation along $\mathbf{u}$

$$\mathbf{k}_\ell = \mathbf{u}\frac{\omega_\ell}{c}, \tag{10.45}$$

with a single polarization

$$\vec{\varepsilon}_\ell = \vec{\varepsilon}. \tag{10.46}$$

We then assume the α_ℓ's have a distribution

$$\alpha_\ell = \frac{K_{qc}}{\omega_\ell - \omega_0 + i\Gamma/2}, \tag{10.47}$$

and we take K_{qc} to be real, for simplicity. We can then calculate explicitly the quasi-classical field (10.44). As in Section 10.2.3, we replace the sum by an integral, using the density of states $\rho(\omega)$. Integration in the complex plane yields

$$\mathbf{E}_{cl}^{(+)}(\mathbf{r},t) = \vec{\varepsilon}\, E_0 \mathrm{H}\left(t - \frac{\mathbf{r}\cdot\mathbf{u}}{c}\right) \exp\left\{-\frac{\Gamma}{2}\left(t - \frac{\mathbf{r}\cdot\mathbf{u}}{c}\right)\right\} \exp\left\{-i\omega_0\left(t - \frac{\mathbf{r}\cdot\mathbf{u}}{c}\right)\right\} \quad (10.48)$$

with

$$E_0 = -i\, 2\pi \rho(\omega_0) \mathcal{E}_{\omega_0}^{(1)} K_{qc}. \quad (10.49)$$

Since $\mathbf{E}_{cl}^{(+)}$ is the eigenvalue of $\hat{\mathbf{E}}^{(+)}$ associated with the radiation state (10.42), the rate of single photodetections (10.11) can be written as

$$w^{(1)}(\mathbf{r},t) = \eta |\mathbf{E}_c(\mathbf{r},t)|^2 = \eta |E_0|^2 \mathrm{H}\left(t - \frac{\mathbf{r}\cdot\mathbf{u}}{c}\right) \exp\left\{-\Gamma\left(t - \frac{\mathbf{r}\cdot\mathbf{u}}{c}\right)\right\}. \quad (10.50)$$

Like (10.24) (with t replaced by $t - \mathbf{r}\cdot\mathbf{u}/c$), Eq. (10.50) suggests the propagation of a wavepacket damped with a time constant Γ^{-1}. However, the quasi-classical wavepacket introduced here differs in many aspects from the one-photon wavepacket of Section 10.2.3. The most striking difference can be seen in Section 10.2.5. Here, we note that the quasi-classical state $|\Psi_{qc}\rangle$ is not an eigenstate of the number of photons operator $\hat{N}$. More precisely, it can be shown that if we were to measure the photon number in such a state, we would find a Poisson distribution. One can readily calculate the average of that distribution, i.e., the average photon number

$$\langle N \rangle_{qc} = \langle \Psi_{qc} | \hat{N} | \Psi_{qc} \rangle = \sum_\ell |\alpha_\ell|^2, \quad (10.51)$$

and its standard deviation

$$\langle \Delta N \rangle_{qc} = \left[\langle \hat{N}^2 \rangle - (\langle \hat{N}\rangle)^2\right]^{1/2} = \left[\langle N \rangle_{qc}\right]^{1/2}. \quad (10.52)$$

Using a method similar to the one yielding Eq. (10.25), we can express (10.51) as

$$\langle N \rangle_{qc} = \int d\omega\, \rho(\omega) \frac{|K_{qc}|^2}{(\omega - \omega_0)^2 + \Gamma^2/4} = \frac{2\pi}{\Gamma} \rho(\omega_0) |K_{qc}|^2. \quad (10.53)$$

It is important to realize that the constant K_{qc} (or equivalently, the average photon number) can be chosen arbitrarily (contrary to constant K_1 in the case of a one-photon wavepacket). A quasi-classical wavepacket thus can be built with any average photon number. In particular, K_{qc} can be chosen small enough to get an average photon number smaller than one. Such a state is the quantum description of a quasi-classical pulse that has been strongly attenuated by a neutral density filter.

Remark. Using the results above, we find that the amplitude E_0 of Eq. (10.49) assumes a form similar to (10.26), with the right-hand side multiplied by $[\langle N \rangle_{\text{qc}}]^{1/2}$

$$E_0 = -i[\langle N \rangle_{\text{qc}}]^{1/2} \mathcal{E}^{(1)}_{\omega_0} [2\pi \, \rho(\omega_0) \Gamma]^{1/2} e^{-i\omega_0 t_0}. \quad (10.54)$$

Taking the same set of top-hat modes as in the remark of Section 10.2.3, we find a rate of photodetection

$$w^{(1)}(\mathbf{r},t) = s \frac{\hbar \omega}{2\varepsilon_0 c} \frac{\Gamma}{S} \langle N \rangle_{\text{qc}} \mathrm{H}\left(t - \frac{\mathbf{r} \cdot \mathbf{u}}{c}\right) \exp\left\{-\Gamma \left(t - \frac{\mathbf{r} \cdot \mathbf{u}}{c}\right)\right\}. \quad (10.55)$$

An integration over the whole section S of the beam, and over time, yields the average number of photoelectrons

$$\iint d^2 S \int dt \, w^{(1)}(\mathbf{r},t) = s \frac{\hbar \omega}{2\varepsilon_0 c} \langle N \rangle_{\text{qc}}. \quad (10.56)$$

For a perfect detector, of sensitivity given by (10.40), the average number of counts is equal to the average number of photon $\langle N \rangle_{\text{qc}}$, as expected.

10.2.5 The Possibility of an Experimental Distinction

We now compare the predictions of quantum optics for a one-photon wavepacket and for a quasi-classical wavepacket. Equations (10.24) and (10.50) show that if we measure the instants of photodetection for wavepackets whose time of emission is known, and build the histogram of the delays between the emission and the photodetection, the results for single-photon wavepackets and quasi-classical wavepackets are similar. Measurements of $w^{(1)}(\mathbf{r},t)$ therefore do not allow us to distinguish between a one-photon wavepacket and a quasi-classical wavepacket. Actually, it is well known that when a distinction between classical light and quantum light is possible, it cannot be observed on single detection signals, but rather on double detection signals [34]. We thus calculate the probability of double detections for both cases.

In the case of a quasi-classical wavepacket of the form (10.42), we again use the fact that it is an eigenstate of $\hat{\mathbf{E}}^{(+)}(\mathbf{r},t)$ and obtain from (10.13)

$$w^{(2)}(\mathbf{r},t; \mathbf{r}',t') = \eta^2 |\mathbf{E}_{\text{cl}}(\mathbf{r},t)|^2 |\mathbf{E}_{\text{c}}(\mathbf{r}',t')|^2. \quad (10.57)$$

The probability of a double detection is the product of the probabilities of the single detections. The detection events are uncorrelated. This is the same result as would be obtained in the semi-classical model of matter-light interaction, for a wavepacket with Fourier components distributed as the α_ℓ's.

Let us now consider the case of a single-photon wavepacket of the form (10.17). We have now

$$\hat{\mathbf{E}}^{(+)}(\mathbf{r},t)|1\rangle = \left[\sum_\ell \vec{\varepsilon}_\ell \mathcal{E}^{(1)}_\ell c_\ell \exp\left\{i\omega_\ell \left(\frac{\mathbf{r} \cdot \mathbf{u}}{c} - t\right)\right\}\right] |0\rangle \quad (10.58)$$

and therefore
$$\hat{\mathbf{E}}^{(+)}(\mathbf{r}',t')\hat{\mathbf{E}}^{(+)}(\mathbf{r},t)|1\rangle = 0 \tag{10.59}$$
since $\hat{a}_\ell |0\rangle = 0$. We conclude that
$$w^{(2)}(\mathbf{r},t;\mathbf{r}',t') = 0. \tag{10.60}$$

The probability of a double detection is thus strictly null in the case of a single-photon wavepacket. This property ("*anticorrelation*") is not surprising if one remembers that the number of photons is a good quantum number, and its value is 1. Since a photodetection amounts to destroying a photon, there is no photon left to allow for a second detection.

In contrast, in a semi-classical wavepacket the number of photons is not a good quantum number, since $|\Psi_{qc}\rangle$ is not an eigenstate of $\hat{N}$, and the probability to have two photons is not null. It is therefore not surprising that one can have two photodetections.

This difference allows one to make an experimental distinction between a true single-photon wavepacket, and a quasi-classical wavepacket, even when attenuated enough that the average number of photons is much less than 1. One can then ask: can such a difference be observed, when one takes into account experimental inefficiencies and noise? We will see in Section 10.3 that it is indeed possible to establish a quantitative criterion that renders the distinction presented above fully operational, leading to practical tests in realistic experiments. But before addressing that question, we will ask, still from a theoretical point of view, whether various kinds of strongly attenuated light beams may exhibit an anomalously small rate of double photodetection, by comparison to what is expected for a classical wave.

10.2.6 Attenuated Continuous Light Beams

In this subsection, we consider the case of a continuous beam emitted by a CW laser, or even a thermal source, attenuated to the point where the average power is so weak that if we insist to describe the beam as made of photons, the average distance between these photons would be large compared to a standard interferometric system (say several meters).

Let us start with the simplest case, the beam emitted by a perfectly stable single-mode laser, of average power P_{Laser}. It is well known [34] that such a beam is well described by a quasi-classical state $|\alpha_{\text{Laser}}\rangle$ of the mode associated with the laser beam. Even in an ideal laser, the complex number α_{Laser} has some fluctuations due to spontaneous emission [38], but the fluctuations of the modulus $|\alpha_{\text{Laser}}|$ can be considered negligible, provided the laser operates well above threshold.

A laser beam has a non-uniform transverse profile (for instance, a Gaussian profile for the fundamental transverse mode), and one should use the corresponding non-uniform modes of the electromagnetic field to correctly describe the quantized field associated with the laser beam. To simplify, we use again the top-hat modes introduced in the remark of Section 10.2.3, with a transverse profile uniform over an area S_{Laser}. The volume of quantization is then $S_{\text{Laser}} \times L$, where L is an arbitrary length along the beam axis, which can be taken as large as necessary. The single-photon amplitude $\mathcal{E}^{(1)}_{\text{Laser}}$ then assumes the value (10.35) with S replaced by S_{Laser}, and the density of modes has the value (10.36). The modulus of α_{Laser} is related to the average number of photons in the quantization volume by

$$\langle N \rangle_{\text{qc}} = |\alpha_{\text{Laser}}|^2 = \frac{P_{\text{Laser}}}{\hbar \omega_{\text{Laser}}} \frac{L}{c}. \tag{10.61}$$

Since $|\alpha_{\text{Laser}}\rangle$ is an eigenstate of the positive-frequency electric-field operator (10.1), the calculation of the single and joint photodetections is trivial (cf. Section 10.2.4). The rate of single photodetections is uniform in the profile, and is equal to

$$w^{(1)}(\mathbf{r},t) = \eta \left[\mathcal{E}^{(1)}_\ell \right]^2 |\alpha_{\text{Laser}}|^2. \tag{10.62}$$

Replacing $\mathcal{E}^{(1)}_\ell$ by its value, and assuming a perfect detector, we obtain

$$w^{(1)}(\mathbf{r},t) = \eta_{\text{perfect}} \frac{\hbar \omega}{2\varepsilon_0 L S} |\alpha_{\text{Laser}}|^2 = |\alpha_{\text{Laser}}|^2 \frac{c}{L}, \tag{10.63}$$

i.e., according to (10.61), the average number of photons per unit time, as it should be.

The density of double detections is also uniform

$$w^{(2)}(\mathbf{r},t; \mathbf{r}',t') = \eta^2 \left[\mathcal{E}^{(1)}_\ell \right]^4 |\alpha_{\text{Laser}}|^4. \tag{10.64}$$

Moreover, we see that

$$w^{(2)}(\mathbf{r},t; \mathbf{r}',t') = w^{(1)}(\mathbf{r},t) \cdot w^{(1)}(\mathbf{r}',t'). \tag{10.65}$$

This means that the detection events are independent from each other. If we take a perfect photodetector that detects every photon, we thus conclude that the photons are randomly distributed in time with a uniform probability density. This property remains true even for an attenuated beam, whatever the level of attenuation, since this only amounts to reducing the magnitude $|\alpha_{\text{Laser}}|$. This property is equivalent to the fact that if one looks for the statistics of photodetections in a given time interval, we expect to find a Poisson distribution.

If now we consider thermal light, it can be considered constituted by a statistical ensemble of quasi-classical states associated with a continuum of modes of the electromagnetic field. Reasoning as in Section 10.2.5, the calculation can be done using the semi-classical model of matter-light interaction, for a classical stochastic field [39]. One can then show, using a standard Cauchy-Schwartz inequality, that the rates of single and double photodetections, calculated according to formulae (10.15) and (10.16), obey the relation

$$w^{(2)}(\mathbf{r},t;\mathbf{r},t) \geq (w^{(1)}(\mathbf{r},t))^2. \tag{10.66}$$

We can thus conclude that such light beams, even strongly attenuated, never lead to a null rate of double detection. The situation is thus explicitly different from what happens with genuine single-photon wavepackets (Section 10.2.5). This conclusion remains valid when we use the criterion that is derived in Section 10.3.3, which applies to real experiments.

10.2.7 Light From a Discharge Lamp

We consider now light emitted by a source constituted of many independent emitters, each emitting one-photon wavepacket, at random times. The light is collimated, and we can thus describe the radiation state as constituted of many independent one-photon wavepackets introduced in Section 10.2.3. We call μ the average number of single photons per unit time, and we consider a time interval T in which we have $N = \mu T$ wavepackets. We will then write the radiation state as

$$|\Psi_N\rangle = |1(t_1)\rangle \otimes |1(t_2)\rangle \cdots \otimes |1(t_N)\rangle. \tag{10.67}$$

In writing this expression, which reflects the fact that the one-photon wavepackets are independent, we assume that the $|1(t_j)\rangle$ states are orthogonal, i.e., the second member of Eq. (10.34) is replaced by $\delta_{jj'}$. This is reasonable if the wavepackets are produced by the same emitter (since then there is a delay between them), or if they are emitted by different emitters with frequencies ω_0 that are not exactly the same because of Doppler effect or inhomogeneous broadening.

The ensemble of the one-photon states $\{|1(t_1)\rangle, \ldots, |1(t_N)\rangle\}$ can then be considered a basis for the Fock space of any combination of such one-photon states. It is then convenient to define creation and destruction operators $\hat{a}^\dagger(t_\ell)$ and $\hat{a}(t_\ell)$ ($\ell = 1, \ldots, N$) such that

$$\hat{a}^\dagger(t_\ell)|0\rangle = |1(t_\ell)\rangle \tag{10.68}$$

and

$$\left[\hat{a}(t_\ell), \hat{a}^\dagger(t_{\ell'})\right] = \delta_{\ell\ell'}. \tag{10.69}$$

The state (10.67) can then be written as

$$|\Psi_N\rangle = \hat{a}^\dagger(t_1)\hat{a}^\dagger(t_2)\ldots\hat{a}^\dagger(t_n)|0\rangle. \tag{10.70}$$

The restriction of the electric-field operator $\hat{\mathbf{E}}^{(+)}(\mathbf{r},t)$ to that space can be written as

$$\hat{\mathbf{E}}_N^{(+)}(\mathbf{r},t) = \vec{\varepsilon} \sum_{\ell=1}^{N} E_{t_\ell}^{(1)}(t)\hat{a}(t_\ell). \tag{10.71}$$

To determine the probability of a single detection per unit time, we need to calculate

$$\hat{\mathbf{E}}_N^{(+)}(0,t)|\Psi_N\rangle = \vec{\varepsilon}\, E_{t_1}^{(1)}(t)|1(t_2)\rangle \otimes |1(t_3)\rangle \otimes \cdots \otimes |1(t_N)\rangle$$
$$+ \vec{\varepsilon}\, E_{t_2}^{(1)}(t)|1(t_1)\rangle \otimes |1(t_3)\rangle \otimes \cdots$$
$$+ \cdots$$
$$= \vec{\varepsilon} \sum_{\ell=1}^N E_{t_\ell}^{(1)} \underbrace{\otimes_{j\neq\ell} |1(t_j)\rangle}_{N-1 \text{ terms}}. \tag{10.72}$$

The state above is a state with $N-1$ photons. Taking its modulus and using (10.24), we obtain

$$w^{(1)}(0,t) = \eta \sum_{\ell=1}^N |E_{t_\ell}^{(1)}(t)|^2$$
$$= \eta |E_0|^2 \sum_{\ell=1}^N \mathrm{H}(t-t_\ell)\exp[-\Gamma(t-t_\ell)]. \tag{10.73}$$

Actually, with such a source we cannot measure the quantity above, since even for an ideal detector we have at most one detection per wavepacket. If we repeat the experiment, and select another interval with N one-photon wavepackets, the distribution of the times $\{t_1, \ldots, t_\ell, t_N\}$ will be different, so that the result after a large number of such experiments is obtained by averaging over each t_ℓ distributed uniformly in the interval $T = N\mu^{-1}$. The result of that averaging is constant in time

$$\overline{w^{(1)}} = \eta |E_0|^2 \mu \Gamma^{-1}. \tag{10.74}$$

Integrating over the whole area of an ideal detector, and reasoning as in Eqs. 10.35–10.40, we obtain an average probability of detection per unit time

$$\frac{\overline{d\mathcal{P}^{(1)}}}{dt} = \mu. \tag{10.75}$$

To evaluate the probability of double detection, we apply the operator $\hat{\mathbf{E}}_N^{(+)}$ to (10.72):

$$\sum_{\ell=1}^N E_{t_\ell}^{(1)}(t) \sum_{p\neq\ell} E_{t_p}^{(1)}(t) \underbrace{\otimes_{j\neq p,\ell} |1(t_j)\rangle}_{N-2 \text{ terms}}. \tag{10.76}$$

In the sum above, the two terms obtained by exchanging ℓ and p are identical, and we can thus write

$$\hat{\mathbf{E}}_N^{(+)}(0,t)\hat{\mathbf{E}}_N^{(+)}(0,t)|\Psi_N\rangle = 2\sum_{\ell=1}^N E_{t_\ell}^{(1)}(t) \sum_{p>\ell} E_{t_p}^{(1)}(t) \otimes_{j\neq p,\ell} |1,t_j\rangle. \tag{10.77}$$

Taking its square modulus, we obtain

$$w^{(2)}(t,t) = 4\eta^2 \sum_{\ell=1}^{N} \sum_{p>\ell} |E^{(1)}(t_\ell)|^2 |E^{(1)}(t_p)|^2. \qquad (10.78)$$

We again average over all t_ℓ and t_p in the interval T, and we obtain

$$\overline{w^{(2)}(t,t)} = 2\frac{N(N-1)}{N^2}\left[\overline{w^{(1)}}\right]^2, \qquad (10.79)$$

where $\overline{w^{(1)}}$ is given in (10.74). If the number of photons is large enough, we have $\overline{w^{(2)}} = 2\left[\overline{w^{(1)}}\right]^2$. The factor 2 is the celebrated Hanbury Brown and Twiss factor.

We thus find that there is no possibility to observe an anticorrelation effect with light emitted from a discharge lamp, even if the one-photon wavepackets are well separated from each other. The reason is that the various wavepackets are emitted at random times independent from each other, and there is a significant probability to have two wavepackets arriving at the same time.

Remark. If we make $N = 1$ in Eq. (10.79), we find $\overline{w^{(2)}} = 0$. It does not mean that if we take a small enough time interval we can expect to observe $\overline{w^{(2)}} = 0$. Indeed, for a source emitting one-photon wavepackets at random times the number N is not strictly fixed, it is in fact distributed according to a Poisson law. A calculation averaging over that distribution would give $\overline{w^{(2)}}(t,t) = 2\left[\overline{w^{(1)}}\right]^2$, whatever the interval.

10.2.8 Conclusion: What is Single-Photon Light?

In this section, we have shown that a genuine single-photon wavepacket, i.e., a one-photon state emitted at a well-known time, exhibits a characteristic behavior, the fact that it cannot be detected jointly by two photodetectors (anticorrelation). Such a behavior is not expected in the case of an attenuated beam from a classical lamp, including the case of a discharge lamp where one has single-photon wavepackets shorter than the average time between the wavepackets. The paradoxical behavior in the latter case is resolved when we realize that the problem is the fact that one has no information about the time when any single-photon wavepacket is emitted. We can thus conclude that it is not enough to have single-photon wavepackets to have single-photon light. We need in addition to know *at which time each single-photon wavepacket is emitted* [40]. This is the case for the two types of sources described below: *heralded single-photon sources* on the one hand, and *on-demand single-photon sources* on the other hand.

10.3 PHOTON PAIRS AS A RESOURCE FOR SINGLE PHOTONS

10.3.1 Introduction

When an atom emits from an excited level, the fluorescent light emitted is a one-photon wavepacket, as can be guessed merely from energy conservation. However, in usual sources, such as discharge lamps, many excited atoms are seen simultaneously by a detector, and their time of excitation is random (Section 10.2.7). The theoretical description of the light then is a mixture of one-photon wavepackets of the form presented in Section 10.2, with random leading-edge times. If one also takes into account the fluctuations of the number of excited atoms, the emitted light can be considered a statistical ensemble of quasi-classical states, and in this situation there is no hope to observe any non-classical effects. To observe non-classical properties in fluorescent light, it is necessary to isolate single-atom emissions, either in space, or in time. This can be done in sources of "heralded" single photons, based on the emission of separated pairs of photons: the first photon then "heralds" the emission of the second one, allowing one to isolate single-atom emission in time.

The production of pairs of photons occurs in many different contexts in physics, including particle physics (e.g., the electron-positron annihilation, producing two γ photons), nuclear physics and atomic physics (through cascade de-excitation between several levels), and non-linear optics (pair production in spontaneous parametric fluorescence). In the latter case, the temporal correlation between the two photons of one pair, a fully quantum property, was observed first in 1970 by Burnham and Weinberg [6] and studied more accurately by Hong *et al.* [41], while the specifically quantum properties of the photon pairs emitted by an atomic cascade were demonstrated in 1974 by Clauser (see Section 10.3.2).

However, it took some time to realize that a very simple way to understand these specifically quantum properties is to consider the quantum state of the light for the second photon only, once the first one has been detected: according to the "projection postulate" of quantum mechanics this second photon is in a state very close to a one-photon state, or more precisely, a one-photon wavepacket with a well-defined leading-edge time (or peak time). One can say that the second photon is "heralded" by the detection of the first [9]. This expression has become popular and is now used as a generic name for such sources.

In this section, we first present and discuss inequalities that apply to "classical" light, i.e., light described by the standard wave model of classical optics, or equivalently light described by the quantum theory as a statistical mixture of quasi-classical states. Such inequalities are derived for the case of an atomic cascade (Section 10.3.2), and for a single photon on a beamsplitter (Section 10.3.3). These inequalities are fully general in the classical context, and since quantum light can contradict them, they delineate a limit beyond which "specifically quantum effects" can be observed. In Section 10.3.4, we present the anticorrelation experiment that allowed us to conclude that our 1986 source was a true single-photon source, and we contrast this result with the one obtained with strongly attenuated classical light pulses.

10.3.2 Non-Classical Properties in an Atomic Cascade

In 1974, John Clauser proposed a scheme to obtain a "model-independent" inequality applying to any classical description of pairs of photons emitted by an atomic radiative cascade [3]. His idea was to "split" simultaneously both the first and second photons of the cascade, by collecting the light emitted on opposite sides of an assembly of excited atoms, and focusing it separately into two beams. The wavelength λ_1 on one side was selected to correspond to that of the first transition of the cascade, and that on the other, λ_2, to the second. The two light beams impinged on beamsplitters, thus creating a total of four beams, between which coincidence rates of photodetections are measured. The semi-classical expression of a coincidence rate between detectors i and j is (see Eq. (10.16))

$$C_{ij} = \eta_i \eta_j T^{-1} \int_{-T/2}^{T/2} \int_{-T/2}^{T/2} \langle I_i(t+t') I_j(t+t'') \rangle dt' \, dt'', \qquad (10.80)$$

where η_i, η_j are the detection efficiencies of the photodetectors, and I_i (respectively I_j) the classical light intensity $I_i = |\mathbf{E}_i^{(+)}(\mathbf{t})|^2$ at photodetector i (respectively j). The time integral bears over the duration T of the run, while the brackets denote a statistical average over many runs.

By using four photomultipliers labeled $\gamma_{1A}, \gamma_{1B}, \gamma_{2A}$, and γ_{2B}, the coincidence rates were monitored between the four combinations: $\gamma_{1A} - \gamma_{1B}, \gamma_{2A} - \gamma_{2B}, \gamma_{1A} - \gamma_{2B}$, and $\gamma_{2A} - \gamma_{1B}$. A diagram of the arrangement is shown in Fig. 10.2. Defining $I_1(t)$ and $I_2(t)$ as the instantaneous light intensities at the $\gamma_{1A} - \gamma_{1B}$ beam splitter with wavelength λ_1, and at the $\gamma_{2A} - \gamma_{2B}$ beam splitter with wavelength λ_2, respectively, it follows directly from the Cauchy-Schwarz inequality that the following inequality holds:

$$\left[\int_{-T/2}^{T/2} \int_{-T/2}^{T/2} \langle I_1(t+t'+\tau_1) I_1(t+t''+\tau_1) \rangle dt' \, dt'' \right]$$
$$\left[\int_{-T/2}^{T/2} \int_{-T/2}^{T/2} \langle I_2(t+t'+\tau_2) I_2(t+t''+\tau_2) \rangle dt' \, dt'' \right]$$
$$\geqslant \left[\int_{-T/2}^{T/2} \int_{-T/2}^{T/2} \langle I_1(t+t'+\tau_1) I_2(t+t''+\tau_2) \rangle dt' \, dt'' \right]^2.$$

Using the definition (10.80) of C_{ij}, this can be written as

$$C_{1A-1B}(0) C_{2A-2B}(0) \geqslant C_{1A-2B}(\tau) C_{1B-2A}(\tau). \qquad (10.81)$$

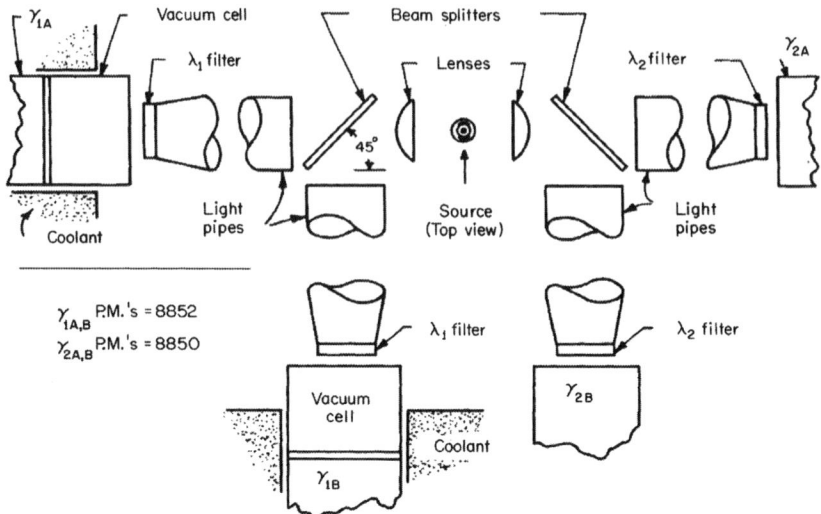

FIGURE 10.2 Schematic diagram of the apparatus used in John Clauser's 1974 experiment.

This simple calculation ignores a possible polarization dependence of the detectors, and the finite photocathode areas, as well as the nonvanishing phototube dark rates (c.f. Chapter 3). However, it can be shown that the above inequality is fully general and holds for these cases as well.

From a quantum point of view, the coincidence rates C_{1A-2B} and C_{2A-1B} are due to the strong temporal correlation between the two photons in each pair emitted by the cascade, and these coincidence rates can reach quite high values. On the other hand, the rates C_{1A-1B} and C_{2A-2B} require random coincidences between photons emitted by different atoms, and for a low-density source, such coincidence rates are much smaller than C_{1A-2B} and C_{2A-1B}. Therefore the above equality can be violated by a large amount, as has been confirmed by the experiment [3].

10.3.3 Anticorrelation for a Single Photon on a Beamsplitter

The main idea of the previous experiment is thus to compare "intra-beam" correlations (auto-correlations), and "inter-beam" correlations (cross-correlations), the first ones being always larger for classical beams, whereas the opposite situation happens for the quantum light emitted by an atomic cascade. This approach, however, does not directly exhibit the anticorrelation behavior that is only associated with a single photon. This is why we introduced the scheme of Fig. 10.3. In that scheme [42], the detection of the first photon of a radiative cascade fires a trigger that generates an electronic gate of duration τ_{gate}, synchronized with, and somewhat longer than, the decay constant Γ^{-1} of the one-photon wavepacket associated with the second photon of the cascade.

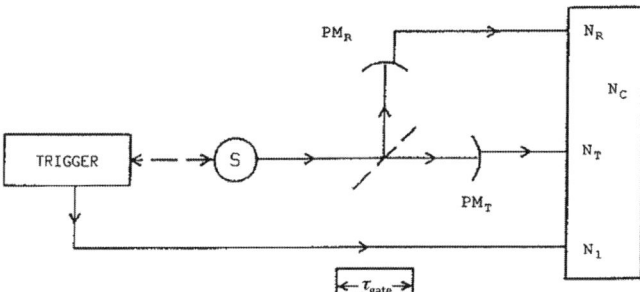

FIGURE 10.3 Experiment to look for an anticorrelation on a beam splitter. The source S emits light pulses that fall on a beam splitter and can be detected in both channels (reflected and transmitted) behind the beam splitter. The detectors are enabled during a gate τ_{gate} synchronized with the light pulses. The rates of single detection (N_R and N_T) and coincidence (N_C) are monitored. If the light pulse contains only one photon, one detection is expected at most, and no coincidence is expected: this is the anticorrelation effect. In sources of heralded single photons based on photon pairs, such as the ones emitted by the radiative cascade of Fig. 10.4, the trigger is activated by the detection of the other photon of the pair.

That single-photon wavepacket is launched toward a beamsplitter with two detectors in the transmitted and reflected legs, and these detectors are enabled only during the gates associated with the trigger, i.e., during the time interval corresponding to the single-photon wavepacket. If both detectors fire during the same gate, a coincidence is recorded. A counting system monitors the triggering events, the detection events, and the coincidences.

Consider an experiment that consists of running the source for a given duration and counting the total number of counts in the transmitted (N_T) or reflected (N_R) channels, the total number of coincidences (N_c), and the total number of gates N_1. We can then estimate the probabilities of single detections per gate,

$$P_R = \frac{N_R}{N_1} \quad \text{and} \quad P_T = \frac{N_T}{N_1}, \tag{10.82}$$

and the probability of a coincidence per gate,

$$P_c = \frac{N_c}{N_1}. \tag{10.83}$$

According to our intuition, we expect P_c to be zero in the ideal case of a one-photon wavepacket, and to be non-zero for a classical light pulse. As in the previous section, this discussion can be rephrased in the context of a comparison between the quantum theory of light and the semi-classical theories of light.

To establish classical inequalities, two equivalent approaches are possible: one is to consider the quantum state "heralded" by the first detection, and to look to the single and coincidence detections on both sides of the beamsplitter; the other one is to look "globally" at the cascade, so that a detection on one side of the beamsplitter is already a coincidence (between the "heralding" and "heralded" photons), whereas clicks on both sides of the beamsplitter will be a "triple" coincidence.

In the first approach, we define

$$\Omega = \tau_{\text{gate}}^{-1} \int_0^{\tau_{\text{gate}}} I_B(t+t')dt'$$

as the time-averaged (classical) intensity impinging on the beamsplitter during the counting window τ_{gate}. For many pulses, one finds

$$P_R = s_R \overline{\Omega}, \quad P_T = s_T \overline{\Omega}, \quad P_c = s_R s_T \overline{\Omega^2},$$

where s_T and s_R are the global detection efficiencies (including the transmission and reflection coefficients of the beamsplitter) of each detector, and the overbar indicates a statistical average over many pulses. From the Cauchy-Schwarz inequality $\overline{\Omega^2} \geq (\overline{\Omega})^2$ one gets $P_c \geq P_R P_T$ or equivalently

$$N_c \geq \frac{N_R N_T}{N_1}.$$

In the second approach we consider the quantity, where ξ is a real variable:

$$F(\xi) = \overline{I_A(t) \int_0^w \int_0^w (\xi + I_B(t+t'))(\xi + I_B(t+t''))dt'\, dt''}$$

$$= \overline{I_A(t) \left(\int_0^w (\xi + I_B(t+t'))dt' \right)^2}$$

$$= \xi^2 w^2 \overline{I_A(t)} + 2\xi w \overline{\int_0^w I_A(t)I_B(t+t')dt'} + \overline{I_A(t) \left(\int_0^w I_B(t+t')dt' \right)^2}.$$

Since $F(\xi) \geq 0$, one obtains the usual Cauchy-Schwarz inequality:

$$\overline{I_A(t)} \times \overline{I_A(t) \left(\int_0^w I_B(t+t')dt' \right)^2} \geq \left(\overline{\int_0^w I_A(t)I_B(t+t')dt'} \right)^2.$$

Reintroducing the appropriate detection-sensitivity factors $s_R s_T$ on both sides, one obtains:

$$N_1 N_c \geq N_R N_T,$$

which is the same as the inequality derived in the first approach. It is usually written

$$\alpha = \frac{P_c}{P_R P_T} = \frac{N_c N_1}{N_R N_T} \geq 1. \tag{10.84}$$

This inequality can be seen either as a property of the "heralded" wavepacket, or as a property of the correlation functions taking into account the "heralding" event, corresponding to $I_A(t)$. Its physical content is very close to the inequality (10.66), and its violation (i.e., $\alpha < 1$), also called "anticorrelation" [5]. As for the antibunching effect [13], the observation of such an anticorrelation is an evidence against the semi-classical theories of light.

Remark. In the limit where τ_{gate} is very small, the inequality (10.84) is strictly equivalent to (10.66), or to $g^{(2)}(0) \geq 1$, where $g^{(2)}(\tau)$ is the usual normalized second-order correlation function [35]. So the condition $\alpha \geq 1$ can be seen as an "integrated" version of $g^{(2)}(0) \geq 1$, over a time window suited to the duration of the single-photon wavepacket. As for the semi-classical inequality $g^{(2)}(0) \geq 1$, or the semi-classical inequality $g^{(2)}(0) \geq g^{(2)}(\tau)$ used in [13], its violation has some relation with sub-Poissonian photon statistics, but no statistics are measured here, only intensity correlation functions. This is why we consider the wording "anticorrelation" well suited to characterize this violation.

10.3.4 The 1986 Anticorrelation Experiment

We have built an experiment corresponding to the scheme of Fig. 10.3, i.e., a setup allowing us to measure the single and coincidence rates on the two sides of a beamsplitter during the opening of gates triggered by events synchronous with the light pulses. This system has been used to study light pulses from a source designed to emit heralded one-photon wavepackets, i.e., based on pairs of photons emitted in a radiative cascade (see Section 10.3.4.1). But we have also used that setup to study strongly attenuated pulses from a classical source. (Section 10.3.4.2.)

10.3.4.1 Heralded One-Photon Pulses From an Atomic Cascade

Our source is composed of atoms excited to the upper level of a two-photon radiative cascade (Fig. 10.4) [4,5]. Each excited atom decays by emission of two photons at different frequencies ν_1 and ν_2. The time intervals between the detections of ν_1 and ν_2 are distributed according to an exponential law, corresponding to the decay time of the intermediate state (lifetime $\tau_s = 4.7$ ns, which is also the time constant Γ^{-1} of the wavepacket describing the heralded single photon ν_2). By choosing the rate of excitation much smaller than $(\tau_s)^{-1}$, we have cascades well separated in time. We use the detection of ν_1 as a trigger for a gate of duration $\tau_{\text{gate}} = 2\tau_s$, corresponding to the scheme of Fig. 10.3. During a gate, the probability of detecting a photon ν_2 coming from the atom that emitted ν_1 is much larger than the probability of detecting a photon ν_2 coming from any other atom in the source. We are then in a situation close to an ideal single-photon pulse, as defined in Section 10.2, and we expect the corresponding anticorrelation behavior on the beamsplitter.

The expected values of the counting rates can be obtained from a straightforward quantum mechanical calculation. Denoting N as the rate of excitation of the cascades, and η_1, η_T, and η_R as the detection efficiencies of

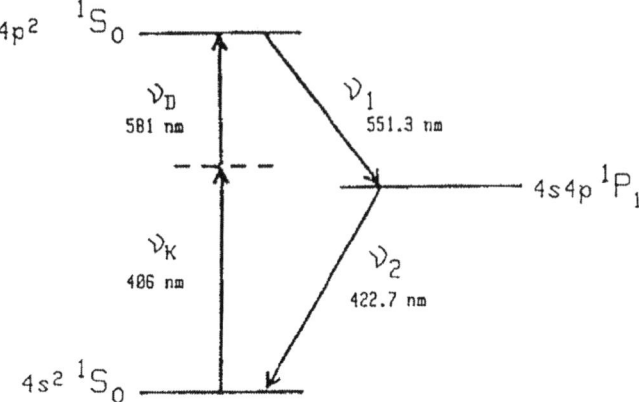

FIGURE 10.4 Radiative cascade in Calcium, used to produce heralded single-photon pulses. The atom is excited to the upper level of the cascade by a resonant two-photon excitation with a Krypton-ion laser and a tunable dye laser. It then re-emits photons ν_1 and ν_2. Detection of photon ν_1 activates the trigger of Fig. 10.3.

photons ν_1 and ν_2 (including the collection solid angles, optics transmissions, and detector efficiencies) we obtain:

$$N_1 = \eta_1 N, \tag{10.85}$$

$$N_T = N_1 \eta_T [f(\tau_{\text{gate}}) + N\tau_{\text{gate}}], \tag{10.86}$$

$$N_R = N_1 \eta_R [f(\tau_{\text{gate}}) + N\tau_{\text{gate}}], \tag{10.87}$$

$$N_c = N_1 \eta_T \eta_R [2f(\tau_{\text{gate}})N\tau_{\text{gate}} + (N\tau_{\text{gate}})^2], \tag{10.88}$$

where $N\tau_{\text{gate}}$ is the probability to have a photon from another atom than the heralding atom, during the gate. The quantity $f(\tau_{\text{gate}})$, very close to 1 in this experiment, is the product of the factor $[1 - \exp(-\tau_{\text{gate}}/\tau_s)]$ (overlap between the gate and the exponential decay) and a factor somewhat greater than 1 that is related to the angular correlation between ν_1 and ν_2 [4,5].

The quantum mechanical prediction for α is

$$\alpha_{QM} = \frac{2f(\tau_{\text{gate}})N\tau_{\text{gate}} + (N\tau_{\text{gate}})^2}{[f(\tau_{\text{gate}}) + N\tau_{\text{gate}}]^2}, \tag{10.89}$$

which is smaller than 1, as expected. The anticorrelation effect is strong (α small compared to 1) if $N\tau_{\text{gate}}$ is much smaller than 1. This condition is easily fulfilled if the cascades are well separated in time, in the average.

Counting electronics, including the gating system, was a critical part of this experiment. The gate τ_{gate} was realized by logical decisions based on the measurement of the time intervals between counts at the various detectors. This

TABLE 5.1 Feeble-Light Interference Experiments. All these Experiments have been Realized with Attenuated Light from a Usual Source

Author	Date	Interferometer	Detector	Photon Flux (s^{-1})	Interferences
Taylor [26]	1909	Diffraction	Photography	10^6	Yes
Dempster et al. [28]	1927	(i) Grating	Photography	10^5	Yes
		(ii) Fabry Perot	Photography	10^5	Yes
Janossy et al. [29]	1957	Michelson interferometer	Photomultiplier	10^5	Yes
Donstov et al. ([30])	1967	Fabry Perot	Image intensifier	10^3	No
Reynolds et al. [31]	1969	Fabry Perot	Image intensifier	10^2	Yes
Bozec et al. [32]	1969	Fabry Perot	Photography	10^2	Yes
Grishaev et al. [33]	1969	Jamin interferometer	Image intensifier	10^3	Yes
Ciamberlini et al. [27]	1994	Diffraction	Image intensifier and CCD	10^5	Yes

TABLE 5.2 Anticorrelation experiment with single-photon pulses from the radiative cascade. The last column corresponds to the expected number of coincidences for $\alpha = 1$. The measured coincidences show a clear anticorrelation effect. These data can be compared to Table 5.3

Trigger Rates	Singles Rates		Duration	Measured Coincidences	Expected Coincidences for $\alpha = 1$
$N_1(s^{-1})$	$N_R(s^{-1})$	$N_T(s^{-1})$	$\theta(s)$	$N_c\theta$	$\frac{N_R N_T}{N_1}\theta$
4720	2.45	3.23	1200	6	25.5
8870	4.55	5.75	17,200	9	50.8
1,21,00	6.21	8.44	14,800	23	64.1
20,400	12.6	17.0	19,200	86	204
36,500	31.0	40.6	13,200	273	456
50,300	47.6	61.9	8400	314	492
67,100	71.5	95.8	3600	291	367

allowed the adjustment of the gates with an accuracy of 0.1 ns. The system also yielded various time-delay spectra, useful for consistency checks.

Table 5.2 shows the measured counting rates for different values of the excitation rate of the cascade. The corresponding values of α have been plotted in Fig. 10.5 as a function of $N\tau_{\text{gate}}$. As expected, the violation of inequality (10.84) increases as $N\tau_{\text{gate}}$ decreases, but the signal decreases simultaneously, and it becomes necessary to accumulate the data for periods of time long enough to achieve a reasonable statistical accuracy. A maximum violation of more than 13 standard deviations has been obtained for a counting time of five hours (second line of Table 5.2). The value of α then is 0.18(6), corresponding to a total number of coincidences of 9, instead of the minimum value of 50 expected for a quasi-classical pulse.

10.3.4.2 Attenuated Classical Pulses

To confirm our arguments experimentally, and to test the photon-counting system, we also studied light from a pulsed light-emitting diode (LED). It produced light pulses with a rise time of 1.5 ns and a fall time about 6 ns. The gates, triggered by the electric pulses driving the photodiode, were 9 ns wide and had an almost complete overlap with the light pulses.

The source was attenuated to a level corresponding to one detection per 1,000 pulses emitted. With a detector quantum efficiency of about 10%, the average energy per pulse can be estimated to be about 0.01 photon. In the context of Table 5.1, this source certainly would have been considered a source of single photons. The results presented in Table 5.3 show that it is definitely not the

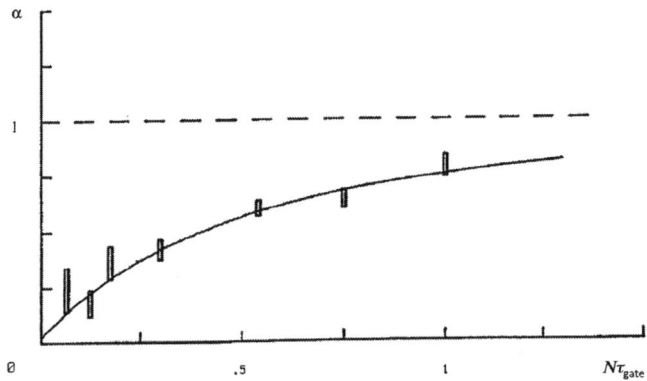

FIGURE 10.5 Correlation parameter α as a function of the excitation rate of the cascade N. The value of α smaller than 1 is the signature of an anticorrelation, corresponding to the one-photon behavior (no classical theory of light can predict a parameter α less than 1). The solid line is the prediction of quantum optics, taking into account the possibility that more than one atom is excited during one gate of duration τ_{gate}: For a single emitter, α would be zero.

TABLE 5.3 Anticorrelation experiment for light pulses from an attenuated photodiode (0.01 Photon/Pulse). The last column corresponds to the expected number of coincidences for $\alpha = 1$. All the measured coincidences are compatible with $\alpha = 1$; there is no evidence of anticorrelation. Note that the singles rates are similar to the ones of Table 5.2

Trigger rates	Singles rates		Duration	Measured coincidences	Expected coincidences for $\alpha = 1$
$N_1(s^{-1})$	$N_{2r}(s^{-1})$	$N_{2f}(s^{-1})$	$\theta(s)$	$N_c\theta$	$\frac{N_R N_T}{N_1}\theta$
4760	3.02	3.76	31200	82	74.5
8880	5.58	7.28	31200	153	143
12,130	7.90	10.2	25,200	157	167
20,400	14.1	20.0	25,200	341	349
35,750	26.4	33.1	12,800	329	313
50,800	44.3	48.6	18,800	840	798
67,600	69.6	72.5	12,800	925	955

case. The quantity α (of inequality (10.84)) is consistently found very close to 1; i.e., no anticorrelation is observed. In fact, the coincidence rate is exactly in agreement with the limit of inequality (10.84).

This experiment thus supports the claim that light emitted by an attenuated classical source does not exhibit one-photon behavior on a beamsplitter, even in the case of very attenuated light pulses with an average energy by pulse much less than the energy of a photon.

10.3.4.3 Conclusion: Anticorrelation as a Characteristic Property of Single Photons

The experiments presented in this subsection confirm that anticorrelation on a beamsplitter is a very clear criterion for discriminating between a one-photon light pulse and a quasi-classical light pulse. A pulse produced by a classical source, even attenuated to a level of 10^{-2} average photon number per pulse, has the behavior expected for a quasi-classical pulse: one observes coincidences in agreement with the inequality (10.84). In contrast, we have been able to produce one-photon pulses for which the number of coincidences was so small that a violation of inequality (10.84) by more than 13 standard deviations was observed. This last result can also be considered as strong experimental evidence against semi-classical theories of light, which never predict a violation of inequality (10.84).

10.4 SINGLE-PHOTON INTERFERENCES

10.4.1 Wave-Particle Duality in Textbooks

Many introductory courses in Quantum Mechanics—whether or not they choose an historical perspective—begin with an "experiment" exhibiting the wave-particle duality of light and matter. This experiment is usually presented by showing an interference pattern, for instance in a Young's slit experiment. Such a phenomenon can be interpreted by invoking a wave that passes through both holes: it is well known that the resulting intensity then depends on the "path difference" Δ, and exhibits a modulation depending on the interference order $p = \Delta/\lambda$, where λ is the wavelength. On the other hand, the "particle" character is usually considered obvious for matter particles such as electrons, neutrons, or atoms, whereas it is actually not obvious for light, as discussed in the previous sections. In the latter case it is therefore useful, before looking for interferences, to present experimental proof that the source S emits well-separated single-photon pulses: if it were not the case, the discussion would be pointless. This is why we have addressed the question of single-photon interferences with the source described in 10.3.4.1.

10.4.2 Interferences with a Single Photon

The quantum theory of light predicts indeed that interferences will happen even with one-photon pulses (see for instance [17] for a detailed calculation). We have thus built a Mach-Zehnder interferometer, keeping the same source and the same beamsplitter as in Fig. 10.3, but removing the detectors on both sides of the beam splitter, and recombining the two beams on a second beam

Chapter | 10 The First Single-Photon Sources

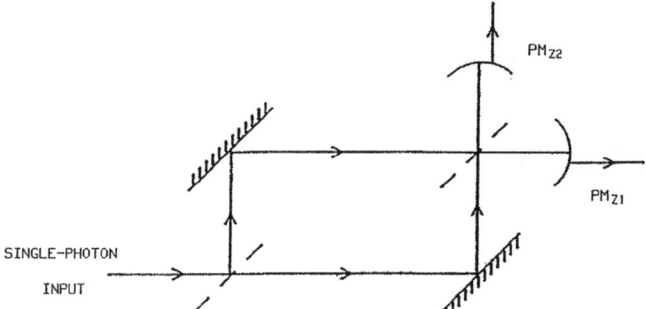

FIGURE 10.6 Single-photon interference experiment. The source and the beamsplitter are similar to Fig. 10.3, but are now configured as a Mach-Zehnder interferometer. The detectors are gated, as in Fig. 10.3, synchronously with the light pulses.

splitter (Fig. 10.6) [5]. The detection rates in the two outputs (1) and (2) are expected to be modulated as a function of the path difference in both arms of the interferometer. To guarantee that we are still working with one-photon pulses, the detectors $PM1$ and $PM2$ are gated synchronously with the pulses, as they were in the experiment of Section 10.3.4.1.

The interferometer has been carefully designed and built to give high-visibility fringes with the beam of large étendue (product of transverse area and solid angle) produced by our source (about 0.5 mm^2 rad^2). The reflecting mirrors and the beam splitters are $\lambda/50$ flat over a 40 mm diameter aperture. A mechanical system driven by piezoelectric transducers permits displacement of the mirrors while keeping their orientation exactly constant: this allows control of the path difference of the interferometer. Preliminary checks with classical light showed a strong modulation of the counting rates of PM_{Z1} and PM_{Z2} when the path difference is modified. For classical pulses shaped as the one-photon pulses from our source, the measured visibility was $V = 98.7(5)\%$, a value very close to the ideal value $V=1$, showing the quality of the interferometer.

Figure 10.7 presents the results obtained by running this interferometer with the one-photon source. The numbers of counts during a given time interval are measured as a function of the path difference. In the first plots, the counting time at each position was 0.01 s, while it was 10 s for the last recordings. This run was performed with the sources in a regime corresponding to an anticorrelation parameter $\alpha = 0.2$, and therefore in the one-photon regime. These recordings clearly show the interference fringes building up "one-photon at a time." When enough data have been accumulated, the signal-to-noise ratio is high enough to allow a measurement of the visibility of the fringes. We repeated such measurements for various regimes of the source, corresponding to the different values of α shown in Fig. 10.5, and observed no deviation from the expected value $V = 98.7$, within the experimental noise, even in a regime where the source emits almost pure one-photon pulses. As predicted by the

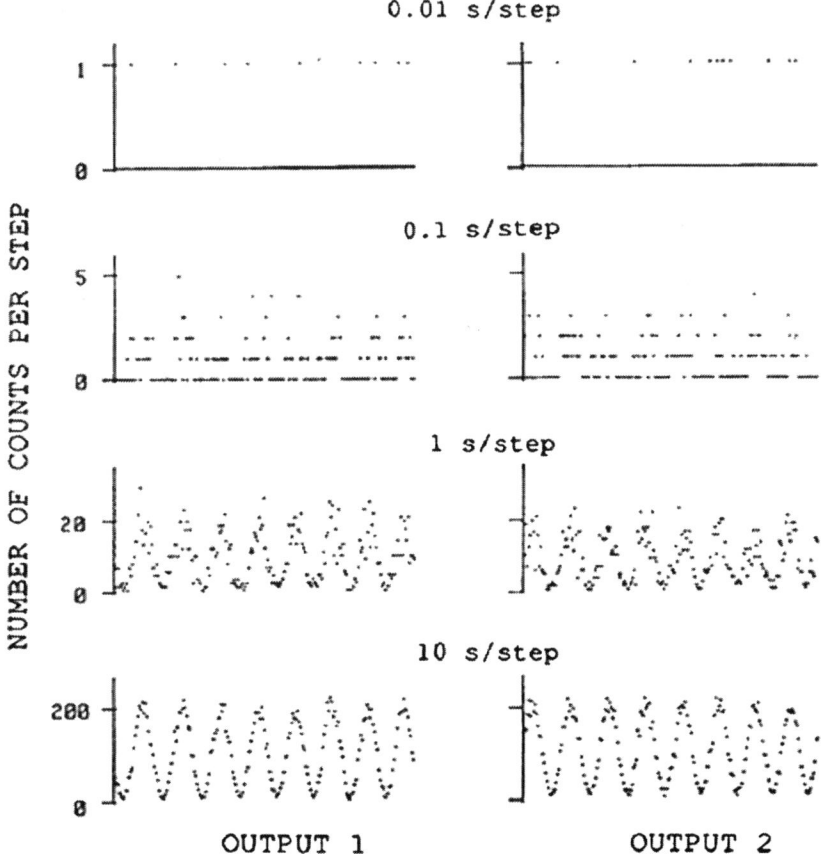

FIGURE 10.7 Number of detected counts in output (1) and (2) as a function of the path difference. The four sets of plots correspond to different counting times at each path difference. This experiment has been realized in the single-photon regime ($\alpha = 0.2$). Note that the interferograms of outputs (1) and (2) are complementary. Original plot for the experiment described in Ref. [5].

quantum theory of light, single-photon pulses do interfere. To our knowledge, this experiment (realized in 1985) was the first of this kind performed with a "fully quantum" light source, a source for which the anticorrelation effect was also directly observed [18].

10.5 FURTHER DEVELOPMENTS

10.5.1 Parametric Sources of Photon Pairs

During the same period as the experiments described above—between the early 1970s and the mid-1980s—another approach to generating photon pairs was developed using parametric fluorescence from $\chi^{(2)}$ crystals, rather than atomic cascades [6,7]. In 1986, Hong & Mandel performed an experiment strongly related to the anticorrelation effect described above, though presented in a different way [8]. Since a full chapter in this book is devoted to such sources, here we only comment that the non-classical features of these photon pairs are similar to the ones described above, but with some notable differences:

- Due to phase-matching conditions, parametric photons are strongly correlated both in their emission times, with a time separation of the order of the inverse of the phase-matching bandwidth, and in their emission directions, due to the conservation of the photon momenta when "splitting" a pump photon into two parametric photons. As a consequence, the heralded photons can be collected with an efficiency orders of magnitude better than in an atomic cascade, and this has been intensively used in experiments.
- A parametric fluorescence experiment is significantly simpler and more reliable than an atomic cascade experiment. Indeed with such sources, a photon-anticorrelation experiment can now be a small and simple table-top experiment that can be done by students in lab work [43].

For these reasons, parametric fluorescence is now widely used to produce heralded single photons, and it is even possible to produce number states in well-defined spatio-temporal modes, and to reconstruct their Wigner functions using quantum homodyne tomography. This has been demonstrated both for one-photon [44] and two-photon Fock states [45]. It should be noted also that parametric photon pairs can be emitted from $\chi^{(3)}$ non-linear effects in optical fibers, rather than $\chi^{(2)}$ non-linear effects in crystals, as described in Chapter 13 of this book.

10.5.2 Other Heralded and "On-Demand" Single-Photon Sources

Many other types of single-photon sources have been proposed and implemented, using quantum dots, single molecules or atoms, possibly in the cavity QED regime, Nitrogen-vacancy centers in diamond, collectively enhanced quantum ensembles, all of which are described elsewhere in this book. Let us emphasize that some of these sources are getting close to being "on-demand" single-photon sources, meaning that the single photon is not only "heralded," but emitted in a "push-button" way at a prescribed time. This can be obtained rather easily from pulsed excitation of a single quantum emitter, but in addition it is desirable that the photon is emitted with a very high efficiency (that is, each "click" gives one and only one photon), and with a perfectly defined spatio-temporal mode (so that, for instance, high-quality quantum tomography of the single photon can be performed). A fully on-demand single-photon source is not yet available, but impressive progress has been achieved during the last 25 years.

10.5.3 "Delayed-Choice" Single-Photon Interference Experiments

To conclude, let us mention some recent developments in single-photon interferences. Following a famous proposal by Wheeler, a very convincing "delayed-choice" interference experiment has been performed by Jacques et al. using a Nitrogen-vacancy (NV) center in diamond as the single-photon source [20]. In this experiment, the "choice" of leaving the interferometer open—and thus observing the "which path" information—or closing the interferometer—and thus observing the interference fringes—is made while the photon is already inside a 50-m long interferometer. In even more recent experiments, it was shown that this choice can be made remotely, by using a second photon entangled with the photon inside the interferometer [21,22]. These experiments demonstrate the impressive control that can be obtained in manipulating single photons, offering more and more possibilities for applications in quantum information and quantum communications.

REFERENCES

[1] M.A. Nielsen and I.L. Chuang, "Quantum Computation and Quantum Information," Cambridge University Press (2010).
[2] S.J. Freedman and J.F. Clauser, "Experimental test of local hidden-variable theories," Phys. Rev. Lett. 28, pp. 938–941, 1972.
[3] J.F. Clauser, "Experimental Distinction Between Quantum and Classical Field-Theoretic Predictions for Photoelectric Effect," Phys. Rev. D 9, 853–860 (1974).
[4] A. Aspect, P. Grangier, and G. Roger, "Experimental Tests of Realistic Local Theories via Bell's Theorem," Phys. Rev. Lett. 47, 460–463 (1981).
[5] P. Grangier, G. Roger, and A. Aspect, "Experimental-Evidence for a Photon Anticorrelation Effect on a Beam Splitter—A New Light on Single-Photon Interferences," Europhys. Lett. 1, 173–179 (1986).
[6] D.C. Burnham and D.L. Weinberg, "Observation of Simultaneity in Parametric Production of Optical Photon Pairs," Phys. Rev. Lett. 25, 84–87 (1970).
[7] S. Friberg, C.K. Hong, and L. Mandel, "Measurement of Time Delays in the Parametric Production of Photon Pairs," Phys. Rev. Lett. 54, 2011–2013 (1985).
[8] C.K. Hong and L. Mandel, "Experimental Realization of a Localized One-Photon State," Phys. Rev. Lett. 56, 58–60 (1986).
[9] D.T. Pegg, R. Loudon, and P.L. Knight, "Correlations in Light Emitted by 3-Level Atoms," Phys. Rev. A 33, 4085–4091 (1986).
[10] B. Lounis and W. Moerner, "Single Photons on Demand from a Single Molecule at Room Temperature," Nature 407, 491–493 (2000).
[11] E.C.G. Sudarshan, "Equivalence of Semiclassical and Quantum Mechanical Descriptions of Statistical Light Beams," Phys. Rev. Lett. 10, 277–279 (1963).
[12] R.J. Glauber, "Coherent and Incoherent States of the Radiation Field," Phys. Rev. 131, 2766 (1963).
[13] H.J. Kimble, M. Dagenais, and L. Mandel, "Photon Anti-Bunching in Resonance Fluorescence," Phys. Rev. Lett. 39, 691–695 (1977).
[14] In his famous Lectures on Physics [46], Feynman cites Wave-Particle duality as the only mystery of quantum mechanics. However, two decades later [47], he emphasizes that up to that point he has missed to recognize the unique feature of entanglement.. and he immediately proposes to use it as a tool for quantum computing.
[15] R. Loudon, "The Quantum Theory of Light," Oxford University Press (2000).
[16] C. Gerry and P. Knight, "Introductory Quantum Optics," Cambridge University Press (2005).

[17] G. Grynberg, A. Aspect, and C. Fabre, "Introduction to Quantum Optics: From the Semi-Classical Approach to Quantized Light," Cambridge University Press (2010).
[18] A modern implementation of such an experiment illustrating wave-particle duality for a single-photon [48] has permitted our collaborators at ENS Cachan to produce a video showing directly the construction of an interference pattern photon by photon, with a single-photon source passing the single-photon test. This video can be found for instance at <http://www.lcf.institutoptique.fr/Alain-Aspect-homepage>.
[19] J.A. Wheeler, "Law Without Law," Princeton University Press (1984).
[20] V. Jacques, E. Wu, F. Grosshans, F. Treussart, P. Grangier, A. Aspect, and J.F. Roch, "Experimental Realization of Wheeler's Delayed-Choice Gedanken Experiment," Science 315, 966–968 (2007).
[21] F. Kaiser, T. Coudreau, P. Milman, D.B. Ostrowsky, and S. Tanzilli, "Entanglement-Enabled Delayed-Choice Experiment," Science 338, 637–640 (2012).
[22] A. Peruzzo, P. Shadbolt, N. Brunner, S. Popescu, and J.L. O'Brien, "A Quantum Delayed-Choice Experiment," Science 338, 634–637 (2012).
[23] A. Einstein, "Generation and Conversion of Light with Regard to a Heuristic Point of View," Annalen Der Physik 17, 132–148 (1905).
[24] Einstein's LichtQuanten was Named Photon Only Two Decades Later [49].
[25] A. Einstein, "On the Evolution of Our Vision on the Nature and Constitution of Radiation," Physikalische Zeitschrift 10, 817–826 (1909).
[26] G.I. Taylor, "Interference Fringes with Feeble Light," Proc. Cambridge Philos. Soc. 15, 114–115 (1910).
[27] C. Ciamberlini and G. Longobardi, "Real-Time Analysis of Diffraction Patterns, at Extremely Low-Light Levels," Opt. Lasers Eng. 21, 317–325 (1994).
[28] A.J. Dempster and H.F. Batho, "Light Quanta and Interference," Phys. Rev. 30, 644–648 (1927).
[29] L. Jánossy and Z. Náray, "The Interference Phenomena of Light at Very Low Intensities," Acta Phys. Hung. 7, 403–425 (1957).
[30] Y.P. Dontsov and A.I. Baz, "Interference Experiments with Statistically Independent Photons," Soviet Phys. JETP-Ussr 25, 1–5 (1967).
[31] G.T. Reynolds, S.K., and D.B. Scarl, "Interference Effects Produced by Single Photons," Nuovo Cimento Della Societa Italiana Di Fisica B-General Physics Relativity Astronomy And Mathematical Physics And Methods 61, 355–364 (1969).
[32] P. Bozec, M. Cagnet, and G. Roger, "Experiments on Interference in a Weak Light," Comptes Rendus Hebdomadaires Des Seances De L Academie Des Sciences Serie B 269, 883 (1969).
[33] I.A. Grishaev, N.N. Naugolny, L.V. Reprints, A.S. Tarasenk, and A.M. Shendero, "Interference Experiment and Photon Statistics for Synchrotron Radiation from Electrons in a Storage Ring," Soviet Phys. JETP-Ussr 32, 16 (1971).
[34] R. Glauber, Quantum Optics and Electronics, in Les Houches Summer School 1964, edited by C. DeWitt, Gordon and Breach, New York (1965).
[35] R.J. Glauber, "The Quantum Theory of Optical Coherence," Phys. Rev. 130, 2529 (1963).
[36] C. Cohen-Tannoudji, J. Dupont-Roc, and G. Grynberg, "Photons and Atoms-Introduction to Quantum Electrodynamics," Wiley-VCH (1997).
[37] L. Mandel and E. Wolf, "Optical Coherence and Quantum Optics," Cambridge University Press (1995).
[38] A.L. Schawlow and C.H. Townes, "Infrared and Optical Masers," Phys. Rev. 112, 1940 (1958).
[39] J. Goodman, "Statistical Optics," Vol. 1, New York, Wiley-Interscience, 567 p. (1985).
[40] This remark was used in the experiment where we observe an interference effect for a single-photon emitted jointly by two atoms flying in opposite directions after a molecular photodissociation [50]. The reason why this observation was successful is that the time of the dissociation, and thus of the leading-edge of the single-photon wave-packet, was precisely known, since the photodissociation was effected by a short laser pulse.
[41] C.K. Hong, Z.Y. Ou, and L. Mandel, "Measurement of Subpicosecond Time Intervals Between 2 Photons by Interference," Phys. Rev. Lett. 59, 2044–2046 (1987).
[42] This scheme is the one that was used in the experiment [5] presented in Section 10.3. A related scheme was proposed independently in [51].

[43] See for instance <http://www.institutoptique.fr/Formation/Ingenieur-Grande-Ecole/Travaux-Pratiques/Physique-quantique-atomique-nanophysique> at Institut d'Optique *Graduate School*.
[44] A.I. Lvovsky, H. Hansen, T. Aichele, O. Benson, J. Mlynek, and S. Schiller, "Quantum State Reconstruction of the Single-Photon Fock State," Phys. Rev. Lett. 87, 050402 (2001).
[45] A. Ourjoumtsev, R. Tualle-Brouri, and P. Grangier, "Quantum Homodyne Tomography of a Two-Photon Fock State," Phys. Rev. Lett. 96 (2006).
[46] R.P. Feynman, "Lectures on Physics," Addison-Wesley (1963).
[47] R.P. Feynman, "Simulating Physics with Computers," Int. J. Theor. Phys. 21, 467–488 (1982).
[48] V. Jacques, E. Wu, T. Toury, F. Treussart, A. Aspect, P. Grangier, and J.F. Roch, "Single-Photon Wavefront-Splitting Interference—An Illustration of the Light Quantum in Action," Eur. Phys. J. D 35, 561–565 (2005).
[49] G.N. Lewis, "The Conservation of Photons," Nature 118, 874–875 (1926).
[50] P. Grangier, A. Aspect, and J. Vigue, "Quantum Interference Effect for 2 Atoms Radiating a Single Photon," Phys. Rev. Lett. 54, 418–421 (1985).
[51] B. Saleh and M. Teich, "Sub-Poisson Light Generation by Selective Deletion from Cascaded Atomic Emissions," Opt. Commun. 52, 429–432 (1985).

Chapter 11

Parametric Down-Conversion

Andreas Christ[*], Alessandro Fedrizzi[†], Hannes Hübel[‡],
Thomas Jennewein[††], and Christine Silberhorn[*]

[*]Applied Physics, University of Paderborn, Warburger Straße 100, 33098 Paderborn, Germany
[†]Centre for Engineered Quantum Systems, Centre for Quantum Computer and Communication Technology, School of Mathematics and Physics, University of Queensland, Brisbane 4072, Australia
[‡]Fysikum, University of Stockholm, Roslagstullsbacken 21, 10691 Stockholm, Sweden
[††]Institute for Quantum Computing and Department of Physics & Astronomy, University of Waterloo, Waterloo, Canada N2L 3G1

Chapter Outline

11.1 Introduction	352
11.2 Single Photons from PDC: Theory	353
11.2.1 Classical Description of PDC	354
11.2.2 Quantum Mechanical Description of PDC	357
11.2.3 Heralding Single Photons from PDC	360
11.2.4 Heralding Pure Single-Photon Fock States	362
11.3 Bulk-Crystal PDC	367
11.3.1 Birefringent Phase-Matching	367
11.3.2 Heralded Single Photons from Triggered PDC	372
11.4 Periodically-Poled Crystal PDC	379
11.4.1 Quasi-Phase-Matching	379
11.4.2 Periodic Poling	383
11.4.3 Optimal Focus Parameters for Heralding Efficiency	384
11.4.4 Number Purity	388
11.4.5 Spectral Purity	390
11.4.6 Non-Uniform Periodic Poling	391
11.5 Waveguide-Crystal PDC	392
11.5.1 History and Experimental Implementations	393
11.5.2 Theory of PDC in Waveguides	394
11.5.3 Heralding Single Photons from PDC in Waveguides	399
11.5.4 Electric Field Modes in Waveguides	401

11.6 Comparison of Experimental Single-Photon
     Sources Using PDC   403
11.7 Overview of the Most Commonly Used
     Nonlinear Materials and Their Properties   404
11.8 Conclusion   404
References   404

11.1 INTRODUCTION

Parametric down-conversion (PDC) is a nonlinear process in which a photon from a strong pump laser is converted into two daughter photons under conservation of energy and momentum. Despite its many contenders discussed elsewhere in this book, PDC is still the most widely used technique for generating single photons. The reasons are that PDC is well understood, simple to implement, and produces photons in well-defined spatio-temporal modes at high rates.

Photon pairs generated in PDC were first observed in the late 1960s and early 1970s [1–4]. Two decades later, two independent groups started to study the coherence properties of the created single photons: Shih and Alley [5] realized that the photon pairs could be used to investigate *fundamental* problems in quantum mechanics, such as John Wheeler's delayed choice Gedankenexperiment, or the Einstein-Podolsky-Rosen paradox. Meanwhile, Ghosh, and Mandel observed non-classical interference effects between down-converted photons [6]. Their work was a direct precursor to a landmark experiment by Hong *et al.* [7], who used non-classical two-photon interference on a beamsplitter to demonstrate that PDC photons were created within at least sub-picosecond time intervals. The so-called Hong-Ou-Mandel dip, a drop in coincident photon detection at the output of the beamsplitter as a function of photon arrival time, firmly established PDC as an experimental tool in quantum optics and has since been the key mechanism for many quantum information processing protocols.

The biggest boost of PDC as a tool was, however, the fact that photon pairs created in this process are naturally entangled. The first bright source of highly entangled photons was demonstrated by Kwiat and co-authors in 1995 [8]. This sparked a host of experiments studying the nature of entanglement, such as Bell tests with space-like separated observers [9], quantum teleportation [10], multi-particle entanglement [11], entanglement-based quantum cryptography [12], and many more.

As these applications evolved, they placed increasing demands on photon sources. The biggest improvement to both pair creation rates and flexibility was due to rapid advances in nonlinear optics, specifically, the development of periodically-poled crystals. Another major development was waveguide technology. Waveguides confine light fields over relatively large distances in optical materials and thus lead to vastly increased PDC generation rates. They also allow for optical integration of sources, optical circuits, and detectors into small-scale devices. Waveguide PDC is however, compared to the more mature bulk-crystal and periodically-poled PDC sources, still in its infancy and it will take some time to eradicate issues such as coupling losses and pump mode suppression.

We will start with a general theoretic description of parametric down-conversion and then move on to the details of heralding photons with PDC. In particular, we will discuss important concepts such as photon purity and phase-matching. The subsequent sections discuss the practical aspects of generating PDC in bulk crystals, in periodically-poled crystals and in waveguides. The reader will find a list of selected experiments at the end of this chapter (see Fig. 11.32). The list is by no means exhaustive, but should give the reader a flavor of the possible variations in single-photon sources based on PDC.

11.2 SINGLE PHOTONS FROM PDC: THEORY

In this section we review the process of parametric down-conversion as a source of single-photon states.

The overall process is sketched in Figure 11.1. A strong pump field is propagating through a medium possessing a $\chi^{(2)}$ nonlinearity. During this interaction one of the pump photons decays into a photon pair, where we label the individual photons signal and trigger. The trigger photon is subsequently detected to herald the presence of its partner the signal photon, effectively generating a source of heralded single photons. Heralded single-photon states are also analyzed and discussed in this chapter, in the context of early (and less efficient) atomic-cascade sources.

This introductory section mainly concentrates on the theory of heralding single photons from PDC and is structured into four parts. In Section 11.2.1

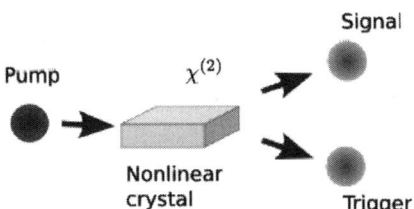

FIGURE 11.1 Single-photon generation using PDC. A pump field is propagating through a medium exhibiting a $\chi^{(2)}$ nonlinearity. During this interaction a pump photon decays into a pair of photons. One of the photons is detected to herald the presence of its partner, effectively creating a heralded single-photon source.

we give an overview of the classical description of PDC and its connection to sum-frequency conversion. In Section 11.2.2 we present a quantum mechanical treatment of PDC and derive the generated two-photon output state. In Section 11.2.3, we elaborate on the process of detecting one photon to herald the presence of its partner and discuss the properties of the heralded signal photon. Finally, in Section 11.2.4 we review the intricacies of heralding pure single-photon Fock states from PDC.

11.2.1 Classical Description of PDC

Although PDC can only be fully described using a quantum mechanical formalisms, it is nevertheless instructional to study a classical form of a nonlinear parametric process like sum-frequency generation. Many of the relations between the classically interacting fields will also hold for the PDC case.

An electro-magnetic field interacting with a dielectric medium will induce a polarization in the material. The normal response of the material is linear, i.e., the incoming electro-magnetic field is not altered in frequency. However at high electric field strengths nonlinear properties cannot be neglected anymore. The polarization P can be expanded in terms of the electro-magnetic field

$$P = \varepsilon_0(\chi^{(1)} E_1 + \chi^{(2)} E_1 E_2 + \chi^{(3)} E_1 E_2 E_3 + \cdots), \tag{11.1}$$

where ε_0 is the vacuum permittivity, $\chi^{(1)}$ is the linear susceptibility, and $\chi^{(2)}, \chi^{(3)}, \ldots$ are the nonlinear susceptibilities of the medium. The nonlinear interaction implies that electro-magnetic waves with different frequencies can interact and, under conservation of energy, frequency conversion can occur. In the $\chi^{(2)}$ process, which is also referred to as three-wave mixing, three electro-magnetic fields interact in a non-centrosymmetric medium and energy can be transferred from one field to another. Since PDC is a three-wave mixing process, the $\chi^{(2)}$ interaction will be discussed in detail here. The higher-order four-wave mixing process with $\chi^{(3)}$ nonlinearities will be discussed in Chapter 12: Four-wave mixing.

The $\chi^{(2)}$ frequency conversion can be roughly divided into two types. The first type has two input fields and produces a single output field. The produced field can have frequencies corresponding to the sum of the input fields, as in sum-frequency generation, or frequencies corresponding to the difference, as in difference-frequency generation. The other type of interaction to which PDC belongs has a single input field that is converted to two output fields, as shown in Fig. 11.1. Historically the resulting fields have been named signal and idler, with the former having the higher frequency. In this book we will refer to the two output fields as *trigger* and *signal*, without specific relation to the frequency. In the quantum mechanical description (Section 11.2.2), PDC gives rise to the spontaneous generation of photon pairs and the presence of a photon in the *trigger* mode heralds the partner photon in the *signal* mode.

Chapter | 11 Parametric Down-Conversion

The susceptibility $\chi^{(2)}$ is a tensor of rank 3 and its components are defined through the nonlinear part of the polarization and the electric field components in Eq. (11.1)

$$P_i^{\text{NL}} = \varepsilon_0 \sum_{j,k=1,2,3} \chi_{ijk} E_j E_k, \tag{11.2}$$

with $i = 1,2,3$. It is customary to assign the cartesian axes x, y, z to the indices values 1, 2, 3. Since the fields E_j and E_k can be permuted without changing the polarization, the 27 elements of the $\chi^{(2)}$ tensor can be reduced to a matrix with only 18 independent elements d_{ij}

$$\begin{pmatrix} P_x \\ P_y \\ P_z \end{pmatrix} = \varepsilon_0 \begin{pmatrix} d_{11} & d_{12} & d_{13} & d_{14} & d_{15} & d_{16} \\ d_{21} & d_{22} & d_{23} & d_{24} & d_{25} & d_{26} \\ d_{31} & d_{32} & d_{33} & d_{34} & d_{35} & d_{36} \end{pmatrix} \times \begin{pmatrix} E_x^2 \\ E_y^2 \\ E_z^2 \\ 2E_y E_z \\ 2E_x E_z \\ 2E_x E_y \end{pmatrix}. \tag{11.3}$$

The d_{ij} coefficients and other optical properties for some well-known nonlinear crystals are summarized in Fig. 11.33.

If the nonlinear interaction couples pump, trigger, and signal fields with the same polarization, it is referred to as a type-0 interaction. If the pump polarization is orthogonal to the signal and trigger polarization the process is labeled type-I PDC, and if the two output fields are orthogonally polarized the interaction is called type-II PDC.

The additional nonlinear part of the polarization can be added to the macroscopic Maxwell's equations to derive the electric fields. Assuming collinear propagation of the waves along the x-direction the following wave equation is obtained

$$\frac{\partial^2 E}{\partial x^2} = -\mu_0 \left(\varepsilon_0 \frac{\partial^2 E}{\partial t^2} + \frac{\partial^2 P^{\text{NL}}}{\partial t^2} \right). \tag{11.4}$$

In the case of sum-frequency generation, assuming monochromatic fields, two incident plane waves in the form of

$$E_{1,2} = A_{1,2}(x) e^{i(\omega_{1,2} t - k_{1,2} x)}, \tag{11.5}$$

with frequencies ω_1 and ω_2, propagation constants k_1 and k_2, and a possible varying amplitude $A_{1,2}(x)$ along the propagation direction will induce a periodic modulation of the polarization at a frequency of $\omega_3 = \omega_1 + \omega_2$. This, in turn, creates a new field E_3 at frequency ω_3. The solution of the new field can be found from Eq. (11.4) by substituting P^{NL} with the input fields

E_1 and E_2 coupled via $\chi^{(2)}$. The electric field amplitude of E_3 along the propagation direction is then found to be

$$\frac{dA_3}{dx} = \frac{-i\omega_3}{2n_3 c} d_{ij} A_1 A_2 e^{i(\Delta k)x}, \tag{11.6}$$

using the slowly varying amplitude approximation $\left(\frac{d^2 E}{dx^2} << k \frac{dE}{dx}\right)$. Similar expressions can be derived for the input fields E_1 and E_2. The nonlinear coefficient d_{ij} is derived from the appropriate matrix element in Eq. (11.3) depending on the polarization of the fields. The wave-vector mismatch Δk is given by

$$\Delta k = k_3 - k_1 - k_2 = \frac{n_3 \omega_3}{c} - \frac{n_1 \omega_1}{c} - \frac{n_2 \omega_2}{c}, \tag{11.7}$$

where $n_j, j = 1,2,3$ are refractive indices of the three propagating fields. The amplitude of the generated field after a propagation distance L can readily be found by integrating Eq. (11.6)

$$A_3(L) = \frac{-d_{ij}\omega_3 A_1 A_2}{2n_3 c} \left(\frac{e^{i\Delta kL} - 1}{\Delta k}\right), \tag{11.8}$$

where we assumed that the fields E_1 and E_2 are constant throughout the interaction (the undepleted-pump approximation). The intensity of the E_3 field after length L is obtained by using the relation $I = \frac{1}{2} n \varepsilon_0 c E E^*$ and results in

$$I_3(L) = \frac{d_{ij}^2 \omega_3^2 I_1 I_2 L^2}{2 n_1 n_2 n_3 c^3 \varepsilon_0} \text{sinc}^2 \left(\frac{\Delta k L}{2}\right). \tag{11.9}$$

The efficiency of the conversion process is strongly dependent on the wave-vector mismatch Δk. If $\Delta k \neq 0$, then the generated field becomes out of phase by π after an interaction length of

$$L_c = \frac{\pi}{\Delta k}, \tag{11.10}$$

which is also called the coherence length. At this point the generated field starts to interfere destructively, in effect lowering the conversion efficiency until it reaches zero at $2L_c$. Therefore, it is paramount that the phase-matching condition $\Delta k = 0$ is fulfilled to obtain optimal frequency conversion. Methods of achieving the phase-matching condition are described in the proceeding sections of this chapter. The above derivation strictly holds only for interactions when at least two fields are present, as in sum-frequency generation. In a PDC interaction, however, only the pump field is present and a purely classical treatment cannot explain the creation of the other two fields. In a quantum mechanical picture employing second-quantization formalism discussed in the following section, the zero-point quantum noise gives rise to a field at ω_1 and ω_2. Through the interaction with the pump, the noise fields at ω_1 and ω_2 are amplified via the nonlinear frequency conversion. If the nonlinear interaction is strong, a macroscopic intensity at ω_1 and ω_2 can be observed.

The classical Hamiltonian for the electric field is given by [13]

$$H_{EM} \propto \int E(r,t) \cdot D(r,t) dr^3. \tag{11.11}$$

By only keeping the 2nd-order nonlinear term in the displacement vector D yields the interaction Hamiltonian for the $\chi^{(2)}$ process

$$H_{\chi^{(2)}} \propto \int \chi^{(2)} E_p(r,t) E_s(r,t) E_t(r,t) dr^3, \tag{11.12}$$

where the subscripts stand for the pump, trigger, and signal fields, respectively.

11.2.2 Quantum Mechanical Description of PDC

Quantizing the electric fields [13] in the Hamiltonian in Eq. (11.12) enables a quantum mechanical description of the process

$$\hat{H}_{PDC} \propto \chi^{(2)} \int_{-\frac{L}{2}}^{\frac{L}{2}} dz\, \hat{E}_p^{(+)}(z,t) \hat{E}_s^{(-)}(z,t) \hat{E}_t^{(-)}(z,t) + h.c., \tag{11.13}$$

where p labels the pump, s the signal, and t the trigger field. We further restrict ourselves to collinear propagation along the z-axis of a crystal of length L, neglecting the transverse degrees of freedom. Their impact will be discussed later in this chapter.

The positive and negative frequency parts of the quantum fields in Eq. (11.13) are defined as

$$\hat{E}_x^{(+)} = \hat{E}_x^{(-)\dagger} = A \int d\omega_x \exp\left[i\left(k_x(\omega_x)z - \omega_x t\right)\right] \hat{a}_x(\omega_x), \tag{11.14}$$

where all constant factors have been merged into the overall constant A and $\hat{a}_x(\omega_x)$ is the photon annihilation operator for a monochromatic frequency ω_x with propagation constant $k(\omega_x)$. The exact nature of the signal and trigger fields is dependent on the applied pump field and the $\chi^{(2)}$-nonlinearity of the crystal. They may be emitted into identical polarizations (type-0/type-I PDC) or into orthogonal polarizations (type-II PDC) [13,14]. Due to the weakness of the nonlinear interaction the incoming pump field must be relatively strong and may be treated as a classical field

$$E_p^{(+)} = E_p^{(-)*} = \int d\omega_p \alpha(\omega_p) \exp\left[i\left(k_p(\omega_p)z - \omega_p t\right)\right]. \tag{11.15}$$

In Eq. (11.15) the function $\alpha(\omega_p)$ describes the spectrum and amplitude of the pump field, which may vary from a delta function $E_p \delta(\omega_p - \omega_c)$ for cw laser sources with central frequency ω_c, to more complex distributions for pulsed laser systems.

We study parametric down-conversion in the Schrödinger picture by calculating the generated state following the presentation in [15]

$$|\psi\rangle_{\text{PDC}} = \exp\left[-\frac{i}{\hbar}\int_{t_0}^{t} dt'\, \hat{H}_{\text{PDC}}(t')\right]|0\rangle. \quad (11.16)$$

We perform a perturbation expansion of Eq. (11.16) to write the PDC process as

$$|\psi\rangle_{\text{PDC}}^{(\text{full})} = |0\rangle - \frac{i}{\hbar}\int_{t_0}^{t} dt'\, \hat{H}_{\text{PDC}}(t')|0\rangle$$
$$+ \left(\frac{i}{\hbar}\right)^2 \int_{t_0}^{t} dt'\, \hat{H}_{\text{PDC}}(t') \int_{t_0}^{t'} dt''\, \hat{H}_{\text{PDC}}(t'')|0\rangle + \cdots, \quad (11.17)$$

where the zero-order term describes the vacuum emission, the first-order term photon-pair emission, the second-order term four-photon emission, and so on. The emission of higher-order photon pairs can be safely ignored as long as the incoming pump field is not too bright and the probability of multi-pair generation is sufficiently small [16,17]. We hence restrict ourselves to the expansion up to first order

$$|\psi\rangle_{\text{PDC}} \approx |0\rangle - \frac{i}{\hbar}\int_{t_0}^{t} dt'\, \hat{H}_{\text{PDC}}(t')|0\rangle, \quad (11.18)$$

which gives us the emitted *not-normalized* two-photon PDC state. Higher-order expansion terms and their effects on the heralding of single photons are discussed in Section 11.2.4.

Combining Eq. (11.13), (11.14), (11.15), and (11.18), we arrive at

$$\int_{t_0}^{t} dt'\, \hat{H}_{\text{PDC}}(t') = B \int_{t_0}^{t} dt' \int_{-\frac{L}{2}}^{\frac{L}{2}} dz \iiint d\omega_p d\omega_s d\omega_t\, \alpha(\omega_p)$$
$$\times \exp\left[-i\left(\omega_p - \omega_s - \omega_t\right)t'\right]$$
$$\times \exp\left[i\left(k_p(\omega_p) - k_s(\omega_s) - k_t(\omega_t)\right)z\right]\hat{a}_s^\dagger(\omega_s)\hat{a}_t^\dagger(\omega_t) + \text{h.c.}, \quad (11.19)$$

where we merged all constants into an overall factor B. Performing the z-integration we obtain

$$\int_{t_0}^{t} dt\, \hat{H}_{\text{PDC}}(t') = B \int_{t_0}^{t} dt' \iiint d\omega_p d\omega_s d\omega_t \alpha(\omega_p)$$
$$\times \exp\left[-i\left(\omega_p - \omega_s - \omega_t\right)t'\right]$$
$$\times L\,\text{sinc}\left[\left(k_p(\omega_p) - k_s(\omega_s) - k_t(\omega_t)\right)\frac{L}{2}\right]\hat{a}_s^\dagger(\omega_s)\hat{a}_t^\dagger(\omega_t) + \text{h.c.}. \quad (11.20)$$

Note that the integration was performed from $-\frac{L}{2}$ to $\frac{L}{2}$, which assumes a pump with no extra phase factors in the center of the crystal. A pump pulse with no additional phase factor at the beginning of the interaction leads to an additional phase factor in the end result.

As a next step we evaluate the time-integration by expanding the bounds to plus and minus infinity. This is justified since we regard the state long before and after the interaction in the crystal. The integration of the time-dependent part then results in a delta function $2\pi\delta(\omega_p - \omega_s - \omega_t)$, and we subsequently take the integral over the pump frequencies ω_p

$$\int_{-\infty}^{\infty} dt\, \hat{H}_{PDC}(t') = 2\pi B \iint d\omega_s\, d\omega_t\, \alpha(\omega_s + \omega_t)$$
$$\times L \operatorname{sinc}\left[\Delta k(\omega_s, \omega_t)\frac{L}{2}\right] \hat{a}_s^\dagger(\omega_s)\hat{a}_t^\dagger(\omega_t) + h.c..\quad (11.21)$$

Here we have introduced the shorthand $\Delta k(\omega_s, \omega_t) = k_p(\omega_s + \omega_t) - k_s(\omega_s) - k_t(\omega_t)$. With Eq. (11.21), we are able to write the generated PDC state as

$$|\psi\rangle_{PDC} = |0\rangle + B' \iint d\omega_s d\omega_t\, \alpha(\omega_s + \omega_t) \operatorname{sinc}\left[\Delta k(\omega_s, \omega_t)\frac{L}{2}\right] \hat{a}_s^\dagger(\omega_s)\hat{a}_t^\dagger(\omega_t)|0\rangle$$
$$= |0\rangle + B' \iint d\omega_s d\omega_t\, \alpha(\omega_s + \omega_t)\Phi(\omega_s, \omega_t)\hat{a}_s^\dagger(\omega_s)\hat{a}_t^\dagger(\omega_t)|0\rangle$$
$$= |0\rangle + B' \iint d\omega_s d\omega_t\, f(\omega_s, \omega_t)\hat{a}_s^\dagger(\omega_s)\hat{a}_t^\dagger(\omega_t)|0\rangle, \quad (11.22)$$

with a scaling factor $B' \propto E_p L$, which depends linearly on the crystal length L and the field amplitude of the pump field.

As described above, PDC creates two-photon states with a given joint-spectral amplitude (JSA) $f(\omega_s, \omega_t)$. The exact shape of $f(\omega_s, \omega_t)$ is defined by the form of the pump distribution $\alpha(\omega_s + \omega_t)$, and by the phase-matching function $\Phi(\omega_s, \omega_t)$ determined by the length and the dispersion of the crystal. An example of all three for PDC pumped by a pulsed laser system is presented in Fig. 11.2.

The pump distribution function $\alpha(\omega_s + \omega_t)$ of Eq. (11.22) reflects the conservation of energy during the process. All pairs created inside satisfy $\omega_p - \omega_s - \omega_t = 0$, whereas the phase-matching function $\Phi(\omega_s, \omega_t)$ depicts

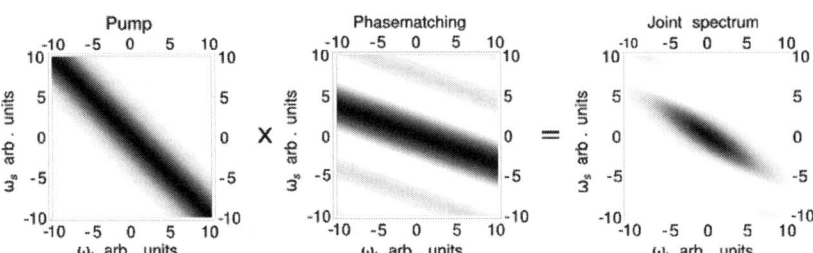

FIGURE 11.2 Exemplary pump spectrum $\alpha(\omega_s + \omega_t)$, phase-matching function $\Phi(\omega_s, \omega_t)$, and joint-spectral amplitude distribution $f(\omega_s, \omega_t)$ of a PDC state.

momentum conservation $k_p(\omega_p) - k_s(\omega_s) - k_t(\omega_t) = 0$. Taken together these functions form the joint-spectral-amplitude distribution of the created photon pairs satisfying both energy and momentum conservation.

11.2.3 Heralding Single Photons from PDC

If the two photons generated by PDC can be efficiently and deterministically separated, for example distinguishing them by their polarizations or their wavelengths, then the detection of one photon heralds the generation of the other. Therefore we have to detect the trigger photon in order to signal the presence of its partner—the signal—from the PDC state in Eq. (11.22). A single-photon avalanche diode (SPAD) with detection efficiency η can be described by the "click" or trigger detection observable $\hat{\pi}_1$ and "no-click" or no detection observable $\hat{\pi}_0$ (see Chapters 2 and 4)

$$\hat{\pi}_0 = \int d\omega \, |0,\omega\rangle \langle 0,\omega|,$$

$$\hat{\pi}_1 = \int d\omega \, \eta \, |1,\omega\rangle \langle 1,\omega|, \qquad (11.23)$$

where $|0,\omega\rangle$ implies vacuum at frequency ω and $|1,\omega\rangle$ a single photon at the same frequency. Since we restrict ourselves to single photons arriving at the detector, higher-photon-number components in $\hat{\pi}_0$ and $\hat{\pi}_1$ are omitted. These SPAD detectors are also known as binary detectors, since they are not able to distinguish the number of photons arriving simultaneously.

The probability of detecting the trigger photon and successfully heralding a single-photon state is directly related to the efficiency of the detector η and the amplitude of the two-photon state created in the down-conversion process B'. While employing efficient detectors is a straightforward method to boost the heralding efficiency, care has to be taken when engineering the PDC process. It might seem straightforward to increase the pump intensity in order to boost the amplitude of the two-photon component in the PDC state. One should however note that at some point the perturbation expansion performed in Eq. (11.18) breaks down and higher-order components have to be considered to correctly model the process.

Using Eqs. (11.22) and (11.23) we calculate the *not-normalized* heralded single-photon state after a successful detection event

$$\begin{aligned}
\rho_s &= \mathrm{tr}\left(\hat{\pi}_1 |\psi\rangle \langle \psi|\right) \\
&= \int d\omega_t'' \, \langle \omega_t''| \left[\eta \iint d\omega_s \omega_t f(\omega_s,\omega_t) \iint d\omega_s' \omega_t' f^*(\omega_s',\omega_t') |\omega_s,\omega_t\rangle \langle \omega_s',\omega_t'|\right] |\omega_t''\rangle \\
&= \eta \iint d\omega_s \, d\omega_s' \left[\int d\omega_t'' f(\omega_s,\omega_t'') f^*(\omega_s',\omega_t'')\right] |\omega_s\rangle \langle \omega_s'| \\
&= \eta \iint d\omega_s \, d\omega_s' \, j(\omega_s,\omega_s') |\omega_s\rangle \langle \omega_s'|. \qquad (11.24)
\end{aligned}$$

Equation (11.24) accurately describes the heralded single-photon state with density matrix $j(\omega_s,\omega'_s)$ that is determined by joint-spectral amplitude $f(\omega_s,\omega_t)$ of the PDC state. $j(\omega_s,\omega'_s)$ already provides some intuition about the purity of the state. If $j(\omega_s,\omega'_s)$ has circular symmetry, and is therefore separable, it can be written as $j(\omega_s,\omega'_s) = f(\omega_s)f^*(\omega'_s)$ and the photon is heralded in a pure state

$$\rho_s = \eta^2 \iint d\omega_s\, d\omega'_s f(\omega_s) f^*(\omega'_s) |\omega_s\rangle \langle\omega'_s|$$
$$= \eta^2 \int d\omega_s f(\omega_s) |\omega_s\rangle \int d\omega'_s f^*(\omega'_s) \langle\omega'_s| = |\psi_s\rangle \langle\psi_s|, \quad (11.25)$$

because pure states always correspond to circular-shaped $j(\omega_s,\omega'_s)$ [18,19].

To obtain further insight into the PDC process we use the Schmidt decomposition theorem [20]. It states that every well-behaved *normalized* two-dimensional function can be decomposed as a sum over a range of positive values κ_k and complete sets of orthonormal functions $g_k(x)$ and $h_k(x)$

$$f(x,y) = \sum_k \kappa_k g(x) h(y), \quad (11.26)$$

with $\sum_k \kappa_k^2 = 1$. Using this decomposition we rewrite the PDC state in Eq. (11.22) as

$$|\psi\rangle_{\text{PDC}} = |0\rangle + b \iint d\omega_s d\omega_t \sum_k \kappa_k g_k(\omega_s) h_k(\omega_t) \hat{a}^\dagger_s(\omega_s) \hat{a}^\dagger_t(\omega_t) |0\rangle, \quad (11.27)$$

where b is used to normalize $B'f(\omega_s,\omega_t)$. Carrying out a basis transformation from the single-frequency modes $\hat{a}^\dagger_s(\omega_s)$ and $\hat{a}^\dagger_t(\omega_t)$ to the broadband frequency modes $\hat{A}^\dagger_k$ and $\hat{B}^\dagger_k$ [21], defined as

$$\hat{A}^\dagger_k = \int d\omega_s\, g_k(\omega_s) \hat{a}^\dagger_s(\omega_s),$$
$$\hat{B}^\dagger_k = \int d\omega_t\, h_k(\omega_t) \hat{a}^\dagger_t(\omega_t), \quad (11.28)$$

the PDC state takes the form of a sum over the broadband modes $\hat{A}^\dagger_k$ and $\hat{B}^\dagger_k$

$$|\psi\rangle_{\text{PDC}} = |0\rangle + b \sum_k \kappa_k \hat{A}^\dagger_k \hat{B}^\dagger_k |0\rangle. \quad (11.29)$$

Hence the photon pair generated during the PDC process with a given frequency distribution $f(\omega_s,\omega_t)$ is emitted into a superposition of strictly correlated broadband frequency modes $\hat{A}^\dagger_k$ and $\hat{B}^\dagger_k$. If one photon is detected in mode $\hat{B}^\dagger_k$ its partner is present in mode $\hat{A}^\dagger_k$, and vice versa.

In a similar manner we expand the detection observables $\hat{\pi}_0$ and $\hat{\pi}_1$ from Eq. (11.23) in the broadband basis of the trigger modes

$$\hat{\pi}_0 = \sum_k |0,h_k(\omega_t)\rangle \langle 0,h_k(\omega_t)|,$$

$$\hat{\pi}_1 = \sum_k \eta |1,h_k(\omega_t)\rangle \langle 1,h_k(\omega_t)|. \quad (11.30)$$

In this broadband formalism, using Eqs. (11.29) and (11.30) the density operator of the heralded single photon can be represented as

$$\rho_s = \sum_k \kappa_k^2 |1,g(\omega_s)\rangle \langle 1,g(\omega_s)|. \quad (11.31)$$

Hence after the detection process the heralded single photon is projected into a statistical mixture of broadband single-photon states with probabilities κ_k^2. The number of κ_k and their respective amplitudes depend on the joint spectrum of the initial PDC state. The reason for this behavior can be attributed to the single-photon detector, which cannot distinguish between the different optical modes. It is impossible to know, in principle, which optical mode was responsible for triggering the detection event. Therefore the heralded single photon is emitted into a statistical mixture of broadband modes.

There exist a variety of different measures to characterize the purity of the heralded single photons. Common are the cooperativity $K = \frac{1}{\sum_k \kappa_k^4}$ [22], which is unity if the signal photon is in a pure state and increases with increased amounts of mixing, or the von Neumann entropy $S = -\sum_k \kappa_k^2 \log_2 \kappa_k^2$ [23], which ranges from zero for a pure signal photon to infinity for rising degrees of impurity.

11.2.4 Heralding Pure Single-Photon Fock States

We now turn our attention to the heralding of pure single photons, or single-photon Fock states, which are vital for various quantum information applications, especially quantum computing. While the heralded signal photons from the PDC process can be delivered in a good approximation of a single-photon state for suitable experimental parameters, the state is not necessarily pure. In order to achieve a pure single-photon state various methods can be pursued.

11.2.4.1 Spectral Purity: Filtering
The simplest and most straightforward approach to create pure Fock states from PDC uses narrowband spectral filtering in the heralding arm [24]. The spectral filter acts as a beamsplitter through which only a single frequency is transmitted

to the detector and all other frequencies are reflected or absorbed by the filter. The general input-output relation of the filter can be written as

$$\hat{a}(\omega) \Rightarrow T(\omega)\hat{a}(\omega) + R(\omega)\hat{b}(\omega), \quad (11.32)$$

where $T(\omega)$ and $R(\omega)$ obey the standard beamsplitter relations. If we assume a delta-function-like transmission of the filter at frequency ω_f, i.e., $T(\omega) = \delta(\omega - \omega_f)$ and condition on the heralding photon passing the filter, the PDC state becomes

$$|\psi\rangle_{\text{PDC}} = \iint d\omega_s d\omega_t f(\omega_s, \omega_t) \delta(\omega_f - \omega_t) \hat{a}_s^\dagger(\omega_s) \hat{a}_t^\dagger(\omega_t) |0\rangle$$
$$= \int d\omega_s f(\omega_s, \omega_f) \hat{a}_s^\dagger(\omega_s) \hat{a}_t^\dagger(\omega_f) |0\rangle. \quad (11.33)$$

After detection of the filtered trigger the heralded signal photon is projected onto the *not*-normalized signal state

$$|\psi\rangle_s = \eta \int d\omega_s f(\omega_s, \omega_f) \hat{a}_s^\dagger(\omega_s) |0\rangle. \quad (11.34)$$

The heralded signal is in a *pure* single-photon state, yet still spans a range of different frequencies, i.e., it is in a broadband single-mode quantum state. Hence the multimode nature of the PDC state is effectively suppressed by filtering the trigger photon down to a single-frequency mode. This is due to the fact that, after the filtering, only a *single* optical frequency mode is impinging on the detector, which leads to a "collapse" of the signal photon into pure, but broadband, single-photon Fock state. However, in practice, care has to be taken when performing the filtering because the filter bandwidth will always have some finite width. This limits the purity of the heralded single-photon state. Additionally standard filters do not feature unit transmissivity. Thus, in this approach to producing pure single-photon states the majority of the trigger photons are absorbed by the filter, severely reducing the rate of successful heralding events. In addition, the losses associated with filtering the trigger may enhance the multi-photon contribution of the PDC state to the heralded signal [25].

11.2.4.2 Spectral Purity: Extended Phase Matching and Group-Velocity Matching

A more sophisticated approach to herald Fock states is to engineer the down-conversion process itself. If the generated PDC state is emitted into only a single-frequency mode, such as $|\psi\rangle_{\text{PDC}} = |0\rangle + \hat{A}^\dagger \hat{B}^\dagger |0\rangle$, then the heralded single photon will be projected into the pure state $\rho_s = \hat{A}^\dagger |0\rangle \langle 0| \hat{A}$.

This configuration can be achieved if the shape of the joint-spectral distribution $f(\omega_s, \omega_t)$ is engineered appropriately. The Schmidt decomposition theorems (Eq. (11.26)) tell us that to obtain a photon pair in a single optical mode, it is necessary for the joint-spectral distribution $f(\omega_s,\omega_t)$ to be factorable as $f(\omega_s,\omega_t) = g(\omega_s)h(\omega_t)$. In other words all κ_k coefficients have to be zero except one. Hence it is necessary to find experimental conditions such that the joint-spectral amplitude distribution from Eq. (11.22)

$$f(\omega_s,\omega_t) = \alpha(\omega_s + \omega_t)\Phi(\omega_s,\omega_t), \quad (11.35)$$

does not feature any frequency correlations between the signal and the trigger.

Consider using a cw pump laser with central frequency ω_c, in which case $f(\omega_s,\omega_t)$ becomes

$$f(\omega_s,\omega_t) = E_p\delta(\omega_s + \omega_t - \omega_c)\Phi(\omega_s,\omega_t). \quad (11.36)$$

In this specific case, the joint-spectral-amplitude distribution is diagonal and continuous in the frequency space (ω_s,ω_t) and hence is not factorable and cannot be written in product form. In fact it is impossible to directly create photon pairs in a single-frequency mode when using a cw laser. Pulsed lasers, however, can achieve the desired effect. Standard pulsed lasers can produce Gaussian-shaped pulses with a given spectral width σ_p and central frequency ω_c, resulting in a JSA is given by

$$\begin{aligned} f(\omega_s,\omega_t) &= A_p \exp\left[-\left(\frac{(\omega_s + \omega_t - \omega_c)^2}{2\sigma_p^2}\right)\right] \Phi(\omega_s,\omega_t) \\ &= A_p \exp\left[-\left(\frac{(\omega_s + \omega_t - \omega_c)^2}{2\sigma_p^2}\right)\right] \operatorname{sinc}\left[\Delta k(\omega_s,\omega_t)L/2\right]. \end{aligned} \quad (11.37)$$

From Eq. (11.37) an analytical expression for the factorability condition can be derived (see [26])

$$\frac{2}{\sigma^2} + \gamma L^2 \left(\frac{1}{v_p} - \frac{1}{v_s}\right)\left(\frac{1}{v_p} - \frac{1}{v_t}\right) = 0, \quad (11.38)$$

where v_i labels the group-velocities of the three interacting fields and $\gamma \approx 0.193$. Therefore the condition for directly generating PDC states that will lead to the heralding of pure single-photon Fock states is to find a crystal in which the pump travels slower than the trigger wave, and travels faster than the signal wave (or vice versa). Engineering the source in this manner has the advantage that no filtering is required, and hence the generated trigger photons arrive at the detector with low loss, resulting in high heralding rates. It is indeed possible to find materials that can fulfill these conditions by generating these waves in different polarizations [26] or directions [27]. Examples will be discussed in the remainder of this section.

11.2.4.3 Photon-Number Purity

Based on the quantum state produced from PDC, we must consider the details of the photon-number distribution of the heralded single-photon states achieved by detecting one mode from a photon pair. Given the strict photon-pair correlations imposed by the PDC process there will be multi-photon terms from the higher-order emissions, however, if the heralding is done by a perfect single-photon detector with ideal coupling efficiency, this scheme would allow to herald perfect single photons.

The photon-number statistics can be derived for the state generated by PDC by applying the PDC Hamiltonian in the approximation of a strong pump mode, acting on two optical modes a_1, a_2 in the form $H_{\text{PDC}} = \varepsilon(a_1^\dagger a_2^\dagger + a_1 a_2)$, where ε is an interaction parameter that includes the nonlinearity, the phase-matching, and the interaction time. The state, including the higher-order photon-number terms with their correct amplitudes, can be estimated by evolution of this Hamiltonian for vacuum inputs

$$\exp(-i H_{\text{PDC}})|0,0\rangle_{a_1,a_2}. \tag{11.39}$$

For $\varepsilon < 1$ this yields the output state in the form

$$|\Psi\rangle_{\text{PDC}} = c_0|00\rangle_{a_1,a_2} + c_1|11\rangle_{a_1,a_2} + c_2|22\rangle_{a_1,a_2} + \cdots, \tag{11.40}$$

with the amplitudes $c_i = \sqrt{1-\varepsilon^2} \cdot \varepsilon^i$ [28]. One of the modes will be the trigger mode, the other the signal. Because the detection process on the trigger photon corresponds, at least in principle, to a photon-number measurement, the remaining photon is no longer in a superposition of number states, as originally given by the PDC. Rather it is a mixture of number states weighted with the appropriate probability. Typically the purity in photon numbers for the heralded single-photon state is characterized by the normalized second-order correlation function $g^{(2)}(t_1, t_2)$, see Chapter 2: Photon statistics and measurements.

The residual photon statistics of the heralded photon will depend on the efficiency of the trigger detection, and on the type of trigger detector, specifically whether it is a click/no-click detector (often referred to as a threshold detector); or a photon-number resolving detector. See Chapters 3–7 for further details on single-photon detectors.

As discussed earlier in this book, the response of a click/no-click detector with single-photon detection efficiency η to two or more photons is given by the combined probability that any number of the incident photons produces a click. For example, with two photons impinging on the detector, $|2\rangle\langle 2|$, a click-event may be generated with probability η^2 in the case that both photons are absorbed, or with probability $\eta(1-\eta)$, that one is absorbed and not the other. Because the latter combination can occur in two different ways, the combined probability for a "click" is $\eta^2 + 2\eta(1-\eta) = 1 - (1-\eta)^2$. Generalizing this approach leads to the click probabilities for a given n-Fock state $P_{\text{BD}}(n) = 1 - (1-\eta)^{(n)}$, with measurement operator

$\hat{\Pi}_{\text{"click"}} = \sum_{n} \left(1 - (1-\eta)^n\right) |n\rangle\langle n|$ (see Chapter 9, and the models described in [29,30]). A photon-number resolving (PNR) detector has a different characteristic, as it can discriminate the number of photons that are detected. For instance, the measurement operator for the detection of exactly one photon has the form $\hat{\Pi}_1 = \sum_{N=1}^{\infty} \binom{N}{1}(1-\eta)^{N-1} \eta |N\rangle\langle N|$, see Refs. [31,30].

In order to determine the photon-number state of the heralded output, these photon-number detection characteristics are used to describe the trigger detector receiving the multi-photon state generated from PDC, shown in Eq. (11.40). For a click/no-click detector this gives a heralded state described by the density matrix:

$$\rho_{\text{heralded1}} = \frac{1}{M} Tr_{a_1}(\hat{\Pi}_{\text{PDC}} |\psi\rangle\langle\psi|_{\text{PDC}}) \tag{11.41}$$

$$= Q|0\rangle\langle 0|_{a_2} + P_1|C_1|^2 |1\rangle\langle 1|_{a_2} + P_2|C_2|^2 |2\rangle\langle 2|_{a_2} + \cdots, \tag{11.42}$$

which describes a mixture of photon-number states, where $P_1, P_2, \ldots$ are the detection probabilities for a "click" in the trigger detector, as defined above for the respective photon-number terms, $C_1, C_2, \ldots$ are the amplitudes of the terms from the PDC state, Q is the probability that the trigger detector does not fire, and $\frac{1}{M}$ is the renormalization factor.

The simulation in Fig. 11.3 shows how $g^{(2)}(0)$ changes with the probability of detecting a trigger photon within the time unit of the system (for example this could be the duration of the pump pulse or the timing window of the detectors), under variation of the interaction strength of the PDC. Essentially this result shows that with a higher pair production through stronger pumping, the photon-number purity suffers severely. One must therefore choose the PDC interaction strength according to what can be tolerated in the specific application [114].

11.2.4.4 Improving the Heralding Rates with Switched PDC

Given the non-deterministic behavior of the PDC interaction, the heralding of a single-photon state will only occur with a low probability. As shown above, the pump strength cannot be increased arbitrarily, because as the probabilities for a trigger detection increase, so will the likelihood of multi-pair emission.

One elegant solution is to combine multiple PDC sources using an array of switches so that the signal output of a PDC source from which a trigger photon was successfully detected can be switched into a single combined output. If enough PDC sources are combined, the chances that at least one PDC process triggers is close to 1, but at the same time each individual heralded photon can be produced with the desired low $g^{(2)}(0)$.

Several possible schemes of this type have been studied, including parallel switches for several spatial modes [32], and switched photon storage in rings or cavities [33]. Generally, all these schemes are challenging to implement, because the additional optical losses caused by currently available switching

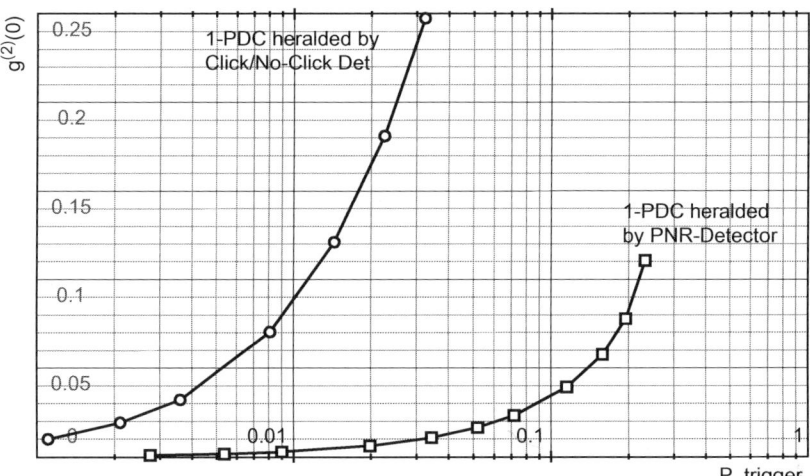

FIGURE 11.3 Single-photon quality $g^{(2)}(0)$ for photons heralded from PDC, versus the probability for a click in the trigger detector. The pump strength of the PDC is varied in the simulation. The assumed detector efficiencies are 65% for the click/no-click detector and 95% for the PNR detector.

technologies outweigh the gains from the combination of several PDC's, as was analyzed in [34].

11.3 BULK-CRYSTAL PDC

In this section we explore parametric down-conversion in *bulk* crystals, i.e., standard nonlinear crystals without poling or embedded waveguides. The vast majority of PDC sources have been realized in bulk crystals, and while they are increasingly being supplanted by periodically-poled structures in continuous-wave down-conversion—due to the higher achievable brightnesses—they are still the system of choice for the generation of multi-photon states in pulsed-pump PDC.

11.3.1 Birefringent Phase-Matching

In Section 11.2.2, we derived the general PDC Hamiltonian. To generate PDC in practice, the three interacting fields have to conserve energy, i.e., the frequencies match

$$\omega_p = \omega_s + \omega_t, \quad (11.43)$$

and conserve momentum *in the crystal*, i.e., the phases match

$$\mathbf{k}_p = \mathbf{k}_s + \mathbf{k}_t, \quad (11.44)$$

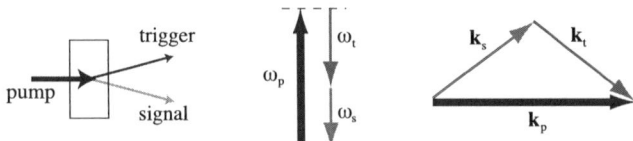

FIGURE 11.4 Energy and momentum conversion between a pump (p), a signal (s), and a trigger field (t) in parametric down-conservation.

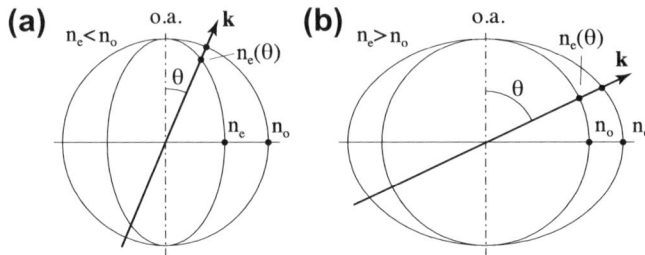

FIGURE 11.5 Refractive index for a light field **k** propagating in birefringent crystals. (a) Negative uniaxial crystal with $n_o > n_e$. (b) Positive uniaxial crystal with $n_e > n_o$. o.a.: optical axis.

as shown schematically in Fig. 11.4. The phase-matching condition Eq. (11.44) is subject to dispersion in the nonlinear medium, through $k = \omega n(\omega)/c$, see also Eq. (11.7). To satisfy Eq. (11.44), the polarizations and wave vectors of the interacting fields must be carefully chosen such that $\omega_p n_p(\omega_p) = \omega_s n_s(\omega_s) + \omega_t n_t(\omega_t)$. In an isotropic bulk crystal however, the normal dispersion ensures that this is not possible while also fulfilling Eq. (11.43).

The solution is to use anisotropic materials, in which fields with different polarizations experience different refractive indices. The plane containing the optical axis and the pump wave vector is called the principal plane, and we denote a light beam polarized orthogonally to that plane the *ordinary* (o) and the beam polarized within that plane the *extraordinary* (e) beam. As can be seen in Fig. 11.5 for a uniaxial crystal, the refractive index for the ordinary beam, n_o, is independent of the field orientation in the crystal. The extraordinary refractive index, shown as an ellipsoid, is dependent on the angle θ between the field vector and the optical axis of the crystal, $n_e(\theta) = (\cos^2(\theta)/n_o^2 + \sin^2(\theta)/n_e^2)^{-1/2}$.

Phase-matching in such a material can be achieved for orthogonally-polarized fields through birefringent phase-matching, which is most commonly done by tuning the angle between the crystal axes and the interacting fields. This technique is also called *critical* phase-matching, because it is quite sensitive to deviations from optimal conditions and thus limits the angular, spectral, and temperature acceptance bandwidth.

Alternatively, the angle θ can be set to 90° and the phase-matching can be achieved by varying the temperature of the crystal, which changes the relative

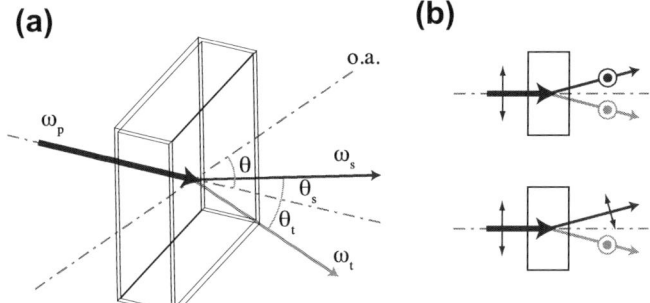

FIGURE 11.6 Parametric down-conversion in a bulk crystal. (a) Angular phase matching. The optical crystal axis (o.a.) spans an angle θ with the pump field. The down-converted fields emerge at angles θ_s and θ_t. (b) PDC in birefringent materials can be achieved via type-I (top) or type-II (bottom) phase matching.

size of the ordinary and extraordinary ellipsoids, Fig. 11.5. This technique is called *non-critical* phase-matching (and sometimes temperature- or 90° phase-matching).

Angular phase matching is schematically shown in Fig. 11.6a. The pump field spans the angles θ_s and θ_t with the two down-converted fields. There are two phase-matching options in this case, type-I down-conversion of a pump beam orthogonally polarized to the two co-polarized down-conversion fields, or type-II down-conversion, where the down-converted fields are orthogonally polarized, see Fig. 11.6b. In a negative uniaxial crystal, type-I phase matching is possible for an extraordinary pump splitting into two ordinary PDC photons, $e \rightarrow o + o$, while type-II PDC can be achieved for $e \rightarrow o + e$. In a positive uniaxial crystal, this situation is inverted, $o \rightarrow e + e$, $o \rightarrow e + o$. The third option, type-0 phase matching where all fields are co-polarized, cannot generally be implemented in bulk materials but is often used in periodically-poled crystals.

One of the challenges of building PDC sources using angle phase-matching is transversal walkoff between the e and o fields. This walkoff restricts the useful length of a bulk crystal and thus limits the photon-pair yield. This can only be avoided by choosing a collinear PDC configuration along one of the crystal axes, which can be achieved by using non-critical phase matching in bulk crystals.

The dispersion relations, i.e., the exact wavelength- and temperature-dependence of the refractive indices, are usually available in the form of empirical *Sellmeier* equations (see [35] for one of the most comprehensive compilations of nonlinear crystal properties). Using these equations, the phase-matching conditions can be solved numerically for a set of target frequencies—which are usually subject to availability of suitable pump lasers and photon detectors. A particularly useful free software suite for this purpose is *SNLO*, written by A. Smith [36]. It contains data for more than 50 commonly used

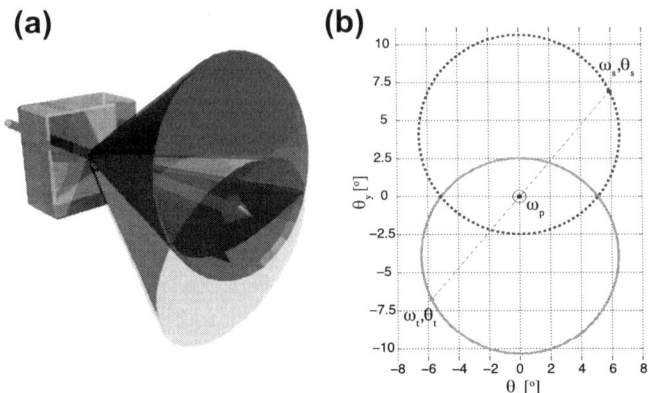

FIGURE 11.7 Non-collinear type-II down-conversion in a BBO crystal, with its optical axis oriented at $\theta = 41°$, pumped at 410 nm.

crystals and allows modeling of most nonlinear optical applications, such as optical parametric oscillators, second-harmonic generation, and parametric down-conversion.

We now present two example solutions for widely used PDC applications. The first is type-II noncollinear PDC in a negative uniaxial β-barium borate (BBO) crystal pumped by a 410 nm laser to generate frequency-degenerate down-conversion at 820 nm. The characteristic emission cones are shown in Fig. 11.7. This geometry is a popular choice for generating polarization-entangled photons, and has had enormous success in experimental quantum information processing since its first demonstration in [8].

The second example is collinear PDC ($\theta_s = \theta_t = 0$) in a biaxial bismuth borate (BiBO) crystal. The BiBO crystal is increasingly replacing BBO due to its higher nonlinearity, as for example demonstrated in [37]. While it is more complicated to calculate phase-matching conditions for biaxial crystals, they are also more versatile. Exemplary tuning curves, for the same pump wavelength of 410 nm, are shown as a function of θ for type-I and type-II phase matching in Fig. 11.8a and b, respectively. These tuning curves show a characteristic difference between type-I and type-II schemes. For frequency-degenerate schemes, where $\omega_s = \omega_t$, the bandwidth of type-I schemes is far broader for a given acceptance angle then for type-II schemes. The bandwidth quickly decreases for type-I as one moves away from degeneracy, whereas it is more or less constant over a wide frequency range for type-II configurations. Furthermore, the type-I configuration does not have a phase-matching solution for angles above the critical angle. A rigorous analysis of non-collinear type-II PDC in BiBO has recently been conducted [38].

Once a crystal has been chosen and a phase-matching solution has been found, the next design challenges are: to optimise (i) the PDC brightness, (ii) the photon heralding efficiency, and (iii) the heralded-photon purity. As we shall

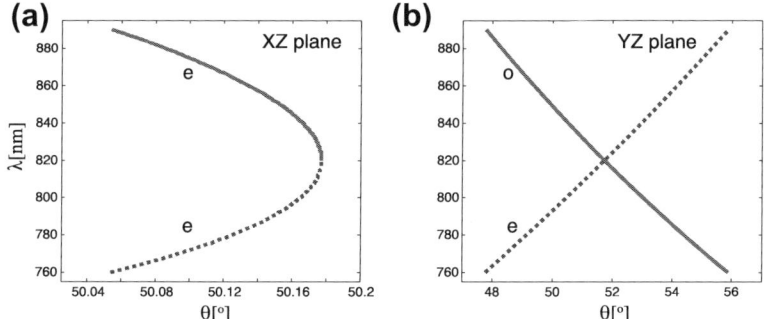

FIGURE 11.8 Tuning curves for PDC in a biaxial BiBO crystal, as a function of optical axis orientation θ, pumped at 410 nm. (a) Type-I phase-matching, $o \to e + e$, with the crystal cut along the X-Z plane. (b) Type-II phase-matching, $e \to o + e$, along the Y-Z plane.

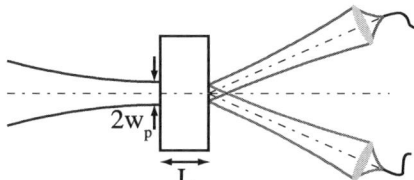

FIGURE 11.9 Fiber coupling non-collinear down-conversion light. The pump beam is focused to a waist size w_p in the crystal of length L. The PDC modes are coupled into two optical fibers, whose mode field diameter is matched to the PDC mode with two lenses.

see, these are intimately interconnected. The source brightness can be defined as either the collected or detected number of photon pairs per unit time and unit pump power. Depending on the application, this is sometimes further divided by the PDC output bandwidth, giving the spectral brightness. Typical values are summarized in Fig. 11.32.

The heralding efficiency is the probability of a PDC photon being in the signal mode conditioned on detection of the trigger photon (see also the formal definition in Section 11.2.3). Both the brightness and the heralding efficiency are primarily controlled by optical mode-matching of the pump and PDC fields, see Fig. 11.9. Early experiments relied mostly on pinholes for spatial filtering and collection of the PDC light, a technique that was soon replaced by the use of single-mode. One early analysis showed that fibre coupling can be optimized for a given spectral range by tuning the (continuous-wave) pump-beam focusing to match the angular spread of the PDC emission to the angular acceptance width of the fibre modes [39]. This work was followed by more sophisticated numerical models for both the continuous-wave [40,41] and—for heralding applications—the pulsed-pump cases [42]. Due to the complicated spatio-temporal form of the two-photon state generated in non-collinear PDC, the calculations must be solved numerically. In the following we briefly outline how this problem can be approached [42].

11.3.1.1 Transverse Two-Photon State in Bulk Crystals

In contrast to the derivation of the collinear case in Section 11.2.2, the exact description of non-collinear PDC includes the transverse components of the interacting fields. The positive and negative components of the field operators (cf. Eq. (11.14)) thus take the form $\hat{E}_i^{(+)}(\mathbf{r}) = \hat{E}_i^{(-)\dagger}(\mathbf{r}) = A \int d^2\mathbf{k}_i d\omega_i \, e^{i(\mathbf{k}_i(\omega_i)\mathbf{r} - \omega_i t)} \hat{a}_i(\mathbf{k}_i, \omega_i)$. The corresponding two-photon state is

$$|\psi\rangle = \int d^2\mathbf{k}_{s\perp} d^2\mathbf{k}_{t\perp} d\omega_s d\omega_t \, \Phi(\mathbf{k}_{s\perp}, \omega_s, \mathbf{k}_{t\perp}, \omega_t) a_s^\dagger(\mathbf{k}_{s\perp}, \omega_s) a_t^\dagger(\mathbf{k}_{t\perp}, \omega_t)|0\rangle, \quad (11.45)$$

with the two-photon amplitude

$$\Phi(\mathbf{k}_{s\perp}, \omega_s, \mathbf{k}_{t\perp}, \omega_t) = \int_0^L dz \, A_p(\mathbf{k}_{s\perp} + \mathbf{k}_{t\perp}, \omega_s + \omega_t) e^{i\Delta k_z(\mathbf{k}_{s\perp} + \mathbf{k}_{t\perp}, \omega_s + \omega_t)z}, \quad (11.46)$$

where A_p represents the pump field. The phase-mismatch Δk_z is now defined through the z components of the wave vectors involved

$$\Delta k_z(\mathbf{k}_{s\perp} + \mathbf{k}_{t\perp}, \omega_s + \omega_t) = k_{pz}(\mathbf{k}_{s\perp} + \mathbf{k}_{t\perp}, \omega_s + \omega_t) \\ - k_{sz}(\mathbf{k}_{s\perp}, \omega_s) - k_{tz}(\mathbf{k}_{t\perp}, \omega_t). \quad (11.47)$$

Equation (11.46) describes the PDC field for all possible output directions, but will only have non-zero amplitude for the propagation directions allowed by the phase-matching conditions. Using these equations one can now tackle the problem of optimizing the key performance parameters of a non-collinear PDC source, for example, by numerically maximizing the overlap of the output wave function with optical fiber modes, as shown in [42]. This treatment also provides a starting point for obtaining conditions for which the joint-spectral PDC amplitude is factorizable, which is the prerequisite for a high heralded-photon purity [115].

11.3.2 Heralded Single Photons from Triggered PDC

A plethora of PDC sources have been demonstrated for the generation of one or more (entangled) photon pairs for experiments in quantum information processing. All of them serve as examples of the creation of heralded single photons via PDC even though they were usually not employed for that purpose.

The creation of *pure* heralded single photons (see theory in Section 11.2.4) via PDC has only recently attracted the focus of in-depth research. In the following we describe two approaches toward this goal. The first investigates how vectorial matching can be used to create pure photons. The second, which can be combined with most other techniques, involves multiplexing downconverters to increase the likelihood of the creation of single pairs while reducing multi-pair emissions.

11.3.2.1 Photon-State Tuning Through Vectorial Matching

In Section 11.2.3 we discussed the theory and necessity of creating photon pairs with a separable joint-spectral amplitude, defined in Eq. (11.35). The goal is to tailor the JSA such that the measurement of the heralding photon does not interfere with the state of the signal photon. In particular, the two photons must not be correlated in frequency. This can be achieved through *group-velocity matching* [26] in one of the following two ways.

Asymmetric JSA
The first is to match the group-velocity of either down-conversion photon to that of the pump field, e.g., $v'_t = v'_p$. This creates a vertical phase-matching function and, consequently, a separable, asymmetric JSA, see Fig. 11.10, which corresponds to emission of one narrowband and one broadband photon. In a bulk crystal, this can be achieved through careful vectorial matching in combination with the proper choice of the spectral properties of the pump laser. However, the necessity of simultaneous phase-matching and group-velocity matching imposes very stringent conditions and can only be achieved for very specific sets of wavelengths in a limited number of nonlinear crystals. This limitation is alleviated by the advent of periodically-poled crystals, which offer more experimental degrees of freedom, as discussed in Section 11.4.

This particular group-velocity matching technique was first demonstrated in an experiment by Mosley *et al.* [43]. The authors employed a frequency-doubled Ti:sapphire laser with 50 fs pulses to pump a pair of 5 mm potassium dihydrogen phosphate (KDP) crystals at 415 nm. For type-II down-conversion in this configuration, the group-velocity of the e-polarized 415 pump field is matched to the o-polarized 830 nm down-conversion photon which theoretically would suffice for the creation of uncorrelated photons. However, the group-velocity matching only holds for a plane wave. In practice, the pump beam has to be focused into the PDC crystal to increase the pair yield and coupling of the down-converted fields into single-mode fibers. This leads to an angular spread of the pump field, which leads to non-ideal group-velocity matching. In [43]

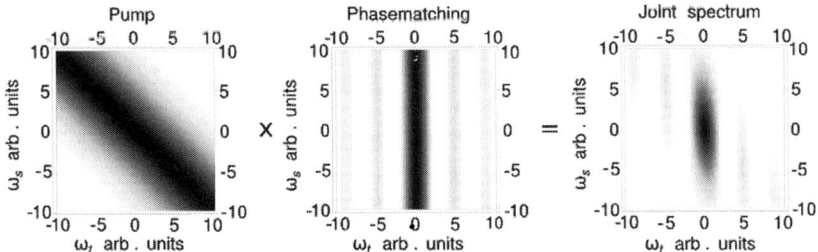

FIGURE 11.10 Group-velocity matching. The condition $v'_t = v'_p$ results in an asymmetric phase-matching function and a vertical, separable joint-spectral amplitude.

this problem was addressed (to some degree) by adjusting the spatial chirp the pump acquired in the frequency-doubling crystal. A detailed analysis of the effects of focusing and spectral pulse shaping can be found in [44].

As a quality benchmark, the authors demonstrated two-photon interference of signal photons created in two independent PDC crystals. This type of experiment was first demonstrated by Hong, Ou, and Mandel [7], and is the basis for many protocols in linear optical quantum information processing. Importantly, the visibility of this two-photon interference is a direct measure of photon-number and spectral purity once the unavoidable spatio-temporal distinguishability in the measurement is accounted for. In [44], the authors achieved an interference visibility of 94.7% for independently heralded photons. The key achievement was that no narrowband spectral filters were used, which usually limit the source brightness in multi-photon schemes.

Symmetric JSA

The second option to attain group-velocity matching is to match the mean group-velocity of the two down-converted fields to the pump group-velocity, $v'_p = (v'_s + v'_t)/2$, see Fig. 11.11. This creates a phase-matching function with orthogonal orientation to the pump function. The resulting JSA is thus roughly circular. In order to achieve perfect separability, the widths of the phase-matching and pump functions must be equal, which can be controlled via crystal length or the pump spectrum, respectively [26].

In either case, one can see that the final JSA is still not perfect—the phase-matching function has a distinct sinc shape, which emanates from the integration of the PDC Hamiltonian over a finite crystal length, see Eq. (11.19). The associated side lobes can not be eliminated through group-velocity matching [24]. One way to reduce this residual distinguishability would be to use frequency filtering, but this defies the main purpose of group-velocity matching. Alternatively, one can manipulate the nonlinearity of a PDC crystal to produce a Gaussian-shaped phase-matching function, a technique which has been

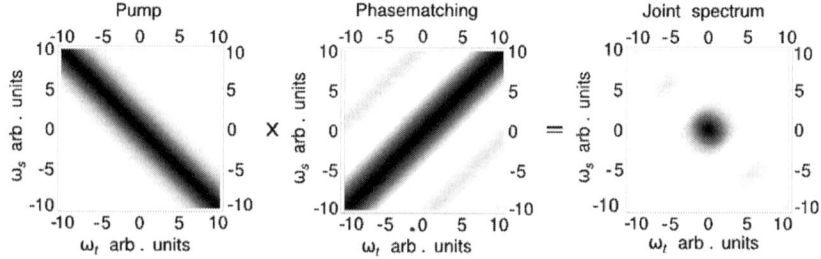

FIGURE 11.11 Group-velocity matching. Setting $v'_p = (v'_s + v'_t)/2$ creates a symmetric phase-matching function and a circular joint-spectral amplitude.

demonstrated for periodically-poled crystals in [45] and will be outlined in Section 11.4.6.

11.3.2.2 Multiplexed PDC Sources

As shown in Section 11.2.4, the number purity of heralded single photons is strongly dependent on the PDC pump power. The simplest method to reduce multi-photon emission and improve this purity is to operate at lower pump power. This however limits the source brightness, which one cannot always afford. Instead, several sources at lower power can be used in parallel and multiplexed into a single output as discussed above.

In the *spatial* multiplexing scheme suggested in [47], and shown schematically in Fig. 11.12a, several down-converters are run in parallel in a photon switchyard. The efficiency for this scheme was simulated in detail in [46]. It is critically limited by the switchyard architecture and the characteristics of the switches, such as switching time and optical loss.

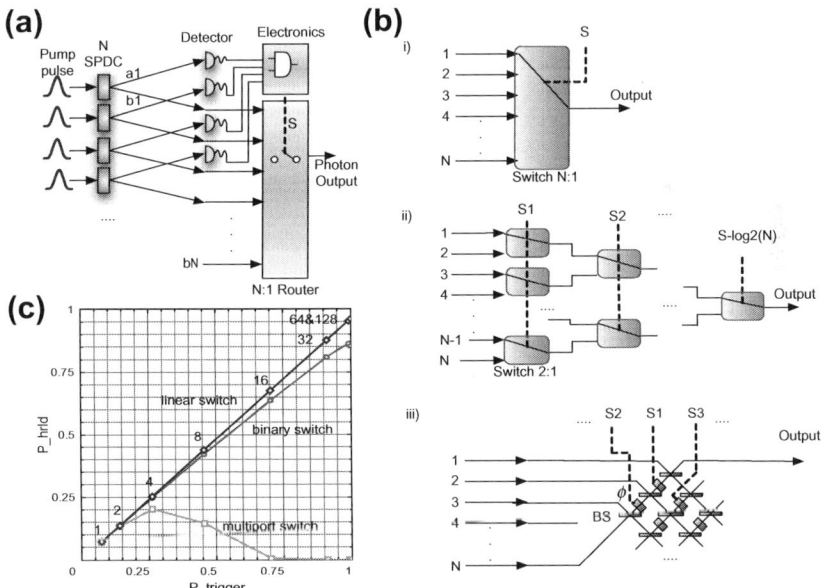

FIGURE 11.12 Heralded photons from spatially multiplexed PDC sources [46]. (a) Tightly synchronized laser pulses pump N PDC sources. One photon from each source is used as a trigger $a_1 \ldots a_N$ while the signal modes $b_1 \ldots b_N$ are routed into the single output mode via an active N:1 switch. (b) Three-photon switchyard configurations are simulated. (i) A single N:1 optical switch; (ii) A succession of 2:1 binary switches; (iii) An all-optical multiport switch. The performance of the multiplexing scheme depends crucially on switching speeds and optical loss in these architectures. (c) Simulation of the three switching schemes, showing the photon heralding efficiency P_{hrld} versus the triggering probability $P_{trigger}$.

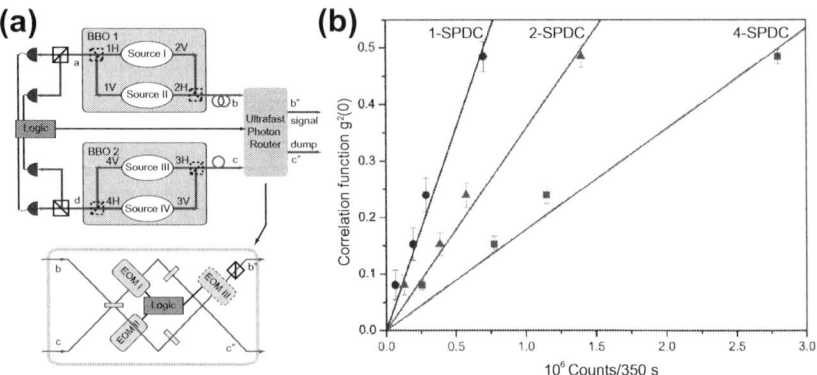

FIGURE 11.13 Demonstration of active PDC multiplexing [34]. (a) Experimental scheme for multiplexing 2 and 4 PDC sources into a common signal mode via fast electro-optical switches (EOM). (b) Correlation function $g^{(2)}(0)$ measured for the multiplexed signal photon. The results show a clear improvement of photon purity for comparable count rates.

Experimentally, this multiplexing scheme has been demonstrated for 2 and 4 PDC sources in [34], see Fig. 11.13. Photon pairs were created in type-II phase-matched BBO crystals. For the 2:1 scheme, one photon of each pair was split off with a polarizing beamsplitter and served as the trigger. The 4:1 scheme was realized with a more sophisticated setup which was based on polarization-entangled photons. The signal photons in both cases were combined with an ultrafast photon router based on electro-optical switches, as shown in Fig. 11.13a. The authors measured the two-photon correlation function $g^{(2)}(0)$ as a benchmark for single-photon purity. The results in Fig. 11.13b show that $g^{(2)}(0)$ for a single PDC source decays quickly toward 1 for high pump powers, where multi-pair emissions dominate. In comparison, the 2:1 and 4:1 multiplexed schemes allow higher pump power and thus higher signal rates for the same $g^{(2)}(0)$, a convincing improvement.

Alternatively, sources can be multiplexed in time. Typically, PDC experiments are pumped with Ti:Sapphire lasers at a fixed pulse repetition rate of 76 MHz, which corresponds to an interval of 13 ns between subsequent pulses. The timing windows for heralded photons can, however, be as small as 1 ns, with the main limitation being the photon detectors and electronics. In this case an experiment could be operated at repetition rates of up to 1 GHz. This would allow operation of a PDC source at lower power, to diminish the number of unwanted higher-order pair emissions, while keeping the signal rate at a reasonable level.

Since the laser repetition rate is usually fixed, it has to be changed externally, as demonstrated in a recent experiment in [48], see Fig. 11.14a. A simple optical delay line, consisting of two beamsplitters and two mirrors, splits off 50% of the frequency-doubled, 76 MHz pump laser and recombines it with the pump mode

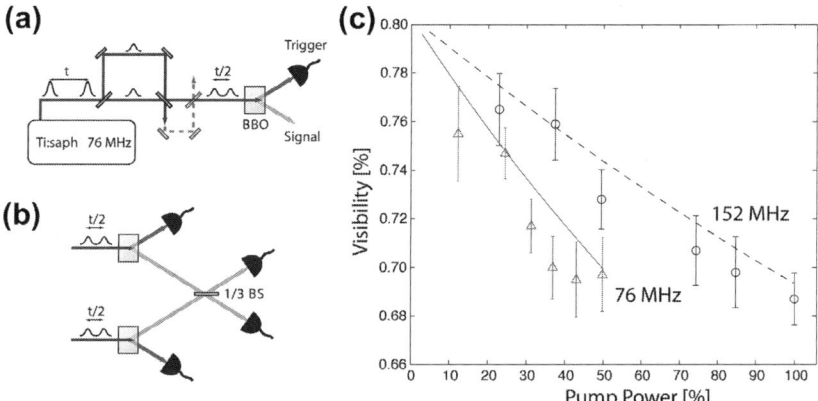

FIGURE 11.14 Temporal multiplexing of PDC sources [48]. (a) The repetition rate of a pump laser can be increased with a simple optical delay line. (b) Two-photon interference on a 1/3 beamsplitter (BS) was used to test the performance of the scheme. (c) The results show that at twice the repetition rate, the interference visibility (which is limited to 80% for this BS) decays by a factor of 2 less with increasing pump power.

after a ≈6.5 ns delay period. The pump repetition rate is doubled to 152 MHz, while the pump power per pulse is reduced to 25%, where 50% of the loss occurs due to the recombination on the second beamsplitter. Usually, this loss can easily be afforded, as state-of-the-art experiments are very efficient and are operated far below the approximately 3–4 W average pump power that can be generated from a frequency-doubled femtosecond Ti:sapph laser. As in [43], this scheme was benchmarked using a two-photon interference experiment between independently generated photons, as shown in Fig. 11.14b. The results in Fig. 11.14c show a clear improvement of the two-photon interference visibility for the doubled repetition rate. The advantage of this temporal multiplexing scheme is that it is simple and can be applied to almost any experiment. The optical delay line can be extended to multiply the repetition rate beyond a factor of 2 without further loss in overall power.

A less complicated alternative for a newly designed experiment would be to employ a pulsed laser with a higher repetition rate. One example would be commercially available picosecond lasers with a repetition rate of 150 MHz. The longer pulse length as compared to femtosecond lasers has some drawbacks, however. Another example are newly available Ti:sapphire femtosecond lasers with up to 1 GHz repetition rate. The overall output power of such a system is considerably lower than for the commonly used 76 MHz Ti:sapphire lasers, but this limitation can be overcome by use of more efficient crystals, as we will see in the next Section 11.4.

Another method to increase the signal to noise ratio of heralded photons, which can be combined with the multiplexing schemes discussed above, is deliberate suppression of unwanted background counts. These background

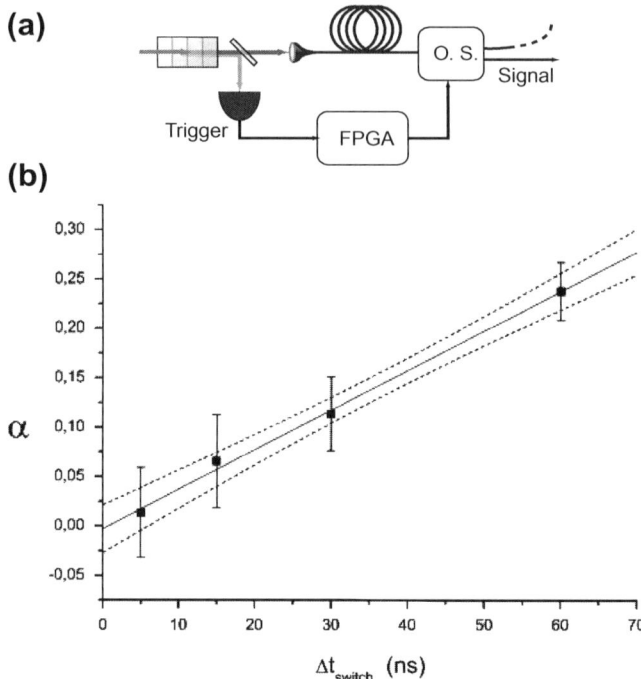

FIGURE 11.15 (a) Scheme for background suppression in heralded PDC [49]. (a) Photon pairs are produced in type-II non-degenerate PDC in a PPLN crystal. The trigger detector synchronizes an optical shutter (O.S.), which blocks the signal line, through a fast field-programmable gate array (FPGA). The 1550 nm signal photon was detected with a InGaAs detector with a timing window of 100 ns. (b) Results. Shorter switching times Δt_{switch} lead to a significant reduction in α, which corresponds to the two-photon correlation function $g^{(2)}(0)$.

counts, either in the heralding or the signal detectors can be intrinsic dark counts caused by the detector, or actual counts induced by background light. The intrinsic dark counts can be minimized by proper choice of detectors (see Chapters 3–7 for an overview).

Extrinsic background counts can be minimized by adding optical shutters to heralded PDC scheme, as demonstrated in [49] and shown in Fig. 11.15. The signal channel in this scheme is blocked until the trigger photon signals the arrival of the signal photon. This technique is especially effective in systems using slow detectors, such as photon-number-resolving transition edge sensors (c.f. Chapter 6). These detectors, due to their high internal jitter, require comparably large coincidence windows ($\approx$100 ns) and operate in a low-count-rate regime. They therefore produce a significant amount of background-induced false-positive coincidence counts, which can be minimized with an optical shutter. The results in Fig. 11.15 show that shutter times below the coincidence timing window increase the single-photon purity.

11.4 PERIODICALLY-POLED CRYSTAL PDC

11.4.1 Quasi-Phase-Matching

Quasi-phase-matching (QPM) was first proposed by Armstrong et al. in 1962 [50] to achieve efficient energy transfer between interacting waves in nonlinear media. It is based on a spatial modulation of the nonlinear properties along the propagation direction. The simplest form of QPM is a periodically alternating orientation of the crystal domain, so that the effective nonlinearity flips between $+d_{\text{bulk}}$ and $-d_{\text{bulk}}$ [50]. In contrast to birefringent phase-matching, the interacting fields still propagate with different phase velocities inside the crystal, but when the accumulated phase mismatch reaches π, the sign of the nonlinear susceptibility is reversed. Rather than starting to interfere destructively the fields at this point start at zero phase difference, which then increases again as the fields propagate until it reaches π again, where the nonlinear susceptibility is reversed once more. This creates a step-wise growth in the output power along the crystal length as shown in Fig. 11.16.

We give here a brief introduction of the principles of QPM, a formal derivation of the QPM can be found in [51]. The longitudinally varying nonlinear susceptibility ($d(z)$) can be expressed as a Fourier series

$$d(z) = d_{\text{bulk}} \sum_m G_m \exp(-ik_m z), \quad (11.48)$$

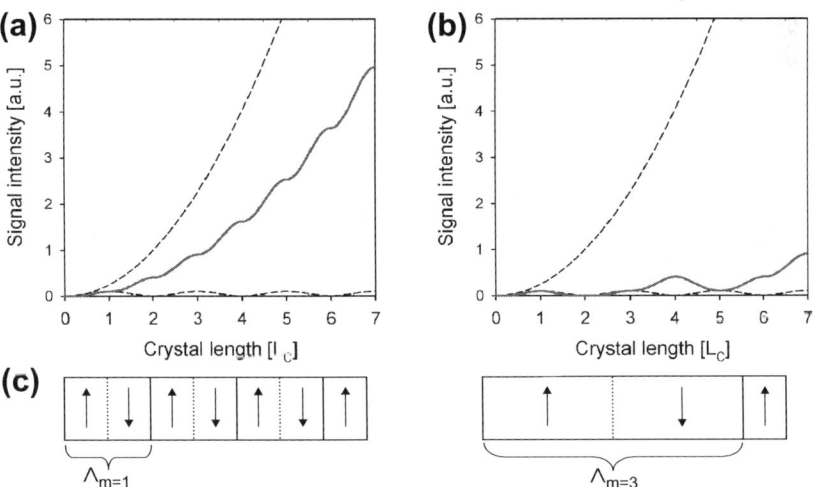

FIGURE 11.16 Solid lines show the effective growth of signal intensity with length in a nonlinear medium (here in units of the coherence length L_c) in the case of first-order QPM (a) and third-order QPM (b). The dashed lines show the signal intensity for perfect birefringent phase-matching and the phase-mismatched case. In (c), the poling structure and periodicity Λ are schematically drawn for the first- and third-order cases.

where d_{bulk} is the nonlinear coefficient of the bulk material, G_m are the Fourier coefficients, $k_m = 2\pi m/\Lambda$ is the grating vector of the m-th Fourier component, and Λ is the spatial period of the modulated structure.

Assuming that only one Fourier component is phase matched and contributes significantly to the PDC process, the integration of the signal amplitude from Eq. (11.9) yields

$$E_s \propto L\, d_Q\, \text{sinc}\left(\Delta k_Q \frac{L}{2}\right), \qquad (11.49)$$

where d_Q is the effective nonlinear coefficient in QPM. For a square-wave modulation of the nonlinearity from $+d_{\text{bulk}}$ to $-d_{\text{bulk}}$ the effective nonlinear coefficient becomes:

$$d_Q = 2\, d_{\text{bulk}}\, \sin(\pi D)/m\pi, \qquad (11.50)$$

where the duty factor $D = l/\Lambda$ is given by the length l of a single reversed domain divided by the period Λ. The value of m determines the order of the QPM. As the efficiency of the nonlinear process is proportional to $1/m^2$, a low-order QPM is desired. However this comes with the price of small periodic structures (Λ). For example, in a first-order QPM (Fig. 11.16) the period Λ has to be twice the coherence length l_c as defined in the bulk medium, see Eq. (11.10). The largest nonlinear coefficient for QPM is obtained for a first-order process with duty cycle of 50%. In this case

$$d_Q = \frac{2}{\pi} d_{\text{bulk}}. \qquad (11.51)$$

The effective nonlinearity in QPM is therefore reduced by at least a factor of $\frac{2}{\pi}$ compared to the value for birefringent phase-matching in the bulk medium.

The Hamiltonian in Eq. (11.21) has the following form for QPM:

$$\int_{t_0}^{t} dt\, \hat{H}_{\text{PDC}}(t') = 2\pi\, B\, d_Q \iint d\omega_s\, d\omega_t\, \alpha(\omega_s + \omega_t)$$
$$\times L\, \text{sinc}\left[\Delta k_Q \frac{L}{2}\right] \hat{a}_s^\dagger(\omega_s)\hat{a}_t^\dagger(\omega_t) + h.c.. \qquad (11.52)$$

The effective wave-vector mismatch due to the quasi-phase-matching is

$$\Delta k_Q = k_p - k_s - k_t - k_m = \Delta k_{\text{bulk}} - k_m, \qquad (11.53)$$

assuming that all wave vectors are collinear with the grating vector. The grating vector is a powerful parameter as it is independent of material properties and can be easily adjusted during crystal fabrication, as shown in the periodic-poling section below. Even if the wave-vector mismatch is non-zero for the bulk material, using quasi-phase-matching a fully phase-matched interaction

($\Delta k_Q = 0$) can be achieved for maximum conversion efficiency. In addition, the phase-matching can be tuned (somewhat) using the temperature dependence of the refractive index, with the wave vector k given by

$$k(\lambda, T) = \frac{n(\lambda, T)\omega}{c}. \tag{11.54}$$

The grating vector k_m is also affected by temperature changes, since the poling period $\Lambda(T)$ depends on the thermal expansion of the medium. The temperature and wavelength dependence of the refractive index can be easily derived using the empirical Sellmeier equations of the medium, e.g., for potassium titanyl phosphate (KTP), the Sellmeier coefficients can be found in Refs. [52,53].

The QPM approach has several advantages over the birefringent phase-matching discussed in the previous chapter:

- Free choice of the trigger and signal wavelengths by appropriate choice of the modulation period Λ.
- For fixed choice of trigger and signal wavelength, the propagation direction of trigger and signal photons can be made collinear with the pump photons, eliminating spatial walk-off effects in the transverse direction.
- With no walk-off, the interaction length inside the nonlinear medium can be very long, increasing the conversion efficiency.
- Free choice of polarization of the pump, trigger, and signal fields. This allows to make use of the largest component of the nonlinear susceptibility tensor to maximize conversion efficiency in a given nonlinear medium.
- Easy tunability of the trigger and signal wavelengths of several nm by controlling the temperature of the nonlinear medium.
- Complex phase-matching conditions can be achieved by using non-periodic modulation of the nonlinear coefficient.

The advantage of QPM is illustrated in Fig. 11.17 where different phase-matching scenarios are realized in the same nonlinear medium, a periodically-poled KTP (ppKTP) crystal. The top row shows data from a simulated type-II phase-matched ppKTP, with signal and trigger wavelengths around 810 nm. The trigger and signal wavelengths for collinear emission are plotted against the poling period, temperature, and pump wavelength. By simply varying the period of the poling, the difference in wavelength can be altered at will around the degeneracy wavelength of 810 nm (set by the pump wavelength) which occurs at $\Lambda = 10.7$ μm. Although not as strong, the additional dependence of the phase-matching on temperature is very helpful in experiments since it allows fine tuning of the trigger and signal wavelengths to match existing filters or atomic absorption lines without interfering with the optical setup. If a tunable pump is available, fine tuning can also be achieved using different pump

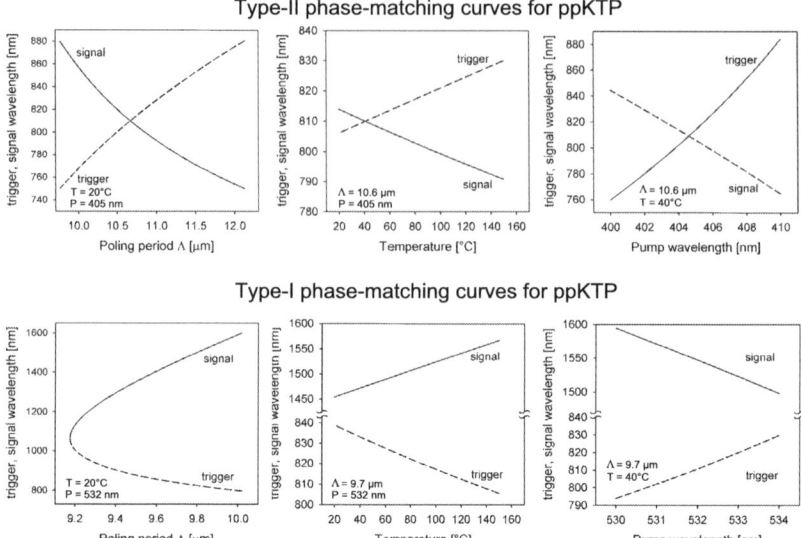

FIGURE 11.17 Simulated phase-matching curves of ppKTP using SLNO (Select Nonlinear Optics (software), by Dr. Arlee Smith, download: http://www.as-photonics.com). The first row shows the phase-matching conditions for a degenerate type-II interaction as functions of three parameters: poling period Λ, temperature, and pump wavelength. The second row shows a non-degenerate type-I interaction.

wavelengths, as shown on the right-most plot. The bottom row of Fig. 11.17 shows a simulation for the same ppKTP crystal but this time phase-matched for type-I with non-degenerate signal and trigger wavelengths. By slightly changing the poling period to $\Lambda = 9.9$ μm, the crystal (at 20°C) produces trigger and signal photons at highly non-degenerate wavelengths. The trigger is found at 810 nm and the signal lies at 1550 nm, a wavelength combination that profits from the high available detection efficiency at 810 nm and low transmission losses at 1550 nm in optical fibers. The same wavelength pair can be obtained from a crystal with period $\Lambda = 9.7$ μm when heated to about 130°C, as shown on the bottom center of Fig. 11.17.

Note that there is no crossing of the trigger and signal wavelengths at the degeneracy point, as in the type-II case. Since in type-I both photons have the same polarization they become indistinguishable and interchangeable at degeneracy. As a result the natural bandwidth of trigger and signal at degeneracy is larger for type-I phase-matching.

Even if the period Λ is fixed after production of the crystal, the exact wavelengths for collinear emission can still be tuned using either the crystal temperature or pump wavelength. It is clear that this technique provides the experimenter with very valuable additional controls when operating or designing an experimental setup.

11.4.2 Periodic Poling

The spatial modulation of the nonlinearity can be realized in ferroelectric crystals by periodically altering the crystal orientation so that the effective nonlinearity alternates between $+d_{bulk}$ and $-d_{bulk}$ in each domain. An early method for generating domain reversal was periodic modulation during crystal growth, but this approach lacked precise control of the domain size over long distances. Later use of lithographic methods improved the control of the domain structure. In this technique, a lithographic mask is used to define the periodic-poling regions after crystal growth. Initially the first periodic poling under lithography was based on in-diffusion of dopants [54] or through ion exchange [55]. However these methods could only produce periodically-poled waveguides because the domains only penetrated a few μm into the substrate. Finally, electric-field poling methods, developed in the early 1990s, enabled the production of bulk periodically-poled structures for high-power applications [56–58].

In electric-field poling the nonlinear medium is lithographically patterned with electrodes. A pulsed electric field is then applied to the electrodes, and if the field is strong enough, spontaneous reversal of the crystal structure occurs. The critical field strength at which poling occurs is called the coercive field. The actual value for the critical field depends on the crystal properties and is on the order of several kV/mm. By reversing the domains only under the electrodes, see Fig. 11.18, a periodic structure with alternating nonlinearity is achieved.

One of the first demonstrations of PDC in a periodically-poled bulk crystal was performed by Mason *et al.* in 2002 [59], where a periodically-poled lithium niobate (ppLN) crystal was used to generate photon pairs at 800–1600 nm. Although waveguide experiments preceded bulk operation in this instance, due to the easier fabrication of periodically-poled structures in small surface layers, the number of PDC experiments based on periodically-poled bulk-crystals grew rapidly thereafter [60–62]. The early experiments were not entirely directed toward the creation of heralded single photons, but many of them already revealed the need for efficient coupling and photon-number purity.

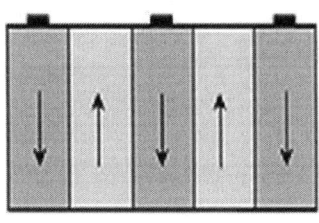
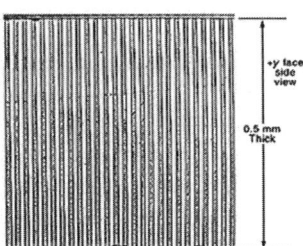

FIGURE 11.18 Left: Domain reversal underneath the electrode region. Right: Uniform 15 μm wide domain structure through a 0.5 mm thick crystal. *Reprinted figure with permission from D.S. Hum and M.M. Fejer, Comptes Rendus Physique 8, 180–198 (2007). Copyright © 2007 Acadmie des sciences. Published by Elsevier Masson SAS. All rights reserved.*

The following sections of this chapter discuss strategies for efficient coupling of the output fields to optical fibers and detail the experimental effort to achieve high photon-number purity and spectral purity with periodically-poled PDC sources.

11.4.3 Optimal Focus Parameters for Heralding Efficiency

In most instances, trigger and signal photons produced inside the nonlinear medium will be coupled to single-mode fibers (SMF), in order to guide them to the analysis and detection stations. In particular, the ability to tune the PDC to collinear emission makes the use of single-mode fibers very attractive since the trigger and signal photons can be coupled very efficiently. The key parameters for a single-photon source are heralding efficiency and the overall rate of heralded photons. The heralding efficiency ($\eta_{s|t}$) is defined as the probability of a signal photon being in the single-mode fiber given the detection of a trigger photon. Since single-photon detectors can have efficiencies ranging from < 1% to above 90%, it is difficult to compare the heralding efficiency when different types of detectors are used. Therefore the detection efficiency has been excluded from the values of the heralding efficiencies stated below for easier comparison.

Several theoretical and experimental investigations have been conducted to find the optimal fiber-coupling configurations [63–67]. In Fig. 11.19, taken from [64], a non-degenerate PDC source for heralded photons is depicted that produces trigger photons at 810 nm and signal photons at 1550 nm. A laser is focused at the center of the ppLN crystal with a focal spot waist size of W_p.

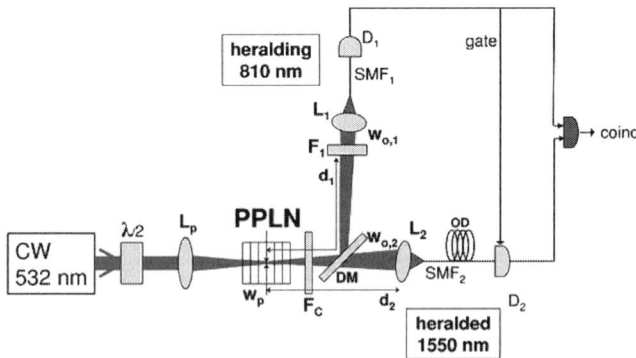

FIGURE 11.19 Setup for production of heralded single photons at 1550 nm. The pump beam is focused to a waist size of W_p inside a 5 mm long ppLN crystal. The fiber cores have waist sizes of $W_{o,1}$ and $W_{o,2}$ for trigger (810 nm) and signal (1550 nm) photons, respectively. In the main text these waists are refereed to as W_t and W_s. *Reprinted figure with permission from S. Castelletto, I.P. Degiovanni, V. Schettini, and A. Migdall, Metrologia, 43, S56-S60, 2006. © Bureau International des Poids et Mesures. Reproduced by permission of IOP Publishing. All rights reserved.*

Likewise the mode accepted by the optical fibers is also imaged at the same location as the pump beam, generating two waists W_t and W_s, for the trigger and signal fiber, respectively. Note that the size of the mode field in the fibers is dependent on the wavelength and therefore W_t and W_s are not necessarily the same.

Measurements of $\eta_{s|t}$ were performed at a fixed $W_p = 144$ μm and a fixed $W_t = 82$ μm for two different signal waists (W_s). Spectral filtering was achieved solely by the geometrical acceptance of the single-mode fibers and a FWHM of ≈2 nm was measured for the trigger photon. For W_s of 158 μm and 197 μm the measured heralding efficiencies were 16% and 21%, respectively. Much higher heralding efficiencies were obtained by using a narrow band filter ($\Delta\lambda = 0.1$ nm) in the trigger path. In this case the efficiency increased nearly to unity when optical losses were removed.

A thorough theoretical model for the case of non-degenerate PDC at 810–1550 nm devised by Ljunggren et al. [63] came to a similar conclusion: when a narrow filter is applied to the trigger field the heralding efficiency can reach up to 100% over a large range of focusing conditions, as shown in Fig. 11.20. Narrow filtering in this context means a reduction of the bandwidth of the photons below the bandwidth limit set by the crystal properties and collection angle of the SMF. The focusing parameters in Fig. 11.20 are not directly given in waist sizes but are defined as $\xi = L/z_R$, the ratio between crystal length L and the Rayleigh-range z_R of a standard Gaussian beam. This representation is helpful as it turns

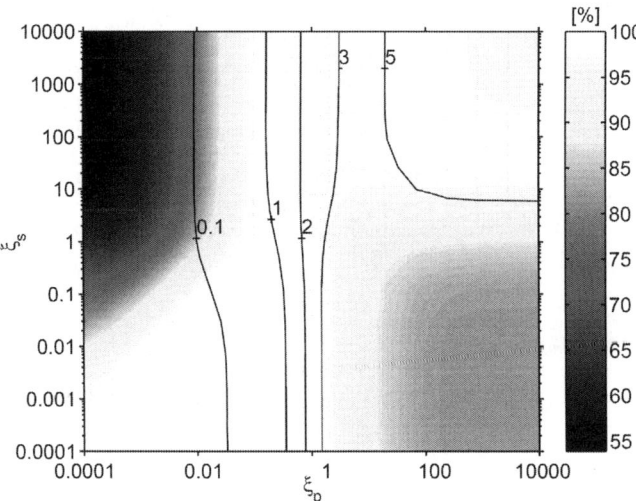

FIGURE 11.20 Plot of maximally achievable heralding efficiencies ($\eta_{s|t}$) as function of focusing parameters for pump and trigger modes (ξ_p, ξ_t). Black lines indicate the optimal signal focusing conditions (ξ_s) in order to maximize $\eta_{s|t}$. *Reprinted figure with permission from D. Ljunggren and M. Tengner, Phys. Rev. A, 72, 062301, 2005. Copyright (2005) by the American Physical Society.*

out that the optimal focusing parameter ξ^{opt} is independent of the crystal length. It is also interesting to see that the exact values of the individual modes is not very important because the maximum for the heralding efficiency is very broad, (note the logarithmic scale for ξ_p and ξ_t). As a rule of thumb, a large heralding efficiency is expected when the ξ's of all three fields are on the order of one, i.e., when the crystal length is about twice the Rayleigh range. Experimental data was also obtained using a 4.5 mm ppKTP crystal phase-matched to 810–1550 nm. A heralding efficiency, excluding all optical losses, of 34% was reported with focusing parameters of $\xi_p = 2.1$, $\xi_t = 3.2$ and $\xi_s = 2.5$.

Another interesting finding of [63] is the dependence of the fiber-coupled photon rate, and hence the heralding rate, on the crystal length, when operating at optimal focusing. With no additional spectral filtering, only the fiber acts as a frequency filter and the coupled photon rate R scales as

$$R(\Delta\lambda_{\text{wide}}) \propto \sqrt{L}, \tag{11.55}$$

whereas in the case of narrow filtering the dependence goes as

$$R(\Delta\lambda_{\text{narrow}}) \propto L\sqrt{L}. \tag{11.56}$$

The reason for the smaller growth of the rate with crystal length, as compared to Eq. (11.49), is the focusing of the pump beam. The angular spread in pump k-vectors generates additional sinc functions that are slightly offset, causing the spectral width to decrease as $1/\sqrt{L}$ with the crystal length.

Another experimental study investigated optimal focusing parameters for the case of degenerate PDC around 810 nm [65]. Absolute coincidence rates and heralding efficiencies were measured for varying pump and signal/trigger waist sizes. The main findings are shown in Fig. 11.21 for a 15 mm long ppKTP crystal. For this crystal the maximum heralding efficiency was measured to be

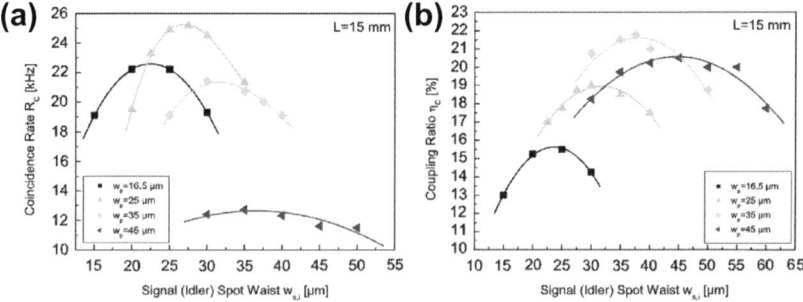

FIGURE 11.21 Coincidence count rates (a) and heralding efficiencies ($\eta_{s|t}$) (b) for a 15 mm long ppKTP for a series of focusing conditions. *Reprinted figure with permission from A. Fedrizzi, T. Herbst, A. Poppe, T. Jennewein, and A. Zeilinger, Opt. Express, 15, 15377-15386, 2007. Copyright (2007) by OSA.*

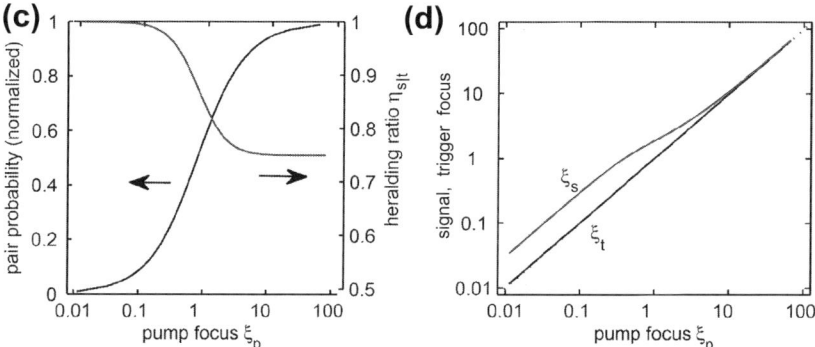

FIGURE 11.22 Simultaneous optimization of the total collection probability and heralding efficiency $\eta_{s|t}$. Panel on the right shows the focusing parameters which result in the best trade-off between collection probability and heralding efficiency. *Reprinted figure with permission from R.S. Bennink, Phys. Rev. A, 81, 053805, 2010. Copyright (2010) by the American Physical Society.*

around 22% for single-mode coupling without any additional filtering of the signal/trigger fields.

The reader should note that the optimal focusing conditions are different for maximal rate and maximal heralding efficiency. The optimum heralding efficiency is achieved with a looser focusing, enabling a higher mode overlap of the signal and trigger modes at the optical fiber. Other crystal lengths, ranging between 10 and 25 mm, were also investigated showing the independence of the focusing parameter ξ^{opt} with crystal length for the signal and trigger fields. The square-root dependence of the fiber-coupled photon rate given in Eq. (11.55) was also verified experimentally in this investigation.

In a recent theoretical work by Bennink [66], the trade-off between maximal pair rate and heralding efficiency was also found. Contrary to [63], Bennink assumed collinear Gaussian spatial modes. The calculation proceeds to find the various probabilities that the trigger and signal fields are emitted into collinear Gaussian modes. Such Gaussian modes can be coupled very efficiently to single-mode optical fibers and therefore the analysis can be regarded as if the trigger and signal fields are fiber coupled. The theoretical results for the heralding efficiency are summarized in Fig. 11.22.

It is evident from the figure that with a strong pump focus ($\xi_p \gg 1$) there will be a high overall pair-coincidence rate. However the trigger and signal might not be emitted into the same modes, and therefore the heralding efficiency will be decreased. In the limit of a weak pump focus ($\xi_p \ll 1$) the trigger and signal fields are in the same mode and have a higher heralding efficiency. No closed forms for ξ_t and ξ_s are given for the maximum values of $\eta_{s|t}$, but the right plot in Fig. 11.22 displays the values of ξ_t and ξ_s, which results in the best trade-off between collection probability and heralding efficiency. Nevertheless the study shows that near unity heralding efficiencies are possible when the pump beam is weakly focused. A recent experimental investigation corroborated these findings by reporting a heralding efficiency of 84% in a 25 mm long ppKTP crystal [68]. This result was achieved with a weakly focused pump ($\omega_p = 200$ μm) and more strongly focused trigger and signal modes ($\omega_{t,s} = 175$ μm). When comparing results, one should note that in [66] the focus parameter is defined as $\xi = \frac{L}{2z_R}$, which is half the size of the definition in [63] for the same focusing condition.

Similar heralding efficiencies were reported in [69,70]. Both these setups are based on a ppKTP crystal phase-matched for creation of photon pairs at 810 nm from a 405 nm pump laser, and both used the results from [65,66] for optimal focusing. A heralding efficiency of 80% was observed in [69]. In [70] a standard telecom fiber (SMF-28), guiding two spatial modes at 810 nm, was used to increase the coupling efficiency of the signal photon. The larger mode field diameter of the SMF-28 fiber increased the heralding efficiency to 87%. Both experiments also employed transition-edge sensors (TES) with $\eta_{DE} > 95\%$ to achieve high heralding efficiencies, including detection.

11.4.4 Number Purity

As outlined in Section 11.2.4, the photon state produced by PDC contains terms with more than one trigger/signal pair. These higher-order pairs lead to a degradation of the purity of the single-photon signal, especially when a non-number resolving detector is used to herald the presence of the signal. Without using the elaborate schemes of switched PDC sources presented at the end of Section 11.2.4 it is possible to approximate a single-photon state using a weakly pumped heralded PDC source. The photon-number purity can be measured using a Hanbury Brown-Twiss setup in the signal arm. Without heralding, the signal photons of a single-mode PDC beam have thermal statistics [71], i.e., it shows bunching of photons with an increased probability that two or more signal photons are detected at the same time. In the heralding case, these measurements are conditioned on the detection of a trigger photon so that the second-order correlation function ($g_c^{(2)}$) of the signal arms becomes [72]

$$g_c^{(2)}(t_1,t_2|t_t) = \frac{\langle \hat{E}_s(t_1)\,\hat{E}_s(t_2)\,\hat{E}_s^\dagger(t_2)\,\hat{E}_s^\dagger(t_1)\rangle_c}{\langle \hat{E}_s(t_1)\,\hat{E}_s^\dagger(t_1)\rangle_c \,\langle \hat{E}_s(t_2)\,\hat{E}_s^\dagger(t_2)\rangle_c}, \quad (11.57)$$

where $\hat{E}_s(t), \hat{E}_s^\dagger(t)$ are the field operators for the signal arm at times t_1 and t_2. The average $\langle \ldots \rangle_c$ is conditioned on the detection of a trigger photon at time t_t. If a PDC source is used for the production of single photons, we are especially interested in the probability of a second signal photon at time $t_2 = \tau$, given that a signal photon at $t_1 = 0$ was heralded by the detection of a trigger photon at $t_t = 0$. In this case the second-order correlation function reduces to: $g_c^{(2)}(0,\tau|0)$.

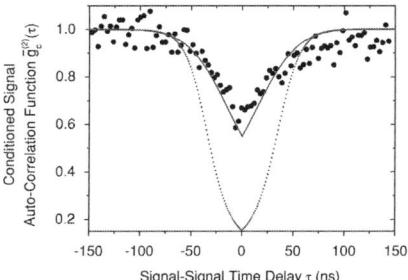

FIGURE 11.23 Measured (dots) and simulated (solid line) second-order coherence function of signal photons conditioned on the detection of a trigger photon. The dotted line represents a simulation without background noise. (For interpretation of the references to color in this figure legend, the reader is referred to the web version of this book.) *Reprinted figure with permission from D. Höckel, L. Koch, and O. Benson, Phys. Rev. A, 83, 013802, 2011. Copyright (2011) by the American Physical Society.*

In Fig. 11.23, this correlation is plotted and clearly shows a dip below unity around $\tau = 0$.

It is thus possible to obtain anti-bunched light from a PDC source, and therefore a suppression of higher photon-numbers, even though the individual signal and trigger fields show thermal (bunching) statistics. This is possible since the heralding uses the very strong non-classical correlations between the two fields (pair generation), which is independent of pump power. On the other hand, the probability to detect two signal photons, which scales quadratically with the pump power (see Eq. (11.40)), can be made arbitrarily small by reducing the pump intensity. It should be noted that the simple heralding presented here approximates a single-photon state with $g_c^{(2)}(0) = 0$ only in the limit of negligible pump power and heralding rate. To obtain single-photon statistics at high rates, one has to revert to the switched PDC scheme as outlined in Section 11.2.4, or some other technique.

Several experimental investigations have observed the antibunching feature of the signal photons when conditioned on the trigger [72–77]. In [72] a cw pumped 10 mm long ppKTP crystal was used to produce trigger and signal photons around 800 nm. A conditioned measurement of the signal second-order correlation function yielded $g_c^{(2)}(0) = 0.7$ at a heralding rate of $\approx 240,000$ s^{-1}. A suppression of higher photon-numbers by 2 orders of magnitude compared with a Poissonian light source ($g_c^{(2)}(0) = 0.01$) was reported by [75], although in this instance the heralding rate was reduced to ≈ 5000 s^{-1}. This experiment is also interesting because it used cavities around the ppKTP crystal to enhance the coherence time of the photons to beyond 140 ns. In this regime the detector jitter is negligible and the shape of the coherence function is given by the temporal extent of the wavepacket alone. The lowest $g_c^{(2)}(0)$ for single-photon sources based on PDC was reported in [77] with a value of 0.005.

11.4.5 Spectral Purity

To have spectrally independent trigger and signal photons, Eq. (11.38) has to be fulfilled, which states that the group-velocity of the pump in the nonlinear medium must lie between the group-velocities of the trigger and signal fields. In the visible wavelength range this requirement is difficult to satisfy since the normal material dispersion at lower (pump) wavelengths results in lower group-velocities for the pump. However, by moving to longer wavelengths the differences in dispersion between pump, trigger, and signal fields become small. It is then possible, by including the dependence of the group-velocity with polarization, to find a solution in which the orthogonally polarized trigger and signal fields have group-velocities above and below the pump [78]. For example, in type-II PDC using KTP, it is possible to match the group-velocities in a pump wavelength range of 650–900 nm, corresponding to degenerate wavelengths of 1300–1800 nm for the trigger and signal fields. In an experiment by Evans *et al.* [79] a pump wavelength of 776 nm was used to produce trigger and signal pairs at 1552 nm in a 20 mm long type-II ppKTP. In addition, the crystal was periodically-poled to yield zero phase mismatch at those wavelengths. To obtain the minimal spectral entanglement a specific width of the pump spectrum had to be chosen, corresponding to a pulse duration of 1.3 ps. Joint-spectral intensities were measured and can be seen in Fig. 11.24a. Analysis of the spectrum yields a spectral Schmidt number of 1.07, indicating a very high spectral purity of the trigger and signal photons.

Spectral purity of trigger and signal photons has also been shown in an experiment using a periodically-poled KDP crystal, quasi-phase-matched for

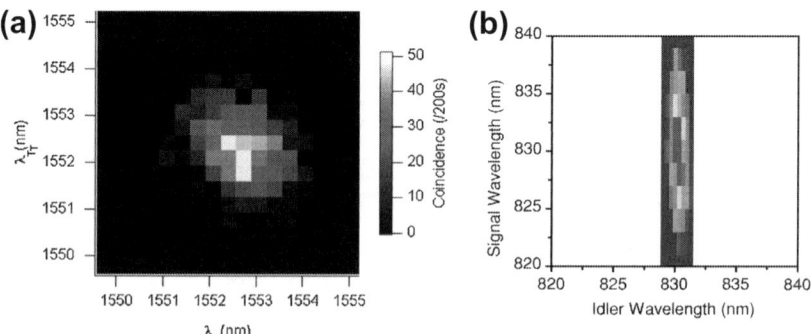

FIGURE 11.24 Joint-spectral intensity of the trigger and signal modes in (a) type-II ppKTP crystal pumped at 776 nm. *Reprinted figure with permission from P.G. Evans, R.S. Bennink, W.P. Grice, T.S. Humble, and J. Schaake, Phys. Rev. Lett., 105, 253601, 2010. Copyright (2010) by the American Physical Society.* and (b) type-II ppKDP crystal pumped at 415 nm. *Reprinted figure with permission from R.-B. Jin, J. Zhang, R. Shimizu, N. Matsuda, Y. Mitsumori, H. Kosaka, and K. Edamatsu., Phys. Rev. A, 83, 031805, 2011. Copyright (2011) by the American Physical Society.*

type-II PDC [80]. The trigger and signal fields were degenerate at 830 nm and their joint-spectral intensity can be seen in Fig. 11.24b. The Schmidt value obtained from the experimental data was 1.03, demonstrating again the high spectral purity.

11.4.6 Non-Uniform Periodic Poling

The versatility of poled structures is further increased through transversal or longitudinal patterning of the quasi-phase-matched poling structure. Transversal patterning can be used to control the spatial component of the PDC biphoton wave function, see e.g., [81,82]. Longitudinal patterning enables the manipulation of the spectral PDC wavefunction, which is arguably more important for the purpose of creating pure heralded single photons. A linearly-chirped poling period $\Lambda(z)$, for example, allows the creation of ultra-broadband single photons for optical coherence tomography [83]. A longitudinal, interleaved superposition of multiple poling periods may lead to quasi-phase-matching solutions for multiple sets of wavelengths [84], or for concurrent type-0, type-I and type-II PDC [85].

Longitudinal domain engineering can be used to address one of the remaining problems not addressed by group-velocity matching. In a standard PDC experiment, wavepackets have a sinc frequency spectrum. As can be seen from Eq. (11.19), this is due to the fact that a crystal has finite length and a rectangular shape—the nonlinear interaction between the pump beam and the crystal is thus turned on abruptly to its full strength when the pump enters the crystal, and remains constant until it is turned off when the pump exits the crystal. In the frequency domain, this temporal step-function transforms into a sinc shape, see Eq. (11.20). This spectral shape causes residual frequency distinguishability even in the presence of group-velocity matching—see the sidelobes in Figs. 11.10 and 11.11—which has a detrimental effect on the purity of heralded PDC photons [24].

To fully reduce these side lobes, the constant nonlinearity of the crystal must be turned into an effective Gaussian function. In [45], a 10 mm ppKTP was longitudinally patterned with discrete sections of increasingly higher-order polings, see Fig. 11.25. The pump beam entering the crystal first encountered a section with poling order $m = 32$, experiencing, according to Eq. (11.50), a very weak effective nonlinearity. The nonlinearity was then increased step by step, peaking at $m = 1$ in the crystal center and then dropping off symmetrically. The theoretic biphoton spectrum obtained with this spectral engineering method, shown in Fig. 11.25b, was confirmed by measuring two-photon interference patterns, which showed a distinctly Gaussian pattern instead of the characteristic triangular pattern of $sinc^2$-shaped bi-photons.

Combining these engineered nonlinearities with group-velocity matching leads to a significant improvement in photon purity without the need for spectral filtering, as shown numerically in [45].

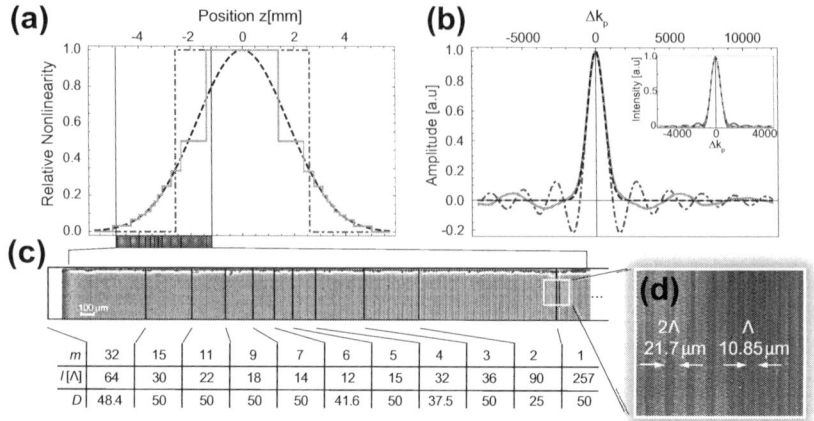

FIGURE 11.25 Spectral biphoton shaping via engineered crystal nonlinearities *(Reprinted figure with permission from A.M. Brańczyk, A. Fedrizzi, T.M. Stace, T.C. Ralph, and A.G. White, Opt. Express, 19, 55–65, 2011. Copyright (2011) by OSA.).* (a) A Gaussian nonlinearity profile (black dashed line) was approximated with discrete crystal sections (gray solid line) of order m. (b) Phase-matching function amplitudes and intensities (inset) for the aperiodically poled crystal (solid line) compared to a ppKTP of the same effective length (dot-dashed line) and target Gaussian profile (dashed line). (c) Magnified image of part of the custom crystal. Vertical lines separate sections with constant nonlinearity, with their poling order, length L and poling duty cycle D. (d) Zoom into the transition from poling order $m = 1$ to $m = 2$.

11.5 WAVEGUIDE-CRYSTAL PDC

After the success of PDC in bulk crystals as a source of heralded single photons (see Section 11.3) and the incorporation of periodic poling for quasi-phase-matching enabling easy tunability of the signal and trigger wavelengths (see Section 11.4) another major development was the inclusion of waveguide technology, as depicted in Fig. 11.26.

The confinement of the electric fields inside a waveguide leads to extremely high field amplitudes in the nonlinear material, increasing the down-conversion rate by several orders of magnitude in comparison to bulk-crystal sources.

FIGURE 11.26 Microscopy of the waveguide used in [86] (color adjusted). The depicted cross-section gives an impression of the tooth-like waveguide structure and enables a rough estimate of its dimensions.

Furthermore the confinement of the signal and trigger fields in the waveguide modes enables efficient coupling to optical fibers. Finally, waveguided PDC allows for the integration into small-scale devices that may include sources, optical circuits, and detectors.

11.5.1 History and Experimental Implementations

The main challenge for the generation of heralded single photons in nonlinear waveguides is the production of high-quality waveguides with low losses and minimal variation over the length of the structure. There are several materials available that are well suited for nonlinear optical waveguides, including dielectric materials such as lithium niobate ($LiNbO_3$ or LN), lithium tantalate ($LiTaO_3$), potassium titanyl phosphate ($KTiOPO_4$ or KTP) [87,88], and potassium niobate ($KNbO_3$) [14]. Nonlinear semiconductor materials include AlGaAs, GaAs, ZnSe, ZnTe, and InP [89–92]. In comparison to dielectric materials, semiconductors feature higher nonlinearities, but also higher loss rates.

The methods to create waveguiding structures inside the nonlinear materials are as varied as the materials, ranging from proton exchange, anneal/proton exchange over ion exchange, to metal-diffused waveguides (Ti-, Rb-, Zn-indiffusion), ion-implantation waveguides, and epitaxial-growth methods. Each method has its inherent advantages and disadvantages. For example ion-exchange waveguides in LN guide only one polarization, but allow burying the waveguide in the material via reverse-proton-exchange processes. To perform type-II PDC in LN one has to resort to metal-diffused LN waveguides, which guide both polarizations but are always located at the surface. There exists an extensive literature on the waveguide production process, with an overview given in [14].

The most commonly used waveguide materials for PDC are LN and KTP, which gained popularity due to their low losses, special dispersion properties, and ease of fabrication. The first photon-pair PDC source using waveguides was presented in 2001 by Tanzilli *et al.* [93] in LN, and already featured down-conversion rates four orders of magnitudes higher than bulk-PDC sources at that time. Soon afterwards the spatial properties of PDC in waveguides were investigated [94]. Since then a lot of improvements were made to this process [95–97]. The generation of type-II PDC in a waveguide yielded signal and trigger states in orthogonal polarizations, such that they could be separated by a polarizing beam splitter and efficient state generation could be accomplished [26,98,99]. Nonetheless, all these sources created frequency-correlated photon pairs, resulting in the heralding of mixed single-photon states. Frequency filtering of photon-pair states for producing pure single-photon Fock states was experimentally investigated in [25].

The first direct production of uncorrelated photon pairs in waveguides based on group-velocity matching [100] was demonstrated in 2011, two years after the initial breakthrough by Mosley *et al.* relying on bulk PDC [43], and ten years after the first demonstration of waveguided PDC [93]. Since then the available generation rates and heralding efficiencies have continuously improved [101]. Modern sources exhibit heralding rates up to 80% and feature high conversion efficiencies [102].

One advantage of PDC in waveguides for heralded Fock-state generation is the fact that it opens up new possibilities to achieve uncorrelated photon-pair generation. For example, Bragg structures placed at the boundaries of the waveguide can be used to modify the dispersion properties inside the waveguides, enabling group-velocity matching in various materials and at wavelengths previously not available due to their unfavorable material parameters [103]. Waveguides also enable the creation of counterpropagating photon pairs, in which one photon travels backward toward the pump [104, 105, 27]. This setup, while technologically challenging, leads to uncorrelated photon-pair emission that is (almost) independent of the dispersion properties of the crystal material.

11.5.2 Theory of PDC in Waveguides

To mathematically describe the PDC process in waveguides we expand the theory presented in Section 11.2 to include the transverse degrees of freedom, as in Section 11.3, while also including the effects of an optical guide.

To calculate the two-photon state emitted by the process of waveguided parametric down-conversion (see Fig. 11.27) we extend the PDC Hamiltonian to include the transverse degrees of freedom

$$\hat{H}_{\text{PDC}}(t) \propto \chi^{(2)} \int_V d^3 r \, \hat{E}_p^{(+)}(\mathbf{r},t) \hat{E}_s^{(-)}(\mathbf{r},t) \hat{E}_t^{(-)}(\mathbf{r},t) + h.c. \quad (11.58)$$

Following the presentation in [86], the boundary conditions imposed by the waveguide on the electric fields define a finite set of transverse-field distributions

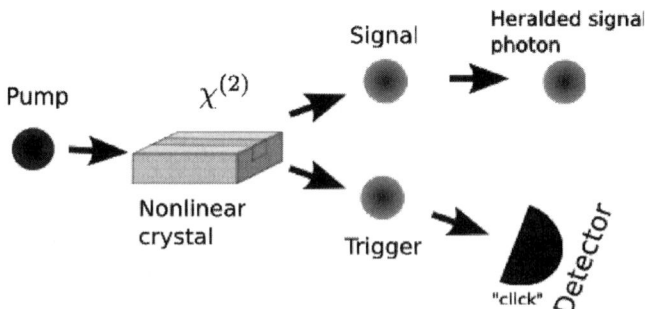

FIGURE 11.27 Single-photon generation using PDC in a nonlinear optical waveguide. The guiding structure gives rise to an enhanced down-conversion rate and a collinear propagation of all involved fields in well-defined spatial modes.

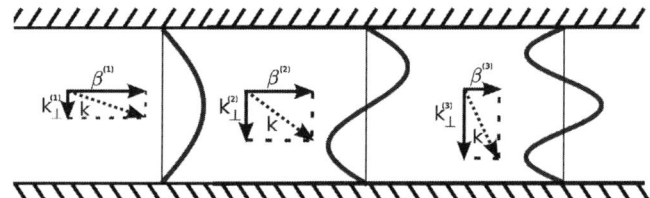

FIGURE 11.28 Schematic spatial modes and k-vectors of an electric field propagating inside a 1D waveguide.

$f^{(k)}(x,y)$ propagating inside the waveguide. This is sketched in Fig. 11.28 for a one-dimensional waveguide. The propagating field in a given spatial mode is defined by its **k**-vector, which is split into the $\beta^{(k)}$ component describing the propagation constant through the material, and $k_\perp^{(k)}$ related to the individual transverse-spatial-mode distribution $f^{(k)}(x,y)$. This is in sharp contrast to down-conversion in bulk crystals. Instead of emitting the signal and trigger fields into a continuous set of spatial modes, the down-converted beams are generated in a finite set of well-defined spatial modes imposed not by the PDC process but by the waveguide parameters. The electric fields inside the channel are therefore of the form

$$\hat{E}_x^{(+)} = \hat{E}_x^{(-)\dagger} = A \sum_k f_x^{(k)}(x,y) \int d\omega_x \exp\left[i\left(\beta_x^{(k)}(\omega_x)z - \omega_x t\right)\right] \hat{a}_x^{(k)}(\omega_x). \tag{11.59}$$

In Eq. (11.59), A collects all constants, the x subscript labels the signal, trigger, or pump field, and the superscript (k) labels the spatial mode. Note that both the spatial field distribution $f_x^{(k)}(x,y)$ and the effective k-vector $\beta_x^{(k)}$ are dissimilar for the signal, trigger, and pump fields due to varying wavelengths and polarizations, which impact the spatial-mode distributions. In Eq. (11.59) we assume fields which are not too broad in frequency, $\Delta\omega \ll \omega_c$, which means we may neglect the frequency dependence in the spatial distribution function about the central wavelength. Again, we treat the strong pump field as a classical wave, which yields the formula

$$E_p^{(+)} = E_p^{(-)*} = A \sum_k f_p^{(k)}(x,y) \int d\omega_p \, \alpha(\omega_p) \exp\left[i\left(\beta_p^{(k)}(\omega_p)z - \omega_p t\right)\right]. \tag{11.60}$$

We use Eqs. (11.18), (11.58), (11.59), and (11.60) to calculate the two-photon output state. By incorporating the three-dimensional structure of all involved fields we finally arrive at the following formula

$$\int_{t_0}^{t} dt' \, \hat{H}_{\text{PDC}}(t') = B \int_{t_0}^{t} dt' \sum_{klm} \underbrace{\iint dx \, dy \, f_p^{(k)}(x,y) f_s^{(l)}(x,y) f_t^{(m)}(x,y)}_{A_{klm}}$$

$$\times \int_{-\frac{L}{2}}^{\frac{L}{2}} dz \iiint d\omega_p d\omega_s d\omega_t \, \alpha(\omega_p)$$

$$\times \exp\left[-i\left(\omega_p - \omega_s - \omega_t\right)t'\right]$$

$$\times \exp\left[i\left(\beta_p^{(k)}(\omega_p) - \beta_s^{(l)}(\omega_s) - \beta_t^{(m)}(\omega_t)\right)z\right]$$

$$\times \hat{a}_s^{(l)\dagger}(\omega_s) \hat{a}_t^{(m)\dagger}(\omega_t) + h.c..$$

(11.61)

Equation (11.61) is very similar to Eq. (11.19) in Section 11.2 neglecting the transverse degree of freedom and describing a collinear interaction. Following the discussion in Section 11.2, we are able to calculate the generated waveguided PDC state as

$$|\psi\rangle_{\text{PDC}} = |0\rangle + B' \sum_{klm} A_{klm} \iint d\omega_s d\omega_t \, \alpha(\omega_s + \omega_t)$$

$$\times \text{sinc}\left[\Delta\beta_{klm}(\omega_s,\omega_t)\frac{L}{2}\right] \hat{a}_s^{(l)\dagger}(\omega_s) \hat{a}_t^{(m)\dagger}(\omega_t) |0\rangle$$

$$= |0\rangle + B' \sum_{klm} A_{klm} \iint d\omega_s d\omega_t \, \alpha(\omega_s + \omega_t)$$

$$\times \Phi_{klm}(\omega_s,\omega_t) \hat{a}_s^{(l)\dagger}(\omega_s) \hat{a}_t^{(m)\dagger}(\omega_t) |0\rangle$$

$$= |0\rangle + B' \sum_{klm} A_{klm} \iint d\omega_s d\omega_t \, f_{klm}(\omega_s,\omega_t) \hat{a}_s^{(l)\dagger}(\omega_s) \hat{a}_t^{(m)\dagger}(\omega_t) |0\rangle ,$$

(11.62)

where all constants have been merged into B'.

The main modification with respect to the simplified model is the appearance of the overlap integral over the three interacting spatial modes, which introduces the coupling constant

$$A_{klm} = \iint dx \, dy \, f_p^{(k)}(x,y) f_s^{(l)}(x,y) f_t^{(m)}(x,y). \quad (11.63)$$

Hence the efficiency of the down-conversion becomes dependent on the spatial shape of the interacting spatial-mode triplet. If the pump, signal, and trigger propagate in similar modes an output state will be generated with high efficiency, but if the signal and trigger modes are sufficiently distinct the photon-pair generation efficiency will be strongly diminished. In addition, the phase-matching

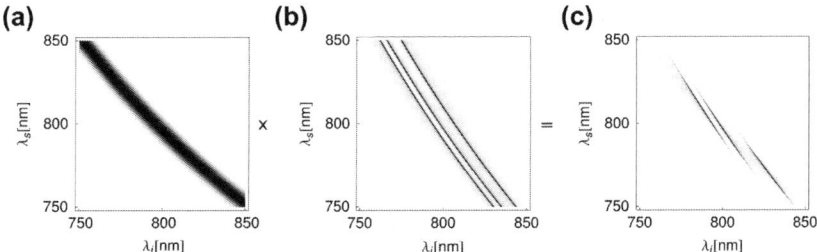

FIGURE 11.29 Sketch of the frequency distribution generated by a spectrally multimode PDC process [86]. (a) Pump distribution $\alpha(\omega_s + \omega_t)$, (b) phase-matching function $\Phi_{klm}(\omega_s,\omega_t)$ for the interacting mode triplets, (c) joint-spectral distribution function $f_{klm}(\omega_s,\omega_t)$. Each triplet of interacting modes exhibits a distinct spectral shape.

function $\Phi_{klm}(\omega_s,\omega_t)$ is now dependent on the spatial modes inside the waveguide, as each interacting mode triplet features a distinct β-vector mismatch

$$\Delta\beta^{(klm)}(\omega_p, \omega_s, \omega_t) = \beta_p^{(k)}(\omega_p) - \beta_s^{(l)}(\omega_s) - \beta_t^{(m)}(\omega_t). \quad (11.64)$$

Different spatial modes lead to a modification of the propagation vectors β_i, which consequently translates the phase-matching function in frequency space. This creates an individual phase-matching function for each interacting spatial-mode triplet, and in effect a different joint-spectral amplitude.

Figure 11.29 sketches, the pump distribution, the multitude of phase-matching functions, and the three distinct spectral shapes generated by the source presented in [86].

Direct measurements of the spatial- and spectral-mode structure in waveguided PDC are presented in [86,106,107]. Figure 11.30 shows the measured spectral distributions of the signal and trigger photons (labeled idler in this picture) of the waveguided KTP down-conversion source presented in [106]. Each peak, labeled A–E corresponds to different spatial-mode triplets (signal, trigger, and pump) propagating inside the material. By measuring the spatial-mode distribution of each peak one can resolve the different spatial modes of the signal and trigger beams, ranging from simple Gaussian distributions to more complicated higher-order structures (see Fig. (11.31)). Details are presented in [106].

Employing waveguides for PDC processes offers several advantages over bulk-PDC sources. It enables control of generated spatial distributions by engineering the design of the waveguiding structure. Whereas bulk PDC emits photon pairs into a large set of spatial modes at different angles, waveguided PDC restricts the two-photon states to well-defined spatial modes ideally suited for high collection efficiencies and coupling into optical fibers. Furthermore the confinement of PDC inside a waveguide leads to an increased overlap of the involved fields A_{klm} and to the restriction of the interaction couplings to

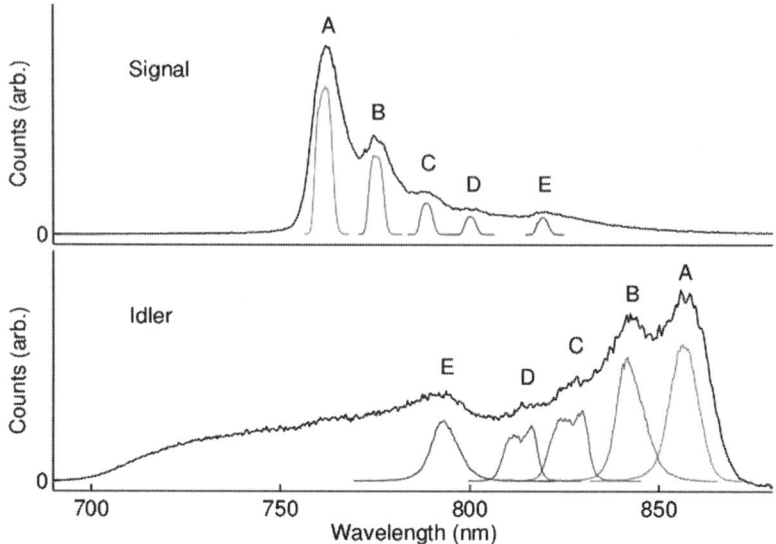

FIGURE 11.30 Spectral distribution of spatial multimode PDC in a nonlinear KTP waveguide [106] (Idler ≡ Trigger). The different peaks A-E stem from distinct interacting pump, signal, and trigger spatial-mode triplets inside the waveguide.

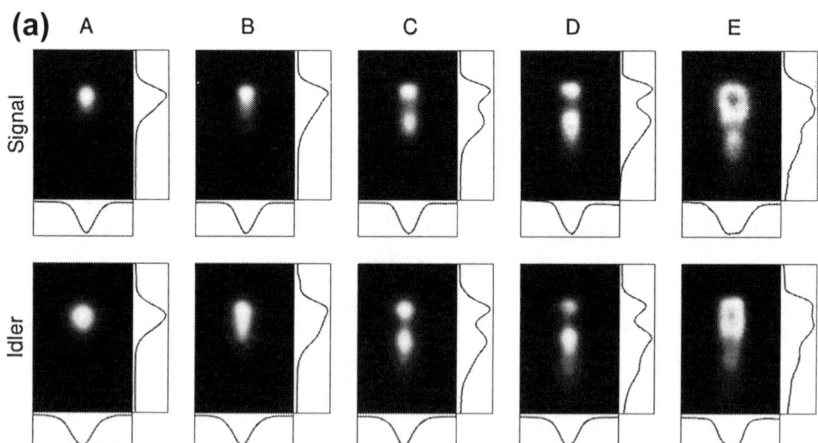

FIGURE 11.31 Spatial mode distribution of the different spectral PDC distributions presented in [106](Idler ≡ Trigger).

a discrete number of modes. This in turn significantly boosts the photon-pair generation probability [108,109]. Finally, the interaction of the three fields is strictly collinear, and this greatly eases the experimental alignment and post-processing of the beams in the laboratory. It should however be noted that waveguides feature a slightly enhanced loss rate in comparison to the surrounding bulk material, which stem from additional scattering losses in the waveguide.

11.5.3 Heralding Single Photons from PDC in Waveguides

If we take into account the spatial degree of freedom for the heralding of single-photon states from waveguided PDC, the heralded signal assumes the form

$$\rho_s = \sum_{l,l'} \iint d\omega_s \, d\omega_{s'} \left[\int d\omega_t \sum_{k,k',m} f_{klm}(\omega_s,\omega_t) f^*_{k'l'm}(\omega_{s'},\omega_{t'}) \right] |\omega_s,l\rangle \langle \omega_{s'},l'|$$

$$= \sum_{l,l'} \iint d\omega_s \, d\omega_{s'} \, j_{l,l'}(\omega_s,\omega_{s'}) |\omega_s,l\rangle \langle \omega_{s'},l'|, \qquad (11.65)$$

where $\omega_s, \omega_{s'}$ label the frequencies of the heralded signal and l, l' its spatial mode. Hence the heralded single-photon states are, in general, prepared in a mixture of different frequencies and spatial modes. Of course this is a highly detrimental effect for the heralding of pure single-photon Fock states. In order to achieve a high purity for the single-photon Fock-state generation we must not only consider the frequency degree of freedom as discussed in Section 11.2.4, but we must also consider the spatial degree of freedom.

In Section 11.2.3 we learned how to cope with mixing effects that stem from frequency correlations. In fact, the impact of the spatial degree of freedom on the heralding of single photons is mathematically almost identical to the spectral degree of freedom. As in the frequency domain, the easiest and most straightforward approach is to apply filtering, for example employing small pinholes or single-mode fibers to purify the trigger photon, which results in the propagation of spatially pure signal states. This, however, greatly reduces the heralding efficiency because a large fraction of the beam is lost.

To study the impact of the spatial modes on the heralded signal in more detail we start by neglecting the frequency properties of the state in Eq. (11.65) and assume monochromatic signal and trigger modes. The heralded signal is transformed to

$$\rho_s = \sum_{k,l} \sum_{k',l'} \sum_m f_{klm} f^*_{k'l'm} |l\rangle \langle l'|, \qquad (11.66)$$

where (k, k') labels the spatial modes of the pump beam, (l, l') the modes of the signal beam and m the trigger modes. From Eq. (11.66) it is evident that the heralded signal photon is, in general, in a mixture of spatial modes, and consequently not in a pure single-photon Fock state.

One method to herald single-spatial-mode single photons is to create the PDC state in only a single trigger mode m, which leads to the heralding of a spatially pure signal photon

$$\begin{aligned}\rho_s &= \sum_{k,l}\sum_{k',l'} f_{klm} f^*_{k'l'm} |l\rangle\langle l'| \\ &= \sum_{k,l} f_{klm} |l\rangle \sum_{k',l'} f^*_{k'l'm} \langle l'| \\ &= |\psi_s\rangle\langle\psi_s|. \end{aligned} \qquad (11.67)$$

Alternatively it is also sufficient to engineer the signal emission into a single mode. In this case the heralded state becomes

$$\rho_s = \underbrace{\sum_{k}\sum_{k'}\sum_{m} f_{klm} f^*_{k'lm}}_{d} |l\rangle\langle l| = d\,|l\rangle\langle l| = |\psi_s\rangle\langle\psi_s|. \qquad (11.68)$$

Engineering single-spatial-mode behavior in a waveguide is an elegant method to cope with spatial-mode effects. If the guiding structure is designed to allow only single-mode propagation of the signal or trigger photon, the heralded state will be pure in this domain and no spatial mixing occurs. This can be achieved by a custom waveguide design where the diameters of the guide geometry are adapted to the signal or trigger field. Note, however, that the pump beam, which has a shorter wavelength then the signal and trigger, might still be able to propagate in a higher spatial mode. This can distort the frequency shape of the heralded photons, as different modes typically also exhibit different spectral properties (see Section 11.5.3). However, this effect can be mediated by a careful coupling of the pump wave into the fundamental waveguide mode.

In between the two extremes of strong filtering and careful engineering of the spatio-spectral structure of PDC there exists a third method to eliminate all but one interacting spatial-mode triplet. Introducing spectral windowing we can exploit the fact that each down-conversion process features a distinct joint-spectral amplitude distribution $f_{klm}(\omega_s, \omega_t)$, for each spatial-mode triplet, as depicted in Fig. 11.29, which is caused by the unique β-vector mismatch between the three interacting beams. Using this characteristic it is possible to spatially purify the down-conversion by placing broadband (and thus low loss) spectral filters in the signal and trigger arm that transmit only one process $f_{klm}(\omega_s,\omega_t)$. Hence we can eliminate the spatial multimode effects without introducing losses to the targeted down-conversion process. Still, the applicability of this approach is closely related to the specific characteristics of the individual source. A more detailed review—featuring experimental data—is presented in [106].

In summary all three approaches lead to the same result. The considered PDC components become single mode in the spatial domain and consequently the heralded signal photon can be represented in the form of Eq. (11.24). The effects from the spatial domain are negated and efficient creation and heralding of pure single-photon Fock states with well-defined spatial-mode profiles is possible.

11.5.4 Electric Field Modes in Waveguides

The main challenge when analyzing the exact spatial structure of PDC in waveguides is the theoretical modeling of the transverse-field distributions defined by the channel. As a first step one has to get access to the refractive-index profile of the waveguide. The easiest approach is to perform microscopy at the input or output facet to get a general impression of the structure, as shown in Fig. 11.26.

This measurement already provides some insight on the waveguide properties, and opens up a route to assess the width and height of the structure. However this approach does not give access to the refractive-index step Δn between the waveguide and the material, which must be estimated (common values are $\Delta n \approx 0.01$). More precise methods have been developed in the context of fabricating classical optical devices, for example M-line spectroscopy, and the inverse WKB method enabling precise access to the to refractive-index distribution of the waveguide [110].

For a given refractive-index distribution inside the waveguide there exist various methods to calculate the guided spatial modes, depending on the geometry of the system. Assuming rectangular or circular waveguides with perfectly conducting edges—the mathematical analog of modes inside an infinitely deep potential well ($\Delta n = \infty$)—is the most simplistic model and has the advantage of yielding an analytic solution. Including the effects of finite refractive-index steps Δn, a completely analytical solution is not possible, but a semi-analytical approach, is presented in [111], which relies on the solution of transcendental equations. As more complex waveguide geometries are considered, such as non-rectangular waveguide structures with slowly varying refractive indices Δn, full numerical theories must be utilized. Mode-solving techniques based on finite element approaches [112] are common, and various software applications are available for this purpose.

11.5.4.1 Analytic Waveguide Model

In this book we restrict ourselves to the discussion of the straightforward analytical model, which gives a fair introduction to the physics involved. In this model we assume a rectangular waveguide of width w and height h with perfectly conducting edges (i.e., the electric fields are zero at the boundaries).

To calculate the electric field propagation in three dimensions, we express the electric field as

$$\begin{aligned} \mathbf{E}(x,y,z) &= \left[\mathbf{e}_x E_x(x) + \mathbf{e}_y E_y(y) + \mathbf{e}_z E_z(z)\right] e^{-i\omega t} \\ E_i(j) &= E_i e^{ik_i j}. \end{aligned} \quad (11.69)$$

Here we define z as the propagation direction of the beam, and the x and y components label the transverse degrees of freedom. The absolute value of the wave vector of a light ray inside a medium, propagating alongside a crystal axis i, is given by

$$|k|^2 = k_x^2 + k_y^2 + k_z^2 = \frac{\omega}{c} n_i(\omega) = \frac{2\pi}{\lambda} n_i(\lambda). \tag{11.70}$$

During the PDC process we are mostly interested in the momentum or wave-vector mismatch between the interacting fields in propagation direction, which we label β and that can be calculated by

$$\beta^2 = |k|^2 - k_x^2 - k_y^2 \tag{11.71}$$

The solutions for an electric field in a infinitely deep potential well are well known

$$E_x(x) = \sin\left(\frac{\nu\pi}{w}x\right),$$
$$E_y(y) = \sin\left(\frac{\mu\pi}{h}y\right), \tag{11.72}$$

where ν and μ label the spatial mode of the standing-wave solution in the waveguide $\nu,\mu = \mathbb{N}^+\setminus\{0\}$, with the corresponding k-vectors

$$k_x = \frac{\nu\pi}{w},$$
$$k_y = \frac{\mu\pi}{w}. \tag{11.73}$$

Using Eq. (11.73) we arrive at the effective propagating wave vector:

$$\beta^2(\lambda) = \left(\frac{2\pi}{\lambda} n_x(\lambda)\right)^2 - \left(\frac{\nu\pi}{w}\right)^2 - \left(\frac{\mu\pi}{h}\right)^2 \tag{11.74}$$

For practical purposes it is more elegant to write this effective k-vector β in the propagation direction as an effective Sellmeier equation

$$n_{\text{eff}}^2(\lambda) = n_x^2(\lambda) - \left(\frac{\lambda\nu}{2w}\right) - \left(\frac{\lambda\mu}{2h}\right). \tag{11.75}$$

Despite the simplicity of this model, it yields quite accurate values for the generated mode distributions and effective k-vectors inside the waveguide, given only the width and height of the system. It is very useful to obtain a qualitative simulation of the process, however, when precise quantitative predictions are required, the semi-analytic approach by Marcuse [111] or finite element methods [112] should be applied.

11.6 COMPARISON OF EXPERIMENTAL SINGLE-PHOTON SOURCES USING PDC

Type	Reference	Crystal	Heralding efficiency	Heralding rate c/s	Wavelength (Bandwidth)	Special remarks
Bulk	Mosley 2008 [43]	KDP	44%	150	t=830 (20), s=830 (3.5)	spectral purity 0.95
Periodically-poled	Ljunggren 2005 [63]	ppKTP	48%	7,200	t=810 (2) s=1550	cw-pumped, SMF-coupling, $g^{(2)}(0)=0.024$
Periodically-poled	Castelletto 2006 [64]	ppLN	48%	-	t=810 (2) s=1550	cw-pumped, SMF-coupling, different focusing conditions
Periodically-poled	Fedrizzi 2007 [65]	ppKTP	22%	53,000	t=810 s=810	cw-pumped, study of different focusing conditions
Periodically-poled	Scholz 2009 [75]	ppKTP	55%	5,000	t=- (3MHz) s=- (3MHz)	in cavity, $g^{(2)}(0)=0.01$
Periodically-poled	Bocquillon 2009 [72]	ppKTP	30%	80,000	t=810 s=810	pulsed pump, $g^{(2)}(0)=0.08$
Periodically-poled	Evans 2010 [79]	ppKTP	19.0%	1000	t=1552 (1) s=1552 (1)	SMF-coupling, spectral Schmidt number 1.07
Periodically-poled	Smith 2012 [69]	ppKTP	80%	~10,000	t=820 (3) s=820	cw-pumped, SMF-coupling, TES detectors
Periodically-poled	Brida 2012 [77]	ppKTP	-	~ 5	t=810 (10) s=1550 (30)	$g^{(2)}(0)=0.005$, optical shutter to supress background
Periodically-poled	Pereira 2013 [68]	ppKTP	84%	11,000	t=810 (0.2) s=810 (0.2)	cw-pumped, SMF-coupling, symmetric heralding efficiency
Periodically-poled	Ramelow 2013 [70]	ppKTP	87%	5,500	t=810 (0.5) s=810 (0.5)	TES detectors
Waveguide	Tanzilli 2001 [93]	ppLN	-	1,500	s=1314 (30)	first experiment in waveguides
Waveguide	U'Ren 2004 [95]	ppKTP	85%	1,270	s=800	-
Waveguide	Zhong 2010 [98]	ppKTP	-	3,000	s=1316 (1.2)	-
Waveguide	Eckstein 2011[100]	ppKTP	60%	-	~1550 (5)	$g^{(2)}(0)=0.05$, Spectral purity=0.8/0.95 (Raw / background correction), mean photon number: 2.5
Waveguide	Krapick 2013 [97]	ppLN	60%	105	t=803 (0.7) s=1575	$g^{(2)}(0) =0.004$
Waveguide	Harder 2013 [102]	ppKTP	80%	-	t=1535 (4) s=1535 (4)	mean photon number: up to 80, spectral purity = 0.86 (raw)

FIGURE 11.32 Overview of selected experimental realisations of heralded single-photon sources using PDC. The table is divided into bulk, periodically-poled and waveguide parts, the references are ordered chronologically, and the name of the nonlinear crystal used is given. The heralding efficiency includes optical loses but no detection efficiencies. The heralding rate is the absolute rate measured, i.e., the raw coincidence rate. The central wavelength of the trigger and signal photons is given with their respective bandwidths in brackets. Additional experimental details and measurements on photon-number purity and spectral purity are summarized under remarks.

11.7 OVERVIEW OF THE MOST COMMONLY USED NONLINEAR MATERIALS AND THEIR PROPERTIES

Crystal	Quasi phase-matching	Waveguide	Nonlinearity	Phase-matching	Damage threshold	Transparency range (nm)
BBO	-	Yes	d_{24} = 0.13 pm/V d_{22} = 2.6 pm/V	Type-II Type-0	5 GW/cm^2	189 - 3500
LBO	-	Yes	d_{24} = -0.67 pm/V	Type-II	9.5 GW/cm^2	160 - 2600
LN	Yes	Yes	d_{33} = 27.0 pm/V	Type-0	0.2 GW/cm^2	420 - 5200
KTP	Yes	Yes	d_{33} = 16.9 pm/V d_{24} = 7.6 pm/V	Type-0 Type-II	2.5 GW/cm^2	350 - 4500
BIBO	-	Yes	d_{11} = 2.5 pm/V	Type-I	0.3 GW/cm^2	286 - 2500
KNbO3	Yes	Yes	d_{32} = -18.3 pm/V	Type-I	0.4 MW/cm^2	400 - 4500
AlGaAs	Yes	Yes	d_{14} = 100 pm/V	Type-II	-	> 1000

FIGURE 11.33 Overview of commonly used nonlinear materials for production of heralded single photons. Quasi-phase-matching and waveguide indicate if these modifications are available for the specific material. Strength of $\chi^{(2)}$-nonlinearities is given together with the type of possible phase-matching (only the most common types are listed). Damage thresholds indicate the maximum intensity for the pump laser. Transparency data list the range over which the material is transparent. Note that nonlinearities and damage thresholds are wavelength dependent. Therefore, the values quoted here should only be taken as approximate.

11.8 CONCLUSION

In conclusion, parametric down-conversion is a mature technology for single-photon sources. The spectral and spatio-temporal profiles of PDC light are extremely well understood and can be tailored to meet the demands of modern applications. Advances in nonlinear optics have made PDC sources very efficient and well suited for on-chip integration, which becomes increasingly important as quantum technology moves from laboratories to real-world applications.

The remaining challenge is to combine the theoretic and engineering advances presented in this chapter into one package that delivers pure heralded single photons on demand. The probabilistic nature of the PDC process is acceptable for some applications, such as quantum key distribution, but less so for others, such as quantum logic gates or quantum metrology. A solution may be provided by combining PDC sources with novel nonlinear techniques such as strong two-photon absorbers similar to the scheme shown in [113], for which several systems are studied. Until such schemes become available, PDC sources can approximate single photons to virtually any desired degree by using multiplexing techniques that necessitate a modest engineering overhead once sources, logic gates, and detectors are all integrated on a single photonic chip.

REFERENCES

[1] D. Magde and H. Mahr, "Study in Ammonium Dihydrogen Phosphate of Spontaneous Parametric Interaction Tunable from 4400 to 16 000 Å," Phys. Rev. Lett. 18, 905–907 (1967).

[2] S.A. Akhmanov, V.V. Fadeev, R.V. Khoklov, and O.N. Chunaev, Sov. Phys. JETP Lett. 6, 85 (1967).
[3] R.L. Byer and S.E. Harris, "Power and Bandwidth of Spontaneous Parametric Emission," Phys. Rev. 168, 1064 (1968).
[4] D.C. Burnham and D.L. Weinberg, "Observation of Simultaneity in Parametric Production of Optical Photon Pairs," Phys. Rev. Lett. 25, 84–87 (1970).
[5] Y.H. Shih and C.O. Alley, "New Type of Einstein-Podolsky-Rosen-Bohm Experiment Using Pairs of Light Quanta Produced by Optical Parametric Down Conversion," Phys. Rev. Lett. 61, 2921–2924 (1988).
[6] R. Ghosh and L. Mandel, "Observation of Nonclassical Effects in the Interference of Two Photons," Phys. Rev. Lett. 59, 1903–1905 (1987).
[7] C.K. Hong, Z.Y. Ou, and L. Mandel, "Measurement of Subpicosecond Time Intervals Between Two Photons by Interference," Phys. Rev. Lett. 59, 2044–2046 (1987).
[8] P.G. Kwiat, K. Mattle, H. Weinfurter, A. Zeilinger, A.V. Sergienko, and Y. Shih, "New High-Intensity Source of Polarization-Entangled Photon Pairs," Phys. Rev. Lett. 75, 4337–4341 (1995).
[9] G. Weihs, T. Jennewein, C. Simon, H. Weinfurter, and A. Zeilinger, "Violation of Bell's Inequality Under Strict Einstein Locality Conditions," Phys. Rev. Lett. 81, 5039–5043 (1998).
[10] D. Bouwmeester, J.W. Pan, K. Mattle, M. Eibl, H. Weinfurter, and A. Zeilinger, "Experimental Quantum Teleportation," Nature 390, 575–579 (1997).
[11] D. Bouwmeester, J.W. Pan, M. Daniell, H. Weinfurter, and A. Zeilinger, "Observation of Three-Photon Greenberger-Horne-Zeilinger Entanglement," Phys. Rev. Lett. 82, 1345–1349 (1999).
[12] T. Jennewein, C. Simon, G. Weihs, H. Weinfurter, and A. Zeilinger, "Quantum Cryptography with Entangled Photons," Phys. Rev. Lett. 84, 4729–4732 (2000).
[13] R.W. Boyd, Nonlinear Optics Second Edition, Academic Press (2003).
[14] T. Suhara and M. Fujimura, Waveguide Nonlinear-Optic Devices (Springer Series in Photonics, V. 11). SpringerVerlag (2003).
[15] W.P. Grice and I.A. Walmsley, "Spectral Information and Distinguishability in Type-II Down-Conversion with a Broadband Pump," Phys. Rev. A 56, 1627 (1997).
[16] W. Wasilewski, C. Radzewicz, R. Frankowski, and K. Banaszek, "Statistics of Multiphoton Events in Spontaneous Parametric Down-Conversion," Phys. Rev. A 78, 033831 (2008).
[17] W. Mauerer, M. Avenhaus, W. Helwig, and C. Silberhorn, "How Colors Influence Numbers: Photon Statistics of Parametric Down-Conversion," Phys. Rev. A 80, 053815 (2009).
[18] K.N. Cassemiro, K. Laiho, and C. Silberhorn, "Accessing the Purity of a Single Photon by the Width of the Hong–Ou–Mandel Interference," New J. Phys. 12, 113052 (2010).
[19] W. Wasilewski, P. Kolenderski, and R. Frankowski, "Spectral Density Matrix of a Single Photon Measured," Phys. Rev. Lett. 99, 123601 (2007).
[20] C.K. Law, I.A. Walmsley, and J.H. Eberly, "Continuous Frequency Entanglement: Effective Finite Hilbert Space and Entropy Control." Phys. Rev. Lett. 84, 5304–5307 (2000).
[21] P.P. Rohde, W. Mauerer, and C. Silberhorn, "Spectral Structure and Decompositions of Optical States, and Their Applications," New J. Phys. 9, 91 (2007).
[22] J.H. Eberly, "Schmidt Analysis of Pure-State Entanglement," Laser Phys. 16, 921–926 (2006).
[23] G. Vidal and R.F. Werner, "Computable Measure of Entanglement," Phys. Rev. A 65, 032314 (2002).
[24] A.M. Brańczyk, T. C. Ralph, W. Helwig, and C. Silberhorn, "Optimized Generation of Heralded Fock States Using Parametric Down-Conversion," New J. Phys. 12, 063001 (2010).
[25] K. Laiho, K.N. Cassemiro, and C. Silberhorn, "Producing High Fidelity Single Photons with Optimal Brightness Via Waveguided Parametric Down-Conversion," Opt. Express 17, 22823–22837 (2009).
[26] A.B. U'Ren, C. Silberhorn, R. Erdmann, K. Banaszek, W.P. Grice, I.A. Walmsley, and M.G. Raymer, "Generation of Pure-State Single-Photon Wavepackets by Conditional

Preparation Based on Spontaneous Parametric Downconversion," Laser Phys. 15, 146–161 (2005).
[27] A. Christ, A. Eckstein, P.J. Mosley, and C. Silberhorn, "Pure Single Photon Generation by Type-I PDC with Backward-Wave Amplification," Opt. Express 17, 3441–3446 (2009).
[28] S.L. Braunstein and P. van Loock, "Quantum Information with Continuous Variables," Rev. Mod. Phys. 77, 513–577 (2005).
[29] P. Kok and S.L. Braunstein, "Postselected Versus Nonpostselected Quantum Teleportation Using Parametric Down-Conversion," Phys. Rev. A 61, 042304 (2000).
[30] P. Kok, W.J. Munro, K. Nemoto, T.C. Ralph, J.P. Dowling, and G.J. Milburn, "Review Article: Linear Optical Quantum Computing," Rev. Mod. Phys. 79, 135 (2007).
[31] H. Lee, U. Yurtsever, P. Kok, G.M. Hockney, C. Adami, S.L. Braunstein, and J.P. Dowling, "Towards Photostatistics from Photon-Number Discriminating Detectors," J. Mod. Opt. 51, 1517–1528 (2004).
[32] P. Kumar, P. Kwiat, A. Migdall, S.W. Nam, J. Vuckovic, and F.N.C. Wong, "Photonic Technologies for Quantum Information Processing," Quantum Inf. Process. 3, 215–231 (2004).
[33] N.A. Peters, K.J. Arnold, A.P. VanDevender, E.R. Jeffrey, R. Rangarajan, O. Hosten, J.T. Barreiro, J.B. Altepeter, and P.G. Kwiat, "Towards a Quasi-Deterministic Single-Photon Source," SPIE 6305, 630507 (2006).
[34] X. Ma, S. Zotter, J. Kofler, T. Jennewein, and A. Zeilinger, "Experimental Generation of Single Photons Via Active Multiplexing," Phys. Rev. A 83, 04 (2011).
[35] V.G. Dmitriev, G.G. Gurzadyan, and D.N. Nikogosyan, Handbook of Nonlinear Optical Crystals, volume 64. Springer, Berlin and New York (1999).
[36] SNLO website. *http://www.as-photonics.com/SNLO.html*.
[37] R. Rangarajan, M. Goggin, and P. Kwiat, "Optimizing Type-I Polarization-Entangled Photons," Opt. Express 17, 18920–18933 (2009).
[38] A. Halevy, E. Megidish, L. Dovrat, H.S. Eisenberg, P. Becker, and L. Bohatý, "The Biaxial Nonlinear Crystal BiB_3O_6 as a Polarization Entangled Photon Source Using Non-Collinear Type-II Parametric Down-Conversion," Opt. Express 19, 20420–20434 (2011).
[39] C. Kurtsiefer, M. Oberparleiter, and H. Weinfurter, "High-Efficiency Entangled Photon Pair Collection in Type-II Parametric Fluorescence," Phys. Rev. A 64, 023802 (2001).
[40] F.A. Bovino, P. Varisco, A.M. Colla, G. Castagnoli, G. Di Giuseppe, and A.V. Sergienko, "Effective Fiber-Coupling of Entangled Photons for Quantum Communication," Opt. Commun. 227, 343–348 (2003).
[41] S. Castelletto, I.P. Degiovanni, A. Migdall, and M. Ware, "On the Measurement of Two-Photon Single-Mode Coupling Efficiency in Parametric Down-Conversion Photon Sources," New J. Phys. 6, 87 (2004).
[42] P. Kolenderski, W. Wasilewski, and K. Banaszek, "Modeling and Optimization of Photon Pair Sources Based on Spontaneous Parametric Down-Conversion," Phys. Rev. A 80, 013811 (2009).
[43] P.J. Mosley, J.S. Lundeen, B.J. Smith, P. Wasylczyk, A.B. U'Ren, C. Silberhorn, and I.A. Walmsley, "Heralded Generation of Ultrafast Single Photons in Pure Quantum States," Phys. Rev. Lett. 100, 133601–133604 (2008).
[44] P.J. Mosley, J.S. Lundeen, B.J. Smith, and I.A. Walmsley, "Conditional Preparation of Single Photons Using Parametric Downconversion: A Recipe for Purity," New J. Phys. 10, 093011 (2008).
[45] A.M. Brańczyk, A. Fedrizzi, T.M. Stace, T.C. Ralph, and A.G. White, "Engineered Optical Nonlinearity for Quantum Light Sources," Opt. Express 19, 55–65 (2011).
[46] T. Jennewein, M. Barbieri, and A.G. White, "Single-Photon Device Requirements for Operating Linear Optics Quantum Computing Outside the Post-Selection Basis," J. Mod. Opt. 58, 276–287 (2011).
[47] A.L. Migdall, D. Branning, and S. Castelletto, "Tailoring Single-Photon and Multiphoton Probabilities of a Single-Photon On-Demand Source," Phys. Rev. A 66, 11 (2002).
[48] M.A. Broome, M.P. Almeida, A. Fedrizzi, and A.G. White, "Reducing Multi-Photon in Pulsed Down-Conversion by Temporal Multiplexing," Opt. Express 19, 22698–22708 (2011).

[49] G. Brida, I.P. Degiovanni, M. Genovese, A. Migdall, F. Piacentini, S.V. Polyakov, and I. Ruo Berchera, "Experimental Realization of a Low-Noise Heralded Single-Photon Source," Opt. Express 19, 1484–1492 (2011).
[50] J.A. Armstrong, N. Bloembergen, J. Ducuing, and P.S. Pershan, "Interactions Between Light Waves in a Nonlinear Dielectric," Phys. Rev. 127, 1918–1939 (1962).
[51] M.M. Fejer, G.A. Magel, D.H. Jundt, and R.L. Byer, "Quasi-Phase-Matched Second Harmonic-Generation: Tuning and Tolerances," IEEE J. Quantum Elect. 28, 2631–2654 (1992).
[52] K. Kato, and E. Takaoka, "Sellmeier and Thermo-Optic Dispersion Formulas for KTP," Appl. Opt. 41, 5040–5044 (2002).
[53] S. Emanueli, and A. Arie, "Temperature-Dependent Dispersion Equations for KTiOPO4 and KTiOAsO4," Appl. Opt. 42, 6661–6665 (2003).
[54] E.J. Lim, M.M. Fejer, and R.L. Byer, "Second-Harmonic Generation of Green Light in Periodically Poled Planar Lithium-Niobate Wave-Guide," Electron. Lett. 25, 174–175 (1989).
[55] C.J. van der Poel, J.D. Bierlein, J.B. Brown, and S. Colak, "Efficient Type-I Blue Second-Harmonic Generation in Periodically Segmented KTiOPO4 Waveguides," Appl. Phys. Lett. 57, 2074–2076 (1990).
[56] M. Yamada, N. Nada, M. Saitoh, and K. Watanabe, "First-Order Quasi-Phase Matched LiNbO$_3$ Wave-Guide Periodically Poled by Applying an External-Field for Efficient Blue Second-Harmonic Generation," Appl. Phys. Lett. 62, 435–436 (1993).
[57] M. Houe, and P.D. Townsend, "An Introduction to Methods of Periodic Poling for Second-Harmonic Generation," J. Phys. D – Appl. Phys. 28, 1747–1763 (1995).
[58] D.S. Hum, and M.M. Fejer, "Quasi-Phasematching," Comptes Rendus Physique 8, 180–198 (2007).
[59] E.J. Mason, M.A. Albota, F. Konig, and F.N.C. Wong, "Efficient Generation of Tunable Photon Pairs at 0.8 and 1.6 μm," Opt. Lett. 27, 2115–2117 (2002).
[60] C.E. Kuklewicz, M. Fiorentino, G. Messin, F.N.C. Wong, and J.H. Shapiro, "High-Flux Source of Polarization-Entangled Photons From a Periodically Poled KTiOPO4 Parametric Down-Converter," Phys. Rev. A 69, 013807 (2004).
[61] M. Fiorentino, G. Messin, C.E. Kuklewicz, F.N.C. Wong, and J.H. Shapiro, "Generation of Ultrabright Tunable Polarization Entanglement Without Spatial, Spectral, or Temporal Constraints," Phys. Rev. A 69, 041801 (2004).
[62] F. Konig, E.J. Mason, F.N.C. Wong, and M.A. Albota, "Efficient and Spectrally Bright Source of Polarization-Entangled Photons," Phys. Rev. A 71, 033805 (2005).
[63] D. Ljunggren, and M. Tengner, "Optimal Focusing for Maximal Collection of Entangled Narrow-Band Photon Pairs into Single-Mode Fibers," Pys. Rev. A 72, 062301 (2005).
[64] S. Castelletto, I.P. Degiovanni, V. Schettini, and A. Migdall, "Optimizing Single-Photon-Source Heralding Efficiency and Detection Efficiency Metrology at 1550 nm Using Periodically Poled Lithium Niobate," Metrologia 43, S56–S60 (2006).
[65] A. Fedrizzi, T. Herbst, A. Poppe, T. Jennewein, and A. Zeilinger, "A Wavelength-Tunable Fiber-Coupled Source of Narrowband Entangled Photons," Opt. Express 15, 15377–15386 (2007).
[66] R.S. Bennink, "Optimal Collinear Gaussian Beams for Spontaneous Parametric Down-Conversion," Phys. Rev. A 81, 053805 (2010).
[67] J.L. Smirr, M. Deconinck, R. Frey, I. Agha, E. Diamanti, and I. Zaquine, "Optimal Photon-Pair Single Mode Coupling in Narrow-Band Spontaneous Parametric Down-Conversion with Arbitrary Pump Profile," J. Opt. Soc. Am. B – Opt. Phys. 30, 288–301 (2013).
[68] M.D. Cunha Pereira, F.E. Becerra, B.L. Glebov, J. Fan, S.W. Nam, and A. Migdall, "Demonstrating Highly Symmetric Single-Mode, Single-Photon Heralding Efficiency in Spontaneous Parametric Downconversion," Opt. Lett. 38, 1609–1611 (2013).
[69] D.H. Smith, G. Gillett, M. de Almeida, C. Branciard, A. Fedrizzi, T.J. Weinhold, A. Lita, B. Calkins, T. Gerrits, S.W. Nam, and A.G. White, "Conclusive Quantum Steering with Superconducting Transition Edge Sensors," Nat. Commun. 3, 625 (2012).
[70] S. Ramelow, A. Mech, M. Giustina, S. Gröblacher, W. Wieczorek, J. Beyer, A. Lita, B. Calkins, T. Gerrits, S.W. Nam, A. Zeilinger, and R. Ursin, "Highly Efficient Heralding of Entangled Single Photons," Opt. Express 21, 6707–6717 (2013).

[71] B. Yurke, and M. Potasek, "Obtainment of Thermal Noise from a Pure Quantum State," Phys. Rev. A 36, 3464–3466 (1987).
[72] E. Bocquillon, C. Couteau, M. Razavi, R. Laflamme, and G. Weihs, "Coherence Measures for Heralded Single-Photon Sources," Phys. Rev. A 79, 035801 (2009).
[73] S. Fasel, O. Alibart, S. Tanzilli, P. Baldi, A. Beveratos, N. Gisin, and H. Zbinden, "High-Quality Asynchronous Heralded Single-Photon Source at Telecom Wavelength," New J. Phys. 6, 163 (2004).
[74] M. Tengner and D. Ljunggren, "Characterization of an Asynchronous Source of Heralded Single Photons Generated at a Wavelength of 1550 nm," arXiv:0706.2985v1, (2007).
[75] M. Scholz, L. Koch, and O. Benson, "Statistics of Narrow-Band Single Photons for Quantum Memories Generated by Ultrabright Cavity-Enhanced Parametric Down-Conversion," Phys. Rev. Lett. 102, 063603 (2009).
[76] D. Höckel, L. Koch, and O. Benson, "Direct Measurement of Heralded Single-Photon Statistics from a Parametric Down-Conversion Source," Phys. Rev. A 83, 013802 (2011).
[77] G. Brida, I.P. Degiovanni, M. Genovese, F. Piacentini, P. Traina, A. Della Frera, A. Tosi, A. Bahgat Shehata, C. Scarcella, A. Gulinatti, M. Ghioni, S.V. Polyakov, A. Migdall, and A. Guidice, "An Extremely Low-Noise Heralded Single-Photon Source: A Breakthrough for Quantum Technologies." App. Phys. Lett. 101, 221112 (2012).
[78] W.P. Grice, A.B. U'Ren, and I.A. Walmsley, "Eliminating Frequency and Space-Time Correlations in Multiphoton States," Phys. Rev. A 64, 063815 (2001).
[79] P.G. Evans, R.S. Bennink, W.P. Grice, T.S. Humble, and J. Schaake, "Bright Source of Spectrally Uncorrelated Polarization-Entangled Photons with Nearly Single-Mode Emission," Phys. Rev. Lett. 105, 253601 (2010).
[80] R.-B. Jin, J. Zhang, R. Shimizu, N. Matsuda, Y. Mitsumori, H. Kosaka, and K. Edamatsu, "High-Visibility Nonclassical Interference Between Intrinsically Pure Heralded Single Photons and Photons from a Weak Coherent Field," Phys. Rev. A 83, 031805 (2011).
[81] J.P. Torres, A. Alexandrescu, S. Carrasco, and L. Torner, "Quasi-Phase-Matching Engineering for Spatial Control of Entangled Two-Photon States," Opt. Lett. 29, 376–378 (2004).
[82] H.Y. Leng, X.Q. Yu, Y.X. Gong, P. Xu, Z.D. Xie, H. Jin, C. Zhang, and S.N. Zhu, "On-Chip Steering of Entangled Photons in Nonlinear Photonic Crystals," Nat. Commun. 2, 429 (2011).
[83] M.B. Nasr, S. Carrasco, B.E.A. Saleh, A.V. Sergienko, M.C. Teich, J.P. Torres, L. Torner, D.S. Hum, and M.M. Fejer, "Ultrabroadband Biphotons Generated Via Chirped Quasi-Phase-Matched Optical Parametric Down-Conversion," Phys. Rev. Lett. 100, 183601 (2008).
[84] H.G. de Chatellus, A.V. Sergienko, B.E.A. Saleh, M.C. Teich, and G. Di Giuseppe, "Non-collinear and Non-degenerate Polarization-Entangled Photon Generation Via Concurrent Type-I Parametric Downconversion in PPLN," Opt. Express 14, 10060–10072 (2006).
[85] M. Pysher, A. Bahabad, P. Peng, A. Arie, and O. Pfister, "Quasi-Phase-Matched Concurrent Nonlinearities in Periodically Poled $KTiOPO_4$ for Quantum Computing Over the Optical Frequency Comb," Opt. Lett. 35, 565–567 (2010).
[86] A. Christ, K. Laiho, A. Eckstein, T. Lauckner, P.J. Mosley, and C. Silberhorn, "Spatial Modes in Waveguided Parametric Down-Conversion," Phys. Rev. A 80, 033829 (2009).
[87] J.D. Bierlein, A. Ferretti, L.H. Brixner, and W.Y. Hsu, "Fabrication and Characterization of Optical Waveguides in KTiOPO4," Appl. Phys. Lett. 50, 1216 (1987).
[88] W.P. Risk, "Fabrication and Characterization of Planar Ion-Exchanged KTiOPO4 Waveguides for Frequency Doubling," Appl. Phys. Lett. 58, 19 (1991).
[89] A.S. Helmy, P. Abolghasem, J.S. Aitchison, B.J. Bijlani, J. Han, B.M. Holmes, D.C. Hutchings, U. Younis, and S.J. Wagner, "Recent Advances in Phase Matching of Second-Order Nonlinearities in Monolithic Semiconductor Waveguides," Laser Photonics Rev. 5, 272–286, 2011.
[90] R. Horn, P. Abolghasem, B.J. Bijlani, D. Kang, A.S. Helmy, and G. Weihs, "Monolithic Source of Photon Pairs," Phys. Rev. Lett. 108, 153605 (2012).
[91] M. Ravaro, Y. Seurin, S. Ducci, G. Leo, V. Berger, A. De Rossi, and G. Assanto, "Nonlinear AlGaAs Waveguide for the Generation of Counterpropagating Twin Photons in the Telecom Range," J. Appl. Phys. 98, 063103 (2005).

[92] L. Lanco, S. Ducci, J.-P. Likforman, X. Marcadet, J.A.W. van Houwelingen, H. Zbinden, G. Leo, and V. Berger, "Semiconductor Waveguide Source of Counterpropagating Twin Photons," Phys. Rev. Lett. 97, 173901 (2006).
[93] S. Tanzilli, H. De Riedmatten, W. Tittel, H. Zbinden, P. Baldi, M. De Micheli, D.B. Ostrowsky, and N. Gisin, "Highly Efficient Photon-Pair Source Using Periodically Poled Lithium Niobate Waveguide," Electron. Lett. 37, 26–28 (2001).
[94] K. Banaszek, A.B. U'Ren, and I.A. Walmsley, "Generation of Correlated Photons in Controlled Spatial Modes by Downconversion in Nonlinear Waveguides," Opt. Lett. 26, 1367 (2001).
[95] A.B. U'Ren, C. Silberhorn, K. Banaszek, and I. Walmsley, "Efficient Conditional Preparation of High-Fidelity Single Photon States for Fiber-Optic Quantum Networks," Phys. Rev. Lett. 93, 093601 (2004).
[96] P. Aboussouan, O. Alibart, D.B. Ostrowsky, P. Baldi, and S. Tanzilli, "High-Visibility Two-Photon Interference at a Telecom Wavelength Using Picosecond-Regime Separated Sources," Phys. Rev. A 81, 021801 (2010).
[97] S. Krapick, H. Herrmann, V. Quiring, B. Brecht, H. Suche, and C. Silberhorn, "An Efficient Integrated Two-Color Source for Heralded Single Photons," New J. Phys. 15, 033010 (2013).
[98] T. Zhong, X. Hu, F.N.C. Wong, K.K. Berggren, T.D. Roberts, and P. Battle, High-Quality Fiber Optic Polarization Entanglement Distribution at 1.3 μm Telecom Wavelength," Opt. Lett. 35, 1392–1394 (2010).
[99] J. Chen, A.J. Pearlman, A. Ling, J. Fan, and A.L. Migdall, "A Versatile Waveguide Source of Photon Pairs for Chip-Scale Quantum Information Processing," Opt. Express 17, 6727–6740 (2009).
[100] A. Eckstein, A. Christ, P.J. Mosley, and C. Silberhorn, "Highly Efficient Single-Pass Source of Pulsed Single-Mode Twin Beams of Light," Phys. Rev. Lett. 106, 013603 (2011).
[101] T. Zhong, F.N.C. Wong, A. Restelli, and J.C. Bienfang, "Efficient Single-Spatial-Mode Periodically-Poled KTiOPO4 Waveguide Source for High-Dimensional Entanglement-Based Quantum Key Distribution," Opt. Express 20, 26868–26877 (2012).
[102] G. Harder, V. Ansari, B. Brecht, T. Dirmeier, C. Marquardt, and C. Silberhorn, "An Optimized Photon Pair Source for Quantum Circuits," Opt. Express 21, 13975–13985 (2013).
[103] J.Ĺ. Svozilǎk, M. Hendrych, A.S. Helmy, and J.P. Torres, "Generation of Paired Photons in a Quantum Separable State in Bragg Reflection Waveguides," Opt. Express 19, 3115–3123 (2011).
[104] M.C. Booth, M. Atatüre, Di Giuseppe, B.E.A. Saleh, A.V. Sergienko, and M.C. Teich, "Counterpropagating Entangled Photons from a Waveguide with Periodic Nonlinearity," Phys. Rev. A 66, 023815 (2002).
[105] M.F. Saleh, B.E.A. Saleh, and M.C. Teich, "Modal, Spectral, and Polarization Entanglement in Guided-Wave Parametric Down-Conversion," Phys. Rev. A (At. Mol. Opt. Phys.) 79, 053842–053852 (2009).
[106] P.J. Mosley, A. Christ, A. Eckstein, and C. Silberhorn, "Direct Measurement of the Spatial-Spectral Structure of Waveguided Parametric Down-Conversion," Phys. Rev. Lett. 103, 233901 (2009).
[107] M. Karpinski, C. Radzewicz, and K. Banaszek, "Experimental Characterization of Three-Wave Mixing in a Multimode Nonlinear KTiOPO$_4$ Waveguide," Appl. Phys. Lett. 94, 181105–181108 (2009).
[108] M. Fiorentino, S.M. Spillane, R.G Beausoleil, T.D. Roberts, P. Battle, and M.W. Munro, "Spontaneous Parametric Down-Conversion in Periodically Poled KTP Waveguides and Bulk Crystals," Opt. Express 15, 7479–7488 (2007).
[109] S.M. Spillane, M. Fiorentino, and R.G. Beausoleil, "Spontaneous Parametric Down Conversion in a Nanophotonic Waveguide," Opt. Express 15, 8770–8780 (2007).
[110] J.M. White, and P.F. Heidrich, "Optical Waveguide Refractive Index Profiles Determined from Measurement of Mode Indices: A Simple Analysis," Appl. Opt. 15, 151–155 (1976).
[111] D. Marcuse, "Theory of Dielectric Optical Waveguides," Academic Press (1974).
[112] T.B. Koch, J.B. Davies, and D. Wickramasinghe, "Finite Element/Finite Difference Propagation Algorithm for Integrated Optical Device," Electron. Lett. 25, 514–516 (1989).
[113] B.C. Jacobs, T.B. Pittman, and J.D. Franson, "Single Photon Source Using Laser Pulses and Two-Photon Absorption," Phys. Rev. A 74, 010303 (2006).

[114] A. Christ, C. Silberhorn, "Limits on the Deterministic Creation of Pure Single-Photon States Using Parametric Down-Conversion." Phys. Rev. A 85, 023829–023835 (2012).

[115] U'Ren, K. Banaszek, and I.A. Walmsley, "Photon engineering for Quantum Information Processing." Quantum Info. Comput. 3, 480–502 (2003).

Chapter 12

Four-Wave Mixing in Single-Mode Optical Fibers

Alex McMillan[*], Yu-Ping Huang[†], Bryn Bell1[*], Alex Clark[‡], Prem Kumar[†] and John Rarity[*]

[*]Photonics Group, Merchant Venturers School of Engineering, University of Bristol, Bristol, BS8 1UB, UK
[†]Center for Photonic Communication and Computing, Department of Electrical Engineering and Computer Science & Department of Physics and Astronomy, Northwestern University, 2145 Sheridan Road, Evanston, Illinois 60208, USA
[‡]Centre for Ultrahigh Bandwidth Devices for Optical Systems (CUDOS), Institute of Photonics and Optical Science (IPOS), School of Physics, University of Sydney, NSW 2006, Australia

---

Chapter Outline

12.1 Introduction	412
12.2 Photon Pair Generation in Optical Fibers	413
12.2.1 Classical Four-Wave Mixing Theory and Phase-Matching Requirements	413
12.2.2 Quantum Theory of Four-Wave Mixing	416
12.2.3 Cross-Polarized Four-Wave Mixing in Birefringent Fibers	419
12.2.4 Raman Scattering	420
12.3 Heralded Single-Photon Sources Based on sFWM	422
12.3.1 Photon-Pair Generation in the Anomalous Dispersion Regime	425
12.3.2 Photonic Crystal Fiber Sources in the Normal Dispersion Regime	427
12.4 Quantum Interference Between Separate Spectrally Filtered Fiber Sources	430
12.5 Intrinsically Pure-State Photons	436
12.5.1 Generation of Spectrally Uncorrelated Two-Photon States Through Group Velocity Matching	436
12.5.2 A Temporal Filtering Approach for Attaining Pure-State Photons	440
12.6 Entangled Photon-Pair Sources	444
12.7 Applications of Fiber Photon Sources—All-Fiber Quantum Logic Gates	454

12.8 Photonic Fusion in Fiber	458
12.9 Conclusion	460
References	461

12.1 INTRODUCTION

In many ongoing experiments in quantum information, probabilistic photon sources based on the generation of photon pairs using nonlinear optics have proven to be an excellent approach, due to their typically high brightness and the wide range of photon properties that can be attained in different nonlinear systems. Photon sources based on spontaneous parametric down-conversion in bulk $\chi^{(2)}$ nonlinear crystals, and more recently waveguides in $\chi^{(2)}$ materials (see Chapter 11), are still the most commonly used, but in recent years there has been much interest in sources based on parametric photon pair generation in optical fibers.

In optical fibers the $\chi^{(3)}$ nonlinear response of the silica glass core can lead to the generation of correlated pairs of photons, known as the signal and idler, through spontaneous four-wave mixing when either a single intense pump field, or pump fields at two distinct wavelengths are propagating in the fiber. This process can only occur efficiently when the constraints of both energy and momentum conservation (i.e., phase-matching) of the optical fields are satisfied. The spectral properties of the photon pairs that can be generated are largely dependent on the dispersion properties of the fiber. When pumping close to the zero-dispersion wavelength of a fiber, depending on the exact choice of pump wavelength, phase-matching can be satisfied for either widely separated signal and idler photon pairs [1], or for signal and idler pairs generated close to the pump wavelength [2].

Even greater flexibility can be achieved for four-wave mixing in photonic crystal fiber, in which the dispersion can be tailored through design of the fiber structure. Pair photon generation has been demonstrated in photonic crystal fiber using a wide range of different pump sources with generated photons covering a considerable wavelength range, from the visible to telecom wavelengths in the near infrared, making them suitable for a wide range of applications [3–7]. Through careful design of the fiber structure it is also possible to influence other properties of the photon pair, in order to generate them on a well-defined polarization axis of the fiber and to avoid undesirable spectral correlations between the signal and idler photons which must otherwise be eliminated through lossy, narrowband spectral filtering [8–10].

Although the generation of photon pairs by four-wave mixing is probabilistic in nature, the photons are always produced simultaneously in pairs, so the detection of one photon can be used to herald the presence of the other. Such fiber-based heralded single-photon sources have been demonstrated with high brightness (generated pair probability) due to the long nonlinear interaction lengths that can be achieved in a waveguide configuration, with the pump beam confined to a small mode area. Furthermore, as the photons can be generated in the fundamental guided mode of the fiber, they can be coupled into standard single-mode fiber with high efficiency. This has led to the development of spliced all-fiber sources with very low optical losses and high stability [5,11,12]. The ease of integration of these photon sources with standard fiber-based optical components has also been crucial for demonstrations of all-fiber quantum logic gates [13].

In addition to heralded single photons, fiber-based photon sources can be designed to generate useful two-photon states. Of particular interest are polarization-entangled signal and idler pairs, which can be generated via various schemes [14–16]. Looking beyond two-photon entangled states, larger entangled cluster states have now been demonstrated by linking the states output from multiple fiber-based photon sources through selective measurement of the output photons [17]. As these states can be used as a resource for universal quantum computation [18], the generation of larger cluster states and investigation into their properties is an area of great future interest.

12.2 PHOTON PAIR GENERATION IN OPTICAL FIBERS

12.2.1 Classical Four-Wave Mixing Theory and Phase-Matching Requirements

Four-wave mixing (FWM) is a third-order nonlinear ($\chi^{(3)}$) optical process that occurs naturally in all optical materials, including fibers, in which an intense, propagating pump field provides amplification for light fields at other wavelengths. In many ways, the process is similar to parametric down-conversion in second-order nonlinear ($\chi^{(2)}$) crystals, which has been widely studied for use as the basis of single-photon sources. FWM results from the coupling of light fields at four distinct wavelengths (in the most general case) due to the polarization response of the transmission medium in which the light is propagating. The polarization response **P** of a material in the presence of light with an electric field component **E** can be described by

$$\mathbf{P} = \epsilon_0 (\chi^{(1)} \cdot \mathbf{E} + \chi^{(2)} : \mathbf{EE} + \chi^{(3)} \vdots \mathbf{EEE} + \cdots), \qquad (12.1)$$

where the $\chi^{(j)}$ terms are the jth order coefficients of a Taylor series expansion of the electric susceptibility of the material [19]. In isotropic materials, such as a silica glass, there will be no $\chi^{(2)}$ term because of symmetry constraints, and the nonlinear response of the material is dominated by the $\chi^{(3)}$ contribution. In addition to effects such as third harmonic generation and self-phase modulation, the $\chi^{(3)}$ nonlinear response allows FWM, where two intense optical fields (pump fields of angular frequency ω_{p1} and ω_{p2}) lose energy and provide gain for two other fields (the so called signal at higher frequency ω_s and the idler at the lower frequency ω_i). Most commonly, the two pump fields are chosen to be degenerate in frequency (ω_p), such that the process can be pumped using a single high-power laser. If a probe beam at the signal or idler wavelength is also launched with the pump it will be amplified (a phase-insensitive process), and result in the generation of light at the other wavelength. When both the signal and idler modes are initially empty, spontaneous four-wave mixing (sFWM) can occur, with the process seeded by quantum vacuum noise. In this case the signal and idler appear in the spectrum as newly generated sidebands, equally spaced in frequency about the central pump peak.

At the quantum level the sFWM process can be regarded as the virtual absorption of two pump (p) photons and subsequent creation of a signal (s) and idler (i) photon pair. As the response time of the $\chi^{(3)}$ Kerr nonlinearity in optical fibers is nearly instantaneous, this allows the detection of one photon of the pair to herald the generation of the other. For a given fiber structure and pump laser, the signal and idler wavelengths that will be generated through sFWM are determined by the energy and phase-matching conditions:

$$2\omega_p = \omega_s + \omega_i, \qquad (12.2)$$

and

$$\kappa = k_s + k_i - 2k_p + 2\gamma P_p. \qquad (12.3)$$

Here κ is the phase-mismatch between the propagation constants of the signal, idler, and pump waves resulting from chromatic dispersion, with $k_j = \frac{n_j \omega_j}{c}$ ($j = s, i, p$) for light with angular frequency ω_j propagating in a medium with refractive index n_j at that frequency. Only when the phase-mismatch is near zero will the average probability of generated signal and idler photon pairs increase along the entire length of the fiber and build up to appreciable levels. The $2\gamma P_p$ term in Eq. (12.3) is the correction to the phase-mismatch due to self-phase modulation of the intense pump pulse, where P_p is the peak pump power and γ is the nonlinear coefficient of the fiber, given by

$$\gamma = \frac{2\pi n_2}{\lambda_p A_{\text{eff}}}. \qquad (12.4)$$

The nonlinear coefficient of a fiber is determined by the pump wavelength λ_p, the effective cross-sectional area of the fiber mode A_{eff} and the nonlinear refractive index of the material n_2. For silica, $n_2 = 2 \times 10^{-20}$ m^2/W [19].

Figure 12.1 shows an example of a phase-matching curve, calculated for a typical fiber using Eqs. (12.2) and (12.3) at different pump powers. Two different regimes of light propagation can be identified on this figure. At short wavelengths the fiber exhibits normal dispersion, where longer wavelength components of a light pulse travel faster than shorter wavelength components. At longer wavelengths the dispersion is anomalous and the situation is reversed. At the boundary between these regions is the zero-dispersion wavelength, where

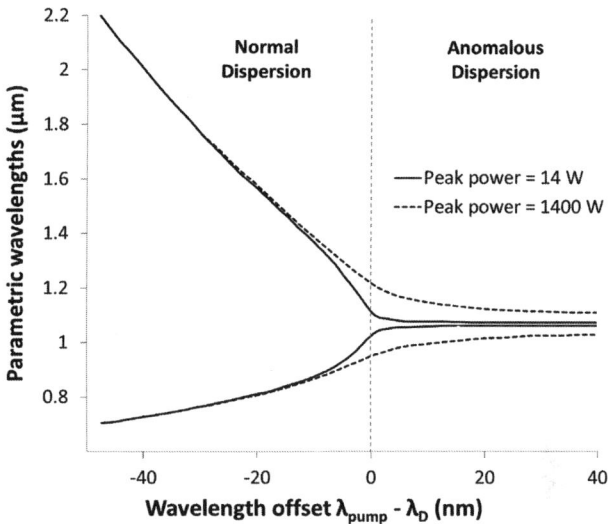

FIGURE 12.1 Calculated phase-matching curves for sFWM in an optical fiber at two different levels of peak pump power. The generated signal and idler wavelengths are shown for pump laser wavelengths λ_{pump} close to the zero-dispersion wavelength λ_D. This phase-matching curve is for a photonic crystal fiber, in which the zero-dispersion wavelength has been shifted toward the visible regime compared to conventional single-mode fibers. *Image adapted from [20].*

the dispersion parameter D of the fiber is zero, such that

$$D = -\frac{2\pi c}{\lambda^2}\frac{d^2k}{d\omega^2} = 0. \tag{12.5}$$

The behavior of the phase-matching conditions for sFWM depends on whether the pump wavelength is above or below the zero-dispersion wavelength. In the anomalous regime, $2k_p > k_s + k_i$, so for signal and idler wavelengths close to the pump the nonlinear self-phase modulation term $+2\gamma P_p$ can be used to compensate for the phase-mismatch. This results in broad sidebands close to the pump wavelength, with their separation highly dependent on the pump power P_p. However, in the normal dispersion regime where $2k_p < k_s + k_i$, the nonlinear term increases the phase-mismatch. Here higher order dispersion terms become important and phase-matching can be achieved close to the zero-dispersion wavelength for large detuning of the signal and idler wavelengths from the pump. In the normal dispersion region the phase-matched signal and idler wavelengths are not strongly affected by the pump power, but can vary considerably in response to small changes in the pump wavelength.

In conventional fibers, the refractive index contrast between the core and cladding that is required for guidance is attained by doping the glass in either the core or cladding regions. As this index contrast is relatively weak, the dispersion properties of conventional fibers are predominantly determined by the inherent material dispersion of bulk silica, with a zero-dispersion wavelength

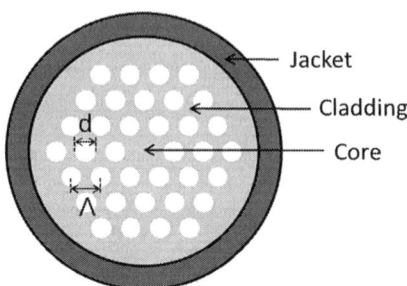

FIGURE 12.2 Schematic representation of a simple photonic crystal fiber design, based on a triangular lattice. The core region is solid silica, while the cladding is formed by a periodic array of air holes. The properties of the fiber are determined by the hole diameter (d) and pitch (Λ).

near 1.3 μm. While it is possible to modify dispersion in conventional fibers with a structured cladding index profile, the extent to which the zero-dispersion wavelength can be moved is limited and it cannot be easily shifted to shorter wavelengths [21].

An alternative approach to sFWM in conventional all-solid fiber is to take advantage of the additional design flexibility available in photonic crystal fiber (PCF). For PCF the fiber is produced from a single material, typically pure silica. In PCF structures designed for photon pair generation through sFWM, the nonlinear interaction occurs in a solid silica core, while the cladding region is formed from a periodic array of air holes in the silica, which run along the length of the fiber (see Fig. 12.2). The presence of the air holes can reduce the effective index of the cladding significantly compared to the solid silica core, leading to stronger guidance than in conventional fibers. Unlike in conventional fibers, the waveguide contribution to dispersion can therefore be of a similar magnitude to the inherent material dispersion, either compensating it or enhancing it depending on the PCF structure.

Physical realizations of various photon sources based on conventional fibers and PCF, and operating with pumping in both the normal and anomalous dispersion regimes are discussed in Section 12.3.

12.2.2 Quantum Theory of Four-Wave Mixing

In order to better understand the properties of the photon-pair state that is generated through sFWM, it is useful to extend the theoretical model described in Section 12.2 to incorporate quantization of the signal and idler modes [22–24]. Commonly a semi-classical approach is taken, in which the bright pump beam is treated using classical nonlinear optics, but its interaction with the signal and idler fields is implemented using photon creation operators $\hat{a}_s^\dagger$ and $\hat{a}_i^\dagger$ acting on an initially empty vacuum state. The nonlinear interaction Hamiltonian associated with the sFWM process (assuming phase-matching is satisfied, and that sFWM occurs in a single temporal, spatial, and polarization mode) is of the form

$$H_{\text{int}} = g(\hat{a}_s^\dagger \hat{a}_i^\dagger + \hat{a}_s \hat{a}_i). \tag{12.6}$$

The first term in Eq. (12.6) corresponds to the creation of a correlated signal and idler pair, achieved through the annihilation of two pump photons (although this is not shown due to the undepleted pump approximation wherein the back effect of sFWM on the pump is neglected; see, e.g., [25]). The second term is the reverse process, by which a signal and idler photon can be converted to two photons at the pump wavelength. The coupling coefficient g determines the strength of the interaction and incorporates several parameters, most notably the intrinsic nonlinear susceptibility of the material $\chi^{(3)}$, the peak power of the pump laser and the effective mode field area of the fiber. The nonlinear response is enhanced for small core fibers where the pump light is tightly confined.

Using the interaction Hamiltonian shown in Eq. (12.6), and assuming for now that the sFWM process described allows generation of signal and idler photons only in the single-mode regime, the generated state is given by

$$|\psi_{\text{out}}\rangle \propto |vac\rangle + \alpha|1,1\rangle_{s,i} + \alpha^2|2,2\rangle_{s,i} + \cdots, \tag{12.7}$$

where $|\alpha|^2$ is the average number of photon pairs generated per temporal mode, $|vac\rangle$ is the vacuum component of the state containing no pairs, and the first and second numbers in each of the other state components denote the number of generated signal and idler photons respectively. Because of the stochastic nature of spontaneous parametric processes, the output signal and idler states prior to measurement consist of a series of terms with increasing photon number and increasing powers of α. However, because the number of photons in the signal and idler modes must be perfectly correlated (in the absence of loss), measurement of either the signal or idler state can be used to determine when the desired number state of photons in the other frequency mode was successfully generated. This can be used to realize heralded single-photon sources, as described in Section 12.3.

The energy and phase-matching requirements identified in Eqs. (12.2) and (12.3) may seem to imply that unique, monochromatic solutions for the signal and idler fields exist for any chosen pump wavelength. While these conditions do define the peak signal and idler wavelengths that are generated, in reality there will always be some finite bandwidth associated with these fields. The requirements for producing spectrally pure-state photons place stringent conditions on the bandwidth of the generated photons, which generally means that they must be spectrally filtered. This is described in detail in Section 12.4. Understanding the factors influencing the gain bandwidth for sFWM is therefore important, as this determines the proportion of the generated photon pairs that will be produced within the desired spectral range.

By using a truncated Taylor expansion of the wavevector terms in the phase-matching equation (Eq. (12.3)) it can be shown [22] that the full-width half-maximum (FWHM) bandwidths of the signal and idler fields, $\Delta\omega_s$ and $\Delta\omega_i$, are given by

$$\Delta\omega_s = \frac{2\pi c}{|\mathcal{N}_s - \mathcal{N}_i|L} + 2\left|\frac{\mathcal{N}_i - \mathcal{N}_p}{\mathcal{N}_s - \mathcal{N}_i}\right|\Delta\omega_p, \qquad (12.8)$$

and

$$\Delta\omega_i = \frac{2\pi c}{|\mathcal{N}_i - \mathcal{N}_s|L} + 2\left|\frac{\mathcal{N}_s - \mathcal{N}_p}{\mathcal{N}_i - \mathcal{N}_s}\right|\Delta\omega_p, \qquad (12.9)$$

where $\Delta\omega_p$ is the FWHM bandwidth of the pump, L is the length of the fiber, and $\mathcal{N}_j$ ($j = s, i, p$) are the group indices for the signal, idler, and pump, given by $\mathcal{N}_j = (\omega_j \frac{\partial n_j}{\partial \omega}\big|_{\omega_j} + n_j)$. The first terms in Eqs. (12.8) and (12.9) show the bandwidth that exists even in the case of a monochromatic pump, due to the finite length of the fiber. As L is increased the tolerance of the sFWM process to phase-mismatch is reduced, and the bandwidth over which signal and idler photons are generated is consequently also reduced.

Working in the interaction picture of quantum mechanics it is also possible to use the Hamiltonian described in Eq. (12.6) to calculate the average number of photon pairs that will be generated by sFWM, which is useful for predicting the attainable brightness of photon sources based on this process. When using a pulsed laser as a pump source, the average number of generated signal photons per pulse $\langle N_s \rangle$ is [22]

$$\langle N_s \rangle = \left(\frac{SIL}{2}\right)^2 \left(\frac{\pi \Delta\omega_p^2}{4\ln(4)}\right)^{3/2} \left(\frac{4\sqrt{2}\pi c}{(\mathcal{N}_s - \mathcal{N}_i)L}\right)\frac{\mathcal{N}_s \mathcal{N}_i}{c^2}, \qquad (12.10)$$

where I is an overlap integral between the spatial mode profiles of the signal, idler, and pump fields and S is a gain coefficient term given by

$$S = \epsilon_0 \chi^{(3)} \frac{E_{p0}^2}{4}\sqrt{\frac{\omega_s \omega_i}{4\epsilon_s \epsilon_i}}, \qquad (12.11)$$

where E_{p0} is the peak electric field amplitude of the pump pulse. The expression for the mean value of generated idler photons per pulse is identical, as expected since the signal and idler photons are always generated together in pairs. As shown in Eq. (12.10), the expectation value for the number of generated pairs depends on the strength of the nonlinearity of the fiber (through $\chi^{(3)}$) and on the square of the peak pump power. The total number of generated pairs increases linearly with the fiber length L. However, when considering only those generated photons that fall within a narrow spectral range, which is small compared to the gain bandwidth (as is often the case for photon sources which rely on narrowband spectral filtering to achieve purity), the increase in fiber length not only increases the total number of generated pairs, but also reduces the gain bandwidth, as shown in Eqs. (12.8) and (12.9). Because of this, the number of useful photon pairs falling within the selected narrow bandwidth grows with the square of the fiber length in this case, as long as the total fiber length does not exceed the pulse walk-off length, the length over which two initial co-propagating pulses become spatially separated [19].

12.2.3 Cross-Polarized Four-Wave Mixing in Birefringent Fibers

In addition to the co-polarized sFWM process described so far, where the pump, signal and idler fields have the same polarization state, phase-matching can also be achieved for cross-polarized spontaneous four-wave mixing. For this process the signal and idler are both generated in a polarization state which is orthogonal to that of the pump. Due to the material symmetry of silica, the relevant nonlinear electric susceptibility coefficient for this process $\chi^{(3)}_{xxyy}$ is only $\frac{1}{3}$ that of the co-polarized process $\chi^{(3)}_{xxxx}$, where x and y define two orthogonal transverse polarization axes of the fiber. The first two terms in the subscript correspond to the polarization state of the pump fields and the third and fourth terms refer to the polarization state of the signal and idler photons [25].

Although the gain for cross-polarized sFWM is inherently lower, in birefringent fibers this process is of considerable interest, as the strength of the birefringence provides an additional parameter that can be used to tailor the fiber dispersion in order to satisfy the phase-matching requirements. A fiber with birefringence exhibits a difference in its refractive index depending on whether light propagates on the fast or slow axis of the fiber. Birefringence in standard index guiding all-solid fibers is usually implemented by the introduction of stress applying elements close to the core, producing an asymmetric refractive index profile across the core region. In strongly birefringent fibers this can provide a difference in refractive index between the fast and slow axis $\delta n \sim 10^{-4}$. In PCF, asymmetry can alternatively be introduced in the geometry of the fiber structure, for instance by slightly collapsing or expanding one of the cladding air holes on either side of the core leading to an elliptical mode field profile [26]. Form-induced birefringence of this type can achieve $\delta n > 10^{-3}$ in the case of a highly elliptical core [27].

The phase-matching requirements shown for co-polarized sFWM in Eq. (12.3) can be modified to describe the cross-polarized sFWM process, in which pump light launched onto one axis of the fiber, generates signal and idler pairs on the orthogonal axis. The effective mode indices for the signal and idler $n_{s,i}$ are modified by an additional term δn to account for propagation on a different axis to the pump, with the sign of the contribution dependent on which axis of the fiber is pumped, such that

$$n_{s,i}(\lambda) = n_p(\lambda) \pm \delta n. \qquad (12.12)$$

The strength of the birefringence can be selected to provide phase-matching for a desired set of operating wavelengths in a similar manner to the $2\gamma P_p$ term in Eq. (12.3) although even in weakly birefringent structures, the magnitude of the δn term is usually far more significant, and can help to achieve phase-matching in both the anomalous and normal dispersion regimes relatively far from the

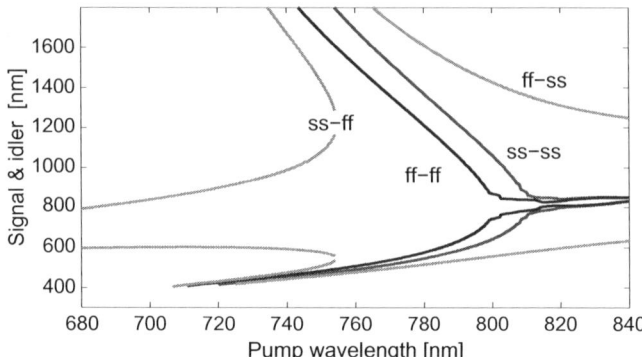

FIGURE 12.3 Phase matching curve for FWM in birefringent PCF. The curves ff–ff and ss–ss represent the phase-matching cases where pump and generated photons are all polarized along the fast or slow axes respectively. In the case of ff–ss and ss–ff, the pump photons are polarized orthogonally to the signal and idler. *Image from [10]*.

zero-dispersion wavelength. As shown on the example phase-matching plot in Fig. 12.3, this allows for generation of signal and idler photons on the fast axis of the fiber when the pump is propagating on the slow axis in the normal dispersion regime (ss → ff), or for generation of photons on the slow axis when pumping on the fast axis in the anomalous dispersion region (ff → ss). The separation of the signal and idler wavelengths from the pump increases with strength of the birefringence. Of particular interest are the turning points on the phase-matching plot where the gradient of either the signal or idler curve become zero. At these points either the signal or idler photons will be generated within a very narrow spectral range, which can be anticipated from Eqs. (12.8) and (12.9). It will be shown in Section 12.5.1 that the states of the generated signal and idler photons become spectrally uncorrelated in these regions, leading to two-photon states with highly desirable properties for use in photon sources.

12.2.4 Raman Scattering

The performance of fiber-based photon source is often limited by the detection of uncorrelated noise photons arising from processes other than the desired sFWM interaction. One such source of noise is the nonlinear shift in the energy of photons, predominantly to longer wavelengths, due to spontaneous Raman scattering. When bright laser light from a pump source propagates in an optical fiber, the Raman effect allows photons at the pump wavelength to be inelastically scattered by the silica glass, resulting in the emission of a phonon, and a photon with a red-shifted frequency. This leads to an increase in the background count rates measured for fiber-based photon sources when the wavelength of the Raman-shifted pump light overlaps with the idler wavelength generated through sFWM.

The shift in frequency associated with Raman scattering depends on the phonon distribution, which is characteristic of the material. Due to the amorphous nature of silica, the energy range of phonons in the material is

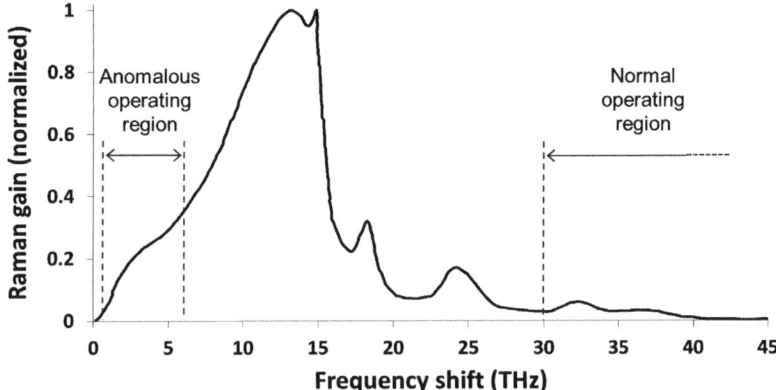

FIGURE 12.4 Normalized gain spectrum for Raman-shifted pump light in silica [19]. The range of idler frequencies (given as an offset from the pump) previously demonstrated for photon sources in the anomalous and normal dispersion regimes of fiber is overlaid (see Sections 12.3.1 and 12.3.2). *Image adapted from [19].*

distributed in a wide and continuous band. The Raman gain peaks for a frequency shift of 13 THz and reduces significantly for larger frequency shifts as well as very close to the original pump wavelength, as shown in Fig. 12.4. For photon sources operating in the anomalous dispersion regime, where the signal and idler wavelengths are constrained to lie close to the pump due to the phase-matching conditions, the Raman contribution can be reduced by minimizing the offset between the pump and idler. For sources pumped in the normal dispersion regime, the signal and idler are generally widely separated, such that the idler wavelength can lie beyond the extent of the Raman gain. In either case, the use of pump pulses with relatively high peak power can help to reduce the effect of Raman noise, as the generation rate of photon pairs through sFWM is proportional to the peak pump power squared, whereas Raman generation increases linearly with the pump intensity and fiber length in the low power regime of interest [28]. However, this effect comes at the price of increased background emission of multiple photon pairs, introducing impurity to the photon-pair state produced. Raman scattering can also be suppressed by cooling of the fiber with liquid nitrogen to reduce the phonon population [29].

Another viable approach to mitigate the Raman scattering effect is to apply on the output photons mode-selective filtering that is capable of discriminating the Raman-scattered photons from the paired photons, even if they are of the same wavelength [30]. It is known that Raman scattering in optical fibers occurs with a time-retarded response to the intense pump fields, while the sFWM occurs in a nearly instantaneous response. As a result, when short-duration pump pulses are applied, the Raman photons will be produced in a mixture of temporal modes different from the photon pairs. The idea is then to construct a temporal-shaped filter such that only a single mode of the photon pairs can pass through. All other modes will be rejected, thereby filtering out most of the Raman photons.

The effectiveness of this method depends on the mode overlap between the photon pairs and the Raman photons. Using this approach, Huang *et al.* found numerically that the quantum-state purity of entangled photons generated in room-temperature fibers can be improved significantly from 82% to 95% in terms of the two-photon Hong-Ou-Mandel interference visibility [30].

12.3 HERALDED SINGLE-PHOTON SOURCES BASED ON sFWM

An ideal single-photon source would emit photons sequentially, such that each output time bin of the source contained one (and only one) photon in a desired pure quantum state. For the case of a pulsed pump, each pulse defines a single time bin which should ideally contain a single output photon. For a CW photon source, the width of the time bins is given by the coherence time of the generated photon state. As is the case with photon sources based on parametric down-conversion in $\chi^{(2)}$ crystals, the advantage of a single-photon source based on sFWM over a weak coherent state arises from the fact that the signal and idler photons are always generated together in correlated pairs. The detection of one photon of the pair can be used to infer the presence of the other, a procedure known as heralding (see Fig. 12.5) [31]. Heralding removes the vacuum contribution to the generated state in which no pairs are generated, which is seen in Eq. (12.7).

While heralding removes the vacuum contribution of the generated state, higher order terms in which multiple photon pairs are generated will still be present due to the probabilistic nature of pair generation. However, if the probability of generating a signal and idler pair in any given time bin is low (when α is small in Eq. (12.7)), the contribution of such higher order terms to the generated state can be made arbitrarily small. The requirement of keeping the pair generation rate low imposes a fundamental limit on the brightness of the source that can be achieved. In the case of a pulsed pump, the pump power is

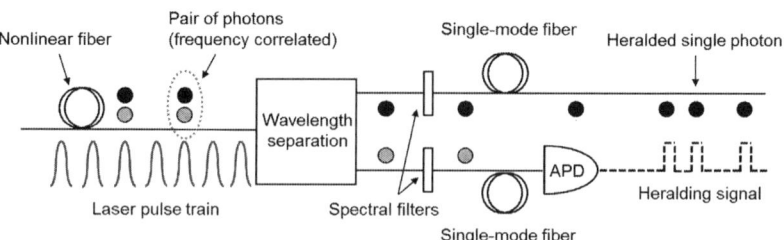

FIGURE 12.5 Typical experimental arrangement of a fiber-based heralded photon source. Pairs of signal and idler photons are generated in a section of nonlinear fiber before being separated by wavelength and filtered to attain spectral purity. After coupling into separate output fibers, one member of the pair is detected using a single-photon avalanche photodiode (SPAD) to herald the presence of the other. *Image from [32]*.

typically chosen to provide a pair generation probability of 0.1 pairs per pulse, or less if high quantum-state purity is desirable. Given this limitation, a high repetition rate pump laser is usually desirable, to maximize the generation rate of photon pairs.

The brightness of the source is commonly defined as the number of heralded single photons output from the source per second, which for the purposes of characterizing such sources is often given as a measured twofold coincidence count rate between signal and idler photons (with units of s^{-1}). The measured coincidence count rate is given by

$$C = r\eta_{opt}^s \eta_{opt}^i \eta_{DE}^s \eta_{DE}^i + C_a, \qquad (12.13)$$

where r is the rate of photon-pair generation in the fiber, $\eta_{DE}^{s,i}$ are the efficiencies of the detectors for the signal and idler photons, and $\eta_{opt}^{s,i}$ are the optical transmission efficiencies for the signal and idler, which incorporates all sources of loss between the point at which the photons are generated and the detectors. C_a are accidental coincidence counts arising from all uncorrelated detection events.

Alternative commonly used definitions of the brightness include the coincidence count rate per unit pump power (s^{-1} mW^{-1}) and per unit bandwidth of the generated photon pairs (s^{-1} mW^{-1} nm^{-1}). However, due to the nonlinear relation between the photon pair generation rate and the pump power, the natural units to measure source brightness should depend on the square of the pump power (s^{-1} mW^{-2} or s^{-1} mW^{-2} nm^{-1}). Such measures aim to highlight both the efficiency of the nonlinear process (through pump power requirements) and the proportion of the generated photon pairs that fall within the useful spectral range (as such sources typically require narrowband spectral filtering to be applied to the signal and idler after generation). As the pump power requirements and bandwidth of the sFWM gain spectrum can be influenced by many factors, such as pump pulse duration and repetition rate, the wavelengths of interest, and properties of the fiber in question, these factors must be taken into account when assessing the brightness of competing sources.

As previously described, there is an inevitable trade-off between the brightness of the source and the detrimental contribution to the output state of multiple pair generation events. The latter can be quantified using the so-called coincidence-to-accidental ratio (CAR). As the signal and idler photons are always generated together in pairs, by selecting the delay between measurement of the signal and idler channels appropriately (to account for any difference in the path length to the detectors for the two photons) the coincident detection rate between photons generated in correlated pairs, on the same pulse of the pump laser, can be observed. In contrast, the accidental rate can be found by selecting the delay between detectors such that coincident detections between the signal photons generated on one pulse of the pump laser and photons at the idler wavelength generated on a subsequent (or preceding) pulse of the pump laser are observed (see Fig. 12.6). These accidental twofold coincidences are a

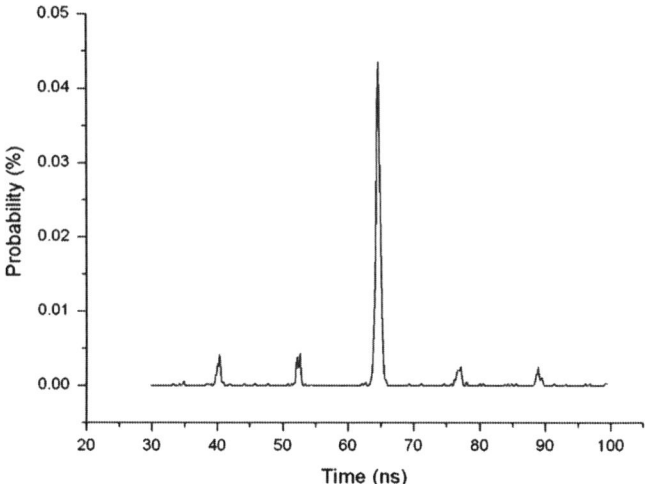

FIGURE 12.6 Typical measured twofold coincident count rate data between signal and idler photons. The histogram shows the probability that a heralded photon will be detected given that a heralded signal was produced, as function of the time delay between the detections. The large central peak arises from the detection of correlated photon pairs, while the smaller satellite peaks are separated by the laser repetition rate and indicate the background uncorrelated accidentals contribution to the count rate. *Image from [3]*.

source of noise in the system which can be present on any pulse of the pump laser (including those in which a heralding signal is generated).

The accidental count rate is due to detections events which are uncorrelated with the heralding signal and can be made up of contributions from several sources. These include residual pump photons which pass through the spectral filtering in the source and reach the detectors, Raman-shifted pump light (primarily at the idler wavelength) and detector dark counts. The other main contribution to the accidentals count rate comes from the generation of additional photon pairs through sFWM that are uncorrelated with the pair that caused the heralding signal. While the other noise sources can be minimized, the accidentals count rate due to multiple pair generation is unavoidable and, when other sources of noise are negligible, limits the CAR to 10 when the source is operated at the typical pair generation rate of 0.1 pairs per pulse.

In addition to the CAR, the heralding efficiency is another important figure of merit for a single-photon source. The heralding efficiency H is the probability that a photon will be output from the source, given that a heralding signal was detected and is given by

$$\eta_{\text{herald}} = \frac{C - C_a}{R_h \eta_{\text{det}}}, \quad (12.14)$$

where C is the measured coincidence count rate for correlated signal and idler photons, C_a is the accidentals count rate, R_h is the measured singles count rate for the heralding photons, and η_{det} is the efficiency of the detector used to measure the output heralded photons. η_{herald} is closely related to the transmission efficiency of the device at the peak output wavelength. However, it may deviate from this value in the case of narrowband filtering, where edge effects can occur due to the profile of the spectral filtering. η_{herald} can also be reduced by uncorrelated noise photons in the heralding channel, leading to an erroneously high singles count rate R_h. To maximize the heralding efficiency it is therefore essential to keep the optical loss in the output photon channel of the source as low as possible. As seen in Eq. (12.13), the brightness of the source will also be improved by minimizing the loss for both the signal and idler photons.

12.3.1 Photon-Pair Generation in the Anomalous Dispersion Regime

Due to the requirement of phase-matching, when pumping in the anomalous dispersion regime of an optical fiber the central wavelengths of the generated signal and idler will depend on the peak power of the pump beam. For generation of single photons in the signal and idler modes, the required peak pump power is relatively low, leading to signal and idler wavelengths separated from the pump wavelength by typically a few tens of nanometers.

The earliest demonstrations of heralded single-photon generation in optical fiber focused on sFWM in the anomalous dispersion regime of dispersion-shifted conventional fibers (DSF), and are reported in [2,33]. By using a 300 m length of fiber with a zero dispersion wavelength shifted to 1535 nm, and pumping using 5 ps pulses at 1536 nm from an optical parametric oscillator, Li *et al.* were able to generate correlated pairs of signal and idler photons at 1532 and 1540 nm respectively (see Fig. 12.7) [33].

This early work highlighted some of the key advantages of fiber-based photon sources. Firstly, despite the relatively weak $\chi^{(3)}$ nonlinear response of silica glass, efficient photon generation can occur at modest pump powers due to the long interaction lengths that can be realized, due to the confinement of the optical mode to the fiber core along the entire fiber length. Secondly, by selecting a fiber with the appropriate dispersion characteristics, phase-matching for the desired signal and idler wavelengths can be achieved without the need for angle and temperature tuning that is often required for photon sources based on parametric down-conversion in bulk $\chi^{(2)}$ crystals. In addition, generating the photon pair directly in the guided mode of a fiber allows the heralded single-photon state to be easily prepared in a single spatial mode. For $\chi^{(2)}$ crystal-based sources, collection of the generated photons into single-mode fiber remains one of the main sources of optical loss, with schemes to improve the collection efficiency often relying on delicate or complex alignment. In the case of optical fiber-based sources, fusion splicing can easily allow sections of conventional fiber to be joined with negligible loss (typically <0.1 dB).

Parallel to the use of standard single-mode fibers, photonic-crystal fibers with relatively large Kerr nonlinearity have also been commonly used to create correlated and entangled photon pairs. An advantage offered by PCFs is that they

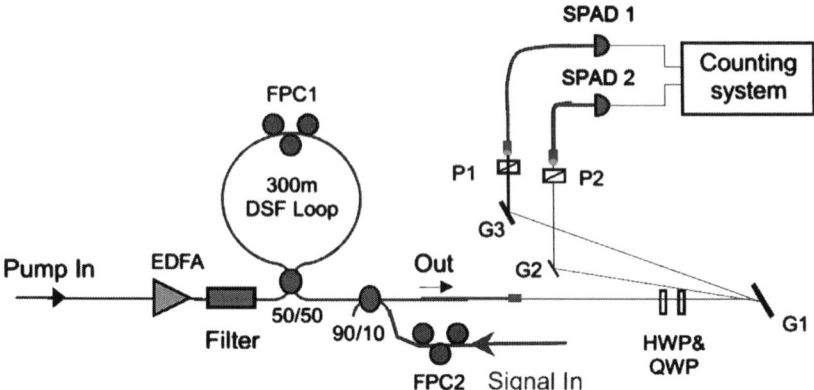

FIGURE 12.7 Schematic representation of a heralded single-photon source operating in the anomalous dispersion regime of a fiber. Pairs of photons were generated at telecoms wavelength by sFWM in a 300 m length of dispersion-shifted fiber. EDFA, erbium-doped fiber amplifier; DSF, dispersion-shifted fiber; FPC, fiber polarization controller; HWP, half-wave plate; QWP, quarter-wave plate; G, grating; P, polarizing beam-splitter; SPAD, single-photon avalanche photodiode. *Image from [33]. Copyright (year of paper) by the American Physical Society.*

can be fabricated having a short zero-dispersion wavelength, so that sFWM can be phase-matched in the 800 nm region of the spectrum (i.e., for wavelengths much shorter than the zero-dispersion wavelength of pure silica) [34,35], leading to the possibility of integration with rubidium-atom quantum memory schemes that have been proposed for the storage of entanglement. Using PCF, Sharping et al. demonstrated photon-pair generation using degenerate pumping at a wavelength of 749 nm, with the signal and idler photons at wavelengths of 761 and 737 nm, respectively [36]. The setup of their experiment was similar to that in Fig. 12.7 but with the DSF Sagnac-loop replaced by a straight-PCF configuration, to eliminate the complexity associated with alignment of the Sagnac interferometer in free-space. In the experiment, linearly polarized, ∼ 3 ps duration pump pulses were launched into a 5.8 m section of PCF such that the plane of polarization was aligned along one of the polarization-mode axes of the fiber. For low pump powers, as the pump pulses propagated through the fiber there was a small probability that a sFWM event would occur. The resulting signal and idler photon pairs co-propagated along with the pump and emerged from the PCF. Because of such a linear configuration, efficient detection of the signal-idler pairs required aggressive filtering of the pump photons by greater than 90 dB. The resulting coincident counts and noncoincidence counts are shown in Fig. 12.8 as functions of the number of pump photons per pulse. It clearly demonstrates that more coincident photons were generated at the signal/idler wavelengths than the accidental coincidences that result from the background photons. The CAR ratio, however, is worse than that reported in

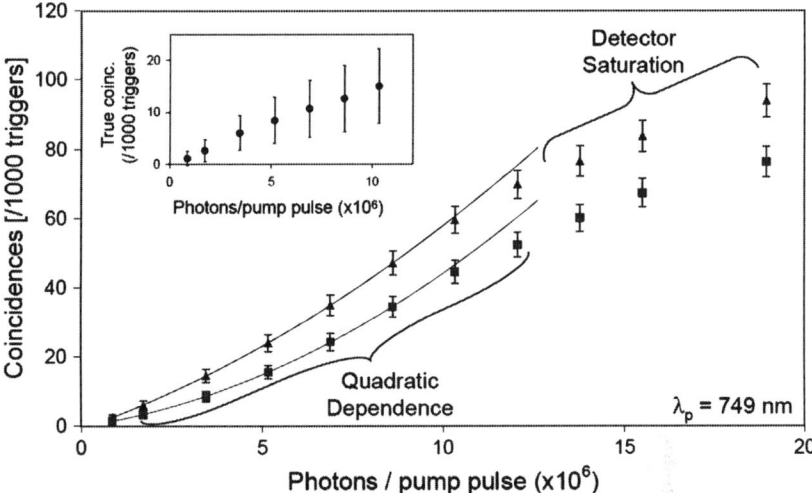

FIGURE 12.8 Plots of total coincidence counts (triangles) and accidental coincidence counts (squares) as a function of the number of pump photons per pulse with the photon counters aligned to detect at 737 and 761 nm wavelengths, respectively. At low pump powers there is a quadratic dependence of the counts on pump power, but as the power is increased the increasing signal and idler count rates start saturating the photon counters. The inset shows true coincidences (the difference between the total coincidence counts and the accidental coincidence counts) as a function of the number of pump photons per pulse. *Image from [36].*

[2,33] because the signal wavelength falls much closer to the peak of the Raman gain curve.

The above work used a degenerate pump to create nondegenerate photon pairs. Fan et al. demonstrated a reverse process, where two pump waves of different colors were used to create degenerate photon pairs [37]. In their experiment, linearly polarized laser pulses of 3 ps at 835 nm, output from a Ti:Sapphire oscillator operating at 80 MHz, drove the sFWM process in a 1 m section of PCF, whose output was collimated onto a high-power diffraction grating. By use of two narrow slits, a pair of linearly polarized signal (837 nm) and idler (833 nm) light pulses was selected with a FWHM bandwidth of 0.5 nm. They were then coupled into a same PCF to create degenerate photon pairs at 834.8 nm. The CAR of such photons was measured to be 8:1 with use of a 50:50 beam-splitter.

12.3.2 Photonic Crystal Fiber Sources in the Normal Dispersion Regime

The flexibility that arises from tailoring the waveguide dispersion in PCF allows the zero-dispersion wavelength (ZDW) of the fiber to be shifted to shorter wavelengths. sFWM can then be pumped at more favorable wavelengths, typically using either picosecond pulsed, high-power fiber lasers at 1064 nm or Ti:Sapphire lasers operating around 750 nm. In both cases the PCF design can be selected to ensure that the pump laser wavelength lies in the normal dispersion region close to the zero-dispersion wavelength. The steep gradient of the phase-matching curves in this region allows the signal and idler wavelengths to be tuned over a wide range through small changes in the pump wavelength. Furthermore, the wide separation of the idler from the pump wavelength allows the idler photons to be generated far from the peak gain of the Raman-shifted pump light, significantly reducing this source of noise.

Using a PCF with a short ZDW, together with a Ti:Sapphire pump laser, means that both the signal and idler can lie within the operating range of high efficiency, silicon-based avalanche photodiodes (500–900 nm), and several such photon sources have been demonstrated to date [4,38,7,39,3]. Operating with these wavelengths can therefore allow higher potential overall efficiencies (combined optical loss and detector efficiency) than can typically be achieved when operating in the anomalous dispersion regime when both the signal and idler photons are in the near infrared range, where detector technology is less well developed. Shifting the ZDW to the shorter wavelengths required for operation in this region is achieved by increasing the size of the holes in the cladding region of the PCF, such that the air filling fraction of the cladding is higher, and by reducing the size of the core, so that the dispersion properties of the fiber approach those of narrow silica strand in air. The smaller core size relative to conventional fiber also results in a reduction in the size of the fundamental guided mode. This enhances the intensity of the guided pump light, enabling high brightness photon sources with modest pump power requirements.

Fulconis et al. [3] demonstrated a source of photon pairs at 587 and 897 nm, pumped using 4 ps pulses from a Ti:Sapphire laser at 708 nm. A PCF with a core diameter of 2 µm, shown in Fig. 12.9, was used to shift the ZDW of the fiber to 715 nm, as required to achieve phase-matching. Due to the high pump intensity resulting from the small mode field diameter of the fiber, a high brightness of 3.2×10^5 s^{-1} detected photon pairs was achieved with an average pump power of 540 µW (1.8 kW peak power), giving a spectral brightness of 4×10^5 s^{-1} mW^{-2} nm^{-1} for the heralded signal photons (see Fig. 12.10). Fan et al. [38] demonstrated a heralded photon source based on a similar PCF structure, generating signal and idler photons at 690.4 and 801.2 nm using a 742 nm pump. A detected pair brightness of 9×10^4 s^{-1} mW^{-2} nm^{-1} was achieved for a CAR near 10. A maximum CAR of 900 was measured for the source at low pump power, but the resulting detected pair brightness was reduced to just 45 s^{-1}. For both of these sources, the overall transmission efficiencies for the signal and idler arms were ∼ 15%, with the implemented free-space spectral filtering and the coupling efficiency into standard single-mode fiber accounting for the majority of the loss.

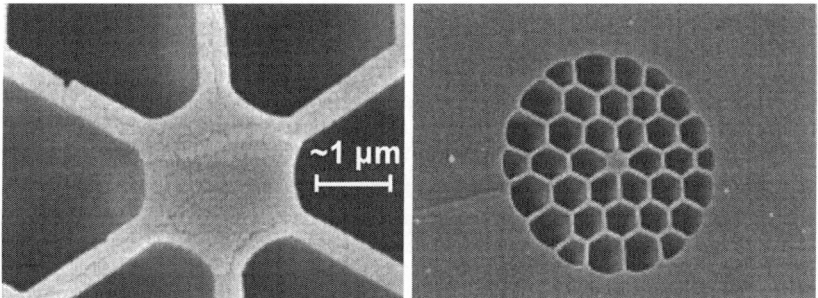

FIGURE 12.9 Scanning electron microscope images showing the cross-section of a highly nonlinear PCF. *Images from [3]*.

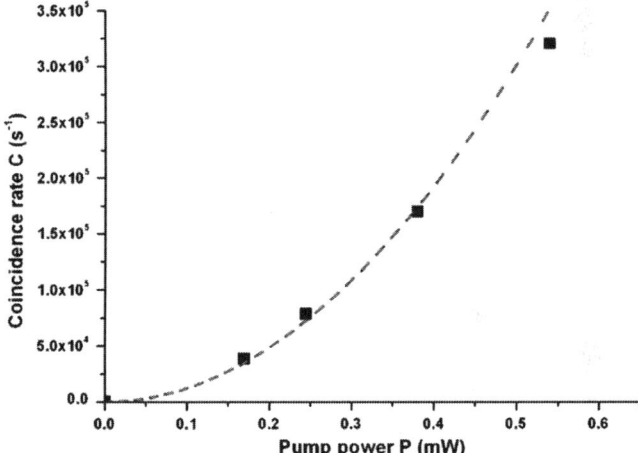

FIGURE 12.10 Measured twofold coincidence count rate as function of average pump power for the photon source described in [3]. *Image from [3]*.

Another interesting approach, with alternative applications, is to generate widely spaced signal and idler photons, with the signal near 800 nm and the corresponding idler photons in the telecommunications band near 1550 nm [1,5,40]. Photons at the idler wavelength can then be transmitted over long distances in optical fibers with minimal loss. At the same time, the wavelength of the signal photons makes them well suited for interfacing with quantum memories, or for processing using quantum logic gates (see Section 12.7), due to the high detection efficiency. This type of source is therefore of considerable interest for many quantum communications applications, such as quantum repeaters [41]. Photon pairs at these wavelengths can be generated using a pump wavelength of 1064 nm, with phase-matching satisfied in the normal dispersion regime of a suitably designed PCF with a ZDW near 1090 nm.

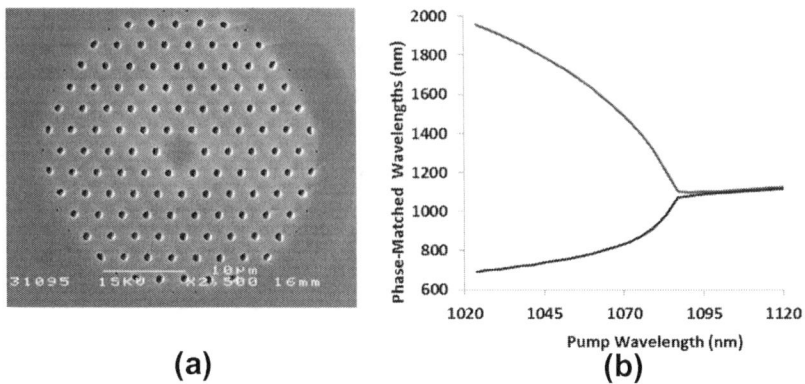

FIGURE 12.11 (a) Scanning electron microscope image of a PCF used to generate signal and idler pairs, widely spaced in frequency as described in [5] and (b) the associated calculated phase-matching curve. *Images from [5].*

The PCF structure required for phase-matching in this type of source has a larger core diameter and lower air filling fraction than for Ti:Sapphire pumped sFWM, as the desired dispersion profile is closer to that of the intrinsic material dispersion. Due to the weaker confinement of the guided mode, and the inherently lower efficiency of sFWM when operating at longer wavelengths, the pump power requirements are typically higher for this type of source. However, the larger core size does allow the PCF to be more easily integrated with conventional single-mode fibers, due to the similarity in mode field profile. Techniques for splicing PCF structures of this type to single-mode fiber have been demonstrated with losses < 1 dB [42]. The lower air filling fraction and larger core also mean that this type of PCF can exhibit endlessly single-mode guidance [43], further improving the compatibility with conventional fibers. McMillan *et al.* [5] demonstrated an all-fiber heralded photon source based on a PCF design of this type (see Fig. 12.11). A picosecond-pulsed pump at 1064 nm generated signal and idler pairs at 802 and 1570 nm respectively. By using low loss, spliced telecom components to separate and spectrally filter the photons, a transmission efficiency of 68% was demonstrated at the idler wavelength, giving a heralding efficiency of 52%.

12.4 QUANTUM INTERFERENCE BETWEEN SEPARATE SPECTRALLY FILTERED FIBER SOURCES

One striking result of the nonclassical nature of single-photon states is the observation of two-photon quantum interference that can occur when two single photons are incident on the two input ports of a beam-splitter. For a 50:50 beam-splitter, when the two input photon states are both pure and indistinguishable the photons will bunch and both exit from the same output port of the beam-splitter, an effect known as Hong-Ou-Mandel (HOM) interference [44] as introduced in Chapter 2. As the HOM effect can be used to implement an effective two-photon interaction, it performs a critical role in many quantum information applications, including quantum logic gates based on the model of linear optical quantum computation (see Section 12.7), and the generation of large entangled cluster states through photonic fusion (see Section 12.8). Achieving high visibility HOM interference between photons from independent fiber-based photon sources is an essential requirement for demonstrating their suitability for use in such applications, and it also serves as a convenient approach for measuring the purity of the photon states generated by such sources.

A general beam-splitter, with reflection and transmission coefficients r and t, can be described by transformation relations between its input modes a_1 and a_2 and output modes b_1 and b_2 in terms of photon creation operators by

$$\hat{a}_1^\dagger = r\hat{b}_1^\dagger + t\hat{b}_2^\dagger, \tag{12.15}$$

and

$$\hat{a}_2^\dagger = t\hat{b}_1^\dagger - r\hat{b}_2^\dagger, \tag{12.16}$$

where the phase-shift seen for the reflected term in Eq. (12.16) is required to satisfy conservation of energy [45]. For an input state of the form $\hat{a}_1^\dagger \hat{a}_2^\dagger |vac\rangle$, where two indistinguishable single-photon states are input, one into each arm of the beam-splitter, the output state after the beam-splitter is given by

$$\begin{aligned} |\psi_{out}\rangle &= (r\hat{b}_1^\dagger + t\hat{b}_2^\dagger)(t\hat{b}_1^\dagger - r\hat{b}_2^\dagger)|vac\rangle \\ &= (rt\hat{b}_1^\dagger \hat{b}_1^\dagger + (t^2 - r^2)\hat{b}_1^\dagger \hat{b}_2^\dagger - rt\hat{b}_2^\dagger \hat{b}_2^\dagger)|vac\rangle. \end{aligned} \tag{12.17}$$

In the case of a 50:50 beam-splitter, $r = t = \frac{1}{\sqrt{2}}$, and the output state of the system is given by

$$|\psi_{out}\rangle = \frac{1}{2}(\hat{b}_1^\dagger \hat{b}_1^\dagger - \hat{b}_2^\dagger \hat{b}_2^\dagger)|vac\rangle = \frac{|2\rangle_{b1}|0\rangle_{b2} - |0\rangle_{b1}|2\rangle_{b2}}{\sqrt{2}}. \tag{12.18}$$

Showing that both photons will emerge from the same output of the beam-splitter (either b_1 or b_2). The cancellation of the $\hat{b}_1^\dagger \hat{b}_2^\dagger$ terms in Eq. (12.17) relies on the output photon modes being indistinguishable, so that the reflected and transmitted terms are equivalent. If the single photons which are input are distinguishable from each other in any degree of freedom then they will not interfere, and will behave like classical particles with each being either transmitted or reflected with 50% probability. HOM interference between single photons can be demonstrated, for example, by introducing a variable delay to the arrival time of one of the photons at the beam-splitter, such that at zero delay the photons are indistinguishable, but at delays greater than their coherence length they are entirely distinguishable.

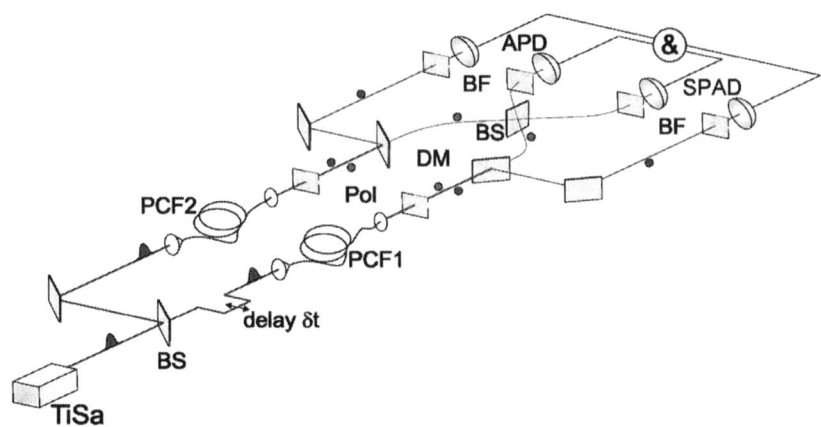

FIGURE 12.12 Experimental setup used to demonstrate HOM interference between two heralded single-photon sources. A time delay δt is incorporated in one of the sources using a retroreflector, to adjust the relative arrival time of the interfering photons at the central beam-splitter. BS, beam-splitter; DM, dichroic mirror; BF, spectral band-pass filter; pol, polarizer; SPAD, single-photon avalanche photodiode. *Image from [10]*.

Figure 12.12 shows a typical experimental setup used to demonstrate HOM interference between two fiber-based pair-photon sources. At the output of the section of nonlinear fiber in each source, the generated photons are separated by wavelength, and one photon of the pair (in this case the idler photon) is detected and used to herald the presence of the other. The heralded single photons are overlapped at a beam-splitter and the coincident counts between the detectors are measured as the variable time delay to the pump pulses for one of the sources is varied. The interference is observed as a dip in the fourfold coincidence count rate between the two heralding and two interfering photons when the signal photons from the two sources are overlapped at the beam-splitter.

The visibility of the interference feature is dependent on the purity of the single-photon states that are generated, and will be reduced by the presence of additional, noninterfering noise photons. If two independent sources are used to generate heralded single photons, then the degree to which the properties of the interfering photons are matched will also influence the interference visibility. For single-photon states to be considered indistinguishable, they must be identical in all degrees of freedom. This includes their spectral properties, temporal mode (relative output time from the source), spatial mode, and polarization state.

Ensuring that the interfering photons are in matching spatial modes can be easily achieved in fiber-based sources, as fibers can be designed such that only a single spatial mode is supported at the desired single-photon wavelength. These sources can also be coupled with low loss to single-mode fused-fiber couplers, which perform an identical operation to free-space beam-splitters, allowing the HOM interaction to take place in an all-fiber configuration. Photons can also readily be generated in a well-defined polarization state in optical fibers by making use of fibers with birefringence. In a birefringent fiber, if linearly polarized laser pump light is launched onto one of the principal fiber axes, its polarization state will be preserved along the fiber length. If the birefringence is sufficiently strong, it will also modify the phase-matching conditions for sFWM, such that the generation of photons at the selected wavelength can only occur on one axis of the fiber. Matching of the polarization state of photons produced from independent sources can be achieved using standard wave plates in a free-space implementation, or with fiber polarization controllers if the photons are delivered in single-mode fiber.

The requirements to realize a heralded photon in a single temporal mode are intimately related to the spectral properties of the state. The coherence time of a generated photon will be determined either by the gain bandwidth of the sFWM process itself, or by the bandwidth of any spectral filtering that has subsequently been applied if this is narrower than the initial unfiltered state. For a photon wave-packet with a Gaussian spectral profile, central wavelength λ_0, and full-width half-maximum bandwidth $\Delta\lambda$, the coherence time τ_{coh} is given by

$$\tau_{coh} = \frac{2\ln 2}{\pi nc} \frac{\lambda_0^2}{\Delta\lambda}, \qquad (12.19)$$

where n is the refractive index of the transmission medium and c is the speed of light in free-space. For a photon source driven by a pulsed pump, the coherence time of the photons should be longer than the pump pulse duration, so that only a single temporal mode is available for the photon pair to be generated in within the timing gate defined by the pump pulse. In that case the desired single-photon state will be pure upon heralding. If the coherence time of the photon is shorter, then due to the limited timing resolution of current realistic photon detectors, the detection of a heralding photon will project its partner into a mixed state.

To meet the required photon coherence time for purity, the heralded photon is often spectrally filtered. For sources pumped with femtosecond laser pulses, the required filtering bandwidth is a few nanometers, which can be readily implemented using standard interference filters. On the other hand, for picosecond pulsed sources the heralded photons may need to be filtered to a bandwidth ~ 0.1 nm, which is more challenging to achieve with low loss. For an experimental setup of the type shown in Fig. 12.12, the maximum visibility of the HOM interference when the photons are overlapped at the beam-splitter is given by [46]

$$V_{max} = \frac{\sqrt{1 + \sigma^2/\sigma_p^2}}{1 + \sigma^2/2\sigma_p^2}, \qquad (12.20)$$

where the signal and idler photons are assumed to be filtered using energy-matched, Gaussian profile spectral filters of bandwidth σ, defined by the filtering function

$$f_j(\omega_j) = \frac{1}{N} e^{(\omega_{j0}-\omega_j)^2/\sigma^2}, \qquad (12.21)$$

where $j = s, i$ denotes the signal or idler, ω_{j0} is the central frequency of the filter and N is a normalizing constant. Similarly, the pump pulse is assumed to be a Gaussian pulse of bandwidth σ_p. As shown in Fig. 12.13, although the maximum interference visibility reaches unity only in the limit when the bandwidth of the spectral filtering ratio tends to zero, visibility of 90% or more is achieved for filtering bandwidths only slightly less than that of the pump bandwidth.

It should be noted that this analysis neglects the effect of group velocity walk-off between the pump and the generated signal and idler, which becomes relevant when dealing with short-duration pump pulses. For the sFWM process described in Section 12.2, phase-matching can be satisfied with the pump close to the ZDW, which coincides with a local minimum in the group velocity of the fiber. For widely separated signal and idler wavelengths the dispersion of the fiber can lead to a significant mismatch between the group velocities of the generated photons with the pump. Pulse walk-off limits the minimum bandwidth of the signal and idler that will be generated and the sFWM gain grows only linearly with the fiber length in this regime. More importantly, walk-off can broaden the emission time of the photon pair beyond the single temporal mode defined by the coherence time of the filtered photon. The walk-off effect is manifested as a timing jitter in the output time of the photons relative to the laser pulses, and the fiber length must be chosen such that this jitter is small in comparison to the coherence length. For photon sources operating in the anomalous dispersion regime of a fiber (see Section 12.3.1), where the generated

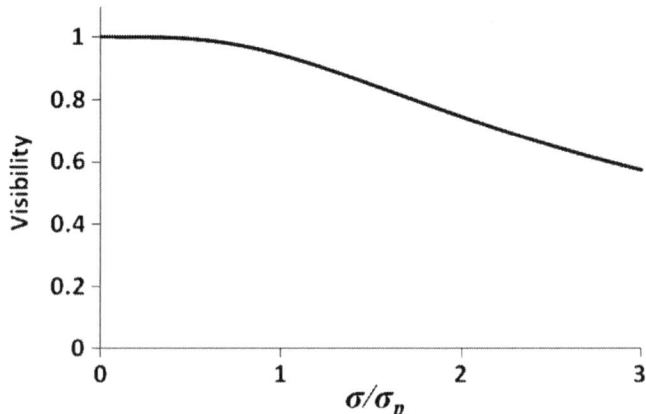

FIGURE 12.13 Theoretical maximum visibility for HOM interference as a function of the spectral filtering bandwidth σ relative to the pump pulse bandwidth σ_p, given by Eq. (12.20). *Image from [32].*

photons may be separated by only tens of nanometers, the limit to the maximum fiber length may be hundreds of meters, while for sources with widely separated signal and idler wavelengths (see Section 12.3.2) the limit can be as little as a few centimeters.

Interference between spectrally filtered photons, generated independently in two separate sections of identical PCF (pumped using the same laser), has been demonstrated with visibility greater than 90% (see Fig. 12.14)) [46]. When operating such sources with low brightness, this interference can be observed in the raw fourfold count rates of the detectors. When the brightness of the source is increased toward 0.1 generated pairs per pulse, the raw visibility is reduced due to the presence of multiple pair generation events in each source, which do not interfere fully. When the average number of pairs per pulse $\bar{n}$ is low, and distinguishability in other degrees of freedom is negligible, the visibility is given by

$$V_{\max} \approx \frac{1 + 8\bar{n}}{1 + 12\bar{n}}. \tag{12.22}$$

The visibility of interference between the sources in the absence of multiple pairs can be determined by measuring the background fourfold count rates due to each source individually, and subtracting these background terms in order to find the net visibility. While the presence of this background term is undesirable for practical applications, measuring the net visibility of interference with higher pump power can be useful for demonstrating the purity of the generated single-photon states. In the future, post-selection using photon-number resolving detectors will allow the contribution from these higher photon-number states to removed.

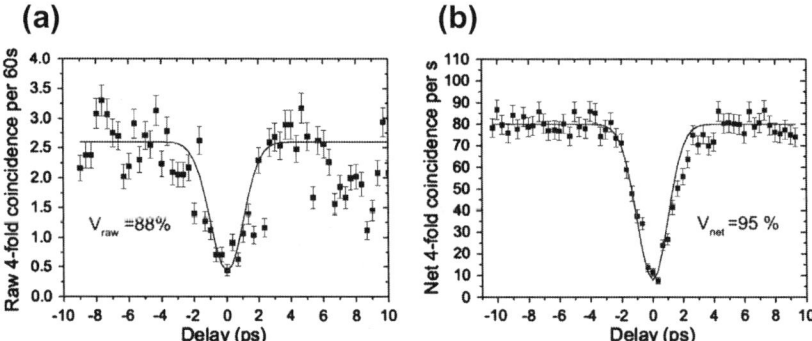

FIGURE 12.14 Experimental demonstration of quantum interference between photons output from separate PCF sources, seen as a dip in the measured fourfold coincidence count rate of signal and idler photons from the two sources (a) with 1 mW average pump power and (b) with 8 mW average pump power and corrected for background from multiphoton emission. *Images from [46]*. ©IOP Publishing Ltd and Deutsche Physikalische Gesellschaft. Published under a CC BY-NC-SA licence.

12.5 INTRINSICALLY PURE-STATE PHOTONS

12.5.1 Generation of Spectrally Uncorrelated Two-Photon States Through Group Velocity Matching

As described in Section 12.4, achieving high visibility quantum interference between independent heralded photon sources requires that the heralded photons be prepared in pure and indistinguishable states. When generating photon pairs through sFWM, the bandwidth of the gain of the process typically allows some variation in the signal and idler wavelengths. This is due to both the finite pump bandwidth, which is necessarily broadened in short pulse duration laser sources, and because of the possibility of generating pairs which are not perfectly phase-matched in a realistic fiber of finite length. Considering only the spectral degree of freedom, the two-photon state generated by sFWM with degenerate pumping can be expressed in the form

$$|\Psi\rangle \propto \iint d\omega_s d\omega_i \, F(\omega_s,\omega_i) \hat{a}_s^\dagger(\omega_s) \hat{a}_i^\dagger(\omega_i) |vac\rangle, \tag{12.23}$$

where $\hat{a}_{s,i}^\dagger(\omega_{s,i})$ are the photon creation operators for signal and idler photons at frequencies ω_s and ω_i respectively, and $|vac\rangle$ is the vacuum state. $F(\omega_s,\omega_i)$ is the joint spectral amplitude function (JSA), which is dependent on both the pump amplitude profile $\alpha(\omega_s + \omega_i)$ and the phase-matching function $\phi(\omega_s,\omega_i)$, with

$$F(\omega_s,\omega_i) = \alpha(\omega_s + \omega_i) \phi(\omega_s,\omega_i). \tag{12.24}$$

The function $\alpha(\omega_s + \omega_i)$ describes the variation in the signal and idler frequencies that are possible due to the finite bandwidth of the pump pulse. The pump is commonly assumed to be Gaussian in profile so that

$$\alpha(\omega_s + \omega_i) = \exp\left(-\frac{(\Delta\omega_s + \Delta\omega_i)^2 \sigma^2}{2}\right), \tag{12.25}$$

where $\Delta\omega_j = \omega_j - \omega_{j0}$ are the detunings of the signal and idler frequencies ω_j away from their peak values ω_{j0}, and σ is related to the full-width half-maximum frequency bandwidth of the pump pulse $\Delta\omega_p$ by $\sigma = 2\sqrt{\ln 4}/\Delta\omega_p$. The phase-matching function is given by [8,10]

$$\phi(\omega_s,\omega_i) = \exp\left(\frac{i\kappa L}{2}\right) \mathrm{sinc}\left(\frac{\kappa L}{2}\right) \tag{12.26}$$

for a fiber of length L, where κ is the phase-mismatch defined in Eq. (12.3).

The JSA is dependent on both the signal and idler frequencies. Because of this the energy and phase-matching requirements generally impose strong correlations in their spectra. When one photon of the pair is detected as a herald, information can in principle be extracted about the state of the other photon. However, since realistic single-photon detectors are unable to resolve

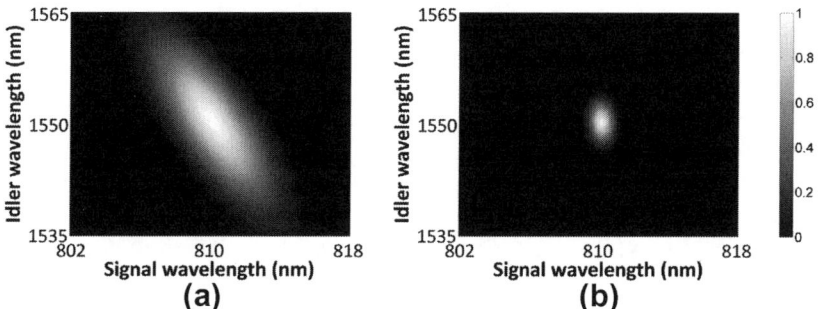

FIGURE 12.15 (a) JSA plot showing correlation between the idler and signal wavelengths and (b) appropriate narrowband spectral filtering can be applied in order to reduce the correlation. *Image from [32]*.

which spectral mode the heralding photon is in, this information is not generally experimentally accessible to the observer of the system with standard single-photon detectors. This projects the heralded single photon into a mixed state, which is a statistical mixture of the available pure states [47].

Figure 12.15a shows the JSA of the two-photon state generated by a typical PCF structure (similar to that shown in Fig. 12.11 used for generating widely separated signal and idler photons pumping in the normal dispersion regime). The undesirable frequency correlation is clearly apparent. As described in Section 12.4, narrowband spectral filtering can be implemented to approximate a single spectral mode, such that a pure heralded single-photon state can be realized, which is the approach taken by all of the photon sources outlined in Sections 12.3.1 and 12.3.2. The effect of spectral filtering can be incorporated into Eq. (12.23) by multiplying the JSA function of the photon pair generated by the fiber by functions $f_s(\omega_s)$ and $f_i(\omega_i)$, describing the spectral profile of the filtering applied to the signal and idler respectively. This reshapes the JSA, and for sufficiently narrowband filtering the correlation between the signal and idler wavelengths is reduced, as demonstrated in Fig. 12.15b. However, implementing narrowband spectral filtering usually introduces significant loss, which reduces both the source brightness and heralding efficiency. This is particularly significant in quantum information experiments involving higher photon numbers, in which multiple heralded photon sources are required to operate simultaneously in order to generate an N-photon state. If the probability of detection of each individual photon is μ, the overall probability of generating the required N-photon state is given by μ^n. This typically limits N to ~ 6 before the measured count rates for experiments become prohibitively low. Additionally, narrowband filtering leads to the majority of the photon pairs generated by sFWM being discarded, significantly increasing the pump power required to achieve the desired source brightness within the operating spectral bandwidth.

An alternative approach to narrowband spectral filtering is to design fiber with a dispersion profile that minimizes the correlation between the generated signal and idler frequencies, and hence allows pure states to be heralded without the need for filtering [8]. From Eq. (12.23) it can be shown that the correlation between the signal and idler photon wavelengths will be eliminated if the JSA function can be factorized such that $F(\omega_s, \omega_i) = F_s(\omega_s) F_i(\omega_i)$, so that the two-photon state can be written as a product of the signal photon state and idler photon state as [48]

$$|\Psi\rangle \propto \left(\int d\omega_s\, F_s(\omega_s) \hat{a}_s^\dagger(\omega_s) \right) \times \left(\int d\omega_i\, F_i(\omega_i) \hat{a}_i^\dagger(\omega_i) \right) |vac\rangle. \quad (12.27)$$

The contribution to the correlation in the JSA from the pump amplitude function $\alpha(\omega_s + \omega_i)$ is due to energy conservation, and the shape of this function always forms an angle of $-45°$ to the signal frequency axis of the JSA plot, with the width of the function determined by the bandwidth of the pump pulse. In contrast, the angle of the phase-matching function on the JSA plot can be rotated by tailoring the dispersion properties of the fiber. By using a linear approximation to the phase-mismatch κ, around the signal and idler frequencies where perfect phase-matching is achieved, the angle of the phase-matching function relative to the signal frequency axis on the JSA plot is found to be [8]

$$\theta_{pm} = -\arctan\left(\frac{\delta_s}{\delta_i} \right), \quad (12.28)$$

with

$$\delta_j = L \left(\left. \frac{dk_p}{d\omega} \right|_{\omega_{p0}} - \left. \frac{dk_j}{d\omega} \right|_{\omega_{j0}} \right) = L \left(\frac{1}{v_p} - \frac{1}{v_j} \right), \quad (12.29)$$

where $j = s, i$ and v_s, v_i and v_p are the group velocities of the signal, idler, and pump.

The form of Eq. (12.28) suggests two possible strategies for generating spectrally uncorrelated two-photon states. In the first, the group velocity of the pump is chosen to lie exactly halfway between those of the signal and idler, which is known as symmetric group velocity matching. The second possible approach, asymmetric group velocity matching, involves matching the group velocity of either the signal or idler field to that of the pump. Both symmetric and asymmetric group velocity matching have been demonstrated using cross-polarized sFWM in birefringent fibers, where the additional flexibility provided by the birefringence enables the group velocity of the pump field on one axis of the fiber to be shifted relative to the group velocity of the generated photon wavelengths on the other axis.

Factorable two-photon states generated by symmetric group velocity have been demonstrated using birefringent PCF [49] and highly birefringent conventional fibers [9,11]. Selecting a fiber with an appropriate birefringence,

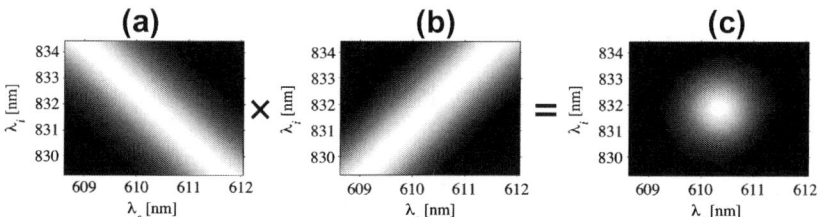

FIGURE 12.16 JSA for symmetric group velocity matching technique. (a) Pump amplitude function; (b) phase-matching function forming an angle of 45° to the signal axis; and (c) resulting JSA with symmetric signal and idler frequency distribution. *Images from [9].*

where the signal and idler group velocities lie on either side of the pump, sets $\theta_{pm} = +45°$. Then, for a given pump bandwidth, the length of the fiber can be selected so that the contribution to the two-photon correlation from the phase-matching function will counteract the correlation due to the pump amplitude function, as shown in Fig. 12.16. Smith *et al.* demonstrated the generation of spectrally uncorrelated signal and idler states of this type at 610 and 830 nm using a 10 cm length of commercial birefringent fiber pumped with femtosecond pulses from a Ti:Sapphire laser at 704 nm, achieving a measured brightness of 100 pairs $s^{-1}mW^{-2}$ [9].

In the case of asymmetric group velocity matching, designing the fiber so that either the signal or idler velocity is matched to that of the pump leads to the phase-matching function being oriented horizontally or vertically respectively on the JSA plot ($\theta_{pm} = 0°$ or $90°$). These conditions are found to coincide with positions on the phase-matching plot where the gradient of either the signal or idler wavelength generated becomes zero with respect to changes in the pump field wavelength. At these points, as long as the bandwidth of the phase-matching function is sufficiently narrow compared to that of the pump amplitude function, the overall two-photon state becomes uncorrelated, although the frequency bandwidths of the signal and idler are not identical in this case. The photon which propagates with same group velocity as the pump is well localized in time, so its partner photon carries all of the timing uncertainty of the pair. As a consequence of the joint energy-time uncertainty of the photon pair, the photon which walks-off in time from the pump pulse exhibits a naturally narrow bandwidth, as shown in Fig. 12.17.

Halder *et al.* [10] demonstrated a heralded single-photon source generating naturally narrowband signal photons at 597 nm by pumping a 40 cm section of birefringent PCF with 0.8 ps pulses at 705 nm from a Ti:Sapphire laser. The measured signal photon bandwidth of 0.13 nm was greatly reduced compared to the bandwidth ($\sim$3 nm) of unfiltered photons generated by sFWM in previous experiments with comparable operating wavelengths [3]. HOM interference was measured between the signal photons generated by two independent, identically prepared sources based on this fiber, showing a raw visibility of 76%.

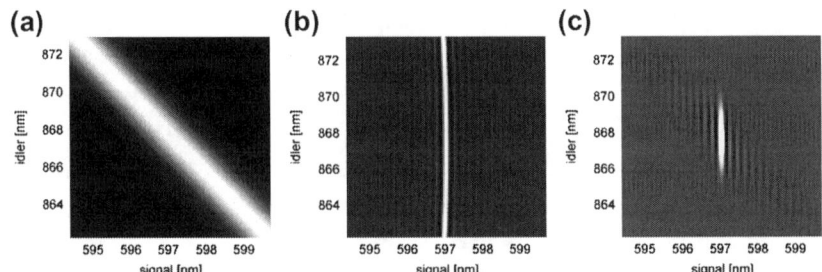

FIGURE 12.17 JSA for the PCF described in [10], designed to generate naturally narrowband signal photons through asymmetric group velocity matching. (a) Envelope of the pump intensity, (b) phase-matching condition of the PCF, and (c) JSA of the created photon pair. *Image from [10]*.

In contrast to spectrally filtered photon sources based on co-polarized sFWM, the visibility of HOM interference is expected to be higher when using a longer fiber length [50]. This means that the brightness and purity of the source could both be improved, indefinitely in principle, by increasing the fiber length. In reality, current limitations in the fiber manufacturing process lead to fluctuations in the fiber structure (and consequently the phase-matching conditions) which typically limit the maximum length to < 1 m.

As with spectrally filtered sources, widely spaced photon-pair generation with the idler wavelength in the telecoms band near 1550 nm is of considerable interest. The generation of naturally narrowband idler photons at 1550 nm has been demonstrated using a 1064 nm picosecond-pulsed pump source operating in the anomalous dispersion regime of a highly birefringent PCF [51]. For PCF with sufficiently strong waveguide dispersion, a second ZDW can be shifted into the transparency window of the fiber in the near infrared, also allowing a factorable two-photon state to be generated by co-polarized sFWM without the need for birefringent phase-matching. This is made possible by the fact the phase-matching curves form closed loops, with turning points on the outer edges of the loops where either the signal or idler become narrowband [8]. This was exploited by Söller et al. to generate widely separated, spectrally factorable signal and idler photons at 514 and 1542 nm from a 771 nm pump source [6].

12.5.2 A Temporal Filtering Approach for Attaining Pure-State Photons

As shown in Section 12.4, applying narrowband spectral filtering to the photon states produced by sFWM can provide a close approximation to a spectrally pure heralded single-photon source, allowing high visibility HOM interference to be achieved. One drawback with the use of narrowband spectral filtering, however, is that the production rate of the heralded photons is low because most usable pairs are filtered out. This problem can not be overcome by simply increasing the pump power in the photon-pair source, since doing so will lead to strong background noise arising from multipair emission. Section 12.5.1 outlined one possible solution to this dilemma—a proposal by Grice et al. [48], who suggested creating the photon pairs in spectrally factorable states so that measuring the signal photon would not disclose any spectral information about the idler. Since the narrowband filters are not needed, this approach can dramatically improve the single-photon production rate. With such motivation, extensive efforts have been made to study the creation of spectrally factorable photon pairs [52,53] and a few experimental demonstrations have been made [54–57]. However, progress has been limited due to the practical difficulty of simultaneously achieving appropriate phase-matching and group velocity matching. More importantly, one has to wait for a relatively long time before a second short-duration photon can be heralded, as in this case the signal (or equivalently the idler) photon must have a bandwidth much narrower than the pump to achieve the spectral factorability. Due to this, the repetition rate of the heralded single-photon creation is restricted, resulting in a relatively low production rate.

To overcome the aforementioned limitations, Huang et al. proposed an approach that does not rely on narrow spectral filtering or spectral factorability of the photon pairs [58]. It is based on time and bandwidth limited detection of the signal photons in a single-mode regime [59], achieved by using a broadband filter and an on/off detector whose measurement window is shorter than the coherence time of the signal photons. To understand the principle of single-mode detection, consider amplitude profiles $f(t)$ and $h(\omega)$ for the time gate and the sequential spectral filter, respectively. The number operator for output photons is given by $\hat{n} = \frac{1}{(2\pi)^2} \int d\omega d\omega' \kappa(\omega,\omega') \hat{a}^\dagger(\omega) \hat{a}(\omega')$ [59,60], where $\hat{a}(\omega)$ is the annihilation operator for the incident photons of angular frequency ω, satisfying

$$[\hat{a}(\omega), \hat{a}^\dagger(\omega')] = 2\pi \delta(\omega - \omega'). \tag{12.30}$$

$$\kappa(\omega,\omega') = \int dt\, h^*(\omega) h(\omega') |f(t)|^2 e^{i(\omega-\omega')t} \tag{12.31}$$

is a Hermitian spectral correlation function, which can be decomposed onto a set of Schmidt modes as

$$\kappa(\omega,\omega') = \sum_{j=0}^{\infty} \chi_j \phi_j^*(\omega) \phi_j(\omega'), \tag{12.32}$$

where $\{\phi_j(\omega)\}$ are the mode functions satisfying

$$\int d\omega\, \phi_j^*(\omega) \phi_k(\omega) = 2\pi \delta_{j,k} \tag{12.33}$$

and $\{\chi_j\}$ are the decomposition coefficients satisfying $1 \geq \chi_0 > \chi_1 > \cdots \geq 0$. Introducing an infinite set of mode operators via ($j = 0, 1, \ldots$)

$$\hat{c}_j = \frac{1}{2\pi} \int d\omega\, \hat{a}(\omega) \phi_j(\omega) \tag{12.34}$$

that satisfy $[\hat{c}_j, \hat{c}_k^\dagger] = \delta_{jk}$, the output photon-number operator for the filtering device can be rewritten as

$$\hat{n} = \sum_{j=0}^{\infty} \chi_j \hat{c}_j^\dagger \hat{c}_j. \tag{12.35}$$

This result indicates that $\{\phi_j(\omega)\}$ have an intuitive physical interpretation: as "eigenmodes" with eigenvalues $\{\chi_j\}$ of the filtering device. In this physical model, the filtering device projects incident photons onto the eigenmodes, each of which are passed with a probability given by the eigenvalues. Specifically, for $\chi_0 \sim 1$ and $\chi_{j \neq 0} \ll 1$ (achievable with an appropriate choice of spectral and temporal filters, as shown below) only the fundamental mode is transmitted while all the other modes are rejected. In this way, truly *single-mode* filtering can be achieved. When combined with a single-photon detector, this can be extended to a single-mode, single-photon detection system. Regardless of the type of spectral and temporal filters used to achieve this kind of single-mode filtering, such a system is capable of separating photons which, even though they may exist in the same spectral band and the same time bin, have different mode structures.

Figure 12.18a shows example plots of χ_0, χ_1, and χ_2 as functions of $c \equiv BT/4$ for a rectangular-shaped spectral filter with bandwidth B and a rectangular-shaped time window of duration T [59,61]. For $c < 1, \chi_0 \approx 1$ is obtained with $\chi_1, \chi_2 \ll 1$, giving rise to approximately single-mode filtering. Note that this behavior is true for any B, as long as $T < 4/B$. In other words, $\{\chi_j\}$ depend only on the product of B and T, rather than on their specific values. Consequently, even a broadband filter can lead to a single-mode measurement over a sufficiently short detection window, and vice-versa. To understand this, consider the case where a detection event announces the arrival of a signal

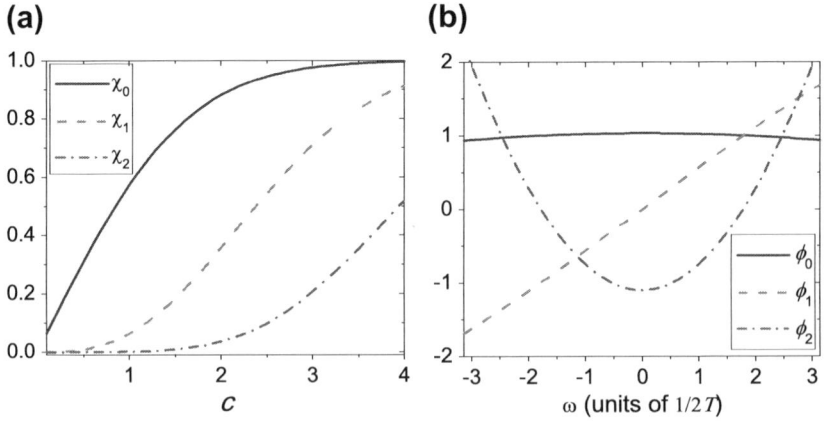

FIGURE 12.18 (a) χ_0, χ_1, and χ_2 as functions of c. (b) Plots of ϕ_0, ϕ_1, and ϕ_2 for $B = \pi/T$ (corresponding to $c = \pi/4$). *Images from [58]. Copyright (year of paper) by the American Physical Society.*

photon at an unknown time within the window T. In the Fourier domain, this corresponds to a detection resolution of $1/T$ in frequency. Given $c < 1$ or $1/T > B/4$, the detector is thus unable to, even in principle, reveal the frequency of the signal photon. Therefore, the signal photon is projected onto a quantum state in a coherent superposition of frequencies within B [62]. This can be seen in Fig. 12.18b, where the fundamental detection mode has a nearly flat profile over the filter band $[-B/2, B/2]$. Lastly, since $T < 4/B$ is required, the pass probability of the fundamental mode will be subunity, but not significantly less than one.

By using single-mode detection, even nonfactorable states can also lead to high-purity heralded single photons without the use of narrowband filters. By using multimode theory to take into account the effect of multiple photon-pair emission, distinctly superior heralding performance was achieved relative to those obtainable via narrow spectral filtering alone or tailoring the phase-matching properties of nonlinear media to obtain spectrally factorable photon-pair states [62]. The theory can also be used to numerically optimize the spectral and temporal filter widths for a variety of photon-pair correlations. The results reveal that while the optimization conditions vary for different correlations, the final performance using optimized filters is always quite similar. This interesting phenomenon suggests that the key to achieving high performance in the heralded creation of single photons is to apply an appropriate measurement scheme to the signal photons [62].

The above theory was validated by a heralded two-photon interference experiment in an all-fiber setup in both multimode ($c > 1$) and single-mode ($c < 1$) regimes [12]. In the experiment, pairs of signal and idler photons were generated in two separate optical fiber spools via sFWM driven by pulsed pumps. By detecting the idler photons created in each spool, the generation of their partner (signal) photons was heralded. To quantify their indistinguishability, the signal photons generated separately from the two spools were mixed on a 50:50 beam-splitter and HOM interference measurements were performed. By changing effective T for the idler photons, heralded generation of single photons in the multimode and single-mode regimes can be realized. The experimental results are shown in Fig. 12.19, where the HOM visibility was found to be quite low (19(2)%) when the signal photons have a temporal length $T > 1/B$ ($c = 3.8$ in this case), owing to the presence of photons with many distinguishable degrees of freedom. However, when $T < 1/B$ ($c = 0.7$ in this case), for which a single-mode detection is effectively realized, a much higher HOM visibility (72(7)%) is obtained. In all cases, the experimental data are in good agreement with predictions of the above multimode-detection theory with careful modeling of multi-pair production, Raman emission, loss, dark-count noise, and the interference between the two arms of a real experimental system, without the need for any fitting parameter. This experiment highlights the single-mode filtering as a viable tool for erasing the quantum distinguishability of single photons.

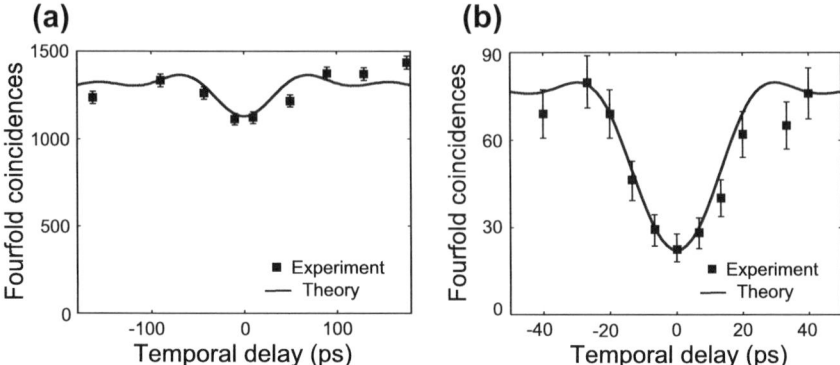

FIGURE 12.19 (a) Fourfold coincidence counts per 20 billion pump pulses recorded as a function of the relative delay between signal photons in the multimode configuration and (b) fourfold coincidence counts per 10 billion pump pulses recorded as a function of the relative delay in the single-mode configuration. In both plots, the error bars are computed following the standard procedure of estimating statistical fluctuations assuming Poisson distributions for the recorded coincidence and single counts. See text for more details on the theory curves in both plots which are overlaid without the use of any fitting parameter. *Images from [12]. Copyright (year of paper) by the American Physical Society.*

12.6 ENTANGLED PHOTON-PAIR SOURCES

The concept of entanglement, in which the quantum states of two or more physically separate systems may be linked, is one of the defining characteristics of quantum mechanics. The pioneering early work of Aspect *et al.* on the nature of entangled photon pairs demonstrated that the state of one photon could not be described independently of the state of the other, meaning that a measurement made on one member of the pair can instantaneously affect the state of its partner [63]. In recent years, efficient, high-fidelity sources of entangled photons have become an essential resource for many quantum information applications, including quantum cryptography and communications protocols [64–66], quantum repeaters [67], quantum memories [68], and schemes for quantum computation [69–71]. An optical fiber-based source of entangled photons can be realized by preparing signal and idler photon pairs through sFWM in such a way that the states of the two photons are correlated in one or more of their degrees of freedom. For many applications, polarization-entangled photons are particularly useful, as the polarization state of light can be easily manipulated and measured using standard linear optical components such as wave plates and polarizers.

The Sagnac-loop configuration shown in Fig. 12.20 can be used to generate entangled photon pairs from a fiber through a counterpropagating scheme [14]. In this scheme, a loop of nonlinear fiber, capable of producing pairs of signal and idler photons by sFWM (as detailed in Section 12.2), is pumped simultaneously in both the clockwise (CW) and counterclockwise (CCW) directions. The

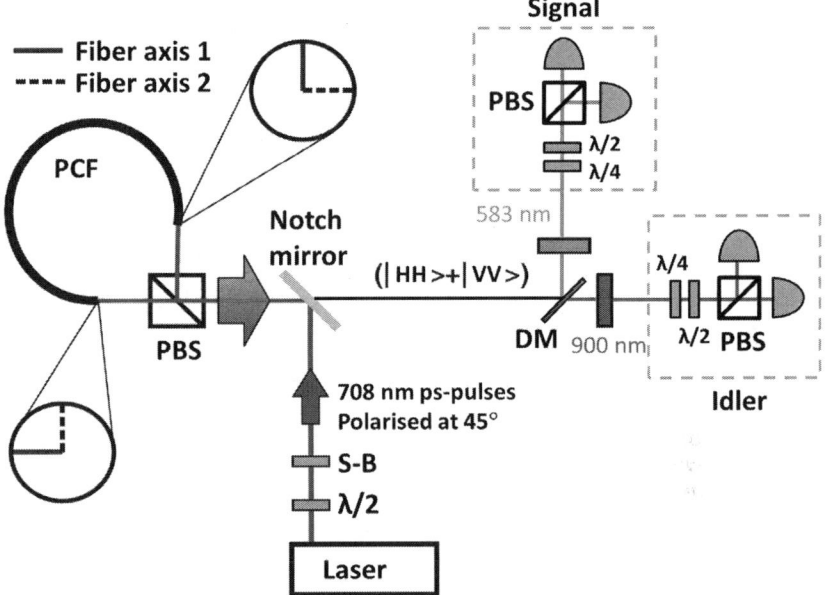

FIGURE 12.20 Schematic representation of a fiber-based entangled photon source setup. A section of PCF forming a Sagnac-loop is pumped in both directions by orthogonally polarized pump pulses, producing entangled pair states when the generated signal and idler pairs are combined onto a single output path. S-B, Soleil-Babinet compensator; PBS, polarizing beam-splitter; DM, dichroic mirror. *Image from [14]. Copyright (year of paper) by the American Physical Society.*

simultaneous pumping is achieved by splitting the pulses from a single pump laser on a polarizing beam-splitter (PBS), and adjusting the polarization state of the laser such that the pump power propagating in both directions through the fiber is balanced. The PBS causes the pump light to be launched into the fiber with a different polarization state for CW and CCW directions, with horizontally polarized (H) pump photons propagating in one direction and vertically polarized (V) pump photons in the other.

If the fiber used for the Sagnac-loop is birefringent, and the fiber is oriented such that the pump photons traveling in both directions are launched with their polarization direction aligned with either the slow or fast birefringent axis of the fiber, then the polarization state of the pump light will remain unchanged as it propagates through the fiber. For a fiber designed to provide phase-matching for co-polarized sFWM, signal and idler photons will be generated predominantly with the same polarization state as the pump. The pump light traveling in one direction around the Sagnac-loop therefore generates H polarized signal and idler photons, while the pump propagating in the opposite direction will generate V polarized signal and idler photons. When the counterpropagating pairs of signal and idler photons meet at the PBS they combine in a coherent superposition, producing the state

$$|\psi\rangle_{out} = \alpha|H_s H_i\rangle + \beta e^{2i\phi_p}|V_s V_i\rangle, \quad (12.36)$$

upon post-selection of the photons by detection. α and β are weighting terms for the probability of generating HH and VV polarized pairs, which are determined by the proportion of pump power propagating in each direction through the loop, and any polarization dependent losses or variations in collection efficiency. The state shown in Eq. (12.36) also contains a relative phase term between photon pairs in the two polarization states, denoted by ϕ_p. This relative phase is determined by the phase difference between the horizontally and vertically polarized components of the pump beam, and can be easily adjusted through the use of a Soleil-Babinet compensator. By setting $\alpha = \beta = \frac{1}{\sqrt{2}}$ and $\phi_p = 0$, the source can be used to generate the maximally entangled $|\Phi^+\rangle$ Bell state. As with the heralded single-photon sources described in Section 12.3, after the entangled photon state is generated, the signal and idler photons can be separated into different output paths using a prism or dichroic mirror, and then filtered in order to attain a spectrally pure state, and to remove the residual pump light and Raman-scattered photons.

To generate the entangled state shown in Eq. (12.36), subwavelength stability of the relative optical path lengths of the HH and VV photon pairs is required to ensure that the phase ϕ_p is not time-varying. One significant advantage of using a Sagnac-loop arrangement for the generation of entangled photon pairs is that any differences in the optical path length of the loop, due to vibrations or temperature variations for example, affect both the CW and CCW paths equally. This arrangement is therefore inherently phase-stable, and ensures that the path lengths for the HH and VV polarized pairs are balanced.

While the use of a birefringent fiber ensures that the polarization state of pump light propagating in the loop is unaffected by environmentally driven polarization fluctuations, care must be taken to ensure that spectral distinguishability is not introduced between the HH and VV photon pairs. As shown in Section 12.2, even for fibers where birefringence is not required to achieve phase-matching, and the signal and idler photon pairs are generated predominately with the same polarization state as the pump, the presence of a strong birefringence can subtly alter the phase-matching conditions on one axis of the fiber relative to the orthogonal axis. This potential source of spectral distinguishability between the HH and VV pairs can be avoided by incorporating a 90° twist in the fiber loop, so that the pump light is launched onto the same birefringent axis of the fiber in both the CW and CCW directions, ensuring that the phase-matching conditions for both are identical.

With the addition of a 90° twist in the fiber, it is also possible to produce a Sagnac-loop-based source with a fiber designed to generate signal and idler pairs through cross-polarized sFWM with birefringent phase-matching. As described in Section 12.5.1, this regime of phase-matching allows for the generation of signal and idler photons which are spectrally uncorrelated. An entangled photon source of this type, was described by Clark *et al.* [50], wherein the Sagnac-loop consisted of a 20 cm section of birefringent PCF, designed to generate signal and idler photons near 600 and 860 nm by pumping in the normal dispersion regime of the fiber with a pulsed Ti:Sapphire laser. The dispersion properties of the PCF were designed such that the requirement of asymmetric group velocity matching between the pump and idler wavelengths was satisfied, allowing the entangled signal and idler photon pairs to be generated in an intrinsically pure state.

To demonstrate that entanglement exists between the generated pairs of signal and idler photons from this type of source it is necessary to look for correlations in their polarization states. The detector setup required to perform this measurement is shown in Fig. 12.21, and consists of a PBS, along with a half-wave plate (HWP) and quarter-wave plate (QWP) in order to select the polarization basis in which each photon will be measured, in both the signal and idler output arms of the source. As the settings of these wave-plates are changed, patterns in the rates of coincident detections from single-photon detectors monitoring the output ports of each PBS can be used to reveal the presence of entanglement.

To test for entanglement, the polarization state measurement of one photon of the pair must be made in one of two nonorthogonal polarization bases. For the detector arrangement shown in Fig. 12.21, the most convenient choices are the computational $\{H/V\}$ basis (where the wave plates are set to leave the photon polarization state unchanged, and H and V polarized photons are directed toward alternate detectors) and the diagonal $\{D/A\}$ basis (where the

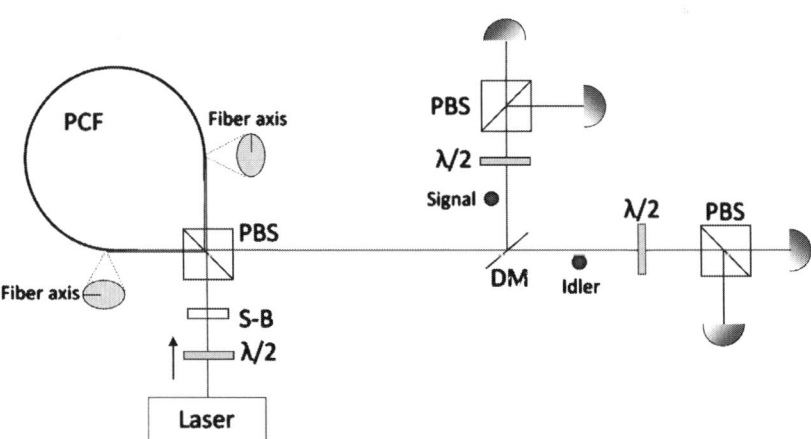

FIGURE 12.21 A setup used to generate entangled photon pairs with a PCF designed for cross-polarized sFWM. The photons can be generated in intrinsically pure spectral states without the need for narrowband spectral filtering. *Image from [50].* ©*IOP Publishing Ltd and Deutsche Physikalische Gesellschaft. Published under a CC BY-NC-SA licence.*

HWP rotates the H and V polarization components by 45° relative to the axis of the PBS). While the measurement of one photon of the pair (for instance, the idler) is performed in one of the two fixed bases, the measurement of the other photon (the signal in this example) is carried out across a variety of basis states by rotating the corresponding HWP. The twofold coincidence count rate between one of the signal arm detectors and one of the idler arm detectors is monitored as the signal arm HWP orientation is varied.

Figure 12.22 demonstrates the coincident count rate fringes that would be expected when measuring an entangled state of the type described in Eq. (12.36). In the $\{H/V\}$ basis (with the idler arm HWP set to 0°) high visibility fringes demonstrate that the signal and idler wavelengths are highly correlated. For this choice of measurement basis, this result would also be observed in the classical regime, as each detector can only receive photons generated in either the CW or CCW direction. The polarization state of the detected signal and idler photons are therefore correlated in this case because they are restricted to a fixed linear polarization state by the requirements of phase-matching and the orientation of the birefringent axes of the fiber. When the idler photon is measured in the $\{D/A\}$ basis (idler arm HWP at 22.5°) fringes in the coincidence count rate of the detected signal and idler photons can only be explained by the coherent superposition of the HH and VV photon pairs at the PBS. Perfect visibility of the coincidence fringes in this second nonorthogonal basis would indicate that the photons were maximally entangled, while a flat coincidence count rate without fringes would show that the polarization state of the photons were uncorrelated. A coincidence fringe visibility exceeding 71% indicates that

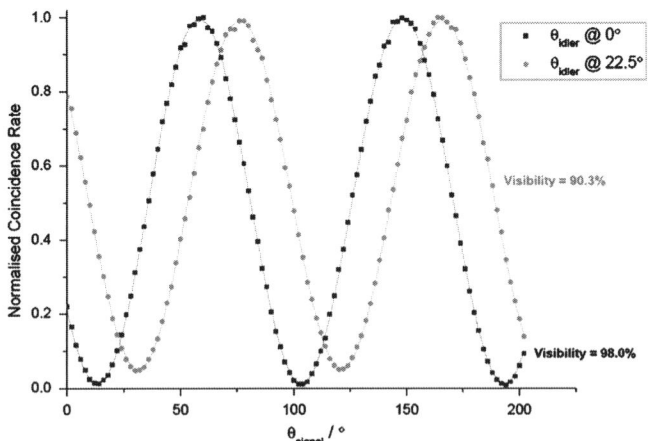

FIGURE 12.22 Measured coincidence count rate data for the entangled photon source described by Clark *et al.* [50]. High visibility fringes can be seen for measurement of the idler in both the {H/V} basis (idler arm HWP set at 0°) and the {D/A} basis (idler arm HWP set at 22.5°), demonstrating the presence of polarization entanglement. *Image from [50].* ©IOP Publishing Ltd and Deutsche Physikalische Gesellschaft. Published under a CC BY-NC-SA licence.

the entanglement of the state is sufficient to violate a Bell inequality [72,73], proving that entanglement was present in the generated state. A more detailed analysis of the entangled state can be made using quantum state tomography measurements to reconstruct the density matrix of the photon pair and determine the fidelity with which the experimentally generated state matches that of an idealized maximally entangled state, such as the $|\Phi^+\rangle$ state [74].

Both of the entangled photon source designs described so far in this section were based on birefringent fiber, in order to avoid environmentally driven fluctuations in the polarization state of the propagating pump light and generated photon pairs. An alternative approach that can be used to generate polarization-entangled photon states is based on the technique of polarization-dependent time delay [75]. Figure 12.23 shows such a scheme using a Sagnac-loop, formed by a 50:50 fiber beam-splitter and a section of dispersion-shifted fiber (DSF). In this scheme, two time-delayed, orthogonally polarized pump pulses are launched into one port of the Sagnac-loop. The pump wavelength is carefully chosen to be in the anomalous dispersion regime of the DSF, so that sFWM is phase-matched in the DSF, with the signal and idler wavelengths closely spaced around the pump wavelength. The signal and idler photons are generated predominantly with the same polarization state as the pump; the probability of cross-polarized sFWM occuring is at least one order of magnitude less [76]. At the output of the Sagnac-loop, the initial time delay is removed by passing the two pulses through a piece of polarization-maintaining fiber (PMF) with proper length ℓ satisfying $\ell = \Delta v_g \tau$, where τ is the initial time delay, and Δv_g is the group-velocity difference between light pulses polarized in the PMF's principle states of polarization. At the end of the PMF, there is no way to tell, not even in principle, which pulse generates the signal/idler photon pair in the

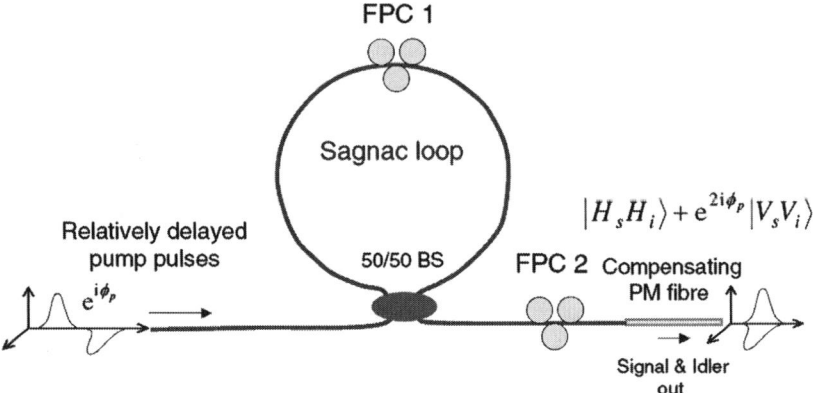

FIGURE 12.23 A simplified schematic of the Sagnac-loop scheme, where most of the experimental details are neglected while putting the emphasis on the gist of the scheme. BS, beam-splitter; FPC, fiber polarization controller; PM, polarization-maintaining. *Image from [16]. ©IOP Publishing Ltd and Deutsche Physikalische Gesellschaft. Published under a CC BY-NC-SA licence.*

overlapped time slot, and as a result the two probability amplitudes $|H_s H_i\rangle$ and $|V_s V_i\rangle$ are added coherently to give the desired polarization-entangled state $|H_s H_i\rangle + e^{2i\phi_p}|V_s V_i\rangle$, where ϕ_p is the phase difference between the original two pump pulses (ϕ_p can be changed by varying the setting of the polarization-dependent time delay circuit). Note that time-bin entanglement can be created from this scheme if, instead of removing the distinguishing timing information using the PMF at the output, one removes the distinguishing polarization information between the two relatively-delayed time slots.

The reason for using a Sagnac-loop in this scheme is technical rather than fundamental. As described in [77], a properly aligned (by adjusting the intra-loop fiber polarization controller (FPC)) Sagnac-loop can function as a total reflector for the pump photons. In this operational mode, $< 0.1\%$ of the incident pump photons get transmitted through the loop together with the signal/idler photons, and the remaining $> 99.9\%$ of pump photons are reflected back (ignoring the propagation loss of the pump inside the Sagnac-loop). This gives at least 30 dB isolation with respect to the pump in the transmitted mode. Since typically a pump rejection ratio > 100 dB is required to reliably detect the trace amount of signal/idler photon pairs [78], placing a free-space double-grating filter at the output of the PMF can provide the additional $\simeq 75$ dB rejection for the pump, satisfying the 100 dB requirement.

A drawback with the setup in Fig. 12.23, however, is that it requires sophisticated phase tracking and locking to maintain a fixed relative phase between the pump pulses. In addition, fiber-birefringence induced polarization fluctuations need to be carefully compensated before sending the photon pairs through the polarization maintaining fiber for accurately removing the time delay. Liang et al. demonstrated an ultra-stable, alignment-free setup that eliminates those needs [15]. A schematic of its experimental setup is shown in Fig. 12.24, where the system's principal axes are defined by a four-port fiber PBS. An optical pump-pulse train is prepared at 45° linearly polarized relative to the principal axes and is injected into a fiber circulator which has the basic property of transmitting incident light between its three ports in a successive fashion: light incident on port one is passed to port two and light that returns back into port two is then passed to port three. To avoid unfavorable polarization rotation of the pump pulses, a polarization-maintaining fiber circulator can be used for bringing the pump pulses to the input port of the PBS. At the output ports of the PBS, each 45° linearly polarized pump pulse is decomposed into horizontally and vertically polarized components P_H and P_V, respectively, with a relative phase ϕ_p equal to zero.

P_V is reflected and guided to a Faraday mirror (FM1) at the start of its itinerary: PBS → FM1 → PBS → FM3 → PBS → FM2 → PBS. A Faraday mirror is a non-reciprocal optical element which produces a reflection with a state-of-polarization (SOP) that is orthogonal to the input SOP. Thus, uncontrolled polarization rotations occurring along the incident path are automatically compensated on the return path. In this setup, P_V becomes

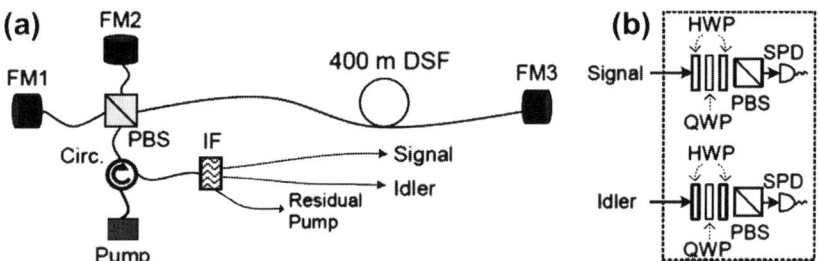

FIGURE 12.24 (a) Schematic setup for an ultra-stable, alignment-free entangled photon source and (b) the polarization control optics and single-photon detectors (SPD) used to verify the generation of entangled photon states. *Images from [15].*

horizontally polarized upon returning to the PBS after reflection from FM1. It then passes through the PBS and enters a 400 m piece of DSF where the sFWM process takes place. Similarly, PH starts its journey PBS → FM2 → PBS → FM3 → PBS → FM1 → PBS by going to FM2. It then arrives back at the PBS vertically polarized and reflects to enter the DSF. A small length difference between the paths to FM1 and FM2 are intentionally designed so that a time delay, which is much greater than the pump pulse duration, is introduced between the arrival times of P_H and P_V at the DSF.

In the DSF, P_H and P_V independently engage in sFWM processes and create signal-idler photon pairs. Note that P_H and P_V experience almost the same insertion loss before entering into the DSF; thus the two sFWM processes have almost the same efficiency. As mentioned above, the signal and idler photon pairs are co-polarized with respect to the pump pulses. With the use of FM3, birefringence-induced polarization fluctuations of the light in the DSF between PBS and FM3 are also automatically compensated. In addition, the polarization states of the two pump pulses and the generated photon pairs are rotated by 90° when they arrive back at the PBS. Hence, P_V and the photon pairs generated from it are directed toward FM2 while P_H and its generated photon pairs travel to FM1. It is easy to see that the information on the generation time of the newly created photon pairs is automatically erased when PH and PV, along with their generated photon pairs, are recombined at the PBS traveling backward through the input port. Upon emerging from the PBS the circulator then directs them away from the pump toward the receivers. With a properly set pump power, the signal and idler photon pairs are emitted in the maximally entangled $|\Phi^+\rangle$ state. It is also important to point out that the pump pulses travel through the DSF twice in this scheme because of FM3. Both P_H and P_V follow a symmetric path through the system. It is this symmetry and the use of Faraday mirrors which automatically compensate any long-term polarization and/or phase fluctuations that enable an ultra-stable and alignment-free source of polarization entanglement. After passing through the circulator, the photon pairs are separated from the residual pump by fiber interference filters, and are then ready for use in various quantum communication applications. Using this Faraday-mirror setup, generation of entangled photon pairs was demonstrated in both the 1310 nm telecom O-band and the 1550 nm C-band with near-unity fidelity [15,79].

All of the above schemes generate polarization-entangled signal and idler photons of different wavelength. Medic et al. demonstrated a source of polarization-entangled photon pairs that is totally degenerate and that deterministically splits each pair of photons into separate single-mode optical fibers [80]. It is built on a "quantum splitter" design for a Sagnac-loop-based source of identical photon pairs [81]. A schematic of the experimental setup is shown in Fig. 12.25, where a dual-frequency, diagonally polarized pump ($|D_p^{(\omega_1,\omega_2)}\rangle$) spontaneously four-wave mixes into two co-polarized, spatially separated, and central-wavelength signal ($|D_s\rangle$) and idler $|D_i\rangle$ photons through $|D_p^{(\omega_1,\omega_2)}\rangle \rightarrow |D_s\rangle|D_i\rangle$, where $|D_j\rangle = (|H_j\rangle) + |V_j\rangle)/\sqrt{2}$, for $j = s, i$. Note that these photon pairs are not polarization entangled. Entanglement generation requires that a distinguishing degree of freedom be coupled to the pump polarization before the sFWM, and that this distinguishing information be subsequently erased, causing orthogonal photon-pair amplitudes to superpose. Consider the diagonally polarized pump $|D_p^{(\omega_1,\omega_2)}\rangle$. After a polarization-dependent time delay (PDD) t', the pump state can be described by $(|H_p^{(\omega_1,\omega_2)}\rangle \otimes |t+t'\rangle + |V_p^{(\omega_1,\omega_2)}\rangle \otimes |t\rangle)/\sqrt{2}$. Here, the PDD is implemented by means of free-space optical delay paths after a polarizing beam-splitter in a Michelson configuration. The degenerate pair production (from reverse HOM interference between the sFWM amplitudes) is then described by

$$|H_p^{(\omega_1,\omega_2)}\rangle \otimes |t+t'\rangle + |V_p^{(\omega_1,\omega_2)}\rangle \otimes |t\rangle) \rightarrow |H_s H_i\rangle \otimes |t+t'\rangle + |V_s V_i\rangle \otimes |t\rangle). \quad (12.37)$$

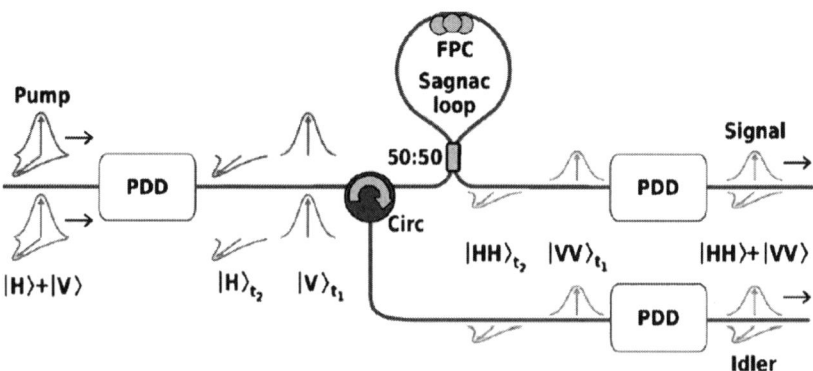

FIGURE 12.25 Setup for generating degenerate, polarization-entangled, and deterministically split photons: PDD, polarization-dependent time delay; FPC, fiber polarization controller; Circ, circulator; $t_2 = t_1 + t'$. Image from [80].

By subjecting these degenerate photons to a second set of complementary PDDs the Bell state $\frac{1}{\sqrt{2}}(|H_s H_i\rangle + |V_s V_i\rangle) \otimes |t\rangle$ is produced.

Finally, Chen *et al.* demonstrated a hybrid scheme that combines the benefits of the Sagnac-loop interferometer and the counterpropagating (CP) schemes. It is called a double-loop (DL) scheme, due to the topological fact that there are two loops intertwined with each other, namely, a PBS loop and a Sagnac-loop (see below). Such a DL scheme can be viewed as a perfect combination of the SL and CP schemes. On one hand, the Sagnac-loop in the DL scheme performs exactly the same way as in the Sagnac-loop scheme, which is to provide pump isolation of around 30 dB. On the other hand, the DL scheme is just a modified CP scheme, with the Sagnac-loop replacing the straight-fiber configuration. The strengths of both the SL and CP schemes are successfully retained in the DL scheme, namely, $\simeq$ 30 dB built-in filtering capability, easy implementation of a CP-like configuration and suppression of the cross-polarized Raman-scattered photons, as well as photons generated by self-phase modulation of the pump.

A simplified schematic of the DL scheme is shown in Fig. 12.26. Just like in the CP scheme, a 45°-polarized pump pulse splits into two equally powered, orthogonally-polarized components after entering the main loop—the PBS loop. The secondary loop, the Sagnac-loop, is configured as a total reflector for both CW and CCW propagating pumps. In this configuration, > 99.9% of each pump pulse (ignoring the propagation loss) is reflected back to its original entrance port (1 → 1, 2 → 2), while < 0.1% of each pump pulse together with its copolarized FWM-generated photon pair gets transmitted (1 → 2, 2 → 1). The FPCs (FPC1 and FPC2) are adjusted such that the following criteria are satisfied:

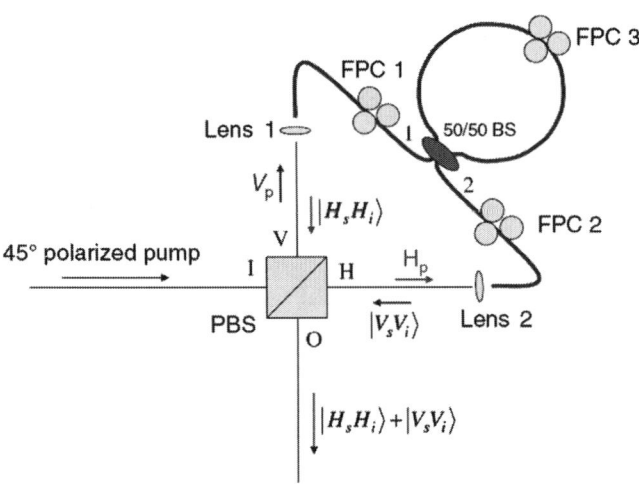

FIGURE 12.26 A simplified schematic of the DL scheme. *Image from [16]. ©IOP Publishing Ltd and Deutsche Physikalische Gesellschaft. Published under a CC BY-NC-SA licence.*

(i) The reflected pump photons maintain their original polarization after their individual round trip (V-port → 1 → 1 → V-port, H-port → 2 → 2 → H-port);

(ii) The transmitted FWM-generated photons, together with the accompanying copolarized pump leakages, regain the initial input polarization after traversing the entire loop (V-port → 1 → 2 → H-port, H-port → 2 → 1 → V-port).

The first criterion is satisfied when all the Sagnac-reflected pump photons go back to the I-port (the input port in Fig. 12.26) of the PBS. The second criterion is satisfied when all the Sagnac-transmitted photons exit from the O-port of the PBS. Due to indistinguishability of the CW and CCW paths, the two coherent FWM-photon amplitudes are thus maximally added to create the desired polarization entanglement, $|H_s H_i\rangle + |V_s V_i\rangle$, with maximum pump isolation provided by the Sagnac-loop. The above seemingly demanding operating conditions were theoretically shown to be feasible. Experimentally, however, it turns out that the first operating criterion can be relatively relaxed without sacrificing the performance of the DL scheme. This is due to the nonoverlapping time slots that the transmitted and reflected pump pulses can be made to occupy when exiting the PBS's O-port. To be more precise, let's denote the light propagation length (including the free-space as well as the fiber) from the PBS's I-port to port 1 of the Sagnac-loop ℓ_1, the Sagnac-loop length L_s, and the light propagation length from port 2 to O-port ℓ_2. The reflected pulse in the CW direction traverses a distance of $2\ell_1 + L_s$, whereas its counterpart in the CCW direction traverses a distance of $2\ell_2 + L_s$. In contrast, the transmitted pulses in both directions traverse the same distance $\ell_1 + L_s + \ell_2$. The pulses' exit-time difference, proportional to their propagation length difference, is thus given by $\Delta t_\pm = \pm |\ell_1 - \ell_2|/c$, where the central (or reference) time slot is the overlapping time slot for the transmitted sFWM photons. As can be clearly seen from the above expression for $\Delta t_\pm$, one can judiciously choose the propagation length difference to be large enough that the leakage photons from the two reflected pump pulses fall outside the detection-time window for the transmitted photons. This essentially means that even though the first operating criterion may not be satisfied, or a real-life PBS may deviate from its ideal performance, the feasibility of the DL scheme is not sacrificed as long as we choose a suitably long $\Delta t_\pm$.

12.7 APPLICATIONS OF FIBER PHOTON SOURCES—ALL-FIBER QUANTUM LOGIC GATES

One application for single and entangled photon states that has generated much interest in recent years is quantum information processing (QIP), with the eventual aim of developing a scalable architecture for quantum computation [82,70]. Photons have many desirable properties that make them well suited for use as qubits in QIP applications. The polarization state of a photon provides a convenient system of two orthogonal quantum eigenstates in which a qubit can be easily encoded, as well as manipulated through the use of wave plates and other birefringent optics. Furthermore, due to their incredibly weak interactions with each other and the environment, qubit states encoded on photons are robust against decoherence. Unfortunately, this weak interaction also makes it extremely challenging to realize the type of two-photon operations that are required for computation, where, for example, the state of one qubit is used to conditionally flip the state of another. Currently there are no known materials with sufficiently strong nonlinearity to implement the required phase-shift of π through cross-phase modulation for one single photon to flip the state of another.

In 2001, a seminal paper by Knill *et al.* showed that an effective two-photon interaction could be realized using only linear optical elements through a combination of nonclassical interference and measurement induced nonlinearity [69]. This approach of linear optical quantum computation facilitated the development of two-photon logic gates, of which the controlled-NOT gate (CNOT) is an important example. This logic gate accepts two input photons—known as the control and target qubits. At the output of the gate, the state of the target qubit is conditionally flipped, based on the state of the control qubit, implementing the required two-photon interaction. The CNOT gate, in combination with easily implemented unitary single qubit operations, has been shown to be a universal set of gates for quantum computation, meaning that any conceivable QIP algorithm can be implemented using just these elements [83].

Early experimental implementations of the CNOT gate [84,85] were based on bulk optical elements, making them unsuitable for scaling up to more complex architectures. They also incorporated multiple interferometers, in order to convert the qubits between path and polarization encoding, and therefore required subwavelength path stability for successful operation. Clark *et al.* reported the first experimental demonstration of an optical fiber-based CNOT gate, which overcame many of these limitations [13].

A schematic of the experimental setup, is shown in Fig. 12.27. The setup includes two PCF-based heralded single-photon sources (see Section 12.3). These generate linearly polarized single photons to be used as the control and target polarization encoded qubits. The qubit values are set through the use of HWPs. The CNOT gate itself consists of three partially polarizing fiber couplers (PPFC). The target and control qubit meet at the first PPFC, which is oriented such that the component has reflectivity of $R_H = \frac{1}{3}$ and $R_V = 1$ for H and V polarized photons respectively. If one or both of the input photons are V polarized then there will be no interaction between them and each will be probabilistically reflected or transmitted independently. Successful operation of the gate requires that a single photon is detected in each of the two output channels. If one or both photons are V polarized then both must be reflected at the first PPFC in order for one photon to be detected in each output mode.

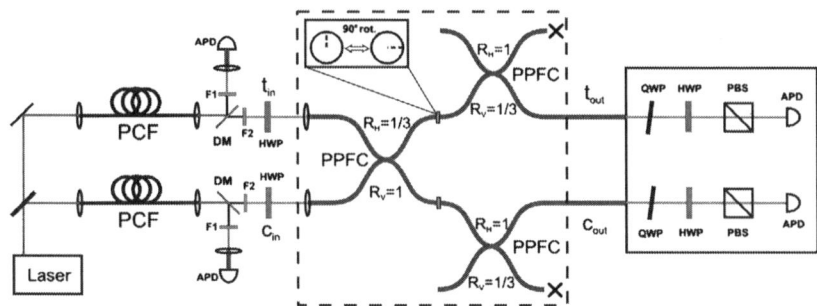

FIGURE 12.27 Experimental setup used to demonstrate an all-fiber CNOT gate. The CNOT gate itself is formed of three partially polarizing fiber couplers (PPFC). Input target and control qubit states (t_{in} and c_{in}) are prepared using two heralded single-photon sources shown on the left-hand side of the figure. The optics shown on the right-hand side of the figure are used to analyze the output states of the target and control qubits after the gate (t_{out} and c_{out}). Image from [13]. Copyright (year of paper) by the American Physical Society.

When the input photons are both H polarized and indistinguishable, nonclassical interference will occur between them at the first PPFC. With probability $\frac{1}{9}$ both photons will be transmitted here, giving a π phase-shift for this output state relative to the other possibilities in which both photons are reflected. This interaction therefore implements a controlled phase-shift when both the control and target photons are H polarized at this PPFC. In combination with single qubit rotations of 45° performed by the HWPs on the target qubit at the input and output of the gate, this implements the required CNOT operation [86]. Failure of the gate to function correctly when two H polarized photons are incident on the PPFC results in one of the two output modes being empty. Successful operation of the gate can be verified through post-selection of events in which output target and control photons are detected in coincidence, which occurs with probability $\frac{1}{9}$ in the absence of losses. The other two PPFCs ensure that the success probability of the gate is uniform for all possible input states. A 90° rotation of the birefringent axes of the optical fiber is also incorporated in the setup at the midpoint of the CNOT gate. This compensates for the polarization-mode dispersion that occurs in the birefringent fiber due to the difference in the propagation constant for H and V polarized photons, which would otherwise lead to decoherence of the polarization state for qubits encoded in a diagonal basis relative to the fiber axes.

In order to evaluate the performance of the CNOT gate, wave plates and a PBS placed before each of the single-photon detectors can be used to determine the probability of finding the control and target qubits in a given polarization state at the output of the gate for each of the four possible combinations of logical input states, as shown in Fig. 12.28. This measurement was performed in both the computational basis $\{H/V\}$, where the photon polarizations were aligned with the principal axes of the PPFCs, and in the nonorthogonal diagonal

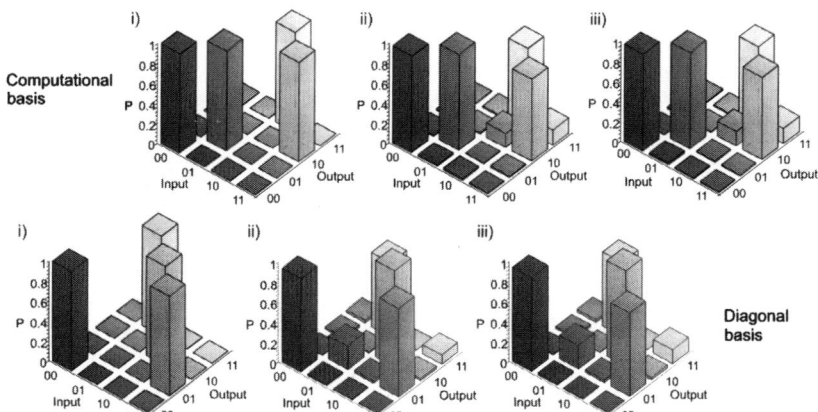

FIGURE 12.28 Truth table results showing the operation performed by the CNOT gate. Results are shown for measurement in both the computational and diagonal bases. The two numbers used to label each input and output state denote the states of the control and target qubits (in that order), with $\{|0\rangle = |V\rangle, |1\rangle = |H\rangle\}$ in the computational basis and $\{|0\rangle = (|H\rangle + |V\rangle)/\sqrt{2}, |1\rangle = (|H\rangle - |V\rangle)/\sqrt{2}\}$ in the diagonal basis. (i) Ideal CNOT gate operation; (ii) experimentally measured values; and (iii) Theoretically modeled predicted values, accounting for imperfect interference visibility. *Image from [13]. Copyright (year of paper) by the American Physical Society.*

basis $\{D/A\}$. Based on these measurements, the classical logical fidelity [87,88] of the all-fiber CNOT gate was calculated to be 0.9 in the $\{H/V\}$ basis and 0.89 in the $\{D/A\}$ basis, and from this the average process fidelity of the gate $\overline{F}$ was found to lie in the range $0.83 \leq \overline{F} \leq 0.91$. It should be noted however that the main limitation on the process fidelity in this case was found to be imperfect quantum interference visibility, resulting from distinguishability between the photons generated by the two independent single-photons sources. Modeling results accounting for this imperfect state preparation showed that the operation of the CNOT gate itself was near-perfect.

This all-fiber implementation of the CNOT gate offers several advantages over bulk optics approaches, including the potential for miniaturization and increased long term stability due to the lack of misalignment issues in a guided wave configuration. In addition, as the gate operates directly on polarization encoded qubits, only a single instance of nonclassical interference is required, without the need for additional classical interferometers. The requirements for the stability of the optical path lengths of the two photons are therefore less stringent, as they only need to be stable to within the coherence length of the photons rather than to a fraction of the photon wavelength. This optical fiber implementation of the CNOT gate is well suited for low loss integration with fiber-based single-photon sources, and paves the way for the development of compact, all-optical fiber QIP circuits.

12.8 PHOTONIC FUSION IN FIBER

Fusion gates are a method for scalably generating large entangled states, which consist of joint measurements applied to pairs of previously unentangled photons [71]. The controlled generation of entanglement is vital to many quantum information applications, including teleportation [89], secret key distribution [90], and most ambitiously quantum computing [82]. Of particular interest is the efficient generation of cluster states, which are a resource for one-way-quantum-computing [91,18]. Once the initial entanglement is generated, the computation can proceed deterministically using a sequence of single qubit measurements.

Figure 12.29a show how overlapping two photons at a PBS can perform a parity measurement. If they are of the same polarization (even parity), they will either both be transmitted or both be reflected, and will leave the beam-splitter in separate modes. This is a successful outcome in that the parity has been measured without gaining any information about the individual polarizations of the photons, and it will leave them in an entangled state. An odd parity outcome leaves both photons at the same output and does not generate entanglement, so the fusion fails 50% of the time. Although this operation is not deterministic, it can be used to build larger states in a scalable way from small entangled states, such as Bell pairs. It is necessary to herald whether the operation has succeeded or failed: in the case of type I fusion (Fig. 12.29a) this is done with a photon-number resolving detector in one of the outputs, so that when the fusion succeeds, one of the photons is detected and destroyed. This allows larger states to be generated in a heralded fashion with some finite probability. Figure 12.29b shows schematically how this can be applied to cluster states—in these pictures, circles represent qubits and bonds represent entanglement between them. Larger states can then be grown in a linear fashion: [71] suggests a strategy of beginning from 5 qubit linear states. Fusing these together, every success adds 4 qubits (because one herald photon is detected and destroyed), while every failed fusion removes 1 qubit. Hence on average the state grows by 1.5 qubits with each fusion.

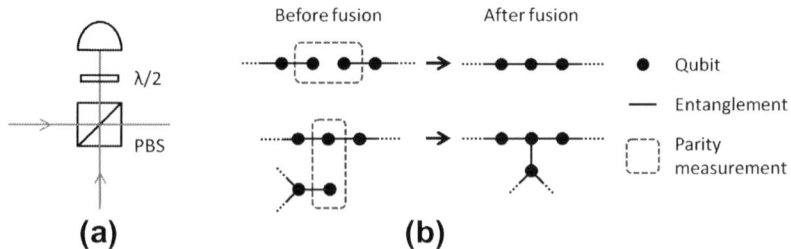

FIGURE 12.29 From [71]. (a) Type I fusion can be used to build long linear cluster states using interference at a PBS. Detection of a photon in one mode heralds the success of the measurement. (b) Linear and 2D cluster geometries produced by fusion operations.

Chapter | 12 Four-Wave Mixing in Single-Mode Optical Fibers

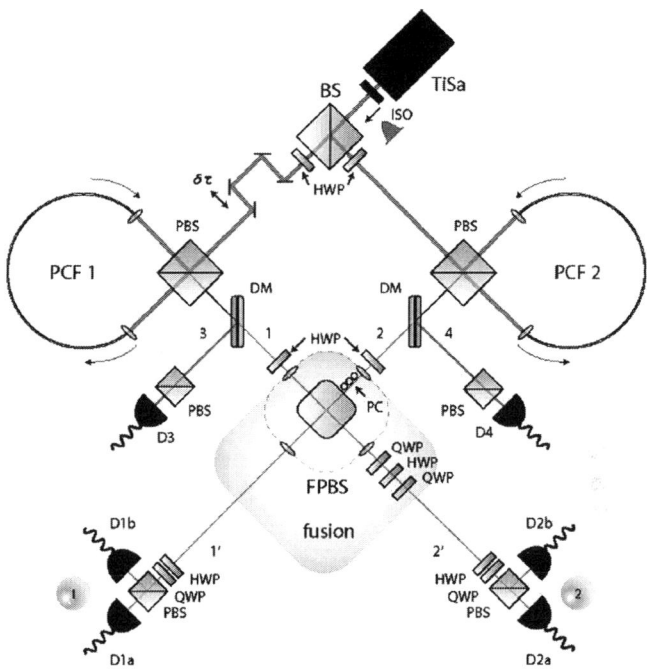

FIGURE 12.30 (a) Type I fusion can be used to build long linear cluster states using interference at a PBS. Detection of a photon in one mode heralds the success of the measurement [71]. (b) Linear and 2D cluster geometries produced by fusion operations. ©IOP Publishing Ltd and Deutsche Physikalische Gesellschaft. Published under a CC BY-NC-SA licence.

Figure 12.30 shows an experimental setup to demonstrate fusion between two photons from separate sources. The input states are prepared using two PCF sources pumped by a pulsed laser. Detections of the idler photons herald the presence of a signal photon from each source. These are both rotated to diagonal polarization, $|D\rangle = \frac{1}{\sqrt{2}}(|H\rangle + |V\rangle)$, so that the state before the fusion gate can be written

$$\frac{1}{2}(|H\rangle_1 + |V\rangle_1)(|H\rangle_2 + |V\rangle_2) = \frac{1}{2}(|HH\rangle_{12} + |HV\rangle_{12} + |VH\rangle_{12} + |VV\rangle_{12}). \quad (12.38)$$

Ideally the fusion operation has the effect of transmitting even parity terms and filtering out odd parity. This can be expressed as

$$|HH\rangle_{12}\langle HH| + |VV\rangle_{12}\langle VV|, \quad (12.39)$$

which, applied to the state in Eq. (12.38), leaves the Bell state $|\Phi^+\rangle$ with 50% success probability:

$$|\Phi^+\rangle_{12} = \frac{1}{\sqrt{2}}(|HH\rangle_{12} + |VV\rangle_{12}). \quad (12.40)$$

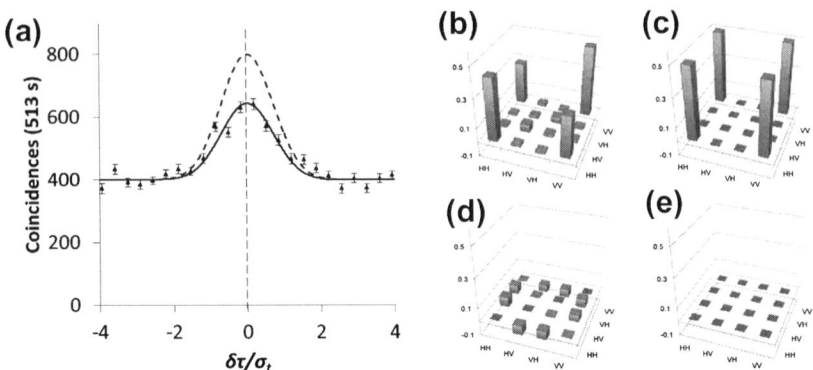

FIGURE 12.31 (a) Increase in DD coincidences when arrival times are matched. The solid line is a fit to the experimental data points and the dashed line shows the ideal case. ©IOP Publishing Ltd and Deutsche Physikalische Gesellschaft. Published under a CC BY-NC-SA licence.

This scheme assumes that the two photons undergoing the fusion are indistinguishable in all degrees of freedom, otherwise they will be left in an incoherent mix of horizontally and vertically polarized pairs [17]. For example, in Fig. 12.30, the arrival time of the photons is matched using a variable delay $\delta\tau$ in one arm. When the output polarization analyzers are set to diagonal, the state should show an increase in the proportion of $|DD\rangle_{12}$ when the photons are indistinguishable—this results from coherent addition of probability amplitudes characteristic of $|\Phi^+\rangle$. This appears as an anti-dip in Fig. 12.31a. The quality of the final state can be characterized more completely with polarization tomography [92,93], as shown in Fig. 12.31b–d.

12.9 CONCLUSION

Spontaneous four-wave mixing in optical fibers is now a well-established approach for realizing probabilistic, nonlinear generation of correlated photon pairs. Signal and idler photons can be generated directly in the guided mode of an optical fiber, allowing them to be easily prepared in both single spatial and polarization modes of a fiber. This makes photon sources based on four-wave mixing well suited for low loss, stable integration with existing single-mode telecommunications fibers. In addition, tight confinement of the interacting guided light fields to the core of a fiber over long lengths allows these sources to demonstrate high brightness in spite of the intrinsically weak nonlinear response of silica.

Sources of heralded single photons and entangled-photon pairs based on four-wave mixing have been demonstrated in both conventional single-mode optical fibers and photonic crystal fibers. Dispersion engineering through design of the fiber structure or the introduction of birefringence can alter the phase-matching conditions for four-wave mixing, leading to considerable flexibility in the signal and idler wavelengths that can be generated. Further, by tailoring the dispersion appropriately it is possible to generate photons in spectrally uncorrelated, intrinsically pure states without the need for lossy narrowband spectral filtering. The wide variety of two-photon states that can be generated through four-wave mixing makes fiber-based photon sources suitable for many applications including quantum communications, quantum metrology, linear optical quantum computation schemes, and the generation of large entangled cluster states.

As with other photon sources based on the probabilistic generation of photon pairs, the main disadvantage with fiber-based photon sources is the stochastic nature of the pair generation process. This will ultimately limit the scalability of quantum information experiments until methods to efficiently multiplex sources, and to detect and correct for multiple photon-pair generation events, can be further developed. In the short term, moving to low loss, all-fiber schemes should lead to improvements in the brightness of these sources, making quantum information experiments involving higher numbers of photons feasible.

REFERENCES

[1] J.G. Rarity, J. Fulconis, J. Duligall, W.J. Wadsworth, and P.St.J. Russell, "Photonic Crystal Fiber Source of Correlated Photon Pairs," Opt. Express 13, 534–544 (2005).
[2] M. Fiorentino, P.L. Voss, J.E. Sharping, and P. Kumar, "All-Fiber Photon-Pair Source for Quantum Communications," IEEE Photonics Technol. Lett. 14, 983–985 (2002).
[3] J. Fulconis, O. Alibart, W.J. Wadsworth, P.St.J. Russell, and J.G. Rarity, "High Brightness Single Mode Source of Correlated Photon Pairs Using a Photonic Crystal Fiber," Opt. Express 13, 7572–7582 (2005).
[4] A. Ling, J. Chen, J. Fan, and A. Migdall, "Mode Expansion and Bragg Filtering for a High-Fidelity Fiber-Based Photon-Pair Source," Opt. Express 17, 21302–21312 (2009).
[5] A.R. McMillan, J. Fulconis, M. Halder, C. Xiong, J.G. Rarity, and W.J. Wadsworth, "Narrowband High-Fidelity All-Fibre Source of Heralded Single Photons at 1570 nm," Opt. Express 17, 6156–6165 (2009).
[6] C. Söller, B. Brecht, P.J. Mosley, L.Y. Zang, A. Podlipensky, N.Y. Joly, P.St.J. Russell, and C. Silberhorn, "Bridging Visible and Telecom Wavelengths with a Single-Mode Broadband Photon Pair Source," Phys. Rev. A 81, 031801 (2010).
[7] J. Fan, A. Migdall, and L. Wang, "Efficient Generation of Correlated Photon Pairs in a Microstructured Fiber," Opt. Lett. 30, 3368–3370 (2005).
[8] K. Garay-Palmett, H.J. McGuinness, O. Cohen, J.S. Lundeen, R. Rangel-Rojo, A.B. U'Ren, M.G. Raymer, C.J. McKinstrie, S. Radic, and I.A. Walmsley, "Photon Pair-State Preparation with Tailored Spectral Properties by Spontaneous Four-Wave Mixing in Photonic-Crystal Fiber," Opt. Express 15, 14870 (2007).
[9] B. J. Smith, P. Mahou, O. Cohen, J.S. Lundeen, and I.A. Walmsley, "Photon Pair Generation in Birefringent Optical Fibers," Opt. Express 17, 23589–23602 (2009).
[10] M. Halder, J. Fulconis, B. Cemlyn, A. Clark, C. Xiong, W.J. Wadsworth, and J.G. Rarity, "Nonclassical 2-photon Interference with Separate Intrinsically Narrowband Fibre Sources," Opt. Express 17, 4670 (2009).
[11] C. Söller, O. Cohen, B.J. Smith, I.A. Walmsley, and C. Silberhorn, "High-Performance Single-Photon Generation with Commercial-Grade Optical Fiber," Phys. Rev. A 83, 031806 (2011).
[12] M. Patel, J.B. Altepeter, Y.-P. Huang, N.N. Oza, and P. Kumar, "Erasing Quantum Distinguishability Via Single-Mode Filtering," Phys. Rev. A 86, 033809 (2012).

[13] A.S. Clark, J. Fulconis, J.G. Rarity, W.J. Wadsworth, and J.L. O'Brien, "All-Optical-Fiber Polarization-Based Quantum Logic Gate," Phys. Rev. A 79, 030303 (2009).

[14] J. Fulconis, O. Alibart, J.L. O'Brien, W.J. Wadsworth, and J.G. Rarity, "Nonclassical Interference and Entanglement Generation Using a Photonic Crystal Fiber Pair Photon Source," Phys. Rev. Lett. 99, 120501 (2007).

[15] C. Liang, K.F. Lee, T. Levin, J. Chen, and P. Kumar, "Ultra Stable All-Fiber Telecom-Band Entangled Photon-Pair Source for Turnkey Quantum Communication Applications," Opt. Express, 14, 6936–6941 (2006).

[16] J. Chen, K.F. Lee, X. Li, P.L Voss, and P. Kumar, "Schemes for Fibre-Based Entanglement Generation in the Telecom Band," New J. Phys. 9, 289 (2007).

[17] B. Bell, A.S. Clark, M.S. Tame, M. Halder, J. Fulconis, W.J. Wadsworth, and J.G. Rarity, "Experimental Characterization of Photonic Fusion Using Fiber Sources," New J. Phys. 14, 023021 (2012).

[18] M.A. Nielsen, "Optical Quantum Computation Using Cluster States," Phys. Rev. Lett. 93, 040503 (2004).

[19] Dawn Hollenbeck, Cyrus D. Cantrell, "Multiple-vibrational-mode model for fiber-optic Raman gain spectrum and response function," J. Opt. Soc. Am. B 19, 2886 (2002).

[20] W. Wadsworth, N. Joly, J. Knight, T. Birks, F. Biancalana, and P. Russell, "Supercontinuum and Four-Wave Mixing with q-switched Pulses in Endlessly Single-Mode Photonic Crystal Fibres," Opt. Express 12, 299–309 (2004).

[21] B. Ainslie and C. Day, "A Review of Single-Mode Fibers with Modified Dispersion Characteristics," J. Lightwave Technol. 4, 967–979 (1986).

[22] O. Alibart, J. Fulconis, G.K.L. Wong, S.G. Murdoch, W.J. Wadsworth, and J.G. Rarity, "Photon Pair Generation Using Four-Wave Mixing in a Microstructured Fibre: Theory Versus Experiment," New J. Phys. 8, 67 (2006).

[23] J. Chen, X. Li, and P. Kumar, "Two-Photon-State Generation Via Four-Wave Mixing in Optical Fibers," Phys. Rev. A 72, 033801 (2005).

[24] Q. Lin, F. Yaman, and G.P. Agrawal, "Photon-Pair Generation in Optical Fibers Through Four-Wave Mixing: Role of Raman Scattering and Pump Polarization," Phys. Rev. A 75, 023803 (2007).

[25] R.W. Boyd, "Nonlinear Optics", 3rd ed., Academic Press (2008).

[26] C. Xiong and W.J. Wadsworth, "Polarized Supercontinuum in Birefringent Photonic Crystal Fibre Pumped at 1064 nm and Application to Tuneable Visible/UV Generation," Opt. Express 16, 2438–2445 (2008).

[27] A. Ortigosa-Blanch, A. Diez, M. Delgado-Pinar, J.L. Cruz, and M.V. Andres, "Ultrahigh Birefringent Nonlinear Microstructured Fiber," IEEE Photon. Tech. Lett. 16, 1667–1669 (2004).

[28] X. Li, P. Voss, J. Chen, K. Lee, and P. Kumar, "Measurement of Co- and Cross-Polarized Raman Spectra In Silica Fiber for Small Detunings," Opt. Express, 13, 2236–2244 (2005).

[29] H. Takesue and K. Inoue, "1.5 μm Band Quantum-Correlated Photon Pair Generation in Dispersion-Shifted Fiber: Suppression of Noise Photons by Cooling Fiber," Opt. Express, 13, 7832–7839 (2005).

[30] Y.-P. Huang and P. Kumar, "Distilling Quantum Entanglement Via Mode-Matched Filtering," Phys. Rev. A 84, 032315 (2011).

[31] C.K. Hong and L. Mandel, "Experimental Realization of a Localized One-Photon State," Phys. Rev. Lett. 56, 58–60 (1986).

[32] A. McMillan, Development of an All-Fibre Source of Heralded Single Photons, PhD Thesis, University of Bath (2011).

[33] X. Li, J. Chen, P. Voss, J. Sharping, and P. Kumar, "All-Fiber Photon-Pair Source for Quantum Communications: Improved Generation of Correlated Photons," Phys. Rev. Lett. 12, 3737–3744 (2004).

[34] J.E. Sharping, M. Fiorentino, A. Coker, P. Kumar, and R.S. Windeler, "Four-Wave Mixing in Microstructure Fiber," Opt. Lett. 26, 1048–1050 (2001).

[35] J.E. Sharping, M. Fiorentino, P. Kumar, and R.S. Windeler, "Optical Parametric Oscillator Based on Four-Wave Mixing in Microstructure Fiber," Opt. Lett. 27, 1675–1677 (2002).

[36] J.E. Sharping, J. Chen, X. Li, and P. kumar, "Quantum Correlated Twin Photons from Microstructure Fiber," Opt. Express 12, 3086–3094 (2004).

[37] J. Fan, A. Dogariu, and L.J. Wang, "Generation of Correlated Photon Pairs in a Microstructure Fiber," Opt. Lett. 30, 1530–1532 (2005).
[38] J. Fan and A. Migdall, "A Broadband High Spectral Brightness Fiber-Based Two-Photon Source," Opt. Express 15, 2915–2920 (2007).
[39] E.A. Goldschmidt, M.D. Eisaman, J. Fan, S.V. Polyakov, and A. Migdall, "Spectrally Bright and Broad Fiber-Based Heralded Single-Photon Source," Phys. Rev. A 78, 013844 (2008).
[40] J.A. Slater, J.-S. Corbeil, S. Virally, F. Bussières, A. Kudlinski, G. Bouwmans, S. Lacroix, N. Godbout, and W. Tittel, "Microstructured Fiber Source of Photon Pairs at Widely Separated Wavelengths," Opt. Lett. 35, 499–501 (2010).
[41] N. Sangouard, C. Simon, H. de Riedmatten, and N. Gisin, "Quantum Repeaters Based on Atomic Ensembles and Linear Optics," Rev. Mod. Phys. 83, 33–80 (2011).
[42] L. Xiao, M.S. Demokan, W. Jin, Y. Wang, and C.-L. Zhao, "Fusion Splicing Photonic Crystal Fibers and Conventional Single-Mode Fibers: Microhole Collapse Effect," J. Lightwave Technol. 25, 3563–3574 (2007).
[43] T.A. Birks, J.C. Knight, and P.St.J. Russell, "Endlessly Single-Mode Photonic Crystal Fiber," Opt. Lett. 22, 961–963 (1997).
[44] C.K. Hong, Z.Y. Ou, and L. Mandel, "Measurement of Subpicosecond Time Intervals Between Two Photons by Interference," Phys. Rev. Lett. 59, 2044–2046 (1987).
[45] Z.Y.J. Ou, "Multi-Photon Quantum Interference," Springer (2007).
[46] J. Fulconis, O. Alibart, W.J. Wadsworth, and J.G. Rarity, "Quantum Interference with Photon Pairs Using Two Micro-Structured Fibres," New J. Phys. 9, 276 (2007).
[47] A.B. U'Ren, C. Silberhorn, R. Erdmann, K. Banaszek, W.P. Grice, I.A. Walmsley, and M.G. Raymer, "Generation of Pure-State Single-Photon Wavepackets by Conditional Preparation Based on Spontaneous Parametric Downconversion," Laser Phys. 15, 146 (2005).
[48] W.P. Grice, A.B. U'ren, and I.A. Walmsley, "Eliminating Frequency and Space-Time Correlations in Multiphoton States," Phys. Rev. A 64, 063815 (2001).
[49] O. Cohen, J.S. Lundeen, B.J. Smith, G. Puentes, P.J. Mosley, and I.A. Walmsley, "Tailored Photon-Pair Generation in Optical Fibers," Phys. Rev. Lett. 102, 123603 (2009).
[50] A. Clark, B. Bell, J. Fulconis, M.M. Halder, B. Cemlyn, O. Alibart, C. Xiong, W.J. Wadsworth, and J.G. Rarity, "Intrinsically Narrowband Pair Photon Generation in Microstructured Fibres," New J. Phys. 13, 065009 (2011).
[51] A.R. McMillan, M. Delgado-Pinar, J.G. Rarity, and W.J. Wadsworth, "Generation of Narrowband 1550 nm Photons in the Anomalous Dispersion Region of a Birefringent PCF," in Proceedings of the International Quantum Electronics Conference and Conference on Lasers and Electro-Optics Pacific Rim 2011, Optical Society of America, p. I591 (2011).
[52] K. Garay-Palmett, H.J. McGuinness, O. Cohen, J.S. Lundeen, R. Rangel-Rojo, A.B. U'ren, M.G. Raymer, C.J. McKinstrie, S. Radic, and I.A. Walmsley, "Photon Pair-State Preparation with Tailored Spectral Properties by Spontaneous Four-Wave Mixing in Photonic-Crystal Fiber," Opt. Express 15, 14870–14886 (2007).
[53] Z.H. Levine, J. Fan, J. Chen, A. Ling, and A. Migdall, "Heralded, Pure-State Single-Photon Source Based on a Potassium Titanyl Phosphate Waveguide," Opt. Express 18, 3708–3718 (2010).
[54] J.S. Neergaard-Nielsen, B.M. Nielsen, H. Takahashi, A.I. Vistnes, and E.S. Polzik, "High Purity Bright Single Photon Source," Opt. Express 15, 7940–7949 (2007).
[55] P.J. Mosley, J.S. Lundeen, B.J. Smith, P. Wasylczyk, A.B. U'Ren, C. Silberhorn, and I.A. Walmsley, "Heralded Generation of Ultrafast Single Photons in Pure Quantum States," Phys. Rev. Lett. 100, 133601 (2008).
[56] O. Cohen, J.S. Lundeen, B.J. Smith, G. Puentes, P.J. Mosley, and I.A. Walmsley, "Tailored Photon-Pair Generation in Optical Fibers," Phys. Rev. Lett. 102, 123603 (2009).
[57] C. Söller, B. Brecht, P.J. Mosley, L.Y. Zang, A. Podlipensky, N.Y. Joly, P.St.J. Russell, and C. Silberhorn, "Bridging Visible and Telecom Wavelengths with a Single-Mode Broadband Photon Pair Source," Phys. Rev. A 81, 031801 (2010).
[58] Y.-P. Huang, J.B. Altepeter, and P. Kumar, "Heralding Single Photons Without Spectral Factorability," Phys. Rev. A 82, 043826 (2010).
[59] D. Splepian and H.O. Pollak, Bell Syst. Tech. J. 40, 43 (1961).
[60] C. Zhu and C.M. Caves, "Photocount Distributions for Continuous-Wave Squeezed Light," Phys. Rev. A 42, 6794–6804 (1990).

[61] M. Sasaki and S. Suzuki, "Multimode Theory of Measurement-Induced Non-gaussian Operation on Wideband Squeezed Light: Analytical Formula," Phys. Rev. A 73, 043807 (2006).
[62] Y.-P. Huang, J.B. Altepeter, and P. Kumar, "Optimized Heralding Schemes for Single Photons," Phys. Rev. A 84, 033844 (2011).
[63] A. Aspect, P. Grangier, and G. Roger, "Experimental Tests of Realistic Local Theories Via Bell's Theorem," Phys. Rev. Lett. 47, 460–463 (1981).
[64] N. Gisin, G. Ribordy, W. Tittel, and H. Zbinden, "Quantum Cryptography," Rev. Mod. Phys. 74, 145 (2002).
[65] N. Gisin and R. Thew, "Quantum Communication," Nat. Photonics 1, 165–171 (2007).
[66] V. Scarani, H. Bechmann-Pasquinucci, N.J. Cerf, M. Dušek, N. Lütkenhaus, and M. Peev, "The Security of Practical Quantum Key Distribution," Rev. Mod. Phys. 81, 1301–1350 (2009).
[67] H.-J. Briegel, W. Dür, J.I. Cirac, and P. Zoller, "Quantum Repeaters: The Role of Imperfect Local Operations in Quantum Communication," Phys. Rev. Lett. 81, 5932–5935 (1998).
[68] M.P. Hedges, J.J. Longdell, Y. Li, and M.J. Sellars, "Efficient Quantum Memory for Light," Nature 465, 1052–1056 (2010).
[69] E. Knill, R. Laflamme, and G.J. Milburn, "A Scheme for Efficient Quantum Computation with Linear Optics," Nature 409, 46–52 (2001).
[70] J.L. O'Brien, "Optical Quantum Computing," Science 318, 1567–1570 (2007).
[71] D.E. Browne and T. Rudolph, "Resource-Efficient Linear Optical Quantum Computation," Phys. Rev. Lett. 95, 010501 (2005).
[72] J.S. Bell, "On the Einstein-Podolsky-Rosen Paradox," Physics 1, 195–200 (1964).
[73] J.F. Clauser, M.A. Horne, A. Shimony, and R.A. Holt, "Proposed Experiment to Test Local Hidden Variable Theories," Phys. Rev. Lett. 23, 880 (1969).
[74] D.F.V. James, P.G. Kwiat, W.J. Munro, and A.G. White, "Measurement of Qubits," Phys. Rev. A 64, 052312 (2001).
[75] X. Li, P.L. Voss, J.E. Sharping, and P. Kumar, "Optical-Fiber Source of Polarization-Entangled Photons in the 1550 nm Telecom Band," Phys. Rev. Lett. 94, 053601 (2005).
[76] X. Li, P. Voss, J. Chen, K. Lee, and P. Kumar, "Measurement of Co- and Cross-Polarized Raman Spectra in Silica Fiber for Small Detunings," Opt. Express 13, 2236–2244 (2005).
[77] D.B. Mortimore, "Fiber Loop Reflectors," J. Lightwave Technol. 6, 1217–1224 (1988).
[78] M. Fiorentino, P. Voss, J.E. Sharping, and P. Kumar, "All-Fiber Photon-Pair Source for Quantum Communications," IEEE Photonic. Technol. Lett. 14, 983–985 (2002).
[79] M.A. Hall, J.B. Altepeter, and P. Kumar, "Drop-in Compatible Entanglement for Optical-Fiber Networks," Opt. Express 17, 14558–14566 (2009).
[80] M. Medic, J.B. Altepeter, M.A. Hall, M. Patel, and P. Kumar, "Fiber-Based Telecommunication-Band Source of Degenerate Entangled Photons," Opt. Lett. 35, 802–804 (2010).
[81] J. Chen, K.F. Lee, and P. Kumar, "Deterministic Quantum Splitter Based on Time-Reversed Hong-ou-Mandel Interference," Phys. Rev. A 76, 031804 (2007).
[82] M.A. Nielsen and I.L. Chuang, "Quantum Computation and Quantum Information," Cambridge University Press, Cambridge, p. 91 (2000).
[83] M. Fox, "Quantum Optics: An Introduction," Oxford University Press (2006).
[84] J.L. O'Brien, G.J. Pryde, A.G. White, T.C. Ralph, and D. Branning, "Demonstration of an All-Optical Quantum Controlled-Not Gate," Nature 426, 264–267 (2003).
[85] J.L. O'Brien, G.J. Pryde, A. Gilchrist, D.F.V. James, N.K. Langford, T.C. Ralph, and A.G. White, "Quantum Process Tomography of a Controlled-Not Gate," Phys. Rev. Lett. 93, 080502 (2004).
[86] A.S. Clark, Quantum Information Processing in Optical Fibres, University of Bristol PhD Thesis (2011).
[87] H.F. Hofmann, "Complementary Classical Fidelities as an Efficient Criterion for the Evaluation of Experimentally Realized Quantum Operations," Phys. Rev. Lett. 94, 160504 (2005).
[88] R. Okamoto, H.F. Hofmann, S. Takeuchi, and K. Sasaki, "Demonstration of an Optical Quantum Controlled-Not Gate Without Path Interference," Phys. Rev. Lett. 95, 210506 (2005).

[89] C.H. Bennett, G. Brassard, C. Crépeau, R. Jozsa, A. Peres, and W.K. Wootters, "Teleporting an Unknown Quantum State Via Dual Classical and Einstein-Podolsky-Rosen Channels," Phys. Rev. Lett. 70, 1895–1899 (1993).
[90] A.K. Ekert, "Quantum Cryptography Based on Bell's Theorem," Phys. Rev. Lett. 67, 661–663, (1991).
[91] R. Raussendorf and H.J. Briegel, "A One-Way Quantum Computer," Phys. Rev. Lett. 86, 5188–5191 (2001).
[92] U. Leonhardt, "Quantum-State Tomography and Discrete Wigner Function," Phys. Rev. Lett. 74, 4101–4105 (1995).
[93] D.F.V. James, P.G. Kwiat, W.J. Munro, and A.G. White, "Measurement of Qubits," Phys. Rev. A 64, 052312 (2001).

Chapter 13

Single Emitters in Isolated Quantum Systems

Glenn S. Solomon[*], Charles Santori[†], and Axel Kuhn[‡]

[*]*Joint Quantum Institute, National Institute of Standards and Technology & University of Maryland, Gaithersburg, MD, USA*
[†]*Hewlett-Packard Laboratories, 1501 Page Mill Rd., Palo Alto, CA, 94304, USA*
[‡]*University of Oxford, Clarendon Laboratory, Parks Road, OX1 3PU, United Kingdom*

Chapter Outline

13.1	**Introduction**	468
13.2	**Single Photons from Atoms and Ions - A. Kuhn**	468
	13.2.1 Emission into Free Space	469
	13.2.2 Cavity-Based Single-Photon Emitters	471
	13.2.3 Photon Coherence, Amplitude, and Phase Control	485
13.3	**Single Photons from Semiconductor Quantum Dots - G. S. Solomon**	492
	13.3.1 Introduction	492
	13.3.2 InAs-Based Quantum-Dot Formation	493
	13.3.3 Exciton Energetics	494
	13.3.4 Optically Accessing Single Quantum Dots	497
	13.3.5 Single Photons From Single Quantum Dots	499
	13.3.6 Weak QD-Cavity Coupling	502
	13.3.7 Quantum-Dot Photon Indistinguishability	505
13.4	**Single Defects in Diamond - C. Santori**	511
	13.4.1 Introduction	511
	13.4.2 The Nitrogen-Vacancy Center	511
	13.4.3 Other Defects	521
	13.4.4 Optical Structures in Diamond	522
	13.4.5 Quantum Communication	525
	13.4.6 Summary	526
13.5	**Future Directions**	526
	References	527

13.1 INTRODUCTION

The interfacing of discrete matter states and photons, the storage and retrieval of single photons, and the mapping of quantum states between distant entities constitute essential building blocks of future quantum communication networks and quantum information processors [1]. Ideally, such systems are composed of individual nodes acting as quantum gates or memories, with optical links between them that allow for the entanglement or teleportation of their quantum states, or for optical quantum information processing using light traveling between the nodes [2]. With individual photons acting as messengers in these networks, substantial efforts are being undertaken that focus on the production and characterization of single photons. Applications which rely on the availability of single photons include quantum cryptography, optical quantum computing, light-matter entanglement, and atom-photon state mapping. All of these have been successfully demonstrated.

For most of these applications, sources of single photons based on discrete, isolated quantum system are ideally suited, given their capability of emitting streams of indistinguishable photons on demand. Examples of a discrete quantum system are a single atom or molecule, or an engineered system like a quantum dot or a color center in a solid-state matrix. These systems are inherently simple and robust, as they can only emit one single photon in a de-excitation process. Here, we discuss both atomic and solid-state sources of single photons on demand. We examine two atomic sources, single atoms and single ions; and, two solid-state sources, strain-induced epitaxial quantum dots and nitrogen-vacancy centers in diamonds. Important fundamental properties of on-demand single-photon sources are analyzed for these four cases, including single-photon purity and indistinguishability. The interactions of these matter states with optical cavities is important for our fundamental understanding as well as the development of improved single-photon emitters. These interactions are examined in the context of cavity-quantum electrodynamic effects.

13.2 SINGLE PHOTONS FROM ATOMS AND IONS - A. KUHN

A large number of atomic species provide simple electronic level structures. Therefore they can be excited in a way that one atom emits exactly one single photon of well-defined frequency and polarization upon either spontaneous or stimulated emission. With all atoms or ions of the same isotope being identical, different photon sources based on the same species can produce indistinguishable photons without further measures, provided the same transitions are used and the electromagnetic environment is identical for all atoms. This makes them ideal candidates for the implementation of large-scale quantum computing networks. However, harnessing their emission into one single optical mode, or holding a single atom or ion trapped in place, are both challenging tasks that are not easy to meet. Here, we first discuss the emission of single photons from freely moving or trapped atoms or ions into free space and then highlight a couple of major achievements, such as the entanglement of remote entities. Second, we introduce the basic principles of cavity-quantum electrodynamics (CQED), and show how to apply these to channel the photon emission into a single mode of the radiation field, with the vacuum field inside the cavity stimulating the process. Third, we elucidate how to determine and control the coherence properties of these photons in the time domain and use that degree of control for information encoding.

13.2.1 Emission into Free Space

In 1977, Kimble *et al.* [3] have been investigating the photon statistics of light emitted from single atoms. Figure 13.1a and b illustrates their setup along with their results. Sodium atoms in an atomic beam are excited on the $2^2 S_{1/2}(F=2, m_F=2) \longrightarrow 3^2 P_{3/2}(F=3, m_F=3)$ transition and behave like an effective two-level system. The fluorescence from the excitation region is collected with a lens of high numerical aperture and then directed to a pair of photon counters in a Hanbury Brown-Twiss arrangement (c.f. Chapter 2) to measure the $g^{(2)}(\tau)$ intensity correlation of the fluorescence light. A minimum around zero detection-time delay is evident. This can be easily understood as the atom is in its ground state immediately after a photon emission has taken place. Hence the probability for emitting another photon drops instantly thereafter to zero and then slowly rises again to the average photon-emission probability. For an ideal single-photon emitter, one would expect to find $g^{(2)}(0) = 0$. However, because of the random distribution of atoms within the atomic beam, a somewhat larger residual signal was found in this experiment.

The situation is different if one analyzes the light emerging from a single trapped emitter, like a single ion in a Paul trap [4]. In this case, a deeper anti-bunching dip where $g^{(2)}(0) = 0$ is observed, see Fig. 13.1c. If the exciting laser is sufficiently intense, the ion is furthermore subject to Rabi oscillations which give rise to the noticeable modulation of the correlation function.

The above experiments [3,4] have in common that an effective two-level system, i.e. atom or ion, is continuously driven. By consequence, one can neither control the exact emission time, nor the number of successive photon emissions, nor the exact time span between those. A straightforward remedy resolving these issues is the application of a pulsed excitation scheme in a three-level atom. For instance, with a laser π-pulse exciting $|e\rangle \longrightarrow |x\rangle$ and the emission taking place upon a transition from $|x\rangle$ to $|g\rangle$, the timing issue may be resolved. Also, no further emissions from the same atom can happen until it is actively brought back into its initial state $|e\rangle$. Monroe [5] and Weinfurter [6] successfully applied such schemes to entangle the polarization of a single photon with the spin of a single ion or atom, respectively. To do so, they realised an excitation scheme leading to two possible final spin states of the atom, $|g_\downarrow\rangle$ and $|g_\uparrow\rangle$ upon emission of either a σ^+ or σ^- polarized photon, thus projecting the whole system into

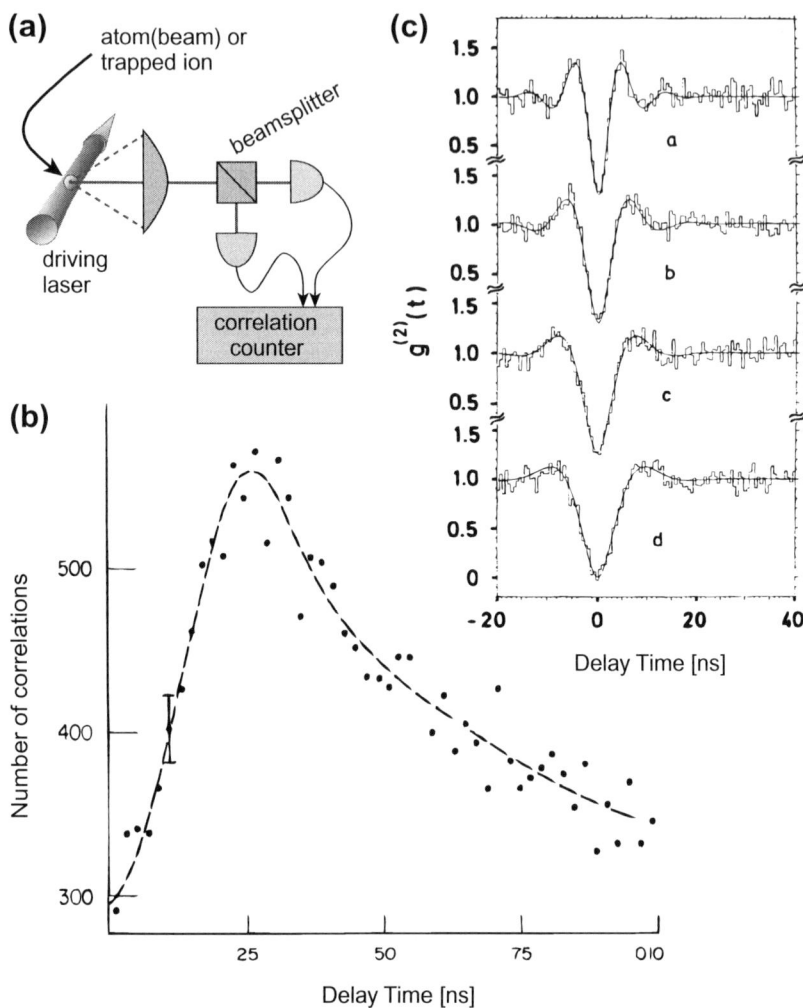

FIGURE 13.1 Intensity correlation function $g^{(2)}(\tau)$ measured using a Hanbury Brown-Twiss arrangement for two-photon coincidence counting (a). Such an arrangement has first been used by Kimble *et al.* [3] to analyse the fluorescence of Sodium atoms in an atomic beam (b), *panel adapted from [3], Copyright (1977) by The American Physical Society.* Thereafter, Diedrich and Walther [4] have been using a similar arrangement to investigate the non-classical photon statistics in the fluorescence of a single trapped ion (c), *panel adapted from [4], Copyright (1987) by The American Physical Society.*

the entangled atom-photon state

$$|\psi^+\rangle = (|g_\downarrow, +\rangle + |g_\uparrow, -\rangle)/\sqrt{2}. \tag{13.1}$$

Projective measurements on pairs of photons emitted from two distant atoms or ions have then been used for entanglement swapping, thus resulting in the entanglement and teleportation of quantum states [7]. Such photon-matter entanglement has a potential advantage of adding memory capabilities to quantum information protocols. In addition, this provides a quantum matter light interface, thereby using different physical media for different purposes in a quantum information application.

This success story impressively demonstrates the inherent potential of using identical atoms or ions as single-photon emitters. The emission spectra are all identical, and coupling different emitters is relatively easy as compared to, e.g. solid-state sources, which need to be tuned individually. However, the spontaneous emission of photons into all directions is an inherent limitation. Even the best collection optics captures at most 25% of the photons [8], with actual experiments reaching overall photon-detection efficiencies of $\approx 5 \times 10^{-4}$ [6]. Combined with the spontaneous character of the emission, these sources are rendered probabilistic and scaling is a serious issue. We shall see in the following that the coupling of atoms or ions to optical cavities is one effective solution to this problem. First, the direction of emission is imposed, with all photons emitted into a single cavity mode, and second, the photon emission is stimulated by the vacuum field of the cavity, with the driving laser allowing for an unprecedented control of the emission process.

13.2.2 Cavity-Based Single-Photon Emitters

In the following, we closely follow, summarize and extend our review published in Contemporary Physics [9] to introduce the concepts, characteristic properties, and major implementations of state-of-the-art single-photon sources based on single atoms or ions in cavities. These have all the potential to meet the requirements of optical quantum computing and quantum networking schemes, namely deterministic single-photon emission with unit efficiency, directed emission into a single spatial mode of the radiation field, indistinguishable photons with immaculate temporal and spatial coherence, and reversible quantum state mapping and entanglement between atoms and photons.

Starting from the elementary principles of cavity-quantum electrodynamics, we discuss how a single quantum system couples to the quantised field within optical resonators. We then show how to exploit these effects to generate single photons on demand in the strong-coupling regime and the bad cavity limit, using either an adiabatic driving technique or a sudden excitation of the emitter. To conclude, we discuss the most prominent experimental achievements and examine the different approaches for obtaining single photons from cavities using either atoms or ions as photon emitters.

13.2.2.1 Atom-Photon Interaction in Resonators

Here we discuss how a single quantum system, which shows discrete energy levels like an individual atom or ion, couples to the quantized modes of the radiation field in a cavity. We introduce the relevant features of cavity-QED and the Jaynes-Cummings model [10,11], and extend these to three-level atoms with two dipole transitions driven by two radiation fields. One of the fields is from a laser, the other is the cavity field coupled to the *atom* (which stands in here for any quantum system showing discrete energy levels). We furthermore explain how the behavior of a coupled atom-cavity system depends on the most relevant cavity parameters, such as the cavity's mode volume and its finesse.

Field quantization in cavity QED: We consider a Fabry-Perot cavity with mirror separation l and reflectivity $\mathcal{R}$. The cavity has a free spectral range $\Delta \omega_{FSR} = 2\pi \times c/(2l)$, and its finesse is defined as $\mathcal{F} = \pi \sqrt{\mathcal{R}}/(1 - \mathcal{R})$. In the vicinity of a resonance, the transmission profile is Lorentzian with a linewidth (FWHM) of $2\kappa = \Delta \omega_{FSR}/\mathcal{F}$, which is twice the decay rate, κ, of the cavity field. Curved mirrors are often used to restrict the cavity eigenmodes to geometrically stable Laguerre-Gaussian or Hermite-Gaussian modes. In most cases, just one of these modes is of interest, characterized by its mode function $\psi_{cav}(\mathbf{r})$ and its resonance frequency ω_{cav}. The state vector can therefore be expressed as a superposition of photon-number states, $|n\rangle$, and for n photons in the mode the energy reads $\hbar \omega_{cav}(n + \frac{1}{2})$. The equal energy spacing allows for an analog treatment of the cavity as an harmonic oscillator. Creation and annihilation operators for a photon, $\hat{a}^\dagger$ and $\hat{a}$, are then used to express the Hamiltonian of the cavity,

$$H_{cav} = \hbar \omega_{cav} \left(\hat{a}^\dagger \hat{a} + \frac{1}{2} \right). \quad (13.2)$$

This Hamiltonian does not account for any losses. In a real cavity, all photon-number states decay until thermal equilibrium with the environment is reached. In the optical domain, the latter corresponds to the vacuum state, $|0\rangle$, with no photons remaining in the cavity.

Two-level atom: We now analyse how the cavity field interacts with a two-level atom with ground state $|g\rangle$ and excited state $|x\rangle$ of energies $\hbar \omega_g$ and $\hbar \omega_x$, respectively, and transition dipole moment μ_{xg}. The Hamiltonian of the atom is

$$H_A = \hbar \omega_g |g\rangle\langle g| + \hbar \omega_x |x\rangle\langle x|. \quad (13.3)$$

The coupling to the field mode of the cavity is expressed by the atom-cavity coupling constant,

$$g(\mathbf{r}) = g_0 \, \psi_{cav}(\mathbf{r}), \quad \text{with } g_0 = \sqrt{(\mu_{xg}^2 \omega_{cav})/(2\hbar \epsilon_0 V)}, \quad (13.4)$$

where V is the mode volume of the cavity. As the atom barely moves during the interaction, we can safely disregard its external degrees of freedom. Furthermore we assume maximum coupling, i.e. $\psi_{cav}(\mathbf{r}_{atom}) = 1$, so that one obtains $g(\mathbf{r}) = g_0$. In a closed system, any change of the atomic state goes hand-in-hand with a corresponding change of the photon number, n. Hence the interaction Hamiltonian of the atom-cavity system reads

$$H_{int} = -\hbar g_0 \left[|x\rangle\langle g| \hat{a} + \hat{a}^\dagger |g\rangle\langle x| \right]. \quad (13.5)$$

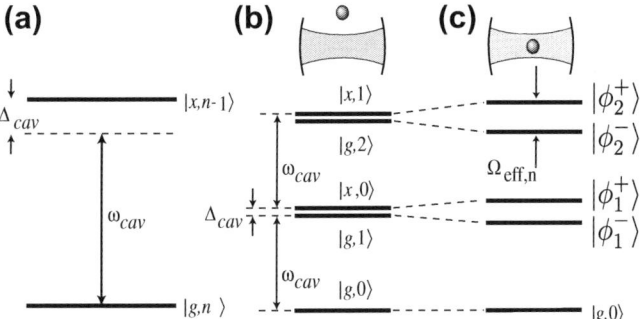

FIGURE 13.2 (a) A two-level atom with ground state $|g\rangle$ and excited state $|x\rangle$ coupled to a cavity containing n photons. In the dressed-level scheme of the combined atom-cavity system with the atom outside (b) or inside (c) the cavity, the state doublets are either split by Δ_{cav} or by the effective Rabi frequency, $\Omega_{\text{eff},n}$, respectively.

For a given excitation number n, the cavity only couples $|g,n\rangle$ and $|x,n-1\rangle$. If the cavity mode is resonant with the atomic transition, the population then oscillates with the Rabi frequency $\Omega_{\text{cav}} = 2g_0\sqrt{n}$ between these states.

The eigenfrequencies of the total Hamiltonian, $H = H_{\text{cav}} + H_A + H_{\text{int}}$, can be found easily. In the rotating wave approximation, they read

$$\omega_n^{\pm} = \omega_{\text{cav}}\left(n + \frac{1}{2}\right) + \frac{1}{2}\left(\Delta_{\text{cav}} \pm \sqrt{4ng_0^2 + \Delta_{\text{cav}}^2}\right), \qquad (13.6)$$

where $\Delta_{\text{cav}} = \omega_x - \omega_g - \omega_{\text{cav}}$ is the detuning between atom and cavity. Figure 13.2 illustrates this level splitting. Two corresponding eigenstates get split by $\Omega_{\text{eff},n} = \sqrt{4ng_0^2 + \Delta_{\text{cav}}^2}$, which is the effective Rabi frequency at which the population oscillates between states $|g,n\rangle$ and $|x,n-1\rangle$. The cavity field stimulates the emission of an excited atom into the cavity, thus de-exciting the atom and increasing the photon number by one. Subsequently, the atom is re-excited by absorbing a photon from the cavity field, and so forth. In particular, an excited atom and a cavity containing no photon are sufficient to start the oscillation between $|x,0\rangle$ and $|g,1\rangle$ at frequency $\sqrt{4g_0^2 + \Delta_{\text{cav}}^2}$. This phenomenon is known as vacuum-Rabi oscillation. On resonance, i.e. for $\Delta_{\text{cav}} = 0$, the oscillation frequency $2g_0$ is therefore called vacuum-Rabi frequency. To summarize, the atom-cavity interaction splits the photon-number states into doublets of non-degenerate dressed states, which are named after Jaynes and Cummings [10,11]. Only the ground state $|g,0\rangle$ is not coupled to other states and is not subject to any energy shift or splitting.

Three-level atom: We now consider an atom with a Λ-type three-level scheme providing transition frequencies $\omega_{xe} = \omega_x - \omega_e$ and $\omega_{xg} = \omega_x - \omega_g$ as depicted in Fig. 13.3. The $|e\rangle \leftrightarrow |x\rangle$ transition is driven by a classical light field

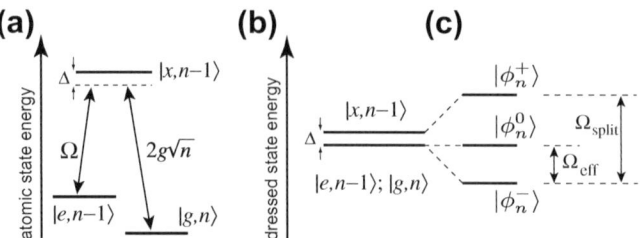

FIGURE 13.3 (a) A three-level atom driven by a classical laser field of Rabi frequency Ω, coupled to a cavity containing n photons. (b) Dressed-level scheme of the combined system without coupling, and (c) for an atom interacting with laser and cavity. The triplet is split by $\Omega_{\text{split}} = \sqrt{4ng_0^2 + \Omega^2 + \Delta^2}$. In the limit of a large detuning Δ, the Raman transition $|e, n-1\rangle \leftrightarrow |g, n\rangle$ is driven at the effective Rabi frequency $\Omega_{\text{eff}} = \frac{1}{2}(\Omega_{\text{split}} - |\Delta|) \approx (4ng_0^2 + \Omega^2)/|4\Delta|$.

of frequency ω_L with Rabi frequency Ω, while a cavity mode with frequency ω_{cav} couples to the $|g\rangle \leftrightarrow |x\rangle$ transition. The respective detunings are defined as $\Delta_L = \omega_{xe} - \omega_L$ and $\Delta_{\text{cav}} = \omega_{xg} - \omega_{\text{cav}}$. Provided the driving laser and the cavity only couple to their respective transitions, the interaction Hamiltonian

$$H_{\text{int}} = \hbar[\Delta_L |e\rangle\langle e| + \Delta_{\text{cav}} |g\rangle\langle g| - \frac{\Omega}{2}(|x\rangle\langle e| + |e\rangle\langle x|)$$
$$- g_0(|x\rangle\langle g|a + a^\dagger |g\rangle\langle x|)] \tag{13.7}$$

determines the behavior of the system. Given an arbitrary excitation number n, this Hamiltonian couples only the three states $|e, n-1\rangle, |x, n-1\rangle, |g, n\rangle$. For this triplet and a Raman-resonant interaction with $\Delta_L = \Delta_{\text{cav}} \equiv \Delta$, the eigenfrequencies of the coupled system read

$$\omega_n^0 = \omega_{\text{cav}}\left(n + \frac{1}{2}\right) \quad \text{and}$$
$$\omega_n^\pm = \omega_{\text{cav}}\left(n + \frac{1}{2}\right) + \frac{1}{2}\left(\Delta \pm \sqrt{4ng_0^2 + \Omega^2 + \Delta^2}\right). \tag{13.8}$$

The previously discussed Jaynes-Cummings doublets are now replaced by triplets,

$$|\phi_n^0\rangle = \cos\Theta |e, n-1\rangle - \sin\Theta |g, n\rangle,$$
$$|\phi_n^+\rangle = \cos\Phi \sin\Theta |e, n-1\rangle - \sin\Phi |x, n-1\rangle$$
$$\quad + \cos\Phi \cos\Theta |g, n\rangle, \tag{13.9}$$
$$|\phi_n^-\rangle = \sin\Phi \sin\Theta |e, n-1\rangle + \cos\Phi |x, n-1\rangle$$
$$\quad + \sin\Phi \cos\Theta |g, n\rangle,$$

where the mixing angles Θ and Φ are given by

$$\tan\Theta = \frac{\Omega}{2g_0\sqrt{n}}, \quad \tan\Phi = \frac{\sqrt{4ng_0^2 + \Omega^2}}{\sqrt{4ng_0^2 + \Omega^2 + \Delta^2} - \Delta}. \quad (13.10)$$

We note that the interaction with the light lifts the degeneracy of the eigenstates. However, $|\phi_n^0\rangle$ is neither subject to an energy shift, nor does the excited atomic state contribute to it. Therefore it's a "dark state" that does not decay by spontaneous emission.

In the limit of vanishing Ω, the states $|\phi_n^\pm\rangle$ correspond to the Jaynes-Cummings doublet and the third eigenstate, $|\phi_n^0\rangle$, coincides with $|e, n-1\rangle$. Also the eigenfrequency ω_n^0 is not affected by Ω or g_0. Therefore transitions between the dark states $|\phi_{n+1}^0\rangle$ and $|\phi_n^0\rangle$ are always in resonance with the cavity.

Cavity-coupling regimes: So far, we have been considering the interaction Hamiltonian and the associated eigenvalues and dressed eigenstates that one obtains whenever a two- or three-level atom is coupled to a cavity. We have been neglecting the atomic polarization decay rate, γ, and the field-decay rate of the cavity, κ (Note that the population decay rate of the atom is 2γ, and the photon loss rate from the cavity is 2κ). It is evident that both relaxation rates result in a damping of a possible vacuum-Rabi oscillation between states $|x, 0\rangle$ and $|g, 1\rangle$. Only in the regime of *strong atom-cavity coupling*, with $g_0 \gg \{\kappa, \gamma\}$, the damping is weak enough so that vacuum-Rabi oscillations can occur. The other extreme is the *bad-cavity regime*, with $\kappa \gg g_0^2/\kappa \gg \gamma$, which results in strong damping and quasi-stationary quantum states of the coupled system if it is continuously driven.

Two properties of the cavity can be used to distinguish between these regimes: First the strength of the atom-cavity coupling, $g_0 \propto 1/\sqrt{V}$ (dependant upon the mode volume of the cavity), and second the finesse $\mathcal{F} = \pi\sqrt{R}/(1-R)$ of the resonator, which depends on the mirror reflectivity R. The finesse gives the mean number of round trips in the cavity before a photon is lost by transmission through one of the cavity mirrors, and it is also identical to the ratio of free spectral range $\Delta\omega_{FSR}$ to cavity linewidth 2κ. To reach strong coupling, a high value of g_0 and therefore a short cavity of small mode volume are normally required. Keeping κ small enough at the same time then calls for a high finesse and a mirror reflectivity $R \geq 99.999\%$.

13.2.2.2 Single-Photon Emission

For the deterministic generation of single photons from atom-cavity systems, all schemes implemented to date rely on the Purcell effect [12]. The spatial mode density inside the cavity is altered substantially, such that the spontaneous emission rate can be either enhanced ($f > 1$) or inhibited ($f < 1$) by the Purcell

factor
$$f = \frac{3Q\lambda^3}{4\pi^2 V},$$
depending on the cavity's mode volume, V, and quality factor, Q. More importantly, the probability of spontaneous emission placing a photon into the cavity is given by $\beta = f/(f+1)$. If the mode volume of the cavity is sufficiently small, the emitter and cavity couple so strongly that $\beta \approx 1$, i.e. emissions into the cavity outweigh spontaneous emissions into free space. A deterministic photon emission into a single field mode is therefore possible with an efficiency close to unity. These effects have first been observed by Carmichael et al. [13] and De Martini et al. [14]. Moreover, with the coherence properties uniquely determined by the parameters of the cavity and the driving process, one should be able to obtain indistinguishable photons from different cavities. Note also that state mapping and entanglement between atomic spin and photon polarization has recently been demonstrated in cavity-based single-photon emitters [15–18]. Additionally the reversibility of the photon generation process, and quantum networking between different cavities has been predicted [19–21], and demonstrated [22–24]. We now introduce different ways of producing single photons from such a system. These include cavity-enhanced spontaneous emission and Raman transitions stimulated by the vacuum field while driven by classical laser pulses. In particular, we discuss a scheme for adiabatic coupling between a single atom and an optical cavity, which is based on a unitary evolution of the coupled atom-cavity system [25,26], and is therefore intrinsically reversible.

For a photon emission from the cavity to take place, it is evident that a finite value of κ is mandatory, otherwise any light would remain trapped between the mirrors. Moreover, as κ is the decay rate of the cavity field, the associated duration of an emitted photon is typically κ^{-1} or more. We also emphasise that γ plays a crucial role in most experimental settings, since it accounts for the spontaneous emission into non-cavity modes, and therefore leads to a reduction of efficiency. The relation of the atom-cavity coupling constant g_0 and the Rabi frequency Ω of the driving field to the two decay rates can be used for marking the difference between three basic classes of single-photon emission schemes from cavity-QED systems.

Cavity-enhanced spontaneous emission: We assume that a sudden excitation process (e.g. a short π pulse with $\Omega \gg \{g_0, \kappa, \gamma\}$, or some internal relaxation cascade from energetically higher states) drives the atom suddenly into its excited state $|x,0\rangle$. From there, a photon gets spontaneously emitted either into the cavity or into free space. To analyse the process, we simply consider an excited two-level atom coupled to an empty cavity. This particular situation is the textbook example of CQED that has been thoroughly analysed in the past. In fact, it was proposed by Purcell [12] and demonstrated by Heinzen et al. [27] and Morin et al. [28] that the spontaneous emission properties of an atom coupled to a cavity are significantly different from those in free space. For an analysis of the atom's behavior, it is sufficient to consider the evolution of the $n = 1$ Jaynes-Cummings doublet under the influence of the atomic polarization decay rate γ and the cavity-field decay rate κ. Non-cavity spontaneous decay of the atom and photon emission through one of the cavity mirrors both lead the system into state $|g,0\rangle$, which does not belong to the $n = 1$ doublet. Therefore we can deal with these decay processes phenomenologically by introducing non-hermitian damping terms into the interaction Hamiltonian,

$$H'_{int} = -\hbar g_0 \left(|x\rangle\langle g|\hat{a} + \hat{a}^\dagger |g\rangle\langle x| \right) - i\hbar\gamma |x\rangle\langle x| - i\hbar\kappa \hat{a}^\dagger \hat{a}. \tag{13.11}$$

Figure 13.4a shows the time evolution of the atom-cavity system when $\kappa > g_0$. The strong damping of the cavity inhibits any vacuum-Rabi oscillation, since the photon is emitted from the cavity before it can be reabsorbed by the atom. Therefore the transient population in state $|g,1\rangle$ is negligible and the adiabatic approximation $\dot{c}_g \approx 0$ can be applied, which gives

$$\frac{d}{dt}c_x = -\gamma c_x - \frac{g_0^2}{\kappa}c_x, \tag{13.12}$$

with the solution

$$c_x(t) = \exp\left(-\left[\gamma + \frac{g_0^2}{\kappa}\right]t\right). \tag{13.13}$$

It is straightforward to see that the ratio of the emission rate into the cavity, g_0^2/κ, to the spontaneous emission probability into free space becomes $g_0^2/(\kappa\gamma) \equiv f$, i.e. the Purcell factor. It equals twice the one-atom cooperativity parameter, C, originally introduced in the context of optical bistability [29]. Hence the photon-emission probability from the cavity reads $P_{Emit} = 2C/(2C+1)$. Note that the atom radiates mainly into the cavity if $g_0^2/\kappa \gg \gamma$. Together with $\kappa \gg g_0$, this condition constitutes the bad-cavity regime.

The other extreme is strong coupling, with $g_0 \gg (\kappa,\gamma)$. In this case vacuum Rabi oscillations between $|x,0\rangle$ and $|g,1\rangle$ occur, with both states decaying at the respective rates γ and κ. Figure 13.4b shows a situation where the atom-cavity coupling, g_0, saturates the $|x,0\rangle \leftrightarrow |g,1\rangle$ transition. On average, the probabilities to find the system in either one of these two states are equal, and therefore the average ratio of the emission probability into the cavity to the spontaneous emission probability into free space is given by κ/γ. The vacuum-Rabi oscillation then gives rise to an amplitude modulation of the photons emitted from the cavity.

Bad-cavity regime: To take the effect of a slow excitation process into account, we consider a Λ-type three-level atom coupled to a cavity. In the bad-cavity regime, $\kappa \gg g_0^2/\kappa \gg \gamma$, the loss of excitation into unwanted modes of the radiation field is small and we may follow Law et al. [30,31]. We assume that the atom's $|e\rangle - |x\rangle$ transition is excited by a pump laser pulse while the atom

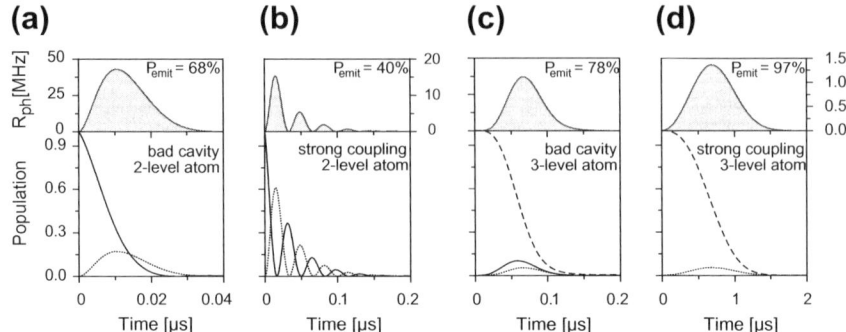

FIGURE 13.4 Evolution of the atomic states and photon emission rate $R_{\text{ph}} = 2\kappa\rho_{gg}$ in different coupling cases. (a) and (b) are for an excited two-level atom coupled to the cavity, showing populations ρ_{xx} (solid) and ρ_{gg} (dotted) of the product states $|x,0\rangle$ and $|g,1\rangle$. (c) and (d) are for a three-level atom-cavity system prepared in $|e,0\rangle$ and exposed to a pump pulse driving $|e\rangle - |x\rangle$ while the cavity couples $|x\rangle$ and $|g\rangle$. The initial-state population ρ_{ee} is dashed. (a) and (c) display the bad-cavity regime with $(g_0,\gamma,\kappa) = 2\pi \times (15,3,20) \times 10^6 \text{ s}^{-1}$, while (b) and (d) depict the strong-coupling case with $(g_0,\gamma,\kappa) = 2\pi \times (15,3,2) \times 10^6 \text{ s}^{-1}$. The pump pulses read $\Omega(t) = g_0 \sin(\pi t/200 \text{ ns})$ in (c), and $\Omega(t) = g_0 \times t/1\,\mu\text{s}$ in (d). No transient population is found in ρ_{xx} in the latter case. The overall photon-emission probability reads always $P_{\text{Emit}} = \int R_{\text{ph}}\,dt$.

emits a photon into the cavity by enhanced spontaneous emission. The cavity-field decay rate κ sets the fastest time scale, while the spontaneous emission rate into the cavity, g_0^2/κ, dominates the incoherent decay of the polarization from the excited atomic state. Provided any decay leads to a loss from the three-level system, the evolution of the wave vector is governed by the non-Hermitian Hamiltonian

$$H'_{\text{int}} = H_{\text{int}} - i\hbar\kappa \hat{a}^\dagger \hat{a} - i\hbar\gamma |x\rangle\langle x|, \tag{13.14}$$

with H_{int} from Eq. (13.7). To simplify the analysis, we take only the vacuum state, $|0\rangle$, and the one-photon state, $|1\rangle$, into account, thus that the state vector reads

$$|\Psi(t)\rangle = c_e(t)|e,0\rangle + c_x(t)|x,0\rangle + c_g(t)|g,1\rangle, \tag{13.15}$$

where c_e, c_x, and c_g are complex amplitudes. Their time evolution is given by the Schrödinger equation, $i\hbar \frac{d}{dt}|\Psi\rangle = H'_{\text{int}}|\Psi\rangle$, which yields

$$\begin{aligned} i\dot{c}_e &= \tfrac{1}{2}\Omega(t)c_x \\ i\dot{c}_x &= \tfrac{1}{2}\Omega(t)c_e + g_0 c_g - i\gamma c_x \\ i\dot{c}_g &= g_0 c_x - i\kappa c_g, \end{aligned} \tag{13.16}$$

with the initial condition $c_e(0) = 1, c_x(0) = c_g(0) = 0$ and $\Omega(0) = 0$. An adiabatic solution of (13.16) is found if the decay is so fast that c_x and c_g are nearly time independent. This allows one to make the approximations $\dot{c}_x = 0$

and $\dot{c}_g = 0$, with the result

$$\begin{aligned}
c_e(t) &\approx \exp\left(-\frac{\alpha}{4}\int_0^t \Omega^2(t')dt'\right), \\
c_x(t) &\approx -\frac{i}{2}\alpha\Omega(t)c_e(t), \\
c_g(t) &\approx -\frac{i}{\kappa}g_0 c_x(t),
\end{aligned} \qquad (13.17)$$

where $\alpha = 2/(2\gamma + 2g_0^2/\kappa)$. Photon emissions from the cavity occur if the system is in $|g,1\rangle$, at the photon-emission rate $R_{ph}(t) = 2\kappa|c_g(t)|^2$. This yields a photon-emission probability of

$$\begin{aligned}
P_{Emit} &= \int R_{ph}(t)dt \\
&= \frac{g_0^2 \alpha}{\kappa}\left[1 - \exp\left(-\frac{\alpha}{2}\int \Omega^2(t)dt\right)\right] \longrightarrow \frac{g_0^2 \alpha}{\kappa}. \qquad (13.18)
\end{aligned}$$

Note that the exponential in Eq. (13.18) vanishes if the area $\int \Omega(t)dt$ of the exciting pump pulse is large enough. In this limit, the overall photon-emission probability does not depend on the shape and amplitude of the pump pulse. With a suitable choice of g_0, α, and κ, high photon-emission probabilities can be reached [31]. Furthermore, as the stationary state of the coupled system depends on $\Omega(t)$, the time envelope of the photon can be controlled to a large extend.

Strong-coupling regime: To study the effect of the exciting laser pulse in the strong-coupling regime, we again consider a Λ-type three-level atom coupled to a cavity. We assume that the strong-coupling condition also applies to the Rabi frequency of the driving field, i.e. $\{g_0, \Omega\} \gg \{\kappa, \gamma\}$. In this case, we can safely neglect the effect of the two damping rates on the time scale of the excitation. We then seek for a method to effectively stimulate a Raman transition between the two ground states that also places a photon into the cavity. For instance, the driving process can be implemented in form of an adiabatic passage (STIRAP process [25,26]) or a far-off-resonant Raman process to avoid any transient population of the excited state, thus reducing losses due to spontaneous emission into free space. An efficiency for photon generation close to unity can be reached this way. Once a photon is placed into the cavity, it gets emitted due to the finite cavity lifetime.

The most promising approach is to implement an adiabatic passage in the optical domain between the two ground states [32,33]. In fact, adiabatic passage methods have been used for coherent population transfer in atoms or molecules for many years. For instance, if a Raman transition is driven by two distinct pulses of variable amplitudes, effects like electromagnetically induced transparency (EIT) [34,35], slow light [36,37], and stimulated Raman scattering by adiabatic passage (STIRAP) [26] are observed. These effects have been demonstrated with classical light fields and have in common that the system's state vector, $|\Psi\rangle$, follows a dark eigenstate, e.g. $|\phi_n^0\rangle$, of the time-dependent interaction Hamiltonian. In principle, the time evolution of the system is completely controlled by the variation of this eigenstate. However, a more detailed analysis [38,32] reveals that the eigenstates must change slowly enough to allow adiabatic following. Only if this condition is met, a three-level atom-cavity system prepared in $|\phi_n^0\rangle$ stays there long enough to control the relative population of $|e, n-1\rangle$ and $|g, n\rangle$ by adjusting the pump Rabi frequency Ω. This is obvious for a system initially prepared in $|e, n-1\rangle$. As can be seen from Eq. (13.9), that state coincides with $|\phi_n^0\rangle$ if the condition $2g_0\sqrt{n} \gg \Omega$ is initially met. Once the system has been prepared in the dark state, the ratio between the populations of the contributing states reads

$$\frac{|\langle e, n-1|\Psi\rangle|^2}{|\langle g, n|\Psi\rangle|^2} = \frac{4ng_0^2}{\Omega^2}. \tag{13.19}$$

As proposed in [25], we assume that an atom in state $|e\rangle$ is placed into an empty cavity, which nonetheless drives the $|g,1\rangle \leftrightarrow |x,0\rangle$ transition with the Vacuum-Rabi frequency $2g_0$. The initial state $|e,0\rangle$ therefore coincides with $|\phi_1^0\rangle$ as long as no pump laser is applied. The atom then gets exposed to a laser pulse coupling the $|e\rangle \leftrightarrow |x\rangle$ transition with a slowly rising amplitude that leads to $\Omega \gg 2g_0$. In turn, the atom-cavity system evolves from $|e,0\rangle$ to $|g,1\rangle$, thus increasing the photon number by one. This scheme can be seen as vacuum-stimulated Raman scattering by adiabatic passage, also known as V-STIRAP. If we assume a cavity decay time, κ^{-1} much longer than the interaction time, a photon is emitted from the cavity with a probability close to unity and with properties uniquely defined by κ, after the system has been excited to $|g,1\rangle$.

In contrast to such an idealized scenario, Fig. 13.4d shows a more realistic situation where a photon is generated and already emitted from the cavity during the excitation process. This is due to the cavity decay time being comparable or shorter than the duration of the exciting laser pulse. Even in this case, no secondary excitations or photon emissions can take place. The system eventually reaches the decoupled state $|g,0\rangle$ once the photon escapes. However, the photon-emission probability is slightly reduced as the non-Hermitian contribution of κ to the interaction Hamiltonian is affecting the dark eigenstate $|\phi_1^0\rangle$ of the Jaynes-Cummings triplet (13.9). It now has a small admixture of $|x,0\rangle$ and hence is weakly affected by spontaneous emission losses [32].

13.2.2.3 Single-Photon Emission from Atoms or Ions in Cavities

Many revolutionary photon generation schemes have recently been demonstrated, such as a single-photon turnstile device based on the Coulomb blockade mechanism in a quantum dot [39], the fluorescence of a single molecule [40,41], or a single color center (Nitrogen vacancy) in diamond [42,43], or the photon emission of a single quantum dot into free space [44–46]. All these schemes emit photons upon an external trigger event. However, the photons are spontaneously emitted into various modes of the radiation field, e.g. into all directions, and they usually show a broad energy distribution. For the same reason, the emission process cannot be described by a Hamiltonian evolution. Hence the process is not reversible, and does not allow for coherent quantum state mapping from source to photon and back. This does not prevent one from using these photons for quantum cryptography and communication, but it represents a major obstacle to many applications in quantum computing or quantum networking. As discussed above most of these limitations can be overcome by cavity-enhanced emission techniques into well-defined modes of the radiation field. Here we focus on these.

Neutral atoms: A straightforward implementation of a cavity-based single-photon source consists of a single atom placed between two cavity mirrors, with a stream of laser pulses traveling perpendicular to the cavity axis to trigger photon emissions. The simplest approach to achieve this is by sending a dilute atomic beam through the cavity, with an average number of atoms in the mode far below one. However, for a thermal beam, the obvious drawback would be an interaction time between atom and cavity far too short to achieve any control of the photon emission time. Hence cold (and therefore slow) atoms are required to overcome this limitation. The author followed this route [47,48], using a magneto-optical trap to cool a cloud of rubidium atoms below 100 μK beneath the cavity. Atoms released from the trap eventually travel through the cavity, either falling from above or being injected from below in an atomic fountain. Atoms enter the cavity randomly, but interact with its mode for 20–200 μs. Within this limited interaction time, between 20 and 200 single-photon emissions can be triggered. Figure 13.5 illustrates this setup, together with the excitation scheme between hyperfine states in ^{87}Rb used to generate single photons by the adiabatic passage technique discussed on Section 13.2.2.2.

Bursts of single photons are emitted from the cavity whenever a single atom passes through the mode of the cavity, and strong antibunching is found in the photon statistics, as shown in Fig. 13.5a. Sub-Poissonian photon statistics are found when conditioning the experiment on the actual presence of an atom in the cavity [49]. In many cases, this is automatically granted—a good example is the characterization of the photons by two-photon interference discussed in section (c.f. Chapter 2). For these experiments, pairs of photons are needed that meet simultaneously at a beamsplitter. With just one source under investigation, this is achieved with a long optical fiber delaying the first photon of a successively emitted pair. With the occurrence of such photon pairs being the precondition to observe any correlation and the probability for successive photon emissions being vanishingly small without atoms, the presence of an atom is actually assured whenever any photon-photon correlations are recorded.

Only lately, refined versions of this type of photon emitter have been realised, with a single atom held in the cavity using a dipole-force trap. McKeever *et al.* [50] managed to hold a single Cs atom in the cavity with a dipole-trapping beam

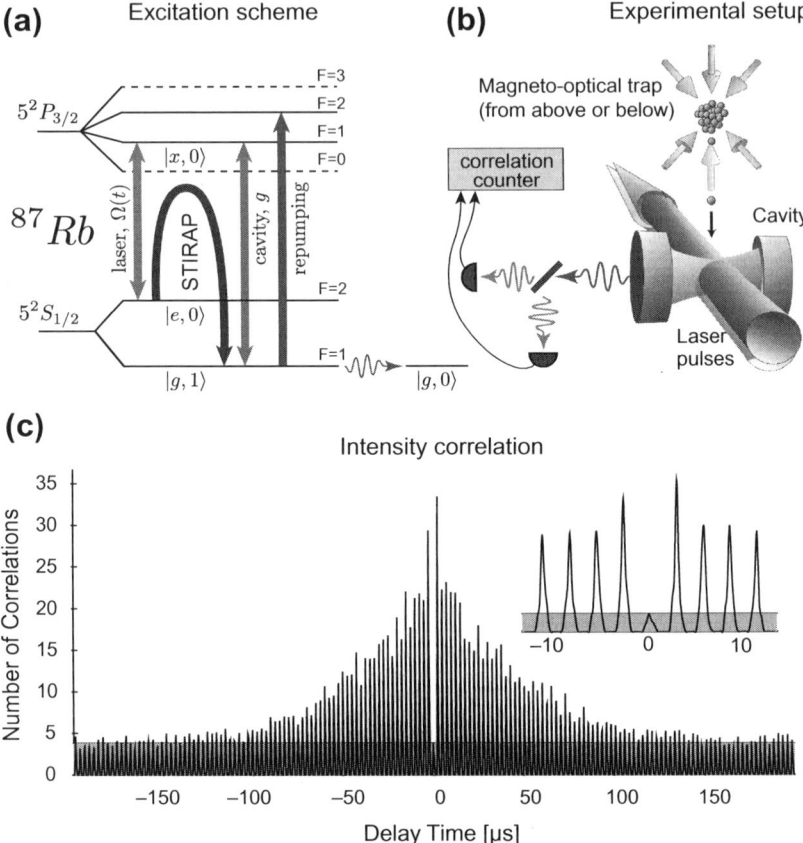

FIGURE 13.5 Single-photon source based on atoms traveling through an optical cavity. (a) Excitation scheme realised in ^{87}Rb for the pulsed single-photon generation. The atomic states labeled $|e\rangle$, $|x\rangle$ and $|g\rangle$ are involved in the Raman process, and the states $|0\rangle$ and $|1\rangle$ denote the photon number in the cavity. (b) A cloud of laser-cooled atoms moves through an optical cavity either from above [47], or from below using an atomic fountain [48]. Laser pulses travel perpendicular to the cavity axis to control the emission process. The light is analysed using a Hanbury Brown & Twiss (HBT) setup with a pair of single-photon avalanche photodiodes. (c) Intensity correlation of the emitted light measured with the HBT setup, with atoms injected using an atomic fountain [48].

running along the cavity axis, while Hijlkema *et al.* [51] use a combination of dipole-trapping beams running perpendicular and along the cavity to catch and hold a single Rb atom in the cavity mode. As illustrated in Fig. 13.6, the trapped atom is in both cases exposed to a sequence of laser pulses alternating between triggering the photon emission, cooling, and repumping the atom to its initial state to repeat the sequence. The atom is trapped, so that the photon statistics are not affected by fluctuations in the atom number and therefore are sub-Poissonian, see Fig. 13.6c. Moreover, with trapping times for single atoms up to a minute, a quasi-continuous bit-stream of photons is obtained.

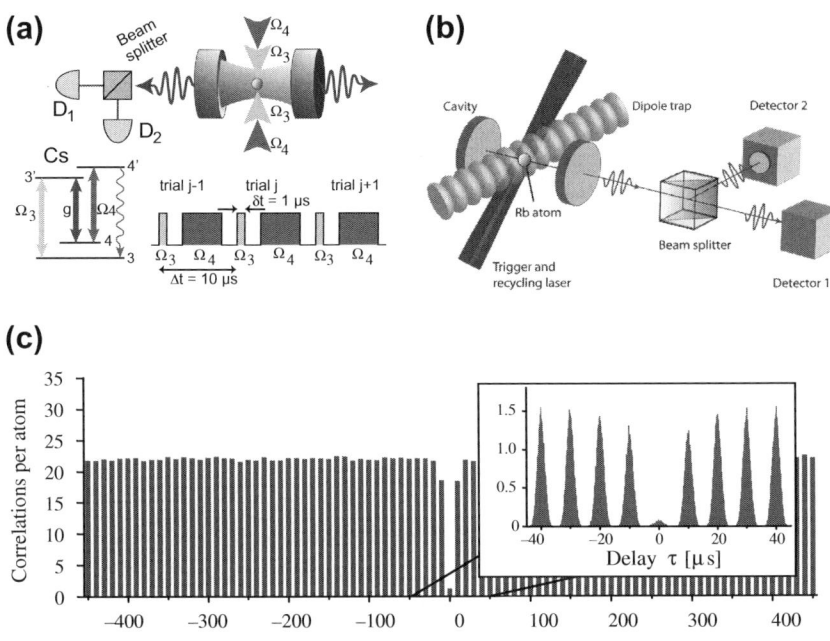

FIGURE 13.6 Atom-cavity systems with a single atom at rest in the cavity mode. (a): The setup by McKeever et al. [50] uses a dipole trap oriented along the cavity axis to hold a single Caesium atom in the cavity. The cavity is symmetric, so that half the photons are directed toward a pair of detectors for analyzing the photon statistics. (b): The Kuhn et al. uses a dipole trap oriented perpendicular to the cavity axis to hold a single Rubidium atom trapped in the cavity [51]. The cavity is asymmetric, and photons emitted through its output coupler are directed to a pair of photon counters to record the second-order correlation function of the photon stream. In both cases, the trapped atom is exposed to a sequence of laser pulses that trigger the photon emission, cool the atom, and re-establish the initial condition by optical pumping. (c): Intensity correlation function of the light emitted from a trapped-atom-cavity system, as found by the Kuhn et al. [51].

The major advantage of using neutral atoms as photon emitters in Fabry-Perot type cavities is that a relatively short cavity (≈ 100 μm) of high finesse (between 10^5 and 10^6) can be used. One thus obtains strong atom-cavity coupling, and the photon generation can be driven either in the steady-state regime or dynamically by V-STIRAP. This allows one to control the coherence properties and the shape of the photons to a large extent, as discussed in Section 13.2.3.2. Photon generation efficiencies as high as 65% have been reported with these systems. Furthermore, based on the excellent coherence properties, first applications such as atom-photon entanglement and atom-photon state mapping [15–18] have recently been demonstrated.

Apart from the above Fabry-Perot type cavities, many other microstructured cavities have been explored in recent years. These often provide a much smaller mode volume and hence boost the atom-cavity coupling strength by about an order of magnitude. However, this goes hand-in-hand with increased cavity losses and thus a much larger cavity linewidth, which might conflict with the desired addressing of individual atomic transitions. Among the most relevant new developments are fiber-tip cavities, which use dielectric Bragg stacks at the tip of an optical fiber as cavity mirrors [52,53]. Due to the small diameter of the fiber, either two fiber tips can be brought very close together, or a single fiber tip can be complemented by a microstructured mirror on a chip to form a high-finesse optical cavity. A slightly different approach are ring-cavities realised in solid state, guiding the light in a whispering gallery mode. An atom can be easily coupled to the evanescent field of the cavity mode, provided it can be brought close to the surface of the substrate. Nice examples are microtoroidal cavities realised at the California Institute of Technology [54,55], and bottle-neck cavities in optical fibers [56]. These cavities have no well-defined mirrors and therefore no output coupler, so one usually arranges for emission into well-defined spatio-temporal modes via evanescent-field coupling to the core of an optical fiber.

Trapped ions: Although neutral-atom systems have their advantages for the generation of single photons, such experiments are sometimes subject to undesired variations in the atom-cavity coupling strength and multi-atom effects. Also trapping times are still limited in the intra-cavity dipole-trapping of single atoms. A possible solution is to use a strongly localized single ion in an optical cavity, as was first demonstrated by Keller *et al.* [57]. In their experiment, an ion is optimally coupled to a well-defined field mode, resulting in the reproducible generation of single-photon pulses with precisely defined timing. The stream of emitted photons is uninterrupted over the storage time of the ion, which, in principle, could last for several days.

The major difficulty in combining an ion trap with a high-finesse optical cavity comes from the dielectric cavity mirrors, which influence the trapping potential if they get too close to the ion. This effect might be detrimental if the mirrors become electrically charged during loading of the ion trap, e.g. by the electron beam used to ionize the atoms. Figure 13.7a shows how this problem has been solved in [57] by shuttling the trapped ion from a spatially separate loading region into the cavity. Nonetheless, the cavity in these experiments is typically more than 10–20 mm long to avoid distortion of the trap. Thus the coupling to the cavity is weak, and although optimized pump pulses were used, the single-photon efficiency in [57] did not exceed 8.0(1.3)%. This is in good accordance with theoretical calculations, which also show that the efficiency can be substantially increased in future experiments by reducing the cavity length. It is important to point out that the low efficiency does not interfere with the singleness of the photons. Hence the $g^{(2)}$ correlation function of the emitted photon stream corresponds to the one depicted in Fig. 13.6c, with $g^{(2)}(0) \to 0$. With an improved ion-cavity setup, Barros *et al.* [60] were able to reach a single-photon emission efficiency of 88(17)% in a cavity of comparable length, using a more favorable mode structure in the near-confocal cavity depicted in

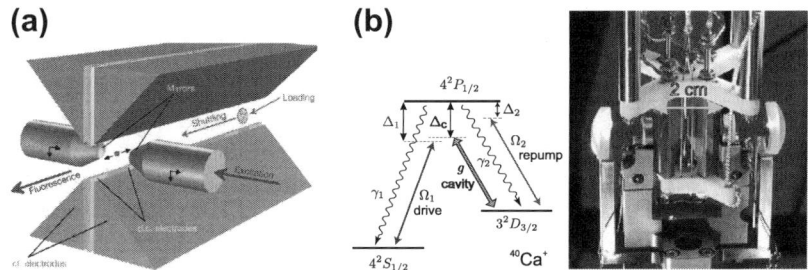

FIGURE 13.7 Arrangement of ion-trap electrodes and cavity in (a) the experiment by Keller *et al.* [57]. The ion is shuttled to the cavity region after loading. Upon excitation of the ion from the side of the cavity, a single photon gets emitted into the cavity mode *(Reprinted by permission from Nature Publishing Group: Nature, Guthöhrlein et al. [58], Copyright 2001)*. The ion-cavity arrangement and excitation scheme in ^{40}Ca$^+$ studied by Russo and Barros *et al.* [59,60] in Innsbruck (b) uses a near-concentric cavity which leads to an increased density of otherwise non-degenerate transverse modes *(Panel b adapted with permission from Springer: Applied Physics B, Russo et al. [59], Copyright 2009)*.

Fig. 13.7b and far-off-resonant Raman transitions between magnetic sublevels of the ion.

13.2.3 Photon Coherence, Amplitude, and Phase Control

The vast majority of single-photon applications do not only rely on the deterministic emission of single photons, but also require them to be indistinguishable from one another. In other words, their mutual coherence is often a key element whenever two or more photons are required simultaneously. The most prominent example to that respect is linear optics quantum computing (LOQC) as proposed by Knill *et al.* [2]. Furthermore, with photons used as information carriers, it is common practice to use their polarization, spatio-temporal mode structure or frequency for encoding classical, or quantum state superpositions. To do so, the capability of shaping photonic modes and controlling their coherence properties is essential. Several of these aspects are discussed in the following.

13.2.3.1 Indistinguishability of Photons

At first glance, one would expect any single-photon emitter that is based on a single quantum system of well-defined level structure to deliver indistinguishable photons of well-defined energy. However, this is often not the case for a large number of reasons. For instance, multiple pathways leading to the desired single-photon emission or the degeneracy of spin states might lead to broadening of the spectral mode or to photons in different polarization states, entangled with the atomic spin [15]. Also spontaneous relaxation cascades within the emitter result in a timing jitter of the last step of the cascade, which is the desired photon emission. Nevertheless, atoms coupled to cavities have been shown to emit nearly indistinguishable photons with well-defined timing. Their coherence properties are normally governed by the dynamics of the Raman process controlling the generation of photons, and—surprisingly—not substantially limited by the properties and lifetime of the cavity mode [61].

Probing photons for indistinguishability is normally done with a two-photon interference experiment of the Hong-Ou-Mandel (HOM) type as discussed in Chapter 2. For two identical photons that arrive simultaneously at different inputs of a 50:50 beamsplitter, they bunch and leave as a photon pair into either one or the other output port. Hence no correlations are found between two detectors that monitor the two outputs. This technique has been well established in connection with photons emitted from spontaneous parametric-down conversion (SPDC) sources, with the correlations between the outputs measured as a function of the *arrival*-time delay between photons.

For the cavity-based emitters discussed here, the situation is somewhat different. The bandwidth of these photons is very narrow, and therefore their coherence time (or length) might be extremely long, i.e. several μs($\approx$100 m). The time resolution of the detectors is normally 3–4 orders of magnitude faster than this photon length, so that the two-photon correlation signal is now determined as a function of the detection-time delay, with the arrival-time delay of the long photons deliberately set to zero [63,62]. This can be regarded as a quantum-homodyne measurement at the single-photon level, with a single local-oscillator photon arriving at one port of a beamsplitter, and a single signal photon arriving at the other port.

Prior to the first photodetection, the two photons arrive simultaneously in the input modes A and B at the beamsplitter and the overall state of the system reads $|1_A 1_B\rangle$. The first photodetection at time t_1 in either output port C or D could have been of either photon, thus the remaining quantum state reduces to

$$|\psi(t_1)\rangle = (|1_A 0_B\rangle \pm |0_A 1_B\rangle)/\sqrt{2}, \quad (13.20)$$

where "+" and "−" correspond to the photodetection in port C and D, respectively. We now assume that the second photodetection takes place $\Delta\tau$ later, at time $t_2 = t_1 + \Delta\tau$, with the input modes A and B having acquired a phase difference $\Delta\phi$ (for whatever reason) during that time span. Hence prior to the second detection, the reduced quantum state has evolved to

$$|\psi(t_2)\rangle = (|1_A 0_B\rangle \pm e^{i\Delta\phi}|0_A 1_B\rangle)/\sqrt{2}. \quad (13.21)$$

By consequence, the probability for the second photon being detected in the same port as the first photon is $P_{\text{same}} = \cos^2(\Delta\phi/2)$, while the probability for the second photon being detected in the respective other beam-splitter port reads

$$P_{\text{other}} = \sin^2(\Delta\phi/2). \quad (13.22)$$

The probability P_{CD} for coincidence counts between the two detectors in the beamsplitter's output ports C and D is therefore proportional to $\sin^2(\Delta\phi/2)$.

Any systematic variation of the phase difference $\Delta\phi$ between the two input modes A and B with time $\Delta\tau$ leads to a characteristic modulation of the coincidence function

$$g_{\text{CD}}^{(2)}(\Delta\tau) = \frac{\langle P_C(t) P_D(t+\Delta\tau)\rangle_t}{\langle P_C\rangle\langle P_D\rangle} \propto \sin^2(\Delta\phi(\Delta\tau)/2). \quad (13.23)$$

A good example is the analysis of two photons of different frequency. We consider one photon of well-defined frequency ω_0 acting as the *local oscillator* arriving at port A at the beamsplitter, and another one of frequency $\omega_0 + \Delta\omega$ which we regard as the *signal photon* arriving simultaneously at port B. Their mutual phase is undefined until the first photodetection at t_1, but then evolves according to $\Delta\phi(\Delta\tau) = \Delta\omega \times \Delta\tau$. In this case, the probability for coincidence counts between the beamsplitter outputs,

$$P_{\text{CD}}(\Delta\tau) \propto \sin^2(\Delta\omega \times \Delta\tau/2), \quad (13.24)$$

oscillates at the difference frequency between local oscillator and signal photon. This phenomenon has been extensively discussed in [61,62] and is also illustrated in Fig. 13.8. Furthermore, the figure shows the effect of random dephasing on the time-resolved correlation function. For photons of 1 µs duration, a 470 ns wide dip is found around $\Delta\tau = 0$. Thereafter, the coincidence probability reaches the same mean value that is found with non-interfering photons of, e.g. different polarization. In this case, we conclude that the dip-width in the coincidence function is identical to the mutual coherence time between the two photons. It is remarkable that it exceeds the decay time of both cavity and atom by one order of magnitude in that particular experiment. This proves that the photon's coherence is to a large extend controlled by the Raman process driving the photon generation, without being limited by the decay channels within the system. We now use a similar system to generate much shorter photons. These have no time to lose their mutual phase relation and the integral two-photon coincidence probability drops to 20% of the reference value found with non-interfering photons. This implies that the photons are nearly indistinguishable and therefore well suited for many-photon interference experiments in linear optics arrangements.

13.2.3.2 Arbitrary Shaping of Amplitude and Phase

From the preceding analysis (see Fig. 13.4) we have seen that the dynamic evolution of the atomic quantum states determines the photon emission probability and thereby also the photon's waveform. This raises the question as to what extent one can arbitrarily shape the photons in time by controlling the envelope of the driving field. This is important for applications such as quantum state mapping, where photon wavepackets symmetric in space and time should allow for a time-reversal of the emission process [19]. Employing photons of soliton-shape for dispersion-free propagation could also help boost quantum communication protocols.

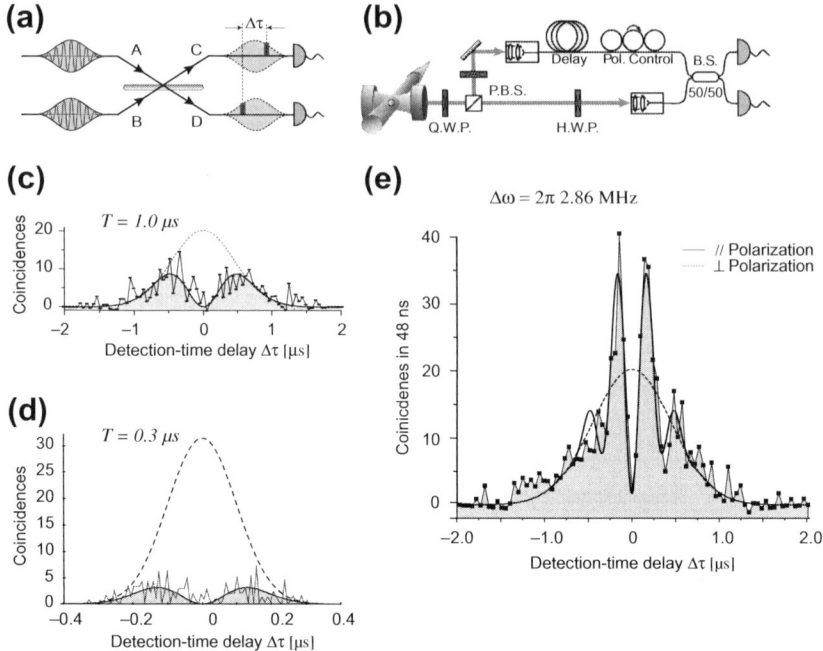

FIGURE 13.8 Time-resolved two-photon interference of photons arriving simultaneously at a beamsplitter (a). With photons emitted successively from one atom-cavity system, this has been achieved using an optical delay line (b). Panel (c) shows the correlation function from [61] for photons of 1.0 µs duration as a function of detection-time delay $\Delta\tau$. A pronounced dip at the origin is found, with the dip-width indicating the photon coherence time. The dotted line shows correlations found if distinguishable photons of perpendicular polarization are used, while the solid line depicts the correlations found if the photon polarization is parallel. Panel (d) shows data from a more recent experiment [48] with photons of 0.3 µs duration. The photons are nearly indistinguishable and the integral two-photon coincidence probability drops to 20% of the reference value found with non-interfering photons. Panel (e) shows data from [61,62] with interfering photons of different frequency. This gives rise to a pronounced oscillation of the coincidence signal as a function of $\Delta\tau$ with the difference frequency $\Delta\omega$.

Photon shaping is normally addressed by solving the Master equation of the atom-photon system, which yields the time-dependent probability amplitudes, and by consequence also the wavefunction of the photon emitted from the cavity [47,57]. Only recently, we have shown [64,21,48] that this analysis can be reversed, giving a unambiguous analytic expression for the time evolution of the driving field in dependence of the desired shape of the photon. This model is not only valid for V-STIRAP in the strong-coupling and bad-cavity regime, but it generally allows control of the coherence and population flow in any Raman process. Designing the driving pulse to obtain photonic wavepackets of any possible desired shape $\psi_{ph}(t)$ is straight forward [64,21]. Starting from the three-level atom discussed on page 9, we consider only the states $|e,0\rangle, |x,0\rangle$, and $|g,1\rangle$ of the $n = 1$ triplet and their corresponding probability amplitudes $\mathbf{c}(t) = [c_e(t), c_x(t), c_g(t)]^T$, with the atom-cavity system initially prepared in $|e,0\rangle$. The Hamiltonian (13.7) and the decay of atomic spin and cavity field at rates γ and κ, respectively, define the Master equation of the system,

$$i\hbar \frac{d}{dt}\mathbf{c}(t) = -\frac{\hbar}{2} \begin{pmatrix} 0 & \Omega(t) & 0 \\ \Omega(t) & 2i\gamma & 2g \\ 0 & 2g & 2i\kappa \end{pmatrix} \mathbf{c}(t). \quad (13.25)$$

The cavity-field decay rate unambiguously links the probability amplitude of $|g,1\rangle$ to the desired wavefunction $\psi_{\text{ph}}(t)$ of the photon. Furthermore, $|g,1\rangle$ only couples to $|x,0\rangle$ with the well-defined atom-cavity coupling g, while the Rabi frequency Ω of the driving laser links $|x,0\rangle$ to $|e,0\rangle$. Hence the time evolution of the probability amplitudes and $\Omega(t)$ can be written as

$$c_g(t) = \psi_{\text{ph}}(t)/\sqrt{2\kappa}, \quad (13.26)$$

$$c_x(t) = -\frac{i}{g}\left[\dot{c}_g(t) + \kappa c_g(t)\right], \quad (13.27)$$

$$\Omega(t) c_e(t) = 2\left[i\dot{c}_x(t) + i\gamma c_x(t) - g c_g(t)\right]. \quad (13.28)$$

We can use the continuity of the system, taking into account the decay of atom polarization and cavity field at rates γ and κ, to get to an independent expression for

$$|c_e(t)|^2 = 1 - |c_x(t)|^2 - |c_g(t)|^2 - \int_0^t dt \left[2\gamma |c_x(t)|^2 + 2\kappa |c_g(t)|^2\right]. \quad (13.29)$$

For an Hamiltonian that does not comprise any detuning and assuming ψ_{ph} to be real, one can easily verify that the probability amplitude $c_x(t)$ is purely imaginary, while $c_e(t)$ and $c_g(t)$ are both real. Hence with the desired photon shape as a starting point, we get analytic expressions for all probability amplitudes. These then yield the Rabi frequency

$$\Omega(t) = \frac{2\left[i\dot{c}_x(t) + i\gamma c_x(t) - g c_g(t)\right]}{\sqrt{1 + c_x^2(t) - c_g^2(t) + \int_0^t dt \left[2\gamma c_x^2(t) - 2\kappa c_g^2(t)\right]}}, \quad (13.30)$$

which is a real function defining the driving pulse required to obtain the desired photon shape.

Figure 13.9 compares some of the results obtained in producing photons of arbitrary wavefunction. The driving laser pulse shown in Fig. 13.9g has been calculated according to Eq. (13.30) to produce the photon shape from Fig. 13.9h. From all the data and calculations, it is obvious that stronger driving is required

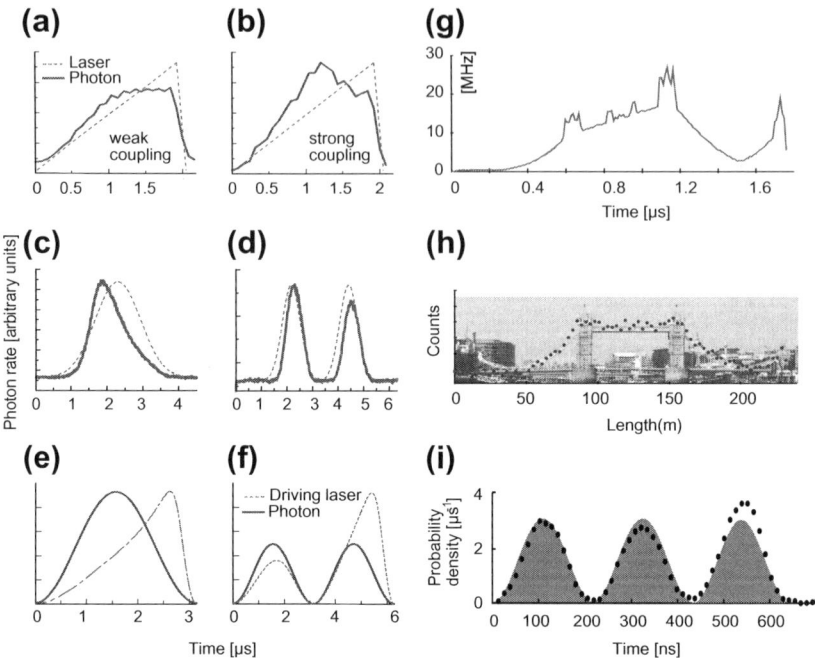

FIGURE 13.9 Photons made to measure: **(a–d)** show photon shapes realised in several experiments and their driving laser pulses. The histogram of the photon-detection time has been recorded for several hundred single-photon emissions. The data shown in **(a + b)** is taken from [47], with neutral atoms falling through a high-finesse cavity acting as photon emitters. The linear increase in Rabi frequency is the same in both cases, and the difference in photon shape is caused by variations in the coupling strength to the cavity. The data shown in **(c + d)** is taken from [57], with a single ion trapped between the cavity mirrors. It shows that the photon shape depends strongly on the driving laser pulse *(Panels c + d adapted with permission from Nature Publishing Group: Nature, M. Keller et al. [57], Copyright 2004)*. **(e + f)** show the Rabi frequency one needs to apply to achieve symmetric single or twin-peak photon pulses with an efficiency close to unity. This is a result from an analytic solution of the problem discussed in [64]. The latter scheme has been applied successfully for generating photons of various arbitrary shapes [48], with examples shown in **(g–i)**.

to counterbalance the depletion of the atom-cavity system. Therefore a very asymmetric driving pulse leads to the emission of photons symmetric in time, and vice versa, as can be seen from comparing Fig. 13.9c and e.

Among the large variety of shapes that have been produced, their possible sub-division into various peaks within separate time-bins is a distinctive feature that allows for time-bin encoding of quantum information. For instance, we recently imprinted different mutual phases on various time bins of multi-peak photons [65], and then successfully retrieved this phase information in a time-resolved quantum-homodyne experiment based on two-photon interference. The latter is illustrated in Fig. 13.10, with subsequently emitted triple-peak photons from the atom-cavity system arriving simultaneously at a beamsplitter. While the mechanism described above is used to subdivide the photons into three peaks of equal amplitude, i.e. three well-separated time bins or temporal modes, we also impose phase changes from one time bin to the next. The latter is accomplished by phase-shifting the driving laser. Eventually, signal photons get emitted from the cavity that are prepared in a W-state with arbitrary relative phases between their constituent temporal modes,

$$|\Psi_{\text{photon}}\rangle = (e^{i\phi_1}|1,0,0\rangle + e^{i\phi_2}|0,1,0\rangle + e^{i\phi_3}|0,0,1\rangle)/\sqrt{3}. \quad (13.31)$$

We may safely set $\phi_1 = 0$, as only relative phases are of any relevance. Every signal photon gets delayed in an optical fiber to arrive simultaneously with a subsequently emitted local-oscillator photon at a beamsplitter. The local-oscillator photon is not subject to phase shifts between its constituents, but is otherwise identical to the signal photon. The time resolution of the photo detectors C and D allows easy identification of whether photons are detected during the first, second, or third peak. The probability for photon-photon correlations is therefore governed by the phase change within the photons– i.e. the probability for correlations between the two detectors monitoring the beam-splitter output depends strongly on the timing of the photodetections. For instance, the coincidence probability $P(C_i, D_j)$ for detector C clicking in time bin i and detector D in time bin j reads

$$P(C_i, D_j) \propto \sin^2((\phi_i - \phi_j)/2). \quad (13.32)$$

We explored this effect [65] with the experiment illustrated in Fig. 13.10, using two types different signal photons. One with no mutual phase shifts, i.e. $\phi_1 = \phi_2 = \phi_3 = 0$, and the other with $\phi_1 = 0, \phi_2 = \pi, \phi_3 = 0$. In the first case, signal and local oscillator photons are identical. By consequence, no correlations between the two detectors arise (apart from a constant background level due to detector noise). In the second case the adjacent time bins within the signal photon are π out of phase. Therefore the probability for correlations between the two detectors increases dramatically if the detectors fire in adjacent time bins, but it stays zero for detections within the same time bin, and for detections occurring in the first and third time bin. These new findings demonstrate nicely that atom-cavity systems give us the capability of fully controlling the temporal evolution of amplitude and phase within single deterministically generated photons. Their characterization with time-resolved Hong-Ou-Mandel interference used for quantum homodyning the photons then reveals these phases again in the photon-photon correlations.

The availability of time bins as an additional degree of freedom to LOQC in an essentially deterministic photon-generation scheme is a big step toward large-scale quantum computing in photonic networks [66]. Arbitrary single-qubit operations on time-bin encoded qubits seem straightforward to implement with phase-coherent optical delay lines and active optical routing to either

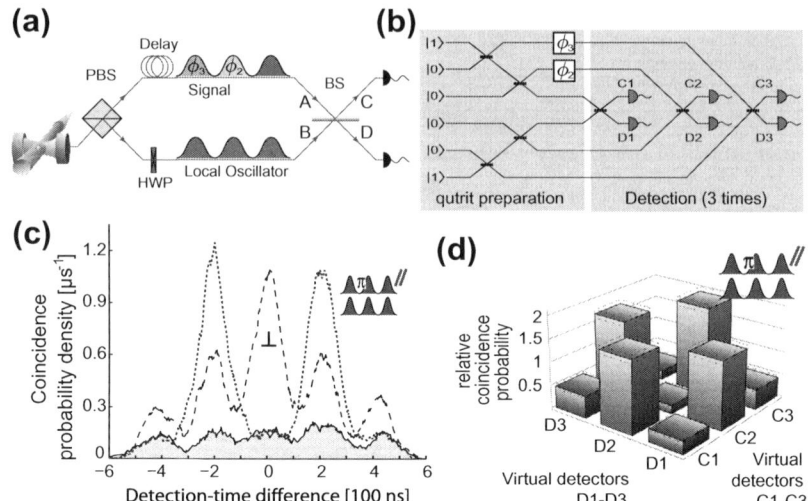

FIGURE 13.10 Qutrits, from [65]: (a) Pairs of triple-peak photons subsequently emitted are delayed such they arrive simultaneously at a beamsplitter. Time-resolved coincidences are then registered between output ports C and D. The signal photon carries mutual phases ϕ_1 and ϕ_2 between peaks, the local oscillator does not. (b) Time-resolved homodyne signal for photons of perpendicular (dashed) and parallel (solid) polarization, with the signal photon having a phase shift in the central time bin of $\phi_1 = \pi$ (dotted). The solid traces result from summing all coincidences found within a 60 ns wide interval around each point of the trace. For some of these data points, the statistical uncertainty is shown. (c) Corresponding virtual circuit if the same experiment was done in the spatial domain. The actual physical system, consisting of one beamsplitter and two detectors, would then correspond to a six-detector setup. All time-resolved photodetections in the real system can be easily associated with the corresponding virtual detectors firing. (d) Relative coincidence probabilities between virtual detectors (diagonal: detections within the same time bin; high columns: detections in successive time bins; outermost columns: detections two time-bins apart).

switch between temporal and spatial modes, or to swap the two time bins. Controlling the atom-photon coupling might also allow the mapping of atomic superposition states to time-binned photons [21,17]; and the long coherence time, combined with fast detectors, makes real-time feedback possible during photon generation.

13.3 SINGLE PHOTONS FROM SEMICONDUCTOR QUANTUM DOTS - G. S. SOLOMON

13.3.1 Introduction

Semiconductor III–V compounds, those made from columns three and five elements of the periodic table, are important for a wide range of optical and electronic devices. For example, optical devices from these materials range from lasers for telecommunications, for chemical sensing, and for cutting tools; a wide variety of light-emitting diodes, such as those used for lighting; and solar cells. While the success of Si-related devices in electronics is partly due to the high-quality interface between Si and it's insulating oxides, III-V devices owe their success in optoelectronics to high-quality interfaces with compounds of different band-gap energies; for example, AlGaAs and GaAs, and InGaAsP-InP. These defect-free, high-quality interfaces allow for quantum confinement, high-efficiency carrier transport through multiple interfaces and relatively low non-radiative recombination rates.

Invariably, new device applications led researchers to push the accepted boundaries of material combinations, and in the late 1980s this included the highly lattice-mismatched material systems of Ge-Si [67] and InGaAs-GaAs [68]. There has been excellent progress in both of these systems, but at the time quantum wells (QWs) of Ge in Si and high alloy compositions of $In_xGa_{1-x}As$ in GaAs showed troublesome thickness and compositional modulations, leading to excessively broad QW emission and dislocation-induced low quantum efficiency. It was realized that for very thin QWs the interfaces remained planar, after which the QW material began to island and eventually form deleterious dislocations. By the late 1980s and early 1990s groups [69,70] began to see this transitional islanding regime as a route to the formation of fully three-dimensionally confined regions, quantum dots (QDs), in the host crystal. In the InAs-GaAs system, QDs of InAs and InGaAs have been developed as an excellent source of discrete single photons. Single-photon emission from these QDs will be discussed in this section.

13.3.2 InAs-Based Quantum-Dot Formation

In homoepitaxial crystal growth of cubic materials the (1 0 0) growth surface is the lowest-energy facet plane [71]. If the growth temperature and flux rates are appropriate, mobile atoms on the surface, called adatoms, attach to the growth surface at atomic step edges of broad flat regions, for instance at kink and ledge sites, or to the edges of flat island nucleation regions. Thus, the crystal growth proceeds by lateral growth at kinks and ledges, or by the expansion of flat, monolayer-scale high islands. Ideally, as one monolayer is filled, new, monolayer-high nucleation sites are created, and the 2D (1 0 0) growth surface propagates. In most cases, the surface is quasi-2D, extending a few monolayers, yet still dominated by lateral growth through adatom attachment. In contrast, during heterogeneous crystal growth, as more adatoms are deposited onto the growing surface, this growth surface can go through structural changes such as changes in surface reconstruction, surface roughness, or reduced abruptness of a heterointerface [72–75]. In the case of InAs on GaAs, the lattice mismatch is 7.2%. Because of the large lattice mismatch, models generally poorly predict the critical thickness for relaxation by dislocation generation and propagation;

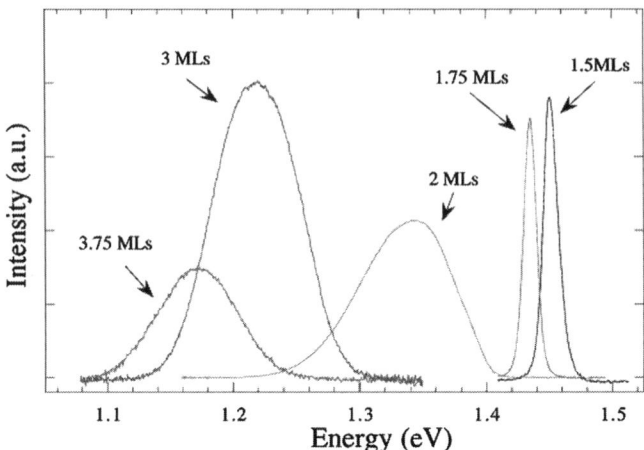

FIGURE 13.11 Photoluminescence (PL) of samples at 8 K containing equivalently planar InAs thickness in GaAs that vary between 1.5 and 3.75 mls. There is a transition from planar growth to quantum island growth between 1.75 and 2 mls. The photoluminescence intensity decrease between 3 and 3.75 mls is due to non-radiative recombination associated with dislocations.

estimates are between 0 and 15 InAs monolayers (mls) depending on the model [76,77]. Because of the similar InAs and GaAs crystal structures, at least one chemisorbed monolayer of InAs can be assumed to be stable on the GaAs substrate. Below the critical thickness or after a chemisorbed layer, a metastable phase can exist. This intermediate growth regime was reported by Stranski and Krastanow in 1938 [78], and is called the Stranski-Krastanow (SK) growth regime. In the SK growth of InAs on GaAs there is a thickness region where the excess strain is partially accommodated by surface islanding. This growth regime is a transitional growth mode between the compliant, planar growth regime that characterizes ideal molecular-beam epitaxy (MBE) growth, and plastically relaxed growth, since as islands grow and merge, the surface area can no longer expand to accommodate the increasing strain energy. It was first observed in semiconductor materials in the Ge-Si system by Eaglesham and Cerullo[79].

Experimentally, as shown in Fig. 13.11, InAs QDs form as islands after a little less than 2 mls of equivalently planar deposition. The islands grow in size and density until about 3.5 mls, where when buried in GaAs, dislocations are observed in transmission-electron microscopy (TEM) [80], and the photoluminescence is diminished [81].

13.3.3 Exciton Energetics

InAs-based islands can be formed dislocation-free, and when subsequently overgrown with a higher band-gap material such as GaAs, high-quality QDs are created. These QDs retain the direct-bandgap character of InAs: conduction-band and valance-band states have aligned minima and maxima in momentum

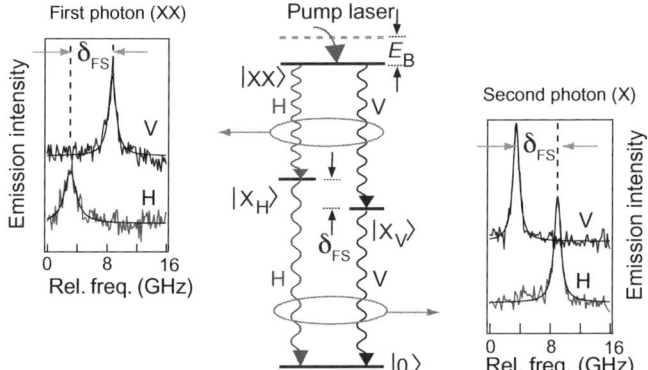

FIGURE 13.12 The decay of the optically active biexciton state, $|XX\rangle$ to the vacuum state, $|0\rangle$, through either of two intermediate single exciton states, $|X_H\rangle, |X_V\rangle$. The single exciton states are ideally degenerate, but are often energetically split by δ_{FS}, leading to linearly polarized optical emission, V,H. The relative energies of the two polarizations are switched between the $|XX\rangle$ decay and the $|X|\rangle$ decay. For InAs QDs, the biexciton binding energy is larger than the exciton binding energy (by E_B), so that in general both $|XX\rangle$ decays (left) are redshifted from the $|X\rangle$ decays (right).

space. In addition, conduction-band electrons and valance-band holes created through an excitation process are naturally localized in the three dimensionally confined QD. The single carrier description ignores the interactions between such localized carriers, so a multi-particle (or local polarization) description— the exciton description, is more appropriate. Many-body interactions described by exciton states are important in many semiconductor processes, but because of three-dimensional localization in QDs, they take on added significance—the single particle, electron-hole description cannot be used in any meaningfully way to described QD states. The canonical excitonic Hamiltonian includes the individual free carrier terms described by the band structure within the single particle approximation and two additional terms: a Coulomb term representing the attractive or repulsive nature of the charged carriers, and an exchange term representing the interactions of the constituent spins with each other and the lattice atoms.

In bulk materials, the excitonic spectrum reflects optical transitions based on the principle quantum numbers [82]. This is seen in ensemble measurements of QDs [83], and in more detailed, single-QD spectroscopy [84]. However, because of localization, multi-excitonic states, particularly the biexciton, is pronounced in QD spectroscopy. The biexciton state ($|XX\rangle$) is a single state composed of two electrons and two holes, where additional interactions make it energetically lower than two single excitons. When excited to the optically active biexciton ground state, the system decays optically through two channels to single exciton states, and these optically decay to the vacuum state. See Fig. 13.12.

The single-particle basis states for excitons are based on the symmetry of the crystal and the QDs. Because of quantum confinement, the heavy hole, with its larger effective mass, is the highest occupied valence energy state at the reciprocal-space zone center (Γ point). The strain in InAs-based QDs breaks the cubic symmetry (Space group $T_d^2 - F43m$) further separating the light- and heavy-hole states (compressive strain). The energy splitting between the two hole states is significant enough so as to neglect the light hole states. The light-heavy hole mixing has been shown theoretically and experimentally to only be on the order of a few percent [85,86]. In addition, the heavy hole-like states also contain a small d-orbital character [87]. While small, the light-heavy hole mixing, and hyperfine interaction of the electron with the nuclei, due to light hole and d-orbital mixing, can relax the selection rules and result in weak but sometime desirable transitions [88]. Thus, the excitons can be constructed from heavy-hole angular moment states $J_h = 3/2, J_{h,z} = \pm 3/2$, and the conduction state electron states $S_e = 1/2, S_{e,z} = \pm 1/2$, where z is aligned to [0 0 1], the growth direction for all QDs discussed here.

Four excitons are formed from the heavy hole and conduction electrons basis states. The [0 0 1] angular momentum projection is ± 2 for two states, and ± 1—the optically active, bright states. Without spin they are degenerate. Using the $\pm|2\rangle$ and $\pm|1\rangle$ states, the exchange Hamiltonian has block diagonal form indicating the dark and bright states do not couple, and furthermore the exchange interaction puts the dark states at lower energy than the bright states by ≈ 150 μeV [89]. Within each block diagonal grouping there is off-diagonal coupling. If the QD has cylindrical symmetry (D_{2d} symmetry), the bright states remain degenerate but the dark states couple, hybridizes into bonding and anti-bonding states [90,91]. However, the usual case is an elliptical in-plane symmetry where the bright states now couple, and the coupling between the dark states can be further altered. The couplings of both the dark and bright states do not produce as large of an energy splitting as the bright-dark separation. This is schematically shown in Fig. 13.13. The optical polarization of these transitions are circular without spin exchange, and the bright state decay remains so with

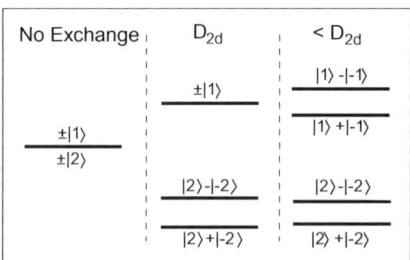

FIGURE 13.13 Schematic diagram of neutral ground state excitons in a QD. With no spin exchange there are four degenerate states, where the two ± 1 angular momentum states are optically active, and the two ± 2 angular momentum states are dark. With the inclusion of spin exchange there is an energy splitting between bright and dark states, and the two dark states couple. If the QD does not have cylindrical symmetry ($<D_{2d}$) the bright state also couples and the optical transitions become linearly polarized (see Fig. 13.12).

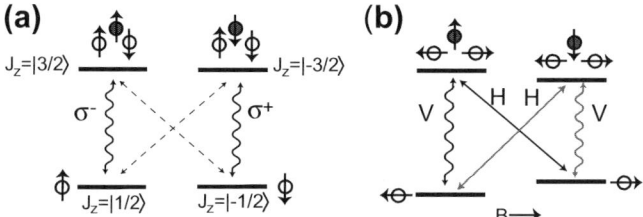

FIGURE 13.14 QD charged exciton decay to a single electron through two circularly polarized transitions (a). Dotted lines indicate ideally forbidden transitions for purely heavy-hole states. (b) In-plane (Voigt configuration) magnetic field splits the electron levels in energy and changes the optical selection rules.

cylindrical symmetry. However, all the neutral exciton transitions are linearly polarized when this symmetry is broken.

The above discussion was for neutral excitons, where the number of electrons and holes are equal; furthermore, it was limited to ground-state excitons and biexcitons. QD emission can result from a variety of excitonic configurations [92,84]. Of the many other types of excitons, the singly charged, ground state exciton, often called a trion, is particularly important. Non-resonantly optically pumping can create excitons, trions and biexcitons. However, without the capture or loss of a charge from the QD, the biexciton—exciton decay cannot yield a trion. The charged exciton can be used in a variety of applications. The energy diagram of the negative trion is shown in Fig. 13.14. For a negative trion with two electrons of opposite spin, there is no spin exchange. The two allowed, optically active transitions are thus circularly polarized. Because the ground state of the decay is an electron or hole, the trion is useful in quantum information, as opposed to the ground state of the neutral exciton manifold which is the vacuum state. When applying an in-plane magnetic field, the degenerate trions are energetically split and the selection rules altered [91]. Thus, the system can be initialized into a particular ground by resonantly or near-resonantly pumping one of the vertically polarized transitions [93]. For instance, in Fig. 13.14, by pumping the lower energy vertical (V) transition, the low-energy electronic ground state can be initialized through the horizontally (H) polarized transition.

13.3.4 Optically Accessing Single Quantum Dots

The ensemble photoluminescence (PL) measurements shown in Fig. 13.11 represent emission from many QDs. To create discrete single photons, individual QD states must be isolated. To resolve individual exciton states, two approaches are used. One uses lithography, typically electron-beam lithography to make etched structures, such as mesas, with dimensions in the 10's to 100's of nanometers. When QD samples are cooled to near liquid He temperatures, the

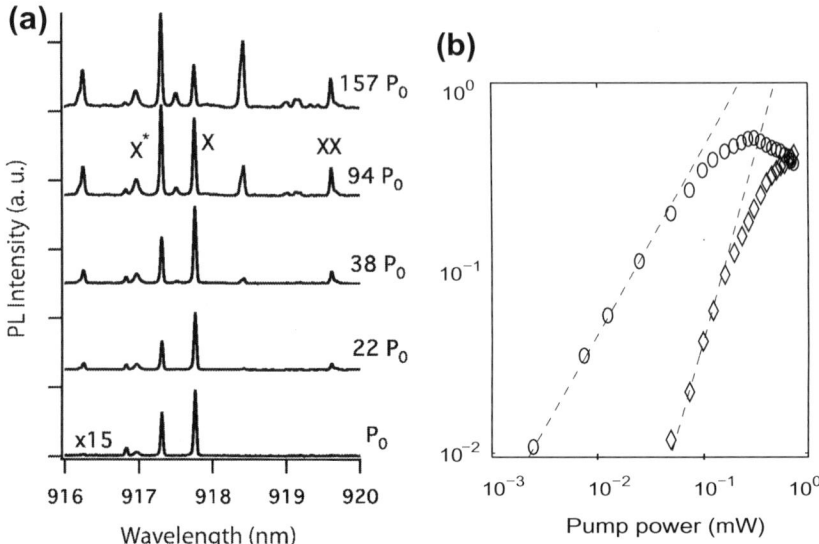

FIGURE 13.15 Photoluminescence of single QD emission with above-bandgap excitation at 4.2 K. (a) The pump power, P_0 is varied and the corresponding spectra are vertically off set for comparison. As the pump power increases the |XX⟩ emission line increases. Notice at higher pump power that the |X⟩ emission intensity begins to decrease. (b) Log-log plot of pump power and PL intensity on a different QD sample. There is a linear rise in |X⟩ emission (circles) and a quadratic rise in the |XX⟩ emission (diamonds) with pump power.

smooth PL peaks in Fig. 13.11 fracture into many small spectral features—evidence of individual QD state decay [94,89,95]. Another approach is to produce samples with dilute QD densities and use near resolution-limited focusing for the PL [96], sometimes with the aid of a solid-immersion lens [97,98]. Often, especially in conjunction with low-mode volume optical cavities, a combination of small structures and low QD density are used. This has historically made the search for QD states aligned both spectrally and spatially to a cavity mode random and time consuming; however, recent improvements using *in situ* optical lithography has made this approach deterministic [99].

In Fig. 13.15 micropost cavities are used in conjunction with low density QD growth for the isolation of single QD states, in a similar manner as in [95,45]. Figure 13.15a shows spectra with above-bandgap excitation at various powers. As the pump power rises the exciton emission intensity rises and the biexciton begins to appear (at $22P_0$). At higher pump powers, the exciton intensity decreases since, on average, the biexcition state population increases in the QD. This is shown in Fig. 13.15b where the log-log plot of intensity versus pump power is shown. The exciton emission saturates and drops, and at the highest powers the biexciton emission intensity is larger than the single exciton emission intensity.

13.3.5 Single Photons From Single Quantum Dots

The lowest-energy excitonic manifold of the quantum dot, containing neutral-exciton, charged-exciton, and biexciton states, has a discrete energy spectrum because of various and strong Coulomb interactions among the confined carriers. Photons result from those excitonic decays having ±1 changes in angular momentum. These include the two bright excitons, the two charged excitons for each of the positively and negatively charged states, and the bright biexciton. Therefore, within the specific energy region of one of these bright states, single photons will be produced. No more than one photon can be produced at a time from the decay of one of these quasi-particles, and on average one photon should be emitted during the radiative lifetime of the state.

Classical sources of light are from thermal sources where the photons in an optical mode are grouped, or bunched; and lasers, where the photon arrive randomly, i.e. a Poissonian distributed. A semiconductor example of a thermal source is a LED, and an example of a Poissonian distributed source is the semiconductor laser. A Poissonian light source can produce on average one photon in a given time, but this is not our interest. For these light sources, no matter how low their intensity there is always a possibility of having more than one photon emitted.[1] Here, we desire a single-photon source where there is only one photon at a time.

The single-photon character of light can be measured using second-order intensity correlation statistics. It is well-developed in Chapter 2. For time difference $\tau = t_2 - t_1$, between times t_1 and t_2, the second-order correlation function, $g^{(2)}(\tau)$ is defined as

$$g^{(2)}(\tau) = \frac{\langle I(t)I(t+\tau)\rangle}{\langle I(t)\rangle\langle I(t+\tau)\rangle} = \frac{\langle E^*(t)E^*(t+\tau)E(t+\tau)E(t)\rangle}{\langle E^*(t)E(t)\rangle\langle E^*(t+\tau)E(t+\tau)\rangle}, \quad (13.33)$$

where $I(t)$ and $E(t)$ are the intensity and electric field at time t, and $E^*(t)$ is the complex conjugate of the field. $\langle \ \rangle$ is the time average. While Eq. (13.33) is second order in intensity, it is fourth order in the field. For single mode fields, the quantum mechanical representation of $g^{(2)}(\tau)$, with photon creation operator, $a^\dagger$ and annihilation operator, a, is

$$g^{(2)}(\tau) = \frac{\langle a^\dagger(t)a^\dagger(t+\tau)a(t+\tau)a(t)\rangle}{\langle a^\dagger(t)a(t)\rangle\langle a^\dagger(t+\tau)a(t+\tau)\rangle}. \quad (13.34)$$

The measurement is made by recording intensity on two detectors as a function of time difference, τ. Note that $g^{(2)}(\tau)$ is not time dependent, but only dependent of the time difference, $t_2 - t_1$. Because detectors have a finite

---

[1] The Poisson number distribution is $P(n) = (\mu^n/n!)e^{-\mu}$, where $P(n)$ is the probability of the distribution containing n photons, and μ is the mean photon number. For a mean photon number, $\mu = 1$ there is a finite probability of having zero or more than one photon.

response time (or more specifically, dead time), this measurement is usually made using a beamsplitter and two detectors. τ can be positive or negative, and results are typically symmetric around $\tau = 0$. For a coherent field such as a laser, there is equal probability of detecting photons at t and $t + \tau$, and thus $g^{(2)}(\tau) = 1$. For thermal light, there is a higher probability of detecting photons in the same or nearby time windows—the source is *bunched*, and $g^{(2)}(0) = 2$. For single photons, coming from the decay of a single two-level system, like a QD exciton, if a photon is detected, then another photon cannot be emitted from the two-level decay until the upper level is repopulated. For pulsed laser excitation, the upper level is populated only once per pulse. For CW excitation, photon emission will on average occur within the radiative decay time. With either excitation technique, the probability of detecting a photon on each of the detectors at $\tau = 0$ is ideally zero, and the single-photon source is called *anti-bunched*. For more details, see Chapter 2.

An example of a $g^{(2)}(\tau)$ measurement made with a cw pump source is shown in Fig. 13.16. The measurement is made on colloidal QDs made from CdSe cores with a surrounding shell of wider band gap ZnS [44]. The measurements were at room temperature. The upper data were made from an ensemble of many QDs, while the lower data were measured on a single CdSe/ZnS QD. The dip in the coincidence counts gives $g^{(2)}(0) < 1$. If $g^{(2)}(0)$ is below 0.5, the emission can only be explained by a process involving a single emitter. However, for an isolated two-level system emitting discrete single photons, $g^{(2)}(0) = 0$. The anti-bunching shown in Fig. 13.16 can be modeled with an exponential function

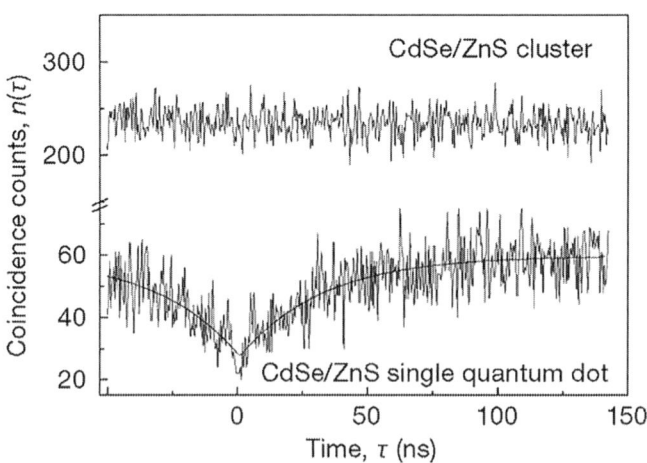

FIGURE 13.16 Second-order correlation counts (unnormalized) of CdSe/ZnS colloidal QDs using cw pumping. (upper data) Measurements on an ensemble of CdSe/ZnS QDs showing no second-order correlation (Poissonian light). (lower data) Measurements on a single QD showing anti-bunching at $\tau = 0$, where τ is the time difference $t_2 - t_1$. From [44]. *Reprinted by permission from Macmillan Publishers Ltd.*

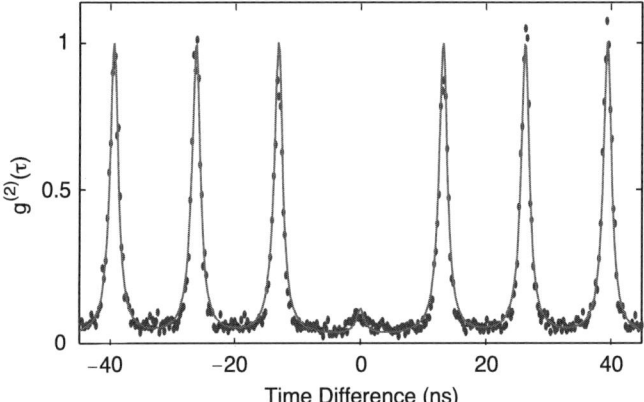

FIGURE 13.17 Second-order correlation measurement of an InAs QD using a 76 MHz repetition rate pump laser. The integrated counts in the time difference, $\tau = 0$ window normalized to nearby peaks is 0.09, which includes the non-zero background floor. *Adapted from [100]*.

of the form $g^{(2)}(\tau) = 1 - e^{-(\Gamma + W_p)\tau}$, where Γ is the radiative recombination rate, and W_p is the pumping rate.

In many fundamental and application-oriented experiments, regulated photons are desired, where by *regulated* we mean the creation of photons synchronized with the pulsed cycle from the pump source. For a regulated single-photon source, we desire just one photon per pulsed cycle. With a pump repetition rate slower than the radiative excitonic decay rate, non-resonant excitation can be used if the excitonic decay is the slowest decay in the total decay process. This generally ensures that only one photon will be emitted from the exciton in each pump cycle. An example of a second-order correlation measurement made using a 76 MHz repetition rate pump laser is shown in Fig. 13.17, where $g^{(2)}(0) \ll 0.5$ is a clear indication of non-classicality, and since $g^{(2)}(0) \approx 0$ this photon emission has very high single-photon purity.

Typically, early measurements of $g^{(2)}(\tau)$ on individual QD photons were made using disk and post microcavities [101,45] where the three-dimensionally confined optical modes improved the collection efficiency because of directional emission and the weak-coupling radiative decay enhancement.[2] Now these three-dimensional cavity—QD systems are used as bright MHz sources of nearly pure single photons [102,103]. Over 80% of the emitted photons are collected and $g^{(2)}(\tau)$ values below 0.15 are obtained [102]. As well, because of improved measurement efficiencies, $g^{(2)}(\tau)$ on single QDs are now often conducted on samples without these strongly confined cavities [100,104], albeit with significantly less collection efficiencies, and often with poorer

---

[2] For a more detailed development of two-level emitter—optical cavity coupling see Section 13.2.2.1 in this chapter.

indistinguishability. As in Figs. 13.16 and 13.17, these experiments can be pumped non-resonantly, which adds flexibility to single-photon sources; for instance, by significantly simplifying experiments and devices, and allowing different systems to be pumped simultaneously [105]. However, non-resonant pumping also induces variation in the initiation of the exciton decay, leading to reduced indistinguishability [106]. Resonant excitation using π-pulses has been demonstrated, showing excellent $g^{(2)}(0)$ (0.012) and high photon count rates (exceeding 200,000 s^{-1}) [104]. While using a solid-immersion lens, the experiments in Ref. [104] do not use a three-dimensional microcavity. They collect ≈6% of the photons from the QD, which can be improved by an order of magnitude with microdisk or micropost cavity [102].

13.3.6 Weak QD-Cavity Coupling

The QD exciton has a large oscillator strength; and modern GaAs-based crystal growth, lithography and etching lead to high quality-factor, low mode volume cavities; creating an ideal system to study cavity-quantum electrodynamics (CQED) in a solid-state environment.

A significant body of successful and intriguing strong-coupling QD–microcavity research exists, where the cavity and two-level system coherently exchange energy. This is detailed in Section 13.2.2.1 of this chapter. The weak-coupling regime of CQED is a significant component of QD-based single-photon sources [95,107]. The enhanced spontaneous emission decay rate increases the single-photon count rate and often improves the photon indistinguishability. Light from the QD decay preferentially couples to the discrete distribution of optical modes present in these cavities. The optical field intensity of these modes is no longer uniform, and in many cases in these small cavity structures the field intensity can be extremely large. These concepts are also developed from an atom/ion perspective in Section 13.2.2.1 of this chapter. Enhanced emission rates and collection efficiency result from orienting the modes of the modified field to the excitonic dipole of the QD. The radiative decay rate, γ is enhanced or suppressed from the isotropic rate, γ_0 by the spontaneous emission rate enhanced factor (the Purcell factor), f, defined as

$$f = \frac{\gamma}{\gamma_0} = \frac{3Q\lambda_c^3}{4\pi^2 n_e^3 V} \left(\frac{\delta\lambda_c^2}{\lambda_c^2 - 4(\lambda_e - \lambda_c)^2} \right) \frac{|\overline{E}(\overline{r})|^2}{|\overline{E}_{\max}|^2} \left(\frac{\overline{d} \cdot \overline{E}(\overline{r})}{|\overline{d}||\overline{E}(\overline{r})|} \right)^2 + \Im.$$

(13.35)

Q is the cavity quality factor, λ_c is the wavelength of the cavity resonance, n_e is the effective index of refraction and V is the cavity mode volume. The three bracketed terms are the spectral, spatial and polarization alignments of the QD exciton with the cavity mode, where $\delta\lambda_c$ and λ_e are the cavity linewidth and excitonic emission wavelength; and $\overline{E}(\overline{r})$ is the electric field at the exciton and $\overline{\mu}$ is the electric dipole of the exciton. Here, we assume that the excitonic linewidth is small compared to λ_c. $\Im$ is the geometric coupling into loss modes. The radiation lost into these modes, $\gamma_0 \Im$ is due to the geometry of these microcavities [95, 108]. Emitted light incident on the microcavity edges at angles smaller than the critical angle for total internal reflection will be lost, as well as light incident on the mirrors at an angle greater than the limit of the angular stopband [108]. $\gamma_0 \Im$ has been shown to be ≈0.3 for small microposts [95]. Besides the wavelength, position and polarization detuning terms, this formulation of f differs from that in Section 13.2.2.1 by the solid-state effective index n_e of the cavity structure, and $\Im$.

The epitaxial growth process used to make strain-induced QDs is well suited for the incorporation of optical microcavities. These microcavities are made by varying combinations of epitaxial crystal growth and processing. For instance, disk microcavities (Fig. 13.18) require a pedestal supporting a disk of typically 2.5–10 µm in diameter. The QDs are centered vertically in the plane of the disk, and are distributed randomly in-plane throughout the disk. Only QDs located near the disk in-plane perimeter couple efficiently to the whispering gallery modes located there. For InAs-based QDs, these layers are made by epitaxial growth and the disk is made by optical lithography. The pedestal layer is usually a high AlAs alloy of AlGaAs so it can be selectively undercut from the GaAs disk. Planar microcavities are also used. They are formed from distributed Bragg reflector (DBR) mirrors separated by a cavity region of $N\lambda/2n$ thickness, where N is an integer. Each DBR mirror is composed of pairs of $\lambda/4n$-thick layers of contrasting refractive index material, so that when many pairs are combined they form a mirror with large effective reflectivity.

FIGURE 13.18 Three different types of microcavities with InAs quantum dots. (Upper Left) Disk microcavity where the cavity modes of interest are whispering gallery modes at the disk perimeter (top view of the mode structure in the insert). (Lower Left) Photonic crystal cavity, where the cavity is formed from three missing air holes (dark regions). *Adapted from [109]*. (Right) Post microcavity formed from alternating AlAs/GaAs layers, lithography and dry etching. The cavity region is the slightly wider, middle gray region. *Reprinted by permission from Macmillan Publishers Ltd: [110]*.

For example, GaAs and AlAs, or different alloys of $Al_xGa_{1-x}As$ are often used, where $n_{GaAs} = 3.5$ and $n_{AlAs} = 2.9$. The quality factors for GaAs disk microcavities depend on the diameter, as well as processing. For 2.5 μm diameter microdisks Q's of 10^5 have been reported in transmission [111]. Often the Q is measured through emission of the QD gain medium, and Q's in the 10^4 range are reported [112,113]. Similarly, the single-mode quality factors of DBR planar microcavities are also $\approx 10^5$ [114]. The planar DBR microcavities have a continuous distribution of modes that can be made into a discrete distribution by providing in-plane optical confinement. This is conveniently accomplished by etching posts from the planar DBR cavity, where the refractive index contrast between the semiconductor and air provides the lateral optical confinement, see Fig. 13.18. As discussed in Section 13.2.2.2.2, the fraction of photons created from the exciton decay that coupled to a single optical mode can be described as

$$\beta = \frac{\gamma}{\gamma_0 + \gamma} = \frac{f}{1+f}. \qquad (13.36)$$

For no enhanced emission coupling $\beta = 0.5$, while if the emitter is completely off-resonance with the mode, so that no coupling occurs, $\beta = 0$.[3] Full three-dimensional mode confinement is important, since within the bad cavity limit, as f increases through larger Q and smaller V the single optical-mode coupling increases, and hence light extraction increases. Besides the disk and post microcavity, the photonic crystal cavity (PCC) is another three-dimensional microcavity in common use (Fig. 13.18). The PCC has extremely small mode volume, on the order of $\frac{1}{2}(\lambda/n)^3$ [115], whereas for the post microcavity, $V \approx 2(\lambda/n)^3$, and for a disk microcavity, $V \approx 5(\lambda/n)^3$. In these three types of microcavities the spontaneous emission enhancement factors are in the range of 4–10 [95,101,115].

As the optical fields in these high f, three-dimensionally confined cavities become increasingly localized, it is increasingly unlikely that a QD will be spatially and spectrally aligned with the cavity mode. Thus, many cavity structures need to be evaluated to find good exciton-cavity alignment. One solution is to build the cavities around the QD and two approaches have been used. Through detailed atomic-force microscopy (AFM) an isolated QD can be found, and lithographic marking made to indicate its location. Then, the cavity can be built around this located QD. This approach is useful when the QD is close to the surface. Otherwise, if the QD is far below the surface, the epitaxial growth front will planarize and QDs below cannot be imaged by AFM. This AFM-based approach has been successfully used to align QDs with PCCs [116]. A more recent approach spatially and spectrally images the QD at the near

---

[3] Other definitions of β are in use. For instance, in [95,107], $\beta \approx 1 - \frac{\gamma_0}{\gamma}$. In this formulation, when there is no decay-rate enhancement, then $\beta = 0$. For enhanced optical-mode coupling, $0 < \beta \leq 1$, and for suppressed mode coupling, $\beta < 0$.

4K measurement temperature after the sample has been covered with optically sensitive resist [99]. After finding exciton emission with good characteristics, they expose the resist, and remove the sample to process post microcavities. Once this lithography is completed, the sample is again cooled to measurement temperature ($\approx 4\ K - 10\ K$) to check the cavity-exciton spectral alignment. A cycle of fine-tuned etching and measurement is repeated until spectral overlap is achieved [99]. Various alternative approaches of patterning QDs in regular arrays by processing the semiconductor sample before QD deposition are also used [117–120], and occasionally this patterning has been incorporated into fully confined microcavities [121].

13.3.7 Quantum-Dot Photon Indistinguishability

Photons emitted from quantum dots can be highly anti-bunched, and with the aid of fully confined microcavities, they can be collected with high efficiency—nearly 80%—and can be delivered with 10^6 repetition rate. Motivated by both fundamental and applied perspectives, we now turn to the question of indistinguishability of the QD photons. While in classical physics identical entities can be distinguishable, in quantum mechanics identical states are fundamentally indistinguishable (see, for example, [122]). Photons are bosonic, and therefore, if a pair of photons are characterized by the same mode of the electromagnetic field, or equivalently if the photons have identical wavepackets, they will coalesce into a single multi-photon state if made to interfere. With the notable exception of the BB84 quantum encryption scheme [123], indistinguishable single photons are a resource critical to the success of many quantum information processing applications [2,19]. The question of indistinguishability is particularly interesting in QDs because of the rich solid-state environment interacting with them.

When two classical fields interfere at a 50-50 beamsplitter, the visibility of the interference seen in the second-order correlation cannot exceed 50% [124]. Since photons follow Bose-Einstein statistics, quantum mechanics says that when identical single photons enter the two ports of a 50/50 beamsplitter, that regardless of their histories, they bunch together, leaving from one of the two output ports as a single two-photon state, rather than each of the two single photons independently choosing an output port (see Chapter 2). As a result, quantum mechanics predicts the probability of joint detection at the two ports is zero, that is, the light is fully anti-bunched. Two-photon bunching was first studied by Hong *et al.* [125]. It is a central concept in quantum optics and has wide applicability, particularly in quantum information science.

In Fig. 13.19 two photons are incident on the two inputs of the beamsplitter (A, B in the figure). Four outcomes are possible and are illustrated in the figure. Of the four, two appear to produce identical results (*tt* and *rr*); however, in each of these two cases the photons leaving each port have different origins: The *tt* and *rr* cases are only identical when the two input photons

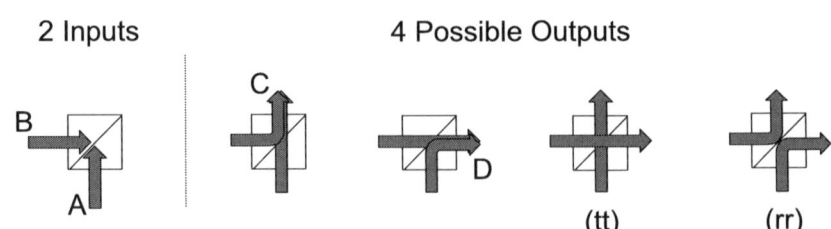

FIGURE 13.19 When two photons are incident on a 50/50 beam splitter (A, B) assuming no losses, there are four possible outcomes. If the two photons are indistinguishable then the (tt) and (rr) cases cancel since transmission and reflection differ by a $\pi/2$ phase shift. The result is the two-photon state, $|\Psi\rangle = \frac{1}{\sqrt{2}}(|2,0\rangle + |0,2\rangle)$.

are identical. Thus, coalescence probability hinges on the indistinguishability of the photons involved. Since the transmission and reflection coefficients differ by a $\pi/2$ phase shift, the tt and rr probabilities cancel for indistinguishable photons, the two single-photon states coalesce into a two-photon state, $|\Psi\rangle = \frac{1}{\sqrt{2}}(|2,0\rangle + |0,2\rangle)$.

The original Hong et al. [125] experiment showed that when pairs of photons produced by parametric-down conversion (PDC) interfere, the reduction in output coincidence detection can be well below 50%, reaching zero for an ideal source of indistinguishable photon pairs. It could be argued that the photons in the PDC experiment have a common history, because they were emitted in the same fluorescence process [126], which motivated efforts to demonstrate indistinguishable photons from physically distinct processes and sources. Indistinguishability measurements of photons from a QD exciton decay was measured by Santori et al. [127]. Because of the enhancement of spontaneous emission decay in the post microcavity, the photon indistinguishability is improved as non-radiative dephasing processes are reduced. The pump laser produces two QD photons from the same excitonic state in the same QD, but separated in time by 2 ns (with a 76 MHz repetition rate). The photons enter a Michelson interferometer with arms of unequal length, see Fig. 13.20a. The difference in arm length is the same 2 ns time difference between the two QD photons. With equal probability, photons can take the short or the long path. There are four options, where photons do not temporally overlap, and one option, represented by peak 3 in Fig. 13.20b where they do overlap. Classically, the probabilities of outcomes represented by peaks 2 and 4 are the same, and differ from peak 3 by a factor of two. However, in Fig. 13.20b the $g^{(2)}(t_2 - t_1)$ at $t_2 - t_1 = 0$ for peak 3 has significantly lower integrated counts than the nearby peaks, 2 and 4. This count suppression is due to the indistinguishability of the two photons. Accounting for imperfections in the measurement system, the two-photon overlap can be as large as 0.8 in [127]. The best results occur with short spontaneous emission lifetimes (90 ps), indicating the importance of the microcavity spontaneous emission enhancement.

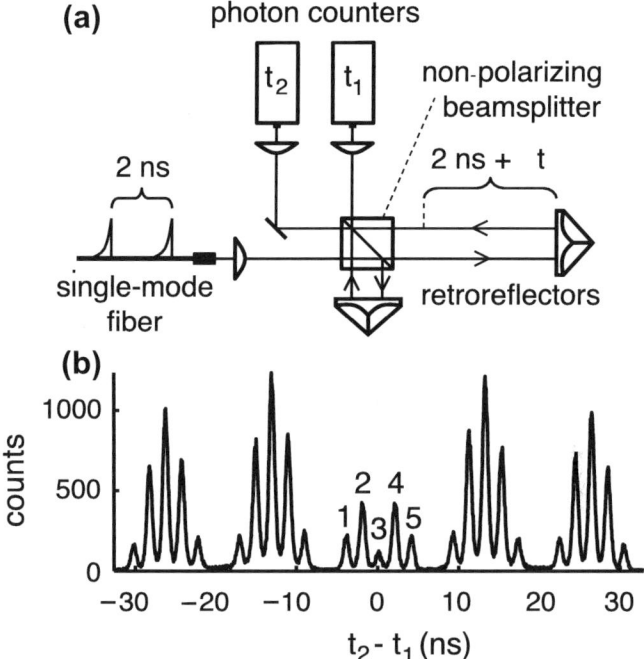

FIGURE 13.20 Two-photon interference from QD exciton photons. (a) Photons separated by 2 ns enter an interferometer, taking one of two paths of different lengths. They reflect at the retroreflectors, again are incident on the beamsplitter, and are detected at photon counters, t_1 and t_2. (b) Five peaks are labeled corresponding to three types of coincidence events. Peak 3 corresponds to the first photon taking the long path and the second photon taking the short path, thus simultaneously arriving at the beamsplitter before exiting the interferometer. The suppression of peak 3 indicates a fraction of the single photons have coalesced into a two-photon state, exiting the same port of the beamsplitter with probability greater than 50%. *From [127].*

The QD photon indistinguishability has also been measured for photons from QD states in different samples. This is fundamentally intriguing because these QD photons share less common history than in the previous single QD experiment, although they still share a common nonresonant pump laser. From an applications perspective, in a distributed quantum information system, photons will likely be emitted from different sources and may even reply on unrelated processes. This means that in many cases photons will require significant manipulation to be made indistinguishable. Recently, Flagg et al. observed the interference of photons emitted by two QDs, where each QD is embedded in separate sample (see Fig. 13.21) [100]. This experiment was subsequently done in the diamond NV center system (see Section 13.4).

Flagg et al. used different types of cavities for each QD sample. One sample is a planar DBR microcavity with 15.5 lower (10 upper) DBR pairs of GaAs and AlAs; the cavity mode is centered at $\lambda = 920$ nm. The other sample is an

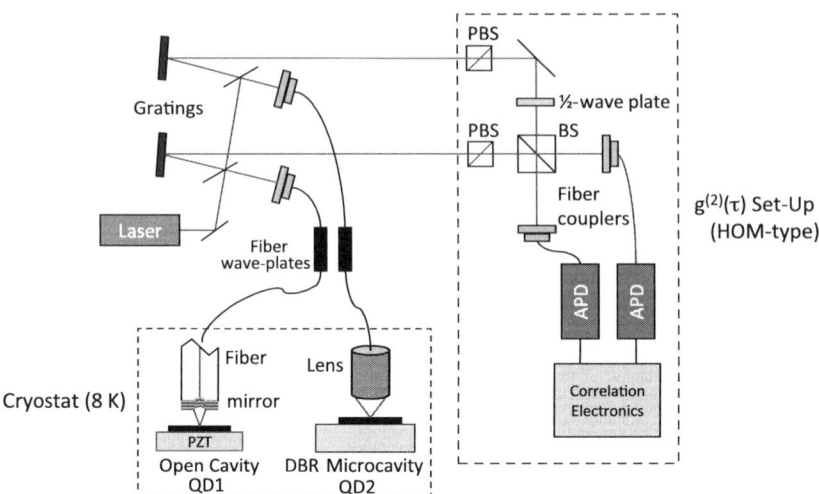

FIGURE 13.21 Schematic of a two-photon interference experiment using InAs QD photons from two different samples [100]. Two cavities at 8 K are used to couple QD photons to the two-photon interference set-up (Hong-Ou Mandel (HOM)-type). A $\frac{\lambda}{2}$ plate rotates the polarization of one input arm to the HOM setup to make the photons completely *distinguishable* as a reference. *Adapted from [100].*

open cavity comprising a lower DBR (35.5 pairs) and an upper external mirror attached to an optical fiber described in [128]. The two types of cavities allow the searching for similar QD states, and the tuning of these states to overlap their emission frequencies. After finding a QD in the fiber-DBR cavity which demonstrated significant anti-bunching and a narrow linewidth, denoted QD1 in Fig. 13.21, the planar DBR microcavity is searched for a second QD, denoted QD2 in Fig. 13.21, whose emission energy is within the ≈10 GHz tuning range of QD1's emission. The fiber cavity sample is glued to a piezoelectric transducer (PZT) so that changing the voltage applied to the PZT strains the sample and tunes the emission energy of the QD [129]. Using the PZT, they align the frequencies of the photons emitted by QD1 and QD2.

Both samples are excited by a common mode-locked laser with a repetition rate of 76 MHz (period ≈13 ns). The emission from the QDs is coupled through optical fibers to gratings and polarizers, and interferes at the beamsplitter of the $g^{(2)}(\tau)$ setup. The spectral and temporal overlaps for the excitonic photons are shown in Fig. 13.22.

The voltage bias on the PZT of the fiber cavity allows for good overlap. The photon fluxes for each QD are adjusted such that their emission intensities are the same; thus the areas under both curves are equal. The coherence times can be extracted from the linewidths; $T_2^{(QD1)} = 580$ ps and $T_2^{(QD2)} = 390$ ps. The linewidths and lifetimes are not externally controlled. The lifetimes are

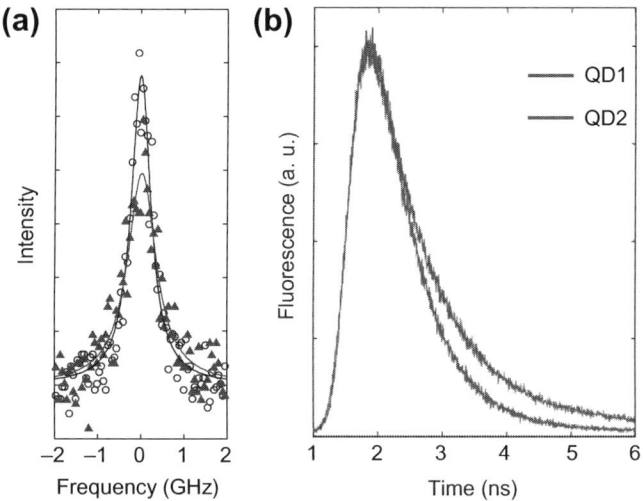

FIGURE 13.22 (a) Spectral alignment of the photons from two QDs at 8 K using a voltage on a piezoelectric transducer to strain-tune one of the QD states. (b) Temporal alignment of the two-photon wavepackets. There is no external tuning in (b). *Adapted from [100]*

$T_1^{(QD1)} = 610$ ps and $T_1^{(QD2)} = 950$ ps. For both QDs, $T_2 < 2T_1$, i.e. the coherence times are not lifetime-limited. Blocking the output from each of the cavities one at a time, the $g^{(2)}(\tau)$ set-up can be used to measure the antibunching properties of the individual photons. The $g^{(2)}(0)$ values for the two QD-photons are below 0.1, indicating quite pure single-photon emission.

Despite differences between the QDs in coherence time and lifetime, their photons still interfere. Figures 13.23a and b show the second-order correlation of the light exiting the two output ports of the interferometer for orthogonal and parallel polarizations, respectively. For parallel polarizations, the height of the $\tau = 0$ peak is lower than that for orthogonal polarizations (completely *distinguishable* photons), indicating that photons from the two different QDs have a non-zero coalescence probability. Figure 13.23c shows a close-up of the center peak for both relative polarizations. While the total coincidence counts in the $\tau = 0$ peak is not influenced by the time response of the detectors, the depth of the dip is reduced.

The central dip in Fig. 13.23c is caused by coalescence of the photons. The probability of coalescence is given by $P_c = \frac{A_\perp - A_\parallel}{A_\perp}$, where $A_{\perp,\parallel}$ is the integrated number of counts in $g^{(2)}_{\perp,\parallel}(\tau)$ during one repetition period around $\tau = 0$. From the data in Fig. 13.23 $P_c = 18\%$. Residual counts in the $\tau = 0$ peak remain because the QDs' coherence time is not lifetime-limited [131]. Though the photons' temporal extent is given by the QD lifetimes, T_1, the time over which they can interfere is given by the coherence times, T_2. Thus, the width of the peaks are determined by T_1, and the width of the dip is determined

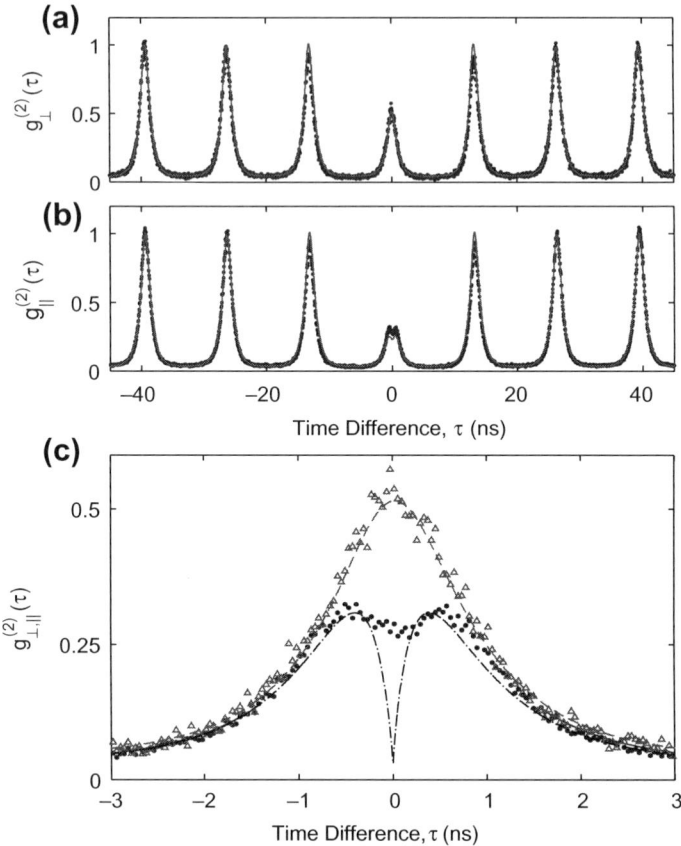

FIGURE 13.23 (a) Correlation of the interference for orthogonal polarizations. (b) Correlation of the interference for parallel polarizations. (c) Close-up of $\tau = 0$ peak for orthogonal (△) and parallel (●) polarizations. The dash curve in Fig. 13.23c is the result of a simulation based on [130] and shows the shape expected if the detectors were infinitely fast and as expected the single-QD $g^{(2)}(\tau)$ goes to zero at $\tau = 0$. From [100]. (For interpretation of the references to color in this figure legend, the reader is referred to the web version of this book.)

by T_2. If the coherence times were lifetime-limited we would have $T_2 = 2T_1$ and the dip would be wide enough to nearly eliminate the $\tau = 0$ peak. Some residual counts would remain because the two-QD lifetimes are different.

One of the differences between the two-photon interference experiments done by Santori *et al.* using photons from a single excitonic state and the experiment done above on photons from two different states is likely the spontaneous emission enhancement in the first example. Using the fully confined cavity, the radiative lifetime is reduced to 90 ps in one case in the single QD experiment, whereas in the two-QD experiment, fully confined cavities of low mode volume are not used and there is negligible spontaneous emission enhancement.

13.4 SINGLE DEFECTS IN DIAMOND - C. SANTORI

13.4.1 Introduction

Diamond contains hundreds of known optically active defects [132], a few of which have been investigated for single-photon generation and for use as spin-based quantum bits (qubits). Here, we begin by describing in detail the optical and spin properties of the most thoroughly studied defect in diamond, the nitrogen-vacancy (NV) center. While the NV center can be used simply as a single-photon source, it is the possibility of obtaining a single-photon source coupled to a long-lived matter qubit that makes this system exceptional, and of great interest for potential quantum networking and computation applications. We also discuss briefly some other known defects, which can act as brighter single-photon sources, but may not be as useful as spin qubits. We then describe current progress on coupling defects to optical structures such as solid-immersion lenses and microcavities, and finally we describe recent work on using single-photon emission, interference and measurement to entangle the spins of spatially separated spins, a potential route toward scalable quantum computing.

13.4.2 The Nitrogen-Vacancy Center

The nitrogen-vacancy center has, by now, been studied in great detail. This must be in part because the NV center is quite common and easily seen in many types of diamond. But more importantly, the NV center is exceptional in providing an electron spin with long-lived coherence that can be individually addressed at room temperature. These properties have enabled a number of pioneering experiments, including experiments on single-photon generation and quantum key distribution [42,43,133], optically detected magnetic resonance of single electronic and nuclear spins [134–140], magnetometry using single spins as probes [141,142], and most recently, remote entanglement of solid-state qubits [143]. In this section, we review the basic properties of the NV center, the first experiments on single-photon generation using this defect, and its capabilities as a spin qubit. Recent experiments involving quantum communication between NV centers are discussed later, in Section 13.4.5.

13.4.2.1 Structure and Formation

While the spectral signature of the NV center has long been known, the structure of the corresponding defect was established in the 1960s and 1970s through irradiation and annealing experiments combined with optical spectroscopy [144]. As shown in Fig. 13.24a, the NV center consists of a substitutional nitrogen atom next to a missing carbon atom (vacancy) in the diamond lattice. The axis connecting the nitrogen atom to the adjacent vacancy can lie along any of the four (1 1 1) crystallographic directions of the diamond lattice. Since either

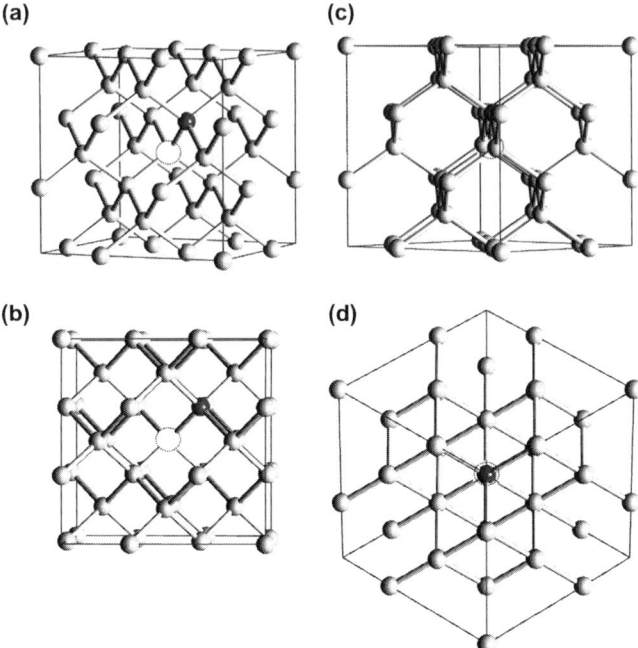

FIGURE 13.24 Structure of the nitrogen-vacancy center in diamond, as seen from (a) a low-symmetry direction, (b) along [1 0 0], (c) along [1 1 0], and (d) along [1 1 1]. The yellow spheres, blue spheres, and dashed circles represent carbon atoms, nitrogen atoms, and vacancies, respectively.

the nitrogen or vacancy can rest at a particular site, a total of eight configurations are allowed. The NV center has trigonal symmetry (point group C_{3v}) with a threefold rotational symmetry about the N-V axis. As discussed below, the NV center has two optical transitions with dipole moments orthogonal to the N-V axis. As a result, the polarization of light emitted by an NV center depends both on the orientation of the N-V axis, and on the direction of the light emission, and hence on the crystal orientation of the polished surface. Figure 13.24b–d illustrates this geometry for diamond samples polished along (1 0 0), (1 1 0), and (1 1 1). For a (1 0 0) surface, the magnitude of the angle between the N-V axis and the surface normal is the same for all orientations, but when collecting light normal to this surface, one will observe preferential polarization along either of two possible directions. For a (1 1 0) surface there are two orientations with the N-V axis parallel to the surface, providing poor optical access, while the other orientations, which are predominantly out-of-plane, have a more favorable collection geometry. For a (1 1 1) surface, there are three N-V orientations with poor optical access, but the fourth orientation has the N-V axis exactly normal to the surface, so that both dipole transitions can be optimally excited, and the emission most efficiently collected.

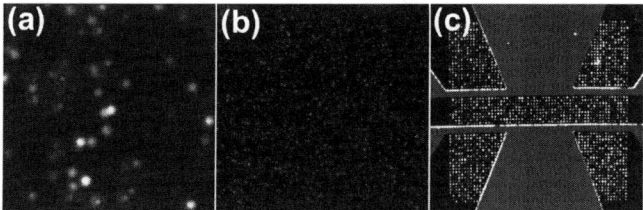

FIGURE 13.25 Confocal microscopy images of NV centers. In (a,b), the dependence of the emission intensity on laser polarization has been encoded into color. (a) NV centers in natural diamond, imaged through a (1 1 1) surface. The four orientations are clearly distinguished through their polarization dependence. Scan size: 12 μm. See Alegre et al., Ref. [145]. (b) NV centers in synthetic diamond grown on a (1 1 0) surface, showing preferential out-of-plane orientation. Scan size: 77 μm. See Edmonds et al., Ref. [146]. (c) NV centers made by nitrogen ion implantation through lithographically defined nano-apertures. *Image courtesy of David Toyli. See also Ref. [154].*

NV centers can be created in several ways. For diamond of a given purity, NV centers created during diamond growth appear to have the best low-temperature optical linewidths and spin coherence properties. Figure 13.25a shows an optical confocal microscopy image of single NV centers in a natural diamond sample. Since the surface of this sample is polished along (1 1 1), the four orientations of NV centers can be uniquely determined through the excitation polarization dependence alone [145]. It has recently been found [146] that in synthetic diamond grown by chemical vapor deposition (CVD) on a (1 1 0) surface, NV centers can be grown preferentially in the two out-of-plane orientations, $[1\,\bar{1}\,\bar{1}]$ and $[\bar{1}\,1\,\bar{1}]$, as shown in Fig. 13.25b. As mentioned above, if the NV center is excited non-resonantly, or at room temperature, the resulting mixture of light emitted from the two dipole transitions can be collected more efficiently for this geometry than for NV centers below a (1 0 0)-polished surface. It has also been shown that NV centers with good properties can be obtained in thin films grown by CVD [147–149].

To obtain a higher density of NV centers, or to obtain a useful density in ultra-pure diamond with nitrogen concentrations in the part-per-billion range, various implantation and irradiation techniques have been developed. For example, nitrogen ions may be implanted with depths ranging from a few nanometers to a few microns, depending on the chosen accelerating voltage. The sample is then annealed in an oxygen-free environment at temperatures ranging typically from 600 to 1000 °C, where vacancies become mobile and can combine with nitrogen impurities to form NV centers [150–152]. The efficiency of converting an implanted nitrogen atom into an NV center is typically only in the $\approx 5\%$ range, but may be increased using co-implantation with carbon to increase the number of vacancies [153]. The ion implantation may be performed through a lithographically defined mask [154], as shown in Fig. 13.25c, or through a scanned aperture [155,156], to control the positions where NV centers will form to within a few tens of nanometers. The optical and spin properties of

NV centers formed by nitrogen ion implantation and annealing are typically degraded, compared with as-grown NV centers in the same material. This is thought to result from interactions between the NV center and other defects in its vicinity resulting from implantation damage. Additional high-temperature annealing steps have been developed to recover better properties [157].

One can also convert existing nitrogen into NV centers using MeV electron irradiation to create vacancies, followed by annealing [144, 158, 159]. While this method, by itself, does not provide control over the position of the NV centers (the electrons can travel for millimeters through the crystal at the energies required to create vacancies), NV centers created by this method can have properties as good as those of NV centers created during crystal growth [160]. Implantation with other particles such as protons, neutrons, helium ions, and gallium ions may also be used [161–164].

None of the methods demonstrated to date can deterministically create a single NV center at a precise location and with a controlled orientation. Even deterministic ion implantation [165, 166] would not be sufficient, since the conversion efficiency from an implanted ion to an NV center is low using existing techniques.

13.4.2.2 Optical Transitions and Level Structure

Our current understanding of the NV center is the result of a number of experiments performed over the last few decades that have revealed the detailed level structure. The electron-spin resonance (ESR) signal from spin-triplet states was first reported by Loubser and van Wyk [167]; they also proposed the six-electron model for the negatively charged NV center that is still considered correct. Later, optically detected magnetic resonance (ODMR) [168] and spectral hole-burning [169] experiments established that the spin-triplet states are the ground states of NV^-, and revealed some information about the excited-state structure.

The basic level structure of NV^- is shown in Fig. 13.26. The ground states, denoted $^3A_2$, are formed from a single orbital state with three electron-spin sublevels. Due to a spin-spin interaction, the $m_s = \pm 1$ spin sublevels are approximately 2.88 GHz higher in energy than the lowest-energy $m_s = 0$ spin sublevel. These ground states are connected by optical transitions to a set of six excited states, denoted ^3E_x and ^3E_y, formed from two orbital states and three spin sublevels. The dipole moments of these optical transitions are orthogonal to the N-V axis (defined here to be the z-axis) and to each other. The spontaneous emission lifetime of these excited states is approximately 12 ns. In addition, the black arrows between the ^3E_x and ^3E_y orbital states represent phonon-assisted population relaxation between them, which becomes faster than the spontaneous decay to the ground states at temperatures above $\approx$20 K [170]. The excited states may also decay through a set of spin-singlet states, denoted $^1A_1$ and ^1E. This decay path affects primarily the $m_s = \pm 1$ spin sublevels of

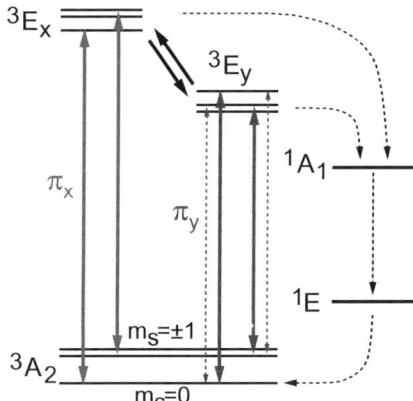

FIGURE 13.26 Schematic energy level diagram of the negatively charged NV center, showing the ground-state spin triplet levels ($^3A_2$), the optically excited triplet levels (^3E_x and ^3E_y), and spin singlet levels ($^1A_1$ and ^1E), along with the π_x- and π_y-polarized optical transitions and additional decay channels.

the excited states [171], and is important for spin polarization and readout of the NV center, as discussed below. This decay path includes a recently discovered optical transition at 1042 nm [172,173]. The lowest-energy singlet state has a relatively long lifetime of ≈200–400 ns, depending on temperature [173,174]. Thus, while the radiative efficiency from the $m_s = 0$ excited states has been estimated to be >0.7 [175], the presence of this fairly long-lived shelving state decreases the overall efficiency of the NV center as a single-photon source.

Figure 13.27 shows typical photoluminescence (PL) spectra from NV centers under various conditions. The spectrum in Fig. 13.27a was obtained from a dense ensemble of NV centers under non-resonant excitation (532 nm) at liquid-helium temperature. The main features that can be seen are the zero-phonon lines (ZPL) from NV^- and NV^0 at 637 nm and 575 nm, respectively, and the associated phonon sidebands (PSB). The dominant emission here is from NV^-, as is often seen in samples with a high concentration of nitrogen impurities. The phonon sideband features arise because of a strong, linear electron-phonon coupling term. Within the Franck-Condon picture, the electronic ground and excited states of the NV center each include a set of vibronic sublevels, corresponding to different motional states of the nuclei near the NV center. If the NV center begins in its electronic excited state and vibronic ground state, and transitions to its electronic ground state and vibronic ground state by emitting a photon (the zero-phonon process), the frequency of the emitted photon is well-defined. On the other hand, if the NV center transitions to an excited vibronic state, one or more phonons are created, and since these phonons have a continuum of energies, the emitted photon has a large spectral bandwidth. The Debye-Waller factor, which measures the fraction of the total

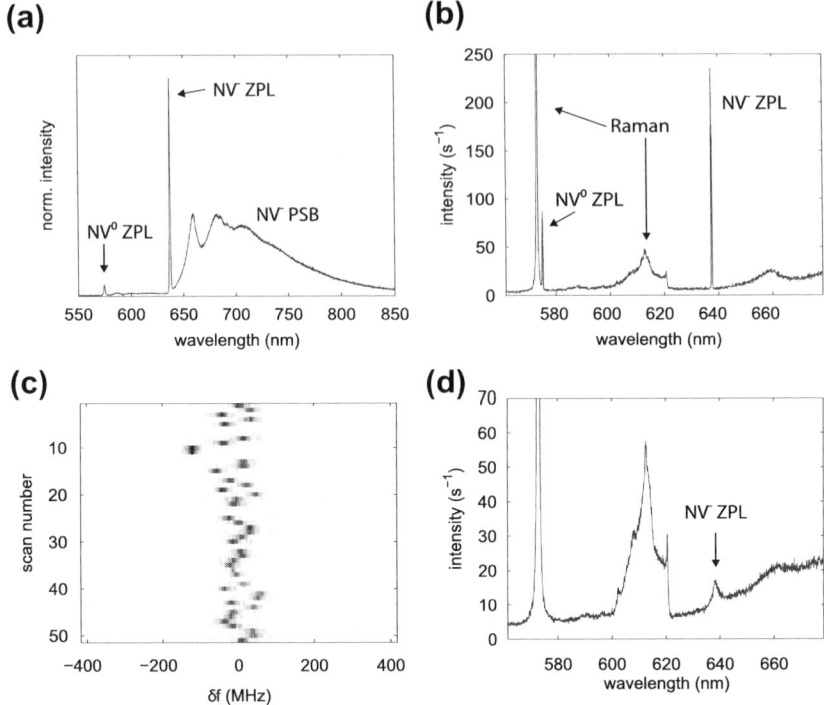

FIGURE 13.27 Typical optical spectra of NV centers: (a) Low-temperature photoluminescence spectrum from a dense ensemble under 532 nm excitation, showing predominantly NV^- emission. Only a few percent of the total emission occurs through the narrow zero-phonon line (ZPL) at 637 nm, with the remainder occurring through broad phonon sidebands (PSB) at longer wavelengths. (b) Low-temperature photoluminescence spectrum of a single NV center under 532 nm excitation, showing both NV^- and NV^0 emission, along with the familiar diamond Raman features of the surrounding crystal. (c) Low-temperature (6 K) photoluminescence excitation spectrum of a single NV center in high-purity diamond containing ≈1 ppb nitrogen, showing a nearly lifetime-broadened single-scan linewidth, with spectral diffusion of ≈100 MHz between scans. In this measurement, an intense 532 nm repump pulse was applied after each scan. Greyness from light to dark represents increased photon emission intensity. (d) Typical room-temperature spectrum from a single NV center, showing a broadened zero-phonon line.

photon emission occurring through the zero-phonon line, is determined by the lattice wavefunction overlap between the ground and excited states, which depends on the strength of the electron-phonon coupling. For NV^-, the Debye-Waller factor is only about 3%, a serious drawback for quantum communication applications requiring spectrally indistinguishable photons.

Figure 13.27b shows a typical PL spectrum obtained from a single NV center under intense (≈1 mW excitation at 532 nm). Even for single NV centers, one typically sees both NV^0 and NV^- emission under such excitation conditions, which cause the charge state to fluctuate rapidly in time. The photo-ionization process driving this charge fluctuation process has recently been studied in detail [176,177]. The ratio between the NV$^-$ and NV^0 ZPL intensities shown here is typical for high-purity diamond with nitrogen concentration in the part-per-billion range.

The most commonly used technique for measuring the low-temperature linewidth of the zero-phonon transitions is photoluminescence excitation (PLE) spectroscopy. In this technique, a laser is scanned across the zero-phonon lines near 637 nm while the photoluminescence intensity through the phonon sidebands is measured. Figure 13.27c shows such a measurement performed on a single NV center in high-purity diamond. In this measurement, each laser scan lasted 10s, and was followed by an intense repump pulse at 532 nm to reset the charge state. This is necessary to reverse photoionization that occurs eventually under resonant excitation. However, the repump pulse is also the main cause of the spectral jumps of the ZPL frequency that are seen to occur from scan to scan. The single-scan linewidth in this sample is below 20 MHz, close to the Fourier transform limit of 13 MHz [178]. In a simple PLE measurement, one usually sees only a single line, corresponding to excitation from the $m_s = 0$ ground state to the $m_s = 0$ spin sublevel of the upper orbital branch in the excited states. The other transitions are normally hidden due to optical pumping effects, but can be revealed by using either optical modulation [179,180] or microwave excitation [180,181] to reverse optical pumping.

At room temperature, the zero-phonon line becomes much broader, as shown in Fig. 13.27d. The phonon sidebands remain qualitatively similar, the most noticeable change being the appearance of anti-Stokes emission on the blue side of the ZPL in emission. Also, at room temperature, phonon-assisted population relaxation between the excited orbital states occurs so rapidly that these states are effectively "averaged" together, as far as the spin properties are concerned [182–184]. For single-photon generation, another consequence is that photons are emitted with random polarization, π_x or π_y, regardless of the excitation polarization.

The excited-state structure at low temperature has been studied in detail, experimentally [144,169,186,187,178,180,171,181,188,189], theoretically [190–194], and computationally [195,196]. Under perfect C$_{3v}$ symmetry, the excited-state energy levels are determined by a combination spin-orbit and spin-spin interactions. These levels, denoted $E_{1,2}$ (twofold degenerate), $E_{x,y}$ (twofold degenerate), A$_1$, and A$_2$, are shown in Fig. 13.28a. However, the random electric fields or strains present even in the best-quality diamond material are typically sufficient to split the energies of the orbital states by $\approx$10 GHz, an amount that is at least comparable to the strength of the spin-orbit and spin-spin interactions. Thus, NV centers with nearly perfect C$_{3v}$ symmetry are usually found only by extensive searching [181,188] or by applying external fields [178,160,185] to cancel the built-in fields. Figure 13.28b shows an example of using an externally applied electric field to change the orbital splitting. Here, the orbital splitting is seen in emission spectroscopy, using a high-resolution grating-based spectrometer and non-resonant excitation at 532 nm. Under such

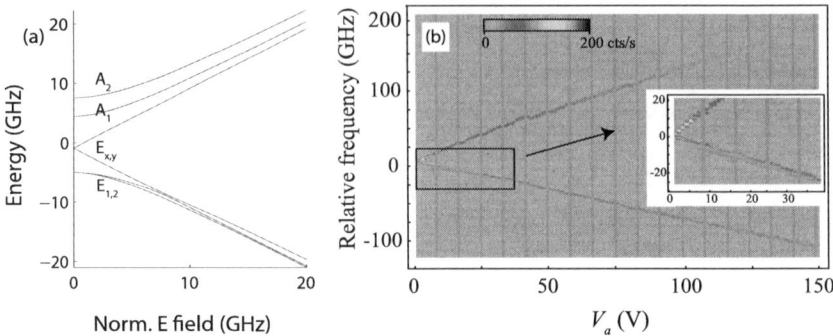

FIGURE 13.28 (a) Theoretical excited-state level structure as a function of electric field, in this case applied normal to the N-V axis and perpendicular to a reflection plane. (b) High-resolution photoluminescence spectra from a single NV center, plotted as a function of voltage applied across microfabricated electrodes on a diamond surface. An orbital splitting of up to 300 GHz is observed. A satellite peak appears in the lower orbital branch at intermediate electric field strengths, corresponding to non-spin-conserving optical transitions. From Acosta et al., Ref. [185]. Copyright 2012 by The American Physical Society.

excitation conditions, the spin of the NV center is polarized predominantly into the $m_s = 0$ state. However, for a range of orbital splittings, non-spin-conserving transitions occur in the lower orbital branch, allowing a satellite emission line to appear, corresponding to a transition to the $m_s = \pm 1$ ground states.

By varying the electric and magnetic fields applied to an NV center, a rich variety of energy level structures can be obtained, which can simultaneously include spin-conserving (cycling) transitions and non-spin-conserving transition. It has been shown that the cycling transitions can be used for single-shot readout of the electron spin [189]. The non-spin-conserving transitions can be used for spin initialization [189], to generate spin-photon entanglement [188], for coherent spin manipulation [297], and for electromagnetically induced transparency [298, 187, 299], with possible applications including quantum memories [200] and magnetometry [299].

13.4.2.3 Single-Photon Generation

The NV center was one of the first solid-state systems to show stable emission of single photons at room temperature. Initial demonstrations of photon antibunching under cw excitation at 532 nm [42] (Fig. 13.29a) or 514 nm [43] were followed by the demonstration of a triggered single-photon source using pulsed laser excitation [133] (Fig.13.29b). In these room-temperature demonstrations, most of the spectral bandwidth, including the phonon sidebands, was collected in order to obtain a high enough total count rate of $\approx 20{,}000$ s^{-1} for pulsed operation. Thus, the bandwidth of the single photons was ≈ 100 nm or more. Nevertheless, a proof-of-principle demonstration of BB84 quantum cryptography was performed successfully using this source [201].

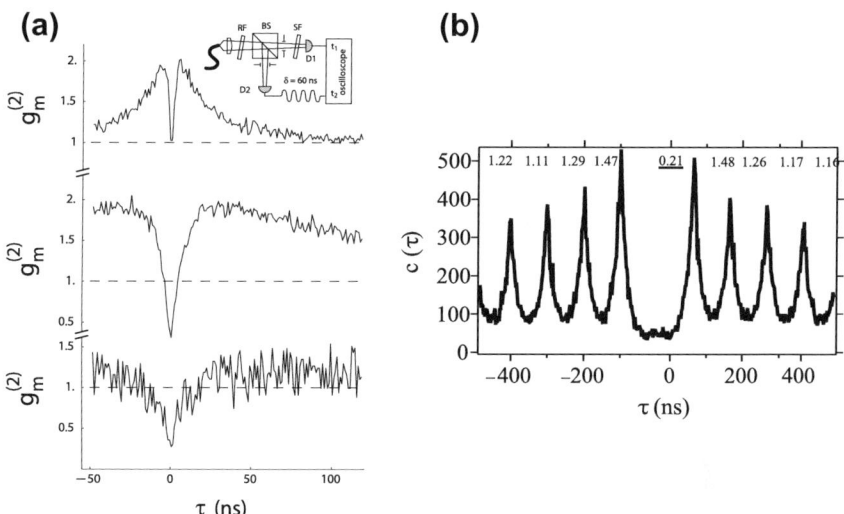

FIGURE 13.29 (a) Early photon antibunching demonstration using an NV center in bulk diamond, excited by a continuous-wave laser, from C. Kurtsiefer *et al.*, Ref. [42]. Copyright 2000 by The American Physical Society. (b) Pulsed single-photon generation from an NV center in a diamond nanocrystal, from Beveratos *et al.*, Ref. [133]. *With kind permission of The European Physical Journal (EPJ)*

As discussed below, nonresonant optical structures have been used to increase the count rate at room temperature by approximately a factor of ten. Diamond nanocrystals also allow for higher collection efficiency since they avoid the problem of total internal reflection at the diamond-air interface. Electrically driven single-photon generation from an NV center in a diamond diode structure has even been demonstrated [260]. The highest reported single-photon count rates from NV centers are now approaching 10^6 s^{-1} [189,202]. Resonant structures can also be used to increase the fraction of light emitted through the narrow zero-phonon line at low temperature to >75% of the total emission [203]. Nevertheless, the NV center appears to have a low overall efficiency, most likely a result of singlet-triplet and charge-state dynamics. Therefore, if one is interested only in generating single photons for applications such as quantum cryptography or linear-optics quantum computation [2], other defects, such as those discussed below, may provide a more ideal two-level structure with correspondingly higher efficiency. The main motivation now for single-photon generation with NV centers is for quantum communication with the long-lived spin states associated with this defect.

13.4.2.4 Spin Properties

Since the first optically detected magnetic resonance (ODMR) experiments were performed on single NV centers [134], it has been realized that the NV

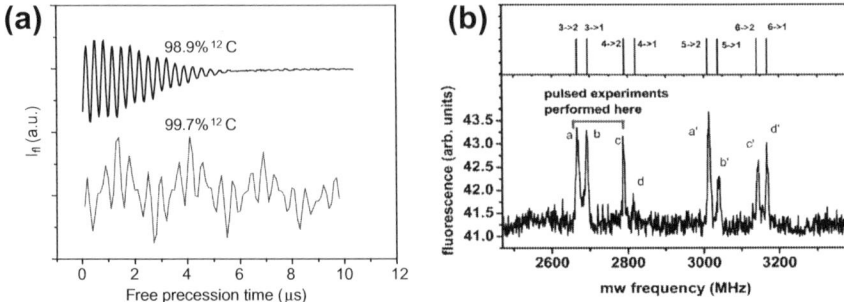

FIGURE 13.30 Optically detected magnetic resonance (ODMR) experiments: (a) Free-induction decay signals from single NV centers in bulk diamond with natural isotopic abundance (top curve), and in isotopically purified diamond (bottom curve). The complicated pattern results from hyperfine coupling between the electronic spin and the nitrogen nuclear spin of the NV center. From G. Balasubramanian et al., Ref. [204]. Reprinted by permission from Macmillan Publishers Ltd: Nature Materials, Copyright 2009. (G. Balasubramanian et al., Nat. Mater. 8, 383–387 (2009).) (b) ODMR spectrum showing hyperfine coupling to a nearest-neighbor ^{13}C nucleus. From Popa et al., Ref. [205]. Copyright 2004 by The American Physical Society.

center offered a special combination of long-lived spin coherence and individual addressability. The NV center is particularly well suited for ODMR at room temperature because, as noted above, non-resonant optical excitation results in a high degree of spin polarization into the $m_s = 0$ ground state, ≈0.8 [174]. Furthermore, because of the long shelving time in the singlet states, the photoluminescence intensity will initially be higher if the NV center begins in the $m_s = 0$ state, so the PL signal can be used for spin-state readout. Recently, it has been shown that higher initialization fidelities, as well as single-shot readout, can be obtained using resonant excitation at low temperature [189].

Because carbon and silicon consist primarily of isotopes with zero nuclear spin (^{12}C has 98.9% natural abundance and ^{28}Si has 92.2% natural abundance), paramagnetic impurities and defects in these materials can potentially have long-lived electron-spin coherence, provided that the concentration of other paramagnetic impurities is sufficiently low. Figure 13.30a, from Ref. [204], shows two examples of free-induction decay signals obtained from NV centers in high-purity diamond, either with natural isotopic abundance, or isotopically enriched. In such an experiment, the spins are first polarized into the $m_s = 0$ state using a laser pulse, typically of ≈1 μs duration. A microwave $\pi/2$ pulse is then applied to create a superposition between the $m_s = 0$ and $m_s = 1$ states. After a variable delay, a second microwave $\pi/2$ pulse is applied, which rotates the state to a new superposition of $m_s = 0$ and $m_s = 1$, with probability amplitudes varying sinusoidally with the delay. A final laser pulse is used to read out the resulting $m_s = 0$ population.

Currently, the best achievable free-induction decay lifetimes measured for single NV centers in isotopically purified materials are $T_2^* \approx 50 - 100$ μs [147]. Using a Hahn echo pulse sequence, in which an additional microwave

π-pulse is applied in between the two $\pi/2$-pulses to cancel magnetic noise that fluctuates on very long timescales, the coherence lifetime can be extended into the millisecond range [204]. Using dynamical decoupling sequences consisting of many microwave pulses, spin coherence lifetimes as long as 0.5 s have been reported [206].

One of the most spectacular achievements using NV centers has been the demonstration of controlled coupling between the electronic spin and one or more nuclear spins. This was first demonstrated using an NV center for which one of the nearest-neighbor carbon atoms is ^{13}C, which has nuclear spin of 1/2 and a 130 MHz hyperfine coupling with the electronic spin of the NV center. As shown in Fig. 13.30b, from Ref. [205], this hyperfine interaction is easily resolved in an ODMR spectrum, and thus it was possible to implement two-qubit gates between the electronic and nuclear spins [135]. More recently, it has been shown that the spin of the NV center can be controllably coupled to more distant ^{13}C nuclear spins [136,137], or to the nuclear spin of the nitrogen atom of the NV center [138–140]. Thus, the NV center can serve as a quantum "register" in which the electronic spin is used to control or detect several longer-lived nuclear spins. Recently, it has been shown that repeated probing of the electronic spin of an NV center at room temperature can allow single-shot readout of the nitrogen nuclear spin [207]. By combining these capabilities with single-photon generation, it is hoped that quantum repeater systems [208] and methods for scalable computation [209] may be developed.

13.4.3 Other Defects

While the nitrogen-vacancy center is quite interesting as a single-photon source coupled to a long-lived matter qubit, its slow radiative decay, small Debye-Waller factor, and apparently low overall efficiency due to shelving dynamics make it less than ideal if one is interested purely in single-photon generation.

A number of other defects in diamond have recently been investigated which, in some cases, have superior properties for single-photon generation. Much of this work has been summarized in two recent review articles [210,211], but here we briefly summarize some recent developments:

- Silicon-vacancy centers: these have an emission line at 738 nm, and are believed to consist of a silicon atom positioned between two vacancies. The first experiment reporting single-photon emission from this defect used a diamond crystal that was implanted with silicon ions and annealed [212]. This experiment reported a short photoluminescence lifetime of 1.2 ns, but very low count rates of $\approx 10^3$ s^{-1}. However, count rates up to 4.8×10^6 s^{-1} were recently reported for silicon-vacancy centers in nanodiamonds grown on an iridium substrate [213]. This is apparently the highest single-photon count rate measured to date for any defect in diamond. The Debye-Waller

factor was ≈0.8, the room-temperature linewidth was 0.7 nm, and the low-temperature linewidth was 0.17 nm.
- Suspected chromium-related defects: defects with emission lines in the 748 – 760 nm range have recently been discovered in diamond nanocrystals grown on sapphire [214], and in single-crystal diamond implanted with chromium and oxygen ions [215]. Single-photon count rates up to 3.2×10^6 s^{-1} were reported [214], with an 11 nm optical linewidth and a 3.7 ns excited-state lifetime.
- Nickel-related or other defects emitting in the near-infrared: Single-photon emission at 802 nm, with a count rate up to 7.5×10^4 s^{-1}, was reported from a defect identified as the NE8 center [216], thought to consist of a nickel atom next to four nitrogen atoms. Pulsed operation has also been achieved [217]. More recently, an unknown defect was found to emit at 734 nm, with a single-photon count rate as high as 1.6×10^6 s^{-1} [218].
- TR12 color center: This defect, thought to consist of an interstitial carbon atom within the diamond lattice, was recently used for single-photon generation at 470 nm wavelength [219]. The defects were produced through high-energy ion implantation of a diamond crystal.

While the study of these numerous defects in diamond continues, several impurities and defects in materials other than diamond have recently shown promise as single-photon emitters, including defects in silicon carbide [220] and zinc oxide [221], and rare-earth ions in crystals (Pr^{3+}:YAG) [222].

13.4.4 Optical Structures in Diamond

Diamond has a fairly high refractive index of $n \approx 2.4$, and for a dipole emitter placed below a planar diamond-air interface, one theoretically expects only ≈4.4% of the emitted light to escape into air when Fresnel losses are included. If immersion oil with $n = 1.5$ is used, the theoretical escape fraction can increase to ≈12%. Of course, the efficiencies are further reduced for actual microscope objectives with limited collection angles. Furthermore, for defects such as the NV center, only a small fraction of the total emission occurs through the zero-phonon line, which is the only part of the emission that can have a linewidth approaching the Fourier transform limit at low temperature, as needed for quantum communications schemes based on optical interference. Thus, there is a strong motivation to place defects such as the NV center into optical structures, both to increase the photon collection efficiency, and to increase the strength of the zero-phonon emission.

Making optical structures in diamond has been difficult because, unlike in III–V semiconductors, it is difficult to grow single-crystal diamond on other materials, and unlike in silicon, one cannot readily obtain a highly uniform, thin layer of diamond bonded to another transparent material with lower refractive index. Nevertheless, several types of optical structures in diamond have been successfully implemented, and some examples are shown in

Fig. 13.31. The first demonstration of spontaneous emission rate enhancement was achieved by coupling NV centers contained inside diamond nanoparticles to gold nanoparticles [225], shown in Fig. 13.31a. In this work, the spontaneous emission rate was increased by a factor up to 9. Other structures that can increase the collection efficiency by a factor of ≈10 without using narrowband resonances include diamond nanopillars [226] made by reactive-ion etching (Fig. 13.31b), and solid-immersion lenses [223,189] made using a focused ion beam (Fig. 13.31c). The solid-immersion lenses have been particularly successful. They do not require tuning, they can be accurately positioned relative to a pre-selected NV center [227,189], and perhaps most importantly, the NV center can be many microns away from any etched surfaces, and thus can have a high degree of spectral stability.

Figure 13.31d–f show examples of all-dielectric (in these cases diamond) microcavity structures that can support modes with high quality factors and small mode volumes; and thus, can be used to selectively enhance the zero-phonon emission. Figure 13.31d shows a microring cavity [228] coupled to a waveguide [224] for efficient light extraction. In such a device, the zero-phonon emission collected from one of the grating couplers was approximately 25 times as bright as a typical zero-phonon signal from an NV center in the unpatterned membrane. This geometry is especially appealing for the long-term goal of building integrating photonic networks [229]. Figure 13.31e shows a two-dimensional photonic crystal cavity [203]. In a similar device, a spontaneous emission enhancement factor as high as 70 was estimated when the cavity was tuned into resonance with an NV center. The fraction of emission occurring through the zero-phonon was estimated to be 75%, compared with ≈3% for bulk diamond. Figure 13.31e shows a nanobeam photonic crystal cavity made using a newly developed angle-etching technique [230]. This approach seems promising for reducing strain and simplifying the fabrication process.

The examples discussed above represent only a small subset of the approaches that have been tried. Other approaches include photonic crystal cavities made from single-crystal diamond grown on a silicon/Ir/YSZ substrate used to enhance emission from SiV centers [231], structures made using various combinations of focused-ion-beam and reactive-ion etching [232–234], structures made from polycrystalline diamond [235,236], gallium-phosphide structures coupled evanescently to bulk diamond [237–239], and diamond pillars coupled to silica microspheres [240], and diamond nanoparticles coupled to silica microspheres [241–243], silica microdisks [244], and gallium phosphide cavities [245–247].

Despite the significant progress that has been made in coupling single defects to cavities with small mode volumes, spectral diffusion of the optical transitions continues to be an obstacle for the use of these structures in quantum networking applications. In interference-based schemes, the visibility scales as $\gamma_{rad}/\gamma_{tot}$, where γ_{rad} is the natural linewidth, and γ_{tot} is the total linewidth including spectral diffusion. In some cases, the increased spectral diffusion linewidth

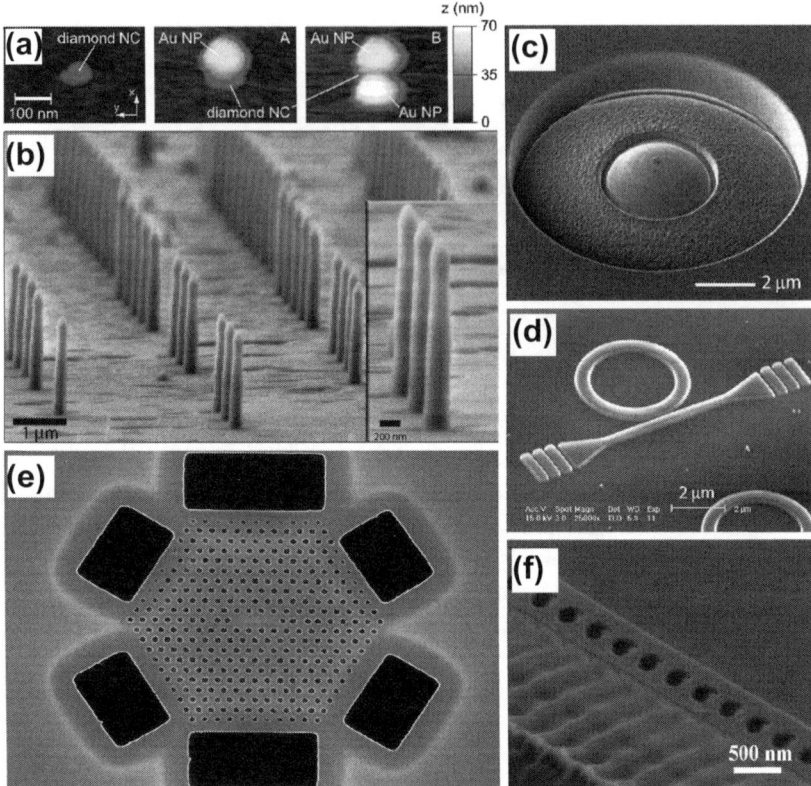

FIGURE 13.31 Optical structures used to enhance NV emission: (a) Gold nanoparticles next to diamond nanoparticles. Reprinted with permission from S. Schietinger, M. Barth, T. Aichele, and O. Benson, Nano Lett. 9, 1694–1698 (2009) [225]. Copyright 2009 American Chemical Society. (b) Diamond nano-pillars. Reprinted by permission from Macmillan Publishers Ltd: Nature Nanotech. T.M. Babinec, B.J.M. Hausmann, M. Khan, Y. Zhang, J.R. Maze, P.R. Hemmer, and M. Loncar, Nat. Nanotechnol. 5, 195–199 (2010) [226], Copyright 2010. (c) Diamond solid-immersion lenses. Courtesy of Sebastian Knauer, J.P. Hadden and Antony C. Stanley-Clarke; see also Ref. [223]. (d) A diamond micro-ring resonator coupled to a waveguide with grating couplers, from Faraon *et al.*, Ref. [224]. (e) A 2D photonic crystal cavity in diamond, from Faraon *et al.*, Ref. [203]. *Copyright 2012 by The American Physical Society.* (f) A nanobeam photonic crystal cavity made by angled reactive-ion etching of diamond. *Reprinted with permission from M.J. Burek, N.P. de Leon, B.J. Shields, B.J. Hausmann, Y. Chu, Q. Quan, A.S. Zibrov, H. Park, M.D. Lukin, and M. Loncar, Nano Lett. 12, 6084–6089 (2012) [230]. Copyright 2012 American Chemical Society.*

caused by proximity to an etched surface can completely offset any gain in collection efficiency. Realizing the full potential of these types of structures will require a combination of materials processing improvements, improved repump methods [176,177] and dynamic compensation [185]. Alternatively, there remains much room to explore structures with larger mode volumes, in which the dipole emitter can be farther from any surfaces.

13.4.5 Quantum Communication

While single NV centers have shown great promise as optically addressable spin qubits, a major challenge in this system (as well as other solid-state defects) is how to connect many qubits together to form scalable networks. Using a photonic network for this purpose could, in principle, allow for nanosecond gate times and micron-scale devices.

At present, numerous schemes have been proposed for quantum-optical networks, including deterministic [19] and nondeterministic (measurement-based) [249,208,250,209] approaches involving single photons. Recent developments suggest that measurement-based approaches may be applied successfully to solid-state defects such as the NV center, if certain technical challenges can be overcome. In Ref. [188], it was shown that if an NV center with nearly ideal C_{3v} symmetry is prepared in its A_2 excited state, the two decay paths, to the $m_s = \pm 1$ ground-state spin levels, emitted photons with orthogonal circular polarization. As a result, the polarization of the emitted photon becomes entangled with the final spin state of the NV center. Separately, as shown in Fig. 13.32, (reprinted from Ref. [248]), two-photon interference was demonstrated for photons emitted by two separate NV centers. Since this can only be observed using the zero-phonon portion of the emitted light, the recent development of solid-immersion lenses in diamond was crucial in obtaining sufficient count rates to perform this experiment. Most recently, two-photon interference from resonantly excited NV centers was used to demonstrate measurement-based entanglement formation between two distant NV centers [248]. The coincidence count rates were such that hours of integration were required to perform this experiment. Nevertheless, this is the first demonstration of its kind in any solid-state quantum system operating at optical frequencies, and represents a significant milestone in the field of solid-state quantum optics.

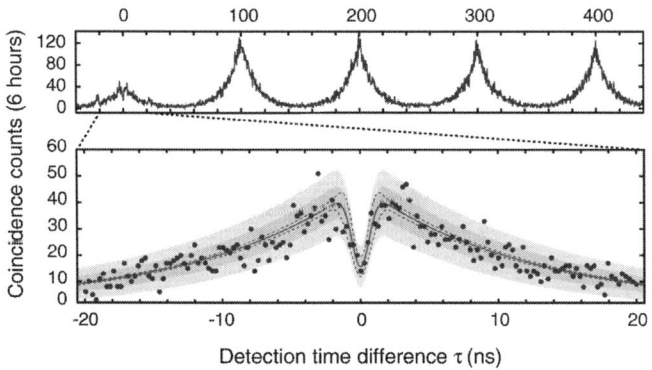

FIGURE 13.32 Demonstration of two-photon interference for light emitted by two different NV centers at low temperature. The drop in coincidences near $\tau = 0$ indicates an interference visibility of 66 ± 10%. From Bernien *et al.*, Ref. [248]. *Copyright 2012 by The American Physical Society.*

Scaling such measurement-based schemes to more than two qubits, or achieving technologically useful entanglement formation rates, will not be easy. As discussed above, the use of resonant cavities to selectively enhance the zero-phonon emission could allow for much higher efficiencies of indistinguishable photon generation. However, resonant structures have their own difficulties: their resonance frequency usually needs to be tuned (experiments so far have used gas condensation inside of a cryostat for tuning), and the spectral stability for NV centers close to an etched structure tends to be degraded. Spectral broadening in turn requires more spectral or temporal filtering to achieve a desired level of photon indistinguishability, and hence lower final count rates. On the other hand, the use of integrated optical networks could allow for stable path lengths (and hence phases) between components. Then, schemes that require only single-photon detection [249,208], but that also require interferometric stability, could be attempted.

13.4.6 Summary

As discussed above, diamond contains a number defects that can be used as optically pumped single-photon sources. Typically, these can operate stably at room temperature, though the spectral linewidth improves with cooling. Some of these defects have yielded single-photon count rates exceeding 10^6 s^{-1}. These may be further improved with the continued development of optical structures and resonators in diamond for enhancing the photon collection efficiency. The NV center is of particular interest because it provides a long-lived spin qubit that is optically addressable at the single-center level. In this system, single-photon emission is a promising route toward quantum communication among a network of spatially separated NV centers, an area of current active research.

13.5 FUTURE DIRECTIONS

We have discussed a variety of ways for producing single photons from simple quantum systems. A large fraction of these photon-production methods lead to on-demand emission of narrowband and indistinguishable photons into a well-defined mode of the radiation field, with efficiencies that can be very close to unity. Therefore these photons are ideal for all-optical quantum computation schemes, as proposed by Knill, Laflamme, and Milburn [2]. These sources are expected to play a significant role in the implementation of quantum networking [19] and quantum communication schemes [251].

The atom- and ion-based sources have already shown to be capable of entangling and mapping quantum states between atoms and photons [15,16]. Processes like entanglement swapping and teleportation between distant atoms or ions, that have first been studied without the aid of cavities [5,252,6,253,254] are beginning to profit enormously from the introduction of cavity-based techniques [24,17,18], as their success-probability scales with the square of the efficiency of the photon generation process. The high efficiency of cavity-based photon sources also opens new routes toward a highly scalable quantum network, which is essential for providing cluster states in one-way quantum computing [255] and for the quantum simulation of complex solid-state systems [256].

The QD system has benefited from a well-established and well-controlled fabrication process used in the semiconductor industry. This sources are highly efficient, showing nearly 80% efficiency in the emission of single photon out of the device [102], and like single atoms and ions, have excellent single-photon purity. Without employing optical cavities, these sources currently show less than adequate indistinguishability properties. However, based on techniques developed for planar microcavity quantum well devices; for instance, the vertical-cavity surface-emitting laser (VCSEL), the indistinguishability properties become very good [45]. Two-photon-number states have been produced from photons from separate QD states, as well as from a QD and PDC states [100,105]. More recently, entanglement of a spin state (from a QD charged exciton) and a photon has been demonstrated by three groups [257–259]. Like the development of photon sources from NV centers in diamond, this system is less mature than the single atoms and ions, yet progress is coming quickly, and we expect that more advanced demonstrations are on the near horizon.

Finally, diamond hosts a number of defects that can serve as stable single-photon emitters even at room temperature. The NV center in diamond is of particular interest since it provides a long-lived, optically addressable electronic spin coupled to a small number of nuclear spins that can serve as an excellent solid-state memory. This capability, and its potential application to quantum networks, motives the continued development and characterization of devices for efficient single-photon emission in this system.

ACKNOWLEDGMENTS

AK acknowledges partial support by the Engineering and Physical Sciences Research Council (QIP IRC and EP/E023568/1), and the Deutsche Forschungsgemeinschaft (Research Unit 635).

GSS acknowledge partial support from NIST and the NSF Physics Frontier Center at the Joint Quantum Institute. GSS would like to thank A. Muller, E. B. Flagg, T. Thomay, Sergey Polyakov and V. Loo for helpful discussions and experimental contributions.

REFERENCES

[1] D.P. DiVincenzo, "Real and Realistic Quantum Computers," Nature 393, 113–114 (1998).
[2] E. Knill, R. Laflamme, G.J. Milburn, "A Scheme for Efficient Quantum Computing with Linear Optics," Nature 409, 46–52 (2001).

[3] H.J. Kimble, M. Dagenais, L. Mandel, "Photon Antibunching in Resonance Fluorescence," Phys. Rev. Lett. 39, 691–695 (1977).
[4] F. Diedrich, H. Walther, "Nonclassical Radiation of a Single Stored Ion," Phys. Rev. Lett. 58, 203–206 (1987).
[5] B.B. Blinov, D.L. Moehring, L.-M. Duan, C. Monroe, "Observation of Entanglement Between a Single Trapped Atom and a Single Photon," Nature 428, 153–157 (2004).
[6] J. Volz, M. Weber, D. Schlenk, W. Rosenfeld, J. Vrana, K. Saucke, C. Kurtsiefer, H. Weinfurter, "Observation of Entanglement of a Single Photon with a Trapped Atom," Phys. Rev. Lett. 96, 030404 (2006).
[7] S. Olmschenk, D.N. Matsukevich, P. Maunz, D. Hayes, L.-M. Duan, C. Monroe, "Quantum Teleportation Between Distant Matter Qubits," Science 323, 486–489 (2009).
[8] T. Bondo, M. Hennrich, T. Legero, G. Rempe, A. Kuhn, "Time-Resolved and State-Selective Detection of Single Freely Falling Atoms," Opt. Commun. 264, 271–277 (2006).
[9] A. Kuhn, D. Ljunggren, "Cavity-Based Single-Photon Sources," Contemp. Phys. 51, 289–313 (2010).
[10] E.T. Jaynes, F.W. Cummings, "Comparison of Quantum and Semiclassical Radiation Theories with Application to the Beam Maser," Proc. IEEE 51, 89–109 (1963).
[11] B.W. Shore, P.L. Knight, "The Jaynes-Cummings Model," J. Mod. Opt. 40, 1195 (1993).
[12] E.M. Purcell, "Spontaneous Emission Probabilities at Radio Frequencies," Phys. Rev. 69, 681 (1946).
[13] H.J. Carmichael, "Photon Antibunching and Squeezing for a Single Atom in a Resonant Cavity," Phys. Rev. Lett. 55, 2790–2793 (1985).
[14] F. De Martini, G. Innocenti, G.R. Jacobovitz, P. Mataloni, "Anomalous Spontaneous Emission Time in a Microscopic Optical Cavity," Phys. Rev. Lett. 59, 2955–2958 (1987).
[15] T. Wilk, S.C. Webster, A. Kuhn, G. Rempe, "Single-Atom Single-Photon Quantum Interface," Science 317, 488 (2007).
[16] B. Weber, H.P. Specht, J. Bochmann, M. Mücke, D.L. Moehring, G. Rempe, "Photon-Photon Entanglement with a Single Trapped Atom," Phys. Rev. Lett. 102, 030501 (2009).
[17] S. Ritter, C. Nölleke, C. Hahn, A. Reiserer, A. Neuzner, M. Uphoff, M. Mücke, E. Figueroa, J. Bochmann, G. Rempe, "An Elementary Quantum Network of Single Atoms in Optical Cavities," Nature 484, 195–200 (2012).
[18] C. Nölleke, A. Neuzner, A. Reiserer, C. Hahn, G. Rempe, S. Ritter, "Efficient Teleportation Between Remote Single-Atom Quantum Memories," Phys. Rev. Lett. 110, 140403 (2013).
[19] J.I. Cirac, P. Zoller, H.J. Kimble, H. Mabuchi, "Quantum State Transfer and Entanglement Distribution Among Distant Nodes in a Quantum Network," Phys. Rev. Lett. 78, 3221–3224 (1997).
[20] D.P. DiVincenzo, "The Physical Implementation of Quantum Computation," Fortschr. Phys. 48, 771 (2000).
[21] J. Dilley, P. Nisbet-Jones, B.W. Shore, A. Kuhn, "Single-Photon Absorption in Coupled Atom-Cavity Systems," Phys. Rev. A 85, 023834 (2012).
[22] A.D. Boozer, A. Boca, R. Miller, T.E. Northup, H.J. Kimble, "Reversible State Transfer Between Light and a Single Trapped Atom," Phys. Rev. Lett. 98, 193601 (2007).
[23] M. Mücke, E. Figueroa, J. Bochmann, C. Hahn, K. Murr, S. Ritter, C. J. Villas-Boas, G. Rempe, "Electromagnetically Induced Transparency with Single Atoms in a Cavity," Nature 465, 755–758 (2010).
[24] H.P. Specht, C. Nölleke, A. Reiserer, M. Uphoff, E. Figueroa, S. Ritter, G. Rempe, "A Single-Atom Quantum Memory," Nature 473, 190 (2011).
[25] A. Kuhn, M. Hennrich, T. Bondo, G. Rempe, "Controlled Generation of Single Photons from a Strongly Coupled Atom-Cavity System," Appl. Phys. B 69, 373–377 (1999).
[26] N.V. Vitanov, M. Fleischhauer, B.W. Shore, K. Bergmann, "Coherent Manipulation of Atoms and Molecules by Sequential Laser Pulses," Adv. At. Mol. Opt. Phys. 46, 55–190 (2001).
[27] D.J. Heinzen, J.J. Childs, J.E. Thomas, M.S. Feld, "Enhanced and Inhibited Spontaneous Emission by Atoms in a Confocal Resonator," Phys. Rev. Lett. 58, 1320–1323 (1987).
[28] S.E. Morin, C.C. Yu, T.W. Mossberg, "Strong Atom-Cavity Coupling Over Large Volumes and the Observation of Subnatural Intracavity Atomic Linewidths," Phys. Rev. Lett. 73, 1489–1492 (1994).
[29] L.A. Lugiato, "Theory of Optical Bistability," in Progress in Optics, edited by E. Wolf, volume XXI, Elsevier Science Publishers, B. V., 71–216 (1984).

[30] C.K. Law, J.H. Eberly, "Arbitrary Control of a Quantum Electromagnetic Field," Phys. Rev. Lett. 76, 1055 (1996).
[31] C.K. Law, H.J. Kimble, "Deterministic Generation of a Bit-Stream of Single-Photon Pulses," J. Mod. Opt. 44, 2067–2074 (1997).
[32] A. Kuhn, G. Rempe, "Optical Cavity QED: Fundamentals and Application as a Single-Photon Light Source," in Experimental Quantum Computation and Information, edited by F. De Martini, C. Monroe, volume 148, IOS-Press, Amsterdam, 37–66 (2002).
[33] A. Kuhn, M. Hennrich, G. Rempe, "Strongly-Coupled Atom-Cavity Systems," in Quantum Information Processing, edited by T. Beth, G. Leuchs, Wiley-VCH, Berlin, 182–195 (2003).
[34] S.E. Harris, "Electromagnetically Induced Transparency with Matched Pulses," Phys. Rev. Lett. 70, 552–555 (1993).
[35] S.E. Harris, "Electromagnetically Induced Transparency," Phys. Today 50, 36 (1997).
[36] L.V. Hau, S.E. Harris, Z. Dutton, C.H. Behroozi, "Light Speed Reduction to 17 metres per second in an Ultracold Atomic Gas," Nature 397, 594–598 (1999).
[37] D.F. Phillips, A. Fleischhauer, A. Mair, R.L. Walsworth, M.D. Lukin, Storage of Light in Atomic Vapor," Phys. Rev. Lett. 86, 783–786 (2001).
[38] A. Messiah, "Quantum Mechanics," volume 2, J. Wiley & Sons, NY (1958).
[39] J. Kim, O. Benson, H. Kan, Y. Yamamoto, "A Single Photon Turnstile Device," Nature 397, 500–503 (1999).
[40] C. Brunel, B. Lounis, P. Tamarat, M. Orrit, "Triggered Source of Single Photons Based on Controlled Single Molecule Fluorescence," Phys. Rev. Lett. 83, 2722–2725 (1999).
[41] B. Lounis, W.E. Moerner, "Single Photons on Demand from a Single Molecule at Room Temperature, Nature 407, 491–493 (2000).
[42] C. Kurtsiefer, S. Mayer, P. Zarda, H. Weinfurter, "Stable Solid-State Source of Single Photons," Phys. Rev. Lett. 85, 290–293 (2000).
[43] R. Brouri, A. Beveratos, J.-P. Poizat, P. Grangier, "Photon Antibunching in the Fluorescence of Individual Color Centers in Diamond," Opt. Lett. 25, 1294–1296 (2000).
[44] P. Michler, A. Imamoğlu, M.D. Mason, P.J. Carson, G.F. Strouse, S.K. Buratto, "Quantum Correlation Among Photons from a Single Quantum Dot at Room Temperature," Nature 406, 968–970 (2000).
[45] C. Santori, M. Pelton, G. Solomon, Y. Dale, Y. Yamamoto, "Triggered Single Photons from a Quantum Dot," Phys. Rev. Lett. 86, 1502–1505 (2001).
[46] Z. Yuan, B.E. Kardynal, R.M. Stevenson, A.J. Shields, C.J. Lobo, K. Cooper, N.S. Beattie, D.A. Ritchie, M. Pepper, "Electrically Driven Single-Photon Source," Science 295, 102–105 (2002).
[47] A. Kuhn, M. Hennrich, G. Rempe, "Deterministic Single-Photon Source for Distributed Quantum Networking," Phys. Rev. Lett. 89, 067901 (2002).
[48] P.B.R. Nisbet-Jones, J. Dilley, D. Ljunggren, A. Kuhn, "Highly Efficient Source for Indistinguishable Single Photons of Controlled Shape," New J. Phys. 13, 103036 (2011).
[49] M. Hennrich, T. Legero, A. Kuhn, G. Rempe, "Photon Statistics of a Non-stationary Periodically Driven Single-Photon Source," New J. Phys. 6, 86 (2004).
[50] J. McKeever, A. Boca, A.D. Boozer, R. Miller, J.R. Buck, A. Kuzmich, H.J. Kimble, "Deterministic Generation of Single Photons from One Atom Trapped in a Cavity," Science 303, 1992–1994 (2004).
[51] M. Hijlkema, B. Weber, H.P. Specht, S.C. Webster, A. Kuhn, G. Rempe, "A Single-Photon Server with Just One Atom," Nat. Phys. 3, 253–255 (2007).
[52] Y. Colombe, T. Steinmetz, G. Dubois, F. Linke, D. Hunger, J. Reichel, Strong Atom-Field Coupling For Bose-Einstein Condensates in an Optical Cavity on a Chip, Nature 450, 272–276 (2007).
[53] M. Trupke, J. Goldwin, B. Darquié, G. Dutier, S. Eriksson, J. Ashmore, E.A. Hinds, "Atom Detection and Photon Production in a Scalable, Open, Optical Microcavity," Phys. Rev. Lett. 99, 063601 (2007).
[54] B. Dayan, A.S. Parkins, T. Aoki, E.P. Ostby, K.J. Vahala, H.J. Kimble, "A Photon Turnstile Dynamically Regulated by One Atom," Science 319, 1062–1065 (2008).
[55] T. Aoki, A.S. Parkins, D.J. Alton, C.A. Regal, B. Dayan, E. Ostby, K.J. Vahala, H.J. Kimble, "Efficient Routing of Single Photons by One Atom and a Microtoroidal Cavity," Phys. Rev. Lett. 102, 083601 (2009).

[56] M. Pöllinger, D. O'Shea, F. Warken, A. Rauschenbeutel, "Ultrahigh-Q Tunable Whispering-Gallery-Mode Microresonator," Phys. Rev. Lett. 103, 053901 (2009).
[57] M. Keller, B. Lange, K. Hayasaka, W. Lange, H. Walther, "Continuous Generation of Single Photons with Controlled Waveform in an Ion-Trap Cavity System," Nature 431, 1075–1078 (2004).
[58] G.R. Guthörlein, M. Keller, K. Hayasaka, W. Lange, H. Walther, "A Single Ion as a Nanoscopic Probe of an Optical Field," Nature 414, 49–51 (2001).
[59] C. Russo, H. Barros, A. Stute, F. Dubin, E. Phillips, T. Monz, T. Northup, C. Becher, T. Salzburger, H. Ritsch, P. Schmidt, R. Blatt, "Raman Spectroscopy of a Single Ion Coupled to a High-Finesse Cavity," Appl. Phys. B 95, 205–212 (2009).
[60] H.G. Barros, A. Stute, T.E. Northup, C. Russo, P.O. Schmidt, R. Blatt, "Deterministic Single-Photon Source from a Single Ion," New J. Phys. 11, 103004 (2009).
[61] T. Legero, T. Wilk, M. Hennrich, G. Rempe, A. Kuhn, "Quantum Beat of Two Single Photons," Phys. Rev. Lett. 93, 070503 (2004).
[62] T. Legero, T. Wilk, A. Kuhn, G. Rempe, "Characterization of Single Photons Using Two-Photon Interference," Adv. At. Mol. Opt. Phys. 53, 253 (2006).
[63] T. Legero, T. Wilk, A. Kuhn, G. Rempe, "Time-Resolved Two-Photon Quantum Interference," Appl. Phys. B 77, 797–802 (2003).
[64] G.S. Vasilev, D. Ljunggren, A. Kuhn, "Single Photons Made-to-Measure," New J. Phys. 12, 063024 (2010).
[65] P.B.R. Nisbet-Jones, J. Dilley, A. Holleczek, O. Barter, A. Kuhn, "Photonic Qubits, Qutrits and Ququads Accurately Prepared and Delivered on Demand," New J. Phys. 15, 053007 (2013).
[66] A.J. Bennett, D.G. Gevaux, Z.L. Yuan, A.J. Shields, P. Atkinson, D.A. Ritchie, "Experimental Position-Time Entanglement with Degenerate Single Photons," Phys. Rev. A 77, 023803 (2008).
[67] D.J. Eaglesham, M. Cerullo, "Low Temperature Growth of Ge on Si(100)," Appl. Phys. Lett. 58, 2276–2279 (1991).
[68] B. Elman, E.S. Koteles, P. Melman, C. Jagannath, J. Lee, D. Dugger, In-situ Measurements of Critical Layer Thickness and Optical Studies of InGaAs Quantum Wells Grown on GaAs Substrates," Appl. Phys. Lett. 55, 1659–1661 (1989).
[69] M. Tabuchi, S. Nodaand, A. Sasaki, "Mesoscopic Structure in Lattice-Mismatched Heteroepitaxial Interface Layers," in Science and Technology of Mesoscopic Structures, edited by S. Namba, C. Hamaguchi, T. Ando, Springer-Verlag, Tokyo, Japan, 379–384 (1992).
[70] D. Leonard, M. Krishnamurthy, C.M. Reaves, S.P. Denbaars, P.M. Petroff, "Direct Formation of Quantum-Sized Dots from Uniform Coherent Islands of InGaAs on GaAs Surfaces," Appl. Phys. Lett. 63, 3203–3205 (1993).
[71] W.A. Tiller, "The Science of Crystallization: Microscopic Interfacial Phenomena," Cambridge University Press, Cambridge, England (1991).
[72] H. Gao, "A Boundary Perturbation Analysis For Elastic Inclusions and Interfaces," Int. J. Solids Struct. 28, 703–725 (1991).
[73] L.B. Freund, F. Jonsdottir, "Instability of a Biaxial Stressed Thin Film on a Substrate Due to Material Diffusion," J. Mech. Phys. Solids 41, 1245–1264 (1993).
[74] B.G. Orr, D. Kessler, C.W. Synder, L. Sander, "A Model for Strain-Induced Roughening and Coherent Island Growth," Europhys. Lett. 19, 33 (1992).
[75] D. Vanderbilt, L.K. Wickham, "Elastic Energies of Coherent Germanium Islands on Silicon," Mat. Res. Symp. Proc. 202, 555 (1991).
[76] J.W. Matthews, A.E. Blakeslee, "Defects in Epitaxial Multilayers. I. Misfit Dislocations," J. Cryst. Growth 27, 118 (1974).
[77] R. People, J.C. Bean, "Calculations of Critical Layer Thickness Versus Lattice Mismatch for GeSi/Si Strain-Layer Heterostructures," Appl. Phys. Lett. 47, 322 (1985).
[78] I.N. Stranski, L. Krastanow, "Abhandlungen der Mathematisch-Naturwissenschaftlichen Klasse IIb," Akademie der Wissenschaften Wien 146, 797–810 (1938).
[79] D.J. Eaglesham, M. Cerullo, "Dislocation-Free Stranski-Krastanow Growth of Ge on Si(100)," Phys. Rev. Lett. 64, 1943–1946 (1990).

[80] S. Guha, A. Madhukar, K.C. Rajkumar, "Onset of Incoherency and Defect Introduction in the Initial Stages of Molecular Beam Epitaxical Growth of Highly Strained $In_xGa_{1-x}As$ on GaAs(100)," Appl. Phys. Lett. 57, 210 (1990).
[81] G.S. Solomon, C. Santori, M. Pelton, J. Vučković, D.F.E. Waks, Y. Yamamoto, "Efficient, Regulated Single Photons from Quantum Dots in Post Microcavities," in Optics of Quantum Dots And Wires, edited by G. W. Bryant, G. S. Solomon, Artech House, Boston, MA, 483–536 (2005).
[82] P.W. Baumeister, "Optical Absorption of Cuprous Oxide," Phys. Rev. 121, 359–362 (1961).
[83] S. Fafard, Z.R. Wasilewski, C.N. Allen, M.S.D. Picard, J.P. McCaffrey, P.G. Piva, "Manipulating the Energy Levels of Semiconductor Quantum Dots," Phys. Rev. B 59, 15368 (1999).
[84] E. Poem, Y. Kodriano, C. Tradonsky, B.D. Gerardot, P.M. Petroff, D. Gershoni, "Radiative Cascades from Charged Semiconductor Quantum Dots," Phys. Rev. B 81, 085306 (2010).
[85] G. Bester, S. Nair, A. Zunger, "Pseudopotential Calculation of the Excitonic Fine Structure of Million-Atom Self-Assembled $In_{1-x}Ga_xAs$-GaAs Quantum Dots," Phys. Rev. B 67, 161306–161310 (2003).
[86] C.Y. Lu, Y. Zhao, A.N. Vamivakas, C. Matthiesen, S. Fält, A. Badolato, M. Atatüre, "Direct Measurement of Spin Dynamics in InAs/GaAs Quantum Dots Using Time-Resolved Resonance Fluorescence," Phys. Rev. B 81, 035332–035337 (2010).
[87] E.A. Chekhovich, M.M. Glazov, A.B. Krysa, M. Hopkinson, P. Senellart, A. Lemaître, M.S. Skolnick, A.I. Tartakovskii, "Element-Sensitive Measurement of the Holenuclear Spin Interaction in Quantum Dots," Nat. Phys. 9, 74–78 (2013).
[88] S.T. Ylmaz, P. Fallahi, A. İmamoğlu, "Quantum-Dot-Spin Single-Photon Interface," Phys. Rev. Lett. 105, 033601–033605 (2010).
[89] M. Bayer, O. Stern, A. Kuther, A. Forchel, "Spectroscopic Study of Dark Excitons in $In_xGa_{1-x}As$ Self-Assembled Quantum Dots by a Magnetic-Field-Induced Symmetry Breaking," Phys. Rev. B 61, 7293–7276 (2000).
[90] V.D. Kulakovskii, G. Bacher, R. Weigand, T. Kmmell, A. Forchel, E. Borovitskaya, K. Leonardi, D. Hommel, "Fine Structure of Biexciton Emission in Symmetric and Asymmetric CdSe/ZnSe Single Quantum Dots," Phys. Rev. Lett. 82, 1780–1783 (1999).
[91] M. Bayer, G. Ortner, O. Stern, A. Kuther, A.A. Gorbunov, A. Forchel, P. Hawrylak, S. Fafard, K. Hinzer, T.L. Reinecke, S.N. Walck, J.P. Reithmaier, F. Klopf, F. Schäfer, "Fine Structure of Neutral and Charged Excitons in Self-Assembled In(Ga)As/Al(Ga)As Quantum Dots," Phys. Rev. B 65, 195315–195338 (2002).
[92] M. Bayer, O. Stern, P. Hawrylak, S. Fafard, A. Forchel, "Hidden Symmetries in the Energy Levels of Excitonic 'Artificial Atoms'," Nature 405, 923–926 (2000).
[93] X. Xu, Y. Wu, B. Sun, Q. Huang, J. Cheng, D.G. Steel, A.S. Bracker, D. Gammon, C. Emary, L.J. Sham, Fast Spin State Initialization in a Singly Charged InAs-GaAs Quantum Dot by Optical Cooling," Phys. Rev. Lett. 99, 097401–097404 (2007).
[94] J.Y. Marzin, J.M.Gérard, A. Izraël, D. Barrier, G. Bastard, "Photoluminescence of Single InAs Quantum Dots Obtained by Self-Organized Growth on GaAs," Phys. Rev. Lett. 73, 716–719 (1994).
[95] G.S. Solomon, M. Pelton, Y. Yamamoto, "Single-Mode Spontaneous Emission from a Single Quantum Dot in a Three-Dimensional Microcavity," Phys. Rev. Lett. 86, 3903–3906 (2001).
[96] A. Muller, W. Fang, J. Lawall, G.S. Solomon, "Emission Spectrum of a Dressed Exciton-Biexciton Complex in a Semiconductor Quantum Dot," Phys. Rev. Lett. 101, 027401–027404 (2008).
[97] Z. Liu, B.B. Goldberg, S.B. Ippolito, A.N. Vamıvakas, M.S. Ünlü, R. Mirin, "Ehigh Resolution, High Collection Efficiency in Numerical Aperture Increasing Lens Microscopy of Individual Quantum Dots," Appl. Phys. Lett. 87, 071905–071907 (2005).
[98] M. Atatüre, J. Dreiser, A. Badolato, A. Imamoglü, "Observation of Faraday Rotation from a Single Confined Spin," Nat. Phys. 3, 101–106 (2007).
[99] A. Dousse, L. Lanco, J. Suffczyński, E. Semenova, A. Miard, A. Lemaître, I. Sagnes, C. Roblin, J. Bloch, P. Senellart, "Controlled Light-Matter Coupling for a Single Quantum Dot Embedded in a Pillar Microcavity Using Far-Field Optical Lithography," Phys. Rev. Lett. 101, 267404–267407 (2008).
[100] E.B. Flagg, A. Muller, S.V. Polyakov, A.M. Ling, G.S. Solomon, Interference of Two Photons from Separate Quantum Dots," Phys. Rev. Lett. 104, 137401–137405 (2010).

[101] P. Michler, A. Kiraz, C. Becher, W.V. Schoenfeld, P.M. Petroff, L. Zhang, E. Hu, A. Imamoğlu, "A Quantum Dot Single-Photon Turnstile Device," Science 290, 2282–2285 (2000).
[102] O. Gazzano, S.M. de Vasconcellos, C. Arnold, A. Nowak, E. Galopin, I. Sagnes, L. Lanco, A. Lemaîtee, P. Senellart, "Bright Solid-State Sources of Indistinguishable Single Photons," Nat. Commun. 4, 1425 (2013).
[103] A. Ulhaq, S. Weiler, S.M. Ulrich, R. Roßbach, M. Jetter, P. Michler, "Cascaded Single-Photon Emission from the Mollow Triplet Sidebands of a Quantum Dot," Nature Photonics 6, 238–242 (2012).
[104] Y.-M. He, Y. He, Y.-J. Wei, D. Wu, M. Atatüre, C. Schneider, S. Höfling, M. Kamp, C.-Y. Lu, J.-W. Pan, "On-demand Semiconductor Single-Photon Source with near-Unity Indistinguishability," Nat. Nanotech. 8, 213–217 (2013).
[105] S.V. Polyakov, A. Muller, E.B. Flagg, A. Ling, N. Borjemscaia, E.V. Keuren, A. Migdall, G.S. Solomon, "Coalescence of Single Photons Emitted by Disparate Single-Photon Sources: The Example of InAs Quantum Dots and Parametric Down-Conversion Sources," Phys. Rev. Lett. 107, 157402–157406 (2011).
[106] E.B. Flagg, S.V. Polyakov, T. Thomay, G.S. Solomon, "Dynamics of Nonclassical Light from a Single Solid-State Quantum Emitter," Phys. Rev. Lett. 109, 163601–163605 (2012).
[107] M. Pelton, C. Santori, J. Vučković, B. Zhang, G.S. Solomon, J. Plant, Y. Yamamoto, "Efficient Source of Single Photons: A Single Quantum Dot in a Micropost Microcavity," Phys. Rev. Lett. 89, 233602–233605 (2002).
[108] M. Pelton, J. Vučković, G.S. Solomon, A. Scherer, Y. Yamamoto, "Three-Dimensionally Confined Modes in Micropost Microcavities: Quality Factors and Purcell Factors," IEEE Q. Electron. (2001).
[109] H. Kim, D. Sridharan, T.C. Shen, G. Solomon, E. Waks, "Magnetic Field Tuning of Two Quantum Dot Spin States to a Photonic Crystal Cavity in the Strong Coupling Regime," Opt. Express 19, 2589–2598 (2011).
[110] J.P. Reithmaier, G. Sęk, A. Löffler, C. Hofmann, S. Kuhn, S. Reitzenstein, L.V. Keldysh, V.D. Kulakovskii, T.L. Reinecke, A. Forchel, "Strong Coupling in a Single Quantum Dot Semiconductor Microcavity System," Nature 432, 197–200 (2004).
[111] K. Srinivasan, O. Painter, "Linear and Nonlinear Optical Spectroscopy of a Strongly Coupled Microdisk Quantum Dot System," Nature 450, 862–866 (2007).
[112] E. Peter, P. Senellart, D. Martrou, A. Lemaître, J. Hours, J.M. Gérard, J. Bloch, "Exciton-Photon Strong-Coupling Regime for a Single Quantum Dot Embedded in a Microcavity," Phys. Rev. Lett. 95, 067401–067404 (2005).
[113] Z.G. Xie, S.G. tzinger, W. Fang, H. Cao, G.S. Solomon, "Influence of a Single Quantum Dot State on the Characteristics of a Microdisk Laser," Phys. Rev. Lett. 98, 117401–117404 (2007).
[114] V. Loo, L. Lanco, A. Lemaître, I. Sagnes, O. Krebs, P. Voisin, P. Senellart, "Quantum Dot-Cavity Strong-Coupling Regime Measured Through Coherent Reflection Spectroscopy in a Very High-Q micropillar," Appl. Phys. Lett. 97, 241110–241112 (2010).
[115] D. Englund, D. Fattal, E. Waks, G. Solomon, B. Zhang, T. Nakaoka, Y. Arakawa, Y. Yamamoto, J.J. Vučković, "Controlling the Spontaneous Emission Rate of Single Quantum Dots in a Two-Dimensional Photonic Crystal," Phys. Rev. Lett. 95, 013904–013907 (2005).
[116] K. Hennessy, A. Badolato, M. Winger, D. Gerace, M. Atatüre, S. Gulde, S. Fält, E.L. Hu, A. Imamoğlu, "Quantum Nature of a Strongly Coupled Single Quantum Dot Cavity System," Nature 445, 896–899 (2007).
[117] S. Kohmoto, H. Nakamura, T. Ishikawa, K. Asakawa, "Site-Controlled Self-Organization of Individual InAs Quantum Dots by Scanning Tunneling Probe-Assisted Nanolithography," Appl. Phys. Lett. 75, 3488–3900 (1999).
[118] D. Chithrani, R.L. Williams, J. Lefebvre, P.J. Poole, G.C. Aers, "Optical Spectroscopy of Single, Site-Selected, InAs-InP Self-Assembled Quantum Dots," Appl. Phys. Lett. 84, 978–980 (2004).
[119] M.H. Baier, S. Watanabe, E. Pelucchi, E. Kapon, "Site-Controlled Self-Organization of Individual InAs Quantum Dots by Scanning Tunneling Probe-Assisted Nanolithography," Appl. Phys. Lett. 84, 1943–1945 (2004).

[120] P. Atkinson, S. Kiravittaya, M. Benyoucef, A. Rastelli, O.G. Schmidt, "Site-Controlled Growth and Luminescence of InAs Quantum Dots Using In Situ Ga-Assisted Deoxidation of Patterned Substrates," Appl. Phys. Lett. 93, 101908–101910 (2008).
[121] Z. Xie, G.S. Solomon, "Spatial Ordering of Quantum Dots in Microdisks," Appl. Phys. Lett. 87, 093106–093108 (2005).
[122] S. Haroche, J.-M. Raimond, "Exploring the Quantum: Atoms, Cavities and Photons," 1st edn., Oxford University Press, New York, NY (2006).
[123] C.H. Bennett, G. Brassard, "Quantum Cryptography: Public Key Distribution and Coin Tossing," Proceedings of the IEEE International Conference on Computers, Systems, and Signal Processing, Bangalore, p. 175 (1984).
[124] R. Loudon, "The Quantum Theory of Light," 3rd edn., Oxford University Press (2000).
[125] C.K. Hong, Z.Y. Ou, L. Mandel, "Measurement of Subpicosecond Time Intervals Between Two Photons by Interference," Phys. Rev. Lett. 59, 2044–2046 (1987).
[126] P. Grangier, "Single Photons Stick Together," Nature 419, 577 (2002).
[127] C. Santori, D. Fattal, J. Vuckovic, G.S. Solomon, Y. Yamamoto, Indistinguishable Photons from a Single-Photon Device," Nature 419, 594–597 (2002).
[128] A. Muller, E.B. Flagg, M. Metcalfe, J. Lawall, G.S. Solomon, Coupling an Epitaxial Quantum Dot to a Fiber-Based External-Mirror Microcavity," Appl. Phys. Lett. 95, 173101 (2009).
[129] S. Seidl, M. Kroner, A. Hogele, K. Karrai, "Effect of Uniaxial Stress on Excitons in a Self-Assembled Quantum Dot," Appl. Phys. Lett. 88, 203113 (2006).
[130] A. Kiraz, M. Atature, A. Imamoglu, "Quantum-Dot Single-Photon Sources: Prospects for Applications in Linear Optics Quantum-Information Processing," Phys. Rev. A 69, 032305 (2004).
[131] J. Bylander, I. Robert-Philip, I. Abram, "Interference and Correlation of Two Independent Photons," Euro. Phys. J. D 22, 295–301 (2003).
[132] A. Zaitsev, "Optical Properties of Diamond: a Data Handbook," Springer (2001).
[133] A. Beveratos, S. Kuhn, R. Brouri, T. Gacoin, J.-P. Poizat, P. Grangier, "Room Temperature Stable Single-Photon Source," Euro. Phys. J. D 18, 191–196 (2002).
[134] A. Gruber, A. Dräbenstedt, C. Tietz, L. Fleury, J. Wratchtrup, C.V. Borczyskowski, "Scanning Confocal Optical Microscopy and Magnetic Resonance on Single Defect Centers," Science 276, 2012–2014 (1997).
[135] F. Jelezko, T. Gaebel, I. Popa, M. Domhan, A. Gruber, J. Wrachtrup, "Observation of Coherent Oscillation of a Single Nuclear Spin and Realization of a Two-Qubit Conditional Quantum Gate," Phys. Rev. Lett. 93, 130501 (2004).
[136] L. Childress, M.V. Gurudev Dutt, J.M. Taylor, A.S. Zibrov, F. Jelezko, J. Wrachtrup, P.R. Hemmer, M.D. Lukin, "Coherent Dynamics of Coupled Electron and Nuclear Spin Qubits in Diamond," Science 314, 281–285 (2006).
[137] M.V. Gurudev Dutt, L. Childress, L. Jiang, E. Togan, J. Maze, F. Jelezko, A.S. Zibrov, P.R. Hemmer, M.D. Lukin, "Quantum Register Based on Individual Electronic and Nuclear Spin Qubits in Diamond," Science 316, 1312–1316 (2007).
[138] R. Hanson, F. Mendoza, R. Epstein, D. Awschalom, "Polarization and Readout of Coupled Single Spins in Diamond," Phys. Rev. Lett. 97, 87601 (2006).
[139] T. Gaebel, M. Domhan, I. Popa, C. Wittmann, P. Neumann, F. Jelezko, J.R. Rabeau, N. Stravrias, A.D. Greentree, S. Prawer, J. Meijer, J. Twamley, P.R. Hemmer, J. Wrachtrup, "Room-Temperature Coherent Coupling of Single Spins in Diamond," Nat. Phys. 2, 408. (2006)
[140] G.D. Fuchs, G. Burkard, P.V. Klimov, D.D. Awschalom, "A Quantum Memory Intrinsic to Single Nitrogen-Vacancy Centres in Diamond," Nat. Phys. 7, 789–793 (2011).
[141] G. Balasubramanian, I.Y. Chan, R. Kolesov, M. Al-Hmoud, J. Tisler, C. Shin, C. Kim, A. Wojcik, P.R. Hemmer, A. Krueger, T. Hanke, A. Leitenstorfer, R. Bratschitsch, F. Jelezko, J. Wrachtrup, "Nanoscale Imaging Magnetometry with Diamond Spins Under Ambient Conditions," Nature (London) 455, 648–651 (2008).
[142] J.R. Maze, P.L. Stanwix, J.S. Hodges, S. Hong, J.M. Taylor, P. Cappellaro, L. Jiang, M.V.G. Dutt, E. Togan, A.S. Zibrov, A. Yacoby, R.L. Walsworth, M.D. Lukin, "Nanoscale Magnetic Sensing with an Individual Electronic Spin in Diamond," Nature (London) 455, 644–647 (2008).

[143] H. Bernien, B. Hensen, W. Pfaff, G. Koolstra, M.S. Blok, L. Robledo, T.H. Taminiau, M. Markham, D.J. Twitchen, L. Childress, R. Hanson, "Heralded Entanglement Between Solid-State Qubits Separated by Three Metres, Nature (London) 497, 86–90 (2013).

[144] G. Davies, M.F. Hamer, "Optical Studies of the 1.945 eV Vibronic Band in Diamond," Proc. R. Soc. London Ser. A 348, 285–298 (1976).

[145] T.P.M. Alegre, C. Santori, G. Medeiros-Ribeiro, R.G. Beausoleil, "Polarization-Selective Excitation of Nitrogen Vacancy Centers in Diamond," Phys. Rev. B 76, 165205 (2007).

[146] A.M. Edmonds, U.F.S. D'Haenens-Johansson, R.J. Cruddace, M.E. Newton, K.-M.C. Fu, C. Santori, R.G. Beausoleil, D.J. Twitchen, M.L. Markham, "Production of Oriented Nitrogen-Vacancy Color Centers in Synthetic Diamond," Phys. Rev. B 86, 035201 (2012).

[147] T. Ishikawa, K.-M.C. Fu, C. Santori, V.M. Acosta, R.G. Beausoleil, H. Watanabe, S. Shikata, K.M. Itoh, "Optical and Spin Coherence Properties of Nitrogen-Vacancy Centers Placed in a 100 nm Thick Isotopically Purified Diamond Layer, Nano Lett. 12, 2083–2087 (2012).

[148] A. Stacey, D.A. Simpson, T.J. Karle, B.C. Gibson, V.M. Acosta, Z. Huang, K.M.C. Fu, C. Santori, R.G. Beausoleil, L.P. McGuinness, et al., "Near-Surface Spectrally Stable Nitrogen Vacancy Centres Engineered in Single Crystal Diamond," Adv. Mater. 24, 3333–3338 (2012).

[149] K. Ohno, F. Joseph Heremans, L.C. Bassett, B.A. Myers, D.M. Toyli, A.C. Bleszynski Jayich, C.J. Palmstrøm, D.D. Awschalom, "Engineering Shallow Spins in Diamond with Nitrogen Delta-Doping," Appl. Phys. Lett. 101, 082413–082413 (2012).

[150] B. Burchard, J. Meijer, I. Popa, T. Gaebel, M. Domhan, C. Wittmann, F. Jelezko, J. Wrachtrup, "Generation of Single Color Centers by Focused Nitrogen Implantation," Appl. Phys. Lett. 87, 261909 (2005).

[151] A. Greentree, P. Olivero, M. Draganski, E. Trajkov, J. Rabeau, P. Reichart, B. Gibson, S. Rubanov, S. Huntington, D. Jamieson, et al., "Critical Components for Diamond-Based Quantum Coherent Devices," J. Phys. Cond. Mat. 18, 825 (2006).

[152] J. Rabeau, P. Reichart, G. Tamanyan, D. Jamieson, S. Prawer, F. Jelezko, T. Gaebel, I. Popa, M. Domhan, J. Wrachtrup, "Implantation of Labelled Single Nitrogen Vacancy Centers in Diamond Using N," Appl. Phys. Lett. 88, 023113 (2006).

[153] B. Naydenov, V. Richter, J. Beck, M. Steiner, P. Neumann, G. Balasubramanian, J. Achard, F. Jelezko, J. Wrachtrup, R. Kalish, "Enhanced Generation of Single Optically Active Spins in Diamond by Ion Implantation," Appl. Phys. Lett. 96, 163108 (2010).

[154] D.M. Toyli, C.D. Weis, G.D. Fuchs, T. Schenkel, D.D. Awschalom, "Chip-Scale Nanofabrication of Single Spins and Spin Arrays in Diamond," Nano Lett. 10, 3168–3172 (2010).

[155] C. Weis, A. Schuh, A. Batra, A. Persaud, I. Rangelow, J. Bokor, C. Lo, S. Cabrini, E. Sideras-Haddad, G. Fuchs, et al., "Single Atom Doping for Quantum Device Development in Diamond and Silicon," J. Vac. Sci. Technol. B: Microelectron. Nanometer Struct. 26, 2596–2600 (2008).

[156] J. Meijer, S. Pezzagna, T. Vogel, B. Burchard, H. Bukow, I. Rangelow, Y. Sarov, H. Wiggers, I. Plümel, F. Jelezko, et al., "Towards the Implanting of Ions and Positioning of Nanoparticles with nm Spatial Resolution," Appl. Phys. A 91, 567–571 (2008).

[157] B. Naydenov, F. Reinhard, A. Lammle, V. Richter, R. Kalish, F. D'Haenens-Johansson, M. Newton, F. Jelezko, J. Wrachtrup, "Increasing the Coherence Time of Single Electron Spins in Diamond by High Temperature Annealing," Appl. Phys. Lett. 97, 242511–242511 (2010).

[158] G. Davies, S.C. Lawson, A.T. Collins, A. Mainwood, S.J. Sharp, "Vacancy-Related Centers in Diamond," Phys. Rev. B 46, 13157–13170 (1992).

[159] V.M. Acosta, E. Bauch, M.P. Ledbetter, C. Santori, K.-M.C. Fu, P.E. Barclay, R.G. Beausoleil, H. Linget, J.F. Roch, F. Treussart, S. Chemerisov, W. Gawlik, D. Budker, "Diamonds with a High Density of Nitrogen-Vacancy Centers for Magnetometry Applications," Phys. Rev. B 80, 115202 (2009).

[160] L. Bassett, F. Heremans, C. Yale, B. Buckley, D. Awschalom, "Electrical Tuning of Single Nitrogen-Vacancy Center Optical Transitions Enhanced by Photoinduced Fields," Phys. Rev. Lett. 107, 266403 (2011).

[161] Y. Mita, Change of Absorption Spectra in Type-Ib Diamond with Heavy Neutron Irradiation," Phys. Rev. B 53, 11360–11364 (1996).

[162] J. Martin, R. Wannemacher, J. Teichert, L. Bischoff, B. Köhler, "Generation and Detection of Fluorescent Color Centers in Diamond with Submicron Resolution," Appl. Phys. Lett. 75, 3096 (1999).
[163] F. Waldermann, P. Olivero, J. Nunn, K. Surmacz, Z. Wang, D. Jaksch, R. Taylor, I. Walmsley, M. Draganski, P. Reichart, A. Greentree, D. Jamieson, S. Prawer, "Creating Diamond Color Centers for Quantum Optical Applications," Diamond Related Mater 16, 1887–1895 (2007).
[164] T. Wee, Y. Tzeng, C. Han, H. Chang, W. Fann, J. Hsu, K. Chen, Y. Yu, "Two-Photon Excited Fluorescence of Nitrogen-Vacancy Centers in Proton-Irradiated Type Ib Diamond," J. Phys. Chem. A 111, 9379–9386 (2007).
[165] T. Schenkel, A. Persaud, S. Park, J. Nilsson, J. Bokor, J. Liddle, R. Keller, D. Schneider, D. Cheng, D. Humphries, "Solid State Quantum Computer Development in Silicon with Single Ion Implantation," J. Appl. Phys. 94, 7017–7024 (2003).
[166] D.N. Jamieson, C. Yang, T. Hopf, S. Hearne, C. Pakes, S. Prawer, M. Mitic, E. Gauja, S. Andresen, F. Hudson, "Controlled Shallow Single-Ion Implantation in Silicon Using an Active Substrate for sub-20-keV ions," Appl. Phys. Lett. 86, 202101–202101 (2005).
[167] J. Loubser, J. Van Wyk, "Electron Spin Resonance in the Study of Diamond," Rep. Prog. Phys. 41, 1201 (1978).
[168] E. Van Oort, N. Manson, M. Glasbeek, "Optically Detected Spin Coherence of the Diamond NV Centre in Its Triplet Ground State," J. Phys. C 21, 4385 (1988).
[169] N.R.S. Reddy, N.B. Manson, E.R. Krausz, "Two-Laser Spectral Hole Burning in a Colour Centre in Diamond," J. Lumin. 38, 46 (1987).
[170] K.-M.C. Fu, C. Santori, P.E. Barclay, L.J. Rogers, N.B. Manson, R.G. Beausoleil, "Observation of the Dynamic Jahn-Teller Effect in the Excited States of Nitrogen-Vacancy Centers in Diamond, Phys. Rev. Lett. 103, 256404 (2009).
[171] A. Batalov, C. Zierl, T. Gaebel, P. Neumann, I. Chan, G. Balasubramanian, P. Hemmer, F. Jelezko, J. Wrachtrup, "Temporal Coherence of Photons Emitted by Single Nitrogen-Vacancy Defect Centers in Diamond Using Optical Rabi-Oscillations," Phys. Rev. Lett. 100, 77401 (2008).
[172] L. Rogers, S. Armstrong, M. Sellars, N. Manson, "Infrared Emission of the NV Centre in Diamond: Zeeman and Uniaxial Stress Studies, New J. Phys. 10, 103024 (2008).
[173] V. Acosta, A. Jarmola, E. Bauch, D. Budker, "Optical Properties of the Nitrogen-Vacancy Singlet Levels in Diamond," Phys. Rev. B 82, 201202 (2010).
[174] L. Robledo, H. Bernien, T. van der Sar, R. Hanson, "Spin Dynamics in the Optical Cycle of Single Nitrogen-Vacancy Centres in Diamond, New J. Phys. 13, 025013 (2011).
[175] F.A. Inam, M.D. Grogan, M. Rollings, T. Gaebel, J.M. Say, C. Bradac, T.A. Birks, W.J. Wadsworth, S. Catelletto, J.R. Rabeau, et al., Emission and Non-radiative Decay of Nanodiamond NV Centers in a Low-Refractive Index Environment, ACS Nano (2013).
[176] K. Beha, A. Batalov, N.B. Manson, R. Bratschitsch, A. Leitenstorfer, "Optimum Photoluminescence Excitation and Recharging Cycle of Single Nitrogen-Vacancy Centers in Ultrapure Diamond," Phys. Rev. Lett. 109, 097404 (2012).
[177] P. Siyushev, H. Pinto, M. Vörös, A. Gali, F. Jelezko, J. Wrachtrup, "Optically Controlled Switching of the Charge State of a Single Nitrogen-Vacancy Center in Diamond at Cryogenic Temperatures," Phys. Rev. Lett. 110, 167402 (2013).
[178] P. Tamarat, T. Gaebel, J. Rabeau, M. Khan, A. Greentree, H. Wilson, L. Hollenberg, S. Prawer, P. Hemmer, F. Jelezko, et al., "Stark Shift Control of Single Optical Centers in Diamond," Phys. Rev. Lett. 97, 83002 (2006).
[179] C. Santori, P. Tamarat, P. Neumann, J. Wrachtrup, D. Fattal, R. Beausoleil, J. Rabeau, P. Olivero, A. Greentree, S. Prawer, et al., "Coherent Population Trapping of Single Spins in Diamond under Optical Excitation," Phys. Rev. Lett. 97, 247401 (2006).
[180] P. Tamarat, N. Manson, J. Harrison, R. McMurtrie, A. Nizovtsev, C. Santori, R. Beausoleil, P. Neumann, T. Gaebel, F. Jelezko, et al., Spin-Flip and Spin-Conserving Optical Transitions of the Nitrogen-Vacancy Centre in Diamond," New J. Phys. 10, 045004 (2008).
[181] A. Batalov, V. Jacques, F. Kaiser, P. Siyushev, P. Neumann, L. Rogers, R. McMurtrie, N. Manson, F. Jelezko, J. Wrachtrup, "Low Temperature Studies of the Excited-State Structure of Negatively Charged Nitrogen-Vacancy Color Centers in Diamond," Phys. Rev. Lett. 102, 195506 (2009).
[182] L. Rogers, R. McMurtrie, M. Sellars, N. Manson, "Time-Averaging Within the Excited State of the Nitrogen-Vacancy Centre in Diamond," New J. Phys. 11, 063007 (2009).

[183] G. Fuchs, V. Dobrovitski, R. Hanson, A. Batra, C. Weis, T. Schenkel, D. Awschalom, "Excited-State Spectroscopy Using Single Spin Manipulation in Diamond," Phys. Rev. Lett. 101, 117601 (2008).
[184] P. Neumann, R. Kolesov, V. Jacques, J. Beck, J. Tisler, A. Batalov, L. Rogers, N. Manson, G. Balasubramanian, F. Jelezko, et al., "Excited-State Spectroscopy of Single NV Defects in Diamond Using Optically Detected Magnetic Resonance, New J. Phys. 11, 013017 (2009).
[185] V.M. Acosta, C. Santori, A. Faraon, Z. Huang, K.-M.C. Fu, A. Stacey, D.A. Simpson, K. Ganesan, S. Tomljenovic-Hanic, A.D. Greentree, S. Prawer, R.G. Beausoleil, "Dynamic Stabilization of the Optical Resonances of Single Nitrogen-Vacancy Centers in Diamond," Phys. Rev. Lett. 108, 206401 (2012).
[186] N. Manson, C. Wei, "Transient Hole Burning in N-V Centre in Diamond," J. Luminescence 58, 158–160 (1994).
[187] C. Santori, D. Fattal, S. Spillane, M. Fiorentino, R. Beausoleil, A. Greentree, P. Olivero, M. Draganski, J. Rabeau, P. Reichart, B. Gibson, S. Rubanov, D. Jamieson, S. Prawer, "Coherent Population Trapping in Diamond NV Centers at Zero Magnetic Field," Optics Express 14, 7986–7993 (2006).
[188] E. Togan, Y. Chu, A.S. Trifonov, L. Jiang, J. Maze, L. Childress, M.V.G. Dutt, A.S. Sorensen, P.R. Hemmer, A.S. Zibrov, M.D. Lukin, "Quantum Entanglement Between an Optical Photon and a Solid-State Spin Qubit," Nature (London) 466, 730–734 (2010).
[189] L. Robledo, L. Childress, H. Bernien, B. Hensen, P.F.A. Alkemade, R. Hanson, "High-Fidelity Projective Read-Out of a Solid-State Spin Quantum Register," Nature (London) 477, 574–578 (2011).
[190] A. Lenef, S. Rand, "Electronic Structure of the NV Center in Diamond: Theory," Phys. Rev. B 53, 13441 (1996).
[191] J. Martin, "Fine Structure of Excited 3E State in Nitrogen-Vacancy Centre of Diamond," J. Luminescence 81, 237–247 (1999).
[192] N. Manson, R. McMurtrie, "Issues Concerning the Nitrogen-Vacancy Center in Diamond," J. Luminescence 127, 98–103 (2007).
[193] J. Maze, A. Gali, E. Togan, Y. Chu, A. Trifonov, E. Kaxiras, M. Lukin, "Properties of Nitrogen-Vacancy Centers in Diamond: The Group Theoretic Approach," New J. Phys. 13, 025025 (2011).
[194] M.W. Doherty, N.B. Manson, P. Delaney, L.C. Hollenberg, "The Negatively Charged Nitrogen-Vacancy Centre in Diamond: The Electronic Solution," New J. Phys. 13, 025019 (2011).
[195] A. Gali, M. Fyta, E. Kaxiras, "Ab Initio Supercell Calculations on Nitrogen-Vacancy Center in Diamond: Electronic structure and hyperfine tensors," Phys. Rev. B 77, 155206 (2008).
[196] F.M. Hossain, M.W. Doherty, H.F. Wilson, L.C. Hollenberg, "Ab Initio Electronic and Optical Properties of the NV^{-} Center in Diamond," Phys. Rev. Lett. 101, 226403 (2008).
[297] C.G. Yale, B.B. Buckley, D.J. Christle, G. Burkard, F.J. Heremans, L.C. Bassett, D.D. Awschalom, "All-Optical Control of a Solid-State Spin Using Coherent Dark States" (2013), preprint arXiv:1302.6638.
[298] P.R. Hemmer, A.V. Turukhin, S.M. Shahriar, J.A. Musser, "Raman-Excited Spin Coherences in Nitrogen-Vacancy Color Centers in Diamond," Opt. Lett. 26, 361–363 (2001).
[299] V.M. Acosta, K. Jensen, C. Santori, D. Budker, R.G. Beausoleil, "Electromagnetically-Induced Transparency in a Diamond Spin Ensemble Enables All-Optical Electromagnetic Field Sensing" (2013), preprint arXiv:1303.6966.
[200] K. Hammerer, A.S. Sørensen, E.S. Polzik, "Quantum Interface Between Light and Atomic Ensembles," Rev. Mod. Phys. 82, 1041 (2010).
[201] A. Beveratos, R. Brouri, T. Gacoin, A. Villing, J.-P. Poizat, P. Grangier, "Single Photon Quantum Cryptography," Phys. Rev. Lett. 89, 187901 (2002).
[202] T. Schröder, F. Gädeke, M.J. Banholzer, O. Benson, "Ultrabright and Efficient Single-Photon Generation Based on Nitrogen-Vacancy Centres in Nanodiamonds on a Solid Immersion Lens," New J. Phys. 13, 055017 (2011).
[203] A. Faraon, C. Santori, Z. Huang, V.M. Acosta, R.G. Beausoleil, "Coupling of Nitrogen-Vacancy Centers to Photonic Crystal Cavities in Monocrystalline Diamond," Phys. Rev. Lett. 109, 033604 (2012).
[204] G. Balasubramanian, P. Neumann, D. Twitchen, M. Markham, R. Kolesov, N. Mizuochi, J. Isoya, J. Achard, J. Beck, J. Tissler, V. Jacques, P.R. Hemmer, F. Jelezko, J. Wrachtrup,

"Ultralong Spin Coherence Time in Isotopically Engineered Diamond, Nature Mater. 8, 383–387 (2009).
[205] I. Popa, T. Gaebel, M. Domhan, C. Wittmann, F. Jelezko, J. Wrachtrup, "Energy Levels and Decoherence Properties of Single Electron and Nuclear Spins in a Defect Center in Diamond," Phys. Rev. B 70, 201203 (2004).
[206] N. Bar-Gill, L.M. Pham, A. Jarmola, D. Budker, R.L. Walsworth, "Solid-State Electronic Spin Coherence Time Approaching One Second" (2012), preprint arXiv:1211.7094.
[207] P. Neumann, J. Beck, M. Steiner, F. Rempp, H. Fedder, P. R. Hemmer, J. Wrachtrup, F. Jelezko, "Single-Shot Readout of a Single Nuclear Spin," Science 329, 542–544 (2010).
[208] L. Childress, J. Taylor, A. Sørensen, M. Lukin, "Fault-Tolerant Quantum Repeaters with Minimal Physical Resources and Implementations Based on Single-Photon Emitters," Phys. Rev. A 72, 52330 (2005).
[209] S. Benjamin, D. Browne, J. Fitzsimons, J. Morton, "Brokered Graph-State Quantum Computation," New J. Phys 8, 141 (2006).
[210] I. Aharonovich, S. Castelletto, D. Simpson, C. Su, A. Greentree, S. Prawer, "Diamond-Based Single-Photon Emitters," Report Progress Phys. 74, 076501 (2011).
[211] S. Pezzagna, D. Rogalla, D. Wildanger, J. Meijer, A. Zaitsev, "Creation and Nature of Optical Centres in Diamond for Single-Photon Emission Overview and Critical Remarks," New J. Phys. 13, 035024 (2011).
[212] C. Wang, C. Kurtsiefer, H. Weinfurter, B. Burchard, "Single Photon Emission from SiV Centres in Diamond Produced by Ion Implantation," J. Phys. B: At. Mol. Opt. Phys. 39, 37 (2006).
[213] E. Neu, D. Steinmetz, J. Riedrich-Möller, S. Gsell, M. Fischer, M. Schreck, C. Becher, "Single Photon Emission from Silicon-Vacancy Colour Centres in Chemical Vapour Deposition Nano-Diamonds on Iridium," New J. Phys. 13, 025012 (2011).
[214] I. Aharonovich, S. Castelletto, D.A. Simpson, A. Stacey, J. McCallum, A.D. Greentree, S. Prawer, "Two-Level Ultrabright Single Photon Emission from Diamond Nanocrystals," Nano Lett. 9, 3191–3195 (2009).
[215] I. Aharonovich, S. Castelletto, B.C. Johnson, J.C. McCallum, D.A. Simpson, A.D. Greentree, S. Prawer, "Chromium Single-Photon Emitters in Diamond Fabricated by Ion Implantation," Phys. Rev. B 81, 121201 (2010).
[216] T. Gaebel, I. Popa, A. Gruber, M. Domhan, F. Jelezko, J. Wrachtrup, "Stable Single-Photon Source in the Near Infrared," New J. Phys. 6, 98 (2004).
[217] E. Wu, J. Rabeau, G. Roger, F. Treussart, H. Zeng, P. Grangier, S. Prawer, J.-F. Roch, "Room Temperature Triggered Single-Photon Source in the Near Infrared," New J. Phys. 9, 434 (2007).
[218] D. Simpson, E. Ampem-Lassen, B. Gibson, S. Trpkovski, F. Hossain, S. Huntington, A. Greentree, L. Hollenberg, S. Prawer, "A Highly Efficient Two Level Diamond Based Single Photon Source," Appl. Phys. Lett. 94, 203107–203107 (2009).
[219] B. Naydenov, R. Kolesov, A. Batalov, J. Meijer, S. Pezzagna, D. Rogalla, F. Jelezko, J. Wrachtrup, "Engineering Single Photon Emitters by Ion Implantation in Diamond," Appl. Phys. Lett. 95, 181109–181109 (2009).
[220] S. Castelletto, B. Johnson, N. Stavrias, T. Umeda, T. Ohshima, "Efficiently Engineered Room Temperature Single Photons in Silicon Carbide," (2012), preprint arXiv:1210.5047.
[221] A.J. Morfa, B.C. Gibson, M. Karg, T.J. Karle, A.D. Greentree, P. Mulvaney, S. Tomljenovic-Hanic, "Single-Photon Emission and Quantum Characterization of Zinc Oxide Defects," Nano Lett. 12, 949–954 (2012).
[222] R. Kolesov, K. Xia, R. Reuter, R. Stöhr, A. Zappe, J. Meijer, P. Hemmer, J. Wrachtrup, "Optical Detection of a Single Rare-Earth Ion in a Crystal," Nat. Commun. 3, 1029 (2012).
[223] J. Hadden, J. Harrison, A. Stanley-Clarke, L. Marseglia, Y.-L. Ho, B. Patton, J. OBrien, J. Rarity, "Strongly Enhanced Photon Collection from Diamond Defect Centers Under Microfabricated Integrated Solid Immersion Lenses," Appl. Phys. Lett. 97, 241901–241901 (2010).
[224] A. Faraon, C. Santori, Z. Huang, K.-M.C. Fu, V.M. Acosta, D. Fattal, R.G. Beausoleil, "Quantum Photonic Devices in Single-Crystal Diamond," New J. Phys. 15, 025010 (2013).
[225] S. Schietinger, M. Barth, T. Aichele, O. Benson, "Plasmon-Enhanced Single Photon Emission from a Nanoassembled Metal-Diamond Hybrid Structure at Room Temperature," Nano Lett. 9, 1694–1698 (2009).

[226] T.M. Babinec, B.J.M. Hausmann, M. Khan, Y. Zhang, J.R. Maze, P.R. Hemmer, M. Loncar, "A Diamond Nanowire Single-Photon Source, Nat. Nanotech. 5, 195–199 (2010).
[227] L. Marseglia, J. Hadden, A. Stanley-Clarke, J. Harrison, B. Patton, Y.-L. Ho, B. Naydenov, F. Jelezko, J. Meijer, P. Dolan, et al., "Nanofabricated Solid Immersion Lenses Registered to Single Emitters in Diamond," Appl. Phys. Lett. 98, 133107–133107 (2011).
[228] A. Faraon, P.E. Barclay, C. Santori, K.-M.C. Fu, R.G. Beausoleil, "Resonant Enhancement of the Zero-Phonon Emission from a Colour Centre in a Diamond Cavity," Nat. Photon. 5, 301–305 (2011).
[229] B.J. Hausmann, B. Shields, Q. Quan, P. Maletinsky, M. McCutcheon, J.T. Choy, T.M. Babinec, A. Kubanek, A. Yacoby, M.D. Lukin, et al., Integrated Diamond Networks for Quantum Nanophotonics," Nano Lett. 12, 1578–1582 (2012).
[230] M.J. Burek, N.P. de Leon, B.J. Shields, B.J. Hausmann, Y. Chu, Q. Quan, A.S. Zibrov, H. Park, M.D. Lukin, M. Loncar, "Free-Standing Mechanical and Photonic Nanostructures in Single-Crystal Diamond," Nano Lett. 12, 6084–6089 (2012).
[231] J. Riedrich-Möller, L. Kipfstuhl, C. Hepp, E. Neu, C. Pauly, F. Mücklich, A. Baur, M. Wandt, S. Wolff, M. Fischer, et al., "One-and Two-Dimensional Photonic Crystal Microcavities in Single Crystal Diamond," Nat. Nanotechnol. 7, 69–74 (2011).
[232] M. Hiscocks, K. Ganesan, B. Gibson, S. Huntington, F. Ladouceur, S. Prawer, "Diamond Waveguides Fabricated by Reactive Ion Etching," Opt. Express 16, 19512–19519 (2008).
[233] I. Bayn, B. Meyler, A. Lahav, J. Salzman, R. Kalish, B.A. Fairchild, S. Prawer, M. Barth, O. Benson, T. Wolf, P. Siyushev, F. Jelezko, J. Wrachtrup, "Processing of Photonic Crystal Nanocavity for Quantum Information in Diamond," Diamond Related Mater. 20, 937–943 (2011).
[234] S. Castelletto, J. Harrison, L. Marseglia, A. Stanley-Clarke, B. Gibson, B. Fairchild, J. Hadden, Y.D. Ho, M. Hiscocks, K. Ganesan, et al., "Diamond-Based Structures to Collect and Guide Light," New J. Phys. 13, 025020 (2011).
[235] C.F. Wang, Y.-S. Choi, J.C. Lee, E.L. Hu, J. Yang, J.E. Butler, "Observation of Whispering Gallery Modes in Nanocrystalline Diamond Microdisks," Appl. Phys. Lett. 90, 081110 (2007a).
[236] C.F. Wang, R. Hanson, D.D. Awschalom, E.L. Hu, T. Feygelson, J. Yang, J.E. Butler, "Fabrication and Characterization of Two-Dimenstional Photonic Crystal Microcavities in Nanocrystalline Diamond," Appl. Phys. Lett. 91, 201112 (2007b).
[237] K. Fu, C. Santori, P. Barclay, I. Aharonovich, S. Prawer, N. Meyer, A. Holm, R. Beausoleil, "Coupling of Nitrogen-Vacancy Centers in Diamond to a GaP Waveguide," Appl. Phys. Lett. 93, 234107 (2008).
[238] P. Barclay, K. Fu, C. Santori, R. Beausoleil, "Chip-Based Microcavities Coupled to Nitrogen-Vacancy Centers in Single Crystal Diamond," Appl. Phys. Lett. 95, 191115 (2009a).
[239] P.E. Barclay, K.-M. Fu, C. Santori, R.G. Beausoleil, "Hybrid Photonic Crystal Cavity and Waveguide for Coupling to Diamond NV-Centers," Opt. Express 17, 9588–9601 (2009b).
[240] M. Larsson, K.N. Dinyari, H. Wang, "Composite Optical Microcavity of Diamond Nanopillar and Silica Microsphere," Nano Lett. 9, 1447–1450 (2009).
[241] Y.-S. Park, A. Cook, H. Wang, "Cavity QED with Diamond Nanocrystals and Silica Microspheres," Nano Lett. 6, 2075–2079 (2006).
[242] S. Schietinger, T. Schroder, O. Benson, "One-by-One Coupling of Single Defect Centers in Nanodiamonds to High-Q Modes of an Optical Microresonator," Nano Lett. 8, 3911–3915 (2008).
[243] S. Schietinger, O. Benson, "Coupling Single NV-Centres to High-Q Whispering Gallery Modes of a Preselected Frequency-Matched Microresonator," J. Phys. B: At. Mol. Opt. Phys. 42, 114001 (2009).
[244] P. Barclay, C. Santori, K. Fu, R. Beausoleil, O. Painter, "Coherent Interference Effects in a Nano-Assembled Diamond NV Center Cavity-QED System," Opt. Express 17, 8081–8097 (2008).
[245] J. Wolters, A.W. Schell, G. Kewes, N. Nusse, M. Schoengen, H. Doscher, T. Hannappel, B. Lochel, M. Barth, O. Benson, "Enhancement of the Zero Phonon Line Emission from a Single Nitrogen Vacancy Center in a Nanodiamond Via Coupling to a Photonic Crystal Cavity," Appl. Phys. Lett. 97, 141108–141108 (2010).

[246] D. Englund, B. Shields, K. Rivoire, F. Hatami, J. Vuckovic, H. Park, M.D. Lukin, "Deterministic Coupling of a Single Nitrogen Vacancy Center to a Photonic Crystal Cavity," Nano Lett. 10, 3922–3926 (2010).
[247] T. Van der Sar, J. Hagemeier, W. Pfaff, E. Heeres, S. Thon, H. Kim, P. Petroff, T. Oosterkamp, D. Bouwmeester, R. Hanson, "Deterministic Nanoassembly of a Coupled Quantum Emitter–Photonic Crystal Cavity System," Appl. Phys. Lett. 98, 193103 (2011).
[248] H. Bernien, L. Childress, L. Robledo, M. Markham, D. Twitchen, R. Hanson, "Two-Photon Quantum Interference from Separate Nitrogen Vacancy Centers in Diamond," Phys. Rev. Lett. 108, 043604 (2012).
[249] C. Cabrillo, J.I. Cirac, P. García-Fernández, P. Zoller, "Creation of Entangled States of Distant Atoms by Interference," Phys. Rev. A 59, 1025–1033 (1999).
[250] S.D. Barrett, P. Kok, "Efficient High-Fidelity Quantum Computation Using Matter Qubits and Linear Optics," Phys. Rev. A 71, 060310(R) (2005).
[251] H.-J. Briegel, W. Dür, J.I. Cirac, P. Zoller, "Quantum Repeaters: The Role of Imperfect Local Operations in Quantum Communication," Phys. Rev. Lett. 81, 5932–5935 (1998).
[252] B. Sun, M.S. Chapman, L. You, "Atom-Photon Entanglement Generation and Distribution," Phys. Rev. A 69, 042316 (2004).
[253] J. Beugnon, M.P.A. Jones, J. Dingjan, B. Darquié, G. Messin, A. Browaeys, P. Grangier, "Quantum Interference Between Two Single Photons Emitted by Independently Trapped Atoms," Nature 440, 779–782 (2006).
[254] P. Maunz, D.L. Moehring, S. Olmschenk, K.C. Younge, D.N. Matsukevich, C. Monroe, "Quantum Interference of Photon Pairs from Two Remote Trapped Atomic Ions," Nat. Phys. 3, 538–541 (2007).
[255] R. Raussendorf, H.-J. Briegel, "A One-Way Quantum Computer," Phys. Rev. Lett. 86, 5188–5191 (2001).
[256] H.P. Büchler, M. Hermele, S.D. Huber, M.P.A. Fisher, P. Zoller, "Atomic Quantum Simulator for Lattice Gauge Theories and Ring Exchange Models," Phys. Rev. Lett. 95, 040402 (2005).
[257] K.D. Greve, L. Yu, P.L. McMahon, J.S. Pelc, C.M. Natarajan, N.Y. Kim, E. Abe, S. Maier, C. Schneider, M. Kamp, S. Höfling, R.H. Hadfield, A. Forchel, M.M. Fejer, Y. Yamamoto, "Quantum-Dot Spinphoton Entanglement Via Frequency Downconversion to Telecom Wavelength," Nature 491, 421425 (2012).
[258] W.B. Gao, P. Fallahi, E. Togan, J. Miguel-Sanchez, A. Imamoğlu, "Observation of Entanglement Between a Quantum Dot Spin and a Single Photon, Nature 491, 426430 (2012).
[259] J.R. Schaibley, A.P. Burgers, G.A. McCracken, L.-M. Duan, P.R. Berman, D.G. Steel, "Demonstration of Quantum Entanglement Between a Single Electron Spin Confined to an InAs Quantum Dot and a Photon," Phys. Rev. Lett. 110, 167401–167405 (2013).
[260] N. Mizuochi, T. Makino, H. Kato, D. Takeuchi, M. Ogura, H. Okushi, M. Nothaft, P. Neumann, A. Gali, F. Jelezko, J. Wrachtrup, and S. Yamasaki, "Electrically Driven Single-Photon Source at Room Temperature in Diamond," Nat. Photonics 6, 299–303 (2012).

Chapter 14

Generation and Storage of Single Photons in Collectively Excited Atomic Ensembles

Bo Zhao and Jian-Wei Pan
Hefei National Laboratory for Physical Sciences at Microscale and Department of Modern Physics, University of Science and Technology of China, Hefei, Anhui, 230026, PR China

Chapter Outline

14.1 Introduction	541
14.2 Basic Concepts	543
14.3 From Heralded to Deterministic Single-Photon Sources	545
14.4 Interference of Photons from Independent Sources	550
14.5 Conclusion and Outlook	555
Appendix	556
A Write Process	556
B Read Process	559
References	560

14.1 INTRODUCTION

Deterministic and storable single-photon sources are of crucial importance to all-optical quantum-information processing. In quantum cryptography, single photons can play a role in establishing unconditional security and high efficiency [1,2]. In linear-optical quantum computation, on-demand and indistinguishable single photons are required to implement measurement-induced quantum logic gates [3,4].

In recent years, different quantum systems have been exploited to realize an on-demand single-photon source, such as quantum dots [5,6], single atoms [7,8], and color centers [9] (see Chapter 14). Such single-emitter-based single-photon sources often require a high-finesse cavity to ensure that photons are emitted predominantly into a well-defined single spatial mode. In this chapter, we introduce a different type of single-photon source that consists of many identical single-photon emitters, i.e. the atomic-ensemble-based single-photon source. Due to the collective enhancement, these sources can be implemented without the use of high-finesse cavity [10]. Moreover, because the process is reversible, atomic ensembles can also provide storage for single photons due to the inherent bandwidth matching. Controllable transfer of quantum states between a flying qubit and a memory is of crucial importance to quantum-information science [11,12]. Therefore, atomic-ensemble-based single-photon sources present a valuable resource for optical quantum-information processing.

The principle of the atomic-ensemble-based single-photon source may be understood as follows [10,13]. Consider an ensemble of three-level atoms in a lambda configuration initialized with all the atoms in one of the ground states. First, the quantum state of a single photon is imprinted into an atomic ensemble using a "write" process. In this process, a single collective excitation is generated probabilistically in an atomic ensemble via spontaneous Raman scattering by applying a weak coherent write light pulse. The successful generation of the collective excitation is indicated by the detection of a corresponding Raman photon. The collective excitation is then coherently stored in the atomic ensemble, which serves as a quantum memory. In the following "read" process, the single excitation is converted into a single photon by applying a strong coherent read pulse after a controllable delay. The spatial mode, bandwidth, and frequency of the single-photon output are determined by the mode-matching, intensity, and frequency of the retrieval light [13–15].

One distinct advantage of atomic-ensemble-based single-photon sources is that the single photon can be stored in the atomic ensemble and be read out when it is needed. The upper limit of the storage time is determined by the coherence time of the atomic memory, which can be on the order of seconds [16]. Because of the long coherence time, a series of reset pulses (optical pulses that pump the atoms to the initial ground state) and write pulses can be applied to the ensemble. The reset and write pulse sequence is stopped by a feedback circuit, once an anti-Stokes photon is detected. In this way, a collective excitation can be stored in the atomic ensemble, subject to a deterministic readout, and thus a deterministic single-photon source can be implemented [17,18]. The same medium can be used as memory that stores single photons [14,15,19,20].

In this chapter we discuss atomic-ensemble-based single-photon sources. In Section 14.2, we introduce the basic concepts of the write and read processes. In Section 14.3, we show how a heralded single-photon source can be converted into a deterministic one. In Section 14.4, we discuss the indistinguishability of single photons generated from independent sources by examining two-photon Hong-Ou-Mandel interference. We conclude with a brief look into the future.

14.2 BASIC CONCEPTS

Let us first consider a Λ-type three-level atomic system. The energy level structure is depicted in Fig. 14.1, where the upper state $|e\rangle$ is the excited state and two lower states $|a\rangle$ and $|b\rangle$ are the ground states used to store the quantum state. Initially all atoms are prepared in one of the ground states, e.g. $|a\rangle$, by optical pumping. In the write process, an off-resonant weak classical laser pulse coupling $|e\rangle$ and $|a\rangle$ is applied to the atomic ensemble. A small number of atoms will be excited to $|e\rangle$ and almost immediately decay. These atoms that decay into $|b\rangle$ emit anti-Stokes photons; this effect is called spontaneous Raman scattering. According to energy conservation, the number of atoms transferred to the $|b\rangle$ state is equal to the number of anti-Stokes photons emitted from the atomic ensemble. If the write pulse is so weak that only one anti-Stokes photon is generated in the mode of interest, then only one atom changes its state, but it is impossible to know which one, even in principle. The excitation rate must be low so that the probability that more than one photon is created is negligible. In this case, if a single anti-Stokes photon is detected, the atomic ensemble is projected into an equally weighted superposition state, a collective excited state, which can be described by

$$|1\rangle_a = S^\dagger |0\rangle_a = \frac{1}{\sqrt{N}} \sum e^{i\mathbf{k}\cdot\mathbf{r}_i} |a \ldots b_i \ldots a\rangle, \quad (14.1)$$

where $S^\dagger = \frac{1}{\sqrt{N}} \sum_{i=1}^{N} e^{i\mathbf{k}\cdot\mathbf{r}_i} |b\rangle_i \langle a|$ is the creation operator of the collective state and $|0\rangle_a = \otimes_i |a\rangle_i$ denotes the atomic vacuum state (see details in Appendix). That is to say, in the write process a quantum state is imprinted into the collective excited state of the atomic ensemble conditional on detecting an anti-Stokes

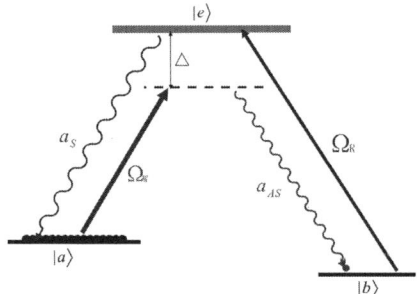

FIGURE 14.1 An illustration of the interaction between the atomic ensemble and light. The excited state $|e\rangle$ and two ground states $|a\rangle$ and $|b\rangle$ form the Λ-type three-level atom. In the write process, an off-resonant write light pulse with Rabi frequency Ω_W and detuning Δ is applied to the atomic ensemble. An anti-Stokes photon is emitted and simultaneously a collective excitation is generated due to spontaneous Raman scattering. In the EIT-based read process, an on-resonance read light pulse with Rabi frequency Ω_R is applied to convert the collective excitation to a Stokes photon.

photon. It should be noted that the two ground states are immune to spontaneous emission, and decoherence of the ensemble as a whole occurs with the rate of a single-atom decoherence.

The collective excitation can be converted into a single photon in the read process. In this process, a strong coherent light pulse, coupling $|e\rangle$ and $|b\rangle$, transfers the collective excitation into a Stokes photon. Since the Stokes photon and the strong read light pulse satisfy the electromagnetically induced transparency (EIT) condition [14,15,19,20], the Stokes photon can propagate through the atomic ensemble with little to no absorption. In the ideal case the excitation stored in the atomic ensemble can be deterministically retrieved (see details in Appendix).

After the write process, the atom-photon state can be described by a two mode squeezed state $|\psi\rangle_{a,\text{AS}} = |0\rangle_a |0\rangle_{\text{AS}} + \sqrt{\chi}|1\rangle_a |1\rangle_{\text{AS}} + \chi |2\rangle_a |2\rangle_{\text{AS}} + O\left(\chi^{\frac{3}{2}}\right)$, where the excitation probability $\chi \ll 1$. After the read process, the collective excitation is converted into a Stokes photon, and thus the photonic state can be described by $|\psi\rangle_{\text{AS},\text{S}} = |0\rangle_{\text{AS}}|0\rangle_{\text{S}} + \sqrt{\chi}|1\rangle_{\text{AS}}|1\rangle_{\text{S}} + \chi|2\rangle_{\text{AS}}|2\rangle_{\text{S}} + O\left(\chi^{\frac{3}{2}}\right)$. The Stokes photon and anti-Stokes photon are non-classically correlated, which leads to a violation of the Cauchy-Schwarz inequality $(g^{(2)}_{\text{AS},\text{S}})^2 \leq g^{(2)}_{\text{AS},\text{AS}} g^{(2)}_{\text{S},\text{S}}$, where $g^{(2)}_{\text{AS},\text{S}}$ is the second-order cross-correlation function, and $g^{(2)}_{\text{S},\text{S}}$ and $g^{(2)}_{\text{AS},\text{AS}}$ are the second-order auto-correlation functions [21]. A straightforward calculation shows $g^{(2)}_{\text{AS},\text{S}} = 1 + 1/\chi$ and $g^{(2)}_{\text{AS},\text{AS}} = g^{(2)}_{\text{S},\text{S}} = 2$. Since $\chi \ll 1$, the Cauchy-Schwarz inequality is violated and the anti-Stokes and Stokes photons are non-classically correlated.

The non-classical correlation between anti-Stokes and Stokes photons allows the preparation of a heralded single-photon source [13–15], in which a single photon can be generated on demand if it is known that there is a collective excitation in the atomic ensemble. The presence of the latter is heralded by the detection of a spontaneous Raman photon in the write process, after which the excitation can be converted into a photon on request. The quality of the heralded single photons can be estimated as follows. Conditional on the detection of an anti-Stokes photon, the Stokes photon may be described by a mixed state $\rho_S = |1\rangle_S\langle 1| + 2\chi |2\rangle_S\langle 2|$. This means the triggered Stokes photon is in a single-photon state that is mixed with a small two-photon component. The normalized auto-correlation of the triggered single photon is $\alpha = \frac{2P_{\text{II}}}{P_{\text{I}}^2} = 4\chi$, where $P_{\text{I}}(P_{\text{II}})$ is the probability of generating one (two) photon(s) per trial (the higher orders are negligible small). For a coherent laser source, $\alpha = 1$, and for an ideal single-photon source $\alpha = 0$. It can be readily seen $\alpha \to 0$ if $\chi \to 0$. Therefore, to obtain a high-quality triggered single photon, the excitation probability must be small. This process is similar to a four-wave-mixing-based single-photon generation (see Chapter 13), but here the "heralded" single photon is stored as an atomic excitation, and its retrieval is a deterministic process. A typical operation of the source would require multiple write attempts, and one read attempt. This source has a number of advantages common to deterministic sources, particularly, it is scalable.

Atomic-ensemble-based single-photon sources have been demonstrated experimentally both in laser-cooled ensembles [13,14,17] and thermal atomic vapors [15]. The cold atomic ensemble has the advantage of higher single-photon quality and longer coherence time; the following discussion focuses on experiments with cold atomic ensembles.

14.3 FROM HERALDED TO DETERMINISTIC SINGLE-PHOTON SOURCES

Heralded single-photon sources are probabilistic. This significantly limits the usefulness of these sources in scalable all-optical quantum-information processing. Thanks to the storable nature of atomic-ensemble-based single-photon sources, it is possible to implement deterministic single-photon sources by means of a feedback circuit. The basic idea is that a single collective excitation is generated in an atomic ensemble by applying a series of subsequent reset and write pulses. The successful generation of a collective excitation is indicated by the detection of a corresponding anti-Stokes Raman photon. This detection is used as a feedback to stop the sequence and to start the next process, i.e. to convert the excitation back into a single photon after a controllable delay. Such a sequence can be taken as a feedforward operation for the deterministic creation of a single photon. In this way, a deterministic single-photon source is obtained. This controllable deterministic single-photon source potentially paves the way for the construction of scalable quantum communication networks and linear-optical quantum circuits. As one example, we present a proof-of-principle demonstration of a deterministic and storable single-photon source [17]. The single-photon quality is preserved while the production rate of single photons is considerably improved by the aid of the repetitive write process and feedback control.

A schematic of the experimental setup used to demonstrate a deterministic single-photon source is shown in Fig. 14.2a. Cold atoms with a Λ-type three-level configuration (two ground states $|a\rangle$, $|b\rangle$ and an excited state $|e\rangle$) collected by a magneto-optical trap (MOT) [22] are used as the medium for quantum memory. In this experiment, more than 10^8 ^{87}Rb atoms are collected by the MOT with an optical depth of about 5 and have a temperature of about 100 μK. To prevent Zeeman splitting, the earth's magnetic field is compensated by three pairs of Helmholtz coils. The two ground states $|a\rangle$, $|b\rangle$ and the excited state $|e\rangle$ are the $|5S_{1/2}, F = 2\rangle$, $|5S_{1/2}, F = 1\rangle$, and $|5P_{1/2}, F = 2\rangle$ states of ^{87}Rb, respectively. The write laser is tuned to the transition from $|5S_{1/2}, F = 2\rangle$ to $|5P_{1/2}, F = 2\rangle$ with a detuning of 10 MHz, and the read laser is locked on-resonance to the transition from $|5S_{1/2}, F = 1\rangle$ to $|5P_{1/2}, F = 2\rangle$. The write and read beams have orthogonal polarizations, and are spatially overlapped

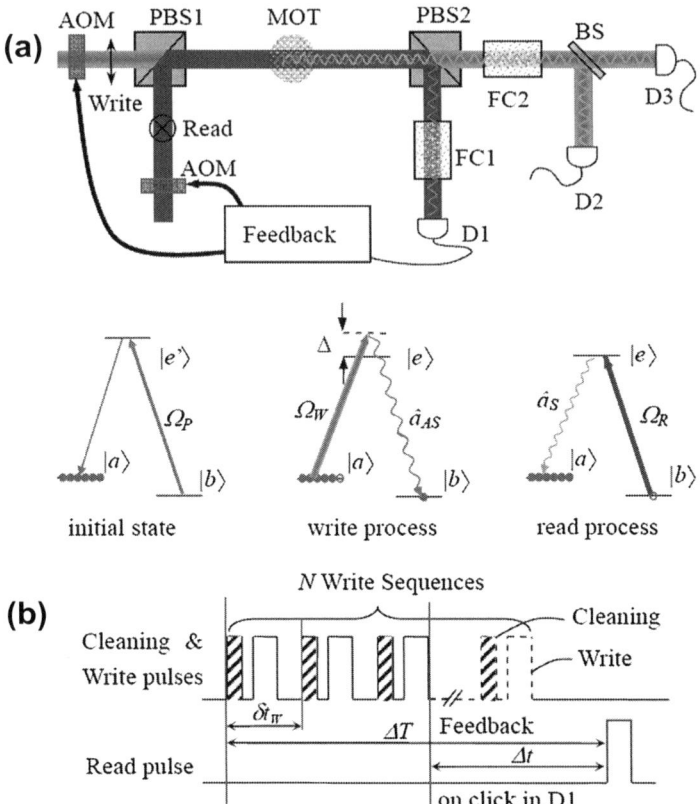

FIGURE 14.2 (a) Illustration of the experimental setup and (b) the time sequence with the feedback circuit for the write and read processes. The atomic ensemble is first prepared in the initial state $|a\rangle$ by applying a reset (cleaning) pump beam resonant with the transition $|b\rangle$ to $|e'\rangle$. A write pulse with the Rabi frequency Ω_W is applied to generate the spin excitation and an accompanying photon of the mode $\hat{a}_{AS}$. After a delay Δt, a read pulse is applied with orthogonal polarization and it spatially overlaps with the write beam in PBS1. The photons, whose polarization is orthogonal to that of the write beam, in the mode $\hat{a}_{AS}$ are extracted from the classical write beam by PBS2 and detected by detector D1. Similarly, the field $\hat{a}_S$ is extracted from the Read beam and detected by detector D2 (or D3). Here, FC1 and FC2 are two filter cells, BS is a 50/50 non-polarizing beam splitter, and AOM1 and AOM2 are two acousto-optic modulators.

on a polarizing beam splitter (PBS1), and then focused into the cold atoms with a beam waist of 35 μm. After passing through the atomic cloud, another polarizing beam splitter (PBS2) separates the classical write (read) beam from the generated anti-Stokes (Stokes) field, and cells filled with ^{87}Rb atoms prepared in state $|5S_{1/2}, F = 2\rangle$ ($|5S_{1/2}, F = 1\rangle$) are used to further suppress the residual light from the read/write pulses.

After switching off both the MOT optical beams and magnetic field, the atoms are first optically pumped to the initial state $|a\rangle$ by a reset pulse. Then a classical write pulse ($\approx 10^4$ photons) with a duration of 100 ns is applied onto the atomic ensemble to produce, via the spontaneous Raman transition $|a\rangle \rightarrow |e\rangle \rightarrow |b\rangle$, a superposition state of the anti-Stokes field and the collective spin, $|\psi\rangle_{a,\text{AS}} = |0\rangle_a|0\rangle_{\text{AS}} + \sqrt{\chi}|1\rangle_a|1\rangle_{\text{AS}} + \chi|2\rangle_a|2\rangle_{\text{AS}} + O\left(\chi^{\frac{3}{2}}\right)$ with a probability of $\chi \ll 1$. After a controllable delay δt, a read pulse with a duration of 75 ns is applied to convert the collective excitation into the Stokes field. The intensity of the read pulse is about 100 times stronger than that of the write pulse.

The cross-correlation between the Stokes and anti-Stokes photons is defined as $g_{\text{S,AS}}^{(2)} = \frac{p_{\text{S,AS}}}{p_S p_{\text{AS}}}$, where $p_{\text{S,AS}} = \chi \eta_{\text{ret}} \eta_{\text{AS}} \eta_S$ is the coincidence probability, η_{ret} is the retrieval efficiency, and η_S (η_{AS}) is the overall detection efficiency of the Stokes (anti-Stokes) channel (including the transmission efficiency of filters and optical components, the coupling efficiency of the fiber couplers, and the detection efficiency of single-photon detectors), and $p_S = \chi \eta_{\text{ret}} \eta_S$ ($p_{\text{AS}} = \chi \eta_{\text{AS}}$) is the probability of detecting a Stokes (anti-Stokes) photon. The anti-Stokes photon is registered by the single-photon detector D1, and the retrieved Stokes photon is registered by detectors D2 and D3. The cross-correlation as a function of p_{AS} is shown in Fig. 14.3a, which clearly demonstrates that the cross-correlation is inversely proportional to the excitation probability. The cross-correlation versus storage time is shown in Fig. 14.3b, which gives a coherence time of the quantum memory of about 12 μs for this implementation. A coherence time of a few tens of microseconds is mainly caused by the residual magnetic field [23].

The single-photon quality is characterized by the conditional auto-correlation of the Stokes photons $\alpha = \frac{P(23|\text{AS})}{P(2|\text{AS})P(3|\text{AS})}$ [24], where $p_{(23|\text{AS})}$ is the coincident probability between single-photon detectors D2 and D3 conditioned on the detection of an anti-Stokes photon on D1, and $p_{(2|\text{AS})}$ ($p_{(3|\text{AS})}$) is the conditional probability on D2 (D3). Figure 14.4a shows α as a function of excitation probability. For small χ, $\alpha = 4\chi$ [13]. $\alpha < 1$ indicates that the triggered Stokes photon approximates a single-photon state. We note that, for $p_{\text{AS}} \rightarrow 0$, $\alpha = 0.057(28)$, which deviates from the ideal value of 0. This offset comes from noise, including the residual leakage of the write and read beams, stray light, and the dark counts of the detectors.

To make a deterministic single-photon source, a number of write pulses per experimental trial and the feedback protocol are applied as discussed above. A realistic time sequence is shown in Fig. 14.2b. In the time interval ΔT, N independent write sequences with a period of δt_W are applied to the atomic ensemble. Each write sequence contains a reset pulse and a write pulse. Once an anti-Stokes photon is detected by D1, the feedback circuit stops further write sequences and enables the read pulse to retrieve the single Stokes photon after a programmable delay Δt. The maximum number of write sequences is determined by the ratio of δt_W and the coherence time of the quantum memory. The feedback protocol enhances the production probability of

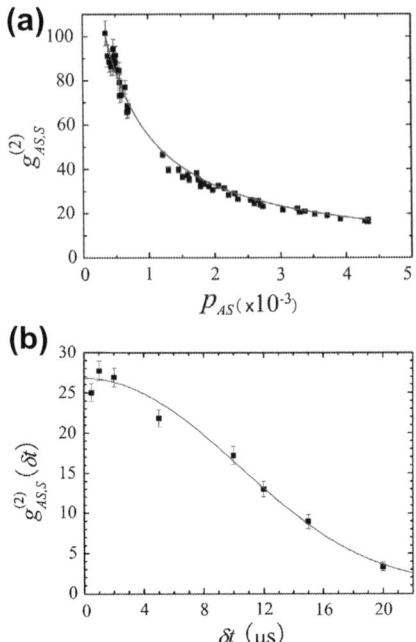

FIGURE 14.3 Intensity correlation functions $g^{(2)}_{S,AS}$ along the excitation probability p_{AS} with $\delta t = 500$ ns (a) and along the time delay δt between read and write pulses with $p_{AS} = 3 \times 10^{-3}$ (b). The observed lifetime is $\tau_C = 12.5 \pm 2.6$ μs.

anti-Stokes photons according to the new excitation probability

$$P_{\text{tot}} = \sum_{j=1}^{N} p_{AS}(1 - p_{AS})^{j-1}. \quad (14.2)$$

And the conditional probability is

$$P_{i/AS} = \sum_{j=0}^{N-1} p_{AS},(1 - p_{AS})^{j-1} P_{i/AS,(\Delta T - j\delta t_W)}, \quad (14.3)$$

with $i = 2, 3$ and 23. $P_{i/AS,(\Delta T - j\delta t_W)}$ is the conditional detection probability after applying the jth write pulse. In this setup, the normalized auto-correlation function is given by $\alpha = \frac{P_{(23|AS)}}{P_{(2|AS)} P_{(3|AS)}}$. This protocol can be executed in different modes. In the first mode, one can fix the retrieve time ΔT. Therefore, the delay Δt varies because the spin excitation is created randomly by one of the write sequences. Single photons are produced at a given time with a high probability, approaching unity if $N \gg 1$. In the second mode, we retrieve the single photons with a fixed delay Δt after a successful write process. In this case, the imprinted

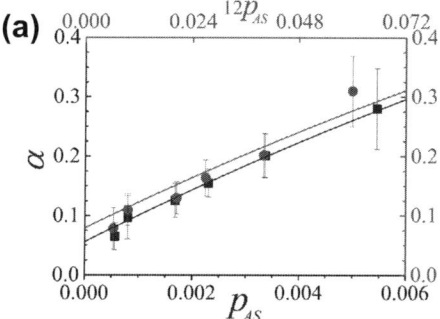

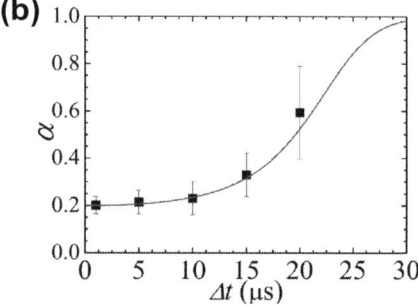

FIGURE 14.4 The auto-correlation parameter as a function of p_{AS} (a) and δt (b). In panel (a), the data in black (square dots) corresponds to the experiment without feedback circuit, in which each write sequence is followed by one read pulse. The data in gray (circular dots) corresponds to the experiment with feedback circuit, in which 12 successive write sequences are followed by one read pulse. The gray curve (upper curve) is the theoretical evaluation taking into account the fitted background of the black dots. In panel (b), 12 write sequences were applied in each trial while measuring.

single excitation can be converted into a single photon at any given time within the coherence time of the quantum memory. The first mode can serve as a deterministic single-photon source and is well suited for all-optical quantum information processing.

In the experiment, $\Delta T = 12.5$ μs and $\delta t_W = 1$ μs, i.e. up to 12 write sequences can be used. The measured auto-correlation parameter α is shown in Fig. 14.4a as a function of the excitation probability p_{AS}. For $N = 1$, α is nearly linear for $p_{AS} \leq 0.006$. For $N = 12$ successive write sequences, α is plotted versus $12 p_{AS}$. Compared with $N = 1$ we can easily see that α is preserved, and the excitation probability is enhanced by the repetitive write sequences. If the coherence time of the spin excitation is long enough to allow many write sequences, the excitation probability can reach unity while preserving the single-photon nature of the output. In this case, the generation efficiency depends only on the retrieval efficiency. In the second experiment, we fix $\delta t_W = 1$ μs and $N = 12$. Figure 14.4b shows the measured as a function of Δt. For each Δt, ΔT varies due to the random creation of the collective excitation by the N write sequences. Again, it is clear that the single-photon nature is well preserved within the coherence time of the quantum memory.

The production rate of the single photons is determined by the coherence time of the quantum memory, the retrieval efficiency, and the duration of the pulses. To enhance the production rate, it is necessary to improve both the coherence time of the quantum memory and the retrieval efficiency. Moreover, by further improving the control circuit, i.e. reducing the period of write pulses due to electronic delays, more write pulses can be applied within the lifetime. A simple estimation shows that with $p_{AS} \sim 0.003$ and a write period of 300 ns, we may obtain a single-photon source with a probability as high as 95% within a coherence time of 300 μs.

14.4 INTERFERENCE OF PHOTONS FROM INDEPENDENT SOURCES

In the linear-optical quantum computing protocol proposed by Knill *et al.* [3], single photons from independent sources interact in a Mach-Zehnder-like interferometer and a single-photon detection is used to implement a controlled phase gate. In the polarization-entanglement-based scheme proposed by Browne *et al.* [4], two photons from independent sources are sent to a Hong-Ou-Mandel (HOM) type interferometer [25] to implement coalescence gates [4]. In both cases, a unity visibility can be achieved only if the photons are indistinguishable [26,27].

The indistinguishability of single photons has been examined in recent years [28–30]. In this section, we show that the two photons generated from two independent atomic ensembles are pure and indistinguishable, as indicated by the high two-photon HOM interference visibility. The two indistinguishable single photons can also be used to generate polarization entanglement by postselection.

The experimental setup is depicted in Fig. 14.5. Atomic ensembles collected by two MOTs that are 0.6 m apart serve as the media for quantum memories and deterministic single-photon sources. The atomic system is the same as the one described in Section 15.3. A write pulse that generates the collective excitation is detuned by 10 MHz and loosely focused to a beam diameter of about 400 μm through the atomic ensemble in the MOT. The emitted anti-Stokes photon is collected into a single-mode optical fiber with the mode diameter of about 100 μm in the interaction region. The optical axis of the anti-Stokes field is tilted by 3° [31] from that of the write beam. As described above, the detection of a single anti-Stokes photon heralds a single collective excitation in the atomic ensemble with complete certainty. After a controllable time delay, a classical read pulse with an orthogonal polarization and spatially mode-matched with the write beam is applied from the opposite direction. The collective excitation in the atomic ensemble is converted into a single-photon excitation in the Stokes field, which propagates in the opposite direction of the anti-Stokes field and is collected into a single-mode fiber. If the retrieval efficiency is unity, the emission of the single Stokes photon is deterministic. Changing from the co-propagating configuration described in the last section to a counter-propagating configuration, and tilting the Stokes and anti-Stokes modes by 3° relative to the classical write and read beams, are convenient ways to separate the quantum fields from the intense classical light.

As shown in Fig. 14.5b, Alice and Bob have similar sources. They each prepare collective spin excitations independently and whichever one finishes the preparation first waits for the other while keeping the collective spin excitation in her/his quantum memory. After both Alice and Bob have finished their preparations, they retrieve the excitations simultaneously at any time within the coherence time of the collective state. Therefore the retrieved photons arrive at the beam splitter contemporarily. The coherence time of the single photons is adjustable and is set to be about 25 ns in this demonstration. This long coherence time is chosen to reduce the impact of a possible small temporal mismatch between Alice and Bob that can arise from the optical path difference. Note that the read and write lasers used for different single-photon sources are completely independent of each other. This guarantees there is no spatial-temporal correlation between the single photons [32].

Compared to a probabilistic photon source, deterministic sources based on atomic ensembles achieve a considerable enhancement of the coincidence rate of single photons coming from Alice and Bob. For instance, consider a probabilistic version of the same experimental setup, in which Alice and Bob operate without the feedback circuit, and instead apply a write pulse and a read pulse in every experimental trial and monitor fourfold coincidences between the anti-Stokes and Stokes photons in the four channels D1, D2, C1, and C2. Assuming that the probability of having an anti-Stokes photon in channel D1 (D2) is p_{AS1} (p_{AS2}), and the corresponding retrieval efficiency of converting the spin excitation to a Stokes photon coupled into channel C1 (C2) is $\eta_{ret1}(\delta t_R)$, the probability of fourfold coincidence is $p_{4c} = p_{AS1}\eta_{ret1}(\delta t_R) p_{AS2}\eta_{ret2}(\delta t_R)$. This can be compared to the feedback-circuit based scheme shown in Fig. 14.5b, where we can apply N write pulses in each trial. Assuming $p_{AS1}, p_{AS2} \ll 1$, the probability of fourfold coincidence can be approximated as $p_{4c} \approx N^2 p_{AS1}\eta_{ret1}(\delta t_R) p_{AS2}\eta_{ret2}(\delta t_R)$, which shows that the probability of fourfold coincidence is enhanced by N^2 for each trial. For this demonstration, $p_{AS1} = p_{AS2} \approx 0.002$ (the corresponding cross-correlation $g_{S,AS}^{(2)} = 30$), $N = 12$, $\tau_C \approx 12$ μs, $\delta t_R = 400$ ns, and $\eta_{ret1}(0) = \eta_{ret2}(0) \approx 8\%$.

The four lasers in Fig. 14.5b are independently frequency-stabilized. The full width at half maximum (FWHM) linewidths of W1 and R1 are about 1 MHz while those of W2 and R2 are about 5 MHz). They are broadened to more than

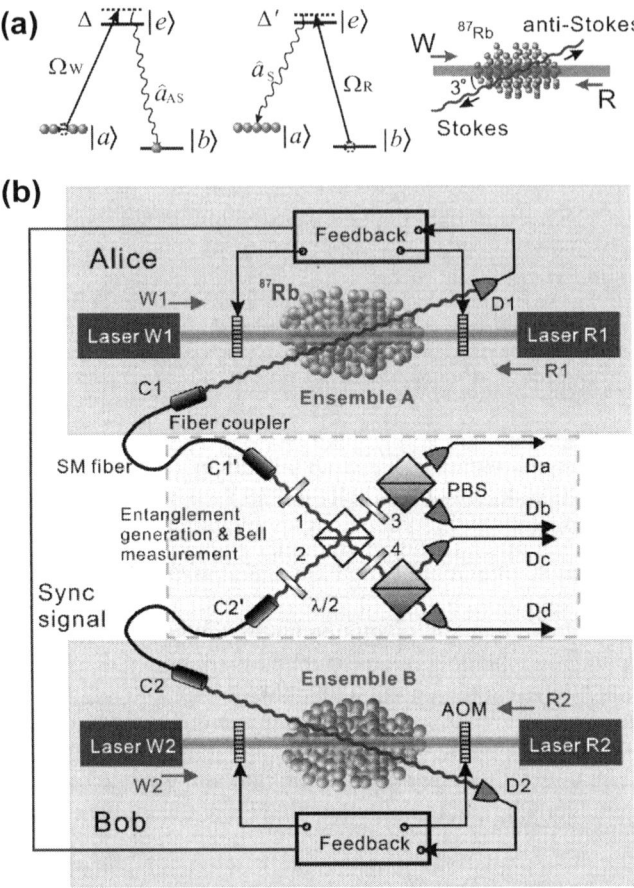

FIGURE 14.5 Illustration of the relevant energy levels of the atoms and arrangement of laser beams (a) and the experimental setup (b). Alice and Bob both keep a single-photon source at two remote locations. Alice applies write pulses continuously until an anti-Stokes photon is registered by detector D1. Then she stops the write pulse, holds on to the spin excitations, sends a synchronization signal to Bob, and waits for his response (This is realized by the feedback circuit and the acousto-optic modulators, AOM). In parallel, Bob prepares a single excitation in the same way as Alice does. After they both agree that each has a collective excitation stored, both simultaneously apply a read pulse to retrieve the spin excitation into a light field $\hat{a}_S$. The two Stokes photons propagate to the place for entanglement generation and a subsequent Bell measurement. Photons overlap at a 50:50 beam splitter (BS) and then will be analyzed by half-wave plates, polarized beam splitters (PBS), and single-photon detectors Da, Db, Dc, and Dd.

20 MHz because the finite bandwidth of AOM limits the shortest pulse width of 40 ns. The linewidth of the retrieved single photons is determined mainly by the linewidth and intensity of the read lasers. In order to verify that the two Stokes photons coming from Alice and Bob are indistinguishable, we overlap them at a non-polarizing beam splitter with the same polarization (horizontal in this case) and measure the two-photon HOM interference. The visibility of the

HOM interference can be expressed in terms of the purity of the single photons [32] $V = Tr(\rho_{S1}\rho_{S2}) = [P(\rho_{S1}) + P(\rho_{S2}) - O(\rho_{S1},\rho_{S2})]/2$, where $\rho_{S1,S2}$ is the reduced density matrix of Stokes photon 1 and 2, $P(\rho_{Si}) = Tr(\rho_{Si}^2) \le 1$ ($i = 1,2$) is the purity, and $O(\rho_{S1},\rho_{S2}) = \|\rho_{S1} - \rho_{S2}\|^2 \ge 0$ is the operational distance between the states of the two photons. It can be easily seen that the visibility gives a lower bound on the mean purity of the single photons. (See Chapter 2 for further discussions on purity, visibility, and indistinguishability.)

We perform HOM interference measurements in both the time domain and the frequency domain. The polarizations of the anti-Stokes photons were set to horizontal with two half-wave plates before they enter the BS, as shown in Fig. 14.5b. The other two half-wave plates after the BS were set to 0°. In the first measurement, fourfold coincidences between detectors D1, D2, Da, and Dd were monitored while changing the time delay between the two read pulses. The excitation probabilities $p_{AS1} = p_{AS2} \approx 0.002$. The coincidence-rate variation with delay is shown in Fig. 14.6. Ideally, there should be complete destructive interference if the wave packets of the two photons overlap perfectly. However, in practice it is difficult to ensure that the two wave packets are identical and perfectly overlapped. We obtained the visibility $V = \frac{C_{\text{plat}} - C_{\text{dip}}}{C_{\text{plat}}} = 80(1)\%$, where C_{plat} is the measured four-photon coincidence rate at the plateau and C_{dip} the rate at the dip. The asymmetry of the profile at a negative delay versus a positive delay shows that the two wave packets are (a) not perfectly identical, (b) not symmetric themselves. Assuming that the HOM dip possesses a Gaussian-type profile, we verify the coherence time to be 25(1) ns FWHM.

In the second measurement, we measured the fourfold coincidence while changing the relative frequency detuning between the two read pulses (Fig. 14.6b, lower panel). The excitation probabilities are $p_{AS1} = p_{AS2} \approx 0.003$. The frequency detuning varies from −30 MHz to 30 MHz limited by the instruments in use. The coincidence rate at the largest detuning was compared to the coincidence rate when the photons impinge the BS with orthogonal polarizations, and the two rates are consistent with each other, thus confirming that the plateau of the HOM dip has been reached. The measured visibility of 82(3)% agrees well with that obtained in the time-domain measurements. The width of the HOM dip is 35(3) MHz FWHM, in agreement with the coherence time of 25 ns. Therefore, the narrow-band nature of the source is directly verified by the HOM dip in the frequency domain. From the visibility, we estimate that the lower bound on the purity of the single photons is about 0.8.

The visibility may be reduced by the imperfect overlap of the two-photon wave packets. Besides, the imperfection of the single-photon sources affects the visibility as well. As discussed in the last section, the quality of the single-photon source can be characterized by the auto-correlation parameter $\alpha = 2P_{II}/P_I^2$, where $P_I (P_{II})$ is the probability of generating one (two) photon(s) from each source (the higher orders are negligible small). If the two single-photon wave packets do not overlap at all then there is no interference, and we obtain the non-correlated coincidence rate $C_{\text{plat}} = \frac{P_I^2}{2} + P_{II}$ between Da and Dd. If they

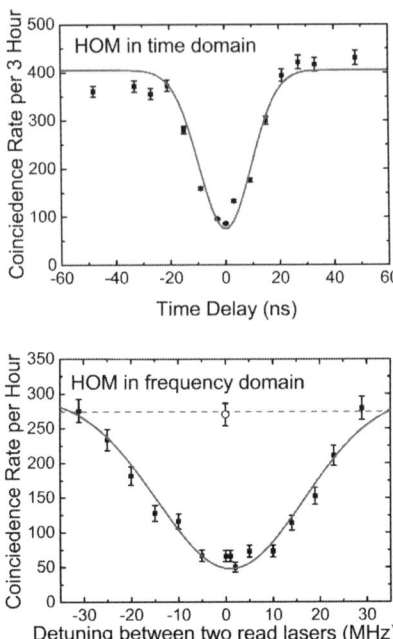

FIGURE 14.6 HOM dips in time domain (upper panel) and frequency domain (lower panel). The open circle in the lower panel was obtained by setting the polarization of the two photons perpendicular to each other and zero detuning between two read lasers, i.e. with fully distinguishable photons. The Gaussian curves that roughly connect the data points are only shown to guide the eye. The dashed line shows the plateau of the dip. Error bars represent statistical errors, which are ±1 standard deviation.

overlap perfectly, there is a destructive interference leading to a coincidence rate $C_{\text{plat}} = P_{\text{II}}$. So the visibility of the HOM dip is $V = 1/(1 + \alpha)$, where $\alpha = 0.12$ for the source prepared later (with the collective excitation retrieved immediately), and 0.17 for the source prepared earlier (it has to wait for the other one). This leads to an average visibility of 87%. In the frequency domain, the average visibility is around 83% because of higher excitation probabilities. In Fig. 14.6, the HOM dip is measured by setting the coincidence window (∼50 ns) larger than the wave-packet length of the single photons (∼25 ns).

Two indistinguishable single photons can be used to generate entanglement by post-selection [33], as illustrated in Fig. 14.5b. Before the BS, half-wave plates set the Stokes photons to orthogonal polarizations (horizontal and vertical). If there is coincidence between output ports 3 and 4, then the state of the two photons will be projected onto a Bell state $|\psi^-\rangle_{12} = \frac{1}{\sqrt{2}}(|H\rangle_1|V\rangle_2 - |V\rangle_1|H\rangle_2)$. With another two half-wave plates and two PBSs after the BS, the entanglement of the two photons can be verified by a Clauser-Horne-Shimony-Holt (CHSH)-type inequality [34], in which local realism implies that $S \leq 2$, with $S = |E(\theta_1, \theta_2) - E(\theta_1, \theta_2') + E(\theta_1', \theta_2) + E(\theta_1', \theta_2')|$. Here $E(\theta_1, \theta_2)$ is the correlation function, where θ_1 and θ_1' (θ_2 and θ_2') are the measured polarization angles of the Stokes photon at port 3 (4). We obtain $S = 2.37(0.07)$, which violates the CHSH inequality by 5 standard deviations. Taking into account the two-photon component in the single-photon sources, a straightforward calculation shows the violation of the CHSH inequality is about 2.3, which is in good agreement with the experimental results.

14.5 CONCLUSION AND OUTLOOK

In this chapter, we have introduced the atomic-ensemble-based single-photon sources. The advantages of this type of source are (1) it does not require strong coupling between light and emitters thanks to the collective enhancement; (2) the linewidth of the photons naturally matches that of an atomic memory implemented in the same physical system, making the emitted single photons readily storable; (3) the emitted photons are narrow-band and highly controllable. With the aid of a feedback circuit, this source can approximate a deterministic single-photon source. Being a key device in any scalable quantum communications network, and for large-scale linear-optical computation, this circuit shows a promising enhancement in excitation probability while preserving the single-photon nature of the output. Synchronized generation of narrow-band single photons from two independent atomic ensembles has been demonstrated. The Hong-Ou-Mandel dip was observed in both the time domain and the frequency domain with a high visibility for independent photons from two separate sources, demonstrating the indistinguishability of these photons.

Atomic-ensemble-based single-photon sources hold promise for all-optical scalable quantum-information processing. Recently, remarkable progresses have been achieved toward quantum-information processing with atomic ensembles and linear optics. For example, DLCZ-type entanglement between two atomic ensembles has been created [35], and functional quantum-repeater nodes with atomic ensembles have been demonstrated [36,37]. The storage of single photons has been realized where heralded single photon emitted from one atomic-ensemble-based single-photon source is stored in a remote atomic quantum memory [14,15,38].

To make atomic-ensemble single-photon sources truly useful in practice, the coherence time and the retrieval efficiency must be significantly improved. In the atomic ensemble, the coherence time of the collective state suffers from the residual magnetic field around the MOT. It has been demonstrated that by using magnetic-field-insensitive states to store the collective excitation and suppressing decoherence due to atomic random motion by reducing the angle between write and anti-Stokes mode, or by using a one-dimensional optical lattice to trap the atoms, millisecond quantum memories can be achieved [39–42]. A recent experiment demonstrated a coherent storage time of 16 s in ultracold atoms for classical states of light [16]. The retrieval efficiency can also be improved, for example, by increasing the optical depth of the atomic ensemble, or by putting the atomic ensemble inside a cavity. Using these techniques a retrieval efficiency of 84 % has been achieved for a storage time of a few hundred nanoseconds [43]. Achieving efficient retrieval and long storage times in a single experiment can be done by combing these two advances, as demonstrated in Ref. [44], where a retrieval efficiency of 73 % and a coherence time of 3 ms are reported.

APPENDIX

A Write Process

Let us consider a pencil-shaped cold atomic ensemble containing N 3-level atoms, with ground states $|a\rangle$, $|b\rangle$ and an excited state $|e\rangle$. We denote the axial direction as z direction and assume the zero point is at the center of the atomic ensemble. Initially, all the atoms are in the ground state $|a\rangle$. The off-resonant classical write pulse coupling the excited state $|e\rangle$ and the ground state $|a\rangle$ is given by $\boldsymbol{E}_W(\boldsymbol{r},t) = \hat{\epsilon}_W E_W(\boldsymbol{r},t) e^{i(\boldsymbol{k}_W \cdot \boldsymbol{r} - \omega_W t)}$+h.c., where $\hat{\epsilon}_W$ is the polarization unit vector, $\omega_W = ck_W$ is the angular frequency of the write light. For simplicity, we assume the write light pulse propagating along the axial direction $\boldsymbol{k}_W = k_W \hat{z}$. The anti-Stokes field coupling the excited state $|e\rangle$ and ground state $|b\rangle$ is quantum mechanically described as $\boldsymbol{E}_{AS}(\boldsymbol{r},t) = \sum_k \hat{\epsilon}_k \varepsilon_k \hat{a}_k e^{i(\boldsymbol{k} \cdot \boldsymbol{r} - \omega_k t)} + \text{h.c.}$, where $\varepsilon_k = \sqrt{\frac{\hbar \omega_k}{2\epsilon_0 V}}$, $\omega_k = ck$, $\hat{\epsilon}_k$ is the polarization unit vector, and $\hat{a}_k$ is the annihilation operator of mode $\boldsymbol{k}$. In the cold atomic ensemble, because of the extremely low temperature and the short pulse length of the write light, we can safely assume the atoms are fixed at certain positions during the write process and denote the coordinate of the ith atom by $\boldsymbol{r}_i$. The total Hamiltonian in the rotating frame is given by

$$H = \sum_{i=1}^{N} \left[\hbar \Delta \sigma_{ee}^i - \hbar \Omega_W(\boldsymbol{r}_i, t) e^{i\boldsymbol{k}_W \cdot \boldsymbol{r}_i} \sigma_{ea}^i + \sum_k \hbar g_k \hat{a}_k e^{i(\boldsymbol{k} \cdot \boldsymbol{r}_i - \Delta \omega_k t)} \sigma_{eb}^i + \text{h. c.} \right], \quad (A.1)$$

where the detuning $\Delta = \omega_{ea} - \omega_W$ and $\Delta \omega_k = \omega_k - \omega_W - \omega_{ba}$, with $\omega_{ea} = \omega_e - \omega_a$ and $\omega_{ba} = \omega_b - \omega_a$ the difference between atomic levels. The spin operators of $\sigma_{lm}^i = |l\rangle_i \langle m|$ $(l,m = e, a, b)$ are the transition operators of ith atom, $\Omega_W(\boldsymbol{r}, t) = \frac{\boldsymbol{d}_{ea} \cdot \hat{\epsilon}_W E_W(\boldsymbol{r},t)}{\hbar}$ with $\boldsymbol{d}_{ea} = \langle e|\boldsymbol{d}|a\rangle$ being the Rabi frequency of the write light, and $g_k = -\frac{\boldsymbol{d}_{eb} \cdot \hat{\epsilon}_k \varepsilon_k}{\hbar}$ with $\boldsymbol{d}_{eb} = \langle e|\boldsymbol{d}|b\rangle$ is the coupling coefficient of each mode of the anti-Stokes light.

If the Rabi frequency of the write light and the linewidth of the excited state are both significantly smaller than the detuning Δ, the upper state $|e\rangle$ can be adiabatically eliminated, and each atom is described by an effective two-level

Chapter | 14 Generation and Storage of Single Photons   557

model. The resulting adiabatic Hamiltonian is given by

$$H = \sum_{i=1}^{N}\left[\frac{\Omega_W(r_i,t)}{\Delta}e^{ik_W \cdot r_i}\sigma_{ba}^i \sum_k \hbar g_k \hat{a}_k^\dagger e^{-i(k \cdot r_i - \Delta \omega_k t)} + \text{h. c.}\right], \quad (A.2)$$

where we have neglected the small AC Stark shift. This adiabatic Hamiltonian describes the spontaneous emission of N atoms from the pseudo excited state $|a\rangle$ to the pseudo ground state $|b\rangle$ where the frequency of the emitted anti-Stokes light is centered at $\omega_S = \omega_W - \omega_{ba}$. The linewidth of the pseudo excited state is $\Gamma' = \frac{\Omega_W^2}{\Delta^2}\Gamma$, with Γ the decay rate from $|e\rangle$ to $|b\rangle$. This Hamiltonian has been extensively investigated in last two decades [45]. The initial stage can be well described by spontaneous emission where the anti-Stokes photon is emitted along all the directions. After a time $1/\Gamma'$, the anti-Stokes light will dominate along the axial direction and enter the superradiance regime. In our case, the interaction time T is determined by the pulse duration of the write beam which is short compared to the lifetime $1/\Gamma'$, and thus we are in the spontaneous emission regime. Therefore we can simply solve the Schrödinger equation by using perturbation theory. To the first order of the perturbation, the atom-light system is described by

$$|\psi\rangle = \left[1 - i\int_0^T H(\tau)\,d\tau\right]|\text{vac}\rangle + o(p) \quad (A.3)$$

with $|\text{vac}\rangle = |0\rangle_a|0\rangle_p$, where $|0\rangle_a = \otimes_i |a\rangle_i$ the atomic vacuum state and is $|0\rangle_p$ the light vacuum. Integrating out τ, we obtain

$$|\psi\rangle = |0\rangle_a|0\rangle_p + \sum_i^N \frac{\Omega_W(r_i)}{\Delta}e^{ik_W \cdot r_i}|a\ldots b_i\ldots a\rangle|\gamma\rangle_i, \quad (A.4)$$

where $|\gamma\rangle_i = -i\int_0^T \sum_k g_k a_k^\dagger e^{-i(k \cdot r_i - \Delta \omega_k t)}|0\rangle_p$ is the spontaneously emitted anti-Stokes light from the ith atom, and we have assumed the Rabi frequency is time independent. It can be easily seen that in the spontaneous emission regime the atoms emit anti-Stokes photons into all the directions independently from each other.

As discussed in standard quantum optics textbooks [46], the spatial wave function of the photon emitted from ith atom can be described by $E_i(\Delta r_i) = \frac{\varepsilon_0}{\Delta r_i}e^{ik_{AS}\Delta r_i}$ where $k_{AS} = \frac{\omega_{AS}}{c}$, ε_0 is the constant proportional to the electro-dipole transition matrix element, $\Delta r_i = |r - r_i|$ is the distance between the ith atom and observation point r. Assume we observe the anti-Stokes light along the axial direction, i.e. z direction, under the paraxial axial approximation $|z - z_i| \gg x^2, y^2, x_i^2, y_i^2$, the wave function on the observation surface may be expressed as

$$E_i(\mathbf{r}) = \frac{\varepsilon_0}{z - z_i} \exp\left[ik_{AS}\left(z - z_i + \frac{x_i^2 + y_i^2}{2(z - z_i)} + \frac{x^2 + y^2}{2(z - z_i)}\right)\right.$$

$$\left. - ik_{AS}\frac{x_i x + y_i y}{z - z_i}\right]$$

$$\approx \frac{\varepsilon_0}{z}\exp(-ik_{AS}z_i)\exp\left[ik_{AS}\left(z + \frac{x_i^2 + y_i^2}{2z}\right.\right.$$

$$\left. + \frac{x^2 + y^2}{2z} - \frac{x_i x + y_i y}{z}\right)\bigg]$$

$$\times \exp\left[ik_{AS}\left(\frac{x_i^2 + y_i^2}{2z^2}z_i + \frac{x^2 + y^2}{2z^2}z_i\right.\right.$$

$$\left. - \frac{x_i x + y_i y}{z^2}z_i\right)\bigg], \qquad (A.5)$$

where $|z_i| \ll z$ is assumed. We define two diffraction angles with $\theta_{W_a} = \frac{1}{k_{AS}W_a}$ and $\theta_L = \left(\frac{1}{k_{AS}L}\right)^{1/2}$, where W_a and L are the waist and length of the atomic ensemble, respectively. It can be readily seen that if the detection angle $\theta \leq \min(\theta_{W_a}, \theta_L)$, all the phase factors which are related to coordinates of the atoms, except $\exp(-ik_{AS}z_i)$, can be safely neglected. Thus the anti-Stokes light on the observation surface can be treated as a single mode and the spatial wave function may be described by $E_i(\mathbf{r}) \approx \zeta_{AS}(\mathbf{r})\exp(-i\mathbf{k}_{AS} \cdot \mathbf{r}_i)$, with $\zeta_{AS}(\mathbf{r}) = \frac{\varepsilon_0}{z}\exp\left[ik_{AS}\left(z + \frac{x^2+y^2}{2z}\right)\right]$ and $\mathbf{k}_{AS} = k_{AS}\hat{z}$ the wave vector of the detected anti-Stokes light. We approximate the detected anti-Stokes photon state by $|\gamma\rangle_i = \sqrt{p}\hat{a}_{AS}^\dagger e^{-i\mathbf{k}_S \cdot \mathbf{r}_i}|0\rangle_p$, where $\hat{a}_{AS}^\dagger$ is a single-mode creation operator, and $p = \frac{\Omega_W^2}{\Delta^2}\Gamma T d\Omega \ll 1$ is the small probability for one atom to scatter one Stokes photon into the detection solid angle $d\Omega$. Substituting $|\gamma\rangle_i$ into Eq. (A.4) we obtain $|\psi\rangle \approx 1 + \sqrt{p}\left(\sum_{i=1}^N e^{i\Delta \mathbf{k} \cdot \mathbf{r}_i}\sigma_{ba}^i\right)\hat{a}_{AS}^\dagger|vac\rangle$, where $\Delta \mathbf{k} = \mathbf{k}_W - \mathbf{k}_{AS}$ is the momentum difference between the write light and the Stokes mode. Defining a bosonic collective-state operator $S^\dagger = \frac{1}{\sqrt{N}}\sum_{i=1}^N e^{i\Delta \mathbf{k} \cdot \mathbf{r}_i}\sigma_{ba}^i$, we have $[S, S^\dagger] \approx 1$. The atom-light system is described by $|\psi\rangle \approx 1 + \sqrt{\chi}S^\dagger \hat{a}_{AS}^\dagger|vac\rangle$, with $\chi = Np$ the probability to detect one Stokes photon in write process. It is easily seen that when an anti-Stokes photon is detected, the atomic ensemble is projected into the collective excited state, or in other words a spin wave is imprinted into the atomic ensemble. The conventional single-mode condition that $F = \frac{A}{\lambda L} \approx 1$ where F is the Fresnel number, $A = \pi W_a^2$, the cross-section area can be obtained by setting the two diffraction angles θ_{W_a} and θ_L are equal. In this case, the detection solid angle can be approximated by λ^2/A. Then we have the total excitation probability $\chi = \frac{\Omega_W^2}{\Delta^2} N\Gamma T \frac{\lambda^2}{A} \sim d_0 \Gamma' T$, where $d_0 = N\sigma_0/A$ with $\sigma_0 = \frac{\lambda^2}{2\pi}$. To ensure we are in the spontaneous Raman scattering regime, we require the excitation probability $\chi \ll 1$. Note that in write process, there is no constructive interference in the forward direction, because when one atom scattering an anti-Stokes photon, it changes to another ground state $|b\rangle$ and thus all the terms in Eq. (A.4) are orthogonal to each other.

B Read Process

In the read process, a strong classical read light is applied to the atomic ensemble to convert the collective excitation into a Stokes photon. The weak Stokes field and the strong read light satisfy the EIT condition [19,20], and thus the Stokes field is not absorbed by the atoms in ground state $|a\rangle$.

Assume the strong classical read light coupling the excited state $|e\rangle$ and $|b\rangle$ ground state is counter-propagating with the write light $k_R = -k_R \hat{z}$. The atom in state $|b\rangle$ is excited by the read light and transferred back to ground state $|a\rangle$ generating an anti-Stokes photon simultaneously. In contrast to the write process, the light emitted from different atoms will interfere with each other, and constructive interference occurs in the direction where mode-match condition is satisfied. The read process can be described by $\frac{1}{\sqrt{N}} \sum_{i=1}^{N} e^{i\Delta k \cdot r_i} |a \ldots b_i \ldots a\rangle \Rightarrow \otimes_i |a\rangle_i E_S(r')$. The spatial wave function of the Stokes field on the observation point can be expressed as

$$E_S(r') = \frac{1}{\sqrt{N}} \sum_{i=1}^{N} e^{i(\Delta k + k_R) \cdot r_i} \frac{\varepsilon_0}{\Delta r_i'} e^{ik_S \Delta r_i'} \tag{B.1}$$

with $\Delta r_i' = |r' - r_i|$, where the atoms are treated as point light sources. Assume we observe Stokes light along the backward direction. Under the paraxial approximation, we can write the Stokes light as

$$E_S(r') \sim \frac{1}{\sqrt{N}} \sum_{i=1}^{N} e^{i(\Delta k + k_R) \cdot r_i} \frac{\varepsilon_0}{|z' - z_i|} e^{-ik_S \cdot r_i}$$
$$\times \exp\left[ik_S \left(|z'| + \frac{x_i^2 + y_i^2}{2|z' - z_i|} + \frac{x'^2 + y'^2}{2|z' - z_i|} - \frac{x_i x' + y_i y'}{|z' - z_i|}\right)\right]. \tag{B.2}$$

It can be readily seen that the once the mode-match condition $k_W - k_{AS} + k_R - k_S = 0$ is satisfied, constructive interference will be observed on the detection surface. The Stokes field can be described by $E_S(r') \approx \sqrt{N} \int dr'' n(r'') \zeta_S(r') = \sqrt{N} \zeta_S(r')$ where $\zeta_S(r') = \frac{\varepsilon_0}{z'} \exp\left[-ik_S \left(z' + \frac{x'^2 + y'^2}{2z'}\right)\right]$ and $n(r)$ is atomic density distribution, and we have assumed the detection angle $\theta' \leq \min(\theta_{W_a}, \theta_L)$.

One can see that the intensity of the Stokes light is proportional to the atomic number N and the detection solid angle. The retrieval efficiency can be estimated by $\eta_{ret} \sim \frac{\Gamma N d\Omega}{\Gamma N d\Omega + \Gamma} = \frac{N d\Omega}{N d\Omega + 1}$, where N is the number of atoms, and $d\Omega$ is the solid angle in which we have constructive interference. Under the single-mode condition $d\Omega \sim \frac{\lambda^2}{A}$, a direct calculation shows that the retrieval efficiency $\eta_{ret} \sim 1 - \frac{1}{d_0}$ is given by the optical depth d_0. Note that taking into account the narrow EIT window, the error in retrieval efficiency scales as $\frac{1}{\sqrt{d_0}}$ [47].

The Stokes field couples the excited state and ground state, while it will not be absorbed since the atom-light system fulfills the EIT condition. In this case the Stokes light propagates in the atomic ensemble slower than the read light. Thus we require the read light pulse is sufficiently long so that all the anti-Stokes light can propagate out of the atomic ensemble.

The read process can be described using the dark-state polariton theory [19,20]. The read light may be described by $E_R(r,t) = \hat{\epsilon}_R E_R(r,t) e^{i(k_R \cdot r - \omega_R t)} +$ h.c., where $\hat{\epsilon}_R$ is the polarization unit vector, $\omega_R = c k_R$ is the frequency of the write light. The Stokes field is approximated by a single-mode light $E_S(r,t) = \hat{\epsilon}_S \hat{a}_S e^{i(k_S \cdot r - \omega_S t)} +$ h.c. The Hamiltonian describing the read process is given by $H = \sum_{i=1}^{N} [-\hbar \Omega_R(t) e^{i(k_R \cdot r_i - \omega_R t)} \sigma_{eb}^i + \hbar g_S \hat{a}_S e^{i(k_S \cdot r_i - \omega_S t)} \sigma_{ea}^i] +$ h. c. with Ω_R the Rabi frequency of the read light and g_S the coupling coefficient, where we have assumed single-photon and two-photon resonance are both satisfied. This Hamiltonian has a series of adiabatic eigenstates with vanishing excited state component, i.e. a dark-state polariton. The simplest dark-state polariton can be described by $|D,1\rangle = \left(cos\theta \hat{a}_S^\dagger - sin\theta S'^\dagger \right) |vac\rangle$, where $tan\theta = \frac{g\sqrt{N}}{\Omega_R(t)}$ and $S'^\dagger = \frac{1}{\sqrt{N}} \sum_{i=1}^{N} e^{i \Delta k' \cdot r_i} \sigma_{ba}^i$ with $\Delta k' = k_R - k_S$. If the Rabi frequency adiabatically changes from 0 to a large value, θ will vary from $\pi/2$ to 0. Consequently, the dark-state polariton will change from the collective excited state to the ground state and simultaneously emit a Stokes photon. Therefore, if the collective state imprinted in the write process $S^\dagger |vac\rangle$ is the same as the collective state $S'^\dagger |vac\rangle$ which can be fully retrieved out during the read process, the retrieve efficiency will reach the maximum. Again we obtain the mode-match condition $k_W - k_{AS} + k_R - k_S = 0$.

After the retrieval process, the whole state of anti-Stokes and Stokes photons can be expressed as $|\psi\rangle \approx 1 + \sqrt{\chi} \hat{a}_{AS}^\dagger \hat{a}_S^\dagger vac\rangle$. It can be easily seen that once there is a photon detected in the anti-Stokes field with a probability of χ, we can obtain a Stokes photon with certainty. In the discussion above, we only expand the perturbation theory to the first order. Taking into account higher-order excitations, the whole state of Stokes and anti-Stokes field may be described a two-mode squeezed state with excitation probability $\chi \ll 1$.

REFERENCES

[1] C.H. Bennett and G. Brassard, in Proceedings of the IEEE International Conference on Computers, systems, and signal processing, IEEE, Bangalore, India, New York (1984).

[2] N. Gisin, G. Ribordy, W. Tittel, and H. Zbinden, "Quantum Cryptography," Rev. Mod. Phys. 74, 145 (2002).
[3] E. Knill, R. Laflamme, and G.J. Milburn, "A Scheme for Efficient Quantum Computation with Linear Optics," Nature 409, 46–52 (2001).
[4] D. Browne and T. Rudolph, "Resource-Efficient Linear Optical Quantum Computation," Phys. Rev. Lett. 95, 010501 (2005).
[5] P. Michler, A. Kiraz, C. Becher, W. Schoenfeld, P. Petroff, L.D. Zhang, E. Hu, and A. Imamoglu, "A Quantum Dot Single-Photon Turnstile Device," Science 290, 2282–2285 (2000).
[6] C. Santori, M. Pelton, G. Solomon, Y. Dale, and Y. Yamamoto, "Triggered Single Photons from a Quantum Dot," Phys. Rev. Lett. 86, 1502–1505 (2001).
[7] A. Kuhn, M. Hennrich, and G. Rempe, "Deterministic Single-Photon Source for Distributed Quantum Networking," Phys. Rev. Lett. 89, 067901 (2002).
[8] J. McKeever, A. Boca, A. Boozer, R. Miller, J. Buck, A. Kuzmich, and H. Kimble, "Deterministic Generation of Single Photons from One Atom Trapped in a Cavity," Science 303, 1992–1994 (2004).
[9] C. Kurtsiefer, S. Mayer, P. Zarda, and H. Weinfurter, "Stable Solid-State Source of Single Photons," Phys. Rev. Lett. 85, 290–293 (2000).
[10] L.-M. Duan, M.D. Lukin, J.I. Cirac, and P. Zoller, "Long-Distance Quantum Communication with Atomic Ensembles and Linear Optics," Nature 414, 413–418 (2001).
[11] P. Kok, W.J. Munro, K. Nemoto, T.C. Ralph, J.P. Dowling, and G.J. Milburn, "Linear Optical Quantum Computing with Photonic Qubits," Rev. Mod. Phys. 79, 135–174 (2007).
[12] J.-W. Pan, Z.-B. Chen, C.-Y. Lu, M. Zukowski, H. Weinfurter, and A. Zeilinger, "Multi-Photon Entanglement and Interferometry," Rev. Mod. Phys. 84, 777–838 (2012).
[13] C.-W. Chou, S.V. Polyakov, A. Kuzmich, and H.J. Kimble, "Single-Photon Generation from Stored Excitation in an Atomic Ensemble," Phys. Rev. Lett. 92, 213601 (2004).
[14] T. Chaneliere, D.N. Matsukevich, S.D. Jenkins, S.-Y. Lan, T.A.B. Kennedy, and A. Kuzmich, "Storage and Retrieval of Single Photons Transmitted Between Remote Quantum Memories," Nature 438, 833–836 (2005).
[15] M.D. Eisaman, A. Andre, F. Massou, M. Fleischhauer, A.S. Zibrov, and M.D. Lukin, "Electromagnetically Induced Transparency with Tunable Single-Photon Pulses," Nature 438, 837 (2005).
[16] Y.O. Dudin, L. Li, and A. Kuzmich, Phys. Rev. A 87, 031801(R) (2013).
[17] S. Chen, Y.-A. Chen, T. Strassel, Z.-S. Yuan, B. Zhao, J. Schmiedmayer, and J.-W. Pan, "Deterministic and Storable Single-Photon Source Based on a Quantum Memory," Phys. Rev. Lett. 97, 173004 (2006).
[18] D.N. Matsukevich, T. Chaneliere, S.D. Jenkins, S.-Y. Lan, T.A.B. Kennedy, and A. Kuzmich, "Deterministic Single Photons via Conditional Quantum Evolution," Phys. Rev. Lett. 97, 013601 (2006).
[19] M. Fleischhauer, A. Imamoglu, and J.P. Marangos, "Electromagnetically Induced Transparency: Optics in Coherent Media," Rev. Mod. Phys. 77, 633–673 (2005).
[20] M. Fleischhauer and M.D. Lukin, "Dark-State Polaritons in Electromagnetically Induced Transparency," Phys. Rev. Lett. 84, 5094–5097 (2000).
[21] D.F. Walls and G.J. Milburn, "Quantum Optics," Springer-Verlag, Heidelberg (1994).
[22] H.J. Metcalf and P. van der Straten, "Laser Cooling and Trapping," Springer-Verlag, New York (1999).
[23] S.V. Polyakov, C.W. Chou, D. Felinto, and H.J. Kimble, Phys. Rev. Lett. 93, 263601 (2004).
[24] P. Grangier, G. Roger, and A. Aspect, "Experimental Evidence for a Photon Anticorrelation Effect on a Beam Splitter: A New Light on Single-Photon Interferences," Europhys. Lett. 1, 173–179 (1986).
[25] C.K. Hong, Z.Y. Ou, and L. Mandel, "Measurement of Subpicosecond Time Intervals Between Two Photons by Interference," Phys. Rev. Lett. 59, 2044–2046 (1987).
[26] Z.Y. Ou, J.-K. Rhee, and L.J. Wang, "Photon Bunching and Multiphoton Interference in Parametric Down-Conversion," Phys. Rev. A 60, 593 (1999).
[27] T. Humble and W. Grice, "Effects of Spectral Entanglement in Polarization-Entanglement Swapping and Type-I Fusion Gates," Phys. Rev. A 77, 022312 (2008).

[28] Z.-S. Yuan, Y.-A. Chen, S. Chen, B. Zhao, M. Koch, T. Strassel, Y. Zhao, G.-J. Zhu, J. Schmiedmayer, and J.-W. Pan, "Synchronized Independent Narrow-Band Single Photons and Efficient Generation of Photonic Entanglement," Phys. Rev. Lett. 98, 180503 (2007).
[29] D. Felinto, C.W. Chou, J. Laurat, E.W. Schomburg, H. de Riedmatten, and H.J. Kimble, "Conditional Control of the Quantum States of Remote Atomic Memories for Quantum Networking," Nature Phys. 2, 844–848 (2006).
[30] T. Chaneliere, D.N. Matsukevich, S.D. Jenkins, S.-Y. Lan, R. Zhao, T.A.B. Kennedy, and A. Kuzmich, "Quantum Interference of Electromagnetic Fields from Remote Quantum Memories," Phys. Rev. Lett. 98, 113602 (2007).
[31] D.A. Braje, V. Balic, S. Goda, G.Y. Yin, and S.E. Harris, "Frequency Mixing Using Electromagnetically Induced Transparency in Cold Atoms," Phys. Rev. Lett. 93, 183601 (2004).
[32] P. Mosley, J.S. Lundeen, B.J. Smith, P. Wasylczyk, A.B. U'Ren, C. Silberhorn, and I.A. Walmsley, "Heralded Generation of Ultrafast Single Photons in Pure Quantum states," Phys. Rev. Lett. 100, 133601 (2008).
[33] Y.H. Shih and C.O. Alley, "New Type of Einstein-Podolsky-Rosen-Bohm Experiment Using Pairs of Light Quanta Produced by Optical Parametric Down Conversion," Phys. Rev. Lett. 61, 2921 (1988).
[34] J.F. Clauser, M. Horne, A. Shimony, and R.A. Holt, "Proposed Experiment to Test Local Hidden-Variable Theories," Phys. Rev. Lett. 23, 880 (1969).
[35] C.W. Chou, H. de Riedmatten, D. Felinto, S.V. Polyakov, S.J. van Enk, and H.J. Kimble, "Measurement-Induced Entanglement for Excitation Stored in Remote Atomic Ensembles," Nature 438, 828–832 (2005).
[36] C.W. Chou, J. Laurat, H. Deng, K.S. Choi, H. de Riedmatten, D. Felinto, and H.J. Kimble, "Functional Quantum Nodes for Entanglement Distribution over Scalable Quantum Networks," Science 316, 1316–1320 (2007).
[37] Z.S. Yuan, Y.A. Chen, B. Zhao, S. Chen, J. Schmiedmayer, and J.W. Pan, "Experimental Demonstration of a BDCZ Quantum Repeater Node," Nature 454, 1098–1101 (2008).
[38] K.S. Choi, H. Deng, J. Laurat, and H.J. Kimble, "Mapping Photonic Entanglement into and out of a Quantum Memory," Nature 452, 67–71 (2008).
[39] B. Zhao, Y.-A. Chen, X.-H. Bao, T. Strassel, C.-S. Chuu, X.-M. Jin, J. Schmiedmayer, Z.-S. Yuan, S. Chen, and J.-W. Pan, "A Millisecond Quantum Memory for Scalable Quantum Networks," Nature Phys. 5, 95–99 (2009).
[40] R. Zhao, Y.O. Dudin, S.D. Jenkins, C.J. Campbell, D.N. Matsukevich, T.A.B. Kennedy, and A. Kuzmich, "Long-Lived Quantum Memory," Nature Phys. 5, 100–104 (2009).
[41] A.G. Radnaev, Y.O. Dudin, R. Zhao, H.H. Jen, S.D. Jenkins, A. Kuzmich, and T.A.B. Kennedy, "A Quantum Memory with Telecom-Wavelength Conversion," Nature Phys. 6, 894–899 (2010).
[42] F. Yang, T. Mandel, C. Lutz, Z.-S. Yuan, and J.-W. Pan, "Transverse Mode Revival of a Light-Compensated Quantum Memory," Phys. Rev. A 83, 063420 (2011).
[43] J. Simon, H. Tanji, J.K. Thompson, and V. Vuletic, "Interfacing Collective Atomic Excitations and Single Photons," Phys. Rev. Lett. 98, 183601 (2007).
[44] X.H. Bao, A. Reingruber, P. Dietrich, J. Rui, A. Dück, T. Strassel, L. Li, N.L. Liu, B. Zhao, and J.W. Pan, "Efficient and Long-Lived Quantum Memory with Cold Atoms Inside a Ring Cavity," Nature Phys. 8, 517–521 (2012).
[45] M.G. Raymer and J. Mostowski, "Stimulated Raman Scattering: Unified Treatment of Spontaneous Initiation and Spatial Propagation," Phys. Rev. A 24, 1980–1993 (1981).
[46] M.O. Scully and M.S. Zubairy, "Quantum Optics," Cambridge University Press (1997).
[47] A. Andre, "Nonclassical States of Light and Atomic Ensembles: Generation and New Applications," PhD Thesis, Harvard University, Graduate School of Arts and Sciences (2005).

Index

A

Active area, 66
Active gating of SPADs, 120–134
Afterpulse probability, 65, 125, 129–130, 132–133, 235, 265, 268
Afterpulsing, 89, 118, 237–238
  in actively gated InGaAs/InP SPADs, 120–122, 133
  in detector arrays, 232–234
  in photomultipliers, 9, 69, 76–82
  in Si–SPADs, 56, 86, 94
  in superconductor–based detectors, 201–203
  probability of, 65–66
All–fiber quantum logic gates, 455–458
Amplifier, 70–71, 85, 127, 129, 131–132, 148, 170, 186, 194–195, 202, 205
  noise properties of, 70–71
Anticorrelation of a single photon on a beam–splitter, 336–339
Atomic–ensemble–based single–photon source, 542–545
  collective excitation, 542
  conditional autocorrelation, 547
  excitation probability, 544–550
  heralded single–photon, 340, 353
  heralded single–photon sources, 316, 333, 388–389, 393–394, 399–402, 545–550
  indistinguishability, 485–487
  post–selection, 435, 446, 544
  storage of photons, 366, 542
  write and read process, 542, 546
Atom–photon interaction in resonators, 471–475
Attenuated laser light, 153, 174, 245, 420
Avalanche photodiode (APD), 85–87, 89, 114, 159–160, *see also* single–photon avalanche photodiode (SPAD)
Avalanche pulse, 90
  discrimination of, 120–134
  trapped charge, 90, 119, 136, 265

B

Bialkali (Sb–Rb–Cs, Sb–K–Cs) photocathodes, 72, 85
Born rule, 284

C

Calibration, 240, 259–263
  correlated–photon–pair calibration method, 262–263
  model of detector's response, 258, 269
  primary standard methods, 260, 262
  substitution method, 261
Cavity–based single–photon emitters, 471–485
  ions, 484–485
  neutral atoms, 481–484
  quantum dots, 471–472
Cavity–coupling regimes, 475
Cavity–quantum electrodynamics, 468–469, 471, 502
Cesium–antimony, 71
$Chi^{(2)}$ nonlinear crystals, 248, 353, 412
Clauser's experiment, 336
Clauser–Horne–Shimony–Holt (CHSH) inequality, 554–555
"Click/no–click" detectors, 40, 52, 56, 79, 192, 318, 219–221, 223–224, 228
Coherent states, 275, 285, 289, 292, 299–300, 326
  collection efficiency, 6, 7, 31, 73, 75–77, 111, 425, 446, 501–502, 519, 522–524, 526
Coincidence–to–accidentals ratio (CAR), 423–424
Continuous–wave (CW) source, 222–226
Controlled–NOT gate (CNOT), 455–458
  all–fiber implementation, 457
  performance, evaluation of, 457
Correlated calibration technique, 267–269, 277–279
Correlation histogram, 39, 50, 267
Crosstalk, 57, 92–93
  electrical crosstalk, 93, 112
  optical crosstalk, 92, 112

563

D

Dark count, 79, 155
  rate of (DCR), 65, 88, 155–156
Dead time, 64–65, 81–82, 86, 94, 96, 110, 112, 157, 219–220, 222–226
Debye–Waller factor, 515–516, 521–522
Detection efficiency (DE), 9–10,12–13, 16, 31, 38–40, 53–56, 58, 60–62, 64–65, 69, 73, 79–82, 85, 87–88, 94, 99, 104, 106, 113–114, 117, 121, 123–126, 128, 130–132, 134, 137, 148–149, 155–157, 163, 165, 188–189, 191, 194, 199–201, 207, 220–221, 224, 232–233, 243, 246–248, 250, 257–259, 261–263, 265, 267–271, 275, 277–279, 283, 291, 298, 303, 360, 382, 384, 429, 547
  assumptions, 288–290
  experimental implementations, 297–310
  photon–number–resolving detectors, 291–293
  silicon–based SPAD, 297–298
  time–multiplexed detector, 236–237
Detector under test (DUT), 261–265, 268–273, 276–278, 289, 298–299
Discharge lamp, light from, 317–318, 320, 331–334
Double–loop (DL) scheme, 453–454
Dynodes, 73–77
  collection efficiency, 73, 75, 77, 111

E

Electromagnetic field, 26, 43, 50, 287–288, 290, 292, 317, 319, 321, 329–330, 354, 505
Entangled photon–pair sources, 444–454
  birefringent fiber (BRF), use of, 446
  coincident count rate, 447
  dispersion–shifted fiber (DSF), 449, 451–452
  double–loop (DL) scheme, 453–454
  polarization–entangled photons, 444–445, 452
  Sagnac–loop configuration, 445–447, 449–450, 452–454
  verification of entanglement, 447–448

F

Field quantization in cavity QED (CQED), 472 , 476, 502
Fill–factor, 93, 106, 108, 135

Four–wave mixing (FWM), 411
  classical four–wave mixing theory, 413–416
  controlled–NOT gate (CNOT) all–fiber implementation, 455–458
  cross–polarized four–wave mixing in birefringent fibers, 419–420
  dispersion–shifted fiber (DSF), 425–426, 449, 451–452
  double–loop (DL) scheme, 453–454
  entangled photon–pair sources, 444–454
  fiber–based heralded photon source, 422–430
    advantages of, 425
    experimental arrangement of, 422
  phase–matching, 413–416
  Sagnac–loop configuration, 445
  spontaneous four–wave mixing (sFWM), 422–430
  symmetric group velocity matching, 438–439
  widely spaced signal and idler photons, 429
  zero–dispersion wavelength (ZDW), 412, 414–416, 419–420, 426–430

G

Geiger mode, 86, 87, 94, 114, 116, 160
  generation efficiency, 31, 33, 396, 549

H

Hanbury Brown–Twiss (HBT), 38, 40, 388, 469
  antibunching, sub–Poissonian behavior, 42–44
  bunching, Poissonian behavior, 42–44
  characterization of single–photon source, 42
  for pulsed sources, 30
  limitations, 42
  Poissonian photon number statistics, 43
  with "click/no–click" detectors, 40–42
  with photon–number–resolving detectors, 39–40
Harmonic subtraction, in gating of SPADs, 131–132
Heralded single–photon sources, 7, 388–389, 393–394, 399–402, 545–550
  accidental count rate, 424
  brightness, 423
  coincidence count rate, 386, 423
  generation efficiency, 396
  heralding efficiency, 384–388

Index

High–order coherence $g(n)$, 53
Hold–off time, 90, 95–96 120, 125, 132–133
Hong–Ou–Mandel (HOM) interference, 46–47, 241, 431–434, 440, 486, 550, 552–553, 554, *see also* two–photon quantum interference
   visibility of, 46–47, 443–444

I

InAs–based quantum–dot formation, 493–494
Indistinguishability, 45–47
   Hong–Ou–Mandel (HOM) interference as a measurement of, 46–47
Infrared SPADs, 113–114

J

Joint spectrum amplitude (JSA), 359, 364, 436–440
   symmetric, 373–374

K

Kronecker delta function, 220

L

Light, 148, 155
   non–classical, 81, 284, 310, 334, 352, 389
   physics of, 1–2, 318–319
Local oscillator, 53, 241, 243, 486–487, 491–492

M

Microlens array, 137, 233
Microwave kinetic–inductance detectors (MKIDs), 204–208
   operating principle, 205–206
Model of a detector, 264
Multialkali (Sb–Na K–Cs) photocathodes, 72–73
Multi–photon emission probability, 31–32, 34, 37
Multiple–quantum–dot detectors, 169–172

N

Nitrogen–vacancy center, 511
   Debye–Waller factor, 515–516, 521
   optical transitions and level structure, 514–518
   optically detected magnetic resonance (ODMR), 519–521
   photoluminescence (PL) spectra, 515–516, 518
   photoluminescence excitation (PLE) spectroscopy, 517
   single–photon generation, 518–519
   spin properties, 519–521
   structure and formation, 511–514
   zero–phonon line, 515–517, 519
Non–classical light, 2–3, 81–82, 284, 334
Nonlinear optics, 394, 416
   second–order effects (parametric up– and down conversion), 369–370
   third–order effects (four–wave mixing), 413, 416–417
Novel semiconductor single–photon detectors, 147

O

On–demand single–photon sources, 17, 333, 347, 468, 541
One–photon wavepacket, 318, 323, 326–328, 331–334, 336–337
Optically detected magnetic resonance (ODMR), 511, 514, 519–521

P

Parametric down–conversion (PDC), 351
   anisotropic materials, 368
   heralded single photons from a triggered PDC source, 372–378
   in bulk crystals, 367
   in waveguides, 401–402
   joint spectrum amplitude (JSA), 374–375
   multiplexed PDC sources, 375–378
   non–critical phase–matching, 369
   nonlinear materials and properties, 404
   optimal focus parameters for heralding efficiency, 384
   periodically–poled crystal, 379–392
   phase matching, 379–382
   photon–state tuning through vectorial matching, 373–375
   quasi–phase–matching (QPM), 379–382
   single–photon sources, comparison of, 403
   spectral purity, 362–363

Periodically poled lithium niobate (PPLN) crystal, 245–248, 378, 383–384
Phase matching, 364–365, 367, 368–371, 373
  in atomic ensembles, 542
  in four–wave mixing, 413–416
  in parametric down–conversion, 343–371
Photocathodes, 72–73
  properties, 74
Photoconductive detector, 148
Photodetector device, 148
Photodetectors, 90, 114, 136–167
Photodiode, 63, 88, 113, 131, 148, 270
Photoelectron emission, 72
Photoluminescence excitation (PLE) spectroscopy, 517
Photomultiplier tube (PMT), 69
  dark–current and signal–current pulse–height distributions, 81
  dynodes, 70, 73–76
  noise properties of amplifier, 70
  photoelectric effect, 70
  photoelectron emission and photocathodes, 72–73
  photon counting, 76–82
  secondary emission, 73–76
  workfunction, 70, 72
Photon, 3, 76–82, 84–85
  definition, 84–85
Photon pairs, 50–52
Photon probability distribution, 41, 48–49, 58
  coherent state, 48–49
  single–photon source, 37–38
  thermal source, 49–50
Photon statistics, 42–44
  coherent state, 43, 47–49
  coincidence to accidental ratio (CAR), 423–424
  Hanbury Brown–Twiss interferometer, 38–42
  high–order coherence $g^{(n)}$, 44–45
  mixed state, 29, 37, 241
  pair source, 50–52
  probability of multi–photon emission, $P(n)$, 30–32, 34, 37–38
  relating $g^{(2)}$ to $P(n)$, 34
  second–order coherence $g^{(2)}$, 32–34
  single–photon source, 4, 5, 9, 316
  thermal source, 49–50
Photonic crystal fiber (PCF), 412, 415–416, 427–430, 461
Photonic fusion, 431
  in fiber, 458–459

Photon–number–resolving (PNR) capability, 9, 12, 56, 159
  definitions of none, some, full, 12–13, 56
Photon–number–resolving (PNR), 39, 172, 218, 232, 291
  multiplexed 10, 13–14
  hybrid PNR–capable detectors, 39–42
  quantum–dot–based detectors, 166–175
  reconstructed POVM elements, 301–305
  transition edge sensor (TES), 195–196, 198–201
Poissonian photon statistics, 42–44
Polarization–maintaining fiber (PMF), 450
Positive–operator-valued measure (POVM), see also Quantum detector tomography, 55, 284–297
  conditioning and regularization, 305–307
  in quantum detector tomography, 288
  measured probabilities relations, 293–294
  optimization problem, 288–289
  phase–sensitivity reconstructed POVM elements, 293–297, 301–305
  reconstruction, 293–297
    robustness of, 307–308
    with phase–sensitivity, 295–297
    without phase–sensitivity, 293–295
Probabilistic generation of photon pairs, 461
Pulsed source, 249–250
Pulsed up–conversion single–photon counting, 245–249

Q

Quantization of electric field, 26–28
Quantum communication, 525
  single–photon detectors, role of, 257, 264
Quantum detector tomography, 259, 286–288
  positive–operator–valued measure (POVM), 284–287, 292
  tomographic techniques, 287–288
Quantum dot, optically gated, field–effect transistor (QDOGFET), 57, 170–175, 177–179
Quantum–dot–based detectors, 166–180
  application to QDOGFET detectors, 177–179
  multiple–quantum–dot detectors, 169–170
  photon–number–resolving detection, 172–175
  single–quantum–dot detectors, 167–169
Quasi–phase–matching (QPM), 379–382, 390–391, 404

Index

R

Radiant power measurements, 261
Raman scattering, 420–422
  phonon distribution, 420–421
Read pulse, 547, 551, 553
Recovery time, 64–65
Refractive index, 368, 381, 401, 419, 433, 503–504, 522
Reset time, 64–65, 170, 188, 191

S

Sagnac–loop configuration, 445
Second–order coherence, 34, 36–37
Second–order coherence $g^{(2)}$, 32–34
  zero–delay value, 33–34, 36, 44, 49
Self–differencing, in gating of SPADs, 129–131
Semiconductor quantum dots, 492–510
  InAs–based quantum–dot formation, 493–494
  optically accessing single quantum dots, 497–499
  quantum–dot photon indistinguishability, 505–510
  single photons from single quantum dots, 499–502
  weak QD–cavity coupling, 502–505
Semiconductor–based detectors for photon counting, 85
Signal–to–noise (SNR), 79, 81, 121, 130, 176–179, 275, 345
Silicon SPAD array detectors, 108
  applications in life sciences, 111
  integration of, 135
  latchless pipeline scheme, 108–109
  pixel access problem, 110
Sine–wave gating SPADs, 127–129
Single defects in diamond, 511, *see also* Nitrogen–vacancy center
  nickel–related or other defects, 522
  optical structures, 514–515
  silicon–vacancy centers, 521–522
  chromium–related defects, 522
  TR12 color center, 522
Single emitters, in isolated quantum systems, 17, 467
  amplitude and phase, arbitrary shaping of, 487–492
  cavity–based single–photon emitters, 471–485

  atom–photon interaction in resonators, 471–475
  cavity–coupling regimes, 475
  cavity–quantum electrodynamics (QED), 468–469, 471–472, 476, 502
  Fabry–Perot type cavities, 472–473
  field quantization in cavity QED, 472–473
  neutral atoms, 481–484
  quantum dots, 499–502
  single–photon emission, 475–485
  strong–coupling regime, 479–480
  three–level atom, 473–475
  trapped ions, 484–485
  two–level atom, 472–473
Single photons from atoms and ions, 468–492
  cavity–based single–photon emitters, 471
  effective two–level system, 469–470
  emission into free space, 469–471
Single–photon avalanche diode (SPAD), 6, 10–11, 16, 56, 63, 65–66, 86–102, 104–124, 127, 129, 131, 134–137, *see also* SPADs, InGaAs/InP, and SPADs Silicon
Single–photon detector, 9, 16, 264–267
  calibration, 267–269
  correlated calibration technique, 267–269, 277
  substitution calibration method, 273–275
  click/no–click detectors, 56
  coincidence peak "shoulder," origin of, 264–265
  comparison of, 273–275
  correlated photon pair and substitution calibration methods, comparison of, 262–263
  detector under test (DUT), 289
  photon-number resolving detectors, 291–293
  photon–counting detector, model of, 84, 260, 267
  radiant power measurements, 261
  transition edge sensors, 275, 286, 297, 378, 388
  trigger detector, 153, 155, 262, 266–267, 278, 365–367, 378
  types of single photon detectors, 56, 65, 121
Single–photon sources, 16–17, 545–550
  comparison, 403

deterministic sources, 7, 15–16, 545–550
  development of, 346–348
  on–demand sources, 347
  probabilistic sources, 545
Single–photon technology, applications of, 4–8
  local realism tests, 554–555
  quantum information, 555
Single–quantum–dot detectors, 167–169
Solid–state detectors, 468, 525
Solid–state photomultipliers, 148–150
Solid–state photomultipliers and visible–light photon counters, 148–166
SPADs, InGaAs/InP, 87, 114–120
  active gating, 120–134
  arrays, 134–135
  discrimination of avalanche signals, 124, 126–127
    by cancellation, 123–125
    by sampling, 122–123
    suppressing gate transient, 121–125
  high–speed periodic gating, 125–127
    comparison of, 132–134
    harmonic subtraction, 131–132
    self–differencing, 129–131
    sine–wave gating, 127–129
    timing resolution in, 117–118
SPADs, Silicon, 115, 117, 119–120, 218, 243, 253
  afterpulsing, 89–90
  avalanche pulse, 66
  circuit principles, 94–98
    active quenching, 95
    bias voltage, 94
    hold–off time, 95
    obtaining accurate photon–timing, 97
    passive quenching, 96
    reset transition, 96
  crosstalk, optical and electrical, 92–93, 112
  dark counts, 79, 155
    rate, 65, 88, 155–156
  deep submicron CMOS, 106–108
  detection efficiency, 291
  fabrication process, 104, 134
  gating, 120–134
  high–voltage, complementary metal–oxide semiconductor (HV–CMOS), 104–106
  microelectronic structure, 93
    absorption and drift region, 93
    avalanche region, 93
    guard–rings, 93
  non–planar, custom process, 102–104
  performance parameters and features, 87–93

dark count rate (DCR), 88
  planar SPAD devices fabricated in custom technology, 98–102
  Poole–Frenkel and trap–assisted tunneling effects, 89
  timing jitter, 90–92
  twilight (transient) regime, 65, 265
Spatially multiplexed detectors, 218–221, 233
  active switching arrangement, 220
  click/no–click (non–PNR) detector, 220
  dead time reduction, 226–231
  dead–time fraction (DTF), 228, 234–236
  passive "detector tree" arrangement, 226–231
  photon–number–resolving detector arrays, 232–234
  with continuous–wave (CW) source, 222–226
  with pulsed source, 34, 220
SQUID, 195–198, 200–201
Standard deep submicron CMOS SPADs, 106–108
Strong–coupling regime, 479–480
Superconducting nanowire single–photon detectors (SNSPDs), 187–194
  detection efficiency, 188–189
  detection process, 189–190
  multiple nanowires, 193–194
  self–heating hotspot, 190
Superconducting tunnel junction detectors, 201–203
  operating principle, 201–203
  quasi–particle trapping structure, 202

T

Three–level atomic system, 143
  lambda–type atomic system, 143
Time–multiplexed detector (TMD), 236
  fiber–loop detectors, 237–241
  non–photon–number–resolving detectors, 172, 232–234
  photon number statistics, 275, 291, 298, 309–310, 365
  weak–homodyne detection, 238, 241–243
Timing jitter, 62, 90, 117
  in actively gated InGaAs/InP SPADs, 125–127
  in single–photon avalanche photodiodes (SPADs), 90–92
  in superconducting nanowire single–photon detectors (SNSPDs), 187–194

in transition–edge sensors (TES), 194–195
in visible light photon counters (VLPCs), 157–159, 165–166
Transition–edge sensors (TES), 194–201
  detection efficiency, 53, 88, 104, 106, 155, 188
  operating principle, 195–199
  photon number resolving capability, 159–161
Two:one photon ratio, 35–37, 48, 51

U

Ultrafast up–conversion, 250–253
  time anticorrelation characteristics, 251–253
  time–frequency Fourier duality, 251–253
Up–conversion-based single–photon counting, 243, 245–249
  energy conservation constraints, 244–245
  frequency translation, 245
  ultrafast up–conversion, 250–253

V

Vertical–cavity surface–emitting laser (VCSEL), 527
Visible light photon counters (VLPC), 148–150, 155–157, 159–160
  calculating impact–ionization, 152–153, 159
  dark count, 155–157
  detection efficiency, 155–157
  device operation, 150–151
  layers grown in degenerate n–type silicon substrate, 150
  photon–number–resolving capability, 159–161
  quantitative model, 161–163
  reducing dark counts and balancing maximum count rates, 164–165
  structure and operation, 150–154
  timing jitter and its reduction, 165–166

W

Waveguide, 401–402
  heralding single photons, 399–401
  single–spatial–mode behavior, 400
  theory of, 394–399
Weak–homodyne detection, 238, 241–243
  local oscillator, 241–242
Wigner function, 242–243, 286–287, 308–310
Write process, 542–544, 548, 556–559
Write pulse, 542–543, 545–548, 550–552, 556

Z

Zero–phonon line, 416, 515–517, 519

Edwards Brothers Malloy
Ann Arbor MI. USA
December 2, 2013